Ganymede

With European Space Agency's (ESA) upcoming JUpiter ICy moons Explorer (JUICE) mission to Jupiter and Ganymede, this book provides a fascinating and timely summary of our current knowledge of Ganymede: the largest moon in the solar system and the only one with an intrinsic magnetic field. Written by a team of multidisciplinary experts spanning geology, space physics and habitability, it provides up-to-date knowledge about Ganymede. The history of its discovery, formation, surface, atmosphere and space environment are discussed in accessible language and supported by the enormous amount of data obtained by Galileo, the Hubble Space Telescope and earlier missions. The latest surface maps of Ganymede are also presented, providing an invaluable reference for graduate students and researchers working in planetary science.

MARTIN VOLWERK is a permanent member of staff at the Space Research Institute of the Austrian Academy of Sciences, Co-Investigator on the JUICE magnetometer and magnetometers on other missions and national Co-Principal Investigator on the Comet Interceptor mission. His 30-year career as a planetary plasma physicist has focused on studying plasma interactions at the Galilean moons, ultra-low frequency waves in planetary magnetospheres and cometary plasma physics.

MELISSA MCGRATH is a senior scientist at the Carl Sagan Science Center, SETI Institute. She has had a distinguished career with over 30 years of experience as a planetary scientist studying Jupiter's four large Galilean satellites: Io, Europa, Ganymede and Callisto. She has worked on numerous NASA missions, including the Hubble Space Telescope and the upcoming JUICE and Europa Clipper missions. She has held leadership positions at NASA, the International Astronomical Union and the American Astronomical Society. She has also served as a scientific editor for *The Astrophysical Journal*.

XIANZHE JIA is a professor at the University of Michigan. He is a space physicist with expertise in combining computer simulations with spacecraft measurements to study magnetic fields and plasma phenomena in planetary environments. He is Co-Investigator on the JUICE and Europa Clipper missions to explore the Jupiter system.

TILMAN SPOHN is a planetary scientist at the DLR Institute of Planetary Research in Berlin, Germany. His work focuses on modelling the interior structure and thermal evolution of planets and icy satellites, with a particular emphasis on their cores and magnetic field generation. He has led instrument teams on the NASA InSight mission to Mars and the ESA Rosetta mission to comet 67P/Churyumov-Gerasimenko. He is an American Geophysical Union Fellow and a recipient of the Runcorn-Florensky medal, awarded by the European Geosciences Union for exceptional contributions to planetology. He is also a member of the Academy of Astronautics and of Academia Europaea.

Cambridge Planetary Science

Series Editors:

Fran Bagenal, David Jewitt, Carl Murray, Jim Bell, Ralph Lorenz, Francis Nimmo, Sara Russell

Books in the Series:

1. Jupiter: The Planet, Satellites and Magnetosphere†
Edited by Bagenal, Dowling and McKinnon
978-0-521-03545-3
2. Meteorites: A Petrologic, Chemical and Isotopic Synthesis †
Hutchison
978-0-521-03539-2
3. The Origin of Chondrules and Chondrites †
Sears
978-1-107-40285-0
4. Planetary Rings †
Esposito
978-1-107-40247-8
5. The Geology of Mars: Evidence from Earth-Based Analogs †
Edited by Chapman
978-0-521-20659-4
6. The Surface of Mars †
Carr
978-0-521-87201-0
7. Volcanism on Io: A Comparison with Earth †
Davies
978-0-521-85003-2
8. Mars: An Introduction to Its Interior, Surface and Atmosphere †
Barlow 978-0-521-85226-5
9. The Martian Surface: Composition, Mineralogy and Physical Properties
Edited by Bell
978-0-521-86698-9
10. Planetary Crusts: Their Composition, Origin and Evolution †
Taylor and McLennan
978-0-521-14201-4
11. Planetary Tectonics †
Edited by Watters and Schultz
978-0-521-74992-3
12. Protoplanetary Dust: Astrophysical and Cosmochemical Perspectives †
Edited by Apai and Lauretta
978-0-521-51772-0
13. Planetary Surface Processes
Melosh
978-0-521-51418-7
14. Titan: Interior, Surface, Atmosphere and Space Environment
Edited by Müller-Wodarg, Griffith, Lellouch and Cravens
978-0-521-19992-6
15. Planetary Rings: A Post-Equinox View (Second edition)
Esposito
978-1-107-02882-1
16. Planetesimals: Early Differentiation and Consequences for Planets
Edited by Elkins-Tanton and Weiss
978-1-107-11848-5
17. Asteroids: Astronomical and Geological Bodies
Burbine
978-1-107-09684-4
18. The Atmosphere and Climate of Mars
Edited by Haberle, Clancy, Forget, Smith and Zurek
978-1-107-01618-7
19. Planetary Ring Systems
Edited by Tiscareno and Murray
978-1-107-11382-4
20. Saturn in the 21st Century
Edited by Baines, Flasar, Krupp and Stallard
978-1-107-10677-2
21. Mercury: The View after MESSENGER
Edited by Solomon, Nittler and Anderson
978-1-107-15445-2
22. Chondrules: Records of Protoplanetary Disk Processes
Edited by Russell, Connolly Jr. and Krot
978-1-108-41801-0
23. Spectroscopy and Photochemistry of Planetary Atmospheres and Ionospheres
Krasnopolsky
978-1-107-14526-9
24. Remote Compositional Analysis: Techniques for Understanding Spectroscopy, Mineralogy, and Geochemistry of Planetary Surfaces
Edited by Bishop, Bell III and Moersch
978-1-107-18620-0
25. Meteoroids: Sources of Meteors on Earth and Beyond
Edited by Ryabova, Asher and Campbell-Brown
978-1-108-42671-8
26. Meteorite Mineralogy
Rubin and Ma
978-1-108-48452-7
27. Vesta and Ceres: Insights into the Dawn of our Solar System
Edited by Marchi, Raymond and Russell
978-1-108-47973-8

† Reissued as a paperback

GANYMEDE

Edited by

MARTIN VOLWERK
Austrian Academy of Sciences

MELISSA MCGRATH
SETI Institute

XIANZHE JIA
University of Michigan

TILMAN SPOHN
DLR Institute of Planetary Research, Berlin

CAMBRIDGE
UNIVERSITY PRESS

CAMBRIDGE
UNIVERSITY PRESS

Shaftesbury Road, Cambridge CB2 8EA, United Kingdom

One Liberty Plaza, 20th Floor, New York, NY 10006, USA

477 Williamstown Road, Port Melbourne, VIC 3207, Australia

314–321, 3rd Floor, Plot 3, Splendor Forum, Jasola District Centre, New Delhi–110025, India

103 Penang Road, #05–06/07, Visioncrest Commercial, Singapore 238467

Cambridge University Press is part of Cambridge University Press & Assessment, a department of the University of Cambridge.

We share the University's mission to contribute to society through the pursuit of education, learning and research at the highest international levels of excellence.

www.cambridge.org
Information on this title: www.cambridge.org/9781108832953

DOI: 10.1017/9781108966474

First published 2025

A catalogue record for this publication is available from the British Library.

A Cataloging-in-Publication data record for this book is available from the Library of Congress.

ISBN 978-1-108-83295-3 Hardback

Additional resources for this publication at www.cambridge.org/ganymede

CONTENTS

The plate section should appear between pages 236 and 237.

Additional resources for this publication at www.cambridge.org/ganymede.

CONTRIBUTORS

CAITLIN AHRENS
NASA Goddard Space Flight Center, Greenbelt, MD, USA

YANN ALIBERT
Space Research & Planetary Sciences, University of Bern, Bern, Switzerland

FRAN BAGENAL
Astrophysical & Planetary Sciences, University of Colorado, Boulder, CO, USA

ROSE-MARIE BALAND
Royal Observatory of Belgium, Brussels, Belgium

AMY BARR
Planetary Science Institute, Tucson, AZ, USA

MICHAEL BLAND
United States Geological Survey, Flagstaff, AZ, USA

BERTRAND BONFOND
LPAP, STAR Institute, University of Liège, Belgium

VERONICA BRAY
lunar and Planetary Laboratory, University of Arizona, Tucson, AZ, USA

MARISSA E. CAMERON
Jet Propulsion Laboratory, California Institute of Technology, Pasadena, CA, USA

GIANLUCA CARNIELLI
Department of Physics, Imperial College London, London, UK

JULIE CASTILLO-ROGEZ
Jet Propulsion Laboratory, California Institute of Technology, Pasadena, CA, USA

JULIAN CHELA-FLORES
The Abdus Salam International Centre for Theoretical Physics Trieste, Italy, and IDEA, Fundación Instituto de Estudios Avanzados, Caracas, Bolivarian Republic of Venezuela

ULRICH CHRISTENSEN
Max Planck Institute for Solar System Research, Göttingen, Germany

GEOFFREY C. COLLINS
Astronomy and Physics Department, Wheaton College, Norton, MA, USA

CHUANFEI DONG
Princeton University, Princeton, USA
Princeton Plasma Physics Laboratory, Princeton, USA

SHAHAB FATEMI
Umeå University, Department of Physics, Umeå, Sweden

MARINA GALAND
Department of Physics, Imperial College London, London, UK

A. GALLI
Space Research and Planetary Sciences, Physics Institute, University of Bern, Bern, Switzerland

HERBERT GUNELL
Umeå University, Department of Physics, Umeå, Sweden

JAMES W. HEAD
Department of Geological Sciences, Brown University, Providence, RI, USA

PAUL HELFENSTEIN
Cornell Center for Astrophysics and Planetary Science, Cornell University, Ithaca, NY, USA

AMANDA HENDRIX
Planetary Science Institute, Tucson, AZ, USA

CHARLES HIBBITTS
Applied Physics Laboratory, Johns Hopkins University, Laurel, MD, USA

SAMUEL M. HOWELL
Jet Propulsion Laboratory, California Institute of Technology, Pasadena, CA, USA

HAUKE HUSSMANN
Institute of Planetary Research, German Aerospace Center (DLR) Berlin, Germany

RALF JAUMANN
Free University of Berlin, Institute of Geological Sciences, Berlin, Germany

XIANHE JIA
Department of Climate and Space Sciences and Engineering, University of Michigan, Ann Arbor, MI, USA

ROBERT E. JOHNSON
University of Virginia, Charlottesville, VA, USA; Physics, New York University, New York, NY, USA

KLÁRA KALOUSOVÁ
Department of Geophysics, Faculty of Mathematics and Physics, Charles University, Prague, Czech Republic

ELKE KERSTEN
Institute of Planetary Research, German Aerospace Center (DLR), Berlin, Germany

KRISHAN K. KHURANA
Department of Earth, Planetary, and Space Sciences, UCLA, Los Angeles, CA, USA

JUN KIMURA
Department of Earth and Space Science, Osaka University, Osaka, Japan

MICHELLE KIRCHOFF
Southwest Research Institute, Boulder, CO, USA

MARGARET G. KIVELSON
Department of Earth, Planetary, and Space Sciences, UCLA, Los Angeles, CA, USA
Department of Climate and Space Sciences and Engineering, University of Michigan, Ann Arbor, MI, USA

VALÉRY LAINEY
IMCCE , Observatoire de Paris, France

FRANÇOIS LEBLANC
LATMOS/IPSL, UPMC Univ. Paris 06 Sorbonne Universités, UVSQ, CNRS, Paris, France

NICOLAS LIGIER
Institut d'Astrophysique Spatiale, Université Paris-Saclay, Orsay, France

ALICE LUCCHETTI
Istituto Nazionale di Astrofisica (INAF), Astronomical Observatory of Padova, Padova, Italy

STEFANO MASSETTI
INAF-IAPS (Institute for Astrophysics and Planetology from Space), Rome, Italy

THOMAS B. MCCORD
Bear Fight Institute, Winthrop, WA, USA

MELISSA MCGRATH
Carl Sagan Science Center, SETI Institute, USA

WILLIAM MCKINNON
Department of Earth, Environmental, and Planetary Sciences, Washington University, St. Louis, MO, USA

PHILIPPA MOLYNEUX
Southwest Research Institute, San Antonio, TX, USA

JEFFREY M. MOORE
NASA Ames Research Center, Mountain View, CA, USA

ALESSANDRO MURA
INAF-IAPS (Institute for Astrophysics and Planetology from Space), Rome, Italy

ERICA M. NATHAN
Department of Geological Sciences, Brown University, Providence, RI, USA

FRANCIS NIMMO
Earth & Planetary Sciences Department, University of Califonia Santa Cruz, CA, USA

APURVA V. OZA
Jet Propulsion Laboratory, California Institute of Technology, Pasadena, CA, USA

PASQUALE PALUMBO
Parthenope University, Department of Applies Sciences, Napoli, Italy

ROBERT T. PAPPALARDO
Jet Propulsion Laboratory, California Institute of Technology, Pasadena, CA, USA

WES PATTERSON
Applied Physics Laboratory, John's Hopkins University, Laurel, MD, USA

CHRISTINA PLAINAKI
Italian Space Agency, Rome, Italy

FRANK POSTBERG
Free University of Berlin, Institute of Geological Sciences, Berlin, Germany

FRANCOIS POULET
Institut d'Astrophysique Spatiale, Univ. Paris Sud/CNRS, Orsay, France

LOUISE M. PROCKTER
Johns Hopkins University Applied Physics Laboratory, Laurel, MD, USA

NICOLAS RAMBAUX
Institut de mécanique céleste et des ephémérides (IMCEE), Paris, France

KURT D. RETHERFORD
Southwest Research Institute, San Antonio, TX, USA

EDGARD G. RIVERA-VALENTÍN
Johns Hopkins University Applied Physics Laboratory, Laurel, MD, USA

THOMAS ROATSCH
Institute of Planetary Research, German Aerospace Center (DLR), Berlin, Germany

COSTANZA ROSSI
Istituto Nazionale di Fisica Nucleare, National Laboratory of Frascati, Frascati, Italy

LORENZ ROTH
Space and Plasma Physics, KTH Royal Institute of Technology, Stockholm, Sweden

TINA RÜCKRIEMEN-BEZ
Institute for Planetary Research, DLR, Berlin, Germany

PAUL M. SCHENK
Lunar and Planetary Institute, Universities Space Research Association, Houston, TX, USA

GERALD SCHUBERT
Department of Earth and Space Sciences, University of California, Los Angeles, CA, USA

YUHITO SHIBAIKE
Physics Institute & NCCR PlanetS, University of Bern, Bern, Switzerland

KRISTA M. SODERLUND
Institute for Geophysics, Jackson School of Geosciences, The University of Texas at Austin, Austin, USA

FRANK SOHL
Institute of Planetary Research, German Aerospace Center (DLR), Berlin, Germany

ANEZINA SOLOMONIDOU
California Institute of Technology, Pasadena, CA, USA and LESIA Observatoire de Paris, CNRS, UPMC Univ., Paris 06, Univ. Paris-Diderot, Meudon, France

CHRISTOPHE SOTIN
Laboratoire de Planétologie et Géodynamique, Nantes University, Nantes, France

JOHN R. SPENCER
Southwest Research Institute, Boulder, CO, USA

TILMAN SPOHN
Institute of Planetary Research, German Aerospace Center (DLR), Berlin, Germany

ALEXANDER STARK
Institute of Planetary Research, German Aerospace Center (DLR), Berlin, Germany

GREGOR STEINBRÜGGE
Department of Geophysics, Stanford University, Stanford, CA, USA; Jet Propulsion Laboratory, California Institute of Technology, Pasadena, CA, USA

KATRIN STEPHAN
Institute of Planetary Research, German Aerospace Center (DLR), Berlin, Germany

GABRIEL TOBIE
Laboratoire de Planétologie et Géodynamique, Université de Nantes, France

TIM VAN HOOLST
Royal Observatory of Belgium, Brussels, Belgium Institute of Astronomy, KU Leuven, Leuven, Belgium

MARTIN VOLWERK
Space Research Institute, Austrian Academy of Sciences, Graz, Austria

AUDREY VORBURGER
Space Research and Planetary Sciences, Physics Institute, University of Bern, Bern, Switzerland

ROLAND WAGNER
Institute of Planetary Research, German Aerospace Center (DLR), Berlin, Germany

LIANG WANG
Princeton University, Princeton, USA
Princeton Plasma Physics Laboratory, Princeton, USA

PETER WURZ
Space Research and Planetary Sciences, Physics Institute, University of Bern, Bern, Switzerland

PHILIPPE ZARKA
LESIA, Observatoire de Paris, CNRS, PSL, Sorbonne Université, Université Paris Cité, Meudon, France

PREFACE

Ganymede is possibly one of the most interesting objects in the solar system: it is the largest moon (even larger than Mercury), has its own internal magnetic field and is an ocean world. All this calls for a book on this intriguing solar system body. The idea for this book came up many years ago, after the publication of the Europa book in 2009, when Melissa and Martin sat together. They decided, however, that at that point the enthusiasm of the publishers would be limited for another 600 page book. And so this idea disappeared into the deskdrawer, to be nearly forgotten, until the launch date of JUICE got closer. In 2019 Melissa and Martin got into contact again and decided that now the time was right for a dedicated book on Ganymede. They invited Xianzhe and Tilman to join the editorial team and the project was started.

Ganymede is the final destination of ESA's JUICE mission, which was launched on 14 April 2023, and expected to arrive at Jupiter in 2031 and entering into an orbit around Ganymede in 2034. With a full set of instruments and a functioning high-gain antenna, there will be an incredible amount of science that can be done invesigating this moon. However, in order to be prepared for the analysis of the data from this mission, and those of NASA's JUNO mission currently flying through Jupiter's system and encountering Ganymede, there needs to be a reference book about what has been done so far. This book should be that reference.

We have tried to collect the best authors to write chapters on a great variety of topics. The chapters are strongly based on the results of the successful NASA mission Galileo, but not limited to that. The authors were asked to make the level suitable as a teaching book for graduate students and a reference book for scientists.

Martin Volwerk, Melissa McGrath,
Xianzhe Jia, and Tilman Spohn
September 2023

Part I

HISTORY, ORIGIN, DYNAMICS

1

Ganymede through the Ages

MARTIN VOLWERK, KATRIN STEPHAN, ROLAND WAGNER, AND APURVA V. OZA

1.1 PROLOGUE

Ganymede is the third Galilean moon of Jupiter and the largest moon in the solar system (even larger than the planet Mercury). Through the history of humankind Ganymede has made a transition from a mythological figure to one of the most interesting objects in the solar system. In this chapter an overview will be given of the development of Ganymede through the various ages: from being a "prince abducted by a god" to being a "habitable colonized world."

Some of this may not be scientifically valid when it is dealing with the mythology around Jupiter and Ganymede in Section 1.2, and with science fiction views of Ganymede in Section 1.6. However, it is part of how Jupiter and Ganymede were and are seen.

The scientifically valid part starts with a discussion of the discovery of this moon in Section 1.3. The optical, ground-based studies of Ganymede including the early development of surface maps discussed in Section 1.4. Space-based observations and spacecraft flybys are discussed in Section 1.5, including the ever-increasing detailedness of images of the moon. The Galileo spacecraft results are only briefly mentioned, as those will be discussed further in the following chapters in this book.

Now please enjoy a trip through the ages covering the discovery and study of Ganymede.

1.2 THE MYTHOLOGY AGE

In many societies Jupiter, as a prominent bright wanderer across the skies, played an important part in their mythologies. Naturally, Ganymede, stemming from Greek mythology, is not mentioned in mythologies from other situations. Nevertheless, two other Jovian mythologies are also mentioned here.

1.2.1 Greek

Ganymede is a Greek mythological figure, mentioned for example in Homer's *Iliad*, and was the son of King Tros, the founder of Troy. He was acclaimed to be the most beautiful of mortal men:

> Now Erichthonius sired Tros, a lord of the Trojans, and Tros, in turn, had three distinguished sons: Ilus, Assaracus and Ganymede radiant as a god, and he was the handsomest mortal man on earth – and so the immortals, awestruck by his beauty, snatched him away to bear the cup of Zeus and pour out wine for all the deathless gods. (Fagles, 1990, p. 511)

Zeus (Jupiter) abducted him, disguised as an eagle, from Mount Ida near Troy (see Fig. 1.1), where Ganymede was herding a flock of sheep. Zeus carried him to Mount Olympus, where he then was made immortal and served as a cup bearer. King Tros did not go uncompensated for his loss and received fine horses from Zeus, which were delivered by Hermes (Mercury):

> "They are the very strain farseeing Zeus gave Tros, payment in full for stealing Ganymede, Tros's son: the purest, strongest breed of all the stallions under the dawn and light of day". (Fagles, 1990, p. 172)

Ganymede is also connected to the constellation Aquarius (Water Carrier), as according to some versions of the story Zeus placed Ganymede in the sky as the constellation Aquarius, next to the constellation Aquila (Eagle).

1.2.2 Chinese

In Chinese astronomy/astrology Jupiter is connected to the element wood (from the Chinese five elements philosophy) and to the Cyan Dragon (chinese: *qing* long), representing the east and spring. As one of the primary colours, cyan represents wood in Chinese philosophy, and Chinese astronomers found that Jupiter was a cyan planet in the skies.

Astronomers studied the heavens and Gan De (~400–340 BCE claimed to be the discoverer of Ganymede, see Section 1.3) and Shi Shen tabulated the orbits of the planets to high precision. Gan De writes in his "Treatise on Jupiter":

> "Every 12 years Jupiter returns to the same position in the sky; every 370 days it disappears in the fire of the Sun in the evening to the west, 30 days later it reappears in the morning to the east."

With the 12-year chronology in China, and the 12 signs of the Chinese zodiac, Jupiter was called "*Sui Xing*," star of the year. However, as its orbit is not exactly 12 years, an idealized Jupiter was entered "*Tai Sui*," the god of the year, who plays an important part in Taoism. With the 12 zodiacal signs and the 5 elements, there is a total of 60 Tai Suis. Interestingly, Tai Sui is a noble's god who only protects the emperor and kings. It is said that every year right under Tai Sui a meat-like growth can be found underground, being the incarnation of the god, which

Figure 1.1 Michelangelo's *Ganymede*. Copy after a lost original (1532) (*Wikimedia Commons* https://commons.wikimedia.org/wiki/File:Andr 37.jpg).

is taboo to dig up. Although it is also said to be a miraculous medicine that can cure all illnesses.

1.2.3 Vedic

The presence of Ganymede as one of Jupiter's abundant satellites may be present in Vedic cosmologies as far back as 800 BCE in the original astronomical text *Vedanga Jyotisha* of the ancient Indian subcontinent (Subbarayappa, 1989, p. 86). The Vedas state that Jupiter is *Brhaspati* (*spati – braha*: spirit – cosmic vastness) who is the guru (*ru – gu*: darkness – dispeller), or teacher, of all devas (gods) (Dalal, 2010).

The Ancient Indian mythology says that the devas would therefore orbit Brhaspati to gain *gnan* (cosmic knowledge), therefore although not explicitly mentioned as one of four bright satellites, Ganymede would indeed be one of 33 students (or satellites) seeking knowledge from Brhaspati.

Brhaspati as a planet is discussed more empirically from naked-eye observations at the turn of the common era in the texts *Aryabhatiya* (Aryabhata; 400 CE), *Romaka* (Latadeva; 500 CE), *Panch Siddhantika* (Varahamhira; 500 CE), *Khandakhadyaka* (Brahmagupta; 600 CE), and *Sidyadhivrddida* (Lalla, 700 CE). Consistent with European literature, no observational claim of a naked-eye detection of Ganymede is made, although speculation of *upagrahon* (satellites) is discussed in the context of Brhaspati as a continuation of the *Vedanga-Jyotisha*.

Thursday, from Proto-Germanic Thor's day (*thonaras daga*) is consistent across Indo-Germanic culture with Greek: *hemera Dios* (day of Zeus), Latin: *Jovis dies* (Jupiter day), and Hindi: *Guruvaar* (Guru/Brhaspati day).

1.3 THE DISCOVERY AGE

It is well known that early civilizations already studied the heavens and planets (see e.g. Kelley and Milone, 2011), sometimes with highly sophisticated techniques. For example, the ancient Babylonians used a "time-velocity" graph to calculate Jupiter's position, as found on cuneiform clay tables from 350 to 50 BCE (Ossendrijver, 2016).

Usually, it is accepted that the four large moons of Jupiter were discovered by Galileo after he obtained his first telescope from the Netherlands. However, there are some old Chinese sources that claim the discovery of Ganymede by a Chinese astronomer (Xi, 1981).

It is said that one of the earliest Chinese astronomers, Gan De, discovered Ganymede 2000 years before Galileo (Hughes, 1982). In the fourth century BCE Gan De studied the heavens, and in particular Jupiter (Chinese: *Sui Xing*, Year Star) and wrote two books on astronomy: *Sui Xing Jing* (*Treatise on Jupiter*) and *Tianwen Xingzhan* (*Astrological Prognostication*) in the "Warring States" period (475–221 BCE). Unfortunately, both books were lost, however, parts of them were incorporated in a later work by Qutan Xida (aka Gautama Siddha): *Kaiyuan Zhan Jing* (*The Kaiyuan Treatise on Astrology*) written between 718 and 726 CE (Xi, 1981). This work quotes from Gan De:

> "In the year of chan yan... Jupiter was in (the Zodaical Division of) Zi, it rose in the morning and went under in the evening together with the Lunar Mansions Xunü, Xü and Wei (Aquarius). It was very large and bright. Apparently, there was a small reddish (*chi*) star appended (*fu*) to its side. This is called 'an alliance' (*'tong meng'*)."

In 1973 the Mawangdui Silk Texts were discovered in Changsha, Hunan (see e.g. Silbergeld, 1982) which contained copies of the works by Gan De and Shi Shen. They stem from before 168 BCE, the year that tomb number three at Mawangdui was sealed. Their writings were gathered under the name *Divination of Five Planets*, and record the motion of Jupiter, Saturn, Venus, and other planets in their orbits between 246 BC and 177 BC (Pankenier, 1992).

The question remains whether it is, indeed, possible to observe Ganymede with the naked eye. The French astronomer Flammarion wrote as a footnote in his book (Flammarion, 1880, p. 536):[1]

> Cette découverte de Galilée montre bien qu'avant lui on n'avait pas observé les satellites de Jupiter. Cependant, d'excellentes vues les ont quelquefois distingués à l'œil nu: cette observation constitue *la plus haute épreuve* que je connaisse pour juge de la portée de la vue humaine.
> This discovery of Galileo shows indeed that before him one had not observed the satellites of Jupiter. Nonetheless, excellent eyesights have sometimes distinguished them with the naked eye: this observation constitutes *the highest challenge* that I know to judge the reach of the human eyesight.

which does not appear in the later English translation by Gore [1907].

[1] https://gallica.bnf.fr/ark:/12148/bpt6k94887w.texteImage.

Admiral William H. Smyth writes (Smyth, 1844, p. 175):

> "Certain esprits fort express surprise that Galileo should have been so gratified by this discovery, since they hold that the satellites of Jupiter are often seen by the naked eye; and they cite the Apennines, and Etna, and the West Indies, and various other fine-climate places, as the spots where such a feat is frequently done."

Xi (1981) mentions that the Beijing Planetarium set up simulations and showed that people with good eyesight would be able to distinguish a 5.5 magnitude moon at an angular distance of 5 arcseconds from a 2 magnitude planet. Thus, seeing Ganymede and Callisto with the naked eye does seem possible.

Now look at the year 1608, in Middelburg, the Netherlands, where either Sacharias Janssen or Hans Lipperhey invented the telescope. The latter patented this instrument. In 1609 Galileo Galilei obtained his first telescope and presented it to the Doge of Venice as a tool for warfare. It was probably the British mathematician/mapmaker Thomas Harriot who first used the telescope to study the heavens (e.g. Roche, 1982), and made a map of the Moon and the first drawings of sunspots and thereby determined the rotation rate of the Sun.

In 1610, then, there is the *Annus Mirabilis*, when Galileo, in March, publishes his *Sidereus Nuncius* (the starry messenger) in which the surface of the Moon, the phases of Venus, and the four large moons of Jupiter are described (see Fig. 1.2). He named the Jovian moons *pianeti Meidici* after his the family who supported him financially, the Medicis.

There are, however, contemporaneous candidates for the discovery of the moons of Jupiter – for example, Simon Mayer (Marius), a student of Tycho Brahe and Johannes Kepler, who independently discovered the four moons. Putting up the timetable of the different discoveries by Galileo and Mayer shows that the latter may have preceded Galileo by a few days in discovering the revolution of the first three moons, but discovered the fourth moon nine days later than Galileo (Johnson, 1930; Pagnini, 1930). Another candidate who is mentioned is Thomas Harriot, who also studied the Jovian system, however this dating of the discovery before Galileo is based on an incorrect interpretation of a date in Harriot's calculations, which was an ephemeris date from Galileo and not an observation (Roche, 1982).

But then there still remains one puzzle: the Frieze in the main hall, Fresoc Casa Marta, Castelfranco Veneto and the painting *The Three Philosophers*, by Giorgio da Castelfranco (Giorgione), which date back to the first decade of the sixteenth century. In the frieze, Jupiter's orbit is portrayed with three concentric circles around the planet, which can be interpreted as the orbits of Io, Europa, and Ganymede. In *The Three Philosophers*, one of the men portrayed shows part of a document that he takes out of his pocket. A detailed analysis of this document shows that Jupiter and possibly the four largest moons are shown. If the interpretation of these two paintings is correct, this would mean that Giorgione would have known about the Galilean satellites about 105 years earlier than Galileo (Keim, 2009).

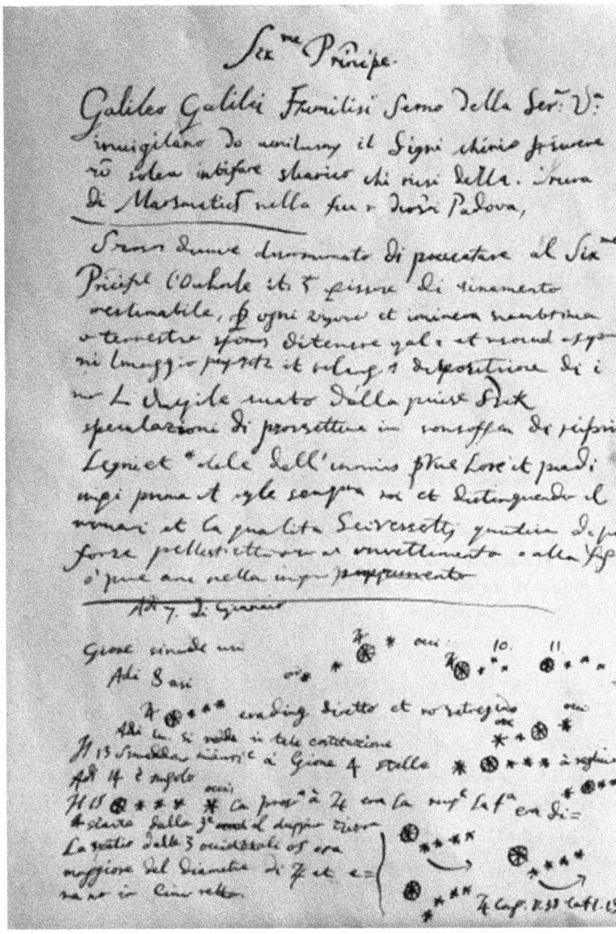

Figure 1.2 Page from *Sidereus Nuncius* by Galileo, showing the four large moons of Jupiter. Credit: University of Michigan Special Collections Library.

1.4 THE TELESCOPE AGE

One of the earliest papers about Ganymede (in NASA's astrophysics data system, ADS) deals with a solar eclipse on Jupiter created by the shadow of the third satellite (Whitmell, 1896). See Fig. 1.3 for an image of Ganymede casting a shadow onto Jupiter. A calculation of the duration of such a solar eclipse is presented for a location *J* at the equator of Jupiter and it is claimed that this point, which experiences a noon eclipse (of 41m duration), also experiences a forenoon and afternoon eclipse (both lasting 21.5m) when *J* is at an angle of $58° \, 0' \, 56''$ with the Jupiter–Sun line at the beginning of the forenoon eclipse. Various other special eclipses are discussed. Of course this is a nice exercise of celestial mechanics and does not give any insights into Ganymede. However, the situation can also be reversed: Ganymede entering Jupiter's shadow.

1.4.1 Ganymede Entering and Exiting Jupiter's Shadow

Fedtke (1942) reports on observations of Ganymede entering the shadow of Jupiter, as seen through the 13″ refractor of the Königsberger Sternwarte on two different occasions. Using the time interval that it took to completely cover Ganymede, an estimated 10 and 9m, the diameter of the moon could be

Figure 1.3 Ganymede throwing a shadow on Jupiter. Image credit and copyright: Damian Peach, Chilescope. (A black and white version of this figure will appear in some formats. For the colour version, please refer to the plate section.)

determined as 6 400 and 5 800 km (official values: $R_G = 2\,634.1$ or $D_G = 5\,268.2$ km). These values contrast with earlier determinations of Ganymede's diameter by Barnard (1895), who used Filar micrometer measures of the diameters of the solar system bodies with the 36″ refractor at the Lick Observatory. Barnard (1895) found a diameter of $3\,558 \pm 54$ miles or $5\,726 \pm 87$ km from an average angular diameter of 1.521″, a value that falls well within diameter determinations between 1829 and 1894 which range from 1.488″ to 1.783″.

A satellite entering the shadow of its host will have less insolation and cool down. Murray and colleagues (1965) studied the 8–14 μm emissions from Ganymede, as well as the optical signal, as Ganymede entered the Jovian shadow on 1963 December 12. Observing the eclipse they found that the cooling or reheating of Ganymede's surface takes less than 15 minutes, with the bulk of the infrared emissions dropping off in less than 10 minutes. This made them conclude that the outermost few millimeters of Ganymede's surface have a thermal inertia as low as that of the Moon, and should be composed of "a fine dielectric powdered material devoid of any gas phase." They also state that it would be interesting to investigate if there is possibly a tenuous atmosphere at Ganymede.

Morrison and colleagues (1971) built further on these results, in which, for the first time, infrared (radiometric) observations at 20 μm wavelength with the, then-new Mauna Kea Observatory in Hawaii were presented. This work provided a more reliable measurement of the thermal inertia of Ganymede's surface than had previously been possible with observations only with a broadband filter at 8–14 μm wavelength. This work was later extended to the other Galilean satellites and Titan (Morrison et al., 1972).

Intermission: Recollections of a Contemporary

Dale Cruikshank narrates:

The discovery of ices on the Galilean satellites is usually credited to Gerard Kuiper. He was the pioneer infrared astronomer in the world, observing many stars and solar system bodies with the near-infrared spectrometer he built with infrared detectors that were developed during World War II and later declassified. He published an abstract in which it was stated that the reflectances of Europa and Ganymede "are most readily explained by assuming that II (Europa) and III (Ganymede) are covered by H_2O snow" (Kuiper, 1957). However, there were no figures shown in this abstract. He also suggested that because Ganymede is darker than Europa, the snow on its surface might be contaminated with silicate dust.

In 1965, Moroz, who pioneered infrared astronomy in the USSR, published spectra of the Galilean satellites and interpreted the Ganymede spectrum as showing H_2O ice. The 1965 paper in Russian was published in English in 1966 (Moroz, 1966).

As the attention of planetary astronomers was drawn to the Galilean satellites by subsequent work, it was always Kuiper who received credit for the discovery of H_2O ice, although he had only published a brief verbal description. Moroz published spectra, and he was very upset and irritated that he did not get credit for the discovery. He expressed this to me during my one-and-a-half years of work with him in Moscow and Crimea, and subsequently in a letter after I returned to the United States.

Kuiper was apparently aware of this tension, and finally published his spectra in 1973 in the Communications of the Lunar and Planetary Laboratory (Kuiper, 1973). Kuiper died later that year. Moroz, who died in 2004, was never satisfied with the situation.

New observations of Ganymede with the infrared spectrometer of Moroz, conducted with the Shajn 2.6 m telescope at the Crimean Astrophysical Observatory confirmed the presence of H_2O ice with higher quality data than Moroz had previously published. Ludmila Gromova, a graduate student at that time (who later married and published under the name Ludmila Zasova), and I, as a visiting American astronomer, together with Moroz published the results (Gromova et al., 1970).

In terms of physical studies of Ganymede, Morrison and colleagues (1971) built on the earlier work of Murray and colleagues (1965), which the authors cite, but for the first time this new work presented infrared (radiometric) observations at 20 μm wavelength. The ability to observe at 20 μm was unique to the then-new Mauna Kea Observatory in Hawaii (elevation 4.2 km), and this paper provided a more reliable measurement of the thermal inertia of Ganymede's surface than had previously been possible with observations only with a broadband filter at 8–14 μm wavelength.

The observations at 20 μm were extended to the other three Galilean satellites and Titan (Morrison et al., 1972).

1.4.2 Ganymede's atmosphere and Surface

After a successful discovery of an atmosphere on Io, Binder and Cruikshank (1964) used photometric measurement taken with the Kitt Peak 36″ telescope to search for an atmosphere on Europa and Ganymede. This is done through searching for an anomalous brightening of the moon after it comes out of eclipse, which can be caused by a frost layer on the surface or a haze over the surface created by a surface temperature drop. Binder and Cruikshank (1966) found at Europa (or JII) a brightness anomaly of 0.03 ± 0.01 magnitudes, which decayed in $\sim$10 minutes. However, for Ganymede (or JIII) no brightening beyond 0.01 magnitudes was measured. This can be caused by the fact that it takes Ganymede longer, $\sim$20 minutes, to exit the shadow of Jupiter and any frost or haze may have sublimed away before it can be noticed as an anomalous brightness.

The presence of (water) frost on the surfaces of the Galilean satellites had already been suggested by Kuiper (1952). Measuring the reflectivity of JII and JIII (Europa and Ganymede) Harris (1961) and Moroz (1966) found that it was lower at 2 μm than at 1 μm, something that is not uncommon for ices (see e.g. Kieffer, 1970).

Using the 60″ McMath telescope at Kitt Peak National Observatory, combined with a scanning Fourier spectrometer, Pilcher and colleagues (1972) measured the reflectivity of the Galilean moons for both the leading and trailing side. They found that the absorption bands in the reflected light, caused by the presence of water frost on the surface of JII and JIII, could be well established, where Ganymede had weaker absorption than Europa. When the geometric albedo of the water frost is assumed to be 0.5 and 0.35 for the underlying material (silicates) they arrive at a frost covering of $\sim$70% of JII and 20–65% of JIII. For a detailed discussion of Ganymede's surface and hydrosphere see Chapters 11 (Ahrens et al.) and 12 (Kalousova et al.) in this book.

It took until 1972 before a successful measurement of Ganymede's atmosphere was made through an occultation of the star SAO 186800 on June 7. The observations showed nonabrupt immersions and emersions of the star, which means that there is a dimming atmosphere surrounding the moon. The surface pressure of the atmosphere was estimated to be greater than 10^{-3} mbar. Through these measurements, also the radius and average density of Ganymede could be determined as $5\,270^{+30}_{-200}$ km and $2.0^{+0.2}_{-0.03}$ g/cm^{-3}, respectively (Carlson et al., 1973).

Combining the results about the water frost on and the atmosphere of Ganymede, led Yung and McElroy (1977) to look at the possibility of an oxygen atmosphere at Ganymede. Water vapour can be dissociated through radiation at wavelengths below 2 800Å, and a series of reactions can take place in Ganymede's atmosphere, leading to the creation of O_2, which in turn can be photo-dissociated again into atomic oxygen that can escape into Jupiter's magnetosphere. They found that there can be an O_2 atmosphere around Ganymede with a surface pressure of 10^{-3} mbar, in agreement with the estimate above. A full discussion of Ganymede's tenuous atmosphere is given in Chapter 15 (Roth et al.) of this book.

After the Pioneer 10 and 11 missions to Jupiter in 1973 (see Section 1.5) interest in planetary satellites got a big push, as evidenced by the review papers by Morrison and Cruikshank (1974) and Mendis and Axford (1974). Sodium emissions from Io were discovered by Brown and Chaffee, Jr. (1974), and Trafton and colleagues (1974) showed that this was not just reflectance from the surface, but that there exists an extended cloud of sodium around Io. Lanzerotti and colleagues (1975) used the vertical spectrograph of the McMath solar telescope to scan the three innermost moons of Jupiter around the Na D I and II lines. They concluded that there was no significant evidence for emission from Ganymede.

Surface reflectance in the 0.5 to 5 μm range can be used to find the temperature of the icy surface of the moons. Using infrared observations of Ganymede, made with the 61″ Catalina Observatory telescope (Fink et al., 1973) and comparing the absorption bands in the reflectance with spectra obtained in the laboratory, Fink and Larson (1975) obtained a surface temperature for Ganymede of 103 ± 10 K. Clark and McCord (1980) studied spectra of Ganymede in the 0.65–2.5 μm band and concluded that the water absorption features are caused by free water on the surface, that is, not bound in minerals. And in a further study, comparing the spectra with laboratory reflectance studies of water frost, water ice, and water and mineral mixtures, Clark (1980) concluded that "the spectra of the icy Galilean satellites are characteristic of water ice (e.g., ice blocks or possibly very large ice crystals $\geq$ 1 cm) or frost on ice rather than pure water frost, and that the decrease in reflectance at visible wavelengths is caused by other mineral grains in the surface." See also Chapters 7 (Pappalardo et al.), 9 (Schenk et al.), 10 (Stephan et al.), and 11 (Ahrens et al.) in this book for more in-depth discussions on these topics.

1.4.3 Ganymede's Surface Maps

With the advancement of telescopes, it became possible to make images of the surface of Ganymede. Cassini, in 1665, claimed he could see markings on the Jovian satellites, although the crude telescopes of that time would actually render this impossible according to Schaeberle and Campbell (1891). Herschel (1797), in 1794–6, noticed brightness changes of the satellites and as this was periodic, he deduced that they rotate around their axis once per revolution (see also Chapter 3 (Stark et al.)). And in 1796, Schröter (1800) observed a dark marking, not unlike a shadow that a satellite would cast on the surface, on the northern hemisphere.

Some drawings with little detail of the surface of Ganymede were made by Schaeberle and Campbell (1891). Better results were obtained by William Pickering, using the Lowell observatory in 1894, who published a plate with 42 drawings of Ganymede at different longitudes, see Fig. 1.4(b) (Pickering, 1900).[2]

Camichel and Lyot (1943)[3] wrote about the difficulty of observing the surface of the Galilean satellites:

[2] http://cdsads.u-strasbg.fr/historical.html.

[3] Note that there are different references to this paper. This is the NASA-ADS reference. On the first page of this paper authorship is given to "MM. Camichel, Gentili et Lyot," and in other papers it is given to "MM. Lyot, Camichel et Gentili."

> *En dehors des passages, les bonnes observations des satellites de Jupiter sont rares. Une seule série importante des dessins de longitudes divers avait été effectuée, en 1894, à l'Observatoire Lowell, par William H. Pickering.*
>
> Apart from the transits, good observations of the Jovian satellites are rare. One single, important series of drawings for different longitudes was carried out by William H. Pickering at the Lowell Observatory in 1894.

In 1941, observations at the Observatoire au Pic du Midi (Camichel and Lyot, 1943) revealed dark and bright patches on the surface of Ganymede. From the full set of observations a map was created showing a dark band along the equator and bright patches at the poles, see Fig. 1.4(f). Based on the recurrence of the observed structures they concluded that Ganymede rotates around an axis perpendicular to its orbital plane with a period practically equal to its revolutionary period. They also found that within a longitude window between 312° and 19° there exists a "vivid brightness" (*vive blancheur*) which does not rotate with the moon, and posed the question whether this is a deposit of highly volatile material that the solar radiation makes disappear.

One year later Danjon (1944) commented on the dearth of observations of the Jovian satellites and that it would be good to add some more in order to study these interesting objects, and presented a set of drawings of Ganymede's and Callisto's surfaces created in 1934. Danjon's observations show similar dark spots as presented by Camichel and Lyot (1943), with two prominent spots on the northern and one prominent spot on the southern hemisphere, see Fig. 1.4(g). Danjon (1944) also discusses the difference between his observations and those of Camichel and Lyot – some spots were not seen and some were at a different longitude. Commenting on the global map, Danjon writes:

> *Il me semble que MM. Lyot, Camichel et Gentili accentuent et schématisent un peut trop, dans le leur, les deux bandes parallèles à l'équateur de Ganymède, à 30° N. et S. Elles n'ont pas cette netteté géometrique sur leurs dessins et nous lisons du reste (page 57 de leur article) qu'elles sont interrompues. Il n'y a pas lieu, à mon avis, de voir là des véritable bandes.*
>
> It looks to me that Messrs. Lyot, Camichel and Gentili accentuate and schematize a bit too much, in their paper, the two parallel bands to Ganymede's equator at 30° N. and S. They do not have this geometrical sharpness as in their drawings and one reads later (page 57 in their paper) that they are interrupted. It is misplaced, in my opinion, to see real bands there.

In the early 1950s further observations of Ganymede were published in *The Strolling Astronomer*. Hare (1951) discusses several observations, for example by Cave (Fig. 1.4d), with smaller telescopes that all agree on brighter polar caps and possibly some darker structures near the equator.

Reese (1951) presented a drawing of Ganymede's surface that he based on observations by Brinckman, Cave, Cragg, Haas, Oberndorfer, and Sandner using telescopes with apertures of 4 to 12″. Assuming a tidally locked rotation of the moon and a zero longitude at superior conjunction, the result is shown in Fig. 1.4(i). This map seems to accord well around the equator with the one from Danjon (1944), but there is definitely less detail near the north and south pole.

Both (1952) praised the map by Reese for its details, but presented a new composite map created by Howe, Fig. 1.4(j), who used the three earlier maps Figs. 1.4(f)–(h), an unpublished map by Howe (1951), and drawings by Barnard and Antoniadi (undated), which is really a milestone in details. There are, of course, differences in the various maps, however, the dark band around the equator seems to be a common feature in all. Also, the two dark spots at longitude 160° to 220° latitude and 0° to 0° N and at 260° to 320° and latitude 0° to 60° N can probably be accepted as real (Both, 1952).

Further information on the surface of Ganymede is given in Chapters 7 (Pappalardo et al.) and 9 (Schenk et al.). The latest composite surface maps are shown in Appendix A (Roatsch et al.).

With the launch of satellites like Pioneer 10 and 11 (1972/73), Voyager 1 and 2 (1977), Galileo (1989), and the Hubble Space Telescope (HST) (1990), whole new windows opened for the exploration of Ganymede in the Space Age. However, the development of ground-based telescopes also continued, with larger and larger mirrors and of course adaptive optics (see e.g. Hart, 2010, and references therein). Dekany and colleagues (2013) show that images of Ganymede taken by the PALM-3000 telescope and the visible fast-frame imaging camera TMAS using adaptive optics, Fig. 1.5 (left) now obtain even better seeing than the HST, Fig. 1.5 (right).

1.5 THE SPACE AGE

1.5.1 The Pioneer Mission

In the beginning of the 70ies of the last century, the first spacecraft were sent to the Jupiter system in a time, when only 12 moons of the largest planet in the solar system were known. These two identical twin spacecrafts were termed Pioneer 10 and 11 (Pioneer F and G, originally Miles, 1973; Hall, 1975). Both had a mass of 258 kg and were equipped with 11 identical instruments, one additional instrument was added to Pioneer 11. Pioneer 10 and 11 were mainly designed for physical measurements, such as studies of cosmic rays, magnetospheric fields, temperatures, the solar wind, and the impact of small meteoroids and interplanetary dust particles, rather than imaging Jupiter and/or its satellites. Both spacecraft rotated several times per minute (at 4.8 rpm) around an axis aligned toward Earth, a configuration not well suited for imaging sensors. The imaging sensor aboard both Pioneer spacecraft is the imaging photopolarimeter (IPP) (Pellicori et al., 1973). It was preferentially used for imaging details and so far unknown features in the Jovian atmosphere.

[4] Reprinted from Murray [1975] with permission from Elsevier.

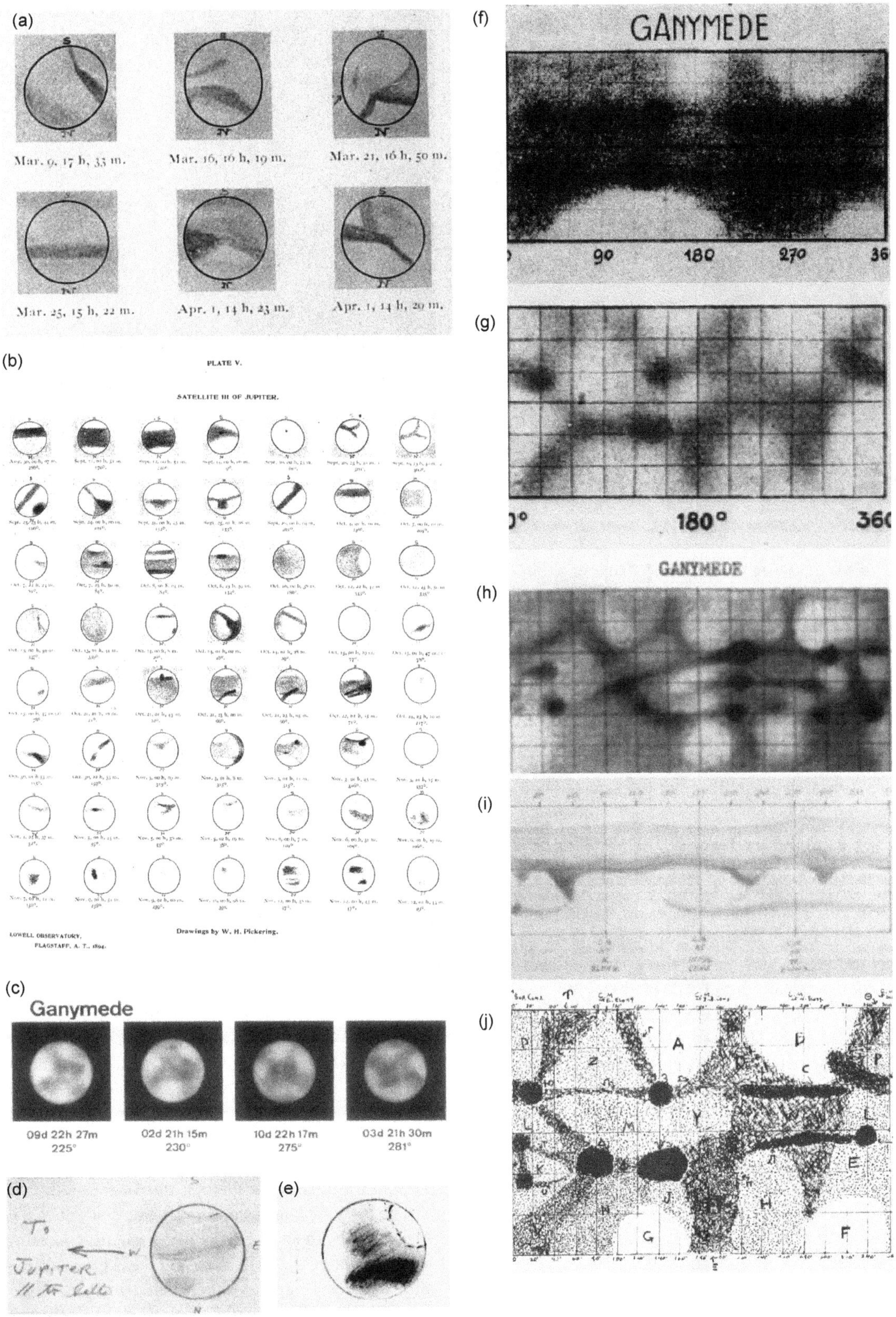

Figure 1.4 Drawings (left) and historical maps (right) of Ganymede's surface. (a): Douglas, A.E. (2006). Drawings of Jupiter's third satellite. *Astronomische Nachrichten*. John Wiley and Sons. Lowell Observatory Archives. (b): Pickering (1900), (c): Murray (1975), reprinted from *Icarus*, 25(3). J.B. Murray, New observations of surface markings on Jupiter's satellites, 397–404, ©1975, with permission from Elsevier.[4] (d): Cave in Hare (1951), (e): Cruikshank (1963), (f): Camichel and Lyot (1943), (g): Danjon (1944), (h): Lyot (1953) (i): Reese (1951), (j): Howe in Both (1952). Note that in the maps on the right −90° latitude is at the top of the image.

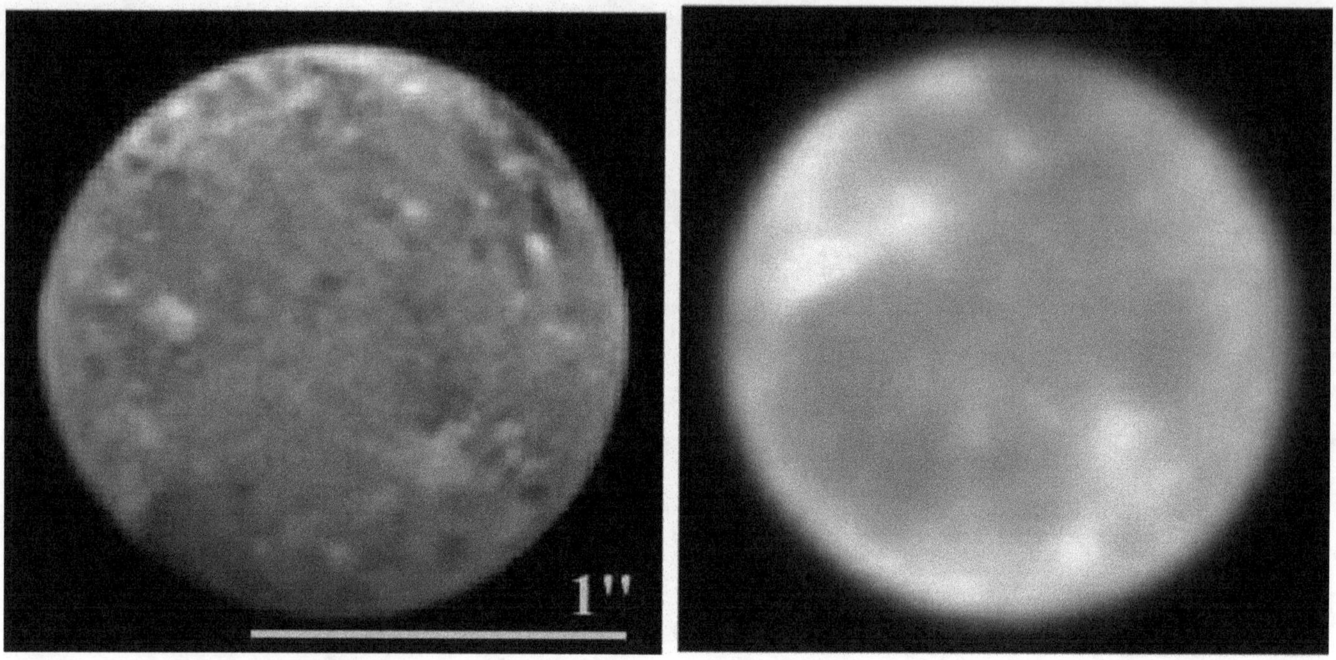

Figure 1.5 Left: PALM-3000 and TMAS false-color image of Ganymede using the Johnson-Cousins BRI filters. The pixel sampling is ~35 km in this image. Right: Hubble Space Telescope false-color image of Ganymede for comparison (NASA image). Figure taken from Dekany and colleagues (2013), a color version can be obtained from the online journal.

Pioneer 10 was launched on 1972 March 2 and provided a series of "firsts": the first spacecraft to fly beyond Mars, the first spacecraft to fly through the main asteroid belt, and the first one to reach the orbit of Jupiter and to continue further into the outer solar system. Pioneer 10 proved for the first time that spacecraft can safely travel through the asteroid belt without being damaged. Only a comparably small number of hits by dust particles were recorded during this passage between July 1972 and February 1973. During its closest approach to Jupiter on 1973 December 4, Pioneer 10 passed the cloud tops at a distance of 130 354 km. The IPP instrument took ~500 images of the Jovian atmosphere at a spatial resolution of ~300 km/pxl. It also took the first images of the three icy Galilean satellites Callisto, Ganymede, and Europa, at that time providing the clearest images of dark and bright surface features on these bodies (see Gehrels, 1977). However, the spatial resolution was not good enough for detailed geoscientific investigations (see Fig. 1.6) compared to standards of more recent missions. The last signal from Pioneer 10 was received on 2003 January 23, from a distance of 12.23 billion km from Earth. The current distance of Pioneer 10 is 18.9 billion km. The spacecraft is heading towards α Tau (Aldebaran), which it will pass in ~2 million years.

Pioneer 11 was launched on 1973 April 6, and passed Jupiter closer than Pioneer 10, at a distance of 42 828 km above the atmosphere. After the flyby, Pioneer 11 continued its journey to Saturn, which it flew by on 1979 September 1, in a year which had seen the first successful flybys of the two Voyager spacecraft at Jupiter. The decision was made by NASA to end the mission on 1995 September 30. Final contact from the spacecraft was received two months later, on 1995 November 24. Pioneer 11 is heading out of the solar system in a direction opposite to that of Pioneer 10. The spacecraft is currently in constellation Scutum and will pass a K-type dwarf star, within 0.8 light years, in 928 000 years. Its current distance (2023 January) is 19.7 billion km from Earth.

1.5.2 The Voyager Mission

In the late 1960s, NASA conceived an Outer Planets Grand Tour to the four large planets Jupiter, Saturn, Uranus, Neptune, as well as to Pluto, the latter at that time still listed as a planet (e.g. Bagenal et al., 2004). The plan was to launch two spacecraft in 1976–7 to reach Jupiter, Saturn, and Pluto, and another dual launch in 1979 to reach Jupiter, Uranus, and Neptune, using gravity assist at each of the four large planets. Constraints in budget severely affected NASA's space program in the early 1970s, however, and the ambitious Grand Tour was cancelled. Then, a modified cheaper mission was redesigned with two Mariner-based spacecraft only, termed Mariner Jupiter Saturn, which eventually became Voyager 1 and 2 in 1977. Due to a planetary alignment of the four large planets, which takes place only once every 176 years, it was possible to send at least Voyager 2 on to Uranus and Neptune, meaning that most of the task of the cancelled Grand Tour could be fulfilled.

The two Voyager spacecraft were identical, each with a mass of 825 kg spacecraft and 11 scientific instruments aboard (Bagenal et al., 2004). Among the 11 scientific instruments, the Voyager Imaging Experiment (ISS) consisted of a wide angle (WA) and a narrow angle (NA) Vidicon camera (Smith et al., 1977). The task of the ISS instrument was to perform detailed investigations of atmospheric features of Jupiter and Saturn, of the rings of Saturn, and imaging of the surfaces of the major Jovian and Saturnian satellites at spatial resolutions of 1 km/pxl or higher (Smith et al., 1977), as shown in Fig. 1.6.

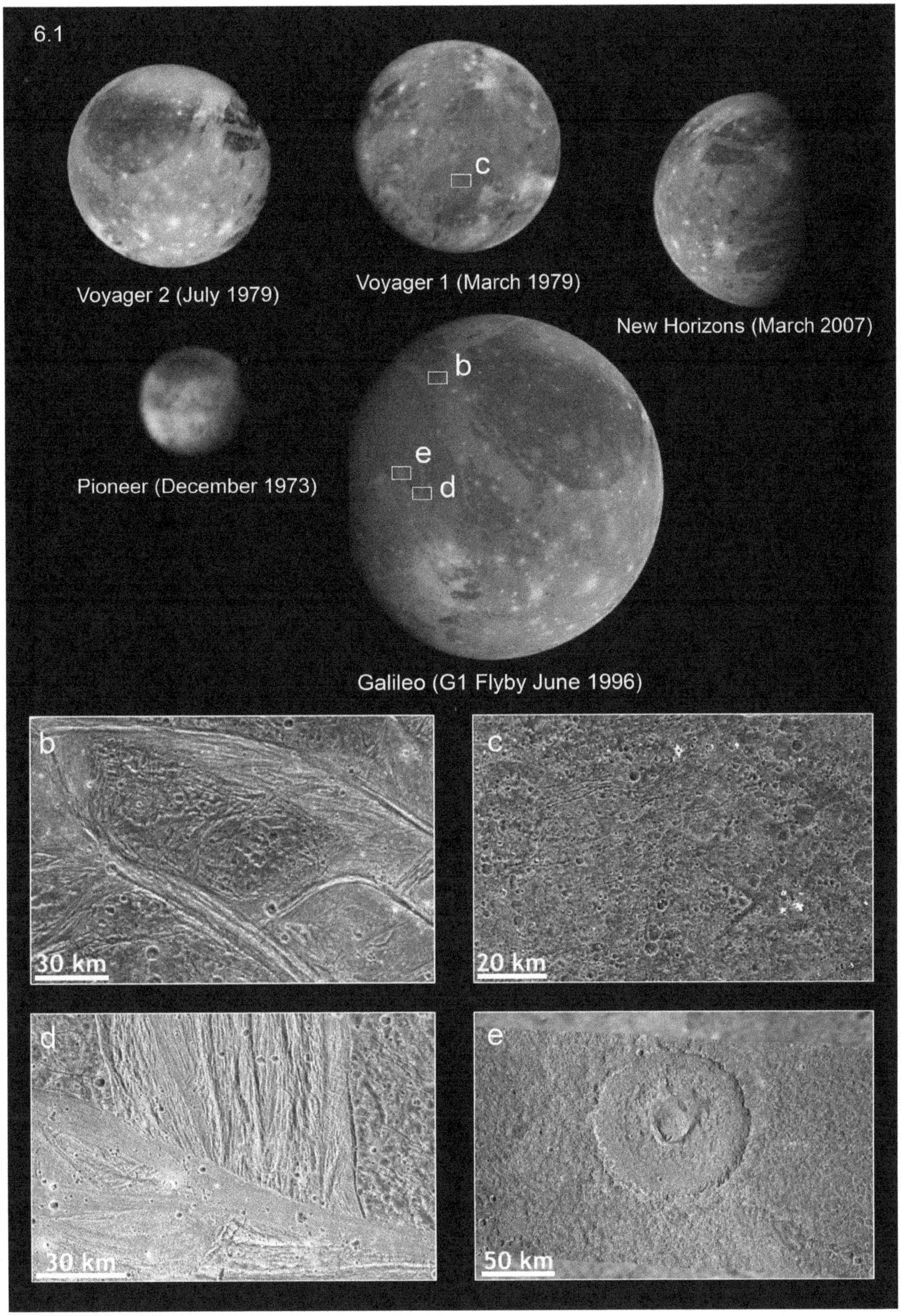

Figure 1.6 Global and detailed views of Ganymede, imaged in various missions to Jupiter and beyond. (a) Global views from the Voyager flybys in 1979; the narrow angle (NA) Vidicon camera aboard Voyager 2 imaged the anti-Jovian hemisphere, the NA camera aboard Voyager 1 the sub-Jovian hemisphere. The large approximately circular dark area is Galileo Regio. The anti-Jovian hemisphere, with the dark terrain of Galileo Regio, was also captured by the photo-polarimeter aboard Pioneer 10 during its flyby in December 1973, which provided the first images of Ganymede and the other Galilean satellites ever taken. These images are about a factor 100 less in spatial resolution compared to global Voyager images. On its cruise to Pluto, the New Horizons spacecraft passed Jupiter and its satellites in March 2007, imaging the leading hemisphere of Ganymede with the LORRI camera. (b–e) Global and detailed images from the Galileo mission (1995–2003). In the first Ganymede flyby (G1) of its orbital tour, the Galileo solid-state camera (SSI) imaged the anti-Jovian hemisphere. Detailed images from Ganymede's major terrain types at higher resolutions (order of 150–500 m/xl) were taken during further close flybys of the Galileo orbiter (see chapters on geology and stratigraphy in this volume). (b) Older, densely cratered dark terrain and younger, bright, tectonically resurfaced terrain (orbit G2). (c) Dark terrain of Nicholson Regio (orbit G28). (d) Tectonic resurfacing in bright terrain of Erech Sulcus (orbit G8). (e) Central pit crater Melkart (orbit G8). The locations of these detailed views are indicated by the labeled rectangles in the global views. Due to the differences in resolution, the sizes of the rectangles do not exactly render the areas shown in the four detailed panels

Voyager 2 was launched first, on 1977 August 20, followed by the launch of Voyager 1 on 1977 September 5. During the Jupiter encounter of Voyager 1 on March 5, 1979 and of Voyager 2 on July 9, 1979, the ISS instrument returned spectacular images not only from Jupiter but especially from the surfaces of the four Galilean satellites Io, Europa, Ganymede, and Callisto, and also of the smaller satellite Amalthea (Smith et al., 1979a, b). The trajectories of the two Voyager spacecraft were chosen such that different hemispheres of the satellites could be imaged during different flybys. Voyager 1 captured the sub-Jovian hemisphere of Ganymede, Voyager 2 the hemisphere facing away from Jupiter. Like Pioneer, the Voyager encounters at Jupiter provided some discovery "firsts" not anticipated prior to Voyager: (1) a faint Jupiter ring; (2) two previously unknown satellites (Metis and Thebe); (3) intense, tidal-driven active volcanism on the innermost Galilean satellite Io; (4) two different terrain types on Ganymede, dark and bright regions; (5) dark densely cratered icy surfaces on Ganymede and Callisto; (6) widespread tectonic features in the bright terrain on Ganymede and on Europa; and (7) surface features suggesting ice volcanism (cryovolcanism) on Europa and, to a lesser degree, on Ganymede (Smith et al., 1979a, b). Subsequent to the two Jupiter encounters, more spectacular images were returned from Saturn by Voyager 1 (1980 November 12) and 2 (1981 August 25), and by Voyager 2 from Uranus (1986 January 24) and Neptune (1989 August 25).

Since 1990 January 1, the Voyager mission has been named the Voyager Interstellar Mission. As of January 2023 Voyager 1 is 159.3 AU or 23.8 billion km from Earth and the farthest-distant man-made object. It is moving into the direction of star AC + 79 3888, also known as Gliese 445, in the Camelopardalis constellation, which it will pass within 1.6 light years in $\sim$40 000 years. Voyager 1 is expected to be out of power in about 2025. Voyager 2 is (as of January 2021) 126.8 AU or 18.9 billion kilometers away from Earth and will pass star Ross 248 within 1.7 light years in $\sim$40 000 years, and Sirius within 4.3 light years in 296 000 years.

1.5.3 The Galileo Mission and Beyond

Most of the knowledge we have about Ganymede today results from NASA's Galileo mission. In 1995, the Galileo spacecraft entered orbit around Jupiter and absolved four close encounters with Ganymede during the primary mission (1995 to 1997). During first flybys, radio-science and other data revealed that Ganymede had a fully differentiated interior with a core, a silicate mantle, and an ice crust (Anderson et al., 1996). The plasma wave experiment and magnetometer data gave evidence of an internally generated magnetic field forming a magnetosphere around Ganymede (Kivelson et al., 2002, and Chapters 13 (Christensen et al.) and 14 (Kivelson et al.) of this book). For the first time, Galileo acquired a large number of UV spectra of Ganymede's surface showing that Ganymede has an extended hydrogen exosphere, SO_2 and O_3 in the surface material through magnetospheric alteration (Hendrix et al., 1999). In addition, near-infrared hyperspectral images (Carlson et al., 1996) enabled for the first time the combined investigation of the satellite's surface composition with geology as seen in Galileo SSI images and revealed the presence of several non-ice compounds on the surface of Ganymede including also radiolytic products of the interaction between Ganymede's surface and magnetosphere (McCord et al., 1998, and Chapters 10 (Stephan) and 15 (Galli et al.)).

Galileo's Europa Mission (GEM) ran from 1997 December 8 to 1999 December 31. Its science objectives were not limited to Europa, but also included analyses of the other satellites, as well as of Jovian fields and particles and atmospheric characteristics. Galileo's Millennium Mission (GMM) ran first from January 2000 through March 2001 and was then extended until the end of mission operations in January 2003. The GMM conducted additional investigations into the dynamics of Ganymede's unique magnetosphere. Galileo's closest approach to Ganymede coincided with that of the Cassini spacecraft, which passed Jupiter in December 2000 within 10 million km in order to pick up speed through a gravity assist on its way to the Saturnian system. Joint Galileo–Cassini observations revealed solar wind effects and magnetospheric dynamics. High-resolution Ganymede images were also taken. Magnetometer field measurements suggest that a salty water layer exists between Ganymede's ice shell and a deep layer of a high-pressure polymorph of ice (Kivelson et al., 2002, and Chapter 14 (Kivelson et al.)). Galileo met its demise in September 2003, when, by design, its trajectory took it on a collision course toward Jupiter and it burned up in the planet's atmosphere.

On its way to Pluto and Charon, the New Horizons spacecraft observed Jupiter's icy satellites, Europa and Ganymede, during its flyby in 2007 and recorded topographic and compositional mapping at visible and infrared wavelengths and that could not be observed properly during the Galileo mission of Ganymede's Jupiter-facing hemisphere (Grundy et al., 2007). The most recent observations of Ganymede's polar regions were made in December 2019, when the Juno spacecraft flew by Ganymede during its 24th orbit of Jupiter at a range between 97,680 to 109,439 kilometers (60,696 to 68,002 miles) (Mura et al., 2020). Juno will continue to observe the Jovian system with multiple rendezvous planned for Ganymede, Europa, and Io through its just extended mission through September 2025, or until the spacecraft's end of life.

In addition to the spacecraft missions, observations by the HST had a major impact on Ganymede science, particularly through UV imaging and spectroscopy. Launched in 1990, HST was placed into orbit around Earth outside the distortion of Earth's atmosphere at about 547 kilometers (340 miles) above the planet's surface and is still in operation with the ability to record images at ultraviolet to near-infrared wavelengths. HST made critical measurements of Ganymede's atmosphere (also Chapter 16 (Roth et al.)) showing that Ganymede also exhibits auroral emission, first observed by Hall et al. (1998) with the Goddard High Resolution Spectrograph (GHRS) on board the HST. These observations constrained the fluxes and spectral shapes of the OI 1356 Å and OI 1304 Å emission lines, which imply a molecular oxygen atmosphere with a column density in the range of $(1–10) \times 10^{14}$ cm^{-2} (Hall et al., 1998).

The line shape of the OI 1356 Å emission is consistent with the radiation being emitted from two circumpolar auroral ovals situated in the north and south polar regions of Ganymede (see also Chapter 18 (Gunell et al.)). Subsequent spatially and spectrally resolved HST observations with the Space Telescope Imaging Spectrograph (STIS) clearly demonstrate the existence of two auroral ovals around Ganymede's magnetic north and south poles as shown in the work of Feldman et al. (2000) and McGrath et al. (2013). In March 2015, using Hubble to study the motion of its aurorae during a time span from 1998 to 2007, the researchers determined that a large saltwater ocean was helping to suppress the interaction between Jupiter's magnetic field and that of Ganymede. The ocean is estimated to be 100 km (60 miles) deep, trapped beneath a 150 km (90 miles) ice crust (Saur et al., 2015).

1.6 THE SCIENCE FICTION AGE

The human exploration of Ganymede has already advanced much further in science fiction, as is to be expected. Human colonization, native Ganymedians and surface conditions that vary from icy, to sub-arctic periodially flooded by tidal action, to being volcanically active and jungle covered, are some of the ideas presented. A short, incomplete discussion of science fiction literature and movies dealing with Ganymede is presented below. The actual status of Ganymede'a habitability is discussed by Chela-Flores in Chapter 20 of this book.

1.6.1 Books

One of the first stories involving Ganymede is the 1938 short story "*Tidal Moon*" by Stanley and Helen Weinbaum, in which, in 2083, human companies are collecting "cree" that is grown by Ganymede's natives, the Nympus, a red moss that turns blue due to the ammonia in Ganymede's atmosphere. On Earth it is used as a combined anaesthetic and medicine. Although it was already known that the moon was tidally locked in its orbit (see Section 1.4.3) in this story the moon has a 6 month rotational period with a sub-arctic climate and large bodies of water. Every three months each part of Ganymede's surface is flooded because of Jupiter's tidal force and only the human settlement "Hydropole" at the south pole stays dry.

The books will be categorized here into two, sometimes overlapping, groups: Ganymede with indigenous inhabitants and Ganymede as a human colony, sometimes used for agriculture.

1.6.1.1 Indigenous Inhabitants

As in *Tidal Moon*, Ganymede has indigenous inhabitants in Isaac Asimov's "*Christmas on Ganymede*" (1941) called the "Ossies" because they resemble OstricheS, who are used as a working force for mining for example wolframite. Introducing the Ossies to Santa Claus causes problems for the human colonists and the mining company.

Not all Ganymedians are collaborating with human visitors. In "*The Counterfeit Man*," a short story by Alan E. Nourse (1963) shapeshifting malicious aliens get onboard the spacecraft returning from Ganymede to Earth, taking over (at least) the captain of the ship after landing on Earth. In Philip K. Dick's "*The Ganymede Takeover*" (1967) the Earth is taken over by Ganymedian worm-like creatures.

Not always are the indigenous inhabitants from Ganymede. In Isaac Asimov's short stories "*Not Final!*" (1940) and "*Victory Unintentional*" (1942) human colonists get into conflicts with the inhabitants of Jupiter. Whereas, in "*The Gentle Giants of Ganymede*" by James P. Hogan (1978) a crashed spacecraft is discovered in which members of an alien race are found, who actually come from a distroyed planet between Mars and Jupiter.

1.6.1.2 Human Colony

In Robert Heinlein's novels Ganymede is often mentioned to be a human colony. In "*Farmer in the Sky*" (1953) there is an agricultural colony, and in "*I Will Fear No Evil*" (1970) a commission proposes to terraform Ganymede. Also in Poul Anderson's "*The Snows of Ganymede*" terraformers are sent to this moon to survey its possibilities, only to be thwarted by an already existing human colony.

After the Earth is devastated by war, the protagonists in "*The Ganymede Club*" (1995) by Charles Sheffield, try to solve a mystery/conspiracy on Ganymede. Therapist Lola Belman meets a patient whose past is a mystery, and finds a dangerous group will stop at nothing to keep Lola from exploring the past and discovering their existence.

In the book/TV series *The Expanse* by James S. A. Corey, Ganymede is colonized and used both as a mining and an agricultural resource. "*Caliban's War*" (2012) describes that on the surface of Ganymede there are agricultural domes on the surface which are illuminated by mirrors in space.

The most rigorous colonization of Ganymede takes place in "*The Marriage of the Living Dark*" (1997), the last book in the "*Chung Kuo*" series by David Wingrove. All Galilean moons are turned into spaceships and leave the solar system for most nearby stars. This way they flee from the warring factions on the Earth where the Chinese great experiment, the Ten Thousand Year Empire of the Han, has failed.

1.6.2 Movies and TV

Ganymede also gets used in TV series and movies As mentioned above, it plays an important part in the book/TV series *The Expanse* (2015–present).

The German (TV) movie *Operation Ganymed* shows the return of remaining astronauts after their multiyear mission to the Jovian system, where they found primitive life on Ganymede. However, they return to a (postnuclear war?) desolate Earth.

On Ganymede a ship was buried under the surface by "the Shadows" in the TV series *Babylon 5* (1993–1999). It was dug up in the episode *Messages from Earth* (1996) leading to dire consequences in the constantly further developing Shadow War.

In the Australian children's science fiction drama *Escape from Jupiter* (1994–95) and its sequel *Return to Jupiter* (1996) Ganymede is home of the parents of the protagonists and a mining colony.

The BBC TV series *Red Dwarf* (1988–present) had Ganymede terraformed and apparently it is one of the more fancy of the Galilean moon colonies.

1.7 EPILOGUE

This is the end of the time travel through the discovery and study of Ganymede. From a mythological figure, it has developed to one of the most interesting objects in our solar system.

A moon with an internal magnetic field (see Chapters 13 (Christensen et al.) and 14 (Kivelson et al.)), with possibly a liquid water layer between two ice layers (Chapter 12 (Kalousova et al.)) embedded in and interacting with the greater Jovian magnetosphere (Chapters 15 (Galli et al.), 18 (Gunell et al.) and 19 (Bonfond et al.)). This is the perfect target for a new space mission: JUICE (Grasset et al., 2013), which was launched on 14 April 2023.

This chapter should have given the reader a taste of what was, and for deeper immersion into the study of Ganymede the following chapters in this book are recommended.

ACKNOWLEDGMENTS

Part of this work was conducted at the Jet Propulsion Laboratory, California Institute of Technology, under contract with NASA.

REFERENCES

Anderson, J. D., Lau, E. L., Sjogren, W. L., Schubert, G. and Moore, W. B. (1996), 'Gravitational constraints on the internal structure of Ganymede', *Nature* **384**, 541–543.

Anderson, P. (1958), *The Snows of Ganymede*, ACE Books, Double Novel Books D-303, New York, USA.

Asimov, I. (October 1941), Not Final, in: Astounding Science Fiction, Stanley Schmidt, USA.

Asimov, I. (January 1942), Christmas on Ganymede, in: Startling Stories, Ned Pines' Standard Magazines, USA.

Asimov, I. (August 1942), Victory Unintentional, in: Super Science Stories, Popular Publications, USA.

Bagenal, F., Dowling, T. E. and McKinnon, W. B. (2004), *Jupiter: The Planet, Satellites and Magnetosphere*, Cambridge University Press, Cambridge.

Barnard, E. E. (1895), 'Filar micrometer measures of the diameters of the four bright satellites of Jupiter, made with the 36-inch equatorial of the Lick Observatory', *Mon. Not. R. Astron. Soc.* **55**, 382–391.

Binder, A. B. and Cruikshank, D. P. (1964), 'Evidence for an atmosphere on Io', *Icarus* **3**, 299–305.

Binder, A. B. and Cruikshank, D. P. (1966), 'Photometric search for atmospheres on Europa and Ganymede', *Icarus* **5**, 7–9.

Both, E. E. (1952), 'Jupiter's satellite Ganymede in 1951–52', *The Strolling Astronomer* **6**, 78–81.

Brown, R. A. and Chaffee, Jr., F. H. (1974), 'High-resolution spectra of sodium emission from Io', *Astrophys. J.* **187**, L125–L126.

Camichel, H. and Lyot, B. (1943), 'Observations planetaire au Pic du Midi, en 1941', *L'Astronomie* **57**, 49–60.

Carlson, R., Smythe, W., Baines, K. et al. (1996), 'Near-infrared spectroscopy and spectral mapping of Jupiter and the Galilean satellites: Results from Galileo's initial orbit', *Science* **274**, 385–388.

Carlson, R. W., Bhattacharyya, J. C., Smith, B. A. et al. (1973), 'An atmosphere on Ganymede from its occultation of SAO 186800 on 7 June 1972', *Science* **182**, 53–55.

Clark, R. N. (1980), 'Ganymede, Europa, Callisto, and Saturn's rings: Compositional analysis from reflectance spectroscopy', *Icarus* **44**, 388–409.

Clark, R. N. and McCord, T. B. (1980), 'The Galilean satellites: New near-infrared spectral reflectance measurements (0.65–2.5 μm) and a 0.325-5 μm summary', *Icarus* **41**, 323–339.

Dalal, R. (2010), *Hinduism: An Alphabetical Guide*, Penguin, India.

Danjon, A. (1944), 'Les satellites de Jupiter', *L'Astronomie* **58**, 33–36.

Dekany, R., Roberts, J., Burruss, R. et al. (2013), 'PALM-3000: Exoplanet adaptive optics for the 5 m Hale telescope', *Astrophys. J.* **776**.

Dick, P. K. (1967), *The Ganymede Takeover*, Ace Books, New York, USA, ISBN: 9780727816870.

Douglass, A. E. (1897), 'Drawings of Jupiter's third satellite', *Astron. Nachr.* **143**, 411–414.

Fagles, R. (1990), *Homer: The Iliad*, Penguin, London.

Fedtke, C. (1942), 'Jupitermond iii (Ganymede)', *Astron. Nachr.* **273**, 6.

Feldman, P. D., McGrath, M. A., Strobel, D. F. et al. (2000), 'HST/STIS ultraviolet imaging of polar aurora on Ganymede', *Astrophys. J.* **535**, 1085–1090.

Fink, U. and Larson, H. P. (1975), 'Temperature dependence of the water-ice spectrum between 1 and 4 microns: Application to Europa, Ganymede and Saturn's Rings', *Icarus* **24**, 411–420.

Fink, U., Dekkers, N. H. and Larson, H. P. (1973), 'Infrared spectra of the Galilean satellites of Jupiter', *Astrophys. J.* **79–179**, L155–L159.

Flammarion, C. (1880), *Astronomie Populaire*, C. Marpon et E. Flammarion, Paris.

Gehrels, T. (1977), *Jupiter*, University of Arizona Press, Arizona.

Grasset, O., Dougherty, M. K., Coustenis, A. et al. (2013), 'JUpiter ICy moons Explorer (JUICE): An ESA mission to orbit Ganymede and to characterise the Jupiter system', *Planet. Space Sci.* **78**, 1–21.

Gromova, L., Moroz, V. I. and Cruikshank, D. P. (1970), 'Spectrum of Ganymede in the region 1–1.7 μ', *Astron. Tsirk.* **569**, 6–8.

Grundy, W. M., Buratti, B. J., Cheng, A. F. et al. (2007), 'New Horizons mapping of Europa and Ganymede', *Science* **318**, 234–237.

Hall, C. F. (1975), 'Pioneer 10 and Pioneer 11', *Science* **188**, 445–446.

Hall, D. T., Feldman, P. D., McGrath, M. A. and Strobel, D. F. (1998), 'The far-ultraviolet oxygen airglow of Europa and Ganymede', *Astrophys. J.* **499**, 475–481.

Hare, E. E. (1951), 'Jupiter 1950 report no. 4', *The Strolling Astronomer* **5**, 8–10.

Harris, D, Pholometry and Colorimetry of Planets and Moons, in G. P. Kuiper and B. M. Middlehurst, eds., 'Planets and Satellites', University of Chicago Press, Chicago, USA, 272–342

Hart, M. (2010), 'Recent advances in astronomical adaptive optics', *Appl. Optics* **49**, D17–D29.

Heinlein, R. (1953). *Farmer in the Sky*, Charles Scribner's Sons, New York, USA, ISBN: 0345324382.

Heinlein, R. (1970). *I Will Fear No Evil*, G. P. Putnam's Sons, New York, USA, ISBN: 978-0441359172.

Hendrix, A. R., Barth, C. A. and Hord, C. W. (1999), 'Ganymede's ozone-like absorber: Observations by the Galileo ultraviolet spectrometer', *J. Geophys. Res.* **104**, 14169–14178.

Herschel, W. (1797), 'Observations of the Changeable Brightness of the Satellites of Jupiter, and of the Variation in Their Apparent Magnitudes; With a Determination of the Time of Their Rotatory Motions on Their Axes. To Which is Added, a Measure of the Diameter of the Second Satellite, and an Estimate of the Comparative Size of All the Four', *Phil. Trans. Royal Soc. London* **87**, 332–352.

Hogan, J. P. (1978), *The Gentle Giants of Ganymede*, Random House Publishing Group, New York, USA, ISBN: 0345323270.

Hughes, D. W. (1982), 'Was Galileo 2,000 years too late?', *Nature* **269**, 199.

James S. E. Corey, Caliban's War, publisher: Orbit, London, UK, ISBN: 9781841499918, 2012

Johnson, J. H. (1930), 'The discovery of the first four satellites of Jupiter', *J. Brit. Astron. Assoc.* **41**, 164–171.

Keim, F. (2009), *Die Entdeckung der Jupitermonde 105 Jahren vor Galileo Galilei*, Peter Lang, Frankfurt.

Kelley, D. H. and Milone, E. F. (2011), *Exploring Ancient Skies*, Springer, New York.

Kieffer, H. (1970), 'Spectral reflectance of CO_2-H_2O frosts', *J. Geophys. Res.* **75**, 501–509.

Kivelson, M. G., Khurana, K. K. and Volwerk, M. (2002), 'The permanent and inductive magnetic moments of Ganymede', *Icarus* **157**, 507–522.

Kuiper, G. P. (1952), *The Atmospheres of the Earth and Planets*, University of Chicago Press, Chicago.

Kuiper, G. P. (1957), 'Infrared observations of planets and satellites', *Astron. J.* **62**, 245.

Kuiper, G. P. (1973), 'Comments on the Galilean satellites', *Comm. LPL* **10**, 28–34.

Lanzerotti, L. J., Robbins, M. F., Tolk, N. H. and Neff, S. H. (1975), 'Scans of Io, Europa, and Ganymede in the NaD region', *Astron. Soc. Pacific* **87**, 449–453.

Lyot, B. (1953), 'L'aspect des planètes au Pic du Midi dans une lunette de 60 cm d'ouverture', *L'Astronomie* **67**, 3–20.

McCord, T. B., Hansen, G. B., Clark, R. N. et al. (1998), 'Non-water-ice constituents in the surface material of the icy Galilean satellites from the Galileo near-infrared mapping spectrometer investigation', *J. Geophys. Res.* **103**, 8603–8626.

McGrath, M. A., Jia, X., Retherford, K. et al. (2013), 'Aurora on Ganymede', *J. Geophys. Res.* **118**, 2043–2054.

Mendis, D. A. and Axford, W. I. (1974), 'Satellites and magnetospheres of the outer planets', *Ann. Rev. Earth Planet. Sci.* **2**, 419–474.

Miles, H. (1973), 'Jupiter probe – Pioneer 10', *J. Brit. Astron. Ass.* **83**, 192–195.

Moroz, V. I. (1966), 'Infrared spectrophotometry of the Moon and the Galilean satellites of Jupiter', *Sov. Astron.* **9**, 999–1006.

Morrison, D. and Cruikshank, D. P. (1974), 'Physical properties of the natural satellites', *Space Sci. Rev.* **15**, 641–739.

Morrison, D., Cruikshank, D. P. and Murphy, R. E. (1972), 'Temperatures of Titan and the Galilean satellites at 20 microns', *Astrophys. J.* **173**, L143–L146.

Morrison, D., Cruikshank, D. P., Martin, T. Z., Berry, J. C. and Shipley, J. P. (1971), 'Thermal inertia of Ganymede from 20-micron eclipse radiometry', *Astrophys. J. Lett.* **167**, L107–L111.

Mura, A., Adriani, A., Sordini, R. et al. (2020), 'Infrared observations of Ganymede from the Jovian infrared auroral mapper on Juno', *J. Geophys. Res.* **125**.

Murray, B. C., Wastphal, J. A. and Wildey, R. L. (1965), 'The eclipse cooling of Ganymede', *Astrophys. J.* **141**, 1590–1592.

Murray, J. B. (1975), 'New observations of surface markings on Jupiter's satellites', *Icarus* **25**, 397–404.

Nourse, A. E. (1967), *The Counterfeit Man*, Scholastic Book Services, USA. ISBN: 0999129031.

Ossendrijver, M. (2016), 'Ancient Babylonian astronomers calculated Jupiter's position from the area under a time-velocity graph', *Science* **351**, 482–484.

Pagnini, P. (1930), 'Galileo and Simon Mayer', *J. Brit. Astron. Assoc.* **41**, 415–422.

Pankenier, D. W. (1992), 'The bamboo annals revisited: Problems of method in using the chronicle as a source for the chronology of early Zhou. Part 2: The congruent mandate chronology in Yi Zhou shu', *Bull. School Orient. Afric. Studies* **55**, 498–510.

Pellicori, S. F., Russell, E. E. and Watts, L. A. (1973), 'Pioneer imaging photopolarimeter optical system', *Appl. Optics* **12**, 1246–1358.

Pickering, W. H. (1900), 'Satellite III', in P. Lowell, ed., "Annals of the Lowell Observatory, Vol. II", Cambridge University Press, Cambridge, pp. 101–126.

Pilcher, C. B., Ridgway, S. T. and McCord, T. B. (1972), 'Galilean satellites: Identification of water frost', *Science* **178**, 1087–1089.

Reese, E. J. (1951), 'Map of Ganymede in 1949', *The Strolling Astronomer* **5**, 1.

Roche, J. J. (1982), 'Harriot, Galileo, and Jupiter's satellites', *Arch. Int. Hist. Sci.* **32**, 9–51.

Saur, J., Duling, S., Roth, L. et al. (2015), 'The search for a subsurface ocean in Ganymede with Hubble Space Telescope observations of its auroral ovals', *J. Geophys. Res.* **120**, 1715–1737.

Schaeberle, J. M. and Campbell, W. W. (1891), 'Observations of markings of Jupiter's third satellite', *Astron. Soc. Pacific* **3**, 359–365.

Schröter (1800), 'Beobachtungen des Saturns, dessen ring und trabanten; des Uranus und des Jupiters, sammt deren trabanten etc. mit einem 10 füssigen Dollond un 13 und 27 füssigen Reflector', *Astron. Jahrbuch für das Jahr 1800* 166–175.

Sheffield, C. (1995). *The Ganymede Club*, Tom Doherty Associates, New York, USA. ISBN: 0312856628.

Silbergeld, J. (1982), 'Mawangdui, excavated materials, and transmitted texts: A cautionary note', *Early China* **8**, 79–92.

Smith, B. A., Briggs, G. A., Danielson, G. E. et al. (1977), 'Voyager imaging experiment', *Space Sci. Rev.* **21**, 103–127.

Smith, B. A., Soderblom, L. A., Beebe, R. et al. (1979a), 'The Galilean satellites and Jupiter: Voyager 2 imaging science results', *Science* **206**, 927–950.

Smith, B. A., Soderblom, L. A., Johnson, T. V. et al. (1979b), 'The Jupiter system through the eyes of Voyager 1', *Science* **204**, 951–957.

Smyth, W. H. (1844), *A Cycle of Celestial Objects, Vol. 1*, J. W. Parker, London.

Subbarayappa, B. V. (1989), 'Indian astronomy: An historical perspective', in S. K. Biswas, D. C. V. Mallik and C. V. Vishveshwara, eds., Cosmic Perspectives, Cambridge University Press, Cambridge, pp. 25–40.

Trafton, L., Parkinson, T. and Macy, Jr., W. (1974), 'The spatial extent of sodium emission around Io', *Astrophys. J.* **190**, L85–L89.

Weinbaum, G. S. and Weinbaum, H. (1938), *Tidal Moon*, in: Thrilling Wonder Stories, Ned Pines' Standard Magazines, USA.

Whitmell, C. T. (1896), 'Solar eclipses on Jupiter due to the third satellite (Ganymede)', *J. Brit. Aston. Ass.* **6**, 424–426.

Wingrove, D. (1997), *The Marriage of the Living Dark*, Doubleday Canada Limited, Toronto, Canada. ISBN: 0385257368.

Xi, Z. (1981), 'The sighting of Jupiter's satellite by Gan De 2000 years before Galileo', *Chin. Astron. Astrophys.* **5**, 242–243.

Yung, Y. L. and McElroy, M. B. (1977), 'Stability of an oxygen atmosphere on Ganymede', *Icarus* **30**, 97–103.

2

Origin of Ganymede and the Galilean Moons

YUHITO SHIBAIKE AND YANN ALIBERT

2.1 INTRODUCTION

The four large moons orbiting around Jupiter, found by Galileo Galilei in 1610, are called Galilean moons. Ganymede is the largest of them, and the largest moon in our solar system. The Galilean system can be seen as a smaller scaled system of planet systems and its origin is potentially also similar to that of planets. We start this chapter with a comparison of the processes of satellite and planet formation.

Ganymede and the Galilean moons are believed to have formed in a gas disc, like planets formed in protoplanetary discs (PPDs). At the end of its formation, Jupiter accreted gas from the circum-stellar disc (CSD) and, as a byproduct, a gas disc formed around the gas planet, which is called the circum-Jovian disc (CJD). The formation processes of the large moons are similar to those of planets; the seeds of the moons grow large by accreting satellitesimals (similar to planetesimals in the case of planets, a few kilometers in size) and/or pebbles (a centimeter to a meter in size). The moons also migrate in the CJD, and the orbits come into resonances. Due to the similarity that dust particles grow larger in gas discs, the physics and technical methods involved in planet formation have also been applied to the formation of the large moons in many ways. However, there are three main differences between them.

- **Small scale** One of the most important differences is the scale. The semimajor axis of Ganymede is about 10^6 km, which is more than 100 times smaller than that of Earth (1 au $=$ 1.5×10^8 km). The orbital period of Ganymede is 7 days, 50 times shorter than that of Earth, suggesting that the timescale of the physical processes of satellite formation is roughly 10 to 100 times shorter than that of the planet formation. The mass ratio between the Sun and Jupiter is about 10^3, and the temperatures of the CJD and CSD are not dissimilar, so the aspect ratio of the CJD is about three times larger than that of the CSD. Due to the short typical length scale of the CJD, the magnetic diffusion is also much faster (i.e., the magnetic Reynolds number is small) and it prevents magnetorotational instability (MRI) in broad regions of the disc (e.g., Fujii et al., 2014; Turner et al., 2014) (see also Section 2.2.2.2).
- **Open system** The CJD is an open system unlike PPDs. In modern views of satellite formation, gas and solids are continuously supplied to the CJD from the CSD and flow into Jupiter. As a result, the continuous inflows of gas and solids dominate the formation of the large moons. This fact is interesting because the timescale of the inflow must be consistent with that of the CSD ($\sim$ 3 Myr), which is much longer than the Kepler timescale inside the CJD, suggesting that the satellite formation is a quasi-steady state phenomenon that continuously repeats. Also, the difference of the timescales makes numerical calculations through the whole satellite formation processes difficult. If there are special structures which prevent the mass flux from the CJD to Jupiter, the system loses the balance between the inflow and outflow and becomes unsteady. Especially an inner cavity of the CJD created by the magnetic field of young Jupiter can solve one of the most severe problems of the satellite formation, a rapid loss of (proto-)satellites from the CJD by migration (see Section 2.2.2.2). The effects of the Sun are also unique to the satellite formation processes. The Hill sphere of Jupiter is distorted into a lemon-like shape and spiral shocks are excited by the Sun in the CJD (e.g., Zhu et al., 2016) (see section 2.2.2.3). However, it is generally considered that the effects of the Sun on Ganymede and the other Galilean moons are limited because their orbits are close enough to Jupiter.
- **Absence of generality** Although the goal of this chapter is to review the origin of Ganymede and the Galilean moons, comparisons to other moon systems and discussion of general satellite formation are important to understanding it. Indeed, in the case of planet formation, recent observations of exoplanets and PPDs brought huge knowledge to planet formation theory. Unfortunately, in the case of satellite formation, there are only two large moon systems, around Jupiter and Saturn, in our solar system and no analogs have been found in exoplanet systems. Therefore, the discussion of satellite formation has not been generalized yet but limited to a unique and historical one about the Jovian and Saturnian systems, which is similar to discussion of the planet formation until 1995 when the first exoplanet was discovered (Mayor and Queloz, 1995).

In this chapter, we focus on the unique history of Ganymede and the Galilean moons, but we note that the discussion of this chapter can be adapted to the formation of Titan, a large moon of Saturn, except for the discussion of the constraints from the observations of the Galilean moons in Section 2.4.3. On the other hand, it is considered that the Saturnian mid-sized moons formed from the materials of ancient massive ring spreading beyond the Roche radius (Crida and Charnoz, 2012). There has been some discussion of applying this ring-related formation scenario to Titan and the Garilean moons, but their mass distribution is difficult to reproduce by the scenario with typical conditions. The origin of the moons around Uranus and Neptune is still controversial: in-situ formation in their circum-planetary discs (CPDs), spreading of the materials of their ancient rings, giant impacts, or captures of bodies after the formation of the planets (e.g., Stevenson, 1984; Agnor and Hamilton, 2006; Canup and Ward, 2006; Crida and Charnoz, 2012; Szulágyi et al., 2018).

We also note that there is a candidate for exomoon, but it is a huge Neptune-size moon around Kepler-1625b. It is considered that the satellite was captured by the planet after its formation phase even if the body is actually an exomoon (Teachey and Kipping, 2018). Dust continuum emission from a CPD around an exoplanet, PDS70c, has been found (Isella et al., 2019; Benisty et al., 2021), and the spectral energy distribution (SED) of PDS70b is best fit to the model assuming the existence of its CPD (Christiaens et al., 2019; Stolker et al., 2020; Wang et al., 2021). However, the estimated mass of the gas accreting planets is $\sim$ 10 M_{J} (Keppler et al., 2018; Müller et al., 2018; Aoyama and Ikoma, 2019; Haffert et al., 2019; Hashimoto et al., 2020), where M_{J} is Jupiter mass, suggesting the CPD is much heavier than the CJD. Thus, it is doubtful that the observed characteristics of the disc can be adapted directly to the discussion of this chapter.

2.2 THE CIRCUM-JOVIAN GAS DISC

During the gas accretion of gas giants, small gas discs (CPDs) form around the planets, because the accreting gas has some angular momentum. Here, we focus on the CJD, which is considered to be the birthplace of Ganymede and the other Galilean moons. First, we review the two classical models of the discs, "minimum mass" and "gas-starved disc" models, in Section 2.2.1 (Lunine and Stevenson, 1982; Canup and Ward, 2002). We then explain more modern views of the discs based on the results of hydrodynamic simulations in Section 2.2.2.

2.2.1 Classical Models of the CJD

The formation of the large moons around Jupiter and Saturn has been investigated a lot, and many CJD/CPD models have been proposed. First, we show the two classical models, which will be the basis of the following discussion of the modern view of the CJD and satellite formation.

2.2.1.1 Minimum Mass Model

Lunine and Stevenson (1982) proposed the "minimum mass" model, which is based on the same concept as the "minimum mass solar nebular" model in the planet formation (Hayashi, 1981). In this model, the CJD is static and isolated from the CSD, and its solid mass is equal to the total mass of the Galilean moons, $M_{\mathrm{T}} \sim 10^{-4}\, M_{\mathrm{J}}$. Therefore, the total mass of the disc is about 1% of Jupiter when the dust-to-gas ratio is equal to the solar composition, 0.01. Some previous works showed that the Jovian (and Saturnian) moons can form in the minimum mass discs (Mosqueira and Estrada, 2003*a*, 2003*b*; Estrada et al., 2009; Mosqueira et al., 2010). This model assumes a low-density region outside the centrifugal radius, $r_{\mathrm{c}} \equiv j^2/(GM_{\mathrm{cp}}) = 15\, R_{\mathrm{J}}$, where j, G, and M_{cp} are the specific angular momentum of the inflowing gas during the formation of the disc, the gravitational constant, and the mass of the central planet, respectively. This assumption means that, when the disc forms, the gas flows inside r_{c} onto the disc, in other words, the outer edge of the gas inflow region is $r_{\mathrm{out}} = r_{\mathrm{c}}$. Thanks to this assumption, Callisto can avoid its internal differentiation by reducing the accretion of solid materials (see also Section 2.4.3.3). Here, the specific angular momentum is assumed as $j = 1/4\Omega_{\mathrm{K,cp}} R_{\mathrm{H}}^2$, where $\Omega_{\mathrm{K,cp}}$ and R_{H} are the Kepler angular velocity of the central planet rotating around the star and the Hill radius, respectively, and the gravitational effect is ignored. They are $\Omega_{\mathrm{K,cp}} = \sqrt{GM_{\mathrm{star}}/a_{\mathrm{cp}}^3}$ and $R_{\mathrm{H}} = (M_{\mathrm{cp}}/M_{\mathrm{star}})^{1/3} a_{\mathrm{cp}}$, where M_{star} and a_{cp} are the mass of the central star and the semimajor axis of the planet, respectively. Here, M_{cp}, M_{star}, and a_{cp} are set as one Jupiter mass, one Solar mass, and the semimajor axis of current Jupiter, respectively. The concept that the formation process of Callisto is different from that of the other Galilean moons is inherited to the subsequent newer formation scenarios (see Section 2.4) (Shibaike et al., 2019; Batygin and Morbidelli, 2020).

However, hydrodynamic simulations both in 2D and 3D suggest that the gas accretion onto CPDs continues even when the gas gap structures open around the discs (e.g., Lubow et al., 1999; Kley et al., 2001). The depth of the gaps created by gas giants are indeed predicted to be shallower than that of the traditional predictions; the gas surface density in the gaps is about 100 times smaller than that of the unperturbed regions (e.g., Kanagawa et al., 2015). It has also been considered that it is difficult to avoid rapid inward migration of the moons induced by the interaction with the massive disc (Miguel and Ida, 2016).

2.2.1.2 Gas-Starved Disc Model

Canup and Ward (2002) proposed the "gas-starved disk" model, which has a smaller gas disc surface density compared to that of the minimum mass model as the name of the model suggests. Although the disc mass is smaller, the gas accretion continues, and it brings sufficient solid material for the satellite formation into the CJD. This continuous supply of gas is consistent with the recent view of the gas accretion process predicted by numerical simulations, but the continuous supply of solid material could be a problem (see Section 2.3). In this model, it is assumed that the gas flows onto the CJD uniformly, and the outer edge of the gas inflow region is the centrifugal radius (i.e., $r_{\mathrm{out}} = r_{\mathrm{c}}$) similar to the minimum mass model. The value of the centrifugal radius is set as $r_{\mathrm{c}} = 30\, R_{\mathrm{J}}$, implicitly including the gravitational effect in the specific angular momentum unlike the minimum mass model. The disc is also assumed to be truncated at $r_{\mathrm{trun}} = 150\, R_{\mathrm{J}}$, considering the shock regions shown in the hydrodynamic simulations (e.g., Lubow et al., 1999; D'Angelo et al., 2002). This is also roughly consistent with an analytical derivation by Martin and Lubow (2011) that the tidal torque from the Sun removes the angular momentum of the gas at the outer region of the disc and makes the disc truncated at $r_{\mathrm{trun}} \approx 0.4\, R_{\mathrm{H}} \approx 300\, R_{\mathrm{J}}$.

Alibert et al. (2005) developed a similar accretion disc model being consistent with the formation process of Jupiter which the original model of Canup and Ward (2002) does not consider. Ward and Canup (2010) made a more detailed analytical model based on Canup and Ward (2002). The satellite formation in the gas-starved disc model has been investigated well, and we will show that in Section 2.4.

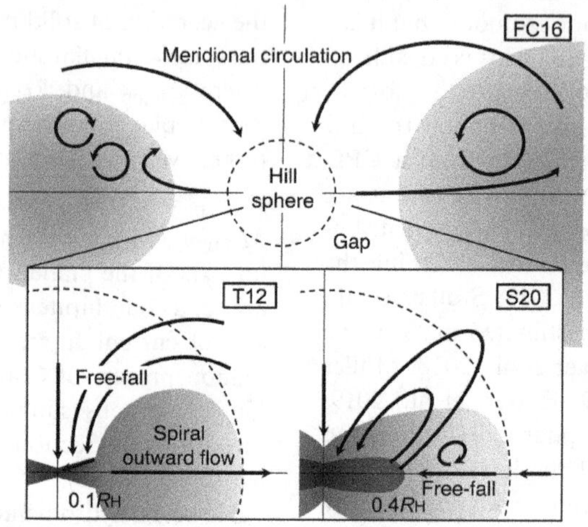

Figure 2.1 Schematic picture of the azimuthally averaged gas structures around the CJD. The upper and lower panels represent the structures shown in the simulations by Fung and Chiang (2016) (FC16, isothermal), and by Tanigawa et al. (2012) (T12, isothermal) and the low opacity case of Schulik et al. (2020) (S20, radiation), respectively. The gray scales and black arrows roughly represent the gas density and the gas flows, respectively.

2.2.2 Modern Views of the CJD

The gas-starved disc model is the basis of the recent CJD/CPD models, but hydrodynamic simulations have revealed more detailed structures of the discs as the performance of computers has improved.

2.2.2.1 Hydrodynamic Simulations

Hydrodynamic simulations have been carried out in many previous works. First, 2D isothermal simulations were calculated in some works (e.g., Lubow et al., 1999), and then a lot of 3D isothermal simulations have been conducted (e.g., Kley et al., 2001). Self-gravitating and/or radiation hydrodynamic calculations resolving inside the Hill sphere have also been calculated (e.g., Ayliffe and Bate, 2009). Ideal and nonideal (including only Ohmic resistivity) magnetohydrodynamic (MHD) simulations have also been conducted by Machida et al. (2006) and Gressel et al. (2013), respectively (see also next).

The gas around the core of Jupiter has an envelope-like structure when the mass of the planet is small, and the envelope becomes flatter as the planet grows heavier. The gas temperature is an important factor because the high temperature supports the gas structure against the gravity. The condition for the formation of the CJD has been investigated by some 3D radiation hydrodynamic simulations (e.g., Ayliffe and Bate, 2009). For example, Schulik et al. (2020) found that the flatness parameter of the envelope/disc starts to decrease (flatten) when the mass of proto-Jupiter reaches around 0.3 M_J, and also the envelope/disc becomes flatter as the opacity is lower.

After the formation of the CJD, the gas basically flows into the Hill sphere thorough the Lagrange points, L_1 and L_2, and is accreted into Jupiter through the CJD in spirals. However, the detailed structures are still controversial. We show a schematic picture of the gas structures in Fig. 2.1. Large structures bridging over the gap named as "meridional circulation" have been found in some hydrodynamic simulations; the gas flows from the edge of the gap to the Hill sphere thorough the high altitude (Morbidelli et al., 2014; Fung and Chiang, 2016; Szulágyi et al., 2021) (the upper part of Fig. 2.1). In the case of isothermal 3D hydrodynamic simulations by Tanigawa et al. (2012) (the left lower part of Fig. 2.1), the gas cannot penetrate inside the Hill sphere (and the shock around its surface) from the mid-plane, but it goes up to the high altitude along the surface of the sphere and then flows down onto the surfaces of the CJD inside about 0.1 R_H ($\approx$ 100 R_J with $M_\mathrm{cp} = M_\mathrm{J}$) with the free-fall speed. The fallen gas turns to Jupiter just after passing through the shock on the surfaces, because the specific momentum of the gas is smaller than that of the local Keplerian rotation. The effective outer edge of the gas inflow area is then $r_\mathrm{out} \approx 0.03\ R_\mathrm{H}$ ($\approx$ 30 R_J), which is near to r_out assumed in Canup and Ward (2002). The gas on the midplane outside about 0.1 R_H flows outward in spirals and goes out from the Hill sphere through L_1 and L_2. It can be transported again to the CJD with the meridional circulation. Inside 0.03 R_H, the gas on the midplane rotates almost Keplerian speed. Although the gas flows have such complex structures, the radial profile of the gas surface density is smooth in the Hill sphere. On the other hand, Schulik et al. (2019, 2020) conducted 3D radiation hydrodynamic simulations where the opacity level is 100 times lower than that of the interstellar medium (ISM) dust (i.e., dust-to-gas ratio is 10^{-4}), which could be a realistic setting due to the filtering of solid particles by the gap (see Section 2.3.1). The simulations showed more complicated gas structures then that of Tanigawa et al. (2012) (the right lower part of Fig. 2.1). Most of the gas can penetrate inside the Hill sphere from the midplane around L_1 and L_2, and it flows inward with the free-fall speed until it collides with another shock around 0.4 R_H from Jupiter. The gas can enter inside the shock from the midplane but is slowed down. The gas then goes up to the high altitude in spirals and falls down to the deeper inside region around 0.2 R_H. The gas density inside the shock is much higher than the outside. In the simulations with the settings of deeper gravitational potential of Jupiter or dust enriched gas, the temperature is higher, and the gas structures are similar to that of Tanigawa et al. (2012) except for the outflows from the Hill sphere through L_1 and L_2 and consistent with other previous 3D radiation hydrodynamic simulations (e.g., Lambrechts et al., 2019). The inner high-density region inside 0.4 R_H appears also in these cases. We note that this robust picture of the inner dense disc inside about 0.4 R_H is consistent with the classical gas-starved disc model by Canup and Ward (2002) (see Section 2.2.1.2).

The large satellites form inside 0.1 R_H in most of the formation scenarios (except for that of Batygin and Morbidelli (2020) where satellites form in about 0.1–0.3 R_H), where the disc is more quiescent than the outside (see Section 2.4). The direction of the gas flow on the midplane is important for the satellite formation, because the outward flow may avoid the loss of the solid materials to Jupiter, but it is still controversial (see Section 2.4.1.1). Beside the two simulations by Tanigawa et al. (2012) and Schulik et al. (2020), Szulágyi (2017) found by 3D radiative hydrodynamic simulations that the gas flows inward inside 10 R_J, outward at 10–100 R_J, and again inward outside 100 R_J, but it depends on the viscosity (see also Drażkowska and Szulágyi (2018)). Nonideal MHD simulations by Gressel et al. (2013) showed that the gas flows inward and outward on the midplane inside and outside of 0.2 R_H, respectively, but the resolution is not enough for the calculations inside about 0.05 R_H.

In the inner regions, however, the temperature is higher than the ice sublimation point in any situations considered in the hydrodynamic simulations, which is not suitable for the formation of the icy moons (see also Section 2.2.3). Thus, the satellites should form later than the situations, when the amount of gas flowing into the Hill sphere has decreased significantly, and the CJD has been cooler. However, it is not clear that the gas structures obtained in the simulations are kept in such later situations. Moreover, the resolutions are not enough in most of the hydrodynamic simulations for investigating the detailed structures of the satellite formation region. Thus, hydrodynamic simulations with longer time ranges and higher resolutions are important future works.

2.2.2.2 *Effects of Magnetic Fields*

Magnetic fields also play important roles in the formation and evolution of the CJD. Young Jupiter has a strong magnetic field and, in addition to that, the magnetic field lines vertically piercing the CSD should be drawn into the CJD with the gas accretion from the CSD to the CJD. Ionization of the CJD large enough for the magnetic field to affect the evolution of the disc is achieved by cosmic ray, stellar X-ray, and radionuclide (Fujii et al., 2011, 2014; Turner et al., 2014). Global 3D MHD simulations of accreting planets have shown that the flow structure of the accreting gas is roughly the same as predicted by hydrodynamical simulations, but it also suggested the existence of the jet structures from around the center of the CJD like in the case of the star formation process (Machida et al., 2006; Gressel et al., 2013). On the other hand, Batygin (2018) argued that the gas flows onto the top of Jupiter and flows down onto the midplane along with the magnetic field of the planet. The magnetic field of the young Jupiter should be stronger than the current magnetic field, and if the planet and the CJD are connected by the magnetic field, the angular momentum is transported from the planet to the disc, resulting in the slowdown of the spin of the planet and formation of the inner cavity (e.g., Takata and Stevenson, 1996; Christensen et al., 2009). The inner cavity is also important for the satellite formation because it can stop their migration (see Section 2.4).

The above works have not investigated the mechanism of the transport of the orbital angular momentum in the CJD. In the case of PPDs, the magneto-rotational instability (MRI) is one of the most important mechanisms (Balbus and Hawley, 1991, 1998). However, the MRI is not maintained in the CJD, because the magnetic diffusion is relatively quick due to the small length scale of the CJD except for the very inner region where thermal ionization occurs (Fujii et al., 2014; Keith and Wardle, 2014; Turner et al., 2014; Fujii et al., 2017). Therefore, the disc mass may be larger than that predicted by the gas-starved disc model, in which a normal strength of viscosity is assumed. We note that the CJD may evolve by the magnetic wind-driven accretion, which can decrease the gas surface density and make the disc laminar (e.g., Bai, 2015).

2.2.2.3 *Other Effects*

There are other factors that can modify the evolution of the CJD. Some 2D hydrodynamic simulations of the disc with radiative cooling show that two spiral shock arms induced by the tidal force from the Sun can transport the angular momentum and cause energy dissipation in the disc (Rivier et al., 2012; Zhu et al., 2016; Chen et al., 2021). The ability of the transport of the angular momentum is consistent with $\alpha \sim 10^{-3}$ in the α disc model. However, this simulation is 2D, and the spiral structures may depend on the dimension of simulations, so future detailed 3D calculations are expected. Photoevaporation plays an important role in the evolution of PPDs, and it can also affect the structures of the CJD (Alexander et al., 2006*a*, 2006*b*; Mitchell and Stewart, 2011). The gap around the disc could be so deep that the disc was exposed to the interstellar far-ultraviolet (FUV) radiation, and the disc could become truncated around the current orbit of Callisto (Oberg et al., 2020).

Unfortunately, the comprehensive understanding of the CJD including all of the above effects (i.e., MRI, disc wind, spiral shocks, and photoevaporation) has not been completed yet. There are still a lot of mysteries in the understanding of the disc.

2.2.3 Simple 1D Disc Model and Constraints Form the Icy Moons

We introduce a simple 1D disc model for the CJD to summarize the discussion. The diffusion equation for the gas surface density of the discs with continuous supplies of gas, Σ_g, is given by:

$$\frac{\partial \Sigma_g}{\partial t} = \frac{1}{r}\frac{\partial}{\partial r}\left[3r^{1/2}\frac{\partial}{\partial r}\left(r^{1/2}\nu\Sigma_g\right)\right] + F_{in}, \tag{2.1}$$

where r, ν, and F_{in} are the distance from the central plane (Jupiter), viscosity, and mass flux of the gas inflow (Shakura and Sunyaev, 1973; Fujii et al., 2014). The viscosity is $\nu = \alpha c_s H_g$, where α is the strength of turbulence, c_s is the sound speed, and $H_g = c_s/\Omega_K$ is the gas scale height where $\Omega_K = \sqrt{GM_{cp}/r^3}$ is the Kepler angular velocity. Here, based on the results of the hydrodynamic simulation by Tanigawa et al. (2012), we model $F_{in} \propto r^{-1}$ for $r < r_{out}$ and $F_{in} = 0$ for $r > r_{out}$ (see Section 2.2.2.1). Then, the stationary solution of the equation can be written as,

$$\Sigma_g = \frac{\dot{M}_g}{2\pi r_{out}}\frac{r^{3/2}}{\nu}\left(-\frac{2}{9}r^{-1/2} + \frac{2}{3}r_{out}r^{-3/2}\right), \tag{2.2}$$

where $\dot{M}_g$ is the gas inflow rate onto the CJD, which is consistent with the (uniform) gas accretion rate inside the disc owing to the conservation of mass in the steady state. We note that if the gas flows only outside of $r < r_{out}$ ($F_{in} = 0$ for $r < r_{out}$), the gas surface density inside r_{out} can be,

$$\Sigma_g = \frac{\dot{M}_g}{3\pi\nu} = \frac{\dot{M}_g\Omega_K}{3\pi\alpha c_s^2}. \tag{2.3}$$

The sound speed depends on the disc temperature, $c_s = \sqrt{k_B T/m_g}$, where k_B, T, and m_g are the Boltzmann constant, disc temperature, and mean molecular mass of gas, respectively. The gas temperature on the mid-plane governed by the viscous heating is given by (Nakamoto and Nakagawa, 1994),

$$T = \left(\frac{3GM_{cp}\dot{M}_g}{8\pi\sigma_{SB}r^3}\right)^{1/4} g, \tag{2.4}$$

where σ_{SB} is the Stefan-Boltzmann constant and,

$$g = \left(\frac{3}{8}\tau + \frac{1}{4.8\tau}\right)^{1/4} \tag{2.5}$$

is a function of the Rosseland mean optical depth $\tau = \kappa\Sigma_g$. In reality, the luminosity from Jupiter is another source of heating, which could be dominant in the outer region of the disc (Canup and Ward, 2002). In principle, the opacity κ depends on the size distribution of solid particles, but we assume $g = 1$ for simplicity as a minimum estimate. The disc temperature is then determined by the gas accretion rate, $\dot{M}_g$, and the gas surface density is regulated by $\dot{M}_g$ and the strength of turbulence, α.

Canup and Ward (2002) assumed that $\dot{M}_g = 0.2\ M_J\ \mathrm{Myr}^{-1}$ and $\alpha = 5 \times 10^{-3}$. In this case, from Eqs. (2.2) and (2.4), the temperature and the gas surface density are the dashed curve in the left panel and the short dashed-dotted curve in the right panel of Figure 2.2, respectively. Shibaike et al. (2017) assumed that $\dot{M}_g = 0.02\ M_J\ \mathrm{Myr}^{-1}$ and $\alpha = 10^{-4}$, which are the dashed-dotted curve in the left panel and the dashed-dotted curve in the right panel of the figure, respectively. In this case, the snow line is around the current orbit of Europa, which is suitable in order to explain the ice mass fractions of the satellites (see Section 2.4.3.2).

On the other hand, 3D hydrodynamic simulations show that if the gas surface density of the CSD is typical, the gas accretion rate is the order of $\dot{M}_g \sim 20\ M_J\ \mathrm{Myr}^{-1}$ even if the gap structure is open (Lambrechts et al., 2019). In this case, the disc temperature around the current orbits of the Galilean moons is much higher than the sublimation temperature of water ice, about 160 K (see the continuous curve in the left panel of Figure 2.2). This is inconsistent with the existence of the icy moons if their solid materials formed in situ. Moreover, such a high accretion rate, by definition, cannot remain for more than $\sim$ 0.01 Myr except if a very strong fraction of this mass is recycled back to the CSD. Therefore, the satellite formation should have occurred in the final phase of the evolution of the CSD, when the gas surface density of the CSD has decreased, and so the gas accretion rate has become lower. On the other hand, there are some research works which proposed the satellite(simal) formation at the outer regions of the CJD where the temperature is low enough even with the high gas accretion rate (Cilibrasi et al., 2018; Drażkowska and Szulágyi, 2018; Batygin and Morbidelli, 2020; Cilibrasi et al., 2020).

2.3 SUPPLY OF SOLID MATERIALS TO THE BIRTHPLACE OF GANYMEDE

In the previous section, we have explained the current understanding of the structure of the CJD, where the large satellites including Ganymede formed. Before we explain how Ganymede formed in the CJD, we have to discuss an important factor in the satellite formation, the supply of solid material to the disc.

2.3.1 Filtering of Pebbles by Gap Structure

In the classical satellite formation models, it is assumed that enough solid material is supplied to the CJD. In the case of the gas-starved disc model, the dust-to-gas mass ratio of the gas inflow is a parameter regulating the total mass of the large satellites of the systems (Canup and Ward, 2006). However, it has been shown that the amount of dust supply to the CJD is not a free parameter but is very limited.

In order to keep the disc temperature cold enough for the formation of the icy moons, the mass flux of the gas accretion from the CSD to CJD needs to be small, which is only possible when the gas surface density of the CSD has decreased enough (see Section 2.2.3). At such a late stage of the CSD evolution, the gap structure has also already formed around the CJD. However, the gap structure filters the pebble-size particles and most of the solid materials cannot enter inside the gap and not be transferred to the CJD (e.g., Paardekooper and Mellema, 2004; Zhu et al., 2012; Homma et al., 2020). Dust particles in the CSD can grow to pebbles (about centimeter or meter size) by collisional growth, but they start to drift toward the Sun as they grow to the size. The negative gas pressure gradient of the CSD supports the gas disc and so the gas rotates around the star with sub-Kepler speed, which becomes the head wind against the particles trying to rotate with Kepler speed. As a result, the particles lose their angular momentum and drift inward. The efficiency of losing angular momentum gets higher when the particles have the size of pebbles. However, at the outer edge of the gap structure, the gas pressure gradient is positive, and pebbles drift outward. Therefore, the pebbles do not enter the gap and pile up at the gas pressure maximum unless their diffusion is so strong that it can pushes pebbles out from the pressure maximum against the piling up mechanism.

An alternative way to transport the materials to the CJD is by changing the size of the particles to be so large that the particles do not suffer the effects of the pressure of the gas or to be so small that the particles can enter the gap coupled with the gas. Some recent works argued that the pebbles piling up at the gas pressure bump can grow to planetesimals by streaming instability and/or mutual collision (Lyra et al., 2009; Ayliffe et al., 2012; Chatterjee and Tan, 2013; Taki et al., 2016; Stammler

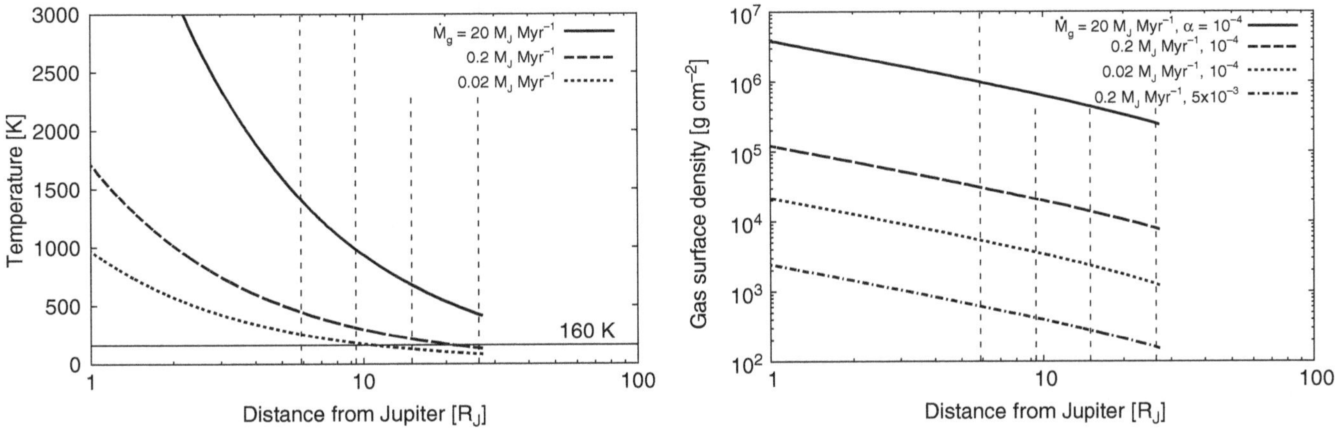

Figure 2.2 Temperature and gas surface density of the CJD. The curves in the left panel represent the mid-plane temperature of the CJD calculated by Eq. (2.4) with the different gas accretion rates. The continuous, dashed, and dotted curves represent the temperature with $\dot{M}_g \sim 20, 0.2, 0.02\ M_J\ \mathrm{Myr}^{-1}$, respectively. The horizontal line is the sublimation temperature of water ice, 160 K. The right panel represents the gas surface density calculated by Eq. (2.2) with $r_{out} = 27\ R_J$. The continuous, dashed, and dotted curves are those with $\dot{M}_g \sim 20, 0.2, 0.02\ M_J\ \mathrm{Myr}^{-1}$, respectively, and with $\alpha = 10^{-4}$. The short dashed-dotted curve is that with $\dot{M}_g = 0.2\ M_J\ \mathrm{Myr}^{-1}$ and $\alpha = 5 \times 10^{-3}$, the same condition with Canup and Ward (2002). The vertical dashed lines in both panels represent the current orbital positions of the Galilean moons.

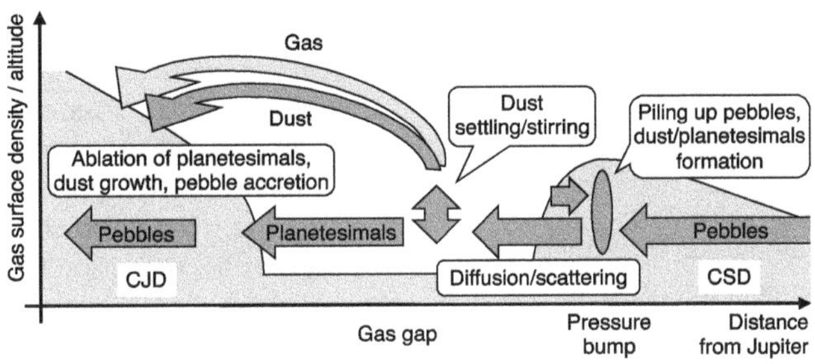

Figure 2.3 Schematic picture of transport of materials to the CJD. The horizontal axis roughly represents the distance from Jupiter. The background black solid curves and the gray shaded regions represent the gas surface density. The solid material of the moons originates at the outer region of the CSD and drift inward as pebbles. The drifting pebbles then pile up at the gas pressure bump created by Jupiter. Planetesimals form from the gathered pebbles by the streaming instability, and their fragmentation also forms small dust particles. Planetesimals scattered by large bodies (such as Saturn) get high eccentricity and penetrate the gap. Small dust particles can also penetrate the gap by their diffusion if it is strong, and they are supplied onto the CJD with the gas inflow from the high altitude, which also depends on their vertical settling and stirring inside the gap. The supplied dust particles grow to pebbles and drift inward as pebbles inside the CJD. The captured planetesimals are also ablated and produce dust particles. They also grow by accreting the drifting pebbles

et al., 2019; Eriksson et al., 2020; Shibaike and Alibert, 2020). At the same time, a lot of small particles can also form as the fragments out of the collisions of the accumulated pebbles and planetesimals (Kobayashi et al., 2012). Therefore, these small dust particles and large planetesimals may be transported to the CJD. We discuss the above two methods of transport in the next two subsections.

2.3.2 Supply of Dust Particles Coupled with Inflowing Gas

The idea of transporting solid materials to the CJD as small dust particles has been assumed implicitly in the classical works (e.g. Canup and Ward, 2002). In most of the formation scenarios, the solid-to-gas mass flux ratio supplied to the CJD is an important parameter which governs the total mass of satellites formed by satellitesimal accretion. Homma et al. (2020), found that most of the dust particles around the CJD can flow onto the disc coupled with inflowing gas if there is enough strength of turbulence to blow up the dust particles to the high altitude where the gas flows toward the CJD as predicted in Tanigawa et al. (2012).

However, when the gap structure exists, in other words, when the large satellites form in the CJD, most of the particles cannot access the root of the gas inflow which is inside the gap. Only small particles can penetrate inside the gap, thanks to the diffusion, and be supplied to the CJD, which should make the dust-to-gas ratio much lower than that expected in the previous satellite formation works (Paardekooper and Mellema, 2006; Paardekooper, 2007; Zhu et al., 2012). The condition for the penetration can be roughly explained as follows. The radial velocity of particle in gas is given by,

$$v_r = v_{drift} + v_{diff}, \tag{2.6}$$

where v_{drift} and v_{diff} are the drift velocity and the diffusion velocity, respectively. The drift velocity is (Adachi et al., 1976; Weidenschilling, 1977),

$$v_{drift} = -2\frac{\mathrm{St}}{\mathrm{St}^2+1}\eta v_K, \tag{2.7}$$

where $\mathrm{St} \equiv \tau_{stop}/\Omega_K$ is the Stokes number of the particle, where τ_{stop} is the stopping time, and $v_K = r\Omega_K$ is the Kepler velocity. The ratio of the pressure gradient force to the gravity of the central planet is

$$\eta = -\frac{1}{2}\left(\frac{H_g}{r}\right)^2 \frac{\partial \ln \rho_g c_s^2}{\partial \ln r}, \tag{2.8}$$

where $\rho_g = \Sigma_g/(\sqrt{2\pi} H_g)$ is the gas density at the mid-plane. At the outer edge of the gap, η is negative. The diffusion velocity is

$$v_{diff} = -\frac{D}{r}\frac{d \ln(\Sigma_d/\Sigma_g)}{d \ln r}. \tag{2.9}$$

The turbulent diffusivity of the particles is $D = \nu/\mathrm{Sc}$, where the Schmidt number is $\mathrm{Sc} = (1 + \mathrm{St}^2)^2/(1 + 4\mathrm{St}^2)$ (Shakura and Sunyaev, 1973; Youdin and Lithwick, 2007). When $|v_{drift}/v_{diff}| < 1$, the particles can penetrate the gap. From the above equations,

$$\left|\frac{v_{drift}}{v_{diff}}\right| \sim \frac{\mathrm{St}}{\alpha}\left(\frac{d \ln \Sigma_g}{d \ln r}\right)\left(\frac{d \ln(\Sigma_d/\Sigma_g)}{d \ln r}\right)^{-1}, \tag{2.10}$$

so the rough condition for the penetration of the particles is $\mathrm{St}/\alpha \ll 1$ (Zhu et al., 2012). On the other hand, the other terms depend on the gas gap structure, which is determined by the mass of the planet, the aspect ratio of the CSD, and so on. When St and α are given, the minimum planet mass which the pebbles cannot penetrate the gap with (i.e., when $v_{drift}/v_{diff}| = 1$) is called "pebble isolation mass" (e.g., Lambrechts et al., 2014; Ataiee et al., 2018). If planets glow larger than their pebble isolation mass, they cannot accrete pebbles anymore (see also Section 2.4.2).

We note that a very recent work, Szulágyi et al. (2021), carried out 3D dust+gas radiative hydrodynamic simulations and found that, even with the gap structure, the meridional circulation drives a strong vertical flow even for mm-sized grains and can deliver them to the CJD with the gas from the high altitude (see Figure 2.1). In addition, the gas easily flows out from the Hill sphere, but the dust is trapped more easily in the planet region. As a result, the accretion rates onto the vicinity (inside $0.5\,R_H$) of Jupiter for the dust can be $\sim 0.01\,M_J$ Myr, and that for the gas is $\sim 0.1\,M_J$ Myr. We also note that the observation of the CPD around PDS70c shows that there are a lot of dust particles inside the CPD (Isella et al., 2019; Benisty et al., 2021). However, in the case of a heavy planet ($\sim 10\,M_J$), its PPD turns eccentric, and the gap edge approaches the CPD and the planet, resulting in an increase of the gas accretion rate (Kley and Dirksen, 2006), which may make it possible to supply a lot of dust with the gas. Therefore, it is difficult to confirm whether this observation of PDS70c supports the argument that plenty of dust can be transported to the CJD or not.

2.3.3 Capture of Planetesimals by the CJD

The idea of the capture of planetesimals by the gas drag from CPDs has been investigated a lot (Fujita et al., 2013; Tanigawa et al., 2014; D'Angelo and Podolak, 2015; Suetsugu and Ohtsuki, 2016; Suetsugu et al., 2016; Suetsugu and Ohtsuki, 2017; Ronnet et al., 2018; Ronnet and Johansen, 2020). In the case of calculations of Ronnet and Johansen (2020), about 23% of the planetesimals in the feeding zone of Jupiter can be captured by the CJD. Suetsugu and Ohtsuki (2017) calculated the capture and subsequently orbital evolution of planetesimals and found that the initial radial distribution of satellitesimals assumed in the satellitesimal accretion scenarios (see Section 2.4.1) could be roughly reproduced by the prograde captured planetesimals. Their surface number density is significantly enhanced at the current location of the Galilean moons. There is also another type of captured planetesimals which take "long-lived" prograde orbits (Suetsugu and Ohtsuki, 2016), which are captured at farther orbits from Jupiter and could be the seeds of the Galilean moons (see Shibaike et al. (2019) and Section 2.4.2). When planetesimals get close to Jupiter, their surface temperature increases, and the ablation (mass loss of the surface material by sublimation and by melting and evaporation) is most vigorous. Some works argued that the ablation (and break-up) of the planetesimals made the CJD dust-laden (D'Angelo and Podolak, 2015; Ronnet et al., 2018). In the pebble accretion scenario by Ronnet and Johansen (2020), the dust particles released by the ablation of the captured planetesimals drift inward as pebbles and they are accreted onto relatively large captured planetesimals which have survived the ablation (see Section 2.4.2).

However, it might be difficult to supply planetesimals to the CJD continuously. According to Ronnet and Johansen (2020), the planetesimals inside the feeding zone have been cleaned up in $\sim 10^4$ yr. Moreover, the gas pressure gradient created by Jupiter and the gravitational interaction with the growing planet itself should push the planetesimals out from Jupiter's feeding zone, rendering the planetesimal capture rate very low (Hayashi et al., 1977; Fujita et al., 2013). On the other hand, the pebbles piled up at the gas pressure bump form planetesimals continuously thanks to the continuous supply of pebbles drifting from the outer region of the CSD to the bump (Shibaike and Alibert, 2020). Massive embryos then from the runaway growth of the planetesimals and can excite the planetesimals' orbits, resulting in the penetration of some planetesimals inside the gap (Kobayashi et al., 2012). Moreover, the outer planet, Saturn, can destabilize the planetesimals and deliver them to the feeding zone, which is a sufficient amount for the formation of the Galilean moons (Ronnet et al., 2018; Ronnet and Johansen, 2020).

2.4 FORMATION OF GANYMEDE AND THE GALILEAN MOONS

In the previous sections, we discussed the possibilities to transport solid materials from the CSD to the circum-Jovian disc. In this section, we show how Ganymede and the other large moons formed from the materials in the disc. First, we explain the "satellitesimal accretion" scenario, which has been considered an orthodox way to form the large moons. Then, we show recent alternative "pebble accretion" scenarios, which may avoid the problems of the satellitesimal accretion scenarios.

2.4.1 Satellitesimal Accretion Scenarios

Satellitesimals, the "satellite version" of planetesimals, have been considered as the building blocks of the large moons. First, the formation process from satellitesimals to the large satellites were investigated. The basic idea of the satellite accretion scenario is as follows. There is always a sufficient number of satellitesimals inside the CJD thanks to the continuous supply of dust particles, resulting in steady satellite formation (Canup and Ward, 2006; Sasaki et al., 2010; Ogihara and Ida, 2012). However, as we discussed in the previous sections, the supply of dust particles is not always guaranteed to be large enough. Moreover, it has been shown that the satellitesimal formation from dust particles in CPDs have many issues similar to the planetesimal formation in PPDs, which were not discussed in the classical satellitesimal accretion research works. In this subsection, we first explain the issues and proposed solutions of the formation of satellitesimals. Then, we show the satellite formation processes starting from the satellitesimals (i.e., satellitesimal accretion and migration) based on classical studies. We also show the schematic pictures of the satellitesimal accretion scenarios in Figure 2.5, including the classical minimum mass model (see Section 2.2.1.1) at the left upper part of the figure.

2.4.1.1 *Satellitesimal Formation in the CJD*

In most of the satellite formation works based on the satellitesimal accretion scenarios, the existence of the satellitesimals has been assumed (Canup and Ward, 2006; Sasaki et al., 2010; Ogihara and Ida, 2012). However, it has been known that dust particles grow to pebbles but drift towards the central stars, and it is difficult to form planetesimals from the pebbles in PPDs except in some special conditions (around snow lines, at gas pressure bumps, etc.). The gas pressure supports the gas and make it rotate slower than the Kepler speed. The pebbles trying to rotate with Kepler speed then receive head wind from the gas and lose their angular momentum, resulting in their radial drift. This mechanism also works for the particles in the CJD. Shibaike et al. (2017) modeled 1D simple CPDs and calculated the simultaneous growth and drift of the dust particles which have single peak mass at each distance from the planet. First, from the conservation of the mass,

$$\dot{M}_{\rm d} = -2\pi r v_r \Sigma_{\rm d}, \tag{2.11}$$

where $v_{\rm r} \approx v_{\rm drift}$, and $v_{\rm diff}$ can be ignored (see Eq. 2.6). The collisional growth under the steady condition is (Sato et al., 2016; Shibaike et al., 2017),

$$v_{\rm r}\frac{dm_{\rm d}}{dr} = 4\pi R_{\rm d}^2 \rho_{\rm d} \Delta v_{\rm dd} \epsilon_{\rm grow} \tag{2.12}$$

where $R_{\rm d}$ is the dust radius. The dust-dust relative velocity (i.e., collision velocity) is

$$\Delta v_{\rm dd} \approx \sqrt{(v_{\rm r}/2)^2 + v_{\rm t}^2}, \tag{2.13}$$

where the turbulent-driven velocity $v_{\rm t} = \sqrt{3\alpha {\rm St}} c_{\rm s}$ (Ormel and Cuzzi, 2007; Sato et al., 2016). The particle density on the midplane is (Youdin and Lithwick, 2007),

$$\rho_d = \frac{\Sigma_{\rm d}}{2\sqrt{\pi}H_{\rm d}} = \frac{\Sigma_{\rm d}}{2\sqrt{\pi}H_{\rm g}}\left(1 + \frac{\rm St}{\alpha}\frac{1+2{\rm St}}{1+{\rm St}}\right)^{1/2}, \tag{2.14}$$

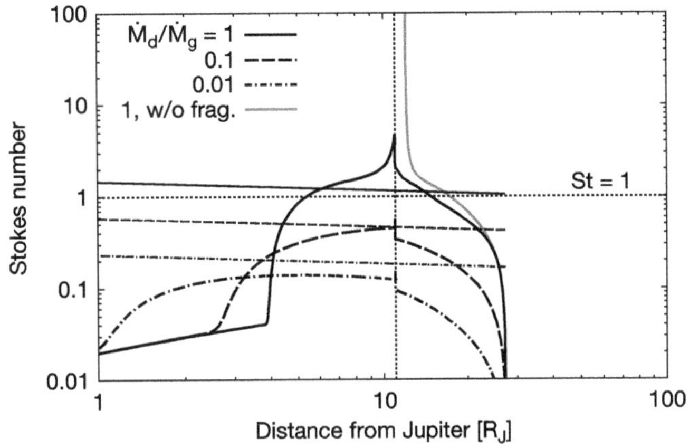

Figure 2.4 Evolution of dust particles in the CJD. The black continuous, dashed, and dashed-dotted curves represent the Stokes number of the dust when $\dot{M}_{\rm d}/\dot{M}_{\rm g} = 1$, 0.1, and 0.01, respectively. The gray curve is the case without the fragmentation (i.e., $\epsilon_{\rm grow}$ is fixed as unity). The thin lines represent the Stokes number estimated by Eq. (2.17). The horizontal and vertical dotted lines represent St = 1 and the position of the snowline, respectively.

where $H_{\rm d}$ is the dust scale height. The Stokes number in the CJD shifts from the Stokes to Newton regimes as the dust grow large,

$${\rm St} = \frac{8}{3C_{\rm D}}\frac{\rho_{\rm int}R_{\rm d}}{\rho_{\rm g}\Delta v_{\rm dg}}\Omega_{\rm K}, \tag{2.15}$$

where $C_{\rm D} \approx 0.5$ (in the Newton regime), $\rho_{\rm int}$, and $\Delta v_{\rm dg}$ are a dimensionless coefficient, the internal density of the particles, and the relative velocity between the particles and gas. The fragmentation occurs when the collision speed is too high. The sticking efficiency is then (Okuzumi et al., 2016),

$$\epsilon_{\rm grow} = \min\left\{1, -\frac{\ln(\Delta v_{\rm dd}/v_{\rm cr})}{\ln 5}\right\}, \tag{2.16}$$

where the critical velocity, $v_{\rm cr}$, for the collision of the rocky and icy particles are about $v_{\rm cr} = 5$ and $50\ {\rm m\ s^{-1}}$, respectively (Wada et al., 2009, 2013).

Figure 2.4 shows the evolution of dust particles in the turbulent CJD calculated by the above equations. The gas surface density is calculated by Eq. (2.2) with $r_{\rm out} = 27\ R_{\rm J}$, $\dot{M}_{\rm g} = 0.02\ M_{\rm J}\ {\rm Myr^{-1}}$, and $\alpha = 10^{-4}$. We assume that all of the dust particles are supplied at $r_{\rm out}$ to carry out the single size calculation. The particles can grow to large icy satellitesimals only when the fragmentation is not considered (i.e., $\epsilon_{\rm grow}$ is fixed as unity) and $x \geq 1$, where $x \equiv \dot{M}_{\rm d}/\dot{M}_{\rm g}$ is the dust-to-gas mass ratio of the accretion inflow, and this condition is virtually unattainable (see also Shibaike et al. (2017)). Furthermore, if the fragmentation is considered, the particles cannot grow to planetesimals but drift inwards even when $x = 1$. The sizes of the particles decrease as they move inward especially inside the snowline because the fragmentation prevents their growth due to the slower critical fragmentation velocity of rocky particles than that of icy particles. The figure also shows that the particles can grow large as the dust-to-gas mass ratio of the inflow is large. When the fragmentation is inefficient ($\epsilon_{\rm grow} = 1$), the Stokes number of the drifting pebbles can be estimated by taking a limit of $r \to 0$ of Eq. (2.12),

$$\mathrm{St} \approx 1.2 \left(\frac{x}{1}\right)^{2/5} \left(\frac{\alpha}{10^{-4}}\right)^{1/5} \left(\frac{T}{160\ \mathrm{K}}\right)^{-2/5} \left(\frac{M_{\mathrm{cp}}}{1\ M_{\mathrm{J}}}\right)^{2/5} \left(\frac{r}{10\ R_{\mathrm{J}}}\right)^{-2/5}, \tag{2.17}$$

where M_{cp} is the mass of the central planet. The drift speed becomes the fastest when the Stokes number of the pebbles is unity (see Eq. 2.7), so if the Stokes number becomes larger than unity, the pebbles start to pile up at that point, and satellitesimals can form there, which is similar to what happens in the planetesimal formation in PPDs. Therefore, Eq. 2.17 shows that the condition for the satellitesimal formation is $x \geq 1$ (if the fragmentation is ignored), and as x is large, the particles can grow large, which is consistent with the results of the calculations. The streaming instability, which is driven by the difference between the velocities of the dust and gas, is one of the most effective ways to avoid the radial drift and form planetesimals in PPDs (Youdin and Goodman, 2005), but it does not work in the CJD because the dust-to-gas density ratio on the mid-plane is much smaller than unity, the critical value for triggering the instability (Johansen and Youdin, 2007; Drażkowska and Dullemond, 2014). Moreover, the internal density of the particles does not affect the results although the fluffy particles may be able to solve the radial drift issue in PPDs (Okuzumi et al., 2012).

If the gas flows outward on the mid-plane in spirals, it may stop the inward drift of pebbles. Although the disc model used in the above discussion is a smooth accretion disc, a previous 3D hydrodynamic simulation showed that the gas on mid-plane at $r \approx 10 - 100\ R_{\mathrm{J}}$ should flow outward Szulágyi (2017). In that case, the dust particles (pebbles) pile up around 85 R_{J} and the dust-to-gas ratio becomes high enough to trigger the streaming instability (Drażkowska and Szulágyi, 2018). Batygin and Morbidelli (2020) argued that satellitesimals form by the gravitational instability at the outer region of the CJD, $r \approx 100 - 300\ R_{\mathrm{J}}$, where the gas on the mid-plane flows outward according to some previous works on the CJD (e.g., Tanigawa et al., 2012; Batygin, 2018). The radial drift of the small particles is stopped by the outward flow and the dust-to-gas ratio increases enough for the occurrence of the gravitational instability (Batygin and Morbidelli, 2020). Based on the above discussion of the outward gas flow on the midplane, recently some scenarios of satellitesimal accretion at the outer region of the CJD have been proposed (see Section 2.4.1.3). Also, the efficiency of the collisional growth of the particles could be higher if the disc is laminar (i.e., turbulence does not occur), for example, with the magnetic wind-driven accretion. The increase of the efficiency actually occurs in the cases of the collisional growth in PPDs (Drażkowska and Dullemond, 2018).

2.4.1.2 Satellitesimal Accretion at the Inner Region of the CJD

The formation process from satellitesimals to satellites is also similar to that from planetesimals to planets. The satellites cannot grow infinitely large because the speed of inward migration (so-called Type I migration) of satellites becomes faster as they become larger (e.g., Tanaka et al., 2002), which governs the final mass of the large moons if there is no structure which can stop the migration (Canup and Ward, 2006). Here we show the estimate of the mass of the moons by Canup and Ward (2006). Thanks to the continuous dust supply, the formation and migration of the satellites repeat until the supply of dust and gas stops. The current satellites are the final generation of the satellites repeating formation and migration. The growth timescale of the (proto-)satellite is,

$$\tau_{\mathrm{acc}} = \frac{M_{\mathrm{s}}}{2\pi r \Delta r x F_{\mathrm{in}}} \tag{2.18}$$

where M_{s}, F_{in}, and $2\pi r \Delta r$ are the (proto-)satellite mass, the gas inflow flux per area, and the annular disc area over which the satellite accumulates the solid material, respectively. We assume that F_{in} and x are uniform and constant. We also note that this equation is given under the assumption that the supplied dust particles are immediately accreted to the satellites, which neglects the inward drift of particles we discussed in the previous section. The annulus width is given by $\Delta r/r \approx 2e$, where e is the characteristic maximum eccentricity of the satellite. The eccentricity is given by $e \approx (H_{\mathrm{g}}/r)(M_{\mathrm{s}}/4\pi r H_{\mathrm{g}} \Sigma_{\mathrm{g}})^{1/5}$, which comes from the balance between the eccentricity damping by density waves and the excitation by gravitational scatterings with other satellitesimals (Ward, 1993). The migration timescale is,

$$\tau_{\mathrm{mig}} = \frac{1}{C_a} \left(\frac{M_{\mathrm{cp}}}{M_{\mathrm{s}}}\right) \left(\frac{M_{\mathrm{cp}}}{r^2 \Sigma_{\mathrm{g}}}\right) \left(\frac{H_{\mathrm{g}}}{r}\right)^2 \tag{2.19}$$

where C_a is the constant of Type I migration order of unity (Tanaka et al., 2002). From the balance between the two timescales, the critical mass of satellites m_{crit} can be given by

$$\frac{m_{\mathrm{crit}}}{M_{\mathrm{cp}}} \approx 5.6 \times 10^{-5} \chi \left(\frac{3.5}{C_a}\right)^{5/9} \left(\frac{H_{\mathrm{g}}/r}{0.1}\right)^{26/9} \left(\frac{r/r_{\mathrm{out}}}{0.5}\right)^{10/9} \left(\frac{\alpha}{3 \times 10^{-3}}\right)^{2/3} \left(\frac{x}{0.01}\right)^{2/3}, \tag{2.20}$$

where χ is order of unity with very weak dependence of the gas accretion rate. Then, the total mass of satellites, M_{T}, is,

$$\frac{M_{\mathrm{T}}}{M_{\mathrm{cp}}} \approx \int_{R_{\mathrm{p}}}^{r_{\mathrm{out}}} \frac{m_{\mathrm{crit}}/M_{\mathrm{cp}}}{\Delta r} dr \sim 2.5 \times 10^{-4} \frac{1}{\chi} \left(\frac{3.5}{C_a}\right)^{4/9} \left(\frac{H_{\mathrm{g}}/r}{0.1}\right)^{10/9} \left(\frac{\alpha}{3 \times 10^{-3}}\right)^{1/3} \left(\frac{x}{0.01}\right)^{1/3}. \tag{2.21}$$

Therefore, the key parameters are α and x, in other words, the efficiency of transport of gas angular momentum and the dust-to-gas mass ratio of the gas inflow onto the CJD, which is similar to the situation of satellitesimal formation. At the same time, the dependence of the parameters is so weak in this scenario that the ratio between the total mass of the Galilean satellites and Jupiter, 10^{-4}, can be explained naturally (provided a sufficient amount of dust is supplied and not lost from the CJD). This ratio is common in the satellite systems in our solar system, the systems of Saturn and Uranus. The right upper part of Figure 2.5 illustrates this scenario. We note that there

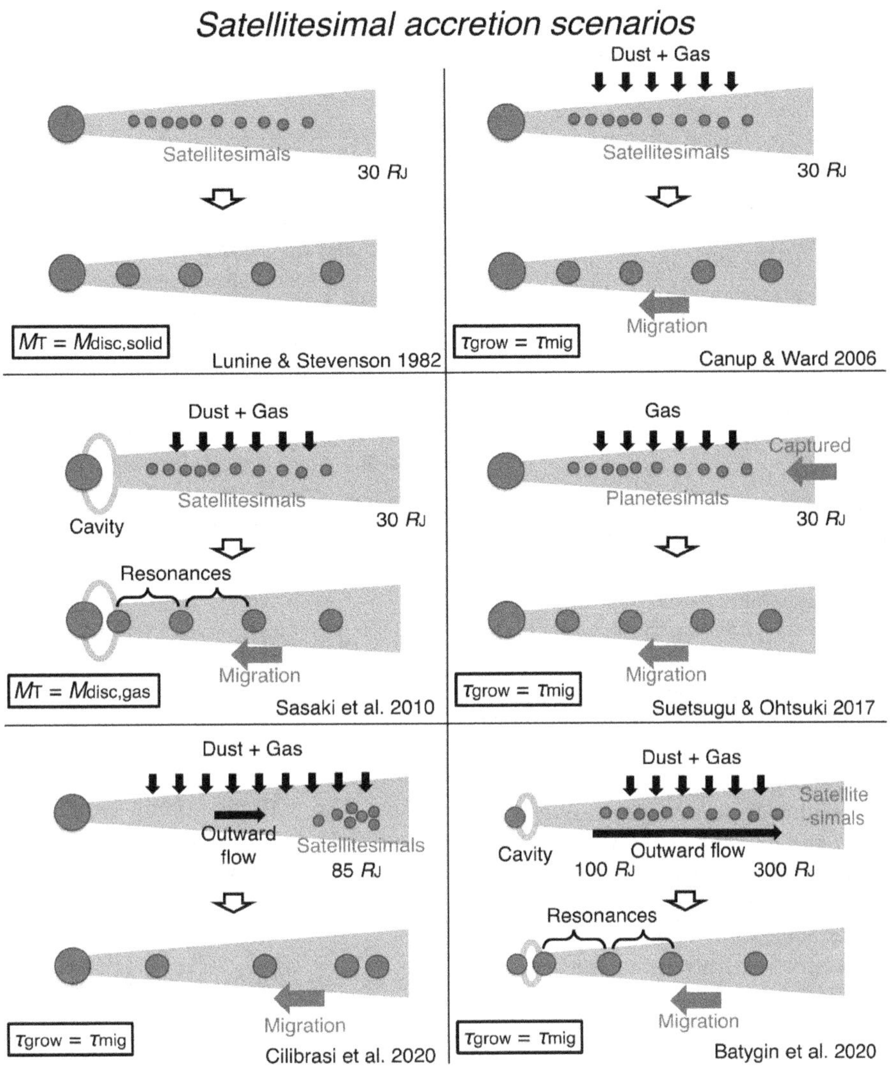

Figure 2.5 Schematic pictures of the satellitesimal accretion scenarios. The black boxes represent how the final mass of the moons is determined in each scenario. The total mass of the moons, the total mass of solid material in the CJD, and the gas disc mass are M_T, $M_{disc,solid}$, and $M_{disc,gas}$, respectively. The timescale of growth and migration of the satellites are τ_{grow} and τ_{mig}, respectively. The representative reference papers are also shown.

is another scenario where the captured planetesimals are used as the alternatives of the satellitesimals (Suetsugu and Ohtsuki, 2017) (see Section 2.3.3), which is shown in the right middle part of the figure.

Some research works invoked an inner cavity of the CJD, a truncation of the gas disc. This is formed by the magnetic coupling between the disc and Jupiter and could stop the migration of the satellites (Ogihara et al., 2010; Liu et al., 2017). The magnetic coupling can also explain the reason the spin rate of Jupiter became sufficiently slower to avoid the rotational breakup (Takata and Stevenson, 1996; Batygin, 2018). If the inner cavity exists, the inward migration of the inner most satellite stops at the edge of the disc. The outer satellites approach inwardly from the outer region, and if the migration speed of the approaching satellites is slow enough, the outer satellites are caught in mean-motion resonances (MMRs) with the next inner satellite one by one. Sasaki et al. (2010) and Ogihara and Ida (2012) performed semi-analytical Monte Carlo simulations and N-body simulations, respectively. They found that four to five similar-mass satellites are likely to remain in the Jovian system trapped in MMRs. This result is very similar to the current orbital characteristics of the Galilean moons; Io-Europa and Europa-Ganymede are in the 2:1 mean motion resonances, but Ganymede-Callisto are not in any resonances (see also Section 2.4.3.1). We note that the condition for maintaining the resonance chain of satellites is not determined by the migration timescale but by the total mass of the CJD (Sasaki et al., 2010), or by the total number of the satellites in the resonance chain (Ogihara and Ida, 2012). When the trapped satellites have the critical values, the inner most satellite is pushed into the central planet by the outer satellites. The left middle part of Figure 2.5 represents this scenario.

2.4.1.3 Satellitesimal Accretion at the Outer Region of the CJD

In the above scenarios, satellitesimal accretion occurs inside about 30 R_J, in other words, around the current orbits of the Galilean moons. However, some recent works argued that the places of the satellitesimal accretion could be more outer regions of the CJD. Cilibrasi et al. (2018) investigated a scenario that the proto-satellites form around 85 R_J, where the dust particles pile up owing to the direction of gas flow on the mid-plane switches from inward to outward there (Drążkowska and Szulágyi, 2018). Thanks to the concentration of the solid materials, a lot of satellites form quickly ($\sim 10^4$ yr) there, but they then migrate inward by Type I migration as they grow larger. The formation and migration of satellites repeatedly occur, and their final generation becomes the Galilean moons. The final mass of the moons is determined by the balance between the growth timescale and the Type I migration timescale, like the classical scenario by Canup and Ward (2006). In order to trigger the streaming instability, Cilibrasi et al. (2018) assumed high dust-to-gas accretion rates onto the CJD (0.03–0.5), resulting in a peak of the distribution of the satellite-to-planet mass ratio locates between 10^{-4} and 10^{-3}, which is heavier than the total mass of the Galilean satellites. However, as we discussed in the previous sections, getting a high value of the dust-to-gas ratio is difficult. The left bottom part of Figure 2.5 represents this scenario.

Much more outer regions could also be the birthplaces of satellitesimals and satellites. Batygin and Morbidelli (2020) argued that $\sim$ 100 km sized satellitesimals are able to form at $r \approx 0.1 - 0.3\ R_H$ by gravitational instability because dust particles can avoid their radial drift owing to the outward gas flow on the mid-plane, and the dust-to-gas ratio can be high. Satellite embryos emerge and grow to larger sizes by accreting the satellitesimals. The embryos also migrate inward, and their growth by satellitesimal accretion stops when they go outside the satellitesimal forming region. The inward migration stops at the inner edge of the disc formed by the magnetic field of Jupiter, which is similar to some of the previous works. The inner three moons, Io, Europa, and Ganymede, are captured into the 2:1 MMRs one by one, and Callisto forms after the gas disappears from the CJD, which is the reason why the satellite is not in any resonances in this scenario. Moreover, Callisto accretes satellitesimals slowly owing to the excitation of their orbital distribution (see also Section 2.4.3). The right bottom part of Figure 2.5 represents this scenario.

2.4.2 Pebble Accretion Scenarios

In the field of the planet formation, the "pebble accretion" scenario has been investigated in many studies (e.g., Ormel and Klahr, 2010; Lambrechts and Johansen, 2012). As we discussed in Section 2.4.1.1, dust particles in PPDs and in CPDs drift inward owing to the headwind they receive from the gas discs rotating with sub-Kepler speed, lose their angular momentum, and drift inward as centimeter or meter sized particles called "pebbles" before they grow larger to planetesimals/satellitesimals. We introduce here satellite formation scenarios which use these pebbles as the building blocks of the moons.

Shibaike et al. (2019) proposed the "slow-pebble-accretion" scenario in which relatively large planetesimals captured by the CJD accrete drifting pebbles on a relatively long timescale, $\sim 10^7$ yr. Basically, the mass of the growing satellites can be calculated by,

$$M_s(t) = \int_{t_{cap}}^{t} \dot{M}_{peb} P_{eff} dt, \tag{2.22}$$

where t_{cap}, $\dot{M}_{peb}$, and P_{eff} are the capture time of the planetesimals, the pebble mass flux in the CJD, and the pebble accretion efficiencies, respectively. From the conservation of mass, the pebble mass flux is equal to the dust accretion rate onto the CJD, $\dot{M}_{peb} = \dot{M}_d$, because a steady state is assumed. The order of magnitude of the (total) satellite-to-planet mass ratio, $M_T/M_{cp}{\sim}10^{-4}$ can be explained very roughly by $M_T/M_{cp}{\sim} \int \dot{M}_g dt/M_{cp} \times x P_{eff} {\sim} 1\times 0.01\times 0.01{\sim}10^{-4}$, when x and P_{eff} are constant. In reality, the pebble accretion efficiency depends on the mass of the satellites and the Stokes number of the drifting pebbles (Liu and Ormel, 2018; Ormel and Liu, 2018),

$$P_{eff} = \left\{ \left(0.32 \sqrt{\frac{\mu_s \Delta v / v_K}{\mathrm{St}\eta^2}} \right)^{-2} + \left(0.39 \frac{\mu_s}{\eta h_{peb}} \right)^{-2} \right\}^{-1/2}, \tag{2.23}$$

where $\mu_s = M_s/M_{cp}$ and $h_{peb} = H_{peb}/r$ are the satellite-to-central planet mass ratio and the pebble aspect ratio, respectively. The first and second terms represent the 2D and 3D regimes of pebble accretion, respectively. The relative velocity between the satellites and pebbles, Δv, is given by the Keplerian shear and the head wind,

$$\Delta v / v_K = 0.52 (\mu_s \mathrm{St})^{1/3} + \eta \left\{ 1 + 5.7 \left(\frac{\mu_s}{\eta^3/\mathrm{St}} \right) \right\}^{-1}. \tag{2.24}$$

Using the above equations, and the approximation of the Stokes number (an improved version of Eq. 2.17), Shibaike et al. (2019) calculated the growth and Type I migration of the seeds of satellites (captured planetesimals) with the key parameters of $x = 0.0026$ and $\alpha = 10^{-4}$. The gas accretion rate is assumed to be $\dot{M}_g = 0.2\ M_J\ \mathrm{Myr}^{-1}$, where $t_{dep} = 3$ Myr. Here, we consider the situations in which the disc has weak turbulence owing to the inactive MRI (Fujii et al., 2014) (see Section 2.2.2.2). It is also assumed that three planetesimals, in other words, the seeds of Io, Europa, and Ganymede, are captured by the CJD one by one and the fourth one, the seed of Callisto, is captured later (Model A) or forms at the gas pressure bump made by Ganymede (Model B). The disc inner cavity assumed to be at the current orbit of Io ($r = 5.9\ R_J$) stops the migration of the innermost body, and the other three (Model A) or two (Model B) embryos are subsequently captured into the MMRs. The final mass of the satellites is regulated by the end of the supply of dust and gas except for Ganymede in Model B whose final mass is determined by its pebble isolation mass. The reasons of the slow pebble accretion in this scenario are as follows. First, the amount of dust supply is small compared to the other scenarios. Second, the dust (and gas) accretion rate onto the CJD has already decreased when the satellites grow large, resulting in the decrease of the pebble accretion efficiency. Figure 2.6 shows the growth tracks of the satellites, which reproduce the

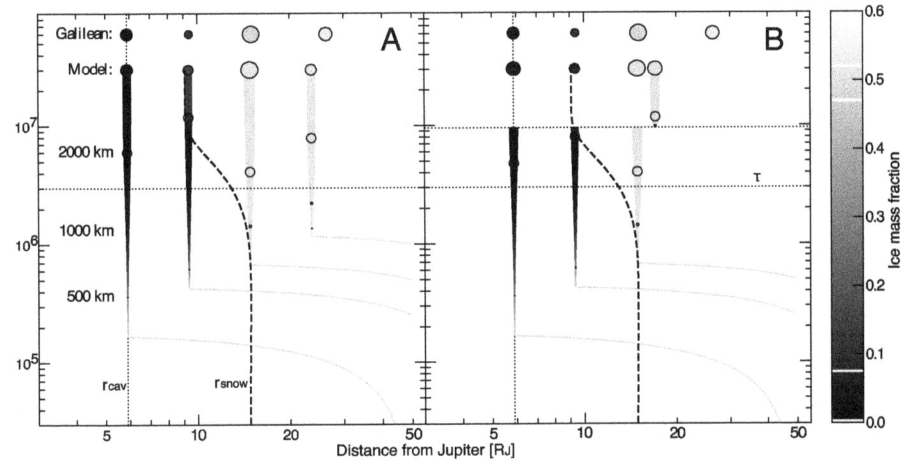

Figure 2.6 Evolution of the Galilean satellites in the slow pebble accretion scenario. The two panels represent the evolution in Models A and B. The sizes and the gray scales of the circles represent the mass and ice mass fractions of the satellites. The curves represent their orbital positions, sizes, and ice mass fractions. The dashed curve is the position of the snowline in this model (r_{snow}). The vertical dotted line is the position of the edge of the inner cavity (r_{cav}), which is assumed to be at the current orbital position of Io. The horizontal dotted lines represent the time when Ganymede reaches its pebble isolation mass (t_{PIM}) and the depletion timescale of the disc (τ_{dep}). The current characteristics of the Galilean moons are also shown. The white bars in the gray scale are also the actual ice mass fractions. This figure is modified from Figures 4 and 6 of (Shibaike et al., 2019).

mass and resonances of the Galilean satellites. This scenario can also explain the ice/rock mass ratio of the Galilean moons and the partial differentiation of Callisto (see next Section). The left part of Figure 2.7 represents this scenario.

On the other hand, Ronnet and Johansen (2020) proposed another pebble accretion scenario which does not rely on the supply of dust particles; pebbles are provided by the continuous ablation of captured planetesimals. A fraction of the captured planetesimals survives and forms several proto-satellites which are large enough to start pebble accretion. It is predicted that the pebble flux inside the CJD is $\dot{M}_{\text{peb}} = 0.003\ M_{\text{J}}\ \text{Myr}^{-1}$ by this process, which is about six times larger than the pebble flux value assumed in Shibaike et al. (2019). Even though the satellites grow faster, they migrate inward before they reach the current sizes of the Galilean moons, suggesting the existence of an inner cavity to stop their migration. After the innermost satellite is stopped at the inner edge of the disc, and the other three satellites are captured into the resonances, they accrete the drifting pebbles and grow at the fixed orbits. In this scenario, Ganymede reaches its pebble isolation mass after ≈ 0.3 Myr. The right part of Figure 2.7 represents this scenario.

We note that Madeira et al. (2021) calculated N-body numerical simulations based on the pebble accretion scenarios. They found that the scattering and mutual collisions of the proto-satellites occur after they are captured into the resonance, which promotes their additional growth. This suggests that more than four seeds of satellites have to grow. In their calculations, the satellites eventually reach the multi-resonant configuration, but Callisto is also in the resonance as predicted in the other previous works.

2.4.3 Constraints from Observations

As discussed in the previous section, many formation scenarios of Ganymede and the Galilean satellites have been proposed. On the other hand, the Jovian system has been the focus of many observational campaigns, from the ground, space telescopes, and space probes. Here we introduce the most important characteristics of Ganymede and the Galilean moons and compare the validity of the proposed formation scenarios in light of these characteristics. We have already discussed the mass of the moons in the previous section, so now we focus on the other characteristics: resonances, ice mass fractions, and internal structures.

2.4.3.1 Orbital Positions and Resonances

First, the Galilean moons are in a very peculiar orbital configuration; Io-Europa and Europa-Ganymede are in the 2:1 MMRs, which constitutes a Laplace resonance. However, the orbit of Callisto is not in any resonances, which implies that there are some differences in the formation process of Callisto compared to the others. There are two classical models to make the resonance chain system; (1) Ganymede migrates fastest because of its largest size and catches up Europa and Io as the gas disc disappears before they fall into Jupiter (Peale and Lee, 2002), and (2) the orbits of the moons evolve with differential tidal expansion by Jupiter after the gas disc disappears (e.g., Yoder, 1979). Peale and Lee (2002) showed that the former way can explain the Laplace relation of the inner three satellites better than the latter. In these two scenarios, the migration of the satellites due to the gravitational interaction with the CJD is not considered. However, as we discussed in the previous sections, the migration timescale of the satellites is much shorter than the growth timescale (and the depletion timescale of the gas disc) in any proposed scenarios. Therefore, it is natural to postulate that there is a mechanism to stop the migration during the formation process. Moreover, if the stopping occurs at the inner edge of the disc, the resonance chain of the inner three Galilean satellites can be explained naturally; the satellites form further out and migrate inward, resulting in the capture into the resonances one by one. This idea has been adopted in both satellitesimal and pebble accretion scenarios (Sasaki et al., 2010; Ogihara and Ida, 2012; Fujii et al., 2017; Shibaike et al., 2019;

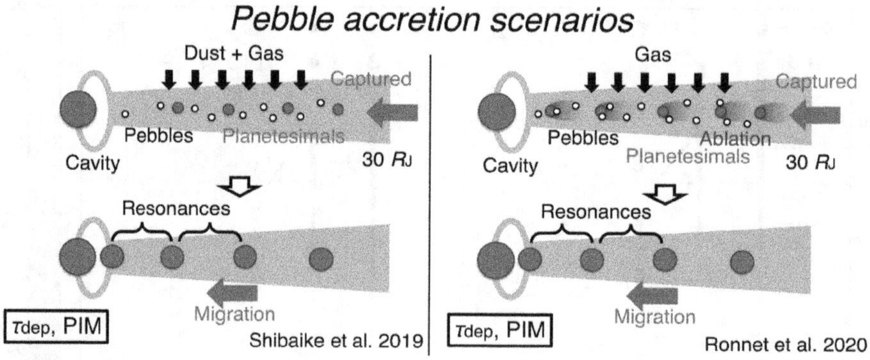

Figure 2.7 Schematic pictures of the proposed pebble accretion scenarios. The final mass of the moons is determined by the depletion timescale of the CJD (and CSD) and the pebble isolation mass of the satellites in both scenarios. The representative reference papers are also shown.

Ronnet and Johansen, 2020; Batygin and Morbidelli, 2020; Madeira et al., 2021).

The halting mechanism is still controversial. The most investigated mechanism is the presence of a disc inner cavity formed by the magnetic field of Jupiter (Takata and Stevenson, 1996; Sasaki et al., 2010; Batygin, 2018; Shibaike et al., 2019). It is generally accepted that the magnetic field of Jupiter was stronger than today, causing magnetospheric accretion and opening a cavity, provided the gas accretion rate is small enough (e.g., Takata and Stevenson, 1996; Christensen et al., 2009). The orbital position of the inner edge of the disc can be estimated from the balance between the torque due to the Jupiter-disc magnetic interaction and the viscous torque of the disc (e.g., Armitage, 2010). The position can then be around 4–5 R_J during the satellite formation (Batygin, 2018; Batygin and Morbidelli, 2020; Ginzburg and Chiang, 2020). Also, the edge moves outward as the gas accretion rate decreases until it reaches the corotation radius (Armitage, 2010), so that Io also migrates together if the satellite has already been trapped there (Ogihara et al., 2010; Liu et al., 2017). The formation and evolution theories of Jupiter show that the planet was larger, and its rotation rate was smaller than the current properties (e.g., Fortney et al., 2011). This suggests that the corotation radius was located around $r \approx 5\ R_J$, which is further out than the current one ($r_{co} = 2.25\ R_J$) but close to the current position of Io (Shibaike et al., 2019).

Incomplete cavities can also stop the migration. Photophoresis is a transfer mechanism whereby gas molecules collide with photo-irradiated dust particles whose motion reflects the surrounding temperature gradient (e.g., Krauss and Wurm, 2005). Arakawa and Shibaike (2019) found that the photophoresis can make a dust-poor region in the inner part of the CJD, which causes MRI only there, and the gas surface density of the inner region decreases. Fujii et al. (2017) found that there is an opacity transition due to the dust sublimation around the orbit of Io, which changes the radial dependence of the temperature and gas surface density distribution, and the gradients are sufficiently steep to stop the migration of the satellites.

The reason why only the orbit of Callisto is not in any resonances is an unsolved issue. If Callisto forms through the same process with the other satellites, Callisto also migrates inward and should be captured into a resonance with Ganymede. Oberg et al. (2020) found that rapid external photoevaporation following the end of gas accretion can stop the migration of Callisto and prevent it from being captured into the resonance. Some other works mentioned the possibility that Callisto escapes from the resonance by dynamical instability after the gas disc has disappeared (Shibaike et al., 2019; Ronnet and Johansen, 2020). In the scenario by Batygin and Morbidelli (2020), the migration of Callisto is caused by the gravitational interaction with neighboring satellitesimals (gravitationally stirring (Safranov, 1969)), but the orbital excitation of the neighboring debris stops the satellitesimal-driven migration before the satellite is captured to the resonance.

2.4.3.2 *Ice Mass Fractions*

The compositional characteristics of the Galilean moons can also constrain the formation history. The ice mass fractions of the satellites increase with their distance from Jupiter; Io is dry, Europa consists of 6–9 wt% ice, while Ganymede and Callisto have ice mass fractions of about 50 wt% (Kuskov and Kronrod, 2005). The temperature of the CJD decreases with the distance from Jupiter in any heating models, so that the variations can be explained by that qualitatively. However, it is difficult to explain that quantitatively, especially the ice mass fraction of Europa, which is rather small compared to those of the outer two moons. Some previous works based on the satellitesimal accretion scenario investigated the ice mass fractions of the satellites and found that the variation may be explained by the appropriate position and migration of the snowline (Canup and Ward, 2009; Sasaki et al., 2010; Ogihara and Ida, 2012). However, Dwyer et al. (2013) found that the rocky and icy satellitesimals must be radially mixed beyond the snowline, which makes it difficult to explain the variations. On the other hand, the pebble accretion scenario should solve this issue because of the size of the icy materials; pebbles are much smaller than satellitesimals and sublimate much more quickly after penetrating the inside of the snowline. Ronnet et al. (2017) found that the balance between the sublimation and drift timescales can explain the small ice fraction of Europa (with fixed snowline). Shibaike et al. (2019) also explained the small fraction by the small amount of accretion of icy pebbles due to the migration of snowline at very final phase of the formation (see Figure 2.6).

Bierson and Nimmo (2020) argued that the ice mass fractions can be explained by the hydrodynamic escape, the vapor loss from the water atmosphere, when the accretion timescale is longer than 0.3 Myr in the satellitesimal accretion scenario by

Canup and Ward (2002). The warm environment at the inner part of the CJD and the deposited accretion heating lead to the formation of water ocean and atmosphere on the early surfaces of Io and Europa. The final density of the satellites strongly depends on the background temperature, in other words, the disc temperature at each orbital position. As a result, Io and Europa lose their water inventories completely and mostly, respectively.

We note that the temperature of the CJD does not significantly depend on the orbital position of Jupiter in the CSD (Heller and Pudritz, 2015).

2.4.3.3 Internal Structures

The internal structures of the moons can also constrain the formation history of Ganymede and the other Galilean moons. The internal structures were inferred by the gravity measurements performed by the Galileo spacecraft and showed that the three inner moons are fully differentiated to iron cores, silicate mantles (Io), and additionally icy outer mantles (Europa and Ganymede), but Callisto is partially differentiated (Anderson et al., 2001; Schubert et al., 2004). First, to avoid the differentiation by the accretion heating, the formation timescale has to be long enough (longer than 0.6 Myr (Barr and Canup, 2008)). In most of the satellitesimal accretion scenarios, the formation timescale is $\sim$ 0.1 Myr or shorter (Canup and Ward, 2006; Sasaki et al., 2010; Cilibrasi et al., 2018). Moreover, the dichotomy between Ganymede and Callisto requires fine-tuned conditions. Since the two satellites have similar mass and compositions, it is natural to expect that they have similar thermal and formation history (Barr and Canup, 2008). The difference of the formation timing after the calcium–aluminium-rich inclusion (CAI) formation could make the dichotomy because the ^{26}Al decay radiogenic heating inside the moons depends on the accreting timing of the materials. In the slow pebble accretion scenario by Shibaike et al. (2019), both Ganymede and Callisto accrete materials slowly enough ($\sim$ 10 Myr) while Callisto forms very late, so that only the interior of Ganymede melts due to the ^{26}Al decay heating. On the other hand, in the scenario by Batygin and Morbidelli (2020), Callisto forms after the gas disc disappears inside the satellitesimal disc with slow accretion timescale.

Alternatively, if both of the interiors of Ganymede and Callisto are not melted during their formation, subsequent processes may be able to melt only the interior of Ganymede: (1) the release of gravitational energy during the differentiation or (2) the Late Heavy Bombardment, the concentration of impacts around 4 Gyr (Friedson and Stevenson, 1983; Barr and Canup, 2010).

We note that the "observed" value of the moment of inertia (MOI) of Callisto is just an estimation from the in-situ gravitational field measurements assuming that Callisto is a simple hydrodynamic two-layer body. Gao and Stevenson (2013) found that the estimated value of MOI of Callisto can also be reproduced even in the case that the interior is fully differentiated but has slight degree 2 distortions away from the hydrostatic shape. We also note that compositional gradients could cause the complete differentiation of Callisto after its formation by preventing efficient transport of radiogenic heating through convection (O'Rourke and Stevenson, 2014).

2.4.3.4 Toward the JUICE Mission and Future Other Observations

As we discussed in the previous sections, many formation scenarios for Ganymede and the Galilean moons have been proposed. Each scenario has advantages and disadvantages for explaining the current characteristics of the system and it is difficult to decide the most accurate one. One missing point which has not been frequently investigated is the history of the (icy) materials that compose the moons. The compositions depend on the thermal history and the trajectory of the icy materials in the CSD and CJD, which may constrain the formation scenarios of the satellites. For example, the icy materials evaporate and then re-condense in the CJD in the classical satellitesimal accretion scenarios and in the recent pebble accretion scenario by Ronnet and Johansen (2020) (Horner et al., 2008). On the other hand, in the slow pebble accretion scenario by Shibaike et al. (2019), the ice condenses at the outer region of the CSD, and the icy materials are transferred to the Jovian systems and accreted to the moons without evaporation through the entire journey. Thus, the icy materials of the Galilean moons might have the same compositional characteristics with the icy dust condensed at the outer region. The characteristics are, for example, the amount of noble gas and the deuterium-to-hydrogen ratio (D/H) and other isotope ratios, and they depend on the surrounding environments during the condensation. The characteristics of these building blocks may have left observable signatures in the current thin atmosphere of the Galilean moons and the water plumes from Europa, which will be observed by the Jupiter Icy Moons Explorer (JUICE) mission led by the European Space Agency (ESA) and by other future observations. Moreover, if the gradient of the ice mass fractions of the Galilean moons is the result of the hydrodynamic escape (see Section 2.4.3.2), Europa would have higher D/H and $^{18}O/^{16}O$ ratios compared to those of Ganymede and Callisto (Bierson and Nimmo, 2020).

2.5 CONCLUSION

We reviewed the origin of Ganymede and the Galilean moons in this chapter, dividing it into the three main processes: (1) the formation of the CJD, (2) the transport of the solid materials from the CSD to the CJD, and (3) the satellite formation in the CJD by the supplied materials.

The CJD is a gas disc that forms around Jupiter as a byproduct of the gas accretion of the planet. Its basic structure and formation process have been revealed by both analytical modelings and hydrodynamic simulations. An envelope forms around the core of Jupiter, and then the envelope becomes a disc as the planet grows heavier. The gas accretion rate then becomes smaller as the gas surface density of the CSD decreases, which makes the CJD colder and allows the survival of the icy materials of the moons. Thus, the icy satellites should form in the later stage of the planet formation, when the gas gap has already been created by Jupiter. It has been revealed that the gas accretes to Jupiter through the CJD in spirals, but the detailed structures (such as gas surface density and direction of the gas flows) and the accretion mechanisms (such as week MRI or shock waves) are still controversial. Future numerical simulations including the effects of magnetic fields, tidal force, and photoevaporation will obtain more detailed characteristics of the CJD. Moreover, observations of CPDs around exoplanets have already started, which will result in a large amount of information about the general characteristics of the discs.

The transport of the solid material of Ganymede and the other moons to the CJD is a newly discussed issue of their formation. The transport dominates the formation processes, but the established theory has not been developed yet. As we discussed above, when the satellite formation starts, the gas gap structure has already been formed, which prevents the transport of pebbles into the CJD. Therefore, the solid materials should be supplied as small dust particles coupled with the gas or as large planetesimals decoupled from the gas. The pebbles trapped at the edge of the gap can form planetesimals by streaming instability and/or mutual collision. At the same time, small fragments by collision of the trapped pebbles and formed planetesimals may also contribute to the supply of solid materials to the CJD. The supply of small particles strongly depends on the gas motion, so that recent and future 3D gas+dust numerical simulations will be important. The planetesimal disc in the CSD can also present a gap structure around Jupiter but another heavy body, such as Saturn, can scatter the planetesimals in the CSD, and later on the planetesimals can be captured by the CJD. The captured planetesimals can take part in the formation of the large moons in different ways: alternatives to the satellitesimals formed in-situ, seeds of the satellites, and a source of dust and pebbles through the ablation.

The formation scenarios of Ganymede and the Galilean satellites have been discussed for a very long time. A lot of scenarios have been proposed, but they can be classified into the two groups following the main accretion processes of solid materials: satellitesimal accretion and pebble accretion. In the satellitesimal accretion scenarios, the fundamental issue is the formation of satellitesimals; dust particles drift inward as pebbles before they grow to satellitesimals by their mutual collisions in the CJD. Some scenarios which can avoid this drift barrier have been proposed. For example, if there is an outward flow on the midplane, it gathers pebbles at one place, and satellitesimals can form there by the streaming instability from the gathered pebbles. Another way is to use the planetesimals captured by the CJD as the alternatives of the satellitesimals. In the pebble accretion scenarios, the drifting pebbles are the primary building blocks of the moons, so that the dust particles do not need to grow large to satellitesimals. Pebbles can also be created by the ablation of the captured planetesimals. The observed characteristics of the large satellites (mass, orbits, ice mass fractions, and internal structures) can contribute to constraining their formation scenarios.

In both scenarios, the migration of the satellites is important. In the classical scenarios, the final mass of the moons is determined by the balance of the growth timescale and migration timescale, which requires a large amount of continuous supplies of solid materials to the CJD. Thus, most recent scenarios consider the cases where the growth time scale is longer than the migration timescale, but the migration stops by the inner edge of the CJD created by the magnetic field of Jupiter. This situation is consistent with the observed resonant orbits of the inner three Galilean moons but not with the orbit of Callisto which is not in any resonances. The ice mass fractions of the moons are roughly consistent with the radial distribution of the temperature of the CJD, but the fine-tuning of the parameters is necessary in any scenarios for reproducing the characteristics quantitatively. The observed characteristics that the inner three moons are fully differentiated but Callisto is only partially differentiated also require fine-tuning in any formation scenarios. In conclusion, the decisive evidence of the one established scenario has not been found yet. Future more detailed observations by JUICE and other missions will bring us more information to find the true origin of Ganymede and the Galilean moons.

REFERENCES

Adachi, I., Hayashi, C. and Nakazawa, K. (1976), 'The gas drag effect on the elliptic motion of a solid body in the primordial solar nebula', *Progress of Theoretical Physics* **56**(6), 1756–1771.

Agnor, C. B. and Hamilton, D. P. (2006), 'Neptune's capture of its moon Triton in a binary–planet gravitational encounter', *Nature* **441**(7090), 192.

Alexander, R. D., Clarke, C. J. and Pringle, J. E. (2006a), 'Photoevaporation of protoplanetary discs – i. hydrodynamic models', *Monthly Notices of the Royal Astronomical Society* **369**(1), 216.

Alexander, R. D., Clarke, C. J. and Pringle, J. E. (2006b), 'Photoevaporation of protoplanetary discs – ii. evolutionary models and observable properties', *Monthly Notices of the Royal Astronomical Society* **369**(1), 229.

Alibert, Y., Mousis, O. and Benz, W. (2005), 'Modeling the Jovian subnebula-i. thermodynamic conditions and migration of protosatellites', *Astronomy & Astrophysics* **439**(3), 1205–1213.

Anderson, J., Jacobson, R. McElrath, T. et al. (2001), 'Shape, mean radius, gravity field, and interior structure of Callisto', *Icarus* **153**(1), 157–161.

Aoyama, Y. and Ikoma, M. (2019), 'Constraining planetary gas accretion rate from hα line width and intensity: Case of pds 70 b and c', *The Astrophysical Journal Letters* **885**(2), L29.

Arakawa, S. and Shibaike, Y. (2019), 'Photophoresis in the circumjovian disk and its impact on the orbital configuration of the Galilean satellites', *Astronomy & Astrophysics* **629**, A106.

Armitage, P. J. (2010), Astrophysics of Planet Formation, Cambridge University Press.

Ataiee, S., Baruteau, C., Alibert, Y. and Benz, W. (2018), 'How much does turbulence change the pebble isolation mass for planet formation?', *Astronomy & Astrophysics* **615**, A110.

Ayliffe, B. A. and Bate, M. R. (2009), 'Gas accretion on to planetary cores: Three-dimensional self-gravitating radiation hydrodynamical calculations', *Monthly Notices of the Royal Astronomical Society* **393**(1), 49–64.

Ayliffe, B. A., Laibe, G., Price, D. J. and Bate, M. R. (2012), 'On the accumulation of planetesimals near disc gaps created by protoplanets', *Monthly Notices of the Royal Astronomical Society* **423**(2), 1450–1462.

Bai, X.-N. (2015), 'Hall effect controlled gas dynamics in protoplanetary disks. ii. full 3d simulations toward the outer disk', *The Astrophysical Journal* **798**(2), 84.

Balbus, S. A. and Hawley, J. F. (1991), 'A powerful local shear instability in weakly magnetized disks. i-linear analysis. ii-nonlinear evolution', *The Astrophysical Journal* **376**, 214–233.

Balbus, S. A. and Hawley, J. F. (1998), 'Instability, turbulence, and enhanced transport in accretion disks', *Reviews of Modern Physics* **70**(1), 1.

Barr, A. C. and Canup, R. M. (2008), 'Constraints on gas giant satellite formation from the interior states of partially differentiated satellites', *Icarus* **198**(1), 163–177.

Barr, A. C. and Canup, R. M. (2010), 'Origin of the Ganymede–Callisto dichotomy by impacts during the late heavy bombardment', *Nature Geoscience* **3**(3), 164.

Batygin, K. (2018), 'On the terminal rotation rates of giant planets', *The Astronomical Journal* **155**(4), 178.

Batygin, K. and Morbidelli, A. (2020), 'Formation of giant planet satellites', *The Astrophysical Journal* **894**(2), 143.

Benisty, M., Bae, J., Facchini, S. et al. (2021), 'A circumplanetary disk around pds70c', *The Astrophysical Journal Letters* **916**(1), L2.

Bierson, C. J. and Nimmo, F. (2020), 'Explaining the Galilean satellites' density gradient by hydrodynamic escape', *The Astrophysical Journal Letters* **897**(2), L43.

Canup, R. M. and Ward, W. R. (2002), 'Formation of the Galilean satellites: Conditions of accretion', *The Astronomical Journal* **124**(6), 3404.

Canup, R. M. and Ward, W. R. (2006), 'A common mass scaling for satellite systems of gaseous planets', *Nature* **441**(7095), 834–839.

Canup, R. M. and Ward, W. R. (2009), 'Origin of Europa and the Galilean satellites', *Europa, University of Arizona Press in collaboration with Lunar and Planetary Institute, Tucson* pp. 59–83.

Chatterjee, S. and Tan, J. C. (2013), 'Inside-out planet formation', *The Astrophysical Journal* **780**(1), 53.

Chen, C., Yang, C.-C., Martin, R. G. and Zhu, Z. (2021), 'The evolution of a circumplanetary disc with a dead zone', *Monthly Notices of the Royal Astronomical Society* **500**(3), 2822–2830.

Christensen, U. R., Holzwarth, V. and Reiners, A. (2009), 'Energy flux determines magnetic field strength of planets and stars', *Nature* **457**(7226), 167.

Christiaens, V., Cantalloube, F., Casassus, S. et al. (2019), 'Evidence for a circumplanetary disk around protoplanet pds 70 b', *The Astrophysical Journal Letters* **877**(2), L33.

Cilibrasi, M., Szulágyi, J., Grimm, S. L. and Mayer, L. [2021], 'An n-body population synthesis framework for the formation of moons around jupiter-like planets', Monthly Notices of the Royal Astronomical Society 504(4), 5455–5474.

Cilibrasi, M., Szulágyi, J., Mayer, L. et al. (2018), 'Satellites form fast & late: A population synthesis for the Galilean moons', *Monthly Notices of the Royal Astronomical Society* **480**(4), 4355–4368.

Crida, A. and Charnoz, S. (2012), 'Formation of regular satellites from ancient massive rings in the solar system', *Science* **338**(6111), 1196–1199.

D'Angelo, G., Henning, T. and Kley, W. (2002), 'Nested-grid calculations of disk-planet interaction', *Astronomy & Astrophysics* **385**, 647–670.

D'Angelo, G. and Podolak, M. (2015), 'Capture and evolution of planetesimals in circumjovian disks', *The Astrophysical Journal* **806**(2), 203.

Drażkowska, J. and Dullemond, C. P. (2014), 'Can dust coagulation trigger streaming instability?', *Astronomy & Astrophysics* **572**, A78.

Drażkowska, J. and Dullemond, C. P. (2018), 'Planetesimal formation during protoplanetary disk buildup', *Astronomy & Astrophysics* **614**, A62.

Drażkowska, J. and Szulágyi, J. (2018), 'Dust evolution and satellitesimal formation in circumplanetary disks', *The Astrophysical Journal* **866**(2), 142.

Dwyer, C., Nimmo, F., Ogihara, M. and Ida, S. (2013), 'The influence of imperfect accretion and radial mixing on ice: Rock ratios in the Galilean satellites', *Icarus* **225**(1), 390–402.

Eriksson, L. E., Johansen, A. and Liu, B. (2020), 'Pebble drift and planetesimal formation in protoplanetary discs with embedded planets', *Astronomy & Astrophysics* **635**, A110.

Estrada, P. R., Mosqueira, I., Lissauer, J., D'Angelo, G. and Cruikshank, D. (2009), 'Formation of Jupiter and conditions for accretion of the Galilean satellites', in R. T. Pappalardo, W. B. McKinnon and K. Khurana, eds., *Europa*, University of Arizona Press, Tucson, p. 27.

Fortney, J. J., Ikoma, M., Nettelmann, N., Guillot, T. and Marley, M. [2011], 'Self-consistent model atmospheres and the cooling of the solar system's giant planets', The Astrophysical Journal 729(1), 32.

Friedson, A. and Stevenson, D. (1983), 'Viscosity of rock-ice mixtures and applications to the evolution of icy satellites', *Icarus* **56**(1), 1–14.

Fujii, Y. I., Kobayashi, H., Takahashi, S. Z. and Gressel, O. (2017), 'Orbital evolution of moons in weakly accreting circumplanetary disks', *The Astronomical Journal* **153**(4), 194.

Fujii, Y. I., Okuzumi, S. and Inutsuka, S.-i. (2011), 'A fast and accurate calculation scheme for ionization degrees in protoplanetary and circumplanetary disks with charged dust grains', *The Astrophysical Journal* **743**(1), 53.

Fujii, Y. I., Okuzumi, S., Tanigawa, T. and Inutsuka, S. (2014), 'On the viability of the magnetorotational instability in circumplanetary disks', *The Astrophysical Journal* **785**(2), 101.

Fujita, T., Ohtsuki, K., Tanigawa, T. and Suetsugu, R. (2013), 'Capture of planetesimals by gas drag from circumplanetary disks', *The Astronomical Journal* **146**(6), 140.

Fung, J. and Chiang, E. (2016), 'Gap opening in 3D: Single-planet gaps', *The Astrophysical Journal* **832**(2), 105.

Gao, P. and Stevenson, D. J. (2013), 'Nonhydrostatic effects and the determination of icy satellites' moment of inertia', *Icarus* **226**(2), 1185–1191.

Ginzburg, S. and Chiang, E. (2020), 'Breaking the centrifugal barrier to giant planet contraction by magnetic disc braking', *Monthly Notices of the Royal Astronomical Society: Letters* **491**(1), L34–L39.

Gressel, O., Nelson, R. P., Turner, N. J. and Ziegler, U. (2013), 'Global hydromagnetic simulations of a planet embedded in a dead zone: Gap opening, gas accretion, and formation of a protoplanetary jet', *The Astrophysical Journal* **779**(1), 59.

Haffert, S., Bohn, A., de Boer, J. et al. (2019), 'Two accreting protoplanets around the young star PDS 70', *Nature Astronomy* **3**(8), 749–754.

Hashimoto, J., Aoyama, Y., Konishi, M. et al. (2020), 'Accretion properties of PDS 70b with muse', *The Astronomical Journal* **159**(5), 222.

Hayashi, C. (1981), 'Structure of the solar nebula, growth and decay of magnetic fields and effects of magnetic and turbulent viscosities on the nebula', *Progress of Theoretical Physics Supplement* **70**, 35–53.

Hayashi, C., Nakazawa, K. and Adachi, I. (1977), 'Long-term behavior of planetesimals and the formation of the planets', *Publications of the Astronomical Society of Japan* **29**, 163–196.

Heller, R. and Pudritz, R. (2015), 'Conditions for water ice lines and Mars-mass exomoons around accreting super-Jovian planets at 1–20 AU from Sun-like stars', *Astronomy & Astrophysics* **578**, A19.

Homma, T., Ohtsuki, K., Maeda, N. et al. (2020), 'Delivery of pebbles from the protoplanetary disk into circumplanetary disks', *The Astrophysical Journal* **903**(2), 98.

Horner, J., Mousis, O., Alibert, Y., Lunine, J. I. and Blanc, M. (2008), 'Constraints from deuterium on the formation of icy bodies in the Jovian system and beyond', *Planetary and Space Science* **56**(12), 1585–1595.

Isella, A., Benisty, M., Teague, R. et al. (2019), 'Detection of continuum submillimeter emission associated with candidate protoplanets', *The Astrophysical Journal Letters* **879**(2), L25.

Johansen, A. and Youdin, A. (2007), 'Protoplanetary disk turbulence driven by the streaming instability: nonlinear saturation and particle concentration', *The Astrophysical Journal* **662**(1), 627.

Kanagawa, K. D., Muto, T., Tanaka, H. et al. (2015), 'Mass estimates of a giant planet in a protoplanetary disk from the gap structures', *The Astrophysical Journal Letters* **806**(1), L15.

Keith, S. L. and Wardle, M. (2014), 'Accretion in giant planet circumplanetary discs', *Monthly Notices of the Royal Astronomical Society* **440**(1), 89.

Keppler, M., Benisty, M., Müller, A. et al. (2018), 'Discovery of a planetary-mass companion within the gap of the transition disk around pds 70', *Astronomy & Astrophysics* **617**, A44.

Kley, W., D'Angelo, G. and Henning, T. (2001), 'Three-dimensional simulations of a planet embedded in a protoplanetary disk', *The Astrophysical Journal* **547**(1), 457.

Kley, W. and Dirksen, G. (2006), 'Disk eccentricity and embedded planets', *Astronomy & Astrophysics* **447**(1), 369–377.

Kobayashi, H., Ormel, C. W. and Ida, S. (2012), 'Rapid formation of saturn after jupiter completion', *The Astrophysical Journal* **756**(1), 70.

Krauss, O. and Wurm, G. (2005), 'Photophoresis and the pile-up of dust in young circumstellar disks', *The Astrophysical Journal* **630**(2), 1088.

Kuskov, O. and Kronrod, V. (2005), 'Models of the internal structure of callisto', *Solar System Research* **39**(4), 283–301.

Lambrechts, M. and Johansen, A. (2012), 'Rapid growth of gas-giant cores by pebble accretion', *Astronomy & Astrophysics* **544**, A32.

Lambrechts, M., Johansen, A. and Morbidelli, A. (2014), 'Separating gas-giant and ice-giant planets by halting pebble accretion', *Astronomy & Astrophysics* **572**, A35.

Lambrechts, M., Lega, E., Nelson, R. P. et al. (2019), 'Quasi-static contraction during runaway gas accretion onto giant planets', *Astronomy & Astrophysics* **630**, A82.

Liu, B. and Ormel, C. W. (2018), 'Catching drifting pebbles-i. enhanced pebble accretion efficiencies for eccentric planets', *Astronomy & Astrophysics* **615**, A138.

Liu, B., Ormel, C. W. and Lin, D. N. (2017), 'Dynamical rearrangement of super-earths during disk dispersal-i. outline of the magnetospheric rebound model', *Astronomy & Astrophysics* **601**, A15.

Lubow, S. H., Seibert, M. and Artymowicz, P. (1999), 'Disk accretion onto high-mass planets', *The Astrophysical Journal* **526**(2), 1001.

Lunine, J. I. and Stevenson, D. J. (1982), 'Formation of the Galilean satellites in a gaseous nebula', *Icarus* **52**(1), 14–39.

Lyra, W., Johansen, A., Klahr, H. and Piskunov, N. (2009), 'Standing on the shoulders of giants: trojan earths and vortex trapping in low mass self-gravitating protoplanetary disks of gas and solids', *Astronomy & Astrophysics* **493**(3), 1125–1139.

Machida, M. N., Inutsuka, S.-i. and Matsumoto, T. (2006), 'Outflows driven by giant protoplanets', *The Astrophysical Journal Letters* **649**(2), L129.

Madeira, G., Izidoro, A. and Winter, S. M. G. (2021), 'Building the Galilean moons system via pebble accretion and migration: A primordial resonant chain', Monthly Notices of the Royal Astronomical Society **504**(2), 1854.

Martin, R. G. and Lubow, S. H. (2011), 'Tidal truncation of circumplanetary discs', *Monthly Notices of the Royal Astronomical Society* **413**(2), 1447.

Mayor, M. and Queloz, D. (1995), 'A Jupiter-mass companion to a solar-type star', *Nature* **378**(6555), 355–359.

Miguel, Y. and Ida, S. (2016), 'A semi-analytical model for exploring Galilean satellites formation from a massive disk', *Icarus* **266**, 1–14.

Mitchell, T. R. and Stewart, G. R. (2011), 'Photoevaporation as a truncation mechanism for circumplanetary disks', *The Astronomical Journal* **142**(5), 168.

Morbidelli, A., Szulágyi, J., Crida, A. et al. (2014), 'Meridional circulation of gas into gaps opened by giant planets in three-dimensional low-viscosity disks', *Icarus* **232**, 266–270.

Mosqueira, I. and Estrada, P. R. (2003 *a*), 'Formation of the regular satellites of giant planets in an extended gaseous nebula i: subnebula model and accretion of satellites', *Icarus* **163**(1), 198–231.

Mosqueira, I. and Estrada, P. R. (2003 *b*), 'Formation of the regular satellites of giant planets in an extended gaseous nebula ii: satellite migration and survival', *Icarus* **163**(1), 232–255.

Mosqueira, I., Estrada, P. and Turrini, D. (2010), 'Planetesimals and satellitesimals: formation of the satellite systems', *Space Science Reviews* **153**(1), 431–446.

Müller, A., Keppler, M., Henning, T., Samland, M., Chauvin, G., Beust, H., Maire, A.-L., Molaverdikhani, K., van Boekel, R., Benisty, M. et al. (2018), 'Orbital and atmospheric characterization of the planet within the gap of the pds 70 transition disk', *Astronomy & Astrophysics* **617**, L2.

Nakamoto, T. and Nakagawa, Y. (1994), 'Formation, early evolution, and gravitational stability of protoplanetary disks', *The Astrophysical Journal* **421**, 640–650.

Oberg, N., Kamp, I., Cazaux, S. and Rab, C. (2020), 'Photoevaporation of the jovian circumplanetary disk-i. explaining the orbit of callisto and the lack of outer regular satellites', *Astronomy & Astrophysics* **638**, A135.

Ogihara, M., Duncan, M. J. and Ida, S. (2010), 'Eccentricity trap: Trapping of resonantly interacting planets near the disk inner edge', *The Astrophysical Journal* **721**(2), 1184.

Ogihara, M. and Ida, S. (2012), 'N-body simulations of satellite formation around giant planets: Origin of orbital configuration of the Galilean moons', *The Astrophysical Journal* **753**(1), 60.

Okuzumi, S., Momose, M., Sirono, S., Kobayashi, H. and Tanaka, H. (2016), 'Sintering-induced dust ring formation in protoplanetary disks: Application to the hl tau disk', *The Astrophysical Journal* **821**(2), 82.

Okuzumi, S., Tanaka, H., Kobayashi, H. and Wada, K. (2012), 'Rapid coagulation of porous dust aggregates outside the snow line: A pathway to successful icy planetesimal formation', *The Astrophysical Journal* **752**(2), 106.

Ormel, C. and Cuzzi, J. (2007), 'Closed-form expressions for particle relative velocities induced by turbulence', *Astronomy & Astrophysics* **466**(2), 413–420.

Ormel, C. and Klahr, H. (2010), 'The effect of gas drag on the growth of protoplanets-analytical expressions for the accretion of small bodies in laminar disks', *Astronomy & Astrophysics* **520**, A43.

Ormel, C. W. and Liu, B. (2018), 'Catching drifting pebbles-ii. a stochastic equation of motion for pebbles', *Astronomy & Astrophysics* **615**, A178.

O'rourke, J. G. and Stevenson, D. J. (2014), 'Stability of ice/rock mixtures with application to a partially differentiated titan', *Icarus* **227**, 67–77.

Paardekooper, S.-J. (2007), 'Dust accretion onto high-mass planets', *Astronomy & Astrophysics* **462**(1), 355–369.

Paardekooper, S.-J. and Mellema, G. (2004), 'Planets opening dust gaps in gas disks', *Astronomy & Astrophysics* **425**(1), L9–L12.

Paardekooper, S.-J. and Mellema, G. (2006), 'Dust flow in gas disks in the presence of embedded planets', *Astronomy & Astrophysics* **453**(3), 1129–1140.

Peale, S. and Lee, M. H. (2002), 'A primordial origin of the laplace relation among the Galilean satellites', *Science* **298**(5593), 593–597.

Rivier, G., Crida, A., Morbidelli, A. and Brouet, Y. (2012), 'Circumplanetary discs as bottlenecks for gas accretion onto giant planets', *Astronomy & Astrophysics* **548**, A116.

Ronnet, T. and Johansen, A. (2020), 'Formation of moon systems around giant planets-capture and ablation of planetesimals as foundation for a pebble accretion scenario', *Astronomy & Astrophysics* **633**, A93.

Ronnet, T., Mousis, O. and Vernazza, P. (2017), 'Pebble accretion at the origin of water in Europa', *The Astrophysical Journal* **845**(2), 92.

Ronnet, T., Mousis, O., Vernazza, P., Lunine, J. I. and Crida, A. (2018), 'Saturn's formation and early evolution at the origin of Jupiter's massive moons', *The Astronomical Journal* **155**(5), 224.

Safranov, V. [1969], 'Evolution of the protoplanetary cloud and formation of the earth and planets', Moscow: Nauka Press (Translated into NASA TT F-677 1972).

Sasaki, T., Stewart, G. R. and Ida, S. (2010), 'Origin of the different architectures of the Jovian and Saturnian satellite systems', *The Astrophysical Journal* **714**(2), 1052.

Sato, T., Okuzumi, S. and Ida, S. (2016), 'On the water delivery to terrestrial embryos by ice pebble accretion', *Astronomy & Astrophysics* **589**, A15.

Schubert, G., Anderson, J., Spohn, T. and McKinnon, W. (2004), 'Interior composition, structure and dynamics of the Galilean satellites', *Jupiter: The planet, satellites and magnetosphere* **1**, 281–306.

Schulik, M., Johansen, A., Bitsch, B. and Lega, E. (2019), 'Global 3d radiation-hydrodynamic simulations of gas accretion: Opacity-dependent growth of saturn-mass planets', *Astronomy & Astrophysics* **632**, A118.

Schulik, M., Johansen, A., Bitsch, B., Lega, E. and Lambrechts, M. (2020), 'On the structure and mass delivery towards circumplanetary discs', *Astronomy & Astrophysics* **642**, A187.

Shakura, N. I. and Sunyaev, R. A. (1973), 'Black holes in binary systems: Observational appearance', *Astronomy & Astrophysics* **24**, 337–355.

Shibaike, Y. and Alibert, Y. (2020), 'Planetesimal formation at the gas pressure bump following a migrating planet: I. basic characteristics of the new formation model', *Astronomy & Astrophysics* **644**, A81.

Shibaike, Y., Okuzumi, S., Sasaki, T. and Ida, S. (2017), 'Satellitesimal formation via collisional dust growth in steady circumplanetary disks', *The Astrophysical Journal* **846**(81), 81.

Shibaike, Y., Ormel, C. W., Ida, S., Okuzumi, S. and Sasaki, T. (2019), 'The Galilean satellites formed slowly from pebbles', *The Astrophysical Journal* **885**(1), 79.

Stammler, S. M., Drażkowska, J., Birnstiel, T., Klahr, H., Dullemond, C. P. and Andrews, S. M. (2019), 'The dsharp rings: Evidence of ongoing planetesimal formation?', *The Astrophysical Journal* **884**(1), L5.

Stevenson, D. J. (1984), Composition, structure and evolution of Uranian and Neptunian satellites, *in* J. T. Bergstralh, ed., 'NASA Conference Publication', Vol. 2330 of *NASA Conference Publication*.

Stolker, T., Marleau, G. D., Cugno, G. et al. (2020), 'Miracles: Atmospheric characterization of directly imaged planets and substellar companions at 4-5 μm. ii. constraints on the mass and radius of the enshrouded planet pds 70 b', *Astronomy & Astrophysics* **644**, A13.

Suetsugu, R. and Ohtsuki, K. (2016), 'Capture of planetesimals by waning circumplanetary gas disks', *The Astrophysical Journal* **820**(2), 128.

Suetsugu, R. and Ohtsuki, K. (2017), 'Distribution of captured planetesimals in circumplanetary gas disks and implications for accretion of regular satellites', *The Astrophysical Journal* **839**(1), 66.

Suetsugu, R., Ohtsuki, K. and Fujita, T. (2016), 'Orbital characteristics of planetesimals captured by circumplanetary gas disks', *The Astronomical Journal* **151**(6), 140.

Szulágyi, J. (2017), 'Effects of the planetary temperature on the circumplanetary disk and on the gap', *The Astrophysical Journal* **842**(2), 103.

Szulágyi, J., Binkert, F. and Surville, C. (2022), 'Meridional circulation of dust and gas in the circumstellar disk: Delivery of solids onto the circumplanetary region', The Astrophysical Journal **924**(1), 1.

Szulágyi, J., Cilibrasi, M. and Mayer, L. (2018), 'In situ formation of icy moons of Uranus and Neptune', *The Astrophysical Journal Letters* **868**(1), L13.

Takata, T. and Stevenson, D. J. (1996), 'Despin mechanism for protogiant planets and ionization state of protogiant planetary disks', *Icarus* **123**(2), 404–421.

Taki, T., Fujimoto, M. and Ida, S. (2016), 'Dust and gas density evolution at a radial pressure bump in protoplanetary disks', *Astronomy & Astrophysics* **591**, A86.

Tanaka, H., Takeuchi, T. and Ward, W. R. (2002), 'Three-dimensional interaction between a planet and an isothermal gaseous disk. i. corotation and lindblad torques and planet migration', *The Astrophysical Journal* **565**(2), 1257.

Tanigawa, T., Maruta, A. and Machida, M. N. (2014), 'Accretion of solid materials onto circumplanetary disks from protoplanetary disks', *The Astrophysical Journal* **784**(2), 109.

Tanigawa, T., Ohtsuki, K. and Machida, M. N. (2012), 'Distribution of accreting gas and angular momentum onto circumplanetary disks', *The Astrophysical Journal* **747**(1), 47.

Teachey, A. and Kipping, D. M. (2018), 'Evidence for a large exomoon orbiting kepler-1625b', *Science Advances* **4**(10), eaav1784.

Turner, N. J., Lee, M. H. and Sano, T. (2014), 'Magnetic coupling in the disks around young gas giant planets', *The Astrophysical Journal* **783**(1), 14.

Wada, K., Tanaka, H., Okuzumi, S. et al. (2013), 'Growth efficiency of dust aggregates through collisions with high mass ratios', *Astronomy & Astrophysics* **559**, A62.

Wada, K., Tanaka, H., Suyama, T., Kimura, H. and Yamamoto, T. (2009), 'Collisional growth conditions for dust aggregates', *The Astrophysical Journal* **702**(2), 1490–1501.

Wang, J., Vigan, A., Lacour, S., Nowak, M., Stolker, T., De Rosa, R., Ginzburg, S., Gao, P., Abuter, R., Amorim, A. et al. [2021], 'Constraining the nature of the pds 70 protoplanets with vlti/gravity', *The Astronomical Journal* **161**(3), 148.

Ward, W. R. (1993), 'Density waves in the solar nebula: Planetesimal velocities', *Icarus* **106**(1), 274–287.

Ward, W. R. and Canup, R. M. (2010), 'Circumplanetary disk formation', *The Astronomical Journal* **140**(5), 1168.

Weidenschilling, S. J. [1977], 'Aerodynamics of solid bodies in the solar nebula', Monthly Notices of the Royal Astronomical Society 180, 57–70.

Yoder, C. F. (1979), 'How tidal heating in io drives the galilean orbital resonance locks', *Nature* **279**(5716), 767.

Youdin, A. N. and Goodman, J. (2005), 'Streaming instabilities in protoplanetary disks', *The Astrophysical Journal* **620**(1), 459.

Youdin, A. N. and Lithwick, Y. (2007), 'Particle stirring in turbulent gas disks: Including orbital oscillations', *Icarus* **192**(2), 588–604.

Zhu, Z., Ju, W. and Stone, J. M. (2016), 'Shock-driven accretion in circumplanetary disks: Observables and satellite formation', *The Astrophysical Journal* **832**(2), 193.

Zhu, Z., Nelson, R. P., Dong, R., Espaillat, C. and Hartmann, L. (2012), 'Dust filtration by planet-induced gap edges: Implications for transitional disks', *The Astrophysical Journal* **755**(1), 6.

3

The Rotation of Ganymede

ALEXANDER STARK, ROSE-MARIE BALAND, TIM VAN HOOLST, NICOLAS RAMBAUX, AND GREGOR STEINBRÜGGE

3.1 INTRODUCTION

The Galilean moon Ganymede is the largest moon in the solar system and orbits Jupiter at an average distance of slightly more than 1 million km. Owing to the strong gravitational field of Jupiter, a large tidal torque is acting on the satellite, which over a few ten thousand years has brought Ganymede to a state of rotational equilibrium, a configuration known as a Cassini state. This rotational state is characterized by a 1:1 spin-orbit resonance, which means that the satellite performs one rotation on its rotational axis as it completes a single orbit about Jupiter. The sidereal rotation period of Ganymede is only slightly longer than a week and amounts to 7 days, 3 hours, and 42.6 minutes (7.15455 days). Ganymede's orbit marginally differs from a circle, with an eccentricity oscillating between 0.0003 and 0.003. Owing to the eccentricity and an asymmetric mass distribution within the satellite, small oscillations around the resonant rotation rate, that is, the longitudinal librations, may occur. In the equilibrium Cassini state, the rotation axis is only slightly offset from the orbital plane normal and lies in a plane spanned by the normal vectors of the orbital and Laplace planes. Ganymede's rotational characteristics are closely tied to its interior structure and rheology. The connecting element is the gravitational torque, which depends on the orbit, the moments of inertia, and Ganymede's Love number k_2 (Van Hoolst et al., 2013).

Hence, measuring Ganymede's rotation can provide insights into the interior of the satellite and thus clues as to its evolution. In particular, the presence of a global subsurface water ocean will be studied further by future spacecraft observations, including the European Space Agency's JUpiter ICy moons Explorer (JUICE) mission (Grasset et al., 2013), through the effects of the ocean on the rotational dynamics. As of today, only limited measurements of Ganymede's rotation are available. In fact, only the presence of the 1:1 spin-orbit resonance is confirmed, and upper bounds for the obliquity and the libration amplitude at the orbital period were obtained by Zubarev et al. (2015) (see Section 3.5.3).

An accurate measurement of rotation is also a prerequisite for the realization of a new reference frame for Ganymede. Such a body-fixed reference frame is needed not only for cartography but also for the planning and georeferencing of observations carried out by spacecraft instruments.

This chapter is organized as follows. In Section 3.2 the Cassini state, that is, a state of rotational equilibrium, is introduced for a solid Ganymede. Next, we provide a description of a resonant rotation model, in which the satellite instantaneously follows all variations in the gravitational torque. Small deviations from the spin-orbit resonance, the physical librations in longitude, are discussed subsequently in Section 3.3. The librational response of Ganymede occurs at the orbital period but long-period librations are also likely to influence the rotational state and are introduced in this section. Moreover, the link between the libration amplitude, that is, the response of the satellite to the gravitational torque, and the interior are highlighted. Section 3.4 is devoted to the variation in orientation of Ganymede's rotational axis. This involves the characterization of the obliquity, that is, the angular separation between the rotational axis and the normal of the orbital plane, as well as the precession of the rotational axis. Again, we describe how these rotational aspects are linked to the interior. In Section 3.5, we outline the rotation model currently in use for Ganymede and discuss the available observational constraints. Finally, the chapter concludes in Section 3.6 with some open questions in the field.

3.2 CASSINI STATE

The Cassini state is a state of rotational equilibrium initially described by Cassini (1693) for the Moon. The famous Cassini laws, which characterize the Cassini state, assert that:

1. The Moon is in a synchronous rotational state, where the rotation period is equal to the orbital period.
2. The rotational axis of the Moon has a constant inclination (called *obliquity*) to the orbit plane and to the ecliptic plane.
3. The rotational axis is always located in the plane formed by the orbit plane normal and the ecliptic plane normal.

The first law is commonly described as the 1:1 spin-orbit resonance. Even before the Pioneer spacecrafts reached Jupiter in the early 1970s, Ganymede was believed to be in such a resonance. This expectation was confirmed by observations carried out by several spacecrafts visiting the Jovian system. However, no direct measurements of the rotation rate were published. The accuracy of such measurements was very limited and the rotation period was only constrained based on the orbital period (Davies et al., 1980; Davies and Katayama, 1981). The spin-orbit resonance is believed to have been formed within the first 10 000 years of Ganymede's evolution and is a result of the balance between the tidal dissipation factor and the permanent equatorial deformation of the body (Goldreich and Peale, 1966). A natural consequence of the spin-orbit resonance is that Ganymede faces the same hemisphere towards Jupiter. However, due to the elliptic orbit of Ganymede and the varying orbital velocity, the x-axis of Ganymede, that is, the long axis of Ganymede's ellipsoid, points exactly towards Jupiter only at the closest (*perijove*) and farthest (*apojove*) point of its orbit

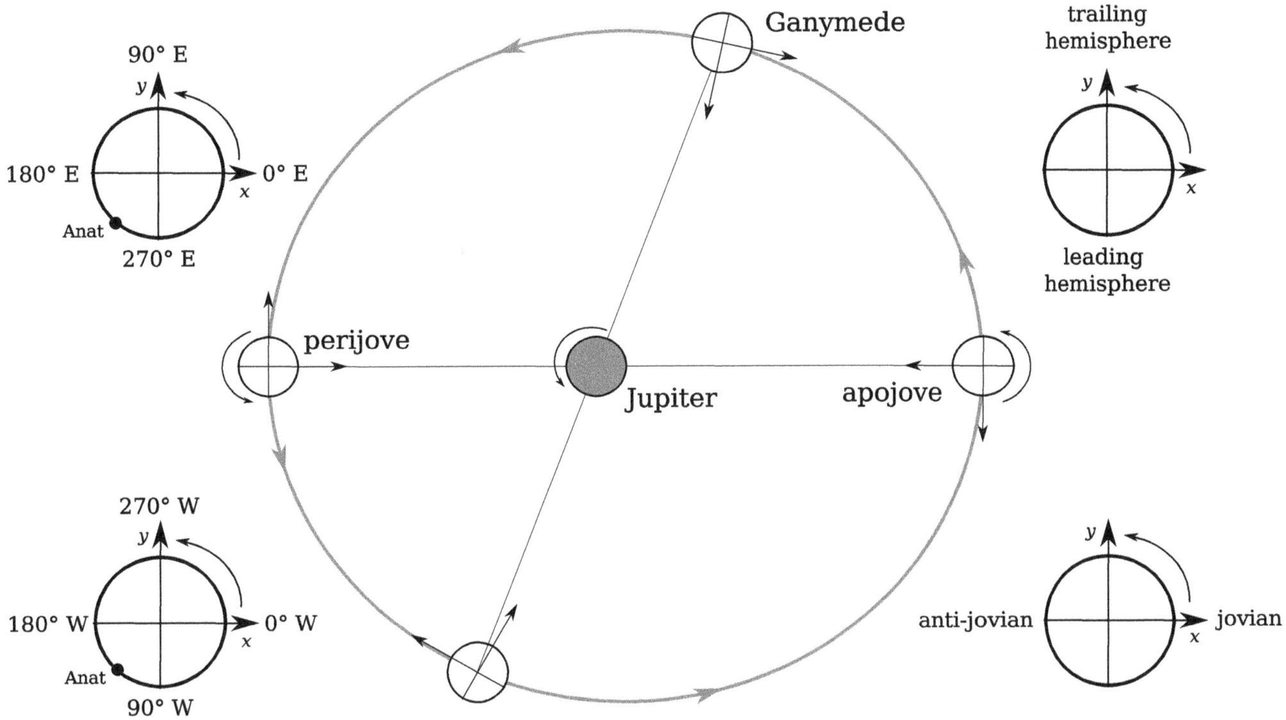

Figure 3.1 Schematic view of the orientation of Ganymede on its orbit around Jupiter. Ganymede's north pole (z-axis) is towards the reader. The crater Anat is used to define the prime meridian (see Section 3.5.2).

(see Fig. 3.1). In fact, at first order in eccentricity, the x-axis points towards the empty focus of the ellipse, and the small angular separation between the direction of the empty focus and Jupiter is commonly known as an *optical libration*. This angle is also enforcing the physical libration in longitude (see Section 3.3) as it determines the dynamics of the gravitational torque (Eq. 3.1). Ganymede orbits Jupiter with the $-y$ direction of its body-fixed frame opposite to the direction of the orbital motion (Fig. 3.1). The corresponding division in *trailing* ($+y$) and *leading* ($-y$) hemispheres is particularly relevant for the interaction between Ganymede's surface and Jupiter's strong magnetic field.

The second and third Cassini laws, initially formulated for Earth's moon, were later generalized by Colombo (1966) for an axisymmetric body and by Peale (1969) for a triaxial body in a p:1 spin-orbit resonance, where p is a half-integer. In these generalized formulations, the ecliptic plane is referred to as the *Laplace plane*, while the plane formed by the normal vectors of the orbit and the Laplace plane is noted as the *Cassini plane*. According to the third law, the Cassini plane also contains the rotation axis. Colombo (1966) and Peale (1969) have shown that there are only two or four stable orientations which correspond to relative equilibria, being extremes of the energy associated with the spin and orientation of the body. These equilibria are nowadays referred to as Cassini state 1 to 4. The number of possible Cassini states is driven by the ratio between the forced precession rate and the free precession rate (see Section 3.4 for further details).

For the Moon, only two Cassini states are possible. But only one of them, Cassini state 2, is prograde like the Moon's rotational direction. This is an equilibrium close to the Laplace plane normal but on the opposite side to the orbit normal. The planet Mercury is another body obedient to Cassini laws. Observations of its small obliquity (Margot et al., 2007; Stark, Oberst, Preusker et al., 2015) revealed that Mercury is in Cassini state 1, that is, close to the orbit normal. Other bodies which are assumed to be in a Cassini state are the Martian moons, Phobos and Deimos, and many of the satellites in the outer solar system. For the Saturnian moon Titan, the analysis of radar images obtained by the Cassini spacecraft confirmed the occupancy of Cassini state 1 (Stiles et al., 2008; Meriggiola et al., 2016) and the unexpectedly large obliquity was explained by the presence of a subsurface ocean (Bills and Nimmo, 2008; Baland et al., 2011; Bills and Nimmo, 2011). The Galilean satellites are also believed to occupy a Cassini state 1 (Henrard and Schwanen, 2004) (see Section 3.4).

3.3 PHYSICAL LIBRATION

As explained in Section 3.2, the stability of the Cassini state that Ganymede occupies does not imply that the rotation is steady. Even if Ganymede were orbiting in a perfect Keplerian orbit, its rotation rate would vary about the mean rotation rate because the long axis of Ganymede is generally not oriented towards Jupiter for an eccentric orbit but towards the empty focus (at first order in the eccentricity, see e.g. Murray and Dermott (1999)). Jupiter therefore exerts a time-variable gravitational torque on Ganymede, which alternatively tends to speed up or slow down the rotation rate. As a result, the direction of the long axis oscillates (or librates) with respect to its orientation at a constant mean rotation rate. Such librations are called 'physical' librations since they are due to a physical restoring force. For a

Keplerian orbit, the longest axis of Ganymede points towards Jupiter only at the pericentre and apocentre.

3.3.1 Diurnal Libration

We first consider the variation in Ganymede's rotation assuming a mean Keplerian orbit. Changes in the rotation rate can be determined from the changes in the angular momentum of Ganymede resulting from the applied torque. For an assumed zero obliquity, the gravitational torque of Jupiter on an entirely solid and rigid Ganymede is

$$\Gamma(t) = \frac{3}{2}(B - A)\frac{GM_J}{r(t)^3}\sin\left(2\theta(t) - 2\nu(t) - 2\omega(t)\right). \tag{3.1}$$

Here, $A < B$ are the two equatorial principal moments of inertia of Ganymede, G is the universal gravitational constant, M_J is the mass of Jupiter, ν is the true anomaly, ω is the argument of pericentre, and r is the distance between the mass centres of Jupiter and Ganymede. The angular momentum of Ganymede is $C\dot{\theta}$, with C being Ganymede's polar moment of inertia and the angle of rotation θ, measured between the long axis of Ganymede and the line of the ascending node. The angular momentum equation

$$C\ddot{\theta}(t) = \Gamma(t) \tag{3.2}$$

connects the applied torque with the change in angular momentum, hence the time derivative of $C\dot{\theta}$. Using Kepler's third law $GM_J = n^2a^3$, where n is the mean motion and a is the semi-major axis, Eq. (3.2) can be written as (e.g. Goldreich and Peale, 1966)

$$C\ddot{\theta}(t) + \frac{3}{2}n^2(B - A)\left(\frac{a}{r(t)}\right)^3 \sin\left(2\theta(t) - 2\nu(t) - 2\omega(t)\right) = 0\,. \tag{3.3}$$

The general solution of Eq. (3.3) can be written as the sum of two distinct librations: a libration with natural frequency σ_{free} and arbitrary amplitude; and forced libration with a period given by the period of the forcing. By averaging Eq. (3.3) over an orbital period, the free frequency σ_{free} can be obtained as (e.g. Comstock and Bills, 2003)

$$\sigma_{\text{free}} = n\sqrt{\frac{3(B - A)}{C}}. \tag{3.4}$$

It is much smaller than the mean motion n since $(B - A)$ is much smaller than C. The ratio $(B - A)/C$ is of the order of $C_{22} = 38.26 \times 10^{-6}$, which has been determined from Galileo fly-bys (Schubert et al., 2004). Since dissipation processes due to tides and libration damp the long-periodic free libration, its amplitude is expected to be close to zero.

For a Keplerian orbit, the gravitational torque by Jupiter has periods equal to the orbital period and its subharmonic periods. The amplitude A_θ of the forced libration $\gamma(t)$, with $\gamma(t) = A_\theta \sin M$ and M the mean anomaly, can be approximated up to the first order in eccentricity e by (e.g. Murray and Dermott, 1999)

$$A_\theta = 2e\frac{\sigma_{\text{free}}^2}{\sigma_{\text{free}}^2 - n^2} \approx -6e\frac{(B - A)}{C}. \tag{3.5}$$

The principal moments of inertia are the only interior structure parameters involved in the amplitude of the rigid librations of Ganymede. The proportionality of the amplitude to the moment of inertia difference $(B - A)$ is due to the gravitational torque. The amplitude is inversely proportional to the polar moment of inertia since that is the inertia for rotational motion. If Ganymede were rigid, and the difference $(B - A)$ known from the gravity field, a libration measurement would inform on the polar moment of inertia C. Assuming that Ganymede's mass distribution and shape can be expressed as a triaxial ellipsoid as a result of its rotation and static tides consistent with a body in hydrostatic equilibrium, the hydrostatic value of C is obtained and the libration amplitude is equal to about 10 m as measured at the equator (angular displacement of 0.85 arcsec, Comstock and Bills (2003)). Amplitudes A_n of librations at subharmonic periods P_{orbit}/n are smaller by a factor of e^n. It is of interest to note that the phase of the libration is such that the libration increases the angular distance of the long axis from the direction to Jupiter (e.g. Tiscareno et al., 2009). The maximum angular deviation of the long axis from the direction to Jupiter without libration is $2e \approx 8.9$ arcmin (with the mean value of the eccentricity $e = 0.0013$).

Besides librations, the gravitational interaction of Ganymede with Jupiter also leads to tides on Ganymede (see Chapter 4). Tidal deformation itself decreases the librational response as part of the energy goes into deforming Ganymede. Without a subsurface ocean, the amplitude of the diurnal libration decreases from the rigid behaviour by less than a metre (Van Hoolst et al., 2013). It is, however, generally agreed that Ganymede has a global subsurface ocean beneath its ice shell, which decouples the libration of the shell from that of the interior below the ocean. Firm support for such an ocean is given by the observed induced magnetic field (Kivelson et al., 2002) and the oscillations of its auroral ovals (Saur et al., 2015). Different layers of Ganymede can then librate with different amplitudes and interact with each other through mutual gravitational torques, pressure torques, viscous torques, and possibly electromagnetic torques. The decoupling between the shell and the interior below the ocean also implies that two different free libration modes exist. One mode describes an in-phase libration of the satellite layers; in the other mode, the shell and interior librate out of phase. What is most crucial in the study of the libration of the ice shell of Ganymede is how the shell responds to tidal forcing. The gravitational interaction of Jupiter with Ganymede leads to an increase in the elastic energy related to tidal deformation and an increase in the gravitational energy related to changing Ganymede's orientation over the libration angle (Goldreich and Mitchell, 2010). For a large satellite such as Ganymede, most of the energy increase will be elastic, in contrast to a small satellite like Enceladus, for which most energy is gravitational. The shape of Ganymede's ice shell conforms to that of the response of the ocean below, and its tidal deformation will be close to that of a layer in hydrostatic equilibrium (see Chapter 6). Ganymede's ice shell is therefore also considered to be soft (Beuthe, 2018), and the libration amplitude will be very close to that of the satellite assumed to have no subsurface ocean. In a model that includes the dominant coupling processes, Van Hoolst et al. (2013) obtained values between 6.5 and 9.5 m (0.5–0.75 arcsec) for a range of interior structure models with a shape consistent with hydrostatic equilibrium.

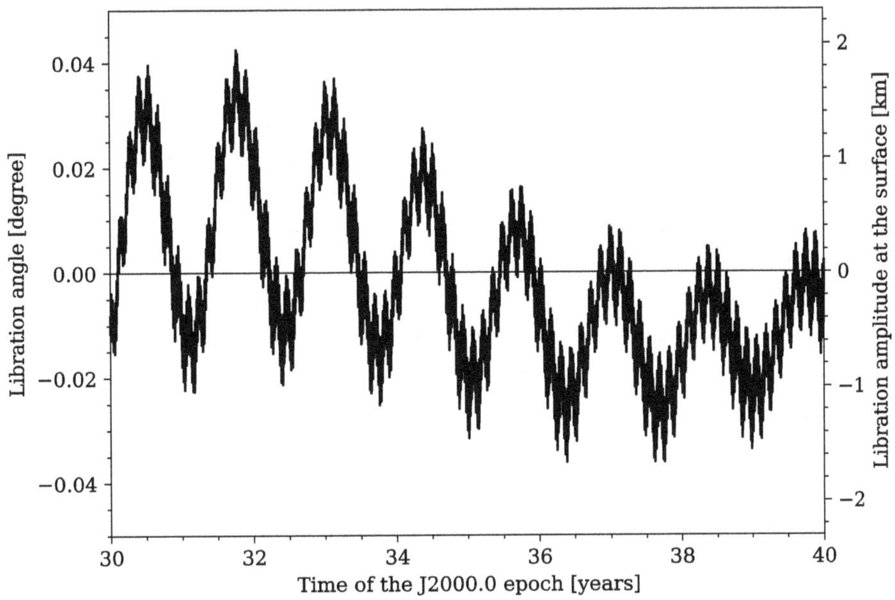

Figure 3.2 Libration angle as obtained by Rambaux et al. (2011) shown for the time period 2030–2040.

3.3.2 Librations due to Orbital Perturbations

Since the orbital motion of Ganymede is not purely Keplerian (Lainey et al., 2006), the gravitational torque of Jupiter on Ganymede differs with respect to that of the Keplerian orbital motion. According to Eq. (3.3), this non-Keplerian motion is the cause of additional librations (Rambaux et al., 2011). Among the variations of the different orbital elements, mainly periodic changes in longitude, which is measured in almost the same plane as the rotation angle as Ganymede's obliquity is very small, lead to those librations. The orbital perturbations are due to gravitational interactions with (1) the Sun, with a frequency equal to Jupiter's orbital frequency, and (2) the other Galilean satellites at different frequencies, most of them much shorter than the orbital frequency. Those frequencies can be expressed as linear combinations of variation rates of the mean longitudes and of the precession rates of the perijoves and nodes of the satellites, in particular Europa, Ganymede, and Callisto. Most of the largest amplitude librations, with amplitudes in the order of 100 m at the equator, have a period of the order of a year or more and are therefore much longer than the orbital period. But short-period perturbations with periods in the order of the orbital period, related to differences in the mean longitudes of pairs of Io, Europa, Ganymede, and Callisto, can also have a significant amplitude. See Rambaux et al. (2011) for a list of the main librations of Ganymede. The librational behaviour as a function of time is given in Fig. 3.2 for an interior model consisting of three layers of homogeneous density: a 50 km thick shell and a 640.8 km thick ocean, both with a density of 1 000 kg/m^3, underlain by a layer with a density of 3 348.95 kg/m^3.

At long periods, much longer than the librational eigenperiods of Ganymede that are between 63 days and 0.9 years (Van Hoolst et al., 2013), the rotation angle of Ganymede closely tracks the orbit longitude variations, and the libration amplitude is well approximated by the amplitude of these variations (Rambaux et al., 2011) (see also Section 3.2). Therefore, Ganymede's long axis is approximately directed towards Jupiter at long periods. This implies that those librations do not contain information on the interior of Ganymede. They are, however, important to account for as they influence the cartographic reference frame for Ganymede. The amplitudes of the short-period librations are smaller than the forcing amplitudes since their periods are smaller than the periods of the two librational eigenmodes (Van Hoolst et al., 2013). The possible proximity of the longest of the shorter period librations (50.14 days) to one of the librational eigenperiods could resonantly increase the amplitude and introduce a dependence on the interior (Rambaux et al., 2011).

3.4 OBLIQUITY AND PRECESSION

Ganymede's obliquity is currently unknown and assumed to be small. Given the current absence of information on the orientation of the Galilean satellites, International Astronomical Union (IAU) orientation and rotation models assume a zero obliquity and an axis of rotation normal to their mean orbital plane (Archinal et al., 2018). A non-zero obliquity induces a small periodic variation in the gravity coefficients C_{21} and S_{21} of a potential expressed in the coordinates of a rotating frame attached to the orbital plane but that was not observed in the Galileo gravity measurements due to the limited measurement accuracy. For a Ganymede obliquity of 0.03° (see Section 3.4.2), the effect on C_{21} and S_{21} is about 10^{-7} and 3×10^{-8}, respectively, and is smaller than the accuracy of 8.6×10^{-7} on C_{22}. These values were provided by Anderson et al. (1996), where the orientation of Ganymede's principal axis was assumed to be known and only the J_2 and C_{22} coefficients were estimated. For Europa and an obliquity of 0.055° (Baland et al., 2012), the effect on C_{21} and S_{21} is 7×10^{-7} and 2×10^{-7}, respectively, and is below the measurement accuracy of 6×10^{-6} and 12×10^{-6} (Anderson et al., 1998), explaining the lack of a definite detection of C_{21} and S_{21}. The recent measurements performed by Zubarev et al. (2015) were compatible with zero

Table 3.1 *Libration forcing and response amplitudes for an interior model consisting of three layers of homogeneous density (see Rambaux et al. (2011) for details).*

Period	Phase	Forcing amplitude	Response amplitude	Response amplitude at equator	Difference forcing and response
575.34359 a	159.808°	0.00309°	0.00309°	141.9 m	0.00000°
137.71359 a	127.748°	0.00280°	0.00280°	128.9 m	0.00000°
30.90433 a	340.031°	0.00078°	0.00078°	35.7 m	0.00000°
26.39823 a	326.086°	0.00089°	0.00089°	40.8 m	0.00000°
11.86294 a	122.630°	0.01324°	0.01325°	609.3 m	0.00001°
5.63872 a	289.313°	0.00097°	0.00098°	44.9 m	0.00000°
1.32951 a	265.581°	0.00547°	0.00594°	272.9 m	0.00046°
1.31980 a	67.453°	0.01048°	0.01138°	523.3 m	0.00090°
1.26630 a	282.887°	0.00867°	0.00952°	437.8 m	0.00085°
50.14182 d	190.809°	0.00083°	0.00489°	225.0 m	0.00406°
12.52325 d	217.225°	0.00540°	−0.00037°	17.2 m	−0.00503°
7.16008 d	36.203°	0.00082°	$< 1''$	<12.8 m	–
7.15559 d	254.678°	0.16379°	−0.00352°	162.0 m	−0.16027°
7.15481 d	56.549°	0.08838°	−0.00190°	87.4 m	−0.08648°
7.05093 d	142.132°	0.06893°	−0.00144°	66.1 m	−0.06749°
6.94928 d	209.587°	0.00081°	$< 1''$	<12.8 m	–
6.26162 d	74.436°	0.01845°	−0.00030°	13.9 m	−0.01815°
4.17442 d	111.597°	0.00130°	$< 1''$	<12.8 m	–
3.52546 d	104.268°	0.00108°	$< 1''$	<12.8 m	–
2.35031 d	66.341°	0.00125°	$< 1''$	<12.8 m	–

Periods are given in years (a) or days (d). The response amplitude for some terms is missing because Rambaux et al. (2011) omits terms below 1 arc second ($< 1''$). The response amplitudes at Ganymede's equator were obtained with the mean radius of 2 631.2 km. The phases of the forcing were initially given in Rambaux et al. (2011) with respect to the J1950.0 epoch and were recomputed here for the J2000.0 epoch.

obliquity within the 0.1° ($\sim$ 5 km at the surface) error bar of the measurement (see Section 3.5.3). The obliquity of Ganymede must, nevertheless, be non-zero.

The obliquity of Ganymede is linked to its interior structure and depends on its precession period. Torque and Hamiltonian approaches coexist in the literature to describe this link. In what follows, we present and discuss recent modelling efforts.

3.4.1 Ganymede as a Solid and Rigid Synchronous Rotator

In the torque approach, the equation governing the precession of the spin axis of a rotating body equates the variations in angular momentum and the applied gravitational torque by the parent body on the flattened rotating body. For an entirely solid and *rigid* synchronous rotator, assuming principal axis rotation and neglecting orbital eccentricity, the angular momentum equation can be written in the form (Ward and Hamilton, 2004; Bills, 2005; Baland et al., 2012)

$$\frac{d\hat{s}}{dt} = \omega_{\text{free}} (\hat{s} \cdot \hat{n}) (\hat{s} \wedge \hat{n}), \tag{3.6}$$

$$\omega_{\text{free}} = \frac{3}{2} n\, MR^2 \frac{(-C_{20} + 2C_{22})}{C} = \frac{3}{2} n \frac{(C - A)}{C}, \tag{3.7}$$

in an inertial reference frame. We will use the Cartesian coordinates (x, y) of the Laplace plane and z along the normal to it, with $\hat{s} = (s_x, s_y, s_z)$ and $\hat{n} = (n_x, n_y, n_z)$ the unit vectors along the rotation axis and the normal to the orbit, respectively. The gravitational torque (here averaged over the orbital period) exerted by Jupiter tends to align the rotation axis with the orbit normal but without ever achieving perfect alignment, because of the continuous orbital precession. As Eq. (3.6) shows, the torque is proportional to $(\hat{s} \wedge \hat{n}) = \sin\varepsilon$ with ε the obliquity. A perfect alignment and a zero obliquity would cancel the torque so that the spin axis could not follow the orbital precession, leading to a non-zero obliquity, in contradiction with the perfect alignment (Bills, 2005).

If the obliquity is small enough that $(\hat{s} \cdot \hat{n}) \simeq 1$, the projection of Eq. (3.6) onto the Laplace plane can be expressed as (Bills, 2005)

$$\frac{dS}{dt} = I\,\omega_{\text{free}}\,(N - S), \tag{3.8}$$

with $S = s_x + Is_y$ and $N = n_x + In_y$, $I = \sqrt{-1}$. The driving non-uniform orbital precession can be expressed in the form of series (e.g. Lainey et al., 2006)

$$N = \sum_j \sin i_j\, e^{I(\dot{\Omega}_j t + \gamma_j - \pi/2)}, \tag{3.9}$$

where i_j are the inclination amplitudes associated with the orbital node precession frequencies $\dot{\Omega}_j$ and the phases γ_j. The solution for the forced precession of the spin axis can then be written as

$$S = \sum_j \sin(i_j + \varepsilon_j)\, e^{I(\dot{\Omega}_j t + \gamma_j - \pi/2)}, \tag{3.10}$$

where, correct up to the first order in ε_j and i_j, the obliquity amplitudes ε_j associated with the frequencies $\dot{\Omega}_j$ are given by

$$\varepsilon_j = -\frac{i_j \dot{\Omega}_j}{(\omega_{\text{free}} + \dot{\Omega}_j)}. \tag{3.11}$$

The frequency ω_{free} of the free spin precession, obtained when $N = 0$, has a period of about 20 years for Ganymede for $C/MR^2 = 0.312$ (Schubert et al., 2004). Free precession is damped by dissipative processes associated to tides and assumed here to be of zero amplitude as no recent or ongoing excitation mechanisms are identified. The time-variable orbital inclination i and obliquity are obtained by (Baland et al., 2011; Bills and Nimmo, 2011)

$$i \simeq \|N\|, \tag{3.12}$$

$$\varepsilon \simeq \|S - N\|. \tag{3.13}$$

The frequency $\dot{\Omega}_1$ (period $= -138$ years, see Table 3.2) is the main precession rate of the orbital plane, with respect to the Laplace plane. When only the first term of the forcing N is considered (e.g. Bills and Ray, 2000), the orbital and spin precessions are uniform, and the orbital inclination $i = i_1 = 0.18°$ and obliquity $\varepsilon = \varepsilon_1 = 0.03°$ are constant over time. The solution for the small uniform obliquity can also be obtained (see e.g. Henrard and Schwanen, 2004; Noyelles et al., 2008 within the Hamiltonian approach) or as one of the four solutions of

$$\omega_{\text{free}} \sin(2\varepsilon) + 2\,\dot{\Omega}\sin(i + \varepsilon) = 0 \tag{3.14}$$

corresponding to the Cassini state 1 (e.g. Ward and Hamilton, 2004; Noyelles, 2010). Equation (3.14) admits four solutions when the free period is much smaller than the forced period, as is the case for Ganymede (20 years versus 138 years) but not for the Moon (81 years versus 19 years), for which only two solutions exist.

Owing to the mutual interactions between the Galilean satellites, the orbital precession is significantly non-uniform and the uniform solution is not representative of the reality. The other terms ($j \geq 2$) of the forcing contribute up to 30% to the orbital inclination and up to 100% to the solution for the obliquity

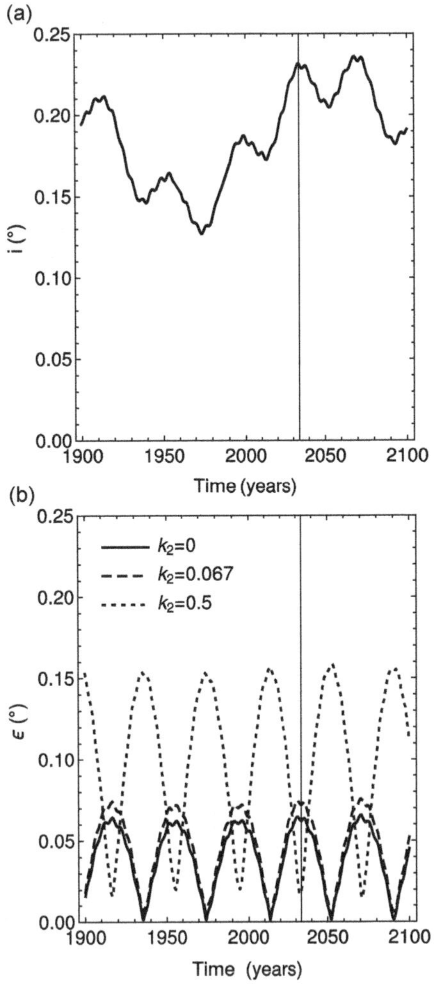

Figure 3.3 Time evolution of (a) the orbital inclination and (b) the obliquity of an entirely solid and rigid ($k_2 = 0$) Ganymede (Baland et al., 2012) and of a solid elastic Ganymede for a $k_2 = 0.067$ and $k_2 = 0.5$ (Moore and Schubert, 2003). The vertical line indicates the epoch of the Ganymede orbital phase of the JUICE mission.

(see Fig. 3.3). The variations in inclination are mainly due to the second and third terms of the forcing in the series expansion of Eq. (3.9), in which the terms are ordered according to decreasing inclination amplitudes i_j with increasing subscript j. The variations in obliquity are mainly due to the third term of the solution ($\varepsilon_3 \simeq -0.03°$). Even though i_3 is one order of magnitude smaller than i_1, $\dot{\Omega}_3$ is about five times larger than $\dot{\Omega}_1$ and a resonant amplification between the forcing period (-30 years) and the free period (20 years) enhances that contribution to the obliquity.

In the entirely solid and rigid case, the link between the obliquity and the interior of Ganymede is therefore given by the parameter ω_{free} and an obliquity measurement would, as libration (see Eq. 3.5), inform on the polar moment of inertia C (the difference $C - A$ is known from the gravitational field) and on the differentiation of the interior. However, since the obliquity varies between 0° and 0.06°, not all times are suitable to infer the polar moment of inertia. The best measuring times are those at which the different terms of the solution combine to provide the maximal obliquity (around 2030, for instance). The

Table 3.2 *Columns 2 to 5: amplitudes, frequencies, periods, and phases (adapted from Lainey et al. (2006), phases at the J1950.0 epoch) of the orbit precession of Ganymede with respect to its Laplace plane.*

j	i_j (deg)	$\dot{\Omega}_j$ (rad/year)	Period (years)	γ_j (deg)	ε_j (deg)	ε_j^{el} for $k_2 = 0.067$ (deg)	ε_j^{el} for $k_2 = 0.5$ (deg)
1	0.18258	−0.0456245	−137.715	52.9125	0.03096	0.03340	0.06812
2	0.04026	−0.0111624	−562.888	203.898	0.00148	0.00158	0.00287
3	−0.01652	−0.207903	−30.2218	−78.2478	−0.03217	−0.03944	0.08596
4	0.00180	1.05936	5.93112	180.311	−0.00139	−0.00141	−0.00156
5	0.00029	−0.529633	−11.8633	−121.96	−0.00071	−0.00065	−0.00042
6	0.00023	0.529663	11.8626	−37.1349	−0.00015	−0.00015	−0.00018
7	0.00021	1.58901	3.95416	122.814	−0.00017	−0.00017	−0.00019
8	0.00016	−9.22056	−0.681432	−89.6335	−0.00016	−0.00016	−0.00016

The Laplace plane has a node of 138.27696° and an inclination of 0.0977821° with respect to the equatorial plane of Jupiter. The obliquity amplitudes of the rigid solid Ganymede given in the 6th column are obtained for $C/MR^2 = 0.312$ (Baland et al., 2012; Schubert et al., 2004). The last two columns correspond to a solid elastic Ganymede for two different values of the tidal Love number k_2.

timing of the JUICE mission would therefore be excellent for measuring the obliquity of Ganymede.

When tidal deformation is included in the modelling of obliquity, the solution for the precession of a solid and *elastic* Ganymede has the same form as in the rigid case, except that the free frequency of precession $\omega_{\text{free}_{el}}$ depends on the material properties through the tidal Love number k_2 (Baland et al., 2016):

$$\varepsilon_j^{el} = -\frac{i_j \dot{\Omega}_j}{(\omega_{\text{free}}^{el} + \dot{\Omega}_j)}, \tag{3.15}$$

$$\omega_{\text{free}}^{el} = \frac{3}{2} n \frac{(C - A) - k_2 MR^2 q_r}{C - k_2 MR^2 q_r/2} \tag{3.16}$$

with $q_r = n^2 R^3/GM$. For a reasonable estimate for the tidal Love number $k_2 = 0.067$ of Ganymede considered to be entirely solid (Moore and Schubert, 2003), the increase of the free period is 6% and does not affect the solution much. For theoretical k_2 values consistent with the presence of a subsurface ocean ($k_2 \simeq 0.5$), the free period increases by about 90%, to about 40 years, leading to a maximal obliquity of 0.16° compared to 0.06° in the rigid case. As a result of the different free period, the third obliquity amplitude changes sign and the maxima of the time-variable obliquity do not occur at the same time. Though this last computation is not self-consistent, it informs on the need to take tidal deformations into consideration and on the precautions to be taken to assess the interior properties from obliquity measurements (different combinations of the polar moment of inertia and of the tidal Love number could lead to the same obliquity at a given time).

3.4.2 Ganymede with Subsurface Ocean

Ganymede most likely has two internal global liquid layers: the liquid water ocean and the iron fluid core. The state of Ganymede's core is not expected to have a noticeable influence on the orientation of the outer ice shell spin axis. The small size, flattening, and tidal deformation of the fluid core have no significant effect on the longitudinal librations (Baland and Van Hoolst, 2010; Van Hoolst et al., 2013). Therefore, Ganymede can be assumed to be made of three layers: a solid ice shell (s), a liquid water ocean (o), and a solid interior (i). The solid interior itself is made up of a high-pressure (HP) ice mantle, a rock mantle, and an iron core.

The precessional behaviour of each layer k is described by an angular momentum equation

$$\frac{d\vec{H}_k}{dt} = \vec{\Gamma}_k \tag{3.17}$$

with the angular momentum written as

$$\vec{H}_k = nC_k\hat{s}_k, \tag{3.18}$$

and $\hat{s}_k = (s_x^k, s_y^k, s_z^k \simeq 1)$ is the unit vector along the rotation axis of layer k. The torque $\vec{\Gamma}_k$ is the sum of an external gravitational torque by Jupiter and of internal torques resulting from gravitational and pressure interactions between the layers. The forced solution for the projection $S_k = s_x^k + Is_y^k$ on the Laplace plane, for layer k, can be expressed by analogy with the rigid case as

$$S_k = \sum_j (i_j + \varepsilon_{j,k}) e^{I(\dot{\Omega}_j t + \gamma_j - \pi/2)}, \tag{3.19}$$

with obliquity amplitudes in the form

$$\varepsilon_{j,k} = -\frac{i_j \, \dot{\Omega}_j \, \chi_k}{(\omega_1 + \dot{\Omega}_j)(\omega_2 + \dot{\Omega}_j)(\omega_3 + \dot{\Omega}_j)}. \tag{3.20}$$

The expressions for χ_k and the $\omega's$ depend on the torques and the properties of the solid layers (deformable or rigid). Under the assumption that the ocean is in hydrostatic equilibrium, the total torque on the liquid layer vanishes and the system reduces to two equations to be solved for $\hat{S}_s$ and $\hat{S}_i$. The obliquity amplitudes can be expressed in the form:

$$\varepsilon_{j,s} = \frac{i_j \dot{\Omega}_j \chi_s}{(\omega_+ + \dot{\Omega}_j)(\omega_- + \dot{\Omega}_j)}, \tag{3.21}$$

$$\varepsilon_{j,i} = \frac{i_j \dot{\Omega}_j \chi_i}{(\omega_+ + \dot{\Omega}_j)(\omega_- + \dot{\Omega}_j)} \tag{3.22}$$

with the two free mode frequencies ω_+ and ω_- which correspond, respectively, to a mode where the shell and the interior precess in and out of phase with each other. The expressions for $\chi_{s,i}$ and the free frequencies are given in Baland et al. (2016) for the cases of rigid or deformable layers. Baland et al. (2012) computed the obliquity for Ganymede with an internal global ocean and solid rigid layers. They found that the period of the in-phase and out-of-phase free periods range between about 15 and 45 years and between 0.5 and 10 years, respectively. As in the solid case, the third obliquity amplitude (forcing period of 30 years) of the shell can be amplified to a significant level, depending on the details on the interior structure (density and size of the different layers), but also the eighth obliquity amplitude can (forcing period of 0.68 years) for a small fraction of the internal structure models considered.

Even in this rather idealized setting (ocean at rest and rigid solid layers), constraining the internal structure from an observed obliquity would be complicated, since many different internal structures could fit the same observation through different combinations of the terms of the solution. The link between the obliquity and the interior structure parameters (density and radius of the layers) is not unique and the degree of complexity increases if tidal deformations (see Baland et al., 2016 for an application to Enceladus) and the hydrodynamical pressure torque at the ocean boundaries ((Dumberry and Wieczorek, 2016) for an application to the Moon and its fluid core (Baland et al., 2019) for an application to Titan and its internal ocean, see also (Boué et al., 2017) for a study of Titan's obliquity within the Hamiltonian approach) are considered. However, the presence of the ocean could be confirmed if the actual internal structure corresponds to a shell obliquity very different from that of a solid rigid Ganymede at the time of measurement, as is the case for Titan (Baland et al., 2011; Bills and Nimmo, 2011). Moreover, as is usual in planetary geodesy, the best constraints on the interior will be obtained from a combination of data, including obliquity.

3.5 CURRENT ROTATION MODEL OF GANYMEDE

In view of the limited observations of Ganymede's rotation, the models currently used for the planning of future observations by spacecraft for example, are derived from the characteristics of Ganymede's orbit. A resonant rotation model, as will be introduced in Section 3.5.1, combined with assumptions on the upper limits of the free libration and precession frequency can provide a working rotation model until more accurate measurements become available.

3.5.1 Resonant Rotation Model

The resonant rotation model is an extreme case where the free libration and precession frequencies, as defined in Eqs. (3.4) and (3.7) respectively, are assumed to be very large, that is, $\omega_{\text{free}} \to \infty$ and $\sigma_{\text{free}} \to \infty$. Such a situation is reached when the body has a very small inertia, that is, $A < B < C \ll MR^2$, where $A < B < C$ are the principal moments of inertia, M is the mass, and R the mean radius of Ganymede. The body then instantaneously responds to the gravitational torque at all forcing frequencies. Such a rotation model, for an inertia-free body, is thus in perfect resonance with the orbital motion and the applied forcing. It is therefore sometimes noted as the 'resonant rotation model' of the body (Stark, Oberst and Hussmann, 2015; Stark et al., 2018). In the resonant rotation model, the rotation axis remains normal to the orbit plane at all times.

Using Keplerian orbital elements, the orientation of the orbital plane with respect to a chosen inertial reference frame can be expressed by the inclination i and the longitude of the ascending node Ω.

For an unperturbed orbit, the orientation of the rotation axis, expressed with right ascension α and declination δ, is directly obtained from $\alpha = \Omega - \pi/2$ and $\delta = \pi/2 - i$. For non-Keplerian perturbed orbits, osculating orbital elements can be used to obtain the evolution of the orbit over time, that is, $\Omega(t)$ and $i(t)$. Hence, the orientation of the rotation axis at time t is given by $\alpha(t) = \Omega(t) - \pi/2$ and $\delta(t) = \pi/2 - i(t)$. The rotation angle W for the inertia-free body is equal to its angular motion along its orbit. Again, for a non-perturbed orbit, W is equal to the true longitude λ, that is, the sum of the argument of pericentre ω and the true anomaly ν. Without perturbation, it holds $\dot{\omega} = 0$ and the mean rotation rate corresponds exactly to the mean motion n. For perturbed orbits, osculating orbital elements are used and the rotation angle at time t is obtained from $W(t) = \lambda(t) = \omega(t) + \nu(t)$. The mean rotation rate is then equal to $\dot{\omega} + \dot{\nu}$ averaged over sufficiently long intervals. In order to get an analytical expression for the resonant rotation model, the osculating orbital elements are decomposed into a secular and periodic part (Stark, Oberst and Hussmann, 2015)

$$x(t) = \sum_{i=0}^{n_s} x_i t^i + \sum_{i=0}^{n_p} A_i \sin(\vartheta_i t + \phi_i), \tag{3.23}$$

where x is an osculating orbital element, n_s is the maximal order of the secular components, and n_p is the number of frequencies in the periodic part. Typically, n_s is chosen such that the secular effects are sufficiently removed and the n_p is constrained by the minimum threshold for the forcing amplitude A_i or by a spectral domain for the forcing frequency ϑ_i. Either an expansion in sine or cosine can be used for the periodic part.

Although the resonant rotation model is an unrealistic extreme case, it can be used to define a rotation model when no definite observations are available and when rotation modelling provides a wide range of possible solutions, as is the case for Ganymede. Moreover, with the information provided by the resonant rotation model the location of the Cassini plane is fully defined and can be used to constrain the rotation axis orientation. However, the resonant rotation model is only as accurate as the ephemeris on which the model is based.

3.5.2 Rotation Model Adopted by the International Astronomical Union (IAU)

The rotation model for Ganymede currently adopted by the Working Group on Cartographic Coordinates and Rotational Elements (WGCCRE) of the IAU (Archinal et al., 2018; Davies et al., 1980) is one example for a resonant rotation model as described in Section 3.5.1. This model was limited to forcing amplitudes larger than 0.01° (Lieske, 1979) and is given with respect to the International Celestial Reference Frame (ICRF). The IAU rotation model reads

$$\alpha(t) = 268.20^\circ - 0.009^\circ\, t/\mathrm{cy} - 0.037^\circ \sin J_4 + 0.431^\circ \sin J_5 + 0.091^\circ \sin J_6 \quad (3.24)$$

$$\delta(t) = 64.57^\circ + 0.003^\circ\, t/\mathrm{cy} - 0.016^\circ \cos J_4 + 0.186^\circ \cos J_5 + 0.039^\circ \cos J_6 \quad (3.25)$$

$$W(t) = 44.064^\circ + 50.3176081^\circ\, t/\mathrm{d} + 0.033^\circ \sin J_4 - 0.389^\circ \sin J_5 - 0.082^\circ \sin J_6 \quad (3.26)$$

with t measured in days since the J2000.0 epoch and $1\,\mathrm{cy} = 100\,\mathrm{a} = 36525\,\mathrm{d}$. The forcing periods are

$$J_4 = 355.80^\circ + 1191.3^\circ\, t/\mathrm{cy} = 355.80^\circ + \frac{360^\circ}{30.2191\,\mathrm{a}}\frac{t}{\,} \quad (3.27)$$

$$J_5 = 119.90^\circ + 262.1^\circ\, t/\mathrm{cy} = 119.90^\circ + \frac{360^\circ}{137.352\,\mathrm{a}}\frac{t}{\,} \quad (3.28)$$

$$J_6 = 229.80^\circ + 64.3^\circ\, t/\mathrm{cy} = 229.80^\circ + \frac{360^\circ}{559.876\,\mathrm{a}}\frac{t}{\,}. \quad (3.29)$$

Here, the right ascension α and declination δ define the orientation of the rotation axis and the rotation angle W denotes the rotation about the rotation axis.

The rotation angle W is measured from the intersection of the body's equator with the ICRF equator and is composed of three parts, the prime meridian constant W_0, the mean rotation rate W_1, and the libration terms, that is, the periodic terms given by $W(t) - W_0 - W_1 t$. For Ganymede, the location of the prime meridian on the surface is defined by stating that the 128° W (232° E) longitude is passing through the centre of the crater Anat (Davies and Katayama, 1981). This definition was made because the sub-Jovian point, that is, the desired location of the prime meridian, was in shadow during the Voyager 2 fly-by and no high-resolution images could be obtained from that area. Anat is 2.9 km in diameter and is located 4.1° south of Ganymede's equator (Fig. 3.4). The orientation of the crater Anat at the J2000.0 epoch with respect to the ICRF, and hence the orientation of Ganymede's body-fixed frame, is given by the prime meridian constant $W_0 = 44.064^\circ$ in Eq. (3.26). This value is measured using geodetic control networks based on Voyager and Galileo images (Davies et al., 1998; Seidelmann et al., 2002) and is, in particular, relevant for cartography (see Appendix A). Ganymede's mean rotation rate of $W_1 = 50.3176081^\circ/\mathrm{d}$ in Eq. (3.26) corresponds to a mean rotation period of 7.15455312 days. Although this value dates back to the pre-Voyager era, it is very accurate when compared with the value obtained with the most recent ephemeris of Ganymede. This is a consequence of the accurate knowledge of Ganymede's orbital period based on Earth-based telescopic observations. However, the rotation of Ganymede is not uniform (see Rambaux et al. 2011 and Section 3.3). As can be observed from Eqs. (3.27) to (3.29), the IAU rotation model features only libration terms which all have periods longer than 30 years. The current knowledge of Ganymede's moments of inertia, and hence of the free libration period, is based on the measurement of the gravity coefficient C_{22} and the assumption of a hydrostatic equilibrium (see Chapter 6). The amplitudes for these *long-period librations* in Eq. (3.26), are plausible, since the free libration period is expected to be in the order of a few to some hundred days and Ganymede can then adapt to variations with periods of several decades (Van Hoolst et al., 2013 and Section 3.3).

Libration forcing terms obtained with the recent ephemeris include amplitudes as large as 0.5 km on Ganymede's surface and periods longer than two years (Table 3.1). These terms are missing in the IAU model, and since their periods are much longer than the expected free libration period they can be adopted without further knowledge of the interior structure of Ganymede.

The orientation of the rotation axis over 1 000 years is outlined in Fig. 3.5(a). Resonant rotation models based on recent ephemeris for Ganymede were obtained by Stark et al. (2018) and show when compared to the IAU model difference as large as 0.05° for the orientation of the rotation axis. Moreover, significant deviations for the obliquity amplitudes in α and δ are expected because the free precession period of Ganymede can be close to the forcing terms in Eqs. (3.24) and (3.25) (see Section 3.4). Even for a solid Ganymede the amplitude for the J_5 forcing term can change by values as large as 0.03°, corresponding to a shift of the rotation axis by about 1.4 km on Ganymede's surface. In the presence of a sub-surface ocean much larger (up to several degrees) obliquity amplitudes are possible (Baland et al., 2011).

3.5.3 Constraints on the Libration and Obliquity from Observations

First measurements of Ganymede's rotation axis orientation were performed with images taken by the two Voyager spacecraft. Davies and Katayama (1981) have established a first control net of Ganymede by means of a photogrammetric analysis of the obtained images. This control net was used to solve for the orientation of the spin axis at the moment of the Voyager 1 (05.03.1979) and Voyager 2 (09.07.1979) fly-bys and the obtained values for the right ascension and declination were $\alpha_0 = 268.56^\circ \pm 0.032^\circ$ and $\delta_0 = 64.64^\circ \pm 0.07^\circ$, respectively. These values were found to be in very good agreement with the prediction based on the orientation of Ganymede's orbital plane. However, as the accuracy of the predicted values was considered better than the measurement error, these control-net-based measurements were not used further.

Earth-based radar observations of Ganymede's spin axis orientations were performed by Margot et al. (2013), where the same technique was used as for measuring Mercury's obliquity (Margot et al., 2007). A precision for the spin axis orientation of $\sim 0.1^\circ$ was quoted but no measurements for the obliquity or spin axis orientation have been reported so far.

By computing a control point network from Voyager 1, Voyager 2, and Galileo images, Zubarev et al. (2015) could not find

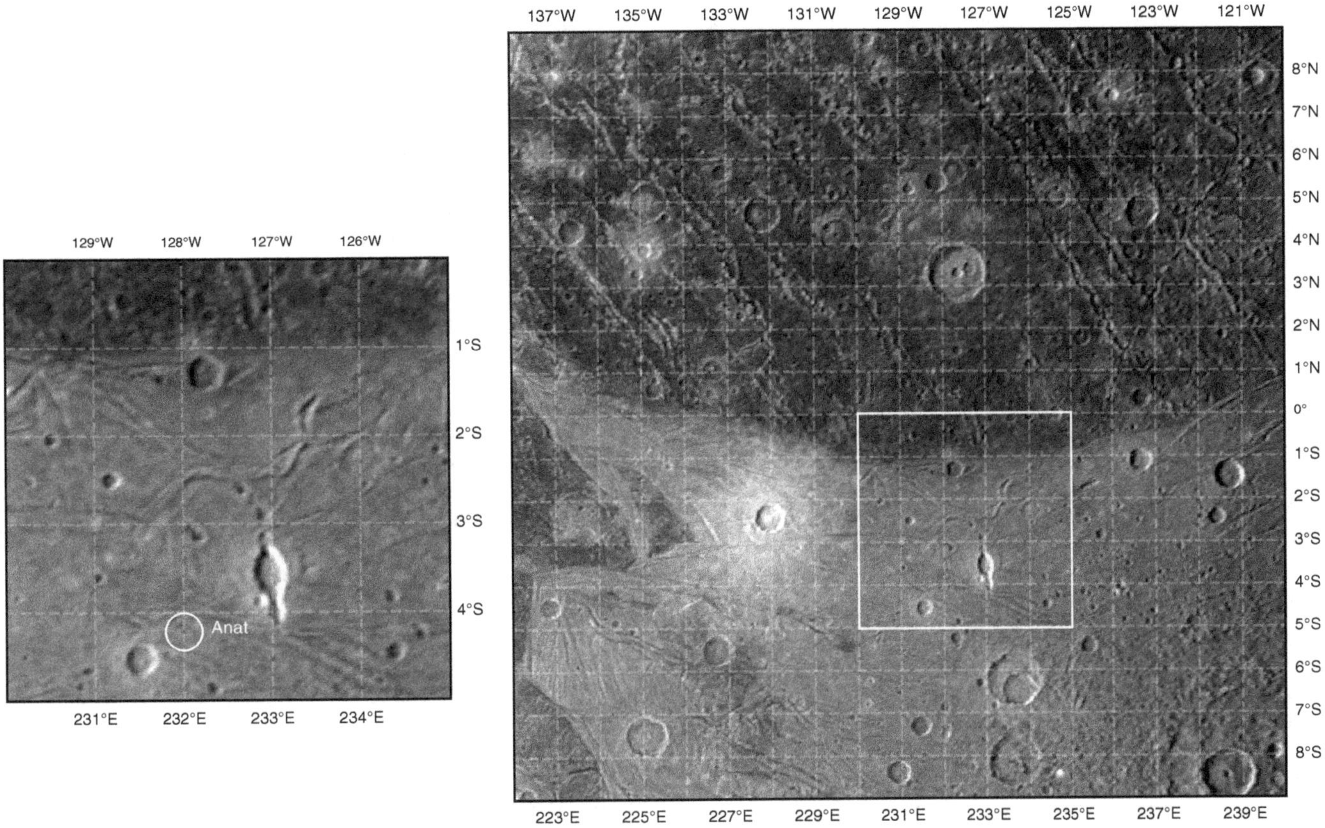

Figure 3.4 Voyager 2 image (c2063714) of the crater Anat. The crater defines the 128° W (232° E) longitude of Ganymede (Davies and Katayama, 1981).

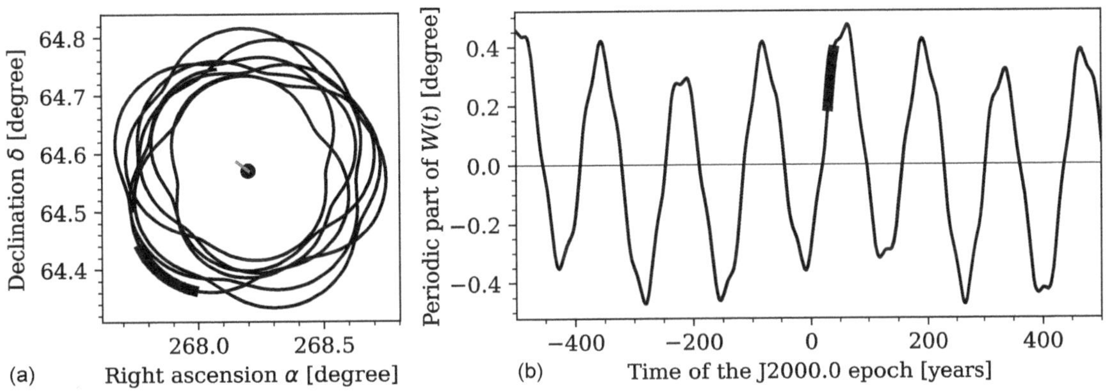

Figure 3.5 Rotation model of Ganymede adopted by the IAU since 1980 (Davies et al., 1980; Archinal et al., 2018). The evolution is shown for 1 000 years, centred at the J2000.0 epoch. The thick black segments denote the time interval between 2030 and 2040: (a) instantaneous orientation of the rotation axis, expressed in right ascension $\alpha(t)$ and declination $\delta(t)$ – the dot denotes the average orientation of the rotation axis at the J2000.0 epoch, while the grey line indicates its secular drift over 500 years; (b) periodic part of the rotation angle $W(t)$, that is, $W(t) - W_1 t$.

a deviation of the rotation state from the assumed synchronous state with zero obliquity. Thereby, the errors for the control points were found to be around 0.1°, that is, ∼ 4.6 km at the surface. This finding rules out large values for the obliquity as were theoretically possible due to resonant amplification (see Section 3.4). Furthermore, Zubarev et al. (2015) tried to solve for the libration amplitude at the orbital period. As a rather broad minimum for the total point errors was found, the analysis allowed only an upper limit of 0.1° to be put forward for the libration amplitude.

3.6 CONCLUSION

Ganymede's proximity to Jupiter (15 Jupiter radii) implies strong tidal forces which have driven its rotational motion into a state of equilibrium corresponding to a synchronous resonance motion. This rotational resonance is confirmed by observations of control points on its surface (Davies and Katayama, 1981; Davies et al., 1998; Zubarev et al., 2015). Ganymede's rotational motion can be described, as a first approximation, by Cassini's laws (Cassini, 1693). However, orbital perturbations and the internal structure of Ganymede will disturb Ganymede's rotational dynamics around this state of equilibrium. Conversely, measuring Ganymede's rotational motion would be a great tool for obtaining information about its interior.

Superimposed on the synchronous rotation, the rotational motion of Ganymede should present an oscillating motion which along the equator is called libration in longitude. The physical libration in longitude corresponds to the dynamical response of the satellite to the gravitational forcing raised by Jupiter and caused by the non-zero eccentricity of its orbit. The amplitude of the libration depends on the moments of inertia of the body and on the presence or absence of an ocean under the ice cover (e.g. Van Hoolst et al., 2013). In addition, Ganymede's orbit is in Laplace resonance, involving a rich and dense orbital frequency spectrum. Certain forcing frequencies could resonate with natural frequencies (which depend on the parameters of the internal structure) and may lead to significant amplitude (e.g. Rambaux et al., 2011). Currently, the amplitude of the physical libration in longitude is limited to a maximum of 0.1° from a study of control points (Zubarev et al., 2015).

As for the libration in longitude, significant orbital perturbations of Ganymede can lead to a pronounced obliquity and precession of the rotation axis. Oscillations of the Ganymede orbital plane can even resonate and induce a greater obliquity than initially expected (e.g. Bills, 2005; Baland et al., 2012). This resonance phenomenon is maintained and even increased in the case of tidal deformations and the presence of a fluid ocean by shifting the proper frequencies as modelled for other satellites and especially Titan (Baland et al., 2011; Boué et al., 2017). However, the link between obliquity and internal structure is not unique, which makes future interpretation of the observations difficult. Currently, the obliquity of Ganymede is unknown and its precessional motion has not been detected.

Currently, the rotation models take into account the orbital perturbations, the gravitational and pressure interaction between the different layers of the interior, and the tidal deformations. Ganymede is certainly composed of an icy shell, an internal ocean, high-pressure ice, a silicate mantle, and a fluid metallic core (which generates its magnetic field) (e.g. Kivelson et al., 2002; Rückriemen et al., 2018). The next step would be a more detailed modelling of the fluid flows induced by the libration, precession, and deformation of tides (Le Bars, 2016). Indeed, the flows in the fluid layers play a key role in the internal dynamics of these bodies and the induced dissipation influences their temporal evolution. In addition, these flows may be at the origin of the generation of the magnetic field observed in Ganymede.

The knowledge of the rotational motion is also important to define and attach a cartographic system to Ganymede's surface for the next observations of the JUICE space mission. In this context, the rotational motion is described according to a series of polynomials and a few periodic terms (Archinal et al., 2018). Future work must address the relation between the theoretical modelling efforts (e.g. Rambaux et al., 2011; Baland et al., 2012; Van Hoolst et al., 2013) and the rotation formalism adopted by the IAU Working Group on Cartographic Coordinates and Rotational Elements (Archinal et al., 2018) and used for the analysis of spacecraft data.

REFERENCES

Anderson, J. D., Lau, E. L., Sjogren, W. L., Schubert, G., and Moore, W. B. (1996), 'Gravitational constraints on the internal structure of Ganymede', *Nature* **384**(6609), 541–3.

Anderson, J. D., Schubert, G., Jacobson, R. A. et al. (1998), 'Europa's differentiated internal structure: inferences from four galileo encounters', *Science* **281**(5385), 2019–22.

Archinal, B. A., Acton, C. H., A'Hearn, M. F. et al. (2018), 'Report of the IAU Working Group on Cartographic Coordinates and Rotational Elements: 2015', *Celestial Mechanics and Dynamical Astronomy* **130**(3), 22.

Baland, R.-M., Coyette, A., and Van Hoolst, T. (2019), 'Coupling between the spin precession and polar motion of a synchronously rotating satellite: application to Titan', *Celestial Mechanics and Dynamical Astronomy* **131**(2), 11.

Baland, R.-M. and Van Hoolst, T. (2010), 'Librations of the Galilean satellites: the influence of global internal liquid layers', *Icarus* **209**(2), 651–64.

Baland, R.-M., Van Hoolst, T., Yseboodt, M., and Karatekin, O. (2011), 'Titan's obliquity as evidence of a subsurface ocean?', *Astronomy and Astrophysics* **530**, A141.

Baland, R.-M., Yseboodt, M., and Van Hoolst, T. (2012), 'Obliquity of the Galilean satellites: the influence of a global internal liquid layer', *Icarus* **220**(2), 435–48.

Baland, R.-M., Yseboodt, M., and Van Hoolst, T. (2016), 'The obliquity of Enceladus', *Icarus* **268**, 12–31.

Beuthe, M. (2018), 'Enceladus's crust as a non-uniform thin shell: I tidal deformations', *Icarus* **302**, 145–174.

Bills, B. G. (2005), 'Free and forced obliquities of the Galilean satellites of Jupiter', *Icarus* **175**(1), 233–47.

Bills, B. G. and Nimmo, F. (2008), 'Proximity of Titan's spin pole to a Cassini state', *AGU Fall Meeting Abstracts*, Vol. 2008, P11D–03.

Bills, B. G. and Nimmo, F. (2011), 'Rotational dynamics and internal structure of Titan', *Icarus* **214**(1), 351–5.

Bills, B. G. and Ray, R. D. (2000), 'Galilean satellite obliquities', *Journal of Geophysical Research* **105**(E12), 29277–82.

Boué, G., Rambaux, N., and Richard, A. (2017), 'Rotation of a rigid satellite with a fluid component: a new light onto Titan's obliquity', *Celestial Mechanics and Dynamical Astronomy* **129**(4), 449–85.

Cassini, G. D. (1693), *De l'origine et du progrès de l'astronomie et de son usage dans la géographie et dans la navigation*, n.p.

Colombo, G. (1966), 'Cassini's second and third laws', *The Astronomical Journal* **71**(9), 891–6.

Comstock, R. L. and Bills, B. G. (2003), 'A solar system survey of forced librations in longitude', *Journal of Geophysical Research: Planets* **108**(5100), E9.

Davies, M. E., Abalakin, V. K., Cross, C. A. et al. (1980), 'Report of the IAU Working Group on Cartographic Coordinates and Rotational Elements of the Planets and Satellites', *Celestial Mechanics* **22**(3), 205–30.

Davies, M. E., Colvin, T. R., Oberst, J. et al. (1998), 'The control networks of the Galilean satellites and implications for global shape', *Icarus* **135**(1), 372–6.

Davies, M. E. and Katayama, F. Y. (1981), 'Coordinates of features on the Galilean satellites', *Journal of Geophysical Research: Space Physics* **86**(A10), 8635–57.

Dumberry, M. and Wieczorek, M. A. (2016), 'The forced precession of the Moon's inner core', *Journal of Geophysical Research (Planets)* **121**(7), 1264–92.

Goldreich, P. M. and Mitchell, J. L. (2010), 'Elastic ice shells of synchronous moons: implications for cracks on Europa and non-synchronous rotation of Titan', *Icarus* **209**(2), 631–8.

Goldreich, P. and Peale, S. (1966), 'Spin-orbit coupling in the solar system', *The Astronomical Journal* **71**(6), 425–38.

Grasset, O., Dougherty, M. K., Coustenis, A. et al. (2013), 'JUpiter ICy moons Explorer (JUICE): an ESA mission to orbit Ganymede and to characterise the Jupiter system', *Planetary and Space Science* **78**, 1–21.

Henrard, J. and Schwanen, G. (2004), 'Rotation of synchronous satellites: application to the Galilean satellites', *Celestial Mechanics and Dynamical Astronomy* **89**(2), 181–200.

Kivelson, M. G., Khurana, K. K., and Volwerk, M. (2002), 'The permanent and inductive magnetic moments of Ganymede', *Icarus* **157**(2), 507–22.

Lainey, V., Duriez, L., and Vienne, A. (2006), 'Synthetic representation of the Galilean satellites' orbital motions from L1 ephemerides', *Astronomy and Astrophysics* **456**(2), 783–8.

Le Bars, M. (2016), 'Flows driven by libration, precession, and tides in planetary cores', *Physical Review Fluids* **1**(6), 060505.

Lieske, J. H. (1979), 'Poles of the Galilean satellites', *Astronomy and Astrophysics* **75**, 158–63.

Margot, J.-L., Padovan, S., Campbell, D., Peale, S., and Ghigo, F. (2013), 'Measurements of the spin states of Europa and Ganymede', *European Planetary Science Congress 2013*, Vol. 8. https://meetingorganizer.copernicus.org/EPSC2013/EPSC2013-393.pdf.

Margot, J. L., Peale, S. J., Jurgens, R. F., Slade, M. A., and Holin, I. V. (2007), 'Large longitude libration of Mercury reveals a molten core', *Science* **316**(5825), 710–14.

Meriggiola, R., Iess, L., Stiles, B. W., Lunine, J. I., and Mitri, G. (2016), 'The rotational dynamics of Titan from Cassini RADAR images', *Icarus* **275**, 183–92.

Moore, W. B. and Schubert, G. (2003), 'The tidal response of Ganymede and Callisto with and without liquid water oceans', *Icarus* **166**(1), 223–6.

Murray, C. D. and Dermott, S. F. (1999), *Solar system dynamics*, Cambridge University Press.

Noyelles, B. (2010), 'Theory of the rotation of Janus and Epimetheus', *Icarus* **207**(2), 887–902.

Noyelles, B., Lemaître, A., and Vienne, A. (2008), 'Titan's rotation: a 3-dimensional theory', *Astronomy and Astrophysics* **478**(3), 959–70.

Peale, S. J. (1969), 'Generalized Cassini's laws', *Astronomical Journal* **74**(3), 483–9.

Rambaux, N., Van Hoolst, T., and Karatekin, O. (2011), 'Librational response of Europa, Ganymede, and Callisto with an ocean for a non-Keplerian orbit', *Astronomy and Astrophysics* **527**, A118.

Rückriemen, T., Breuer, D., and Spohn, T. (2018), 'Top-down freezing in a Fe–FeS core and Ganymede's present-day magnetic field', *Icarus* **307**, 172–96.

Saur, J., Duling, S., Roth, L. et al. (2015), 'The search for a subsurface ocean in Ganymede with Hubble Space Telescope observations of its auroral ovals', *Journal of Geophysical Research: Space Physics* **120**(3), 1715–37.

Schubert, G., Anderson, J. D., Spohn, T., and McKinnon, W. B. (2004), 'Interior composition, structure and dynamics of the Galilean satellites', in F. Bagenal, T. E. Dowling, and W. B. McKinnon, eds., *Jupiter: the planet, satellites and magnetosphere*, Vol. 1, Cambridge University Press, pp. 281–306.

Seidelmann, P. K., Abalakin, V. K., Bursa, M. et al. (2002), 'Report of the IAU/IAG Working Group on Cartographic Coordinates and Rotational Elements of the Planets and Satellites: 2000', *Celestial Mechanics and Dynamical Astronomy* **82**(1), 83–111.

Stark, A., Hussmann, H., Steinbrügge, G., Oberst, J., and Roatsch, T. (2018), 'Resonant rotation parameters of the Galilean satellites', *EGU General Assembly Conference*, Vol. 20. https://meetingorganizer.copernicus.org/EGU2018/EGU2018-10488.pdf.

Stark, A., Oberst, J., and Hussmann, H. (2015), 'Mercury's resonant rotation from secular orbital elements', *Celestial Mechanics and Dynamical Astronomy* **123**(3), 263–77.

Stark, A., Oberst, J., Preusker, F. et al. (2015), 'First MESSENGER orbital observations of Mercury's librations', *Geophysical Research Letters* **42**(19), 7881–9.

Stiles, B. W., Kirk, R. L., Lorenz, R. D. et al. (2008), 'Determining Titan's spin state from Cassini RADAR images', *The Astronomical Journal* **135**(5), 1669–80.

Tiscareno, M. S., Thomas, P. C., and Burns, J. A. (2009), 'The rotation of Janus and Epimetheus', *Icarus* **204**(1), 254–61.

Van Hoolst, T., Baland, R.-M., and Trinh, A. (2013), 'On the librations and tides of large icy satellites', *Icarus* **226**(1), 299–315.

Ward, W. R. and Hamilton, D. P. (2004), 'Tilting Saturn; I. analytic model', *The Astronomical Journal* **128**(5), 2501–9.

Zubarev, A., Nadezhdina, I., Oberst, J., Hussmann, H. and Stark, A. (2015), 'New Ganymede control point network and global shape model', *Planetary and Space Science* **117**, 246–9.

4

Tidal Deformation and Tidal Evolution of Ganymede

GREGOR STEINBRÜGGE, HAUKE HUSSMANN, GABRIEL TOBIE, JULIE CASTILLO-ROGEZ, VALÉRY LAINEY, AND ALEXANDER STARK

4.1 INTRODUCTION

Being bonded by the Laplace resonance, a 4:2:1 orbital resonance with its inner neighbors Io and Europa, Ganymede is in a unique state. Within the Jovian system, the comparably small satellites have been heavily influenced by the gravity of their central planet at the origin of significant tidal forces. Very different worlds emerged from the mutual interactions between the Galilean satellites. Although Ganymede's size and ice-to-rock ratio are similar to those of Callisto, both satellites have undergone different geologic histories. Callisto's surface is heavily cratered and dark while Ganymede's surface displays heavily cratered regions mixed with a brighter "grooved" terrain that has been tectonically modified with surface ages of 4.2 billion years for the older terrain and 3.6 billion years for the younger terrain, respectively (Zahnle et al., 2003). Ganymede is further unique among solar system satellites by being the largest and the only one that generates its own magnetic field (see Chapter 13). The moon's large size leads to a unique hydrosphere, most likely composed of a water ocean sandwiched between an outer shell of ice I and a deeper layer composed of high-pressure water ice polymorphs (see Chapter 12). Ganymede's high degree of differentiation and geological contrast compared to its neighbor Callisto represents a long-standing puzzle whose explanation lies in the formation and evolution of the Galilean satellites and of the Laplace resonance (e.g. Kirk and Stevenson, 1987; Showman and Malhotra, 1997; Mosqueira and Estrada, 2003). While tidal heating does not significantly contribute to the heat budget of Ganymede at present, that process might have played a critical role in the moon's history. The exact mechanism and chronology of the Laplace resonance' emplacement and evolution are still not well understood and it will require the upcoming missions to the Jovian system and ground-based observations to shed some light on the Jovian system's enigmatic past.

While the tidal fluctuations on Ganymede, due to its small orbital eccentricity, are smaller than for its inner neighbors Europa and Io, they still lead to meter-scale surface deformation if a subsurface ocean is present underneath the icy crust (Moore and Schubert, 2003). A deep ocean is suggested by the detection of an induced magnetic field by the *Galileo* mission (Kivelson et al., 2002). The pattern of the tidal-driven deformation depends on the moon's orbital parameters, rotation state, interior structure, and thermal state. The eccentricity, which is more variable for Ganymede than for the other Galilean satellites, is well constrained but the obliquity has not yet been measured. Also, Ganymede's interior structure is still largely unconstrained, especially the structure of the hydrosphere (see Chapter 6). Hence, determining the tidal amplitude and phase lag would help retire key gaps in our understanding of Ganymede's subsurface. Therefore, measuring the expression of Jupiter's tidal force exerted on Ganymede is a critical investigation in the future exploration of Ganymede and will in turn increase our understanding of the formation and evolution of the Galilean satellites.

In this chapter, we start with an overview of the tidal interactions within the Jovian system (Section 4.2), treating the fundamentals of tidal potentials, tidal heating, orbital, and coupled thermal-orbital evolution. We then describe Ganymede's tidal behavior at present in the context of its interior parameters, based on our current knowledge (Section 4.3). Finally, Section 4.4 explores the tidal evolution of Ganymede, the influence of tidal heating on its internal differentiation, and the implications for the formation of its hydrosphere and surface geology.

4.2 TIDAL INTERACTIONS WITHIN THE JOVIAN SYSTEM

4.2.1 The Tidal Potential

Whenever a moon is subject to the gravitational force of its central planet, a reciprocal tidal force is exerted. Starting with the two-body problem of classical mechanics, the gravitational acceleration can be expressed as the gradient of the gravitational potential $\vec{a} = \nabla(GM/d)$, where G is the gravitational constant, M is the mass of the planet, and d the distance between the centers of the respective masses of the two bodies. If the satellite were a point mass, no tide would arise. However, if we consider the satellite as a finite spherical body and put ourselves in a coordinate frame centered on the satellite's center of mass O, we can examine the gravitational pull on an arbitrary point P within the satellite (Figure 4.1). The gravitational force of the planet $\vec{F_c}$ has to be compensated for by the centrifugal force $-\vec{F_c}$ at the center of mass as the satellite revolves about the planet. All points in our reference frame follow the motion of O and hence the centrifugal force is the same at each point on the satellite as it describes the same motion as the center of mass. In contrast, the gravitational force scales with $1/d^2$ and is consequently slightly weaker at the far side than at the near side of the satellite. At the point P, we therefore obtain an effective tidal force $\vec{F} - \vec{F_c}$. In terms of gravitational potential and as a function of the zenith angle ϕ, we can express the gravitational pull as $GM \cos\phi/d^2$. To obtain a zero pull at the center of mass, we have to subtract GM/d such that we obtain the potential

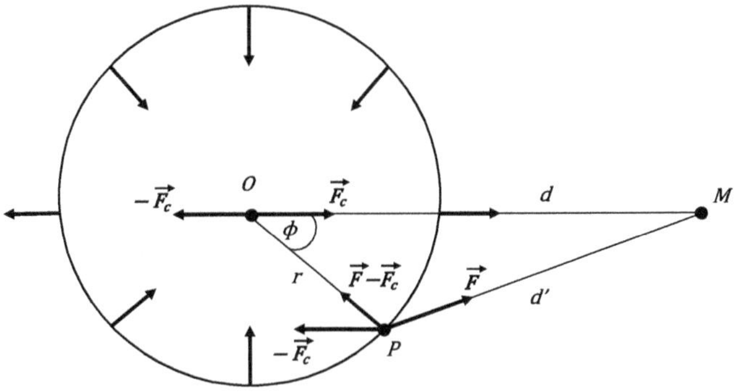

Figure 4.1 Schematic view of the forces acting on a satellite. At the center of mass, the centrifugal force F_c and the gravitational force F raised by the planet of Mass M cancel out. However, since the gravitational force raised by the planet is dependent on the distance, effective tidal forces arise.

$$V = GM\left(\frac{1}{d'} - \frac{1}{d} - \frac{r\cos\phi}{d^2}\right). \tag{4.1}$$

Using the relation $d'^2 = d^2 + r^2 - 2rd\cos\phi$, d' can be expressed as

$$\frac{1}{d'} = \frac{1}{d}\left[1 - 2\frac{r}{d}\cos\phi + \left(\frac{r}{d}\right)^2\right]^{-1/2}. \tag{4.2}$$

Turning this equation into a series expansion in $\frac{r}{d}$ then yields the general form of the tidal potential (Kaula, 1964), as a sum of associated Legendre polynomials P_l^0 of degree l and order 0.

$$V = \frac{GM}{d}\left[-\frac{1}{2}\left(\frac{r}{d}\right)^2 + \frac{3}{2}\left(\frac{r}{d}\right)^2\cos^2\phi - \ldots\right] \tag{4.3}$$

$$= \frac{GM}{d}\sum_{l=2}^{\infty}\left(\frac{r}{d}\right)^l P_l^0(\cos\phi). \tag{4.4}$$

The potential described by Equation (4.4) contains two components. The static component leads to a static tidal bulge similar to the deformation resulting from the rotation of a body. However, if the satellite's orbit is eccentric or if the satellite is not in synchronous rotation, then the resulting deformation will contain a periodic component. In a Keplerian orbit, Equation (4.4) can be expressed as a function of colatitude and longitude by the Kaula formalism, which then results in

$$V = \frac{GM}{d}\sum_{l=2}^{\infty}\left(\frac{r}{d}\right)^l \sum_{m=0}^{l}\frac{(m-l)!}{(m+l)!}(2-\delta_{0m})P_l^m(\cos\theta)$$
$$\cdot\sum_{p=0}^{l}F_{lmp}(i)\sum_{q=-\infty}^{\infty}G_{lpq}(e)S_{lmpq}(\omega, M, \Omega, \Theta) \tag{4.5}$$

with δ_{0m} being the Kronecker delta and S_{lmpq} capturing the constraints arising from the spin-orbit resonance (for details, see Kaula, 1964). $F_{lmp}(i)$ and $G_{lmp}(e)$ are functions dependent on the inclination and eccentricity and are tabulated in Kaula (1964). For a body trapped in a 1:1 resonance and with a mean motion n, the resulting expression up to the second degree, first order eccentricity e and first order obliquity i leads to the tidal potential as a function of longitude ϕ and colatitude θ (= 90° − latitude) (e.g. Jara-Orué and Vermeersen, 2016).

$$\Phi_2(r,\theta,\phi,t) = r^2n^2[-\frac{3}{2}eP_2^0(\cos\theta)\cos nt$$
$$+\frac{1}{4}eP_2^2(\cos\theta)(3\cos nt\cos 2\phi + 4\sin M\sin 2\phi)$$
$$+\sin i\, P_2^1(\cos\theta)\cos\phi\,\sin(\varpi + nt)]. \tag{4.6}$$

We use Φ here instead of V to indicate that the static part of the potential has been dropped. The index 2 means that the potential has been developed up to the second degree, that is, $l = 2$. ϖ is the argument of pericenter measured with respect to Ganymede's ascending node. The associated Legendre polynomials of degree and order 2 are defined by

$$P_2^0(\cos\theta) = \frac{1}{2}\left(3\cos^2\theta - 1\right)$$
$$P_2^1(\cos\theta) = -3\sin\theta\cos\theta$$
$$P_2^2(\cos\theta) = 3\sin^2\theta.$$

For bodies in a Keplerian orbit with small eccentricity and obliquity, Equation (4.6) is a good approximation to Equation (4.4). However, as will be examined in detail in Section 4.3 for Ganymede, this only applies if the orbital elements are averaged over long timescales (i.e. thousands of years).

The tidal potential leads to a periodic deformation of Ganymede with a tidal bulge being raised around the sub-jovian and anti-jovian point with tidal amplitudes being strongest at the equator (compare to Section 4.3.1). The reaction of a body to the tidal potential is commonly parameterized in terms of the complex and dimensionless tidal Love-Shida numbers h, k, and l. The Love number h expresses the radial deformation u_r of the body $u_r = h\Phi/g$, with g being the gravitational surface acceleration, and k describes the gravitational distortion as a secondary potential $\Phi_{int} = k\Phi_{ext}$. The tidal Love number l describes the lateral displacement $u_t = l\nabla\Phi/g$. From Equation (4.5) it can be seen that the tidal potential decreases with $(r/d)^l$. Therefore, only the degree-two term is commonly considered, and the corresponding tidal Love number is indexed by the same number (e.g. k_2 for the potential of order 2). Tidal Love numbers are complex numbers with an amplitude and a phase. The latter describes the temporal difference between the acting force and the viscoelastic reaction of the satellite.

4.2.2 Tidal Heating

For viscoelastic bodies, the deformation induced by the tidal potential leads to friction in the internal layers and consequently to the dissipation of mechanical energy. The amount of tidal dissipation depends on the amplitude of the perturbing potential but also on the density and the mechanical properties of each internal layer. For ice, these properties have been experimentally studied, for example by Goldsby and Kohlstedt (2001) and Durham et al. (2001). The presence of liquid layers (water ocean, metallic core) has a strong impact on the global deformation and on the way energy is dissipated in the adjacent layers.

Another critical property is the Maxwell time τ defined as the ratio between the viscosity η and the rigidity μ. While the Maxwell model is in many ways a simplified model (see Section 4.3.2 on rheology), the Maxwell time is characteristic of the amount of energy a material dissipates. If the Maxwell time is significantly larger than the periodic forcing frequency, then the deformation is mostly elastic and only a very small amount of energy is dissipated. In the other end member case of a Maxwell time significantly smaller than the forcing period, the material approaches a fluidlike behavior, and viscoelastic dissipation is strongly reduced.

Even if the layer is fully fluid, dissipation other than viscoelastic dissipation, such as solid/liquid friction, turbulent friction, and resonant waves, may occur, similarly to dissipation mechanisms acting on the Earth's ocean (e.g. Egbert and Ray, 2000). However, the occurrence of oceanic dissipation strongly depends on the ocean configuration (e.g. thickness, stratification, geometry of boundary layers), the coupling with the icy shell, and the physical processes at play, which are not well constrained. Hence, the level of fluid dissipation that can be achieved in ice-covered oceans is still uncertain (e.g. Tyler, 2008, 2014; Chen et al., 2014; Beuthe, 2016; Matsuyama et al., 2018; Hay et al., 2020; Rovira-Navarro et al., 2020).

In solid layers, the energy dissipation is maximized when the Maxwell frequency matches the forcing frequency. In the Maxwell model, a peak of dissipation is reached when $\omega\eta/\mu = 1$ (e.g. Tobie et al., 2003). Ice has a rigidity of about 3–3.6 GPa (e.g. Sinha, 1989; Sotin et al., 1998) and a highly temperature-dependent viscosity. The latter can range between 10^{13} and 10^{15} Pa.s near the melting point, which, for the low-stress regime expected in icy shells, is mostly determined by the grain size of ice crystals (Budd and Jacka, 1989; Montagnat and Duval, 2000; Durham et al., 2001; Goldsby and Kohlstedt, 2001). At Ganymede's orbital frequency ($n = 1.02 \times 10^{-5}$ rad.s^{-1}), a peak of dissipation would be reached for $\eta = 3 - 3.5 \times 10^{14}$ Pa.s, similar to the viscosity value expected at the ocean/ice interface (Durham et al., 2001; Goldsby and Kohlstedt, 2001; Sotin and Tobie, 2004). As the amount of dissipated energy scales with ω^5 and e^2, the amplitude of tidal heating in Ganymede's ice shell is much smaller than in Europa's shell (e.g. Tobie et al., 2005). Hence, while tidal heating significantly contributes to the thermal budget of Europa, it is less significant for Ganymede at present. The relative contribution of tidal heating within the ice shell to Ganymede's heat budget at present is about 1 mW/m^2 or 5–15% of the total heat flux (e.g. Chen et al., 2014). However, it might have influenced the moon's past evolution (Bland et al., 2009).

Tidal heating is regulated by several competing effects. If the eccentricity remains high from being continuously re-excited by a mean-motion resonance with other moons, then increased tidal heating could lead to melting of the ice shell, hence decreasing the volume of ice subject to tidal forcing. This reduces the heat budget and eventually leads to an equilibrium. This equilibrium state has been investigated for Europa, Ganymede, and Callisto. While it is unclear whether the ice shells of the Galilean satellites are in thermal equilibrium, Hussmann et al. (2002) argued that, in the case of heating or cooling Europa's ice shell by 20 K, and considering the latent heat and heat capacity of ice, the transition from a thin (order of km) to a thick ice shell (order of tens of km) or vice versa can occur within 1–2 million years. Similar timescales would be expected for Ganymede and the ice shell should therefore be at least close to equilibrium.

4.2.3 Orbital Evolution

The tidal interactions between satellites and their planet have further effects on the orbital evolution of the satellites, as the energy dissipation by internal friction in the planet causes an angular momentum being transferred from the planet to its satellites. If the planet rotates faster than a satellite revolves about it, the tidal bulge of the planet will be ahead of the axis connecting the planet and satellite and hence cause a tidal torque. Consequences of that torque are an increase in the semi-major axis and eccentricity of the satellite and simultaneously a slight despinning of the planet. In a similar way, the satellite's rotation period is decreased by the tidal interaction with the planet: If the satellite's spin period is smaller than its orbital period, for instance shortly after formation of the satellite before spin-orbit synchronization, the tidal bulge on the satellite is slightly ahead of the direction to the planet. The resulting torque causes the satellite's spin to decrease. Assuming initial spin periods of 10 hours for the satellites and homogeneous viscous bodies, the timescale for reaching the 1:1 synchronous spin-state can be estimated to a few tens of kyr for Europa and Ganymede (Hussmann et al., 2010). Under these assumptions, Ganymede despun slightly faster than Europa due to its greater size, which overcompensates its greater distance from Jupiter. For Io, the closest Galilean moon, the despinning timescale is only a few thousand years. For Callisto, despinning occurs on the order of millions of years due to the greater distance as compared to Ganymede. For all Galilean moons, these timescales are much smaller than the age of the solar system, which is consistent with the currently observed synchronous rotation states. The second effect is that the planetary tidal bulge leading the satellite's motion exerts a gravitational pull, effectively accelerating the satellite and hence increasing its semi-major axis. Since this gravitational pull is greater when the satellite is at periapsis, the increase in velocity raises the apoapsis altitude and thereby also increases the eccentricity. This latter effect is, however, counteracted by dissipation within the satellite. Eccentricity damping timescales $e/\dot{e}$ are on the order of a few Myr for Io and several hundred Myr for Europa and Ganymede. Thus their orbits should become circular over time. The fact that Io and Europa still have significant orbital eccentricities is a consequence of the resonant forcing. The forcing of Ganymede's eccentricity by Europa, however, is small compared to Ganymede's free eccentricity. The latter is the eccentricity that would remain without any resonant perturbation. Ganymede's free eccentricity of 0.0015, which is damped by tidal interaction with Jupiter on long timescales, might be a remnant of a previous Laplace-like resonance passage (Showman et al., 1997). In summary, for most satellites the eccentricity decreases in response to tidal forcing, leading to the circularization of their orbits. This important feedback mechanism leads to a rapid decrease of tidal dissipation in the satellites, unless the eccentricity is forced by a resonance with one or several other satellites.

In the Galilean satellite system, nonzero eccentricities are maintained by the Laplace resonance, which is a mean motion resonance that obeys the relation $n_1 - 3n_2 + 2n_3 = 0$, where n_1, n_2, and n_3 are the mean motions of Io, Europa, and Ganymede, respectively. The mean motions of Io and Europa obey the relation $n_1 - 2n_2 = \nu$, whose present-day values are $n_1 = 203$ deg/day, $n_2 = 101$ deg/day, and $\nu = 0.74$ deg/day (e.g. Yoder,

1979; Greenberg, 1987). A ν of zero means that Io completes two orbits in the same time Europa orbits Jupiter once; however, the slight offset leads to the longitude of conjunction between these moons to slowly drift over time. The conjunction of Europa and Ganymede also drifts slowly over time exactly at the same rate, allowing the angle between the two conjunctions to remain the same. The stability of the Laplace resonance hinges on the extreme tidal dissipation in Io (Yoder, 1979). In an initial state where Io formed inside Europa's orbit, the initial eccentricities of both moons would have been damped very quickly. Subsequently, their orbits would have moved outwards due to tidal torques raised by Jupiter. The rate at which the orbits are receding from the primary will depend on the semi-major axis of the satellite and on the dissipation rate in Jupiter. However, dissipation in the satellites counteracts these effects and would decrease both the eccentricity and semi-major axis. Thus, it depends on the ratio of dissipation inside the planet and in the satellite whether the moons move outwards and at what rate. Before being in resonance with Europa, Io's eccentricity would be quickly damped and Io would then move outwards at a fast rate exceeding the one of Europa. Owing to its smaller semi-major axis and greater mass, Io would drift faster until approaching the 2:1 commensurability with Europa. Yoder (1979) argues that once this state was reached, Io's eccentricity increased; however, the resulting tidal dissipation led to damping the eccentricity and semi-major axis, thus counterbalancing the effect of Jupiter's tides and forcing Io further outwards. Hence, an equilibrium could be established, and Io and Europa migrated outwards together, until Europa's orbit encountered the 2:1 commensurability with Ganymede. At that stage, Io's dissipation could increase even further within the three-body resonance, and Jupiter's torque forced the satellites into their resonance locks (Yoder and Peale, 1981). The resulting resonance is stable, as a further increase in ν would decrease Io's eccentricity and hence Io's tidal heating, driving ν back towards equilibrium (Greenberg, 1987). However, this theory hinges on Jupiter being subject to high tidal dissipation and it has been argued that the theory by Yoder and Peale (1981) is not compatible with the estimated dissipation rate in Jupiter (Greenberg, 1982). Nevertheless, the physical mechanisms behind tidal dissipation inside giant planets have been reinvestigated in light of the tidal dissipation inside stars. Recent results suggest that tides inside Jupiter could be significant while highly dependent on tidal frequency (Fuller et al., 2016; Lainey et al., 2020). In particular, the resonance locking theory predicts that moons involved in such a dynamical state will expand their orbit at a specific rate to preserve the excitation of proper modes inside the stably stratified interior of their primary. While extremely attractive, the hypothesis of the resonant lock mechanism currently at work within Jupiter has not been demonstrated yet.

4.2.4 Thermal-Orbital Evolution in the Laplace Resonance

As described in the previous section, the three satellites Io, Europa, and Ganymede are locked in the Laplace resonance. Consequently, the orbital eccentricities of Io and Europa are forced, resulting in substantial tidal heating of these satellites on geologic timescales. In the current configuration, Ganymede's eccentricity is dominated by its free contribution and tidal heating does not play a significant role at present. However, Ganymede's surface record (e.g. crater relaxation, groove formation, and flexure of furrows and grooves) suggests that the satellite has gone through phases of intense heating around 3–4 Ga ago, which correspond to surface heat fluxes ranging from 50 to 140 mW/m^2 (Nimmo et al., 2002; Bland and Showman, 2007). In contrast, the maximum radiogenic heating rate shortly after formation of the satellite has been estimated to be about 30 mW/m^2 based on a chondritic composition of the silicate core (Bland et al., 2017) and it has been decreasing since then. Thus, radiogenic heating is by far insufficient to explain the degree of internal activity inferred from Ganymede's geology. While tidal heating might be a possible explanation, Ganymede's eccentricity would need to be significantly enhanced with respect to its current state. However, due to the coupling of Io's thermal evolution to its orbital state, Io might run through cyclic changes in eccentricity and thus heating rate (Ojakangas and Stevenson, 1986; Fischer and Spohn, 1990; Hussmann and Spohn, 2004). With the satellites being locked in resonance, this would have consequences for Europa's and Ganymede's thermal-orbital states as well. The orbital eccentricity of Ganymede could oscillate, peaking at about 0.01, for which tidal heating is about the same order as radiogenic heating. Whether this effect was coupled with variations in Ganymede's ice layer thickness still remains to be investigated. In the aforementioned models, Ganymede is only included in the orbital equations and not with a full time-dependent thermal model, hence neglecting changes in its dissipation rate. Another possible consequence of intense tidal heating could arise during the formation of the Laplace resonance while the satellites could go through different Laplace-like resonance locks (Malhotra, 1991; Showman et al., 1997; Showman and Malhotra, 1997). In these models, Ganymede's eccentricity could reach values up to about 0.02, which could introduce a large peak in the satellite's heating rate, depending on the assumed total amount of dissipation in Jupiter (parameterized by Jupiter's quality factor Q_{jup}). The enhanced tidal dissipation in Ganymede could lead to a positive feedback between the rate of tidal heating and the temperature of the ice such that a thermal runaway can occur, which could explain Ganymede's volume expansion and geological activity in the past (see also Bland et al., 2009). While a quantitative assessment of Ganymede's thermal-orbital evolution remains challenging considering the many uncertainties in the parameters and their time-dependence, tidal heating most likely was a significant contributor to Ganymede's heat budget in the past.

4.3 GANYMEDE'S INTERIOR FROM TIDAL MONITORING

Ganymede's present-day tides are driven by its interior structure and its orbital parameters. The orbital period is 7.155 days, which corresponds to a mean motion n of 1.0164×10^{-5} rad/s[1]. Its mean orbital eccentricity is $e = 0.0015$, however subject to considerable variations, and oscillates from values close to zero up to eccentricities of 0.003 on a scale of about 100 years (Figure 4.2). While the eccentricities of Io and Europa are also variable, Ganymede's variability is the highest relative variation of all Galilean satellites as its absolute eccentricity value is the lowest.

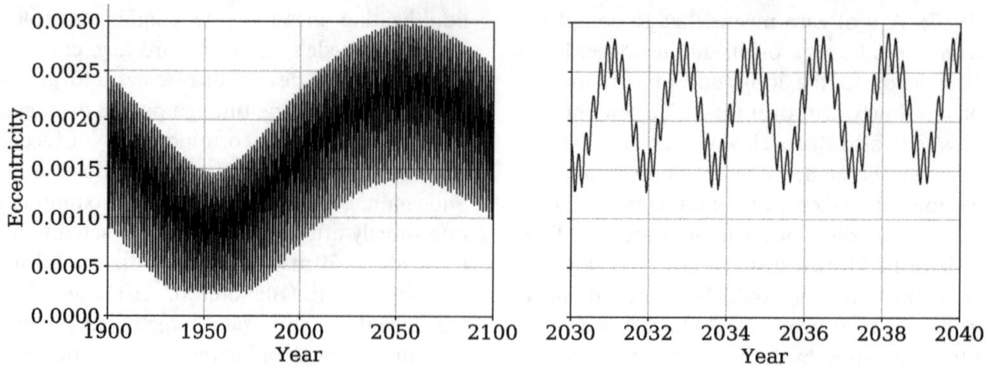

Figure 4.2 Variation of Ganymede's eccentricity as a function of time. The ephemerides have been evaluated and averaged over one orbital period.

The obliquity of Ganymede has not been measured yet but is likely small. However, it cannot be exactly zero as Ganymede's orbit has a periodically varying inclination towards the equatorial plane of Jupiter. If the obliquity would be zero for a certain moment, the precessional torque would vanish and the spin axis would stop following the precession of the orbit. This implies that the obliquity would be ultimately increasing again and not remain zero. The variations in the orbital inclination are small but still larger than the obliquity variations if the rotation state were forced into a general Cassini state by tidal dissipation (Bills, 2005).

As the eccentricity leads to a tidal bulge migrating in longitude, a nonzero obliquity leads to a migration in latitude. The possible range of Ganymede's obliquity could be derived from its orbital evolution but depends on the interior structure. The presence of a global ocean could lead to resonant amplifications for some periodic terms resulting in obliquity values of the order of a few degrees (Baland et al., 2012). However, as the obliquity was not detectable by the Galileo spacecraft, and also not in Earth-based radar observations (Margot et al., 2013), with a reported detection limit of 0.1 degrees, the present-day value is likely to be smaller than this.

4.3.1 Tidal Potential and Radial/Lateral Surface Deformation on Ganymede

The amplitude of the radial surface deformation of Ganymede depends on its interior structure. For a typical interior consisting of a metallic core, a silicate mantle, and an ocean sandwiched between a high-pressure ice layer and an outer icy crust (see Chapter 6), the h_2 Love number is between 1 and 1.5. Using the tidal potential as given in Equation 4.6 with the mean eccentricity of 0.0015, the radial tidal deformation of Ganymede would be of the order of 3 to 4 m in amplitude (Figure 4.3), leading to a peak-to-peak amplitude of up to ~7 m. The tidal response is most sensitive to the thickness and rheology of the ice shell. Generally, the influence of the rigidity is greater for thicker shells, reducing the tidal amplitude; however, even a very rigid outer ice shell ($\mu > 10$ GPa) would reduce the peak-to-peak amplitude by no more than 2 m (Moore and Schubert, 2003). The better constrained ice density and ocean density also have a bearing on the tidal Love numbers k_2, and to a lesser extent on h_2. Both parameters increase with increasing ocean densities in particular (Jara-Orué and Vermeersen, 2016).

Without an ocean, the tidal amplitude would be between 10 and 20 cm. A hypothetical exception would be if a very dissipative high-pressure ice layer is present (Kamata et al., 2016). In this case, the high-pressure ice would be characterized by a low viscosity, leading to tidal amplitudes in the meter scale, however the ice would still be solid. According to Moore and Schubert (2003) (Figure 4.3), the viscosity in the high-pressure ice layer needs to be lower than 10^{12} Pa.s in order for Ganymede to exhibit a tidal amplitude mimicking that of an ocean. For such a low viscosity, the convective heat transfer through the high-pressure ice would be very efficient, rapidly cooling that layer far from its melting temperature (Kalousová et al., 2018). Hence, a very low viscosity high-pressure ice layer appears unlikely.

The pattern of maximum deformation is characterized by two tidal bulges at the sub- and anti-jovian hemispheres with maximums offset about 30 degrees to the east and west. This pattern might further be tilted in case the spin axis is tilted with respect to the orbital plane. As noted by Jara-Orué and Vermeersen (2016), the effect of the obliquity on the tidal deformation pattern is significantly more pronounced on Ganymede than it would be, for example, on Europa. This is a consequence of the comparably small eccentricity of 0.0015 in average, when compared to a conceivable obliquity value in the order of 0.01. However, with the variation in the eccentricity (Figure 4.2) the relative impact of the obliquity will change over time. As noted, the obliquity of Ganymede has not been measured yet and at the same time simulations provide a rather wide range of possible obliquities (Baland et al., 2012). This is due to possible resonant amplification of Ganymede's response to variation of the orbital inclination on some timescales (see Chapter 3). In order to visualize the effect of obliquity tides in Figure 4.3, we chose an obliquity of 0.032 degrees, which corresponds to an entirely solid interior for Ganymede (Baland et al., 2012).

In addition to the radial deformation, the tidal response includes a lateral component characterized by the tidal Love number l_2. Ganymede's tidal Love number l_2 would be of the order of 0.2 to 0.4 for a typical interior model consisting of an 800 km iron core, 1 080 km silicate mantle, 450 km high-pressure ice layer, and 300 km outer ice/ocean layer (Steinbrügge et al., 2015). The corresponding lateral amplitudes would be about a factor three smaller than the radial displacement. Consequently, this effect is difficult to measure and is thus less often considered. However, as noted in the context of Europa (Hussmann et al., 2011), a lander could potentially measure linear

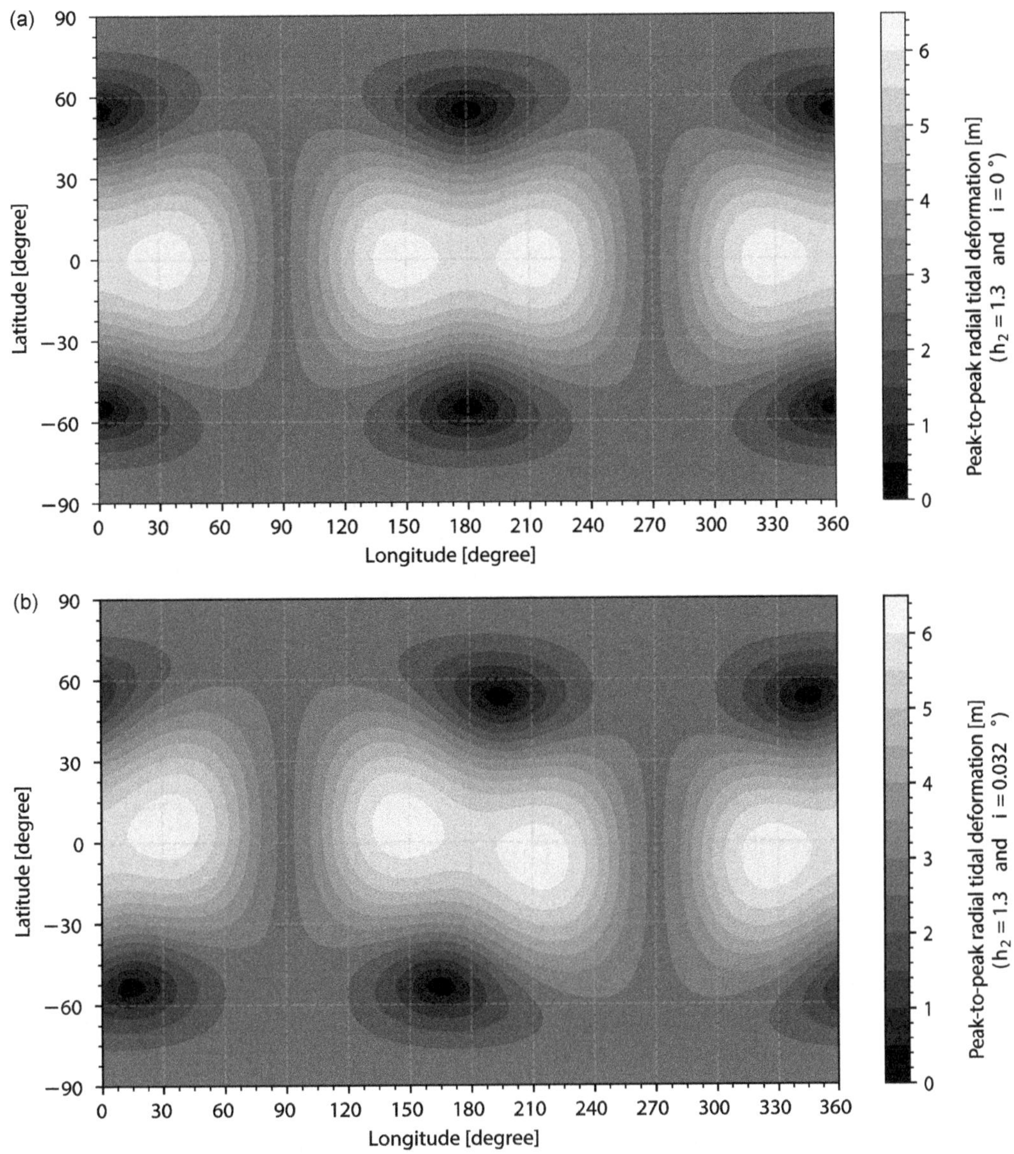

Figure 4.3 Radial tidal deformation of Ganymede's surface assuming a tidal Love number h_2 of 1.3 with (a) zero obliquity and (b) an obliquity of 0.032°. The plots show the tidal deformation for an eccentricity of 0.0015.

combinations of Love numbers, including l_2. The latitudinal factor can be expressed by $1 + k_2 - l_2$, and the strain factor by $2h_2 - 6l_2$ (Zürn and Wilhelm, 1984; Hussmann et al., 2011). The former parameter characterizes variations in the elevation angle of stars when observed from the surface. The strain factor can be obtained from measurements with strain-meters of the distance variations between surface landmarks.

4.3.2 Impact of the Rheology and Thermal State on Ganymede Tides

As discussed in Section 4.2.2, the tidal response of Ganymede depends on both the elastic and the viscous properties of the satellite interior. To first order, the presence of layers with near-zero shear modulus and viscosity (ocean, metallic core) controls the tidal amplitude. In the absence of an internal ocean, the tidal response amplitude, described by the Love numbers, is reduced by about one order of magnitude (Moore and Schubert, 2003) (Figure 4.4). The influence of the rigidity of the solid layers on the tidal response is of second order. For example, reducing the average shear modulus of the outer shell by one order of magnitude would result in a tidal amplitude increase of about 15–30%, depending on the thickness of the ice shell (Moore and Schubert, 2003). As indicated in Section 4.2.2, there is no physical reason to change the shear modulus by one order of magnitude, as this physical quantity is relatively well known for ice. Based on calculations performed for Titan (Mitri et al., 2014), the Love number is expected to vary by 4–6% for shear modulus ranging between 2 and 4 GPa, for ice shell thicknesses comprised between 50 and 100 km.

The viscosity of the ice is poorly understood and could vary significantly depending on the thermal state of the interior. For values larger than μ/ω, the viscosity's influence on the tidal response is minor. Below that Maxwell value, the viscosity has

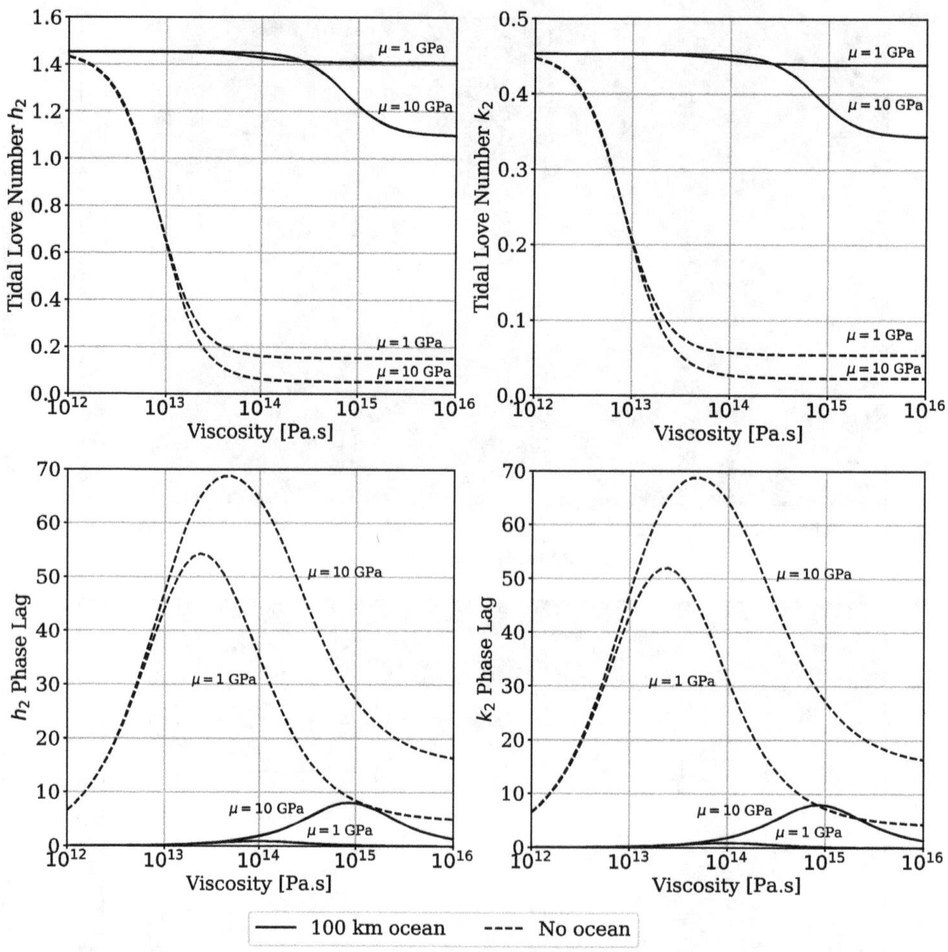

Figure 4.4 Tidal Love numbers h_2 and k_2 and their respective phase lags as a function of viscosity for a weak ice shell ($\mu = 1$ GPa) and a strong shell ($\mu = 10$ GPa). Shown is the case for the interior structure model with a 800 km iron core, 1 000 km mantle, 450 km high-pressure ice shell, 200 km ocean, and 100 km ice I shell from Steinbrügge et al. (2015) with a total hydrosphere thickness of 300 km. The densities of the layers are 5 900, 3 080, 1 310, and 1 000 kg.m^{-2}, respectively. The ice-I shell and ocean are assumed to have equal density for simplicity.

a significant impact on the tidal response, resulting in increased tidal amplitude and phase lag (Moore and Schubert, 2003; Hussmann et al., 2016). Precise monitoring of the tidal phase lag may provide key information on the average viscosity in the outer ice shell (Hussmann et al., 2016) and thus on the thermal state of that layer. A large phase lag would be indicative of a low viscosity and thus a convective ice shell, while a small phase lag would suggest a high viscosity that would hint at a conductive shell.

Nevertheless, the relationships between phase lag, viscosity, and thermal state are not straightforward and depend on several assumptions on material rheology. First, the relationship between the phase lag or dissipation function, Q, and the viscosity depends on the assumed rheological model. The Maxwell model, which is classically used (e.g. Moore and Schubert, 2003; Hussmann et al., 2016), provides a good estimate of the viscoelastic tidal response when the Maxwell time (i.e. η/μ) is close to the tidal forcing period. However, it strongly underestimates viscous dissipation and phase lag (or overestimates Q) for forcing periods much smaller than the Maxwell time (e.g. Sotin et al., 2009). The Andrade rheology, which is consistent with existing experimental data on silicate minerals and ices (Castillo-Rogez et al., 2011; Efroimsky and Makarov, 2013), provides a much better description of anelastic response on a wide range of viscosity and tidal frequency (Castillo-Rogez et al., 2011; Mitri et al., 2014; Tobie et al., 2019).

4.3.3 Tidal Heating Distribution from the Ice Shell to the Core

As illustrated by Figure 4.5, for some examples of density and rheological profiles taken from Vance et al. (2018), tidal heating per volume is largest in the outer ice shell due to the mechanical decoupling by the internal ocean which leads to larger tides in the shell than in the mantle and due to the difference in shear modulus and viscosity values. In the deep interior, the tidal heating rate in both the high-pressure ice mantle and the rocky mantle is much smaller than the radiogenic heating rate and therefore has no influence on the thermal budget, although the lower viscosity in the high-pressure ice still leads to a higher heating rate than in the mantle. In the outer ice shell, the volumetric tidal heating rate ($\sim 10^{-8}$ W/m^3) is comparable to the volumetric radiogenic heating rate in the rocky mantle. However, once integrated throughout the entire ice shell, this heating rate corresponds to a surface heat flux barely reaching 1 mW/m^2, which remains small compared to the heat flux due to radiogenic heating in the rocky mantle and the latent heat of

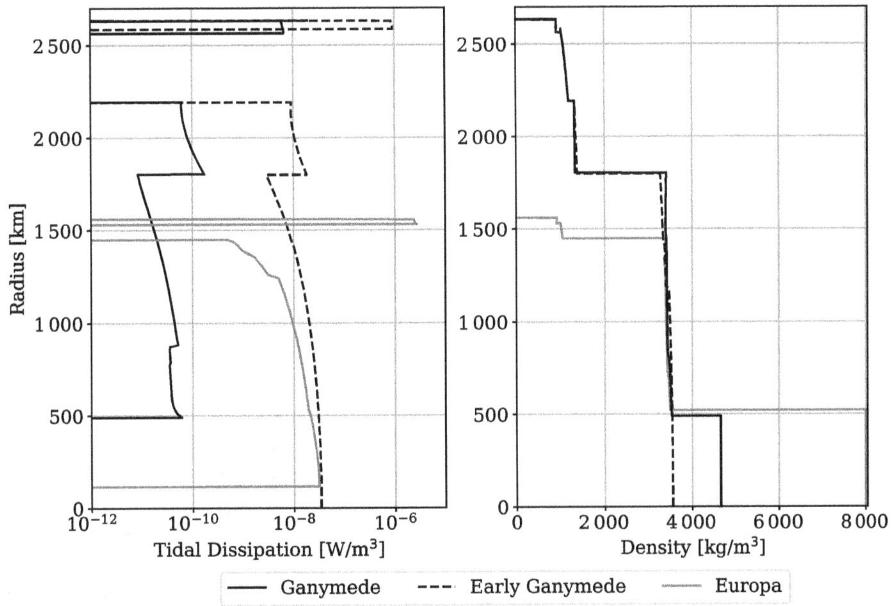

Figure 4.5 Comparison of tidal heating in Europa, Ganymede, and early Ganymede for three representative density and elastic parameter profiles taken from Vance et al. (2018), assuming the same viscosity parameters for the present-day Europa and Ganymede model (10^{14} Pa.s for low-pressure and high-pressure water ice, 10^{20} Pa.s for rock) and a less viscous undifferentiated rock core (10^{18} Pa.s) and 10 times higher eccentricity for an early Ganymede.

ocean crystallization estimated at 5–10 mW.m^{-2}. For an average present-day eccentricity of 0.0015, the tidal heating rate in the outer shell of Ganymede is about two orders of magnitude smaller than on Europa (Figure 4.5).

As proposed by Malhotra (1991) and Showman and Malhotra (1997), Ganymede may have experienced periods with enhanced eccentricity up to about 0.02 when entering the Laplace resonance. During these periods, tidal heating in the ice shell would be comparable to the present-day value for Europa and thus may result in a strong increase in surface heat flux, potentially explaining the past high heat flux (> 40–50 mW.m^{-2} inferred from crater relaxations (Bland et al., 2017)).

As composition can significantly influence the viscoelastic and thermodynamic properties of planetary materials, there is a need to better quantify the abundance and behavior of non-ice materials in icy moons. In particular, gas clathrate hydrates and salt hydrates could significantly change the thermal profile and the viscosity structure of the outer shell if they are abundant (e.g. Bland et al., 2017; Kamata et al., 2019; Kalousová and Sotin, 2020). While the partial pressure of gas in solution was considered too small for clathrate hydrates to form in the Galilean moons, recent formation models suggest these bodies accreted with a larger fraction of cometary and carbonaceous chondrite material that could lead to high partial pressures of CO_2, in particular (e.g. Néri et al., 2020). By affecting the thermal and viscous structure of the icy shell, gas and salt hydrates may modify the distribution of tidal dissipation, especially if their abundance varies spatially. Depending on the level of heterogeneity, detailed tidal monitoring may reveal local or regional variations in the tidal amplitude and phase lag.

The salt content of the hydrosphere also significantly affects the phase diagram of ice and thus the structure of the hydrosphere and the density stratification in the ocean (Vance et al., 2018). The thickness and thermal state of the high-pressure ice mantle forming at the base of the ocean are conditioned by the thickness of the outer ice shell and the composition of the ocean. The density evolution with depth across the ocean has an effect on the overall tidal response measured at the surface, as well as on the level of tidal heating in the high-pressure ice layer. Modeling the hydrosphere structure in a self-consistent manner, including experimental constraints from phase diagrams and thermodynamic properties of aqueous solutions (Vance et al., 2018; Journaux et al., 2020), is essential in order to more accurately predict its tidal response.

Another unknown is the state of the rocky mantle following internal differentiation. Formation from carbonaceous chondrite material implies the rock is already aqueous altered to a large extent at the time of accretion and iron partitioned between the silicates, magnetite, and pyrrhotite. On the other hand, formation from dry material (e.g. ordinary chondrite, cometary material) implies that only partial hydration would occur during differentiation as particle sinking in a body of that size is expected to proceed on a faster scale than aqueous alteration processes like serpentinization as has been proposed for Ceres (Castillo-Rogez and Bland, 2021). The fraction of water trapped in the rocky mantle can then influence the long-term evolution of that layer, for example by shifting its solidus toward lower temperatures. Furthermore, aqueous alteration led to the leaching of major elements that contribute to the ocean salinity, including potassium, thus depriving the rocky mantle from a major heat source (Castillo-Rogez and Lunine, 2010). Lastly, organic matter has been suggested to be more abundant in icy moons than previously considered (Néri et al., 2020) and could influence the thermal state and viscoelastic properties of the rock. However, rigorous internal evolution models accounting for the effect of tens of wt.% of organic matter remain to be developed.

At present, tidal dissipation is expected to be very small in the rocky part ($< 10^{-10}$ W/m^{-3}), but may have reached values comparable to the radiogenic heating in the past, and even possibly higher values if the rocky interior contained large fractions of hydrated minerals and organic matter, which are much more deformable than dry olivine minerals (Hilairet et al., 2007). Future modeling and experimental efforts are required to assess the possible amplification of tidal heating in this kind of organic-rich, hydrated rocky mantle.

4.4 TIDAL EVOLUTION OF GANYMEDE

4.4.1 Short-Term Evolution and Present State

The first determination of tidal dissipation within the Jovian system using astrometry was performed during the early 20th century (de Sitter, 1928). While numerous estimations were later published, most models suffered from a simplistic or incomplete dynamical representation of the orbits (Lainey et al., 2004). Benefiting from numerical simulations, the first accurate estimation of tides inside Jupiter and Io was given by Lainey et al. (2009). Assuming a constant quality factor Q for Jupiter, they found that Io's orbit is decaying, while Ganymede's and Europa's orbits are expanding. The fitted dissipation values correspond to orbital acceleration values $\dot{n}/n$ of $(0.14 \pm 0.01) \times 10^{-10} yr^{-1}$, $-(0.43 \pm 0.10) \times 10^{-10} yr^{-1}$, and $-(1.57 \pm 0.27) \times 10^{-10} yr^{-1}$ (formal error bars, 1σ) for Io, Europa, and Ganymede, respectively. These accelerations represent a shift in the satellite orbital positions of respectively 55 km, 2 125 km, and 2 365 km over the 116 years considered. Using such numerical values, the study of the Laplace resonance angle suggests that the Laplace resonance is evolving, which could be the consequence of tidal cycles within Io (Ojakangas and Stevenson, 1986). However, the former results assumed a constant Jovian Q value. How far such an approximation is reliable remains an open question (Fuller et al., 2016).

4.4.2 Influence on the Formation and Evolution of the Metallic Core

The mechanism and timing of the formation of Ganymede's metallic core and the processes enabling the long-term occurrence of a self-excited dynamo are not fully understood. This is in part due to the large uncertainties in the metallic core density (5 500 to 8 000 kg.m^{-3}) and size (600 to >1 000 km). The density depends in large part on the fraction of S in the core (as FeS), and to a lesser extent on other volatile elements such as O and C, which depends on the evolution of the mantle. In particular, thermal metamorphism of FeS originally in association with other minerals can lead to the release of sulfur-based volatiles at temperatures between 600 to >1 000 K and pressures up to 6 GPa. Hence, the overall thermal evolution of the mantle can influence the amount of sulfur trapped in the core.

Kimura et al. (2009) argue that tidal heating had little effect on the formation of the core. Scott et al. (2002) suggested that, following the global melting of the volatile phase and resulting phase of water–rock interaction, metal-enriched particles could have sunk faster than metal-depleted particles and sedimented to form a low-density core of pyrrhotite. Hence, that core differentiated at a relatively colder temperature (around 500°C) than if it had formed from the partial melting of silicates (above 1 000°C). Scott et al. (2002) assume that this pyrrhotite core could eventually reach a temperature of 1 000°C Celsius, above which a dynamo onset may be possible, but they do not explicitly model that step in Ganymede's thermal evolution. Covering a broad parametric space, Bland et al. (2008) found that tidal heating in the overlaying silicate mantle would not be sufficient to maintain a dynamo inside Ganymede. These authors then attributed the long-lived dynamo to other factors, in particular compositional (low <3% or high >21% sulfur content) and rapid cooling of the silicate mantle. They also suggested as an alternative explanation that the metallic core could have separated from the rock late, so that it could still be in the process of freezing. The low-temperature differentiation model proposed by Scott et al. (2002) could also lead to delayed melting and cooling of the core while ensuring a high content in sulfur. On the other hand, Rückriemen et al. (2018) found that Ganymede could maintain a liquid Fe-FeS layer until the present day for a broad range of sulfur content, provided that the rocky mantle viscosity is of the order of 10^{19}–10^{22} Pa.s. Successful models tend to have rocky mantles depleted in heat sources (subchondritic abundances of long-lived radioisotope, little/no tidal dissipation). Lastly, these authors pointed out that future studies should aim to broaden the range of possible compositions for Ganymede's core by accounting for the likely presence of light elements such as oxygen and carbon.

As explained in Section 4.3.3, tidal heating has a negligible influence on the interior heat budget if the eccentricity is not at least 10 times larger than the present-day value. Even during periods of enhanced eccentricity, tidal heating remains moderate and of the same level as radiogenic heating. Tidal heating alone would probably not be sufficient to melt the rocky interior and drive the formation of the metallic core, but it could represent a complementary heat source that accelerated the melting process. In particular, the presence of partial melting (metals and/or rocks) is expected to reduce the shear modulus and viscosity (Schmeling et al., 2012), leading to an increase in tidal deformation and hence a local amplification of tidal heating. Although tidal heating was likely moderate during most of Ganymede's evolution, it might have engendered heat pulses during specific periods of evolution and may have played a key role in the differentiation process.

4.4.3 Implications for the Hydrosphere and Surface Geology

Two-thirds of Ganymede's surface is covered by bright grooved terrain which is evidence of past geologic activity (see Chapter 7). This grooved terrain is commonly attributed to an extension of the crust. The grooves are characterized by a regular periodicity, which implies a high heat flow, at least temporarily. Extensional instabilities are a viable mechanism for explaining the regular structural periodicity found in the grooved terrain of Ganymede but require temperature gradients of 35 K/km, which is not consistent with radioactive decay heating alone (Kirk and Stevenson, 1987; Mueller and McKinnon, 1988). Old grooved terrain could have been created as a result of phase changes, which increased Ganymede's volume by about 6–7% (Squyres, 1980). Geological observations of crater relaxation (Bland et al., 2017) and flexural uplift (Nimmo et al., 2002) suggest at least one episode of high heat flux (see Section 4.2.4), which these authors attribute to a transient heating event. The model by Malhotra (1991) predicted that such a high eccentricity phase would happen if the dissipation factor of Jupiter is a few 10^5, which appears consistent with observations (e.g. Lainey et al., 2009). Future astrometric observations from the ground as well as from upcoming space missions (Dirkx et al., 2017) will provide key constraints on the dissipation factor of Jupiter and on the past geological activity that will allow for testing this scenario and the consequences for the thermal evolution of its water-rich interior.

Previous studies have attempted to explain the geophysical and geological dichotomy between Ganymede and Callisto. The latter's relatively high moment of inertia (about 0.355, Anderson et al. (2001)) suggests the moon is only partially differentiated. However, in the absence of high-resolution gravity data, the inversion of the degree-two gravity coefficients into an estimate of the moment of inertia generally assumes that the body is in hydrostatic equilibrium via use of the Radau–Darwin approximation. In practice, the presence of non-hydrostatic anomalies, which have been predicted by Mueller and McKinnon (1988), can invalidate the applicability of the Radau-Darwin approximation, and even more so if the body is a slow rotator like Callisto (Gao and Stevenson, 2013). It is thus possible that Callisto could be more differentiated than suggested by its apparent moment of inertia, but it is not possible to be more accurate with the data available. In the case of Ganymede, accretional heating and radiogenic heating were likely sufficient to trigger the melting of ice on a global scale. It is also likely that tidal heating played a key role in enabling Ganymede's advanced differentiation state, in comparison to Callisto, and would have also contributed to global expansion and tectonic activity (Showman et al., 1997; Bland et al., 2009). These authors show that tidal heating could help preserve a thin ice shell (5–20 km) as long as Ganymede remained in a Laplace-like resonance. This follows the prediction of Malhotra (1991) who suggested that Ganymede could have gone through a phase of high eccentricity excitation that would have enhanced tidal heating when the Laplace resonance was established. Lastly, Barr and Canup (2010) showed that heat produced from an intense impacting event (the late heavy bombardment period) could lead by itself to the differentiation of rock from ice in Ganymede, without invoking tidal heating. Those impacts would be more energetic at Ganymede than Callisto because of the gravitational focusing effect of Jupiter, leading to only partial separation of the rock phase from the ice in Callisto.

4.5 SUMMARY AND OPEN QUESTIONS

Ganymede is the largest satellite and the largest ocean world in the solar system, containing about 40 wt.% H_2O. Being bonded to its inner neighbors Europa and Io by the Laplace resonance, Ganymede is subject to forced eccentricity and consecutive tidal forces. As the internal ocean decouples the icy crust from the deeper interior, the tidal deformation amplitudes can reach up to 7 m peak-to-peak deformation at the surface assuming the mean eccentricity value of 0.0015. The exact tidal pattern and amplitude depend on the obliquity, which is currently unknown, and the interior structure of the moon, which is partially constrained. The dominant control on the tidal amplitude comes from the hydrosphere structure, in particular the thickness of the outer ice shell, the ocean density and the rheology of the outer ice shell, and the high-pressure ice shell. If the ice responds viscoelastically to the tidal forcing, a phase lag in the shell's tidal response is expected. It could reach up to about 8 degrees for an elastic ice shell in presence of a subsurface ocean. Viscoelasticity leads to tidal heating via friction and the volumetric heating rate is most likely of the order of 10^{-8} W/m^3 and hence less than the amount of radiogenic heating in the silicate mantle. While today's eccentricity is small with an average value of 0.0015, it might have been higher in the past, leading to enhanced tidal heating, global melting, and expansion of the outer icy shell. Despite the fact that a pulse of internal heat does not necessarily mean a large heat flux because it may take time for the internal heat to dissipate to the surface, the past heat flux could have reached tens to hundreds of mW/m^2, driving geological activity.

The main open questions related to Ganymede's tidal deformation and tidal evolution are (1) What is the tidal response of Ganymede at present and does it provide independent evidence (from the magnetometer data) for a subsurface ocean? (2) What is the origin of the Ganymede–Callisto dichotomy and are the eccentricity tides key to its explanation? (3) What is the viscoelastic response of the ice shell and how much energy is dissipated?

Future measurements of Ganymede's and Europa's tidal responses will significantly advance our knowledge about the satellite and the thermal-orbital evolution of the Galilean satellites (Grasset et al., 2013). The European Space Agency's Jupiter Icy moons Explorer (JUICE) is equipped with a highly capable suite of geophysical instruments. After multiple flybys of the icy moons Europa, Callisto, and Ganymede, the JUICE spacecraft will be inserted into a 280-day orbit around Ganymede: 150 days in Elliptical Orbit (GEO) followed by 130 days Circular Orbit at 500 km (GCO-500).

During the low altitude phase (GCO-500), the tidal Love numbers k_2 and h_2 will be accurately measured by the radio science experiment 3GM (Iess, 2013) and the Ganymede Laser Altimeter (GALA) (Hussmann et al., 2019), respectively. Using precise radio tracking, 3GM will be capable of determining the Love number k_2 with an accuracy of $\sim 10^{-4}$ and of retrieving the obliquity and the libration at orbital period, with a level of uncertainty of 1 and 2 μrad, respectively (Cappuccio et al., 2020). The GALA instrument will determine the amplitude of the tide-induced radial displacement of Ganymede's surface using the crossover technique with an accuracy better than 2% (Steinbrügge et al., 2015) and possibly better if an even lower altitude phase is considered in an extended mission. Depending on the level of dissipation inside Ganymede, both measurements may quantify the tidal phase lag. Moreover, depending on the level of heterogeneity in the outer ice shell, deviations from the standard degree-two response may be revealed, which will put constraints on the thermo-compositional state of the ice shell.

In complement to the time monitoring of the gravity field and surface displacements during the orbital phase, an accurate measurement of the current Laplace resonance state by astrometric measurements should also be made, which will put constraints on the dissipative state of the moons. If the orbit of Europa and Ganymede will be strongly constrained by radio science data using both JUICE and Europa Clipper data, the monitoring of Io's orbit may require extra effort (Dirkx et al., 2017). In that respect, an observation campaign of star occultations by Io could be extremely useful (Morgado et al., 2019).

In combination with magnetic induction experiments, subsurface radar sounding, and in situ compositional investigations, the tidal monitoring will provide key constraints on the hydrosphere structure, in particular the thickness of the outer ice shell and the density of the ocean, which is essential to constrain the dynamical evolution of this water-rich world and better understand the astrobiological potential of its subsurface ocean.

REFERENCES

Anderson, J., Jacobson, R., McElrath, T. et al. (2001), "Shape, mean radius, gravity field, and interior structure of Callisto," *Icarus* **153**, 157–161.

Baland, R.-M., Yseboodt, M., and Van Hoolst, T. (2012), "Obliquity of the Galilean satellites: The influence of a global internal liquid layer," *Icarus* **220**(2), 435–448.

Barr, A. and Canup, R. (2010), "Origin of the Ganymede–Callisto dichotomy by impacts during the late heavy bombardment," *Nature Geoscience* **3**, 164–167.

Beuthe, M. (2016), "Crustal control of dissipative ocean tides in Enceladus and other icy moons," *Icarus* **280**, 278–299.

Bills, B. G. (2005), "Free and forced obliquities of the Galilean satellites of Jupiter," *Icarus* **175**(1), 233–247.

Bland, M. T. and Showman, A. P. (2007), "The formation of Ganymede's grooved terrain: Numerical modeling of extensional necking instabilities," *Icarus* **189**(2), 439–456.

Bland, M. T., Showman, A. P., and Tobie, G. (2008), "The production of Ganymede's magnetic field," *Icarus* **198**(2), 384–399.

Bland, M. T., Showman, A. P., and Tobie, G. (2009), "The orbital–thermal evolution and global expansion of Ganymede," *Icarus* **200**(1), 207–221.

Bland, M. T., Singer, K. N., McKinnon, W. B., and Schenk, P. M. (2017), "Viscous relaxation of Ganymede's impact craters: Constraints on heat flux," *Icarus* **296**, 275–288.

Budd, W. F. and Jacka, T. (1989), "A review of ice rheology for ice sheet modelling," *Cold Regions Science and Technology* **16**(2), 107–144.

Cappuccio, P., Hickey, A., Durante, D. et al. (2020), "Ganymede's gravity, tides and rotational state from JUICE's 3GM experiment simulation," *Planetary and Space Science* **187**, 104902.

Castillo-Rogez, J. and Bland, P. (2021), *Ceres' Internal Evolution*, Cambridge: Cambridge University Press.

Castillo-Rogez, J. C., Efroimsky, M., and Lainey, V. (2011), "The tidal history of Iapetus: Spin dynamics in the light of a refined dissipation model," *Journal of Geophysical Research: Planets* **116**(E9).

Castillo-Rogez, J. C. and Lunine, J. I. (2010), "Evolution of Titan's rocky core constrained by Cassini observations," *Geophysical Research Letters* **37**(20).

Chen, E., Nimmo, F., and Glatzmaier, G. (2014), "Tidal heating in icy satellite oceans," *Icarus* **229**, 11–30.

de Sitter, W. (1928), "Orbital elements determining the longitude of Jupiter's satellites, derived from observations," *Annalen van de Sterrewacht te Leiden* **16**, B1–B96.

Dirkx, D., Gurvits, L., Lainey, V. et al. (2017), "On the contribution of PRIDE-JUICE to Jovian system ephemerides," *Planetary and Space Science* **147**, 14–27.

Durham, W. B., Stern, L. A., and Kirby, S. H. (2001), "Rheology of ice i at low stress and elevated confining pressure," *Journal of Geophysical Research: Solid Earth* **106**(B6), 11031–11042.

Efroimsky, M. and Makarov, V. V. (2013), "Tidal friction and tidal lagging: Applicability limitations of a popular formula for the tidal torque," *Astronomy & Astrophysics* **764**(1), 26.

Egbert, G. and Ray, R. (2000), "Significant dissipation of tidal energy in the deep ocean inferred from satellite altimeter data," *Nature* **405**(6788), 775–778.

Fischer, H. J. and Spohn, T. (1990), "Thermal-orbital histories of viscoelastic models of Io (J1)," *Icarus* **83**(1), 39–65.

Fuller, J., Luan, J., and Quataert, E. (2016), "Resonance locking as the source of rapid tidal migration in the Jupiter and Saturn moon systems," *Monthly Notices of the Royal Astronomical Society* **458**(4), 3867–3879.

Gao, P. and Stevenson, D. J. (2013), "Nonhydrostatic effects and the determination of icy satellites' moment of inertia," *Icarus* **226**(2), 1185–1191.

Goldsby, D. and Kohlstedt, D. (2001), "Superplastic deformation of ice: Experimental observations," *Journal of Geophysical Research: Solid Earth* **106**(B6), 11017–11030.

Grasset, O., Dougherty, M., Coustenis, A. et al. (2013), "Jupiter icy moons explorer (JUICE): An ESA mission to orbit Ganymede and to characterise the Jupiter system," *Planetary and Space Science* **78**, 1–21.

Greenberg, R. (1982), "Orbital evolution of the Galilean satellites," *Satellites of Jupiter* pp. 65–92.

Greenberg, R. (1987), "Galilean satellites: Evolutionary paths in deep resonance," *Icarus* **70**(2), 334–347.

Hay, H. C. F. C., Trinh, A., and Matsuyama, I. (2020), "Powering the Galilean satellites with moon-moon tides," *Geophysical Research Letters* **47**(15), e2020GL088317.

Hilairet, N., Reynard, B., Wang, Y. et al. (2007), "High-pressure creep of serpentine, interseismic deformation, and initiation of subduction," *Science* **318**(5858), 1910–1913.

Hussmann, H., Choblet, G., Lainey, V. et al. (2010), "Implications of rotation, orbital states, energy sources, and heat transport for internal processes in icy satellites," *Space Science Reviews* **153**(1–4), 317–348.

Hussmann, H., Lingenauber, K., Kallenbach, R. et al. (2019), "The Ganymede laser altimeter (GALA): Key objectives, instrument design, and performance," *CEAS Space Journal* **11**(4), 381–390.

Hussmann, H., Shoji, D., Steinbrügge, G., Stark, A., and Sohl, F. (2016), "Constraints on dissipation in the deep interiors of Ganymede and Europa from tidal phase-lags," *Celestial Mechanics and Dynamical Astronomy* **126**(1), 131–144.

Hussmann, H., Sohl, F., and Oberst, J. (2011), "Measuring tidal deformations at Europa's surface," *Advances in Space Research* **48**, 718–724.

Hussmann, H. and Spohn, T. (2004), "Thermal-orbital evolution of Io and Europa," *Icarus* **171**(2), 391–410.

Hussmann, H., Spohn, T., and Wieczerkowski, K. (2002), "Thermal equilibrium states of Europa's ice shell: Implications for internal ocean thickness and surface heat flow," *Icarus* **156**(1), 143–151.

Iess, L. (2013), "3GM: Gravity and geophysics of Jupiter and the Galilean moons." In *European Planetary Science Congress*. https://meetingorganizer.copernicus.org/EPSC2013/EPSC2013-491.pdf. "European Planetary Science Congress," pp. EPSC2013–491.

Jara-Orué, H. and Vermeersen, B. (2016), "Tides on Jupiter's moon Ganymede and their relation to its internal structure," *Netherlands Journal of Geosciences – Geologie en Mijnbouw* **95**(2), 191–201.

Journaux, B., Kalousová, K., Sotin, C. et al. (2020), "Large ocean worlds with high-pressure ices," *Space Science Reviews* **216**(1), 1–36.

Kalousová, K. and Sotin, C. (2020), "The insulating effect of methane clathrate crust on Titan's thermal evolution," *Geophysical Research Letters* **47**(13), e2020GL087481.

Kalousová, K., Sotin, C., Choblet, G., Tobie, G., and Grasset, O. (2018), "Two-phase convection in Ganymede's high-pressure ice layer: Implications for its geological evolution," *Icarus* **299**, 133–147.

Kamata, S., Kimura, J., Matsumoto, K. et al. (2016), "Tidal deformation of Ganymede: Sensitivity of love numbers on the interior structure," *Journal of Geophysical Research: Planets* **121**(7), 1362–1375.

Kamata, S., Nimmo, F., Sekine, Y. et al. (2019), "Pluto's ocean is capped and insulated by gas hydrates," *Nature Geoscience* **12**(6), 407–410.

Kaula, W. M. (1964), "Tidal dissipation by solid friction and the resulting orbital evolution," *Reviews of Geophysics* **2**(4), 661–685.

Kimura, J., Nakagawa, T., and Kurita, K. (2009), "Size and compositional constraints of Ganymede's metallic core for driving an active dynamo," *Icraus* **202**(1), 216–224.

Kirk, R. L. and Stevenson, D. J. (1987), "Thermal evolution of a differentiated Ganymede and implications for surface features," *Icarus* **69**(1), 91–134.

Kivelson, M., Khurana, K., and Volwerk, M. (2002), "The permanent and inductive magnetic moments of Ganymede," *Icarus* **157**(2), 507–522.

Lainey, V., Arlot, J.-E., Karatekin, Ö., and van Hoolst, T. (2009), "Strong tidal dissipation in Io and Jupiter from astrometric observations," *Nature* **459**(7249), 957–959.

Lainey, V., Casajus, L. G., Fuller, J. et al. (2020), "Resonance locking in giant planets indicated by the rapid orbital expansion of Titan," *Nature Astronomy* **4**, 1053–1058.

Lainey, V., Duriez, L., and Vienne, A. (2004), "New accurate ephemerides for the Galilean satellites of Jupiter. I. Numerical integration of elaborated equations of motion," *Astronomy & Astrophysics* **420**, 1171–1183.

Malhotra, R. (1991), "Tidal origin of the Laplace resonance and the resurfacing of Ganymede," *Icarus* **94**(2), 399–412.

Margot, J.-L., Padovan, S., Campbell, D., Peale, S., and Ghigo, F. (2013), "Measurements of the spin states of Europa and Ganymede." In *Proceedings of the 45th Division for Planetary Sciences Meeting*. Denver, CO: American Astronomical Society.

Matsuyama, I., Beuthe, M., Hay, H. C., Nimmo, F., and Kamata, S. (2018), "Ocean tidal heating in icy satellites with solid shells," *Icarus* **312**, 208–230.

Mitri, G., Meriggiola, R., Hayes, A. et al. (2014), "Shape, topography, gravity anomalies and tidal deformation of Titan," *Icarus* **236**, 169–177.

Montagnat, M. and Duval, P. (2000), "Rate controlling processes in the creep of polar ice, influence of grain boundary migration associated with recrystallization," *Earth and Planetary Science Letters* **183**(1–2), 179–186.

Moore, W. B. and Schubert, G. (2003), "The tidal response of Ganymede and Callisto with and without liquid water oceans," *Icarus* **166**(1), 223–226.

Morgado, B., Benedetti-Rossi, G., Gomes-Júnior, A. R. et al. (2019), "First stellar occultation by the Galilean moon Europa and upcoming events between 2019 and 2021," *Astronomy & Astrophysics* **626**, L4.

Mosqueira, I. and Estrada, P. R. (2003), "Formation of the regular satellites of giant planets in an extended gaseous nebula i: subnebula model and accretion of satellites," *Icarus* **163**(1), 198–231.

Mueller, S. and McKinnon, W. B. (1988), "Three-layered models of Ganymede and Callisto: Compositions, structures, and aspects of evolution," *Icarus* **76**(3), 437–464.

Néri, A., Guyot, F., Reynard, B., and Sotin, C. (2020), "A carbonaceous chondrite and cometary origin for icy moons of Jupiter and Saturn," *Earth and Planetary Science Letters* **530**, 115920.

Nimmo, F., Pappalardo, R. T., and Giese, B. (2002), "Effective elastic thickness and heat flux estimates on Ganymede," *Geophysical Research Letters* **29**(7), 62–1–62–4.

Ojakangas, G. W. and Stevenson, D. J. (1986), "Episodic volcanism of tidally heated satellites with application to Io," *Icarus* **66**(2), 341–358.

Rovira-Navarro, M., Gerkema, T., Maas, L. R. et al. (2020), "Tides in subsurface oceans with meridional varying thickness," *Icarus* **343**, 113711.

Rückriemen, T., Breuer, D., and Spohn, T. (2018), "Top-down freezing in a Fe–FeS core and Ganymede's present-day magnetic field," *Icarus* **307**, 172–196.

Schmeling, H., Kruse, J. P., and Richard, G. (2012), "Effective shear and bulk viscosity of partially molten rock based on elastic moduli theory of a fluid filled poroelastic medium," *Geophysical Journal International* **190**(3), 1571–1578.

Scott, H. P., Williams, Q., and Ryerson, F. J. (2002), "Experimental constraints on the chemical evolution of large icy satellites," *Earth and Planetary Science Letters* **203**(1), 399–412.

Showman, A. P. and Malhotra, R. (1997), "Tidal evolution into the Laplace resonance and the resurfacing of Ganymede," *Icarus* **127**(1), 93–111.

Showman, A. P., Stevenson, D. J., and Malhotra, R. (1997), "Coupled orbital and thermal evolution of Ganymede," *Icarus* **129**(2), 367–383.

Sinha, N. K. (1989), "Elasticity of natural types of polycrystalline ice," *Cold Regions Science and Technology* **17**(2), 127–135.

Sotin, C., Grasset, O., and Beauchesne, S. (1998), "Thermodynamic properties of high pressure ices: Implications for the dynamics and internal structure of large icy satellites." In B. Schmitt, C. De Bergh, and M. Festou (eds.), *Solar System Ices*. Dordrecht: Springer, pp. 79–96.

Sotin, C. and Tobie, G. (2004), "Internal structure and dynamics of the large icy satellites," *Comptes Rendus Physique* **5**(7), 769–780.

Sotin, C., Tobie, G., Wahr, J. et al. (2009), "Tides and tidal heating on Europa," *Europa* **11**.

Squyres, S. W. (1980), "Surface temperatures and retention of H_2O frost on Ganymede and Callisto," *Icarus* **44**(2), 502–510.

Steinbrügge, G., Stark, A., Hussmann, H., Sohl, F., and Oberst, J. (2015), "Measuring tidal deformations by laser altimetry: A performance model for the Ganymede laser altimeter," *Planetary and Space Science* **117**, 184–191.

Tobie, G., Choblet, G., and Sotin, C. (2003), "Tidally heated convection: Constraints on Europa's ice shell thickness," *Journal of Geophysical Research: Planets* **108**(E11).

Tobie, G., Grasset, O., Dumoulin, C., and Mocquet, A. (2019), "Tidal response of rocky and ice-rich exoplanets," *Astronomy & Astrophysics* **630**, A70.

Tobie, G., Mocquet, A., and Sotin, C. (2005), "Tidal dissipation within large icy satellites: Applications to Europa and Titan," *Icarus* **177**(2), 534–549.

Tyler, R. (2014), "Comparative estimates of the heat generated by ocean tides on icy satellites in the outer solar system," *Icarus* **243**, 358–385.

Tyler, R. H. (2008), "Strong ocean tidal flow and heating on moons of the outer planets," *Nature* **456**(7223), 770–772.

Vance, S. D., Panning, M. P., Stähler, S. et al. (2018), "Geophysical investigations of habitability in ice-covered ocean worlds," *Journal of Geophysical Research: Planets* **123**(1), 180–205.

Yoder, C. F. (1979), "How tidal heating in Io drives the galilean orbital resonance locks," *Nature* **279**(5716), 767–770.

Yoder, C. F. and Peale, S. J. (1981), "The tides of Io," *Icarus* **47**(1), 1–35.

Zahnle, K., Schenk, P., Levison, H., and Dones, L. (2003), "Cratering rates in the outer solar system," *Icarus* **163**(2), 263–289.

Zürn, W. and Wilhelm, H. (1984), "Tides of the solid earth," *Geophysik der festen Erde, des Mondes und der Planeten; Landolt-Börnstein Neue Serie* **2**, 280–299.

5

Ganymede's Stratigraphy and Geologic Evolution

RALF JAUMANN, KATRIN STEPHAN, ROLAND WAGNER, HAUKE HUSSMANN, PASQUALE PALUMBO, ALICE LUCCHETTI, FRANK POSTBERG, THOMAS ROATSCH, AND ALEXANDER STARK

5.1 INTRODUCTION

The observed surface features on the Jovian satellites Ganymede, Europa, and Callisto owe their appearance to the unique behaviour of H_2O ice. Although ice behaves as a brittle material at the cold temperatures in the outer solar system, the amount of energy necessary to bring it close to its melting point is lower than in the inner solar system. As rheology of ice is complicated with respect to elastic, brittle, and viscous properties (e.g. Petrenko and Whitworth, 1999; Beeman et al., 1988; Durham and Stern, 2001), each of the icy bodies in the outer solar system experienced variable geological histories driven by both internal and external forces.

On one hand, exogenic modification is primarily caused by impacts of asteroids and comets as well as micrometeorite bombardment (affecting essentially all airless solar system surfaces). Also UV radiation and partly energetic particles originating from Jupiter's magnetospheres modify the physical and chemical properties of surface materials. On the other hand, endogenic modification includes tectonic resurfacing, as well as icy volcanism (cryovolcanism) of several sorts. Many of the outer icy satellite surfaces display evidence of tectonic activity, although to different degrees. Most of these satellites show predominately extensional deformation, with some exhibiting strike-slip faulting, whereas contractional tectonism apparently plays a minor role. In particular, Europa's surface is pervasively tectonized, covered with a great variety of tectonic features (Fig. 5.1), some of which are unique to the outer solar system. Ganymede and Callisto exhibit geologically old heavily cratered terrain that has been cut and pulled apart by normal faulting. On Ganymede, these faults form a global network.

Our current understanding of Ganymede is primarily based on eight fly-by encounters from the Voyager spacecrafts. Latterly, the Galileo spacecraft, building on the kilometre-scale resolution images first obtained during the Voyager encounters, continued to fill in the gaps with similar resolution imaging. However, variation in imaging quality and viewing geometry prevents basic geological observations for about a third of the surface. Only a few small areas, about 2% of the surface, were imaged by Galileo at tens to hundreds of metres per pixel (m/px).

The specific tectonics of Ganymede has implications for surface deformation and the role of liquid water within Ganymede's ice shell. Heavily cratered dark-appearing terrain covers one-third of the surface of Ganymede. The other two-thirds are dominated by tectonically modified structures known as grooves and occur as bright-appearing terrain. Although, high-resolution a images of parallel lineaments cutting across the surface of bright-appearing terrain are available, the style of resurfacing, either tectonic and/or cryovolcanic, is still unsolved (Pappalardo et al., 2004). Extensional tectonism erasing the rims of impact craters is observed in some areas (Pappalardo and Collins, 2005), while not all bright-appearing terrain is tectonized to such an extent. Smooth topographically low areas associated with local flow-like features are the best evidence so far for some form of cryovolcanism (Schenk et al., 2001). Other smooth regions on Ganymede suggest that they formed through crustal separation and spreading, as bands on Europa (Head et al., 2002).

Grooves on Ganymede's surface form a global, interconnected recording of an intense period of activity at some point in the middle of Ganymede's history. One possible explanation for this pulse of activity is that it is tied to Ganymede's capture into the Laplace orbital resonance with neighbouring Europa and Io (Showman and Malhotra, 1997). Europa and Io have been resurfaced much more recently, thus Ganymede may be the only place showing evidence for the early and particularly mid-term geological history in the Jovian system.

The presence of ancient dark-appearing terrain and the younger bright-appearing terrain provides a record of impact cratering throughout Ganymede's history (Fig. 5.1). The oldest of Ganymede's impact structures are palimpsests, low-relief albedo features thought to have formed while the Ganymede crust was warmer, the ice depth to a subsurface ocean was less, and relaxation was the major crustal reaction. Central smooth intercrater regions reflect either extensive impact melt pools or exposure of the subsurface ocean as a result of the impactor fully penetrating the ice crust (Passey and Shoemaker, 1982). Palimpsests and other old crater forms in the dark-appearing terrain have little or no topographic relief due to significant viscous relaxation, indicating significantly higher heat flows in the past (Dombard and McKinnon, 2006). While Ganymede's heat flow decreases with time, the depth to its subsurface ocean increased, resulting in a reduction in the rate of viscous relaxation, and impact craters resemble more recognizable crater morphologies (e.g. Schenk et al., 2004; Bray et al., 2008). Studies of the variation in crater morphology between Ganymede's dark-appearing terrain, bright-appearing terrain, and the Earth's moon thus allow us to develop our understanding of the cratering process in rock, ice, and ice–rock mixes. Comparisons with craters on Europa help us to understand and separate out the effect of the layering of a stiff lithosphere over a fluid layer at some depth within the target body. Ganymede's impact history can serve as a bridge between

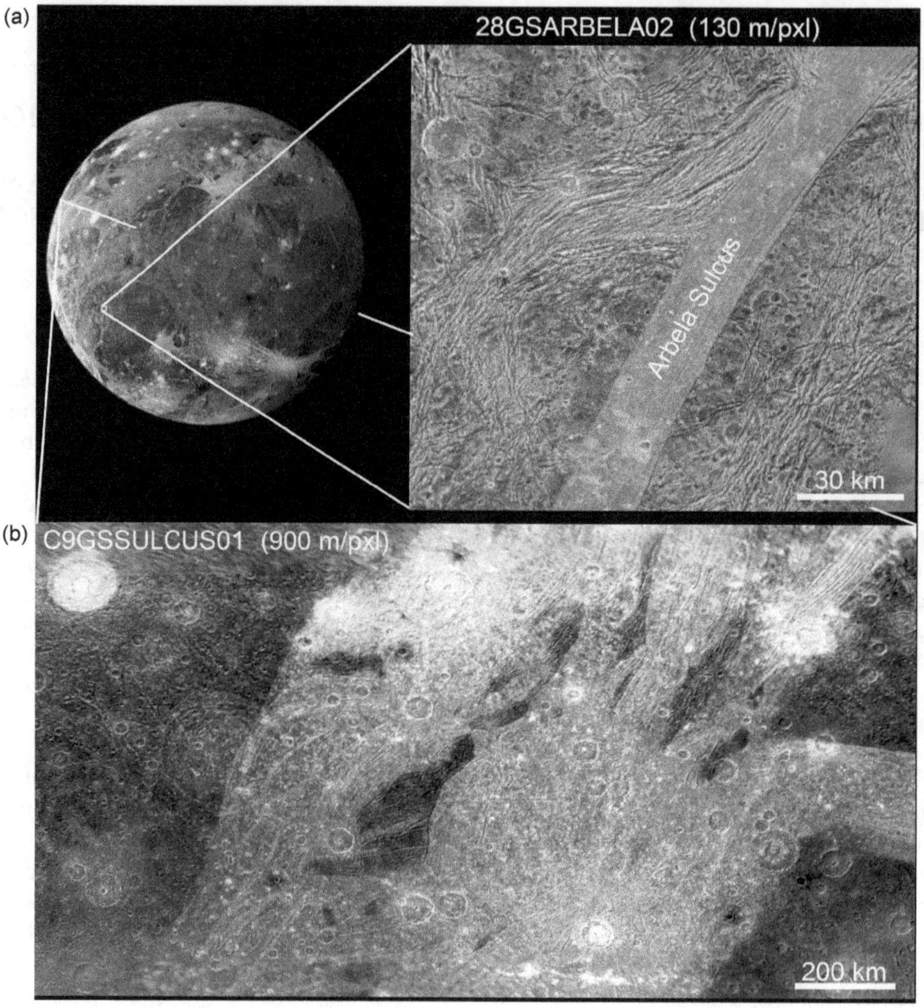

Figure 5.1 Ganymede's surface is characterized by old, dark densely cratered plains and by younger, bright, and more ice-rich, tectonically resurfaced terrain: (a) bright, smooth bands (Galileo 28GSARBELA02, 130 m/px) indicate lithospheric spreading, involving extension as well as strike-slip movements, as, for example, in Arbela Sulcus (AS); (b) bright-appearing terrain formed at the expense of dark-appearing terrain, mostly through extensional tectonism (Galileo C9GSSULCUS, 900 m/px).

thick-lithosphere geologic features that are more easily interpretable from terrestrial experience and the more difficult to interpret thin-lithosphere features found on places like Europa and the south pole of Enceladus.

5.2 GANYMEDE'S GEOLOGY AND SURFACE EVOLUTION

Ganymede, the largest satellite in the solar system, shows a mixture of geologic features (Fig. 5.2) which are morphometrically somewhat in between the geology of its inner neighbour Europa and its outer neighbour Callisto. Ganymede's surface is a mix of old and young terrain, ancient impact basins, fresh craters, landscapes dominated by tectonics, icy volcanism, and slow degradation by space weathering. Therefore it serves as an archetype for understanding several icy satellite surface and evolutional processes throughout the outer solar system. In addition to craters, dark-appearing terrain also displays hemisphere-scale sets of linear or curved troughs and furrows termed 'fossae', in general 6–20 km wide, which are probably the remnants of vast multi-ring impact basins, now broken up by subsequent bright-appearing terrain tectonism (Smith et al., 1979) with furrows on a global scale (Hirata et al., 2020). This type of terrain appears dark due to the addition of a non-water ice contaminant that appears to be concentrated at the surface by a variety of processes including sublimation, sputtering, and mass wasting (Prockter et al., 1998). At higher image resolution, dark-appearing terrain is shown to be more heterogeneous in albedo. Dark deposits on crater floors and furrow scarps, together with the bright ice exposed on steep scarps, reveal that the dark material is a comparably thin veneer on relatively clean ice (Prockter et al., 1998) (Fig. 5.2).

In general, dark-appearing terrain is in the order of 4 billion years old (Neukum et al., 1998; Zahnle et al., 1998, 2003; Schenk et al., 2004). In dark-appearing terrain, smooth units in some parts less densely cratered, were thought to represent cryovolcanic flows, extruded as icy slushes or slurries with darker, rockier components incorporated (Croft and Goudreau, 1987; Allison and Clifford, 1987; Croft et al., 1990; Murchie et al., 1990; Schenk, 1995; Head et al., 1998; Prockter et al., 1998; Pappalardo et al., 2004). On the other hand, earlier

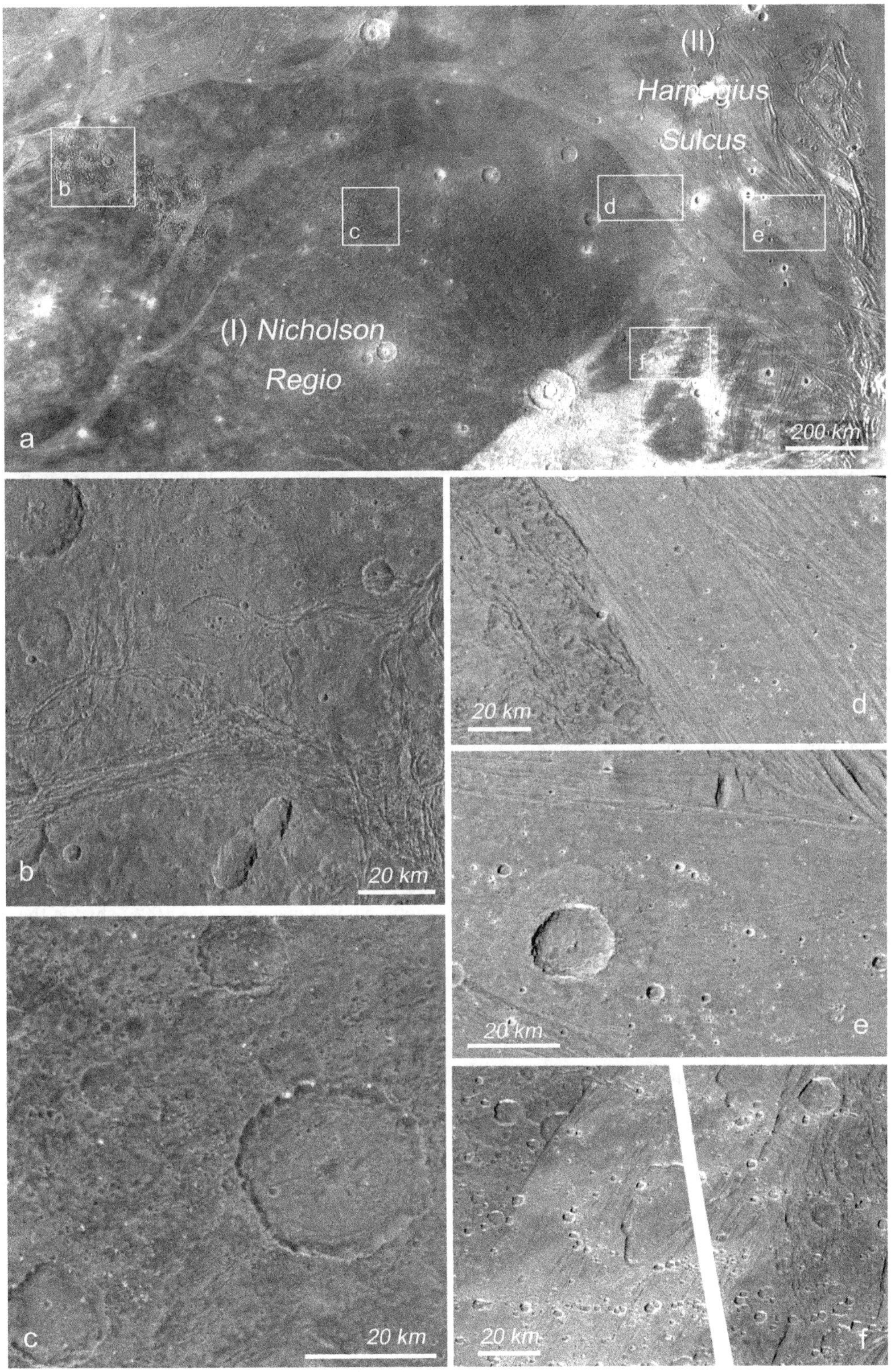

Figure 5.2 Type localities Nicholsonian and Harpagian: (a) context image of the dark-appearing terrain of Nicholson Regio and bright-appearing terrain Harpagia Sulcus, respectively, with the white frames indicating the location of close-up views onto (b) the ancient heavily cratered dark-appearing terrain, (c) its transition into the tectonically resurfaced, and (d) partly very smooth terrain, which (e) sometimes shows local possibly cryovolcanic surface features.

interpretations of high-resolution Galileo solid state imaging (SSI) images revealed that dark, smooth units originated from mass wasting processes along slopes rather than cryovolcanism (Prockter et al., 1998). The observations provide arguments for both hypotheses and more measurements are needed to identify the formation process.

Bright-appearing terrain separates the dark units by broad, up to several hundred kilometres wide, linear, or curved tectonic grooves termed 'sulci' (Pappalardo et al., 2004; Prockter et al., 1998, 2010). Specifically, grooves are regional-scale, linear to curvilinear, morphotectonic structures that represent the brittle deformation of the bright-appearing terrain (Fig. 5.2). The bright-appearing terrain units formed predominantly at the expense of dark-appearing terrain through a process termed 'tectonic resurfacing', causing the partial or total transformation of dark-appearing terrain into bright-appearing terrain (e.g. Pappalardo et al., 1998). However, some bright-appearing terrain is possibly formed from ocean material owing to dike intrusion or exposure (Pizzi et al., 2017; Howell and Pappalardo, 2018). Bright-appearing terrain has a lower superimposed crater density than dark-appearing terrain and therefore is younger (e.g. Neukum et al., 1998; Zahnle et al., 1998, 2003; Schenk et al., 2004). Generally, grooved terrain represents rifts created by extensional stress and strike-slip kinematics (Pappalardo et al., 2004). Several caldera-like, scalloped depressions termed 'paterae' found in the bright-appearing terrain represent probable volcanic vents (Pappalardo et al., 2004). Ridged deposits in one of the largest of such paterae were interpreted as cryovolcanic flows (Head et al., 1998). The presence of an internal magnetic field created by a dynamo in Ganymede's core (e.g. Schubert et al., 2004) could be an indication that tectonism, which deforms the old dark-appearing terrain and creates the bright-appearing terrain units, has lasted a long time, in the order of more than 2 billion years (e.g. Prockter et al., 1998; Collins et al., 1998a, 1998b, 2013; Pappalardo et al., 1998, 2004; Head et al., 2002; Schenk et al., 2003, 2004).

As mentioned, dark smooth units have been hypothesized to represent either cryovolcanic flows (Pappalardo et al., 2004) or mass wasting deposits (Prockter et al., 1998). The smoothest units also exhibit some degree of tectonics, implying that cryovolcanism and tectonic deformation are closely linked (Head et al., 2002). The driving mechanism for the formation of grooves is uncertain but may be tied to the internal forces and the history of the orbital evolution of the Galilean satellite system (Showman et al., 1997).

Ganymede is unique in having the widest range of crater morphologies compared to any other body (except Callisto) in the solar system (Shoemaker et al., 1982; Murchie et al., 1986; Murchie and Head, 1989; Wagner et al., 1995; Neukum, 1997; Prockter et al., 1998; Zahnle et al., 2003; Pappalardo et al., 2004; Schenk et al., 2004). With increasing size, crater forms become more complex and feature (1) central pits, (2) large central domes, (3) pedestals (outward-facing scarps in the continuous ejecta), or (4) palimpsests (bright, flat pancake-shaped patches) which show concentric morphology such as ridges at higher resolution (Guest et al., 1988; Lucchitta et al., 1992; Jones et al., 2003; Pappalardo et al., 2004; Schenk et al., 2004). Ductile material in the icy subsurface beneath a brittle crust or shell is thought to be responsible for these crater forms (e.g. Prockter et al., 1998; Schenk et al., 2004).

The subdued character of Ganymede's oldest impact craters implies a steep thermal gradient in Ganymede's early history. More recent impact structures reflect a thicker and stiffer elastic lithosphere (e.g. Shoemaker et al., 1982). This suggests that there was a warmer shallow subsurface ocean early in Ganymede's history.

Ganymede was the first object in the solar system where an impact generated ballistic dust exosphere was detected. In 1996, during a series of close fly-bys of the Galileo spacecraft, the onboard dust detector recorded a series of impact events (Krüger et al., 1999). The detected particle size distribution and their dependence on altitude (Fig. 5.5) are consistent with predictions from impact ejecta models (e.g. Krivov et al., 2003) in which the most efficient impactor population consists of micro-meteorites. These are relatively large interplanetary dust particles with typical sizes of 1–100 μm (Divine, 1993) and typical speeds of a few 10 km/s. They can liberate orders of magnitude more mass than the impacting particle itself but at much lower speeds (in the order of hundreds of m/s). Owing to the high escape speed of Ganymede, nearly all ejected particles eventually fall back to the surface on ballistic trajectories, contributing to the regolith which is assumed to cover the moon's surface.

The Galileo mission revealed remarkable evidence of surface degradation and mass movements on Ganymede's surface. Since all Galilean satellites orbit Jupiter within its strong magnetosphere, sputter ablation of water ice by impacting magnetospheric particles is expected to be a prominent degradation process on the Galilean satellites (Clark and Johnson, 1996). The potential effect of sputtering, however, decreases substantially with increasing distance from Jupiter. Hence, the efficiency of sputtering on water ice is 20 times less at Ganymede than at Europa (Johnson, 1990). In addition, Ganymede's own magnetic field (Kivelson et al., 1996) shields the equatorial region between $\pm\sim40°$ latitude from major effects of sputtering. Hence, in the polar regions sputtering is expected to be the most important process and may significantly affect surface morphology and albedo (Spencer, 1987; Hillier et al., 1996; Johnson, 1997). In the equatorial region, however, sublimation and impact processes are expected to be dominant (Prockter et al., 1998; Moore et al., 1999), however, new studies (Ligier et al., 2019; Carnielli et al., 2020; de Kleer et al., 2021) point out that sputtering seems to also significantly affect the leading/trailing hemisphere. Shoemaker et al. (1982) pointed out that the ice-rich crater rays on Ganymede, which Passey and Shoemaker (1982) calculate have a probable survival age of »1 Ga, are probably not more than a few metres thick and so may be taken as an additional indicator that the ice sputter–ablation rate on Ganymede may be low (i.e. less than a few metres per Ga) (Moore et al., 1999).

Locally, if sputtering rates of water ice are greater than the rate of thermal sublimation, re-homogenization of the surface material would prevent the segregation of dark material into bright frost and dark lag deposit (Spencer, 1987; Prockter et al., 1998). But, instead, strong evidence for a more dominant thermal segregation of water ice and dark non-ice material on Ganymede at sub-kilometre scales has been recognized in higher-resolution Galileo images (Prockter et al., 1998). Water ice predominantly sublimates at sun-facing and thus warmer slopes facing the equator, whereas frost redeposits again on

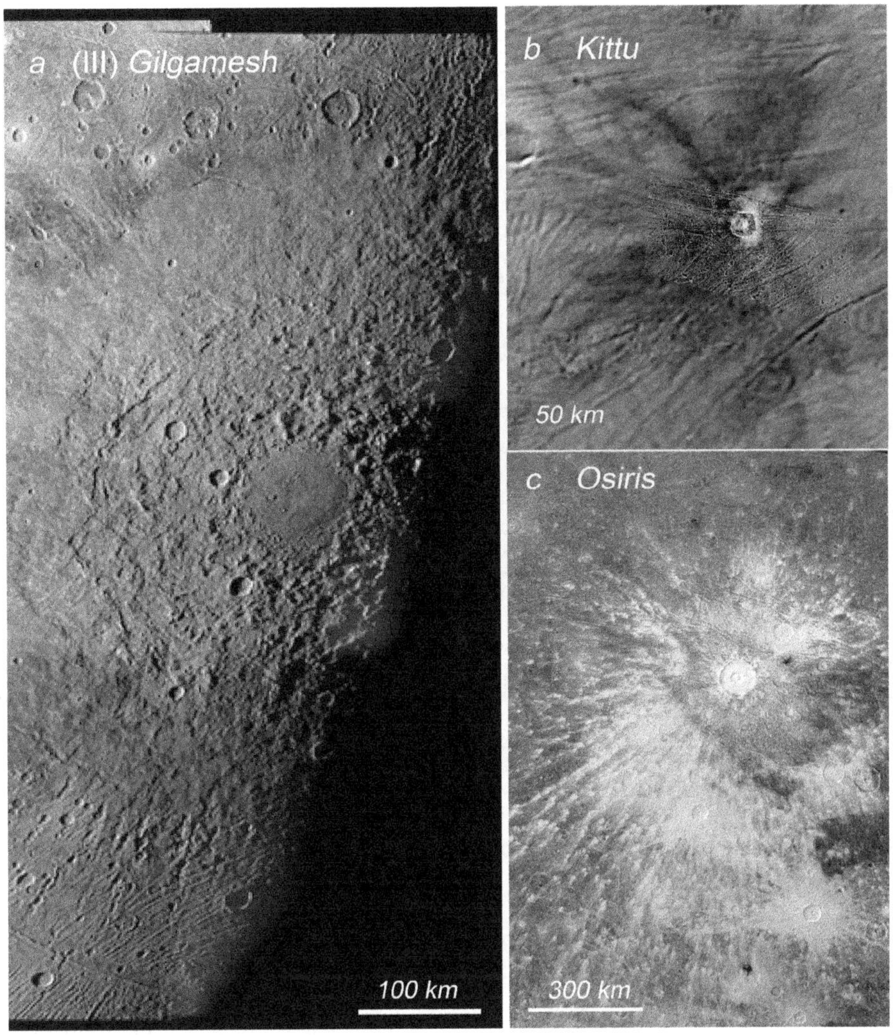

Figure 5.3 Type locality for the Gilgameshan: (a) the ~150 km large Gilgamesh basin (63°S/ 235°E); (b) the 15 km large dark ray crater Kittu (0°/335°W); and (c) the 107 km large bright ray crater Osiris (38°S/194°E).

the less sunny and thus colder pole-facing slopes (Prockter et al., 1998). Purves and Pilcher (1980) and Squyres (1980) imply that frost should accumulate at Ganymede's poles at the cost of the equatorial regions, where exposed water ice should entirely sublimate away during diurnal temperature variations. A sublimate-driven redistribution of water ice is one of the processes discussed so far, which could explain the formation of Ganymede's so called 'polar caps' (see Stephan et al., 2003 and Galli et al., 2023, this issue chapter). As water ice sublimes the dark non-ice component of the surface, material remains behind to form a refractory-rich sublimation lag deposit. The lag may eventually become thick enough to suppress any further sublimation from underneath (Prockter et al.,1998).

While no significant effect on the topography could be observed in Ganymede's dark terrains (Prockter et al., 1998), Galileo images show a rather 'softened' surface at small scales (<100 m/px). This is contrary to what can be seen at Callisto, which has a surface dominated by sublimation processes (Moore et al., 1999, 2004), visible in the rounded crests of crater rims and furrows (Prockter et al., 1998). On Ganymede, the rain of micro-meteorites is thought to be responsible for the peculiar appearance of the surface, which counteracts thermal segregation by mixing the surface material, including the addition of possible meteoritic impactor material (Prockter et al.,1998).

Both sublimation and impact gardening together presumably play an important role in the formation and evolution of Ganymede's surficial regolith layer, a thin layer of dark veneer overlying an ice-rich substrate. These processes in combination might also explain the presence of significant slumping or sliding of surface material at a variety of scales including gravity-induced movement of 'dry' rocky debris along steep slopes (Moore et al., 1999, 2004). Downslope movement of material on Ganymede at the smallest observable scale visible by Galileo SSI consists of numerous dark streaks on steep slopes (Pappalardo et al., 1998; Prockter et al., 1998; Moore et al., 1999). These dark streaks are seen on all terrains observed at resolutions greater than 100 m/px, although they are less prevalent on bright-appearing terrains. The streaks always appear superimposed on other materials, implying they are superficial deposits.

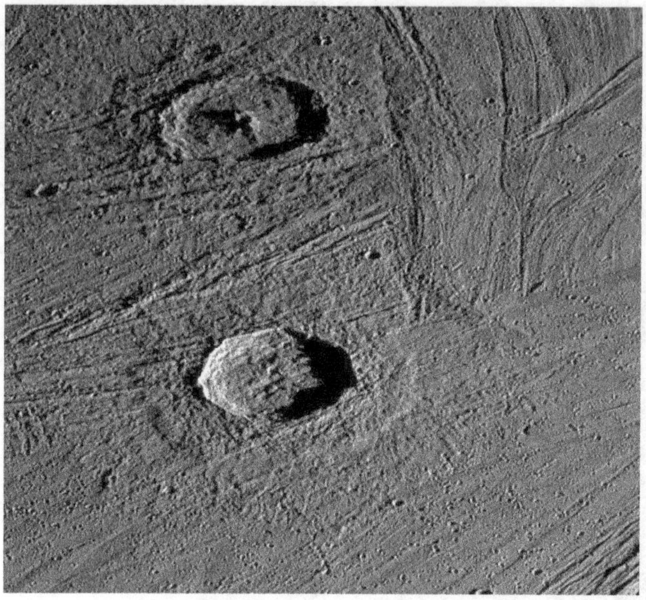

Figure 5.4 The crater Gula (top), about 40 km in diameter, and Achelous (bottom), about 30 km in diameter, in the grooved terrain of Ganymede with ejecta 'pedestals' (image resolution 180 m/px).

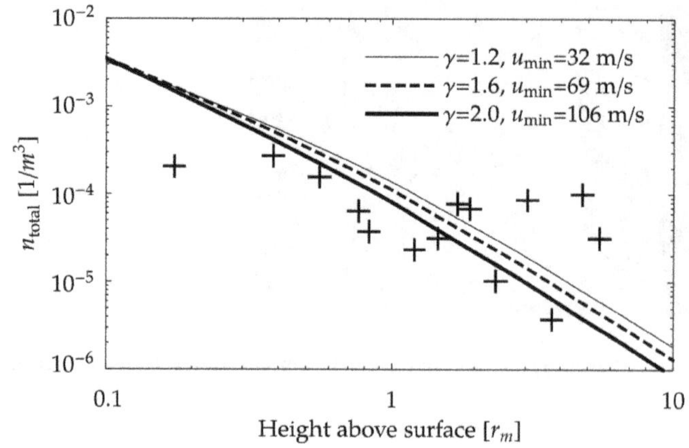

Figure 5.5 Number density of ejecta larger than 10^{-16} kg detected by the Galileo dust detector (Krüger et al., 1999, 2000) depending on altitude given in Ganymede radii (r_m = 2 634 km). The different lines represent fits for different model parameters (Krivov et al., 2003) where γ represents the porosity of the surface (1 = highly porous, 2 = compact) and u_{min} the minimum velocity of the dust ejecta speed distribution. *Source*: from Szalay, S. R., Poppe, R. A., Agarwal, J., Britt, D., Belskaya, I., Horányi, M., Nakamura, T., Sachse, M., Spahn, F., (2018). Dust Phenomena Relating to Airless Bodies. *Space Sci Rev* 214:98, Springer Nature.

5.3 AGES AND STRATIGRAPHIC RELATIONS

Crater statistics and relative ages.
In addition to applying the rules of superposition for geologic units, measurements of the crater size–frequency distribution (CSFD) of craters provide a highly effective tool to extract ages for units with a statistically significant number of superimposed impact craters. The method has been described extensively in the landmark paper by Arvidson et al. (1979) and ever since in a series of summary papers or book chapters (e.g. Chapman and McKinnon, 1986; Neukum and Ivanov, 1994; Neukum et al., 2001; Werner and Ivanov, 2015). In the following, we briefly introduce the procedure and the currently available cratering chronology models for the Galilean satellites for stratigraphic purposes (see Kirchoff, Barr, Bland et al., Chapter 8, this volume).

Crater size–frequency distributions are graphically represented in log-log plots of crater frequency (log N) versus crater diameter (log D). Using crater counts in geologic units for stratigraphy and the derivation of surface ages is based on two main premises: (1) a geologic unit with a higher density or frequency of superimposed craters is relatively older than a unit with a lower frequency and (2) a CSFD measured in a geologic unit represents the image of the size–frequency distribution (SFD) of the impactor population whose members created the craters since the unit has formed. In this (ideal) case, the CSFD reflects the so-called production distribution. The slopes of CSFDs in a log N versus log D graphical representation are diameter-dependent and, in general, vary from a (cumulative) slope index of −3 to −1 in the range between the smallest (1 km or less) and largest (hundreds of kilometres) impact craters measured in planetary images (e.g. Chapman and McKinnon, 1986; Neukum and Ivanov, 1994; Werner and Ivanov, 2015). Thus, production distributions can be approximated by straight lines, depending on the respective slope in the diameter range measured.

A 'standard' production distribution polynomial (termed 'production function' (PF)) can be fit to measured CSFDs to obtain the relative age of a geologic unit from the cumulative frequency at a reference diameter (e.g. Neukum and Ivanov, 1994; Neukum, 1997; Neukum et al., 1998; Neukum et al., 2001; Marchi et al., 2009; Le Feuvre and Wieczorek, 2011; Hiesinger et al., 2016; see also Kirchoff, Barr, Bland et al., Chapter 8, this volume). However, the use of CSFDs for stratigraphic studies based on the simple relationship that the frequency of impact craters in a geologic unit is directly related to the age of the unit, as first formulated by Öpik (1960), is compromised by a number of issues or caveats which have to be taken into account: (1) the issue of saturation/equilibrium; (2) contamination with secondary craters; (3) the SFD of the potential impactor population; (4) time-dependence of the impactor SFD; (5) apex-antapex asymmetries of CSFDs; and (6) geological processing affecting the CSFD of a geologic unit. In the following, we briefly summarize these issues (see also Kirchoff, Barr, Bland et al., Chapter 8, this volume).

Impact chronology models for the Galilean satellites of Jupiter.
Absolute ages of geologic units on the icy Galilean satellites of Jupiter, derived from crater size–frequency measurements, can only be inferred if the crater-forming rate by members of a candidate impactor family, or more than one potential impactor families, is known. Here, we introduce the two currently used impact chronology models lunar-like or lunar-derived chronology model (LDM) and comet-derived chronology model (CDM) which are treated in more detail by Kirchoff, Barr, Bland et al. in Chapter 6 (this volume): lunar-like or lunar-derived chronology model (LDM) and comet-derived chronology model (CDM). A summary of these models is also given in Dones et al. (2009), Burger et al. (2010), and Kirchoff et al. (2018), among others. The historically oldest model developed after the Voyager fly-bys was a lunar-like or

lunar-derived chronology model (LDM), with a steep, exponential decline of the impact and crater-forming rate prior to ~3 Ga ago and a constant impact and cratering rate since that time. This model was advocated for the Jovian moons by, for example, Shoemaker and Wolf (1982), Neukum and Pozio (1983), and Neukum (1985). A lunar-like model for bodies other than the Moon is based on the premise that impact and cratering rates had a similar time dependence as on the Moon, whose cratering chronology could well be correlated with radiometric ages from Apollo (USA/NASA) and Luna (USSR) surface materials (Neukum and Ivanov, 1994; Neukum et al., 2001, and references therein). The notion of a steep exponential decay has been challenged by the idea of the so-called terminal lunar cataclysm or Late Heavy Bombardment (LHB), a strong peak in the impact rate inferred from a clustering in radiometric ages around ~3.9 Ga by Tera et al. (1974). This variant of the lunar-derived model has also been termed Nice model chronology (Dones et al., 2009, and references therein). Whether such an LHB peak occurred and whether the bombardment rate showed a sawtooth-line time dependence on the Moon or on other terrestrial planets (and asteroids) is still an unsolved question (pro: Strom et al., 2005; Morbidelli et al., 2012, 2018; contra: Baldwin, 2006; Michael et al., 2018).

An LDM for the Galilean satellites of Jupiter (and of Saturn) with a smooth exponential decay in the bombardment rate prior to ~3 Ga ago was favoured by Neukum (1985, 1997), Neukum and Pozio (1983), and Neukum et al. (1998). Based on similarities in the shapes of CSFDs between the Moon and Jupiter's icy satellite Ganymede in their measurements, the authors concluded that most craters were formed by bodies stemming from the asteroid main belt. Their model is calibrated on the assumption that the impact event which created the Gilgamesh basin on Ganymede happened 3.8 Ga ago, like when Orientale, the youngest lunar basin, was formed. Using this fixed point in time, the lunar chronology model can be transferred to Ganymede and to the other Galilean satellites. The authors also concluded that the shapes of CSFDs on Ganymede and Callisto indicate not only that the craters were formed mainly by asteroids but that these were preferentially captured into planetocentric orbits about Jupiter and impacted at impact velocities comparably low to those of heliocentric impactors (Horedt and Neukum 1984a, 1984b; Burger et al., 2010). Predominantly planetocentric impactors would also explain the lack of apex–antapex asymmetries in crater frequencies for strictly non-synchronously rotating bodies. Neukum (1985, 1997) and Neukum et al. (1998) did not offer a dynamical mechanism of how asteroids can be captured into planetocentric orbits about Jupiter to impact the Jovian satellites, therefore the validity of the LDM, based on asteroidal impacts, remains an open issue.

Shoemaker and Wolf (1982) first presented a similar lunar-like chronology model, based on impacts by comets and asteroids. Later, the lunar-like time dependence of bombardment rates was dropped by Shoemaker (1995), who extrapolated current impact rates preferentially for comets back in time. Zahnle et al. (1998, 2003) carried out similar estimations and came forward with a *comet-derived chronology model* (CDM), with impacts by comets from heliocentric orbits and a bombardment rate mainly constant back into earliest times. According to Zahnle et al. (1998, 2003), the contribution of main-belt asteroids (MBAs) to the cratering of the Jovian satellites is negligible at present time, as well as the contribution of nearly isotropic comets (NICs) from the Oort cloud (long-period comets (LPCs) and Halley-type comets (HTCs)). The primary impactors are ecliptic comets (ECs) or Jupiter-family comets (JFCs). The uncertainties of their impact rates, however, are high and on the order of a factor of 5 (Zahnle et al., 1998) or a factor of 3 (Zahnle et al., 2003). Using updated estimates of JFC impact rates, absolute model ages (AMAs) from CSFD measurements based on Zahnle et al. (2003) are higher than in their original paper (Zahnle et al., 1998).

The SFDs of JFCs and bodies from the Kuiper belt are difficult to reconcile with CSFDs measured on the surfaces of Ganymede, especially for measurements by Neukum (1985), Neukum and Pozio (1983), and Neukum et al. (1998). These measurements are also in disagreement with results by, for example, Strom et al. (1981), who advocated CSFDs on Ganymede to be different from lunar CSFDs. What can be concluded from the shape of these CSFDs with some certainty, however, is that the satellites of Jupiter were predominantly bombarded by members of a collisional evolved impactor family (Burger et al., 2010, and references therein). Main-belt asteroids are collisional evolved, as verified from astronomical observations (Ivanov et al., 2002, and references therein), and their collisional evolution can be modelled (Durda et al., 1998). If preferentially cometary objects were responsible for the cratering of the satellites of Jupiter, these bodies should have had a collisional history to produce a shape of their SFD like that of the MBAs. Collisional evolution has been discussed for Kuiper belt objects (KBOs) and trans-Neptunian objects (TNOs), but the shape of their SFD is unknown since the smallest KBOs to be generally observable are several tens of kilometres across – producing craters with diameters in the order of ~100 km or more. Improved observation techniques as well as the New Horizons Mission, which flew by Pluto and its moons in July 2015, and the KBO Arrokoth in January 2019 will help to extend the known SFD of these bodies to smaller diameters in the order of at least several kilometres.

Absolute model ages derived from the two chronology models by Neukum et al. (1998) and Zahnle et al. (1998, 2003) are comparable for surfaces older than 3.5–4 Ga but differ greatly for younger surfaces. Since the cratering chronology scenario for the Galilean satellites is only known with high uncertainties, it is recommended that both models are applied in satellite stratigraphy and chronology, as used by the Galileo SSI Team.

Relative and absolute model ages (AMAs) of geologic units on Ganymede.

Relative (crater frequencies) and absolute model ages (AMAs) were measured on Ganymede's major geologic units and on several impact features. In the following, AMAs are listed for the LDM chronology (Neukum, 1997) and for the CDM chronology, but only AMAs based on the cratering rates in the updated paper of Zahnle et al. (2003) are used. Cumulative crater distributions in the dark, densely cratered plains which represent the oldest units on Ganymede (Pappalardo et al., 2004; Patterson et al., 2010; Collins et al., 2013) were first measured in Voyager data and were shown to continue to smaller diameters measured in high-resolution Galileo SSI data without major changes in cumulative slope, indicating no major resurfacing events had taken place, except in specific localities (Neukum, 1997; Neukum et al., 1998). The largest areas of measurement used for crater counts are in the two most prominent dark-appearing terrains of Galileo Regio and Nicholson Regio (Fig. 5.2). Their cumulative frequencies for crater diameters D $\geq$ 1 km are within a range of $\sim$3 10^{-2} to 10^{-2} per km^2 (Neukum, 1997; Neukum et al., 1998; Prockter et al., 1998) which translate into a range in LDM AMAs of 4.13–3.97 Ga (average error is in the order of $\pm$0.05 Ga; Neukum, 1997; Neukum et al., 1998); the corresponding range for CDM AMAs is 4.38–3.01 Ga (with an upper limit of 4.56 Ga and lower limit of 1.38 Ga; Zahnle et al., 2003). In some dark units, lower crater frequencies imply resurfacing by mass wasting and accumulation of material in topographic lows (Prockter et al., 1998).

Collins et al. (2013) associate these oldest dark units with the lowest time-stratigraphic system or chronologic period which they termed the Nicholsonian System/Period. The time-stratigraphic systems established for Ganymede by Collins et al. (2013) in the context of the two cratering chronology function curves are shown in Fig. 5.6. Measurements of CSFDs in Nicholson Regio render the average AMA of dark material in this area and are within the range listed above. The cumulative crater frequency for D $\geq$ 1 km is 1.96 10^{-2}, corresponding to an LDM AMA of 4.07 $\pm$ 0.05 Ga, and a CDM AMA of 4.1 Ga (upper limit 4.55 Ga; lower limit 2.31 Ga) (Wagner et al., 2017, 2018). This measurement, however, does not represent the base of the Nicholsonian System since this oldest system is not defined by a stratigraphic marker horizon (Patterson et al., 2010; Collins et al., 2013). The variability of crater frequencies and AMAs in the old dark materials, and hence potential small-scale geologic resurfacing processes in these units, can be studied in detail in the future (years 2030>) since several areas of interest were selected and suggested for higher-resolution imaging with the JANUS camera aboard the European Space Agency (ESA) JUpiter ICy moons Explorer (JUICE) Jupiter and Ganymede Orbiter (Stephan et al., 2021).

The bright, mostly heavily tectonized materials were classified into the Harpagian System/Period (Collins et al., 2013). The marker horizon of the Harpagian a bright unit preferentially smooth at Voyager resolution in Harpagia Sulcus (Fig. 5.2), close to the dark cratered materials of Nicholson Regio. It has a cumulative frequency for D $\geq$ 1 km of 4.34 $\cdot$ 10^{-3} (Wagner et al., 2017, 2018). This corresponds to an LDM AMA of 3.85 $\pm$ 0.05 Ga, and a CDM AMA of 1.71 Ga (upper limit 3.44 Ga; lower limit 0.66 Ga) (Fig. 5.6).

Owing to the intense tectonic resurfacing over time, CSFDs in bright units show some scatter. In the Uruk Sulcus region on the anti-jovian hemisphere, crater frequencies and AMAs higher (older) than the Harpagian base were found in units mapped in the SSI target area G1GSURUSUL01 and its Voyager context (Neukum, 1997; Neukum et al., 1998). From a cumulative frequency for D $\geq$1 km of 1.9 10^{-3}, an LDM AMA of 3.95 $\pm$ 0.05 Ga could be derived, versus a CDM AMA of 2.83 (upper limit 4.31; lower limit 1.26 Ga). Bright grooved material at the G1GSURUSUL01 SSI target area is even older, with a cumulative frequency for D $\geq$ 1 km of 1.9 $\cdot$ 10^{-2}, corresponding to an LDM AMA of 4.06 $\pm$ 0.05 Ga, versus a CMD AMA of 3.97 Ga (upper limit 4.55 Ga; lower limit 2.25 Ga). These units would be classified as pre-Harpagian/Nicholsonian in the time-stratigraphic system by Collins et al. (2013). These frequencies and AMAs in the G1GSURUSUL01 target area, however, must be treated with some caution when comparing them with ages in Harpagia Sulcus because the SSI Uruk Sulcus target area is closer to the apex point than the Harpagia Sulcus region. These differences could be an effect of a possible apex–antapex variation in crater frequency rather than an actual difference in age (Schenk et al., 2004). The Galilean satellites rotate non-synchronously, and such satellites are expected to show a pronounced apex–antapex asymmetry in crater frequencies for craters predominantly formed by impactors from heliocentric orbits (Passey and Shoemaker, 1982; Shoemaker and Wolf, 1982; Horedt and Neukum, 1984a, 1984b). Recent studies using both Voyager and Galileo SSI images showed that a pronounced apex–antapex asymmetry is less and only occurs in the younger, light, tectonized terrain (Schenk et al., 2004). Two explanations exist for an apex–antapex asymmetry to be absent or to be little pronounced: planetocentric impactors which are inferred to be main source of impact craters in the chronology model by Neukum (1997) and Neukum et al. (1998) are difficult to generate as a primary source for reasons of orbital dynamics (Horedt and Neukum, 1984b; Zahnle et al., 2001); non-synchronous rotation for an extended period in early time, prior to, for example, 2–3 Ga, is an explanation which is preferred by, Zahnle et al. (1998, 2001) and Schenk et al. (2004), among others. Non-synchronous rotation of a satellite needs the decoupling of an ice shell from a liquid interior, such as water oceans, which are thought to be present in the icy Galilean satellites (e.g. Schubert et al., 2004; Hussmann et al., 2010; Sohl et al., 2010). Similarly, high ages were derived in the bright material of Xibalba Sulcus in the C9GSSULCUS01 SSI target area which is located near the apex point (Stephan et al., 2003; Stephan, 2006). The comparably high age of the bright units in Harpagia Sulcus or Uruk and Xibalba Sulcus in the LDM by Neukum (1997) and Neukum et al. (1998) is in strong contradiction to modelling the thermal evolution of Ganymede with time. In the thermal models by Kirk and Stevenson (1987) and Mueller and McKinnon (1988), the tectonic processes that formed the bright grooved and smooth units could date back to $\sim$2 Ga, much later than the $\sim$3.8 Ga inferred in the LDM by Neukum (1997) and Neukum et al. (1998). Absolute model ages of $\sim$2 Ga and less are much more in agreement with the CDM by Zahnle et al. (2003).

Crater counts in the highly tectonized bright (grooved or smooth (subdued)) units based on Voyager images are difficult since the areas of measurements in the individual structural cells (e.g. Pappalardo et al., 2004; Patterson et al., 2010; Prockter et al., 2010) are small and statistical uncertainties high. However, there is still some potential to use these lower-resolved Voyager images and the comparably small number of Galileo SSI target areas for a detailed study and reassessment of the range of crater distributions and ages in the bright units. These results will then provide a good base for a much more complete investigation based on JANUS camera images of the JUICE mission and selected high-resolution targets (Stephan et al., 2021).

Gilgamesh (Fig. 5.3a) is the only well-preserved impact basin on Ganymede and was formed mainly in bright-appearing terrain (Pappalardo et al., 2004; Schenk et al., 2004). Gilgamesh is the base horizon of the Gilgameshan System/Period in the

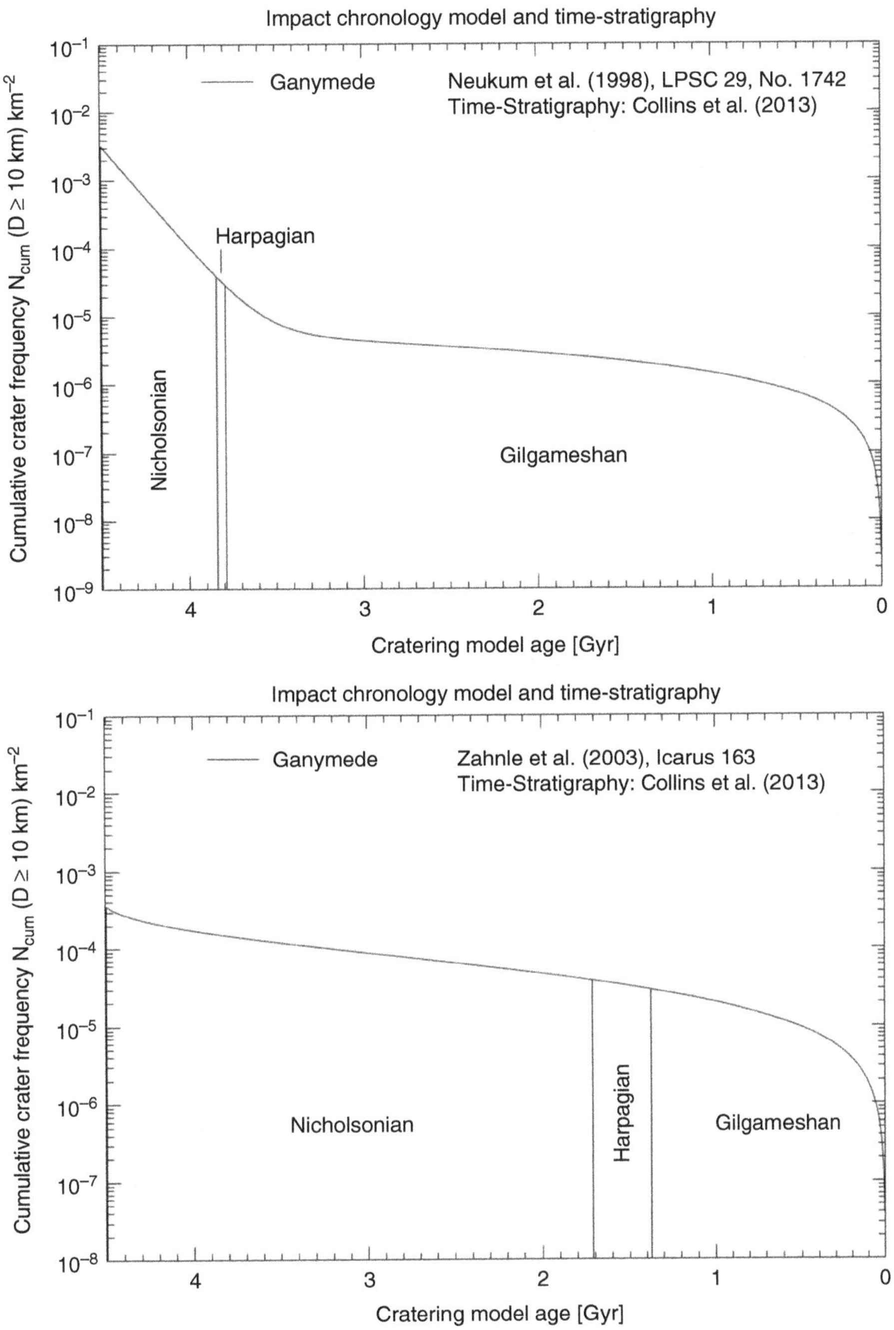

Figure 5.6 Time-stratigraphic systems (= chronologic periods) of Ganymede based on the global geologic map by Collins et al. (2013) in the two impact chronology models: (a) lunar-derived chronology model (LDM) (Neukum, 1997; Neukum et al., 1998) and (b) comet-derived chronology model (CDM) (Zahnle et al., 2003). Both graphs are shown for a cumulative frequency for craters equal to or greater than 10 km diameter.

Collins et al. (2013) time-stratigraphic scheme. In the LDM, Gilgamesh was used as a fixed time step for establishing the LDM chronology for Ganymede and the Galilean satellites (Neukum and Pozio, 1983; Neukum, 1997). The CDM AMA of Gilgamesh is 1.37 Ga (upper limit 3.0 Ga; lower limit 0.51 Ga). Additional measurements in a higher-resolution mosaic from Galileo's orbit E12 (SSI target area 12ESGLGMSH01) are impactable due to the extreme high sun of this image scene. Fig. 5.6a shows that the Harpagian Period is unreasonably short in the LDM compared to Gilgameshan and Nicholsonian, in the order of 10–20 My only. If Gilgamesh represents the largest final impact event which took place towards the end of bright-appearing terrain formation, the tectonic resurfacing period which created most of the bright grooved or smooth (subdued) materials may have lasted more than ~300 Ma according to the CDM (Fig. 5.6b).

Individual impact craters, classified according to the various crater morphologies on Ganymede (Schenk et al., 2004), are difficult to use for crater statistics due to the insufficient spatial resolution and the mostly small areas covered by the Voyager images. Only a relatively small number of impact features could be targeted for high-resolution imaging by the

Galileo SSI camera. These impact features, however, show a wide range in ages. For the following examples, only the two AMAs are cited, for brevity. With two exceptions, palimpsests are abundant mostly in the dark units (Pappalardo et al., 2004). Buto Facula, a palimpsest located in Marius Regio (observed in orbit G8 (area G8GSBUTOFC01)), has a derived AMA of 3.82 ± 0.05 Ga (LDM). According to CDM, recalculated based on Zahnle et al. (2003), the AMA yield is 1.64 Ga (upper limit 3.37 Ga; lower limit 0.63 Ga) (Wagner et al., 1999). Zakar (SSI target area G7GSPALIMP01), one palimpsest superimposed on bright units, was shown to have a similar age. Neith, an impact feature classified either as a pene-palimpsest as an anomalous dome crater which is located at a dark-appearing terrain/bright-appearing terrain boundary (Schenk et al., 2004, and references therein), shows AMAs of 3.91 ± 0.05 Ga (LDM) and 2.35 Ga (CDM; upper limit 4.04 Ga; lower limit 0.98 Ga) (Wagner et al., 1999). Melkart, a prominent central pit and dome crater in Marius Regio with remarkable spectral properties, could have been formed 3.78 ± 0.05 Ga (LDM) or 1.31 Ga ago (CDM; upper limit 2.92 Ga; lower limit 0.49 Ga) (Wagner et al., 1999; Stephan et al., 2003; Stephan, 2006).
Owing to the limited number of high-resolution images of impact features in Galileo SSI data, and due to the high uncertainties in the AMAs of both impact chronologies, it is difficult to draw conclusions about the absolute time in which these specific craters were formed or about potential changes in rheologic properties of the Ganymede surface with time. In the LDM scenario, these impact features are mostly old features formed between 3.95 and 3.75 Ga ago (a time span of ~200 Ma), while in the CDM scenario they may have formed between 2.35 and 1.31 Ga ago, over a time span of ~1 Ga. Ganymede features a considerable number of bright and dark ray craters which could be shown to have characteristic and very specific spectral properties (Stephan et al., 2003; Pappalardo et al., 2004; Schenk et al., 2004 Stephan, 2006). These ray craters could be used as marker horizons to define a youngest time-stratigraphic system.

Osiris (Fig. 5.3c) is a prominent 107 km large bright ray crater (Schenk et al., 2004; Stephan, 2006). Wagner et al. (2018) used Voyager 2 images to reinvestigate if dating Osiris would be possible. The superimposed crater frequency, however, can only be estimated with a high degree of uncertainty since the spatial resolution is not sufficient. The derived AMAs for Osiris from this estimation (Wagner et al., 2018, updating older AMAs by Wagner et al., 1999) are in much better agreement with Osiris's pristine morphological appearance, its high albedo, and the fact that its extended ray system extensively covers pre-existing geologic units, placing Osiris in a more recent time in Ganymede's history. The estimated AMAs, based on the existence of a single small crater close to the Voyager 2 image resolution limit are 744 ± 350 Ma (LDM), or 59 Ma (CDM; upper limit 175 Ma; lower limit 20 Ma). Wagner et al. (2018) suggested that an Osirian System/Period be introduced into Ganymede's time-stratigraphic scheme. Osiris will be a major target for the JANUS camera aboard JUICE and these data will be used to constrain the age of Osiris, as well as for other prominent ray craters (Stephan et al., 2021).

Kittu is a spectrally remarkable dark ray crater observed in Galileo SSI target areas G7GSKITTU_01 and_02 (Stephan et al., 2003; Stephan, 2006). The SSI image covering a part of Kittu is not dense enough to perform a reliable CSFD measurement; therefore only a maximum age can be estimated based on the spatial resolution of the image and a possible smallest crater diameter. This procedure yields a maximum AMA of 734 Ma (LDM) and 58 Ma (CDM; upper limit 173 Ma; lower limit 20 Ma). In the high northern latitudes of the sub-jovian hemisphere, two craters with similar diameters but different morphology and ejecta patterns are emplaced in proximity. Both craters were imaged by Galileo SSI (target area G7GSACHELS01) and are shown in Fig. 5.4. Gula is a central-peak crater, while Achelous is a so-called pedestal crater, with bright ejecta rays which are in the Voyager data (the SSI target area was imaged at low sun). Both craters were redated recently (Wagner et al., 2019), updating older values by Wagner et al. (1999). Gula is relatively old with an AMA of 3.92 ± 0.04 Ga (LDM) and 3.1 Ga (CDM; upper limit 4.41 Ga; lower limit 1.44 Ga). The bright ray crater Achelous could be as young as 175 ± 135 Ma (LDM) or 20 Ma (CDM; upper limit 59 Ma; lower limit 6.6 Ga).

5.4 TOPOGRAPHY

Topographic data for Ganymede have been obtained based on the first detailed imaging acquired by the Voyager 1 and 2 spacecraft. Photoclinometry was used to obtain elevation models for bright- and dark-appearing terrain, as well as for specific features. The bright-appearing terrain was found to be consisting of subparallel linear grooves typically 3–10 km apart with a typical height of 300–400 m reaching a value up to 700 m (Squyres, 1981). Slopes were found to be very gentle, showing primarily subdued topography. Results obtained for Ganymede's south polar region were consistent with later Galileo data sets. Based on photogrammetric analysis of stereo imaging data, Giese et al. (1998) derived a digital terrain model (DTM) of Uruk Sulcus. Located near the equator, Uruk Sulcus showed prominent wavelengths ranging from 4 to 6 km and wave amplitudes reaching some hundreds of metres, 500 m at the highest. The topographic wave trains are not symmetric in shape, showing different slopes (e.g. 15° on one side and 4° on the other). The steepest slopes measured are 19°. Both concave and convex slopes exist. Several other bright-appearing terrain regions show similar characteristics (e.g. Patel et al. 1999). The similarities between the earlier analyses of completely different regions suggest that grooved terrain is formed by the same mechanisms operating globally on Ganymede. The grooved terrain has been interpreted as a result of the necking instabilities that are related to phases of global extension (e.g. Dombard and McKinnon, 2001; Bland et al., 2007). Tectonism appears to be the dominant geological process that has shaped the terrain. A broad wavelength of deformation may be the result of ductile necking of the lithosphere, while a finer scale of deformation probably reflects faulting of the brittle near surface (Pappalardo et al., 1998; Patel et al., 1999).

The furrow systems of Ganymede's dark-appearing terrain were investigated in a few locations, utilizing stereo imaging of Galileo data. Unfortunately, overlapping regions suitable for photogrammetric stereo analysis were very sparse in the Ganymede Galileo mission data set (Figs. 5.7 and 5.8). The most prominent example is the Zu furrow, located in Galileo Regio. The DTM obtained covers a patch of 63 × 102

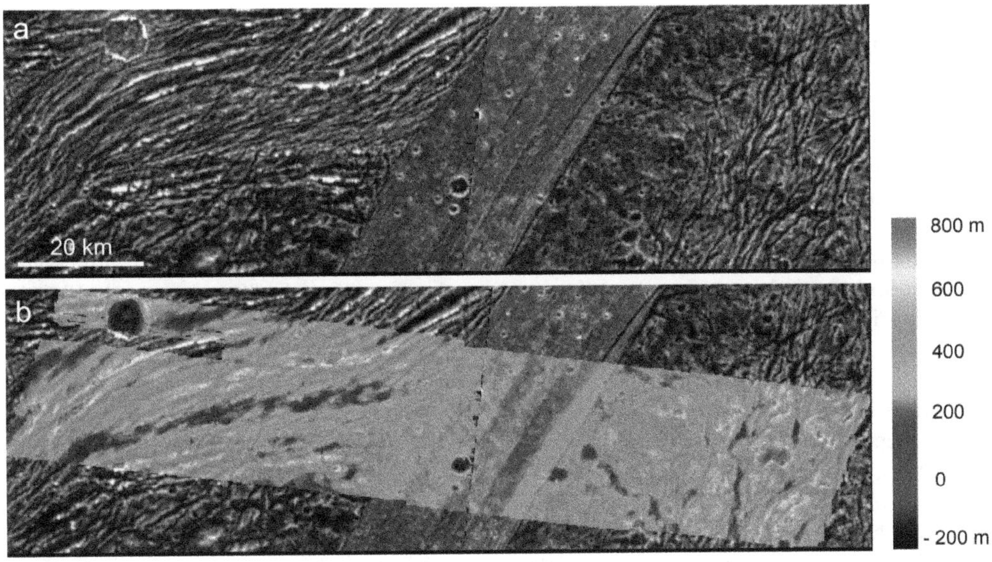

Figure 5.7 (a) High resolution image (34 m/px) and (b) topographic data of Arbela Sulcus (15°S/347°E) derived from stereo images. Arbela Sulcus exhibits small-scale lineations and straight margins indicative of tectonic deformation.
Note that the eastern margin of Arbela is elevated relative to local depressions in the adjacent dark-appearing terrain, arguing against a primary origin by cryovolcanic flooding. The topographic model has a horizontal resolution of 600 m and a vertical resolution of 15–25 m. North is to the top. (A black and white version of this figure will appear in some formats. For the colour version, please refer to the plate section.)
Source: adapted from Giese et al. (2001).

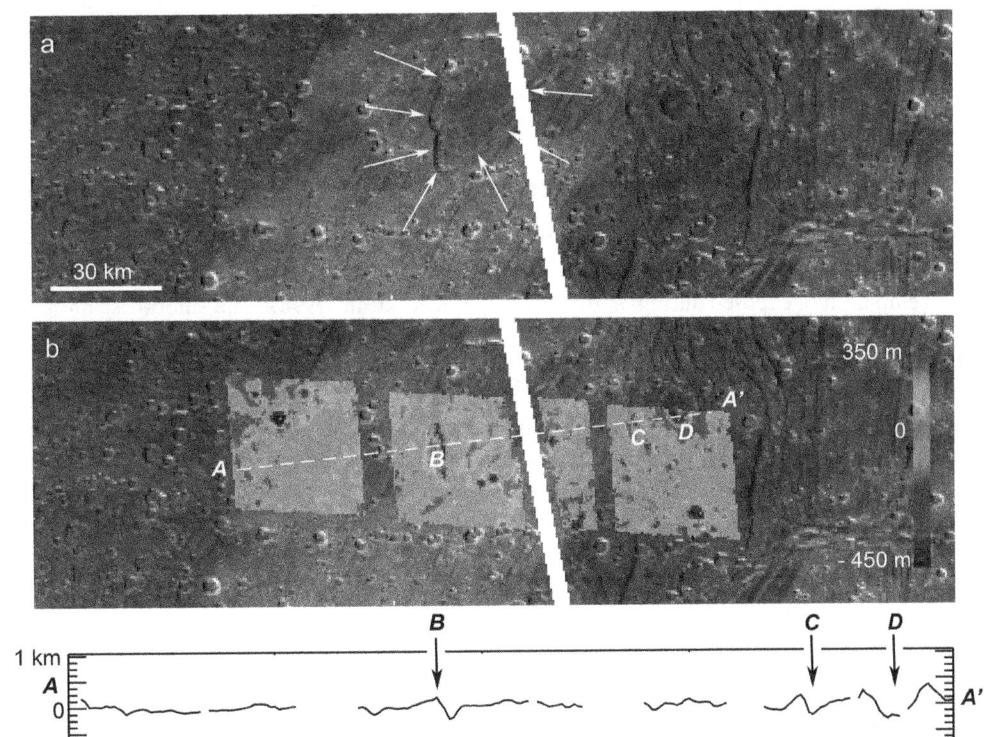

Figure 5.8 (a) Galileo observation G28GSCALDRA01 (180 m/px) covering 20 × 40 km near Hammamat Patera (ID #1.1) at 24°S/42°E (arrows) surrounded by bright-appearing terrain with (b) the DTM-derived topographic information (Giese et al., 2017). The DTM derived from stereo image analysis exhibits a spatial resolution of 600 m with a horizontal resolution of about 1 km and vertical precision of 20–40 m with (c) topographic profiles. The profiles are marked as dashed lines A to A'. Additional surface features such as prominent grooves (B, C, and D) and impact craters can be well recognized in the DTM and profile. (A black and white version of this figure will appear in some formats. For the colour version, please refer to the plate section.)

km; this region includes furrows, ridges, plateaus, isolated knobs, and large degraded impact structures (Giese et al., 1998; Zubarev et al., 2017). Furrows and ridges extend over tens of kilometres, with widths between a few kilometres up to 10 km, and heights from hundreds of metres up to 1 km; local slopes reach 30° (Giese et al., 1998). Unlike the topography in Uruk Sulcus, the furrows and ridges in Galileo Regio are not aligned and are remnants of ancient multi-ringed structures. As compared with other bright-appearing terrains, the surface of Galileo Regio is rougher and has a shorter characteristic correlation length. Overall variations in elevation are greater and slopes are generally steeper (Giese et al., 1998). Several processes are interpreted to have occurred within Galileo Regio. These include tectonic deformation, mass wasting, sublimation, resurfacing by impact ejecta, and possibly cryovolcanism and isostatic adjustment (Prockter et al., 1998). Arbela Sulcus is another band of grooved terrain that could be analysed in a single DTM, along with the adjacent dark-appearing terrain (e.g. Prockter et al., 2010). It was found that large parts of the band stand ~100 m lower than the surrounding dark-appearing terrain. Arbela Sulcus features a pronounced smoothness as compared with the adjoining terrain. Heat induced to Ganymede's icy lithosphere would increase isostatic adjustment for pre-existing topography like at Tiamat Sulcus. Topography with a shorter wavelength relaxes faster than long-wavelength topography with increasing interior heat (Singer et al., 2018). In particular, impact structures show viscous relaxation (e.g. Passey and Shoemaker, 1982; Dombard and McKinnon, 2006; Schenk, 2010; Bland et al., 2017; Singer et al., 2018).

Oberst et al. (1999) found a strong correlation between albedo and elevation. In both bright-appearing (Uruk Sulcus) and dark-appearing (Galileo Regio) terrain regions, the bright material is found on topographic highs, whereas dark material is in depressions. The relative abundance of the dark material increases with lower elevation. The authors suggest that while the bright material was emplaced or exposed during or after formation of the terrain in its present location, the dark material was eroded and redistributed downslope.

Squyres (1980) identified dome-shaped features from photoclinometric analysis of Voyager images. The most prominent dome has a diameter of roughly 250 km and an elevation of 2–2.5 km above the surrounding plains. The formation mechanisms and the role of cryovolcanism in the creation of dome-shaped features on Ganymede remain unclear. Further topographic data, for example from the JUICE mission, will provide invaluable new data sets for addressing that question.

5.5 GEOLOGICAL HINTS FOR A SUBSURFACE OCEAN

Impact as a candidate process to reach a subsurface ocean can be ruled out. During an impact, a transient cavity forms with a depth approximately $\frac{1}{3}$ to $\frac{1}{4}$ of the transient crater diameter (Melosh and Ivanov, 1999). Impact craters with diameters of 10–20 km have formed in substantial number over the past ~3 Ga on Ganymede and are common on both terrain types. These would require only kilometre-sized projectiles, which could only reach depths of a few kilometres; this is not enough to penetrate to the top of Ganymede's subsurface ocean, which is estimated to be located at a depth of approximately 170–200 km (Pappalardo et al., 2004; Schubert et al., 2004). The formation of impact basins like Gilgamesh, Ganymede's largest impact structure, which has an assumed nominal crater rim diameter of ~600 km and a transient excavation depth of 150–200 km (Melosh and Ivanov, 1999; Schenk et al., 2004), is a rare event in the history of the evolution of the Jovian system. Gilgamesh is in the order of ~1–3.8 Ga old, depending on the cratering chronology model applied (Neukum et al., 1998; Zahnle et al., 1998, 2003).

Ganymede's lithosphere overlying a plastic asthenosphere was originally estimated to be ~10 km thick, however, the topography of the younger Gilgamesh basin implies a thicker crust of ~75–100 km (McKinnon and Melosh, 1980; Schenk et al., 2004). Nimmo et al. (2002) used Galileo SSI stereo images to measure the elastic thickness in tectonized areas, which is an indicator of how deep tectonic structures can extend until the brittle-to-ductile layer is reached. The authors found a value of ~1–2 km at the time of groove formation (Nimmo et al., 2002). Indeed, gravity and pressure of the overlaying crust will close fractures at that depth. The morphology of the post-groove Gilgamesh basin reveals, however, that Ganymede's brittle layer has thickened considerably since the time when the bright grooved terrain formed (e.g. Schenk et al., 2004). In conclusion, the probability of dragging any material down to the convective sublayer is negligible. In the bright-appearing terrain, regional and high-resolution-scale images from the Galileo SSI camera revealed that intense tectonism was the dominant process in creating and shaping Ganymede's bright-appearing terrain, while cryovolcanism played a comparably minor role (Pappalardo et al., 2004). Several scalloped paterae are found in the bright-appearing terrain, representing probable volcanic vents (Pappalardo et al., 2004). Ridged deposits in one of the largest of such paterae were interpreted as cryovolcanic flows (Head et al., 1998). The smoothest units, as found in several localities in bright-appearing terrain, could have been created by the extrusion of low-viscosity cryomagmatic materials (Giese et al., 1998; Schenk et al., 2001). The smoothest units also exhibit some degree of tectonism, suggesting that cryovolcanism and tectonic deformation are closely linked (Head et al., 2002). Material in dark smooth patches was likely created by sublimation and degradation processes rather than by cryovolcanism (e.g. Moore and Schubert, 2003).

5.6 CONCLUSION

Ganymede shows a tremendous diversity of surface features which differ significantly in their specific evolutionary paths. Each of the Jovian moons exhibits its own fascinating geologic history – formed by the competition and combination of external and internal processes. Their origins and evolutions are influenced by factors such as density, temperature, composition (volatile compounds), stage of differentiation, volcanism, tectonism, the rheological reaction of ice and salts to stress, tidal effects, and interactions with the Jovian magnetosphere and space. These interactions are still recorded in the present surface geology. The record of geological processes spans from possible cryovolcanism through widespread tectonism to surface degradation and impact cratering. The Galileo mission has allowed us to describe for the first time the global geology of the moons. But it was not possible, except in a very few cases, to study regional and local geology, as most of the data are of low or medium resolution (less than 1% of the surface was observed at a resolution greater than 100 m/px).

Thus, several questions remain unsolved: What are the relative roles of tectonism and volcanism in shaping the terrains? What does the distribution of craters on Ganymede tell us about the evolution of the impactor population in the Jovian system through time? How is the geological evolution related to the impact, tectonic, and cryovolcanic history? What is the age of certain geological units on Ganymede, and how will these findings contribute to our understanding of the origin of the Jupiter system? What is the rheological reaction of ices and ice/salt/clathrate mixtures with regards to tectonic stress? To what extent are surfaces altered by cosmic weathering and what are the major exogenic surface alteration processes (micrometeorites, radiation, charged particles)?

The ESA JUICE mission will fully observe Ganymede at a medium resolution (total coverage at a few 100 m/px) and will provide at least 50 times more high-resolution imaging than Galileo. It will study the satellites' geological history using global, regional, and local mapping at different levels of resolution, as well as stereo imaging on regional and local scales. Coloured maps down to 200 m/px and four-colour coverage for selected large areas, up to 50 m/px using spectral filters from about 350 nm to 1000 nm, will be acquired to constrain mineralogical/chemical constituents of the near-surface layers and to correlate geologic features with compositional variations. The JUICE mission will also acquire detailed topographic profiles of tectonic features, grooved terrain, craters, and cryovolcanic features using laser altimetry which, combined with imaging data, will enable the identification of dynamical processes that cause internal evolution and near-surface tectonics.

The JUICE mission will use radar sounding on Ganymede's subsurface to assess any linkage of the ocean to the surface and the dynamics of the near-surface ice layers by detecting compositional or phase boundaries between different intra-ice horizons and unconformities. The stratigraphic and structural data acquired will also provide information on crust formation and its possible destruction by deformational processes. Ganymede is a densely cratered object with a record of large impact structures, including multi-ring structures, which implies an old surface age (e.g. Pappalardo et al., 2004). The widest range in crater morphology on any planet or satellite is found on Ganymede (e.g. Pappalardo et al., 2004). JUICE will significantly improve the current estimate of Ganymede's surface ages by measuring crater distributions from nearly global image coverage at 200–400 m/px resolutions as well as sufficiently high-resolution target areas (5–50 m/px) and by monitoring the surfaces on a timescale in the order of hundreds of days up to three years to identify potentially newly formed craters.

ACKNOWLEDGEMENTS

The authors are grateful for the experience gained with the JUICE project. A portion of this work was performed at the Free University of Berlin. Pasquale Palumbo and Alice Lucchetti were supported for this activity by the Italian Space Agency (ASI) under the ASI-INAF contract 2018-25-HH.0. We also thank the two anonymous reviewers for their very helpful comments.

REFERENCES

Allison, M. L., and Clifford, S. M. (1987). Ice-covered water volcanism on Ganymede. *Journal of Geophysical Research* 92, 7865–76.

Arvidson, R., Boyce, J., Chapman, C. et al. (1979). Standard techniques for presentation and analysis of crater size-frequency data. *Icarus* 37, 467–74.

Baldwin, R. B. (2006). Was there ever a Terminal Lunar Cataclysm? With lunar viscosity arguments. *Icarus* 184, 308–18.

Beeman, M., Durham, W. B., and Kirby, S. H. (1988). Friction of ice. *Journal of Geophysical Research* 93, 7625–33.

Bland, M. T., and Showman, A. P. (2007). The formation of Ganymede's grooved terrain: Numerical modeling of extensional necking instabilities. *Icarus* 189, 439–56.

Bland, M. T., Singer, K. N., McKinnon, W. B., Schenk, P. M. (2017). Viscous relaxation of Ganymede's impact craters: Constraints on heat flux. *Icarus* 296, 275–88, doi: https://doi.org/10.1016/j.icarus.2017.06.012.

Bray, V. J., Collins, G. S., Morgan, J. V., and Schenk, P. M. (2008). The effect of target properties on crater morphology: Comparison of central peak craters on the Moon and Ganymede. *Meteoritics and Planetary Science* 43, 1979–92.

Burger, M. H., Wagner, R., Jaumann, R., and Cassidy, T. A. (2010). Effects of the external environment on icy satellites. *Space Science Reviews* 153, 349–74.

Carnielli G., Galand, M., Leblanc, L. et al. (2020). Simulations of ion sputtering at Ganymede. *Icarus* 351, 113918, https://doi.org/10.1016/j.icarus.2020.113918.

Chapman, C. R., and McKinnon, W. B. (1986). Cratering of planetary satellites. In: J. A. Burns and M. S. Matthews (Eds.), *Satellites*. Tucson: University of Arizona Press, pp. 492–580.

Clark, B. E., and Johnson, R. E. (1996). Interplanetary weathering: Surface erosion in outer space, *Eos, Transactions, American Geophysical Union*, 77(15), 141–5.

Collins, G. C., Head, J. W., and Pappalardo, R. T. (1998a). Formation of Ganymede grooved terrain by sequential extensional episodes: Implications of Galileo observations for regional stratigraphy. *Icarus* 135(1), 345–59.

Collins, G. C., Head, J. W., and Pappalardo, R. T. (1998b). The role of extensional instability in creating Ganymede grooved terrain: Insights from Galileo high-resolution stereoimaging. *Geophysical Research Letters* 25, 233–6.

Collins, G. C., Patterson, G. W., Head, J. W. et al. (2013). *Global geologic map of Ganymede*. Scientific Investigations Map 3237. Reston, VA: US Geological Survey.

Croft, S., Cassachia, R., and Strom, R. (1990). *Geologic map of the Tiamat Sulcus quadrangle of Ganymede*. Map I-1548 (Jg-9). Reston, VA: US Geological Survey.

Croft, S. K., and Goudreau B .N. (1987). Tectonism and volcanism in Ganymede's dark terrain. *Lunar and Planetary Science Conference Abstracts* XVIII, 209–10.

de Kleer, K., Butler, B., de Pater, I. et al. (2021). Ganymede's surface properties from millimeter and infrared thermal emission. *The Planetary Science Journal* 2(5), http://doi.org/10.3847/PSJ/abcbf4.

Divine, N. (1993). Five populations of interplanetary meteoroids. *Journal of Geophysical Research* 98, 17029–48, http://doi.org/10.1029/93JE01203.

Dombard, A. J., and McKinnon, W. B. (2001). Formation of grooved terrain on Ganymede: Extensional instability mediated by cold, superplastic creep. *Icarus* 154, 321–36.

Dombard, A. J., and McKinnon, W. B. (2006). Elastoviscoplastic relaxation of impact crater topography with application to Ganymede and Callisto. *Journal of Geophysical Research* 111, https://doi.org/10.1029/2005JE002445.

Dones, L., Chapman, C. R., McKinnon, W. B. et al. (2009). Icy satellites of Saturn: Impact cratering and age determination. In: M. K. Dougherty, L. W. Esposito, and S. M. Krimigis (Eds.), *Saturn from Cassini-Huygens*. Dordrecht: Springer, pp. 613–35.

Durda, D. D., Greenberg, R., and Jedicke, R. (1998). Collisional models and scaling laws: A new interpretation of the shape of the Main-Belt Asteroid size distribution. *Icarus* 135, 431–40.

Durham, W. B., and Stern, L. A. (2001). Rheological properties of water ice: Applications to satellites of the outer planets. *Annual Review of Earth and Planetary Sciences* 29, 295–330. https://doi.org/10.1146/annurev.earth.29.1.295.

Giese, B., Hauber, E., and Hussmann, H. (2017). On the formation of caldera-like features on Ganymede: Implications from Galileo-g28 images. *Lunar and Planetary Science Conference Abstracts* XLVIII, Abstract No. 274.

Giese, B., Oberst, J., Roatsch, T. et al. (1998). The local topography of Uruk Sulcus and Galileo Regio obtained from stereo images. *Icarus* 135, 303–16.

Guest, J. E., Bianchi, R., and Greeley, R. (1988). *Geologic map of the Uruk Sulcus quadrangle of Ganymede*. Map No. 1934. Reston, VA: US Geological Survey.

Head, J. W., Pappalardo, R. T., Collins, G. et al. (2002). Evidence for Europa-like tectonic resurfacing styles on Ganymede. *Geophysical Research Letters* 29, 4–1.

Head, J. W., Pappalardo, R. T., Kay, J. et al. (1998). Cryovolcanism on Ganymede: Evidence in bright terrain from Galileo Solid State Imaging data. *Lunar and Planetary Science Conference Abstracts* XXXIX, Abstract No. 1666.

Hiesinger, H., Marchi, S., Schmedemann, N. et al. (2016). Cratering on Ceres: Implications for its crust and evolution. *Science* 353, https://doi.org/10./1126/science.aaf4759.

Hillier, J., Helfenstein, P., and Veverka, J. (1996). Latitude variations of the polar caps on Ganymede. *Icarus* 124, 308–17.

Hirata, N., Suetsugu, R., and Ohtsuki, K. (2020). A global system of furrows on Ganymede indicative of their creation in a single impact event. *Icarus* 352, 113941, http://doi.org/10.1016/j.icarus.2020.113941.

Horedt, G. P., and Neukum, G. (1984a). Cratering rate over the surface of a synchronous satellite. *Icarus* 60, 710–17.

Horedt, G. P., and Neukum, G. (1984b). Planetocentric versus heliocentric impacts in the Jovian and Saturnian satellite system. *Journal of Geophysical Research* 89, 10405–10.

Howell, S. M., and Pappalardo, R. T. (2018). Band formation and ocean-surface interaction on Europa and Ganymede. *Geophysical Research Letters*, 45, 4701–9. https://doi.org/10.1029/2018GL077594.

Hussmann, H., Choblet, G., Lainey, V. et al. (2010). Implications of rotation, orbital states, energy sources, and heat transport for internal processes in icy satellites. *Space Science Reviews* 153, 317–48.

Ivanov, B. A., Neukum, G., Bottke, W. F. Jr., and Hartmann, W. K. (2002). The comparison of size-frequency distributions of impact craters and asteroids and the planetary cratering rate. In: W. Bottke, A. Cellino, P. Paolicchi, and R. P. Binzel (Eds.), *Asteroids III*. Tucson: University of Arizona Press, pp. 89–101.

Johnson, R. E. (1997). Polar 'caps' on Ganymede and Io revisited, *Icarus*, 128(2), 469–71.

Johnson, R. E. (1990). Plasma erosion and modification of natural surfaces and atmospheres. In R. E. Johnson, *Energetic Charged-Particle Interactions with Atmospheres and Surfaces*. Berlin and Heidelberg: Springer, pp. 137–203, https://doi.org/10.1007/978-3-642-48375-2_4.

Jones, K. B., Head, J. W., Pappalardo, R. T., and Moore, J. M. (2003). Morphology and origin of palimpsests on Ganymede based on Galileo observations. *Icarus* 164(1), 197–212.

Kirchoff, M. R., Bierhaus, E. B., Dones, L. et al. (2018). Cratering histories in the Saturnian system. In: P. M. Schenk, R. N. Clark, C. J. A. Howett, A. J. Verbiscer, and J. H. Hunter (Eds.), Enceladus and the icy moons of Saturn. Tucson: University of Arizona Press, pp. 267–84.

Kirk, R. L., and Stevenson, D. J. (1987). Thermal evolution of a differentiated Ganymede and implications for surface features. *Icarus* 69, 91–134.

Kivelson, M. G., Khurana, K. K., Russell, C. T. et al. (1996). Discovery of Ganymede's magnetic field by the Galileo spacecraft. *Nature* 384(6609), 537–41.

Krivov, A. V., Sremcevic, M., Spahn, F., Dikarev, V. V., and Kholshevnikov, K. V. (2003). Impact-generated dust clouds around planetary satellites: Spherically symmetric case. *Planetary and Space Science* 51, 251–69.

Krüger, H., Krivov, A., and Grün, E. (2000). Dust cloud of Ganymede maintained by hypervelocity impacts of interplanetary micrometeoroids. *Planetary and Space Science* 48, 1457–71, http://doi.org/10.1016/S0032-0633(00)00092-1.

Krüger, H., Krivov, A. V., Hamilton, D. P., and Gru¨n, E. (1999). Detection of an impact-generated dust cloud around Ganymede. *Nature* 399, 558–60, http://doi.org/10.1038/21136.

Le Feuvre, M., and Wieczorek, M. A. (2011). Nonuniform cratering of the Moon and a revised crater chronology of the inner solar system. *Icarus* 214, 1–20.

Ligier, N., Paranicas, C., Carter, J. et al. (2019). Surface composition and properties of Ganymede: Updates from ground-based observations with the near-infrared imaging spectrometer. *Icarus*, 333, 496–515.

Lucchitta, B. K., Barnes, C. W., and Glotfelty, M. F. (1992). *Geologic map of the Memphis Facula quadrangle (Jg-7) of Ganymede*. Map No. 2289. Reston, VA: US Geological Survey.

Marchi, S., Mottola, S., Cremonese, G., Massironi, M., and Martellato, E. (2009). A new chronology for the Moon and Mercury. *The Astronomical Journal* 137, 4936–48.

McKinnon, W. B., and Melosh, H. J. (1980). Evolution of planetary lithospheres: Evidence from multiringed structures on Ganymede and Callisto. *Icarus* 44, 454–71.

Melosh, H. J., and Ivanov, B. A. (1999). Impact crater collapse. *Annual Review of Earth and Planetary Sciences* 27, 385–415.

Michael, G. G. (2013). Planetary surface dating from crater size-frequency measurements: Multiple resurfacing episodes and differential isochron fitting. *Icarus* 226, 885–90.

Michael, G., Basilevsky, A., and Neukum, G. (2018). On the history of the early meteoritic bombardment of the Moon: Was there a terminal lunar cataclysm? *Icarus* 302, 80–103.

Moore, J. M., Asphaug, E., Morrison, D. et al. (1999). Mass movement and landform degradation on the icy Galilean satellites: Results of the Galileo Nominal Mission, *Icarus* 140, 294–312.

Moore, J. M., Chapman, C. R., Bierhaus, E. B. et al. (2004). Callisto. In F. Bagenal, T. Dowling, and W. B. McKinnon (Eds.), *Jupiter. The planet, satellites, and magnetosphere*. Cambridge: Cambridge University Press, pp. 397–426.

Moore, W. B., and Schubert, G. (2003). The tidal response of Ganymede and Callisto with and without liquid water oceans, *Icarus* 166, 223–26.

Morbidelli, A., Marchi, S., Bottke, W. F., and Kring, D. A. (2012). A sawtooth-like timeline for the first billion years of lunar bombardment. *Earth and Planetary Science Letters* 355–6, 144–51.

Morbidelli, A., Nesvorny, D., Laurenz, V. et al. (2018). The timeline of the lunar bombardment: Revisited. *Icarus* 305, 262–76.

Mueller, S., and McKinnon, W. B. (1988). Three-layered models of Ganymede and Callisto: Compositions, structures, and aspects of evolution. *Icarus* 76, 437–64.

Murchie, S. L., and Head, J. W. (1989). *Geologic map of the Philus Sulcus (Jg-4) quadrangle of Ganymede*. Map No. 1966. Reston, VA: US Geological Survey.

Murchie, S. L., Head, J. W., Helfenstein, P., and Plescia, J. B. (1986). Terrain types and local scalestratigraphy of grooved terrain on Ganymede. In: Lunar Planet. Sci. Conf.Proceedings, pp. 17.

Murchie, S. L., Head, J. W., and Plescia, J. B. (1990). Tectonic and volcanic evolution of dark terrain and its implications for the internal structure of Ganymede. *Journal of Geophysical Research* 95(B7), 10743–68.

Neukum, G. (1985). Cratering records of the satellites of Jupiter and Saturn. *Advances in Space Research* 5(8), 107–16.

Neukum, G. (1997). Bombardment history of the Jovian system. In: C. Barbieri, J. H. Rahe, and T. V. Johnson (Eds.), *The three Galileos: The man, the spacecraft, the telescope*. Dordrecht: Kluwer, pp. 201–12.

Neukum, G., and Dietzel, H. (1971). On the development of the crater population on the Moon with time under meteoroid and solar wind bombardment. *Earth and Planetary Science Letters* 12, 59–66.

Neukum, G., and Ivanov, B. A. (1994). Crater size distributions and impact probabilities on Earth from lunar, terrestrial-type planets, and asteroid cratering data. In: T. Gehrels (Ed.), *Hazards due to comets and asteroids*. Tucson: University of Arizona Press, pp. 359–416.

Neukum, G., Ivanov, B. A., and Hartmann, W. K. (2001). Cratering records in the inner solar system in relation to the lunar reference system. *Space Science Reviews* 96, 55–86.

Neukum, G., and Pozio, S. (1983). The cratering record of Ganymede. *Lunar and Planetary Science Conference Abstracts* XV, Abstract No. 601–602.

Neukum, G., Wagner, R., Wolf, U. et al. (1998). Cratering record in the Jovian system and the derivation of absolute ages. *Lunar and Planetary Science Conference Abstracts* XXIX, Abstract No. 1742.

Nimmo, F., Pappalardo, R. T., and Giese B., (2002). Effective elastic thickness and heat flux estimates on Ganymede. *Geophysical Research Letters* 29(7), https://doi.org/10.1029/2001GL013976.

Oberst, J., Schreiner, B., Giese, B. et al. (1999). The distribution of bright and dark material on Ganymede in relationship to surface elevation and slopes. *Icarus* 140, 283–93.

Öpik, E. J. (1960). The lunar surface as an impact counter. *Monthly Notices of the Royal Astronomical Society* 120, 404–11.

Pappalardo, R. T., and Collins, G. C. (2005). Strained craters on Ganymede. *Journal of Structural Geology* 27, 827–38.

Pappalardo, R. T., Collins, G. C., Head III, J. W. et al. (2004). Geology of Ganymede. In F. Bagenal, T. Dowling, and W. B. McKinnon (Eds.), *Jupiter: The planet, satellites, and magnetosphere*. Cambridge: Cambridge University Press, pp. 363–96.

Pappalardo, R. T., Head, J. W., Collins, G. C. et al. (1998). Grooved terrain on Ganymede: First results from Galileo high-resolution imaging, *Icarus* 135, 276–302.

Passey, Q. R., and Shoemaker, E. M. (1982). Craters and basins on Ganymede and Callisto: Morphological indicators of crustal evolution. In D. Morrison (Ed.), *Satellites of Jupiter*. Tuscon: University of Arizona Press, pp. 379–434.

Patel, J. G., Pappalardo, R. T., Head, J. W. et al. (1999). Topographic wavelengths of Ganymede groove lanes from Fourier analysis of Galileo images. *Journal of Geophysical Research* 104(E10), 24057–74.

Patterson, G. W., Collins, G. C., Head, J. W. et al. (2010). Global geologic mapping of Ganymede. *Icarus* 207, 845–67.

Petrenko V. F., and Whitworth R. W. (1999). *Physics of ice*. Oxford: Oxford University Press.

Pizzi, A., Di Domenica, A., Komatsu, G. et al. (2017). Spreading vs. rifting as modes of extensional tectonics on the globally expanded Ganymede. *Icarus* 288, 148–59. https://doi.org/10.1016/j.icarus.2017.01.034.

Prockter, L. M., Head, J. W., Pappalardo, R. T. et al. (1998). Dark terrain on Ganymede: Geological mapping and interpretation of Galileo Regio at high resolution. *Icarus* 135, 317–44.

Prockter, L. M., Lopes, R. M. C., Giese, B. et al. (2010). Characteristics of icy surfaces. *Space Science Reviews* 153, 63–111.

Purves, N. G., and. Pilcher, C. B. (1980). Thermal migration of water on the Galilean satellites. *Icarus* 43(1), 51–55.

Richardson, J. E. (2009). Cratering saturation and equilibrium: A new model looks at an old problem. *Icarus* 204, 697–715.

Schenk, P. M. (1995). The geology of Callisto. *Journal of Geophysical Research* 100, 19,023–40.

Schenk, P. M. (2010). The timing of viscous relaxation on Ganymede. *Lunar and Planetary Science Conference* 41, 2083.

Schenk, P. M., Chapman, C. R., Zahnle, K., and Moore, J. M. (2004). Ages and interiors: The cratering record of the Galilean satellites. In F. Bagenal, T. Dowling, and W. B. McKinnon (Eds.), *Jupiter: The planet, satellites, and magnetosphere*. Cambridge: Cambridge University Press, pp. 427–56.

Schenk, P. M., McKinnon, W. B., Gwynn, D., and Moore, J, M. (2001). Flooding of Ganymede's bright terrains by low-viscosity water-ice lavas. *Nature* 410, 57–60.

Schenk, P. M., and Moore, J. M. (1995). Volcanic constructs on Ganymede and Enceladus: Topographic evidence from stereo images and photoclinometry. *Journal of Geophysical Research* 100(E9), 19,009–22.

Schubert, G., Anderson, J. D., Spohn, T., and McKinnon, W. B. (2004). Interior composition, structure, and dynamics of the Galilean satellites. In F. Bagenal, T. Dowling, and W. B. McKinnon (Eds.), *Jupiter: The planet, satellites, and magnetosphere*. Cambridge: Cambridge University Press, pp. 281–306.

Shoemaker, E. M. (1995). The collision of Comet Shoemaker-Levy 9 with Jupiter. *Geophysical Research Letters* 22, 1555–636.

Shoemaker, E. M., Lucchitta, B. K., Wilhelms, D. E., Plescia, J. B., and Squyres, S. W. (1982). The geology of Ganymede. In D. Morrison (Ed.), *Satellites of Jupiter*. Tuscon: University of Arizona Press, pp. 435–520.

Shoemaker, E. M., and Wolf, R. F. (1982) Cratering time scale for the Galilean satellites. In D. Morrison (Ed.), *Satellites of Jupiter*. Tuscon: University of Arizona Press, pp. 277–339.

Showman, A. P. Malhotra, R. (1997). Tidal evolution into the Laplace resonance and the resurfacing of Ganymede. *Icarus* 127, 93–111.

Showman, A. P., Stevenson, D. J., Malhotra, R. (1997) Coupled orbital and thermal evolution of Ganymede. *Icarus* 129, 367–83.

Singer, K. N., Bland, M. T., Schenk, P. M., and McKinnon, W. B. (2018) Relaxed impact craters on Ganymede: Regional variation and high heat flows. *Icarus* 306, 214–24, https://doi.org/10.1016/j.icarus.2018.01.012.

Smith, B. A., Soderblom, L. A. Beebe, R. et al. (1979). The Galilean satellites and Jupiter: Voyager 2 imaging science results. *Science* 206, 927–50.

Sohl, F., Choukroun, M., Kargel, J. et al. (2010) Subsurface water oceans on icy satellites: Chemical composition and exchange processes. *Space Science Reviews* 153, 485–510.

Spencer, J. R. (1987) Thermal segregation of water ice on the Galilean satellites, *Icarus*, 69(2), 297–313.

Stephan, K. (2006) Chemisch-physikalische Zusammensetzung der Ganymedoberfläche. PhD Dissertation (in German), Freie Universität Berlin, Germany, http://refubium.fu-berlin.de/handle/fub188/1446?show=full.

Stephan, K., Jaumann, R., Wagner, R. (2013). Geology of Icy Bodies. In: Gudipati, M., Castillo-Rogez, J. (eds.) *The Science of Solar System Ices*. Astrophysics and Space Science Library, vol. 356. Springer, New York, NY. https://doi.org/10.1007/978-1-4614-3076-6_10.

Stephan, K., Jaumann, R., Wagner, R., Hibbits, C. A., and Hansen, G. B. (2003) Ganymede craters: Relationships between spectral properties and crater retention age. *Lunar and Planetary Science Conference Abstracts* XXXIV, Abstract No. 1687 [CD-Rom].

Stephan, K., Roatsch, T., Tosi, F. et al. (2021). Regions of interest on Ganymede's and Callisto's surfaces as potential targets for ESA's JUICE mission. *Planetary and Space Science* 105324, http://doi.org/10.1016/j.pss.2021.105324.

Strom, R. G., Malhotra, R., Ito, T., Yoshida, F., and Kring, D. A. (2005) The origin of planetary impactors in the inner solar system. *Science* 309, 1847–50.

Strom, R. G., Woronow, A., and Gurnis, M. (1981). Crater populations on Ganymede and Callisto. *Journal of Geophysical Research* 86(A10), 8659–74.

Squyres, S. W. (1980). Surface temperatures and retention of H_2O frost on Ganymede and Callisto, *Icarus*, 44, 502–10.

Squyres, S. W. (1981). The topography of Ganymede's grooved terrain. *Icarus* 46, 156–68.

Szalay, S. R., Poppe, R. A., Agarwal, J. et al. (2018). Dust phenomena relating to airless bodies. *Space Science Reviews* 214(98), http://doi.org/10.1007/s11214-018-0527-0.

Tera, F., Papanastassiou, D. A., and Wasserburg, G. J. (1974). Isotopic evidence for a terminal lunar cataclysm. *Earth and Planetary Science Letters* 22, 1–21.

Wagner, R. J., Jaumann, R., and Neukum, G. P. (1995). Geologic map of the Etana Region (Jg-1) of Ganymed. Map No. 2497. Reston, VA: US Geological Survey.

Wagner, R. J., Schmedemann, N., Stephan, K. et al. (2017). Ganymede's stratigraphy and crater distributions in Voyager and Galileo images: Results from the antijovian hemisphere. *Bulletin of the American Astronomical Society*, Abstract No. 214.06.

Wagner, R. J., Schmedemann, N., Werner, S. C. et al. (2019). Ray craters on Ganymede as time-stratigraphic markers. *Lunar and Planetary Science Conference Abstracts* XL, Abstract No. 1849.

Wagner, R. J., Stephan, K., Schmedemann, N. et al. (2018). The large bright ray crater Osiris on Ganymde: Its age, role as a potential time-stratigraphic marker, and target for detailed imaging by the JUICE/JANUS Camera. *Europlanet Science Congress (EPSC)*, Abstract No. EPSC2018-855.

Wagner, R., Wolf, U., Neukum, G., and the Galileo SSI Team (1999) Ages of individual craters on the Galilean satellites Ganymede and Callisto. *Lunar and Planetary Science Conference Abstracts* XXX, Abstract 1818.

Werner, S. C., and Ivanov, B. A. (2015). Exogenic dynamics, cratering, and surface ages. In: T. Spohn (Ed.), *Treatise on Geophysics, Vol. 10: Planets and Moons*. Amsterdam: Elsevier, pp. 327–365.

Zahnle, K., Dones, L., and Levison, H. F. (1998). Cratering rates on the Galilean satellites, *Icarus* 136, 202–22.

Zahnle, K., Schenk, P., Levison, H., and Dones, L. (2003). Cratering rates in the outer solar system, *Icarus* 163, 263–89.

Zahnle, K., Schenk, P., Sobieszczyk, S., Dones, L., and Levison, H. F. (2001). Differential cratering of synchronously rotating satellites by ecliptic comets. *Icarus* 153, 111–29.

Zunarev, A., Nadezhdina, I., Brusnikin, E., Giese, B., and Oberst, J. (2017). A search for Ganymede stereo images and 3D mapping opportunities. *Planetary and Space Science* 146, 40–54. http://dx.doi.org/10.1016/j.pss.2017.07.021.

Part II

GEOLOGY, SURFACE, INTERIOR

6

Internal Structure of Ganymede

HAUKE HUSSMANN, GERALD SCHUBERT, GREGOR STEINBRÜGGE, FRANK SOHL, AND JUN KIMURA

6.1 INTRODUCTION

Jupiter's moon Ganymede is the largest satellite in the solar system. With a mean radius of 2 631.2 +/-1.7km, it exceeds Mercury in size, albeit not in mass due to its lighter constituents as compared to the iron-rich innermost planet. As a typical icy satellite of the outer solar system, Ganymede's bulk structure consists of about 60% rock and metal and 40% H_2O ice by mass. Even in its outer part the internal pressure reaches a certain threshold – the triple-point of water ice – at which the low-pressure ice phase, ice-I, is not stable anymore. At pressures beyond the triple-point, and depending on temperature, the H_2O phase would be either in a liquid state or in a dense high-pressure solid state. The first case suggests the possibility of a global internal subsurface ocean on Ganymede. Both theoretical modeling (Spohn and Schubert, 2003) and data from the Galileo mission (Kivelson et al., 2002) provide evidence that Ganymede may contain a huge reservoir of liquid water in its interior.

Besides the possibility of providing a stable liquid water environment potentially conducive for life to evolve, Ganymede is unique in several ways regarding its interior: (1) It is the only satellite in the solar system for which a self-sustained magnetic field that is believed to be generated inside a liquid iron core was confirmed by magnetometer data (Kivelson et al., 1996; Schubert et al., 1996). Actually, apart from the giant planets in which magnetic fields are generated in a different way, Ganymede is one among only three planetary bodies generating a present-day magnetic dipole field, with Mercury and the Earth being the other two. Hence, Ganymede creates its own magnetosphere, which is embedded in Jupiter's strong magnetic field. (2) Among all planetary bodies for which gravitational field data were acquired, Ganymede has the smallest moment of inertia factor, implying that Ganymede's interior is strongly condensed and highly differentiated (Anderson et al., 1996). Thus, the metal, rock, and H_2O phases are well separated into an iron-rich core, a silicate mantle, and an outer H_2O layer consisting of high-pressure ices, an internal ocean, and an outermost ice-I layer. A typical interior structure model consistent with the moment of inertia, total mass, and radius of the satellite is shown in Fig. 6.1. Ganymede's strong differentiation is even more remarkable when compared to its outer sibling Callisto, which is similar with respect to total mass, size, and bulk composition. However, with its moment of inertia factor being considerably larger, Callisto is not fully differentiated and followed a completely different evolutionary path as compared to Ganymede (Anderson, Jacobson, McElrath et al., 2001). (3) The latter fact is also evident in the satellites' surface records. Whereas Callisto remained almost inactive over its entire history, Ganymede displays surface terrain that was intensely modified by tectonic activity and/or cryovolcanism (Pappalardo et al., 2004). Crater morphologies, in particular the relaxation state of crater floors, suggest the presence of weak material, that is, ice close to its melting point over certain periods in the satellite's past (Bland et al., 2017). The amount of energy involved in such processes and the resulting heating of the interior is quite remarkable. Whether states of high orbital eccentricity during formation or in the Laplace resonance are a key factor for Ganymede's active phases (Showman et al., 1997a) remains uncertain.

In this chapter, we summarize the knowledge of the internal structure of Ganymede based mainly on the gravitational field collected by the Galileo spacecraft. We briefly discuss the context and methods for inferring the internal structure of satellites in Section 6.2. The acquired relevant data, both from the ground and space missions, are summarized in Section 6.3. Interior structure models that are consistent with the data and with our current knowledge of planetary formation and evolution are discussed in Section 6.4. Implications for the evolution of Ganymede and considerations from comparative planetology are given in Section 6.5 ending with open questions for future missions.

6.2 METHODOLOGY

There are several ways and different aspects of how knowledge of the internal structure of a planetary body can be gained. The most direct approach is the use of seismometers emplaced on the surface of a body. By recording seismic waves and using inversion techniques to determine the propagation of waves inside the body, layers with different density and/or rheology can be detected. However, other than the Earth, seismometers have only been placed on the Moon, with the Apollo missions, and recently on Mars, with NASA's InSight mission. For Ganymede and other planetary bodies, seismic data are still lacking. Thus, the most important technique is the determination of the gravity field of the body by Doppler tracking and ranging by a spacecraft at close flybys or in orbit. For Ganymede, this technique was a key element of the Galileo mission. Measurements of the body's shape can be used in combination with gravity data to assess whether the body is in hydrostatic equilibrium. Other indirect evidence on Ganymede's internal structure comes from the detection of a self-sustained magnetic field that is generated inside the satellite's metallic core. Furthermore, an

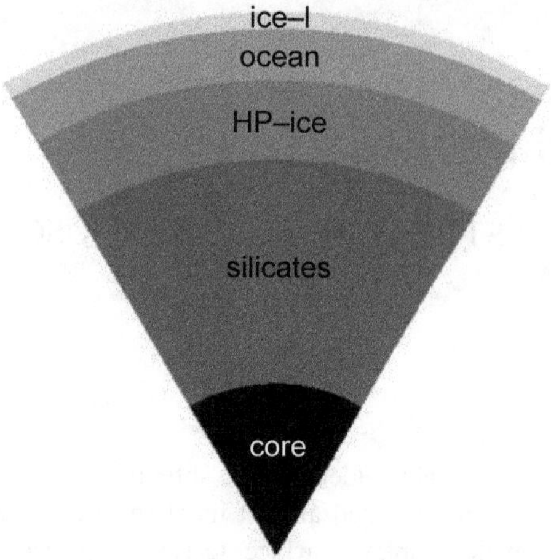

Figure 6.1 A typical five-layer model of Ganymede's interior consisting of an iron core, a silicate mantle, a high-pressure ice layer, an internal water ocean, and an outermost ice-I layer.

induced magnetic field in response to the external Jovian field was detected and interpreted as being generated inside a global water ocean below the outer icy crust. Cosmochemical arguments and compositional data are important to assess in which environment Ganymede accreted and which main constituents could be expected inside the moon. Also, the rotational state of a planetary body, the orientation of the axis with respect to the orbital plane, its precession period, and possible physical librations are indicative of the internal density distribution. In the following, we describe the main aspects and techniques leading to our current knowledge of the internal structure of Ganymede.

6.2.1 Formation and Composition

The formation of Ganymede within the Jovian sub-nebula is described by Shibaike and Alibert (Chapter 2, this volume). From the thermal gradient inside that nebula it can be assessed that volatiles, in particular H_2O ice and to a lesser extent other volatile components, played a major role in Ganymede's accretion process besides the silicate and metal components (e.g., Schubert et al., 1986; Kirk and Stevenson, 1987; Mueller and McKinnon, 1988). The satellite's mean density of 1 940 kg/m^3 shows that a large fraction of Ganymede consists of volatiles with densities much lower than typical rock densities of 3 500 kg/m^3. From spectroscopic measurements, it is evident that H_2O ice is by far the main constituent of Ganymede's surface. Assuming water ice and typical silicate rock and metal densities, we can estimate that about 35 wt.-% of Ganymede's bulk structure is H_2O and 65 wt.-% is silicate rock and metal, mainly iron. Whereas we can estimate these mass fractions from cosmochemical arguments and the mean density, we cannot say whether these components have separated into a layered structure based on these data alone.

6.2.2 Gravitational Fields

Important constraints on the interior of planetary bodies can be obtained from the determination of their gravitational fields. Tracking the change of velocity of a spacecraft in orbit or during flybys allows us to infer local density variations. Analogous to the global topography, the gravitational potential is often expressed in spherical harmonics with zero degrees solely depending on the GM, gravitational constant times mass, of the body, while higher degrees express subsequently finer local variations. The gravitational potential U can be expressed in spherical coordinates as a function of radial distance r from the planetary center, longitude λ, and latitude ϕ by

$$U(r,\lambda,\phi) = \frac{GM}{r}\sum_{n=0}^{\infty}\sum_{m=0}^{n}\left(\frac{R}{r}\right)^n P_{nm}(\sin\phi) \times [C_{nm}\cos(m\lambda) + S_{nm}\sin(m\lambda)], \tag{6.1}$$

with R being the radius of the body. P_{nm} are the associated Legendre polynomials, and the higher moments C_{nm} and S_{nm} contain the information on the shape of the gravitational field, respectively the body's mass distribution. If the coordinate system is properly chosen, that is, the origin coincides with the body's center of mass, the z-axis is aligned with the rotational axis, and the x, y, and z axes are aligned with the principal moments of inertia $C_{10} = C_{11} = S_{11} = 0$ as well as $C_{21} = S_{21} = 0$. The zonal coefficients ($m = 0$) are commonly written by $C_{n0} = -J_n$. Due to the limited number of flybys from the Galileo mission, in addition to GM, only the quadrupole terms ($n = 2$) could be determined for Ganymede. For degree-2, Eq. 6.1 can be significantly simplified. Using the above relations and the associated Legendre polynomials $P_{20}(x) = 3/2\sin^2 x - 1/2$ and $P_{22}(x) = 3\cos^2 x$, we obtain for degree-2:

$$U(r,\lambda,\phi) = \frac{GM}{r}\left(1 - \left(\frac{R}{r}\right)^2 J_2\left(\frac{3}{2}\sin^2\phi - \frac{1}{2}\right) + \left(\frac{R}{r}\right)^2 3C_{22}\cos^2\phi\cos(2\lambda)\right). \tag{6.2}$$

In hydrostatic equilibrium, the Radau–Darwin relation can be applied, which relates the degree-2 gravity coefficients to the dimensionless moment of inertia factor $MoI = C/MR^2$, where C is the principal moment about the rotational axis. For a homogeneous sphere, a value of $2/5 = 0.4$ is expected, with smaller values indicating higher degrees of differentiation. With $MoI = 0.3105 \pm 0.0028$ (Anderson et al., 1996), Ganymede's moment of inertia factor is significantly smaller than 0.4, thus indicating the satellite's high degree of differentiation.

For a layered model planet with constant density shells, two equations – one for the total mass (or equivalently mean density) and one for the moment of inertia – have to be satisfied simultaneously. Using reasonable assumptions for the densities of the layers and different layer thicknesses, two quantities (e.g., silicate shell density and thickness) can be determined. For a number of layers exceeding two, the solution is not unique and possible model solutions span a certain parameter space (e.g., Anderson et al., 1996; Sohl et al., 2002), providing a set of consistent interior structure models. The results of the Galileo gravity flybys are shown in Section 6.3 and their implications in Section 6.4.2.

6.2.3 Shape

Like all major satellites in the solar system, Ganymede is locked in a 1:1 spin-orbit coupling with respect to its central planet Jupiter. On long timescales, such a synchronously rotating satellite will assume an equilibrium shape with – apart from libration effects – its long axis a pointing towards Jupiter, the polar axis c being the shortest axis due to rotational oblateness, and with the in-orbit axis b ranging in-between. From equilibrium theory (Hubbard and Anderson, 1978; Dermott, 1979; Zharkov et al., 1985), the expected shape can be predicted for both homogeneous and differentiated satellites. However, whether the satellite has actually assumed equilibrium depends on its thermal state, which has a strong influence on material parameters, for example rigidity and viscosity. These parameters are frequency-dependent and would tend to evolve towards equilibrium on long timescales, that is, on infinite excitation period. The latter is described by the fluid Love numbers k_f and h_f, where the term "fluid" refers to the completely relaxed state of the satellite. k_f describes the gravitational potential and h_f the satellite's static shape. Both numbers depend on the density distribution inside the satellite but due to the infinite timescale not on rheologic parameters. From several Galileo flybys, the potential fluid Love number of second degree and order k_f was determined for Ganymede as $k_f = 0.804 \pm 0.018$ (Schubert et al., 2004) corresponding to $h_f = 1 + k_f = 1.804 \pm 0.018$. These values are consistent with a highly differentiated satellite and the corresponding axes can be computed (e.g., Zharkov et al., 1985) from

$$a = R\left(1 - \tfrac{7}{18} q_t h_f\right), \tag{6.3}$$

$$b = R\left(1 - \tfrac{1}{9} q_t h_f\right), \tag{6.4}$$

$$c = R\left(1 - \tfrac{5}{18} q_t h_f\right), \tag{6.5}$$

where $q_t = 3R^3\omega^2/(GM)$, with R, ω, and GM being Ganymede's radius, rotation frequency, and mass times the gravitational constant, respectively. For Ganymede, the obtained axes are $a = 2\,633.40$ km, $b = 2\,632.04$ km, and $c = 2\,631.59$ km. The differences with its maximum of $a - c = 1.8$ km are in the order of a few kilometers, which could be detected from imaging and laser altimetry data in orbit. Zubarev et al. (2015) have determined Ganymede's shape from a control-point network using Galileo imaging data and find that the derived triaxial shape is consistent with equilibrium theory and earlier determinations by Davies et al. (1998). However, more accurate data would be needed to identify additional topographic variations and an exact orientation of the ellipsoid and its time variability due to librations.

6.2.4 Tides and Ocean

The induced magnetic field of Ganymede indicates a subsurface ocean present underneath the outer ice-I shell. Since a liquid layer in the interior decouples the outer crust from the rest of the interior, an ocean would significantly enhance the tidal response of Ganymede on its diurnal cycle of 7.15 days (period of revolution about Jupiter). However, due to the likely presence of high-pressure ice layers underneath the ocean (see Kalousová, Soderlund, and Solomonidou, Chapter 12, this volume), the situation for Ganymede can be more complex than for other icy satellites like Europa or Enceladus where the ocean is embedded between ice and a silicate core. In the absence of an ocean, the tidal response will be smaller for almost all conceivable cases (Moore and Schubert, 2003). However, theoretically the possibility exists of a warm high-pressure ice layer with an almost fluidlike viscosity, which could emulate the role of a liquid layer by sufficiently decoupling the outer crust to allow for an enhanced tidal response (Kamata et al., 2016). Thus, care has to be taken in interpreting future measurements of Ganymede's tidal distortion.

A further complexity is Ganymede's orbital eccentricity, which is varying from values close to zero up to eccentricities of approximately 0.003 on a timescale of approximately 50 years with a mean value of 0.0015 (see Steinbrügge, Hussmann, Tobie et al., Chapter 4, this volume). Thus, the actual values at times of the tidal measurements have to be taken into account, which are also oscillating on a shorter timescale of about 1.3 years. In addition to the amplitude, the tidal response of Ganymede could contain a phase-lag dependent on the amount of dissipation in the ice and/or silicate layers. In the presence of a subsurface ocean, Ganymede's radial deformation due to diurnal tides can exceed 7 m and would thus be accessible to altimetry measurements from orbit. Without an ocean, the tidal deformation would be a few tens of cm only (Moore and Schubert, 2003). For details and implications on tides, we refer the reader to (Chapter 4, this volume) and note that measuring the tidal variations of the Love numbers k_2 and h_2 not only would confirm or disprove a subsurface ocean on Ganymede independently from induced magnetic field measurements but could also constrain the outermost ice thickness and mechanical parameters of Ganymede's interior. As described by Wahr et al. (2006) in application to Europa, best constraints on Ganymede's ice thickness can be gained from analyzing the linear combination $1 + k_2 - h_2$, the so-called diminishing factor, instead of each Love number separately. Combining gravity field and altimetry measurements would thus be the best way to obtain such constraints complementary to the assessment of the induced magnetic field.

6.2.5 Magnetic Field

After two close flybys at closest approach altitudes of 838 and 264 km, respectively, the Galileo spacecraft discovered a magnetic field associated with Ganymede (Kivelson et al., 1996). The data were consistent with a dipole field tilted about 10° with respect to Ganymede's rotational axis and an equatorial field strength of approximately 750 nT. The dipole field is consistent with a self-sustained dynamo in Ganymede's core. Thus the detection of such a field has important implications for interior structure models. Interpretations suggest that Ganymede is separated into a silicate core surrounding an inner iron or iron-sulfur core that could be partially (or entirely) liquid (Schubert et al., 1996). Convective currents inside the liquid part of the core would generate the magnetic dipole field. The magnetic field and its generation are discussed in detail by Christensen, Rückriemen-Bez, and Schubert (Chapter 13, this volume).

In addition to the self-sustained dipole field generated in Ganymede's deep interior, an induced component of the field in response to the time-variable strength and direction of the external Jovian magnetosphere was detected or is at least consistent with the magnetic field data (Kivelson et al., 2004; Kivelson et al., 2002). Such an induced field would be generated in a global electrically conducting layer relatively close to the surface and can be explained by a globally encircling subsurface H_2O ocean containing salts. The ocean would be located between an ice-I layer of about 100 km thickness and high-pressure ice layers at the bottom. A clear separation of the induced field component from the dipole field would be one of the goals for future orbiters around Ganymede (see also Kivelson, Bagenal, and Jia, Chapter 14, this volume).

The important implications of the magnetic field detection for interior structure models are (1) that the self-sustained field is consistent with the highly differentiated state with an iron core being separated from the silicates and ices; (2) that possible core densities are roughly in the range of 5 000 to 8 000 kg/m^3, depending on the amount of volatiles (e.g., FeS) inside the core; (3) that the entire core or an outer shell is probably liquid, driving the currents that generate the dipole field; (4) that the temperature and/or compositional gradients are strong enough to drive convection in the core's liquid phase, and (5) that the induced field component is consistent with a salty liquid water layer located between ice-I and high-pressure ices in Ganymede's hydrosphere. The latter in particular has important implications for assessing the potential habitability of Ganymede.

6.3 OBSERVATIONAL DATA

6.3.1 Galileo Data

Most relevant for Ganymede's internal structure are measurements obtained from tracking the Galileo spacecraft at close approaches during Ganymede flybys. For the data analysis in particular, the first two flybys, one almost equatorial pass at 835 km at the closest approach and a near-polar pass at 261 km altitude, had been used (Anderson et al., 1996; Anderson, Jacobson, Lau et al., 2001; Schubert et al., 2004). The flyby configuration was ideal for independently determining J_2 from a polar pass and C_{22} from an equatorial flyby. However, the data analysis led to unphysical derived parameters without invoking the hydrostatic constraint of $J_2 = 10/3C_{22}$. Subsequently, the assumption of hydrostatic equilibrium was used to derive Ganymede's moment of inertia. This of course introduces a complete correlation between these two quadrupole moments and the tidal and rotational component of the gravity field could not be determined independently. Thus our current knowledge on Ganymede's internal structure is based on the strong assumption of hydrostatic equilibrium. Later analysis of Galileo tracking data for the Europa flybys yielded updates of the parameters (Gomez Casajus et al., 2021). For Ganymede, this analysis is still ongoing. On June 7, 2021, the Juno spacecraft had successfully performed a Ganymede flyby at an altitude of 1 050 km as part of the Juno extended mission. The results from Doppler tracking may thus be used to further constrain the internal structure of Ganymede (Gomez Casajus et al., 2022). By adding this new flyby to the data set, an analysis without invoking the assumption of hydrostatic equilibrium might be possible. The confirmation (or disproof) of the hydrostatic ratio between Ganymede's J_2 and C_{22} would be an important scientific result for Juno and future missions. The results that are the basis for interior structure analysis for Ganymede shown here are based on Galileo data only and are summarized in Table 6.1.

Table 6.1 *Results from the Galileo mission.*

GM	9887.83 ± 0.03 km^{3}s^{-2}
mass, M	1.4818×10^{23} kg
mean radius, R	2631.2 ± 1.7 km
J_2	$(127.53 \pm 2.9) \times 10^{-6}$
C_{22}	$(38.26 \pm 0.87) \times 10^{-6}$
fluid Love number, k_f	0.804 ± 0.018
moment of inertia factor	0.3115 ± 0.0028
mean density, ρ	$1\,942.0 \pm 4.8$ kg/m^3

Source: Schubert et al. (2004).

6.3.2 Post-Galileo Observations

Telescopic observations for the understanding of Ganymede's interior structure have been made after the Galileo mission. Information related to the existence of the subsurface ocean came from the auroral oval by Hubble Space Telescope (HST) observations. With Ganymede being exposed to the time-periodic plasma and magnetic fields of Jupiter, Ganymede's aurora is expected to fluctuate with time. An approximately 10° tilt between Ganymede's orbital plane and Jupiter's magnetic equator due to Jupiter's rotation and dipole axis misalignment is responsible for the periodic variations. Therefore, Ganymede's magnetic latitude oscillates ±10°, in Jupiter's synodic rotation period of 10.5 hours. This time-varying magnetic field generates the oscillations in Ganymede's auroral oval. Saur et al. (2015) analyzed the response of Ganymede's auroral oval to the time-varying electric field in Jupiter's magnetosphere around Ganymede using 135.6 nm auroral images acquired by the Space Telescope Imaging Spectrograph (STIS) of HST and found that the amplitude of the observed oscillations was slight (2.0° ± 1.3°). Since the oscillation amplitude of the auroral oval in absence of an ocean is predicted to be 5.8° ± 1.3°, it is interpreted that the time-varying electric field induces a magnetic field in the electrically conductive ocean, which reduces the expected oscillation of the oval. In other words, it is concluded that the small oscillation amplitude is due to induction in the salty subsurface ocean within Ganymede. Together with the measurements of the induced magnetic field by the Galileo spacecraft, it was found that the electrical conductivity of the ocean is 0.09 S/m or more, which corresponds to a minimum salinity of 0.9 grams of $MgSO_4$ per kilogram of ocean water at depths of 150 to 250 km. However, time-varying components contain stochastic variability, which need to be investigated in further studies. In addition, further HST observations will provide a more detailed picture of the time variability and will be improved by observations from the UV spectrograph onboard the European Space Agency (ESA) JUpiter ICy moons Explorer (JUICE) spacecraft planned to orbit around Ganymede.

6.4 INTERIOR STRUCTURE MODELS

6.4.1 Early Models

In the pre-Voyager era, interior structure models were entirely based on ground-based data providing measurements of the masses and radii as well as surface spectroscopy of the Galilean satellites. On Ganymede (as well as Europa and Callisto), H_2O ice was identified as the major component on the surface. Based on theoretical modeling of satellite evolution in the outer solar system, Ganymede was considered as either homogeneous or partially differentiated. From the Voyager data, which provided the first detailed images of the satellites' surfaces, the striking difference between Ganymede and Callisto with respect to their thermal evolution became apparent. Consequently, interior structure models considered Callisto as undifferentiated in contrast to Ganymede where the rock and ice phases were believed to have separated from each other. Accretional models started with homogenous ice–rock mixtures, which then separated into an outer ice layer (by accretional heating), a shell of undifferentiated primordial ice–rock mixture, and a rock core by radiogenic heating gradually warming up the moon's deep interior (Schubert et al., 1981; McKinnon and Parmentier, 1986; Schubert et al., 1986). Whereas a precise gravity field determination of the Galilean moons was not possible with the too distant Voyager trajectories, this was a major objective of the Galileo mission entering Jupiter's orbit in 1995. Only after obtaining the quadrupole moments of Ganymede's gravity field from close flybys and after the detection of its magnetic dipole field the highly differentiated state became apparent, suggesting full separation of the interior into an iron core, silicate mantle, and outer H_2O shell.

6.4.2 Current Models

As discussed in Sections 6.2 and 6.3, Ganymede's gravitational field provides insight into its interior. In addition, the presence of an induced magnetic field suggests the existence of an electrically conducting layer, that is, the subsurface ocean, and the intrinsic magnetic field indicates a metallic core (see Section 6.2.5). Here, we briefly summarize Galileo and post-Galileo understandings of Ganymede's interior structure.

From the radio Doppler measurements of the gravity field by the Galileo spacecraft (Anderson et al., 1996), the moment of inertia of Ganymede has been determined based on the assumption that Ganymede is in hydrostatic equilibrium. The dimensionless value of the moment of inertial (MoI) factor of 0.3115 ± 0.0028 shows a considerable degree of differentiation in Ganymede's interior. For the intrinsic dipole magnetic field discovered by the magnetometer measurements of the Galileo spacecraft, it is considered that the most likely source is a dynamo in a liquid or partially liquid metallic core (Schubert et al., 1996). This supports the inference that Ganymede has a metallic core independently of the constraints from the MoI. In addition, the Galileo spacecraft found an induced dipole magnetic field, which is connected to the variation in the Jovian magnetic field near Ganymede and whose moment is an order of magnitude weaker than the intrinsic one. The existence of such an induced field indicates that Ganymede has a subsurface salty water ocean. If the ocean has an electrical conductivity comparable to or higher than that of terrestrial sea water, ocean models favor a depth in the order of 150 km and a thickness of at least 10 km (Kivelson et al., 2002), although the interpretation of the magnetometer data is not unambiguous (see Section 6.2.5 and Kalousová, Soderlund, and Solomonidou, Chapter 12, this volume).

Using the bulk density and the MoI, as model constraints Anderson et al. (1996) calculated the areas of solutions for the permissible variations in the core radius, density of the rocky mantle, and thickness of the water layer, with constant densities for each layer. A typical example of a five-layered Ganymede is shown in Fig. 6.2 using the hydrostatic equation

$$dP = -g(r)\rho(r)dr, \tag{6.6}$$

where P, g, ρ, and r are pressure, gravitational acceleration, density, and radial distance from the planetary center, respectively. The gravity measurement and constraints from the MoI cannot distinguish between liquid water and ice because their densities are similar. The density of the metallic core is uncertain because of its unknown composition, two end-member possibilities for the metallic core are considered, an Fe core of density 8 000 kg m^{-3} and a eutectic Fe-FeS core of density 5150 kg m^{-3}. Depending on the assumed sulfur concentration in the core, the density can attain values of 6 597, 6 105, and 5 825 kg m^{-3} for sulfur concentrations of 7, 13, and 19 wt %, respectively (Rückriemen et al., 2015). A pure FeS core would have a density of about 5 330 kg m^{-3} (Sohl et al., 2002). For an assumed rocky mantle density of 3 300 kg m^{-3}, the possible radius of the Fe core is in a range between 0.15 and 0.4 Ganymede radii, assuming an outer H_2O ice-liquid shell density from 1 000 to 1 350 kg m^{-3}. An Fe-FeS core radius would be in a range between 0.2 and 0.5 Ganymede radii for the same assumed H_2O densities. For a typical mean density of 1 100 kg m^{-3} for the H_2O shell, the core radius is 0.22 Ganymede radii for the pure Fe core and 0.30 Ganymede radii for an Fe-FeS core.

Kuskov and Kronrod (2001) constructed interior models assuming a layered structure consisting of an outer water-ice shell, a three-layer silicate mantle, and an Fe-FeS core. For the silicate mantle, the Na_2O-TiO_2-CaO-FeO-MgO-Al_2O_3-SiO_2 system based on ordinary and carbonaceous chondrites was considered and they used a realistic ice equation of state for the outer shell. The structure was investigated using thermodynamic calculations of phase relations and physical properties of minerals and a Monte Carlo inversion procedure for constructing model solutions. The bulk composition, density distribution in the mantle, and the radii of the metallic core under the geophysical constraints were estimated (in terms of satisfying the constraints of the bulk density and the moment of inertia). Their model suggests that the bulk composition of the rock–iron core is similar to L and LL chondrites, while H and C chondrite compositions may be excluded. For such rock–iron core composition, the radius of the metallic core can be estimated to be 580 to 650 km for a pure Fe core and 820 to 900 km, that is, 0.33 to 0.34 of Ganymede's total radius, for a eutectic Fe-FeS core. The thickness of the H_2O column is estimated to be 890 to 920 km (0.34 R). The Fe/Si ratio of the Ganymede rock–iron core, which they evaluated for an LL chondrite composition, is relatively small and less than the solar value. It strongly depends on the iron-silicate fraction in the sub-Jovian nebula and may be due to the depletion of iron during the formation of Ganymede. As the circum-Jovian disk cooled, the

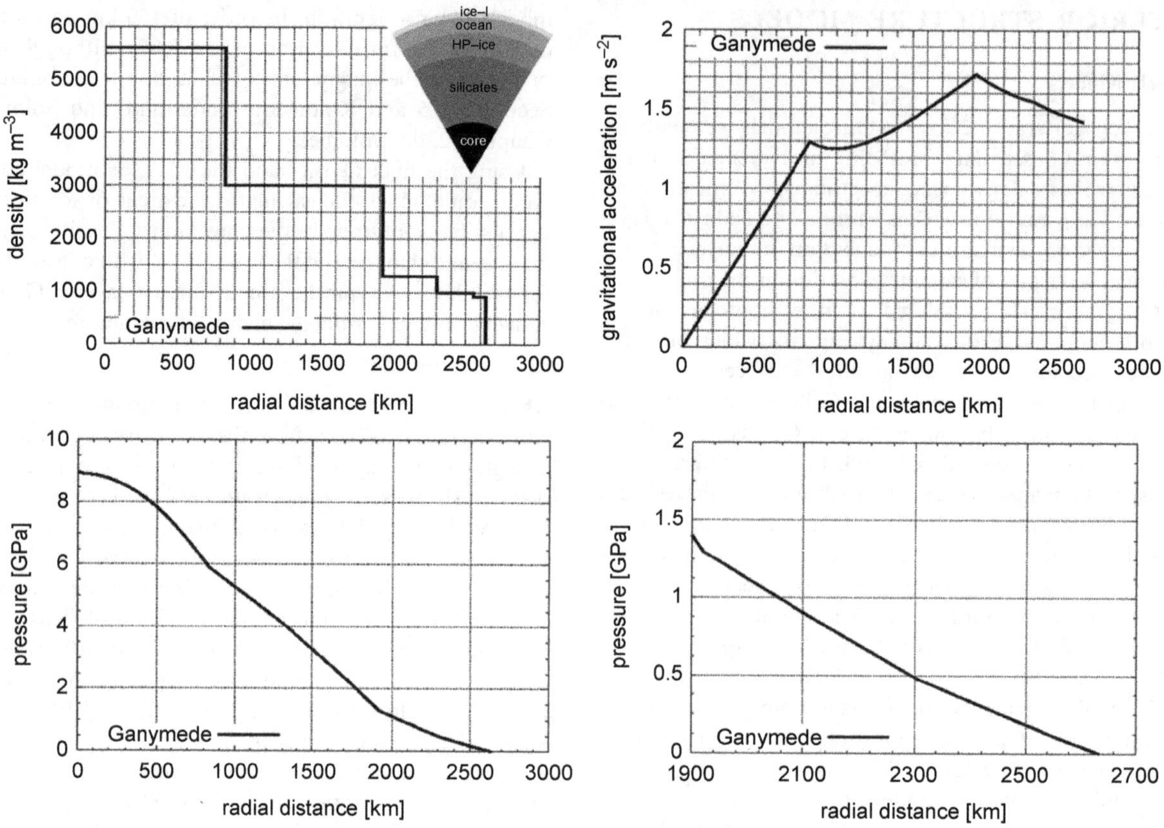

Figure 6.2 Density, gravitational acceleration, and pressure as a function of radial distance from the center for a typical Ganymede five-layer model consisting of an Fe-FeS core, silicate mantle, high-pressure ice layer, internal ocean, and ice-I layer. The model is consistent with Ganymede's moment of inertia and mean density and radius. The lower right figure shows the pressure in the upper part within the ice and ocean layers.

iron particles condensed from the gas first and partly accreted to Jupiter with the gas. However, the details of this process and physical and chemical conditions have not yet been clarified, thus further research is needed.

Sohl et al. (2002) used a layered interior structure consisting of the ice shell, the rocky mantle, and the metallic core assuming chemically homogeneous layers and investigated the admissible range of the structure to satisfy the bulk density and the MoI factor for thermal-mechanical equilibrium. They assume that Ganymede's metallic core is composed of iron and sulfur and the rocky mantle is made of pure olivine, based on an analogy to the Earth's upper mantle. Although the possible core radius depends on the relative proportions of the major light elements and is not well constrained, it is constrained by the moment of inertia to a range of 600 km for pure iron cores and 1 150 km for FeS cores. In addition, their model suggests that Fe/Si ratios of Ganymede have to be significantly superchondritic (Fe/Si = 2 to 5) in order to fit mass-MoI data with simplified mineralogical models. Comparing their estimates for the inner Galilean moons, Io and Europa, of Fe/Si = 0.7 to 1.6, they note that such a difference in the Fe/Si ratios is puzzling. If the concentration of refractory elements decreases and the concentration of volatile elements increases with increasing distance from Jupiter, an inverse gradient in the Fe/Si ratio would be expected. This suggests the formation of chemically different regions in the Jovian sub-nebula and a different condensation mechanism than in the solar nebula. The thickness of the H_2O layer ranges from 1 000 km to 1 100 km above the rocky mantle. Due to the high pressure values at the bottom of the water layer, high-pressure phase ice should exist below a depth of 150 km. Therefore, the water layer is subdivided into ice Ih, ice III, ice V, and ice VI as the pressure increases. If a liquid water layer exists, it should be located between ice I and the high-pressure ice layer, as shown by the magnetic data.

Interior structure models were also coupled to ocean chemistry, which has a significant impact on the melting temperatures of the respective ice phases (Nakamura and Ohtani, 2011). Vance et al. (2014) have included thermodynamics of magnesium sulfate oceans in interior structure models of Ganymede, finding possible ice-I thicknesses ranging between 13 and 144 km. Depending on ocean salinity, some of the high-pressure ice phases become bouyantly unstable and might thus float in a saline ocean. For details on Ganymede's hydrosphere, see Kalousová, Soderlund, and Solomonidou (Chapter 12, this volume).

Sohl et al. (2002) also considered compressibility in their models, showing that even for the largest moon, Ganymede, these effects on density increasing with depth are relatively small (see Fig. 6.3). This shows that errors introduced in models with constant-density assumptions are very small for the Galilean moons.

The metallic core is also not well constrained in size and structure. Owing to the lack of observational constraints, it is unknown whether it is divided into an inner and outer core similar to the Earth's core. Based on the consistency of the magnetometer data with the numerical simulations of dynamo motion in the core, and assuming a composition of the Fe-FeS system, the structure within the core and its evolution would

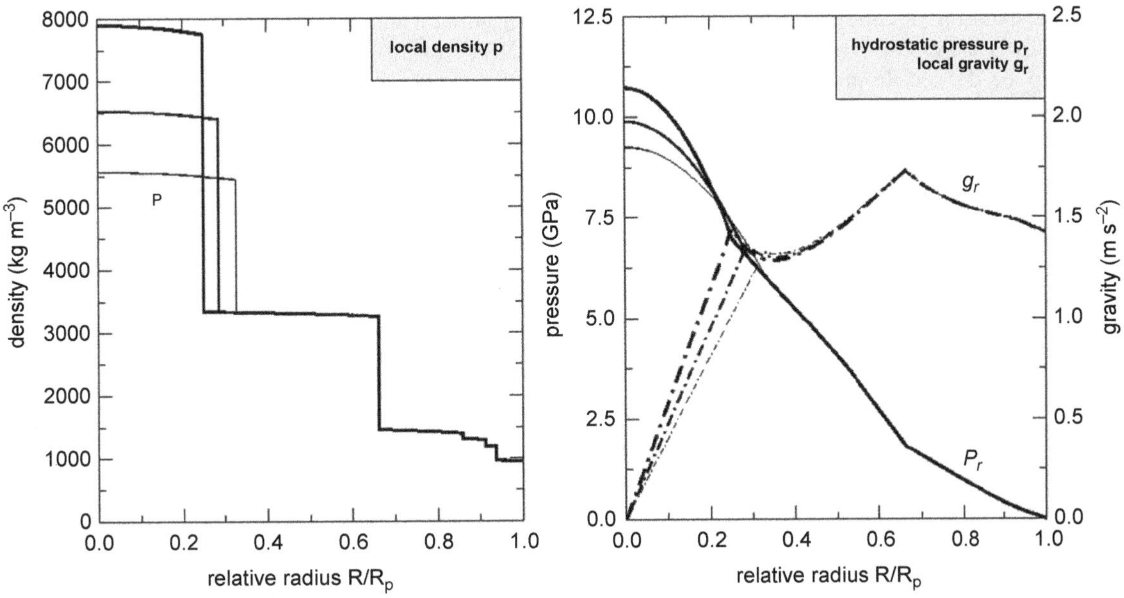

Figure 6.3 Compressible interior structure models showing the moderate self-compaction of Ganymede's density layers. Three models with different core densities (bold curve, 100 wt% Fe; semibold curve, 50 wt% Fe and 50 wt% FeS; light curve, 100 wt% FeS) that are consistent with moment of inertia and mean density are shown.
Source: adapted from Sohl et al. (2002).

strongly depend on the bulk sulfur content (e.g., Fei et al., 1997; Hauck et al., 2006; Bland et al., 2008; Kimura et al., 2009; Zhan and Schubert, 2012; Christensen, 2015; Rückriemen et al., 2015). If the sulfur content is less than the eutectic composition, denser solid Fe forms at the core–mantle boundary at which the iron starts to crystallize. Sinking Fe particles (iron snow) would remelt at depth driving chemical convection. The stable snow zone grows with time. Therefore, when it reaches the center, a solid-Fe inner core and a S-rich liquid outer core would form. It should be noted that if the dynamo is generated by the iron-snow regime the existence of a solid inner core is less likely (e.g. Christensen, 2015; Rückriemen et al., 2015).

On the other hand, if the sulfur content is above the eutectic composition, solid FeS will crystalize within the core. Depending on the scenario assumed, the FeS may crystallize at the center (Hauck et al., 2006) or at the core–mantle boundary (Rückriemen et al., 2015). Over time, a layer forms where solid FeS rises (FeS floating) and the floating zone grows towards the core–mantle boundary until a solid FeS outer layer forms at the core-mantle boundary because solid FeS will be less dense than the residual liquid. Owing to subsequent crystallization, the solid core grows in thickness towards the center. The evolution of the core is essentially governed not only by the sulfur content but also by the release of latent heat by the crystallization, heat transport processes, and the actual temperature distribution (see also Christensen, Rückriemen-Bez, and Schubert (Chapter 13, this volume).

6.5 SYNTHESIS

From the current gravity field determination, a dimensionless moment of inertia factor between 0.3087 and 0.3143 can be considered taking into account uncertainties in the measurements. This introduces further ambiguity in the interior structure models and the corresponding compositional constraints. Again, it should be noted that these values rely on the assumption of hydrostatic equilibrium. Further flybys by Juno and/or future missions, in particular the JUICE mission, which will be inserted into a polar Ganymede orbit, will provide definite answers to whether Ganymede is in a hydrostatic state.

For the moon's thermal evolution, the silicate mass fraction is a key factor due to its constraints on the amount of long-lived radiogenic isotopes. However, the radiogenic heat production rate depends both on uncertainties in the exact composition of the silicates and on the overall mass fraction. The latter will depend on the MoI factor and on the specific densities for the individual layers. With the MoI range as given above, the H_2O layer can have a thickness from about 600 to 900 km (Deschamps and Sotin, 2001; Sohl et al., 2002), which corresponds to an approximate silicate mass fraction (including metal) of 56 to 67 wt.-%. Assuming compositions ranging in heat production between the one of CI chondrites (lowest) and LL chondrites (highest), the total amount of radiogenic heating lies between 4.2 and 5.6 × 10^{11} W. The corresponding heat flux out of the silicate mantle ranges from about 9 to 11 mW m^{-2} and is thus far better constrained than the total heat production. The reason for this is that high total heating rates correspond to a larger surface area of the silicate shell while smaller heating rates correspond to a smaller surface area. Here, we have considered the radiogenic heat flux only. If we include other heat sources, for instance secular cooling, the heat flux out of the silicate shell would even exceed the values mentioned above. The heat flux out of the silicate shell has important consequences on the evolution of Ganymede's hydrosphere, which are further discussed by Kalousová, Soderlund, and Solomonidou (Chapter 12, this volume).

Looking at the Galilean satellite system as a whole, the amount of volatiles is increasing with greater distance from Jupiter, which can be derived from the respective densities. Likewise, the amount of thermal activity, evident in the satellites' surface records, decreases with distance, which can be understood in the context of strongly decreasing tidal heating with distance from Jupiter. However, the peak of activity in Ganymede's history, reaching possible surface heat fluxes of several tens of mW/m^2 or even exceeding 100 mW/m^2 based on observed relaxation of craters and the formation of grooves and furrows (Bland et al., 2017), remains not fully explained with conventional heat sources, especially if we compare this to Callisto's evolution. Whereas radiogenic heating is insufficient to provide such an amount of energy, tidal heating and the process of differentiation might be candidates. If ocean formation and subsequent differentiation in Ganymede proceeded within about 100 Myr, a total amount of energy of about 4×10^{13} W could be released (Hussmann et al., 2009). However, we do not know how long the process of differentiation lasted. Furthermore, it is unclear whether the separation of ice and rock took place simultaneously with the separation of metal from silicates leading to a fully differentiated state. This will largely depend on the initial temperature profiles and corresponding viscosities of the ice and rock mixtures. In the scenario by Friedson and Stevenson (1983), high viscosities in the ice–rock mixture would impede convection and thus the removal of heat generated by radioactive decay. The satellite would heat up and a runaway process leading to melting of the ice and complete separation of rock and ice would occur on Ganymede. In later stages of evolution, the separation of metal and rock would occur according to the different scenarios described at the end of Section 6.4.2. Late differentiation, however, would also be conceivable. Kimura and Kurita (2011) assumed that a primitive Ganymede (or Callisto, which has a slightly smaller rocky content) had an undifferentiated hydrous core. Subsequently, it is heated by long-lived radioisotopes, and after about several hundred million to one billion years the temperature in deep regions of the hydrous core attains the dehydration point. Owing to further radiogenic heating, the temperature reaches the melting point of the metal component (e.g., the eutectic point of FeS-Fe_3O_4 is 1 250 K) and the radiogenic heating may lead to the formation of a metallic core long after Ganymede's formation (up to 1 Ga after satellite formation). However, this hypothesis requires a relatively low melting point of the metallic component. If Ganymede's metallic component has a higher melting point, such "late differentiation" will probably not be achieved and alternative heat sources are required to separate the metal component from the rock.

Tidal heating in the current orbital configuration is significantly smaller than the radiogenic heating rate. However, peaks in the orbital eccentricity during formation of the Laplace resonance (Showman et al., 1997b) or oscillations in the Laplace resonance (Hussmann and Spohn, 2004) might significantly enhance that value. With Callisto not being locked in resonance, these scenarios are attractive because they provide a significant difference in the evolutions of the two satellites. An alternative explanation for the Ganymede–Callisto dichotomy could be the different impact rates about 700 Myr after formation, during the period of late heavy bombardment. In that scenario, sufficient energy would be provided to Ganymede, triggering a complete differentiation process (Barr and Canup, 2010).

It is unclear when and how the differentiation between silicate and iron took place. It largely depends on the initial temperature profile whether the iron and rock have separated simultaneously with the ice phase. The main questions with respect to the internal structure of Ganymede that remain and should be addressed by future missions like the Juno extended mission, Europa Clipper, and JUICE are the following:

- How is Ganymede's field generated and why are Europa and Io lacking such fields?
- Does Ganymede have an inner solid core?
- How can we explain the full differentiation of Ganymede in comparison to Callisto?
- How thick is the outer ice layer and can we unambiguously confirm the presence of a subsurface ocean on Ganymede?
- How can Ganymede's active geologic history be explained? Is it related to the orbital configuration or do other factors play a major role?
- Do Ganymede's gravitational field and shape correspond to hydrostatic equilibrium?
- What are the tidal amplitudes on Ganymede and are they consistent with a subsurface ocean?
- What is the salinity and composition of a subsurface ocean on Ganymede?
- How active is the high-pressure ice layer?
- Are chemical compounds transported from the silicate layer into the ocean?
- Is Ganymede a habitable world?

It should be noted that data from space missions should be complemented by Earth-based observations and laboratory data. Many questions regarding phase transitions and physical parameters under conditions relevant for Ganymede's evolution and interior can be investigated experimentally, which would improve our theoretical models in preparation for the next missions to the Jupiter system.

The analysis of data from Juno's first Ganymede flyby is currently ongoing (Gomez Casajus et al., 2022) and ESA's JUICE mission was successfully launched in 2023. It is equipped with a geophysical instrument suite highly capable of investigating the interior structure of Ganymede in orbit around the moon. The instruments relevant for geophysical investigations include a radio science package, an accelerometer, a subsurface radar, a magnetometer and plasma wave instrument, a laser altimeter, and a camera system (Grasset et al., 2013). First data from flybys and orbit around Ganymede will arrive in the first half of the 2030s, according to current mission planning.

REFERENCES

Anderson, J. D., E. L. Lau, W. L. Sjogren, G. Schubert, and W. B. Moore (1996). Gravitational constraints on the internal structure of Ganymede. *Nature* 384, 541–543.

Anderson, J. D., R. A. Jacobson, T. P. McElrath et al. (2001). Shape, mean radius, gravity field, and interior structure of Callisto. *Icarus* 153, 157–161.

Anderson, J. D., R. A. Jacobson, E. L. Lau et al. (2001). Shape, mean radius, gravity field and interior structure of Ganymede. *AAS/Division for Planetary Sciences Meeting Abstracts* No. 33. p.35.09.

Barr, A. C. and R. M. Canup (2010). Origin of the Ganymede–Callisto dichotomy by impacts during the late heavy bombardment. *Nature Geoscience* 3, 164–167.

Bland, M. T., Showman, A. P., and Tobie, G. (2008). The production of Ganymede's magnetic field. *Icarus* 198, 384–399.

Bland, M. T., K. N. Singer, W. B. McKinnon, and P. M. Schenk (2017). Viscous relaxation of Ganymede's impact craters: constraints on heat flux. *Icarus* 296, 275–288.

Christensen, U. R. (2015). Iron snow dynamo models for Ganymede. *Icarus* 247, 248–259.

Davies, M. E., T. R. Colvin, J. Oberst et al. (1998). The control networks of the Galilean satellites and implications for global shape. *Icarus* 135, 372–376.

Dermott, S. F. (1979). Shapes and gravitational moments of satellites and asteroids. *Icarus* 37, 575–586.

Deschamps, F. and C. Sotin (2001). Thermal convection in the outer shell of large icy satellites. *Journal of Geophysical Research* 106, 5107–5121.

Fei, Y., Bertka, C. M., and Finger, L. W. (1997). High-pressure iron-sulfur compound, Fe_3S_2, and melting relations in the Fe-FeS system. *Science* 275, 1621–1623.

Friedson, A. J. and D. J. Stevenson (1983). Viscosity of rock–ice mixtures and applications to the evolution of icy satellites. *Icarus* 56, 1–14.

Gomez Casajus, L., M. Zannoni, D. Modenini et al. (2021). Updated Europa gravity field and interior structure from a reanalysis of Galileo tracking data. *Icarus* 358, 114187.

Gomez Casajus, L., M. et al. (2022), Gravity Field of Ganymede After the Juno Extended Mission. *Geophysical Research Letters* 49, doi: 10.1029/2022GL099475.

Grasset, O., M. K. Dougherty, A. Coustenis et al. (2013). JUpiter ICy moons Explorer (JUICE): an ESA mission to orbit Ganymede and to characterise the Jupiter system. *Planetary and Space Science* 78, doi: 10.1016/j.pss.2012.12.002.

Hauck, S. A., Aurnou, J. M., and Dombard, A. J. (2006). Sulfur's impact on core evolution and magnetic field generation on Ganymede. *Journal of Geophysical Research* doi:10.1029/2005JE002557.

Hubbard, W. B. and J. D. Anderson (1978). Possible flyby measurements of Galilean satellite interior structure. *Icarus* 33, 336–341.

Hussmann, H., G. Choblet, V. Lainey et al. (2009). Implications of rotation, orbital states, energy sources, and heat transport for internal processes in icy satellites. *Space Science Reviews* 153, 317–348.

Hussmann, H. and T. Spohn (2004). Thermal-orbital evolution of Io and Europa. *Icarus* 171, 391–410.

Kamata, S., J. Kimura, K. Matsumoto et al. (2016). Tidal deformation of Ganymede: sensitivity of Love numbers on the interior structure. *Journal of Geophysical Research: Planets* 121, 13662–13675.

Kimura, J. and K. Kurita (2011). EPSC-DPS Joint Meeting 2011, held 2–7 October 2011 in Nantes, France, p. 358.

Kimura, J., T. Nakagawa, and K. Kurita (2009). Size and compositional constraints of Ganymede's metallic core for driving an active dynamo. *Icarus* 202, 216–224.

Kirk, R. L. and D. J. Stevenson (1987). Thermal evolution of a differentiated Ganymede and implications for surface features. *Icarus* 69, 91–134.

Kivelson, M. G., F. Bagenal, W. S. Kurth, F. M. Neubauer, C. Paranicas, and J. Saur (2004). Magnetospheric interactions with satellites. In F., Bagenal, T. Dowling, and W. B. McKinnon (eds.) *Jupiter: The Planet, Satellites and Magnetosphere*. Cambridge: Cambridge University Press, pp.

Kivelson, M. G., K. K. Khurana, C. T. Russell et al. (1996). Discovery of Ganymede's magnetic field by the Galileo spacecraft. *Nature* 384, 537–541.

Kivelson, M. G., K. K. Khurana, and M. Volwerk (2002). The permanent and inductive magnetic moments on Ganymede. *Icarus* 157, 507–522.

Kuskov, O. and V. A. Kronrod (2001). Core sizes and internal structure of Earth's and Jupiter's satellites. *Icarus* 151, 204–227.

McKinnon, W. B. and E. M. Parmentier (1986). Ganymede and Callisto. In J. A. Burns and M. S. Matthews (eds.) *Satellites*. Tucson: University of Arizona Press, pp. 718–763.

Moore, W. B. and G. Schubert (2003). The tidal response of Ganymede and Callisto with and without liquid water oceans. *Icarus* 166, 223–226.

Mueller, and W. B. McKinnon (1988). Three-layered models of Ganymede and Callisto: compositions, structures, and aspects of evolution. *Icarus* 76, 437–464.

Nakamura, R. and E. Ohtani (2011). The high-pressure phase relation of the $MgSO_4$-H_2O system and its implication for the internal structure of Ganymede. *Icarus* 211, 648–654.

Pappalardo, R. T., G. C. Collins, J. W. Head III et al. (2004). Geology of Ganymede. In F., Bagenal, T. Dowling, and W. B. McKinnon (eds.) *Jupiter: The Planet, Satellites and Magnetosphere*. Cambridge: Cambridge University Press, pp.

Rückriemen, T., D. Breuer, and T. Spohn (2015). The Fe-snow regime in Ganymede's core: a deep seated dynamo below a stable snow zone. *Journal of Geophysical Research* 120, 1095–118.

Saur, J., S. Duling, L. Roth et al. (2015). The search for a subsurface ocean in Ganymede with Hubble Space Telescope observations of its auroral ovals. *Journal of Geophysical Research* 120, 1715–1737.

Schubert, G., J. D. Anderson, T. Spohn, and W. B. McKinnon (2004). Interior composition, structure and dynamics of the Galilean satellites. In F., Bagenal, T. Dowling, and W. B. McKinnon (eds.) *Jupiter: The Planet, Satellites and Magnetosphere*. Cambridge: Cambridge University Press, pp. 281–306.

Schubert, G., T. Spohn, and R. T. Reynolds (1986). Thermal histories, compositions and internal structures of the moons of the Solar System. In J. A. Burns and M. S. Matthews (eds.) *Satellites*. Tucson: University of Arizona Press, pp. 224–292.

Schubert, G., D. J. Stevenson, and K. Ellsworth (1981). Internal structures of the Galilean Satellites. *Icarus* 47, 46–59.

Schubert, G., K. Zhang, M. G. Kivelson, and J. D. Anderson (1996). The magnetic field and internal structure of Ganymede. *Nature* 384, 544–545.

Showman, A. P., D. J. Stevenson, and R. Malhotra (1997a). Coupled orbital and thermal evolution of Ganymede. *Icarus* 129, 367–383.

Showman, A. P., D. J. Stevenson, and R. Malhotra (1997b). Tidal evolution into the Laplace resonance and the resurfacing of Ganymede. *Icarus* 127, 93–111.

Sohl, F., T. Spohn, D. Breuer, and K. Nagel (2002). Implications from Galileo observations on the interior structure and chemistry of the Galilean satellites. *Icarus* 157, 104–119.

Spohn, T. and G. Schubert (2003). Oceans in the icy Galilean satellites of Jupiter? *Icarus* 161, 456–467.

Vance, S., M. Bouffard, M. Choukroun, and C. Sotin (2014). Ganymede's internal structure including thermodynamics of magnesium sulfate oceans in contact with ice. *Planetary and Space Science* 96, 62–70.

Wahr, J. M., M. T. Zuber, D. E. Smith, and J. I. Lunine (2006). Tides on Europa, and the thickness of Europa's icy shell. *Journal of Geophysical Research* 111, E12005, doi: 10.1029/2006JE002729.

Zhan, X. and G. Schubert (2012). Powering Ganymede's dynamo. *Journal of Geophysical Research* 117, doi: 10.1029/2012JE004052.

Zharkov, V. N., V. V. Leontjev, and A. V. Kozenko (1985). Models, figures, and gravitational moments of the Galilean satellites of Jupiter and icy satellites of Saturn. *Icarus* 61, 92–100.

Zubarev, A., I. Nadezhdina, J. Oberst, H. Hussmann, and A. Stark (2015). New Ganymede control point network and global shape model. *Planetary and Space Science* 117, 246–249.

7

Ganymede's Geology

ROBERT T. PAPPALARDO, MARISSA E. CAMERON, GEOFFREY C. COLLINS, JAMES W. HEAD, PAUL HELFENSTEIN, SAMUEL M. HOWELL, ALICE LUCCHETTI, THOMAS B. MCCORD, JEFFREY M. MOORE, ERICA M. NATHAN, LOUISE M. PROCKTER, COSTANZA ROSSI, PAUL M. SCHENK, AND JOHN R. SPENCER

7.1 INTRODUCTION

This chapter reviews the current state of knowledge about Ganymede's geology based largely on Galileo spacecraft results, updating the earlier summary by Pappalardo et al. (2004). Aspects of Ganymede's geology are discussed elsewhere in this volume, for example the stratigraphy and geologic evolution (Chapter 5), cratering record (Chapter 8), topography (Chapter 9), composition (Chapters 10 and 11), hydrosphere (Chapter 12), and links to tidal evolution (Chapter 4). Surface geology serves to constrain the properties and history of the surface, lithosphere, and deeper interior and the evolution of the satellite as a whole (e.g., Patterson et al., 2010).

Ganymede's surface is divided into two principal terrain types: relatively old regions of dark terrain and younger cross-cutting lanes of light, typically grooved terrain. Ganymede's surface consists of 35% ancient dark terrain, which is relatively heavily cratered, and 64% light terrain, which is more recent and heavily tectonized (Shoemaker et al., 1982; Schenk et al., 2001; Patterson et al. 2010; Figure 7.1). The remaining 0.5% of the surface is composed of tectonized reticulate terrain, which does not fit neatly into either the dark or the light terrain categories. Age relations show that Ganymede's geologic history can be divided into three distinct phases (Collins et al., 2013): the period before the formation of light materials (the Nicholsonian period), during light material formation (Harpagian), and after light material formation (Gilgameshan).

Frosts caps extend over ±40° latitude in a thin shroud, hinting of interactions with the magnetospheric particles. Impact features of diverse forms and ages are clues to surface and interior history. Multispectral evidence confirms that light terrain is ice-rich, while dark terrain contains a greater fraction of non-ice materials. Though Ganymede's surface and lithosphere are primarily water ice, the geology of light terrain is reminiscent of terrestrial rift zones.

7.2 DARK TERRAIN

Dark terrain bears witness to processes that have affected Ganymede since its earliest surface evolution (Figure 7.2). Crater density estimates suggest an average dark terrain age >4 Gyr (Neukum, 1997; Zahnle et al., 2003), though it is uncertain how long the Nicholsonian period lasted (Collins et al., 2013). The majority of dark terrain is heavily cratered, occurring as polygonal regions bounded by light terrain, commonly containing ~10 km wide furrow sets (Smith, Soderblom, Beebe et al., 1979; Smith, Soderblom, Johnson et al., 1979).

Dark terrain is heterogeneous in albedo at decameter scales, probably due to thermally driven segregation of ice and non-ice surface components (Spencer, 1987a). Geological mapping from high-resolution Galileo images (Prockter et al., 1998; Prockter et al., 2000) shows subunits including bright high-standing massifs interpreted to be furrow and crater rims or their remnants; intermediate-albedo hummocky-textured plains material interpreted to be degraded impact ejecta; and low-albedo material in topographic lows interpreted to be mass-wasted deposits.

Ganymede's dark terrain is likely composed of a relatively thin and dark lag deposit (Spencer, 1987a; Moore et al., 1999; Oberst et al., 1999; Moore et al., 2004; Johnson et al., 2004) overlying brighter icy material (Prockter et al., 1998). Some zones of dark terrain are also pervasively tectonized (Patel et al., 1999; Prockter et al., 2000), and commonly dark terrain structures share a similar structural fabric to adjacent light terrain swaths (Patterson et al., 2010; Collins et al., 2013); these may represent precursor units to tectonically deformed light terrain (see Section 7.3.1.1).

7.2.1 Furrow Systems

Ganymede's furrow sets are its oldest recognizable surface structures (Passey and Shoemaker, 1982). Individual furrows are linear to curvilinear troughs with generally flat floors and raised rims, are typically ~6 to 20 km wide and 10–100 km long (Passey and Shoemaker, 1982; Smith, Soderblom, Beebe et al., 1979; Shoemaker et al., 1982), and are spaced approximately ~50 km apart.

On the basis of morphology and planform, along with their similarity to impact-formed multi-ringed structures on Europa (Moore et al., 1998; Kadel et al., 2000; Moore et al., 2001), furrows are generally accepted to be fault-bounded troughs formed in response to large impacts into a relatively thin lithosphere early in Ganymede's history, an epoch when Ganymede's surface may have had a higher thermal gradient, perhaps overlying a more ductile, or even liquid, substrate (McKinnon and Melosh, 1980; Schenk et al., 2004). Individual furrows likely formed during basin collapse, resulting from radially inward asthenospheric flow with accompanying brittle failure of the overlying lithosphere (McKinnon and Melosh, 1980;

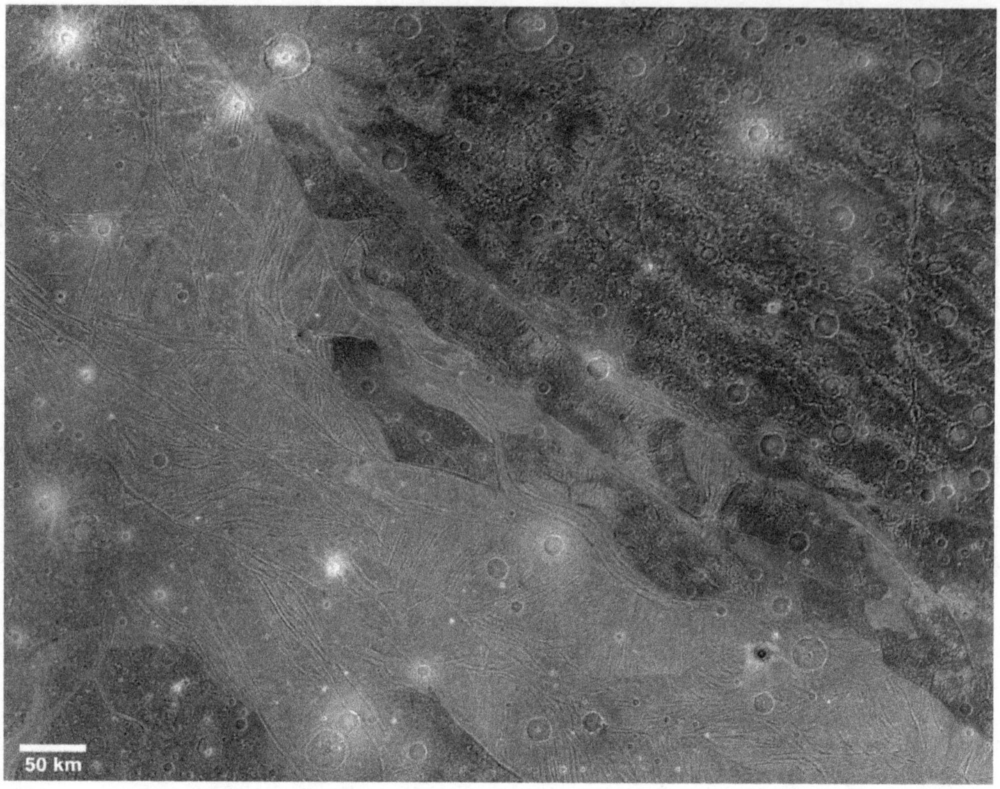

Figure 7.1 Boundary between the furrowed dark terrain of Galileo Regio (upper right) and the light terrain of Uruk Sulcus (lower left), as imaged at ~1 km/pixel (mosaic of Voyager 2 images 0452-, 0458-, 0461-, 0464-, 0470-, and 0473J2-001). Furrows of the Lakhmu Fossae system trend subparallel to the northern boundary of Uruk Sulcus, while the Zu Fossae cut them obliquely. A variety of crater forms are visible, including bright ray, central pit, and dark floor craters.
North is to the top in this figure (and in other figures, unless indicated).

Melosh, 1982). Furrows may have been reactivated and modified by later endogenic activity (Murchie et al., 1990). Hirata et al. (2020) propose that furrows resulted from a single, giant impact centered on 20°S 180°W, which may have induced melting in Ganymede's interior, potentially explaining the difference between Callisto's and Ganymede's evolution.

7.2.1.1 Tectonism

Two primary types of tectonic deformation have been identified in Ganymede's dark terrain: furrow formation and related tectonics (Section 7.2.1) and dark terrain fracturing. Fracturing and/or reactivation of preexisting structures within dark terrain is primarily concentrated near the boundaries of dark terrain regions (Shoemaker et al., 1982; Murchie et al., 1986; Prockter et al., 2000; Patterson et al., 2010; Collins et al., 2013).

Small-scale tectonic features in dark terrain have been focused by preexisting structures (Asphaug et al., 1998; Prockter et al., 2000). In furrowed regions, there is a greater areal density of fractures within older broad furrows, suggesting that younger tectonism may reactivate older structures. Some craters have acted as zones of weakness that focused tectonic deformation, as in the Nicholson Regio area (Figure 7.2).

7.2.1.2 Mass Wasting

Galileo images provide strong evidence that mass wasting has been an important dark terrain modification process (Prockter et al., 1998; Moore et al., 1999) but with lesser degradation compared to Callisto (Moore et al., 1999; Chuang and Greeley, 2000). Evidence for downslope movement of Ganymede's dark material includes low albedo streaks on otherwise bright slopes, with an inverted V-shape characteristic of mass wasting deposits, and accompanying aprons that are likely talus piles. The correlation of low albedo material with local topographic lows (Oberst et al., 1999) suggests that loose dark material has collected in furrow floors and other low-lying areas. This style of mass wasting is consistent with gravity-driven dry sliding or slumping in areas of oversteepened slopes (Moore et al., 1999).

7.2.1.3 Sublimation

High-resolution images of Galileo Regio show a concentration of dark material on equator-facing slopes, with brighter slopes facing away from the direction of solar illumination, inferred to result from the sublimation and deposition of surface volatiles, specifically H_2O (Spencer, 1987a; Spencer et al., 1998; Prockter et al., 1998). Condensation of water vapor as frost may occur on relatively cold pole-facing slopes and topographic highs. In addition, as dark material is shed off of these slopes, a relatively high albedo, ice-rich surface is revealed. These observations suggest that sublimation has been an important process for the redistribution of ices at sub-kilometer spatial scales (Spencer, 1987a; Moore et al., 1996). Sublimation of interstitial volatile ices material may trigger mass wasting and contribute

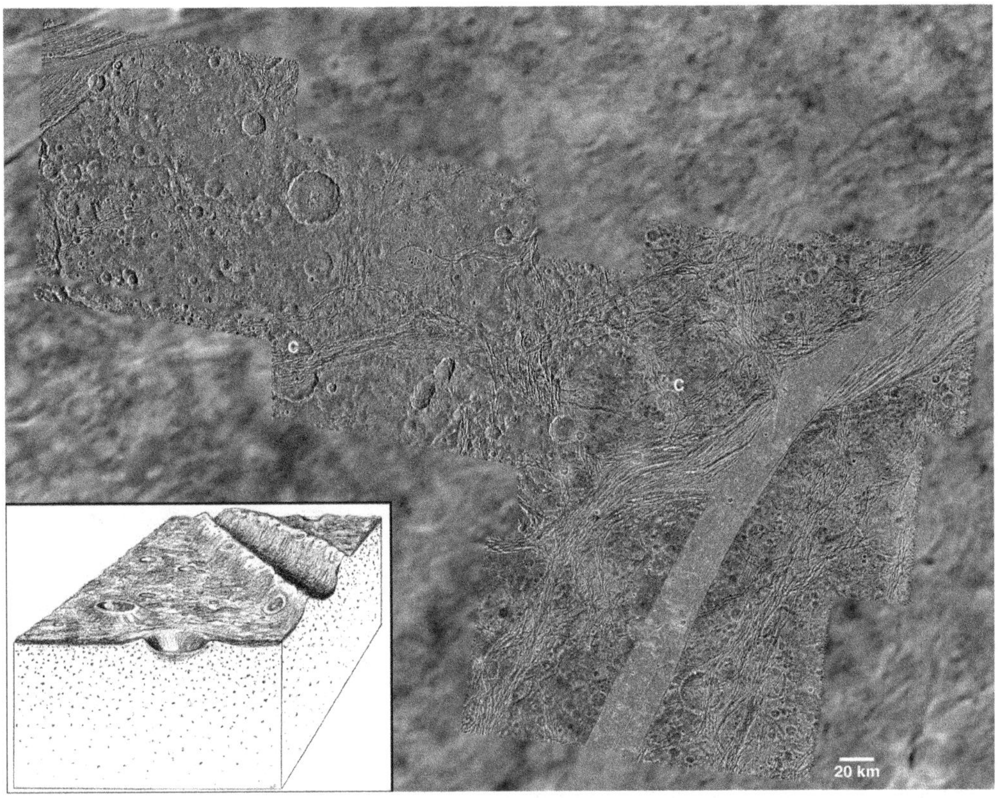

Figure 7.2 A portion of Nicholson Regio centered near 14° S, 353° W. Here the dark terrain is straddled by light grooved terrain and shows evidence of heavy tectonic disruption. The NE-trending swath of smooth light terrain in the southeast is Arbela Sulcus, and grooved swaths in the far northwest belong to the Mysia Sulci. The Galileo images reveal that irregular bright patches observed in the Voyager 1 images (background; USGS 2020) are generally tectonically disrupted areas and patches of impact ejecta. Galileo data are a mosaic of observations G7GSNICHOL01 (176 m/pixel, center and left) and 28GSARBELA02 (133 m/pixel, right). Examples of craters focusing tectonic deformation are at (C). Inset: Schematic interpretation of dark terrain derived from Galileo observations, in which the low albedo material is proposed to be a lag deposit of dark silicate material overlying cleaner ice; impacts and furrow formation through the surface lag result in brighter topographic features, and the lag deposit may be concentrated in topographic lows by mass wasting and thermal segregation (from Prockter et al., 1998).

significantly to topography degradation at local scales, analogous to Callisto's erosional style (Moore et al., 2004).

7.2.1.4 Sputtering

Ganymede's magnetic field deflects most charged particles in the satellite's equatorial regions but funnels plasma onto the polar regions (Kivelson et al., 1998; Kivelson et al., 2004). The energy imparted to Ganymede's surface by impinging charged particles from Jupiter's magnetosphere will cause sputtering, the ejection and redistribution of molecules (Johnson et al., 2004). The presence of extreme albedo variations within dark terrain implies that sublimation occurs at a significantly higher rate than sputtering, the latter process tending to re-homogenize the surface (Spencer, 1987a). This is consistent with estimates of particle fluxes and ice-sputtering rates suggesting an erosion rate of $\sim$80 cm Gyr^{-1} in the equatorial region, which is two orders of magnitude less than the expected sublimation rate (Cooper et al., 2001). Sputter ablation will be further reduced by the presence of a refractory lag deposit, and impact gardening would distribute sputter-produced chemical products throughout the regolith. Thus, sputtering is unlikely to be a major influence on dark terrain surface morphology, at least in Ganymede's nonpolar regions (cf. Section 7.4).

7.2.2 Model of Dark Terrain Formation

Prockter et al. (1998) suggest that the surface of dark terrain is composed of a thin dark veneer of silicate-rich lag, overlying a cleaner, icier substrate (Figure 7.2, inset). In this hypothesis, the substrate has a small amount of a low albedo component heterogeneously admixed, emplaced by projectile infall during the late stages of Ganymede's accretion. High albedo, topographically elevated units, such as crater and furrow rims, are inferred to be outcrops of icy substrate from which the dark lag deposit has sloughed off and/or which have trapped a bright frost veneer. In the warmest regions and local topographic lows, volatiles have been removed and the non-ice material has concentrated. Large impacts, including palimpsests (Section 7.5), have penetrated the lag-enriched surface to the ice-rich crust below, mixing and excavating the impacted material to create relatively bright ejecta.

7.3 LIGHT TERRAIN

Light terrain (alternatively "light material" or "bright terrain") crosscuts dark terrain in a patchwork of swaths 10s to 100s of kilometers wide, covering 64% of Ganymede's surface. Half of the area of light terrain is dominated by tectonically formed grooves, while much of the remaining area exhibits more subtle lineaments (Patterson et al., 2010). The material making up light terrain predominantly occurs in depressions. These topographic lows may have been infilled by extrusion of relatively clean (silicate-poor) liquid water, warm ice, or icy slush that hardened as it cooled to ambient temperatures (Golombek and Allison, 1981; Golombek, 1982; Parmentier et al., 1982; Shoemaker et al., 1982; Squyres, 1982; Allison and Clifford, 1987), or they may have been depressed as the lithosphere was stretched and faulted (Howell and Pappalardo, 2018). The nominal cratering age of light terrain is ~2 Gyr, though large uncertainties in the impact flux through time imply that grooved terrain may have been emplaced any time from ~400 Myr to >4 Gyr ago (Zahnle et al., 1998; Zahnle et al., 2003; Schenk et al., 2004). The beginning and end of light terrain formation defines the Harpagian period (Collins et al., 2013), but it is currently unknown how long this period lasted.

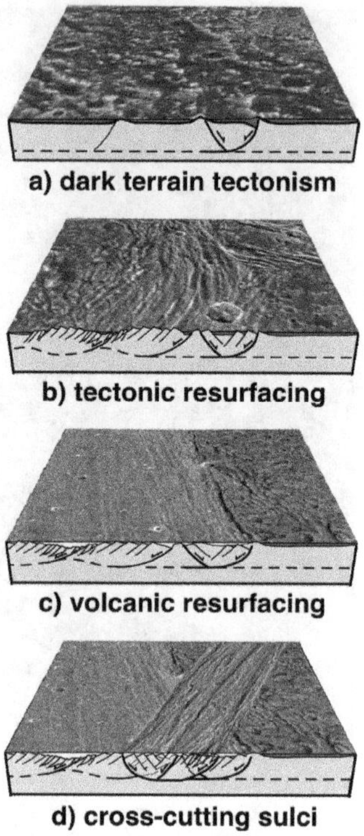

Figure 7.3 Summary model for the formation and evolution of light terrain, including grooved terrain (Pappalardo et al., 2004). (a) Normal faults inferred to define furrows in dark terrain may be reactivated to focus later light terrain deformation. (b) Some grooved light terrain may form by tectonic disruption of the preexisting surface, without concurrent cryovolcanism. (c) Some light terrain may form by a combination of tectonism and cryovolcanism which brightens and smooths the surface. (d) Light terrain swaths, or "sulci," can cross-cut one another, dissecting the preexisting surface into a polygonal patchwork.

7.3.1 Light Terrain Characteristics

7.3.1.1 Regional-Scale Analyses

Light terrain on Ganymede's surface is visually dominated by a network of crosscutting tectonic grooves (Figure 7.3), and is often referred to as "grooved terrain," even though some of the individual areas of light terrain lack prominent grooves. Grooved terrain comprises alternating ridges and troughs with a length >100 km (Shoemaker et al., 1982). The largest-scale grooves have a generally regular mean spacing of ~8 km, derived from Fourier analysis of photometric profiles across groove sets in Voyager images (Grimm and Squyres, 1985). Grooves of an individual set commonly trend subparallel to the long axis of the elongate cell of light terrain in which they occur, but in regions of complexly crosscutting set they can also trend obliquely to a polygon's long axis (Figure 7.1). More prominent grooves commonly bound individual polygonal cells, while structures within a groove set are typically less prominent (Golombek and Allison, 1981).

In some locations, there appears to be an inheritance of structural patterns at the boundaries between dark and grooved terrain (Figure 7.3a), suggesting that prominent grooves have utilized preexisting zones of structural weakness, specifically furrows, in dark terrain (Murchie et al., 1986; Patterson et al., 2010; Collins et al., 2013). At the northern boundary of Uruk Sulcus (Figure 7.1), prominent structures within light terrain trend subparallel to furrows of similar morphology, degree of curvature, and spacing within adjacent dark terrain of Galileo Regio. Other locations called reticulate terrain (Shoemaker et al., 1982; Patterson et al., 2010; Collins et al., 2013) show a style of modification of both dark and light terrain in which two dominant sets of nearly orthogonal intersecting grooves occur.

7.3.1.2 High-Resolution Analyses

High-resolution Galileo images (Figure 7.4) reveal the extent of albedo heterogeneity in grooved terrain (Pappalardo et al., 1998). Stereo imaging shows that the bright, intermediate, and dark materials generally correspond to ridge peaks, slopes, and troughs, respectively (Pappalardo et al., 1998; Oberst et al., 1999). The close correlation of albedo to topography implies that profiles across albedo striping or shading (at low or high incidence angle, respectively) reveal the wavelength characteristics of topography. Fourier analysis on several sites imaged at high resolution by Galileo reveals multiple small-scale wavelengths (~1 to a few km), generally superimposed on longer wavelength topography (~5 to 10 km) (Patel et al., 1999). This result is confirmed by limited high-resolution stereo imaging within Uruk Sulcus that indicates that topographic amplitudes reach ~500 m and the resolved (long wavelength) slopes are ≤19° (Giese et al., 1998); superimposed smaller-scale ridges are probably ≤200 m high (Pappalardo et al., 1998).

7.3.1.3 Styles of Extension

Light terrain is dominated by various expressions of extensional strain and tectonism (Figure 7.4), as described in detail by Pappalardo et al. (1998). At high resolution, some grooved terrain has a morphology resembling flat-floored graben and intervening flat-topped horst ridges (Figure 7.4a). Offset has occurred by cumulative displacement along several of the subkilometer-scale fractures or faults that are pervasive across these grooved terrains; such horst-and-graben style faulting is typically expressed on stratigraphically lower light terrain units, so it may have been more common in the early Harpagian

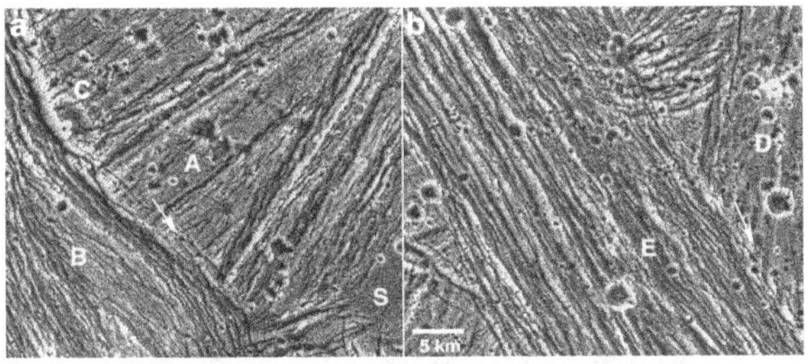

Figure 7.4 Grooved light terrain within Uruk Sulcus (from Galileo observation G1GSURUSUL01, 74 m/pixel). (a) Boundary between older grooved terrain unit A with horst-and-graben-like morphology, and younger unit B with tilt-block-style morphology. Boundary relationships suggest that unit B has formed through tectonic dismemberment of unit A, exemplified by incipient structures (e.g., at arrow), and by an apparent cut crater immediately south of C. Smooth material at S suggests the possibility of an earlier stage of cryovolcanic resurfacing. (b) Unit E with tilt-block-style morphology has apparently formed at the expense of older grooved units, including D. Incipient structures associated with unit E have cut the flanking terrains (e.g., at arrow).

(Collins et al., 1998a). Bland and McKinnon (2015) employ finite-element simulations to model grooved terrain structures, assuming ~10% strain, uniform material properties, and symmetric pure shear boundary conditions. Their results produce zones of symmetric faulting, quite reminiscent of these lineated horst-and-graben-like structures.

In Uruk Sulcus (Figure 7.4b) and elsewhere, narrow triangular ridges and V-shaped troughs with no topographic break in between are suggestive of tilt-block-style normal faulting (also termed "bookshelf" or "domino-style"), where arrays of normal faults dip in the same direction and fault motion induces back-tilting of the surface, forming blocks that are commonly asymmetric in cross section (Figure 7.3b–d). Tilt-block normal faulting creates regularly spaced ridges and troughs, with the faulted layer thickness several times the fault spacing, based on terrestrial examples and sandbox experiments (Mandl, 1987; Vendeville et al., 1987). Bland et al. (2010) find that grooved terrain deformation depends on both thermal gradient and the rate of strain weakening, potentially accounting for a range of grooved terrain styles within close proximity on Ganymede's surface.

Domino-style faulting is commonly associated with at least tens of percent of extensional strain (Wernicke and Burchfiel, 1982). Based on the geometry of faults in high resolution, strain estimates indicate that low-relief groove terrain has experienced only a few percent strain, high-relief grooves typically experienced 25–50% strain, and some localized areas have experienced >100% strain (Collins et al., 1998b; Pappalardo and Collins, 2005; Collins, 2006). Bland and Showman (2007) found that when strain exceeds ~10–20%, instability growth rates decrease, limiting the deformation resulting from extensional instability. Bland and McKinnon (2015) have confirmed modest regional strain of 10–15% and much larger strain of 30–60% at local scale.

Extensional necking of Ganymede's lithosphere can explain the topography of groove sets dominated by tilt-block faulting (Collins et al., 1998b; Pappalardo et al., 1998; Dombard and McKinnon, 2001). Where ductile material underlies a brittle lithosphere and the decrease in viscosity with depth is rapid, an extensional necking instability can occur in which the stretched lithosphere evolves into a series of pinches and swells, as first modeled for the Basin and Range Province of the western United States (Fletcher and Hallet, 1983). Modeling indicates that the multiple topographic wavelengths of Ganymede's grooved terrain topography likely formed by extensional instability in a rheological lithosphere with a surface heat flux 20–120 mW m^{-2}, and with maximum instability growth rates occurring at temperature gradients $\geq$45 K km^{-1} and strain rates $\sim10^{-13}$–10^{-16} s^{-1} (Dombard and McKinnon, 2001; Bland and Showman, 2007; Hammond and Barr, 2013). Including the effect of strain weakening in the models can generate more realistic groove topography at a somewhat lower thermal gradient of ~10 K km^{-1} (Bland et al., 2010).

The high degree of strain in zones of tilt-block normal faulting suggests that preexisting surface features may be altered beyond recognition through "tectonic resurfacing," without requiring cryovolcanism to erase preexisting terrain (Head et al., 1997; Patel et al., 1999; Pappalardo et al., 2004; Pizzi et al., 2017; Bland and McKinnon, 2018; Howell and Pappalardo, 2018). Support for this hypothesis is provided by extensionally strained craters, where preexisting crater structures such as central peaks or pits are absent, suggesting that strains ~10s% may render preexisting topography unrecognizable (Pappalardo and Collins, 2005). Tectonic resurfacing in this context differs from analogous resurfacing on Earth, in that Ganymede's ice shell is not likely to sustain plate-tectonic-like activity (Howell and Pappalardo, 2019). Modeling suggests that tectonic resurfacing is enabled where localized high heat flux has viscously relaxed initial topography (Bland and McKinnon, 2018).

Lithospheric separation and spreading are well-documented on neighboring Europa (Schenk and McKinnon, 1989; Prockter et al., 2002), and a spreading-analog model for grooved terrain was suggested based on Voyager observations of central ridges and regions of apparent axial symmetry within grooved terrain (Lucchitta, 1980). However, little evidence has been found for cut and separated older features, or large-scale horizontal shear, expected to accompany lateral motion (McKinnon and Parmentier, 1986). The portion of Arbela Sulcus imaged by Galileo (Figure 7.2) presents the intriguing possibility that limited lithospheric spreading and associated shear might have occurred on Ganymede (Head et al., 2002; Pizzi et al., 2017). However, reconstruction is imperfect, leaving ambiguous conclusions regarding spreading of Arbela Sulcus and other "plank-like" lanes of smooth light terrain, such as at Perrine Regio and Anshar Sulcus (Collins et al., 2001).

Howell and Pappalardo (2018) find that the surface morphology of extensionally deformed terrain varies as a function of strain and lithosphere strength, with faulting dominating at low strains and spreading dominating at high strains (Figure 7.5). They find that significant surface disruption may occur at low strains, but extensional deformation on Ganymede is unlikely to expose fresh materials from deep within the ice shell or from the ice-ocean interface. The transition between spreading- and rifting-dominated deformation within the ice shell may indicate spatial or temporal variability in ice shell structure and state.

Despite plentiful evidence for extensional strain on Ganymede, there is little evidence of widespread contractional

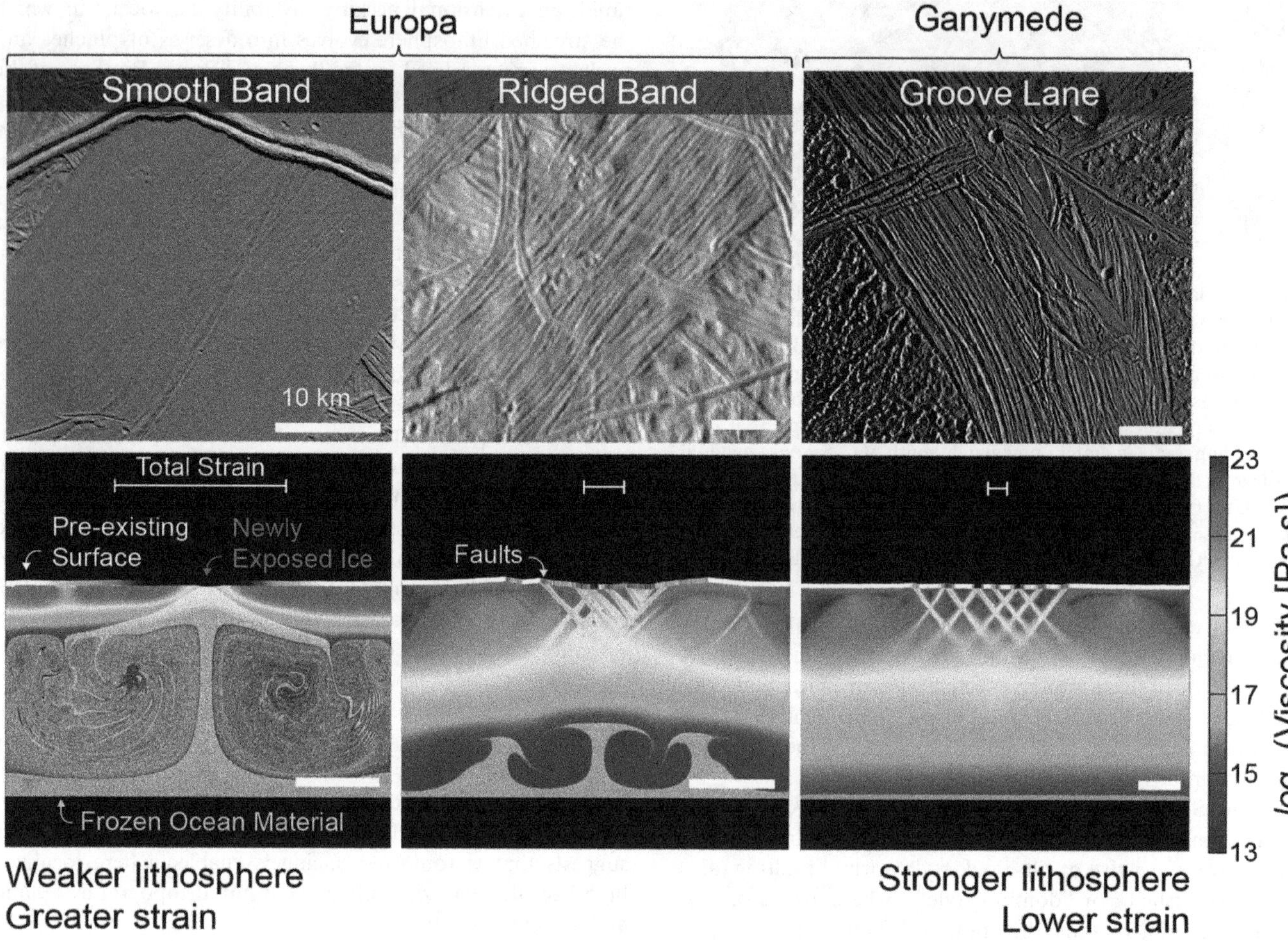

Figure 7.5 Galileo spacecraft images of Europa compared to Ganymede, showing extensional landform morphology in map view (top row) and associated numerical models in cross section (bottom row), modified from Howell and Pappalardo (2018). From left to right, the images are of Europa smooth band Thynia Linea (Galileo observation E17ESTHYLIN01, 40 m/pixel), a ridged band in Europa's northern leading hemisphere (E15ESREGMAP02, 234 m/pixel), and Ganymede groove lane Tiamat Sulcus (G8GSTIAMAT01, 494 m/pixel). From left to right, models increase in icy lithosphere strength and decrease in total strain. For the Europa smooth band, selected models exhibit symmetrical axial yielding with central axis for high regional strain, whereas the Europa ridged band model exhibits widespread asymmetrical faulting. Ganymede's Tiamat Sulcus is the most tectonized, exhibiting large-scale rifting at lower modeled strain than the Europa examples. Model outputs are colored by viscosity and show ice shell material derived in part from ocean material (white dots within ductile ice). The model surfaces are colored white where the original surface has not been significantly disrupted and black where the model surface strain is the greatest and fresh interior ice is exposed. (A black and white version of this figure will appear in some formats. For the color version, please refer to the plate section.)

strain (thrust faults or folds), with the exception of limited transpressional structures (see Section 7.3.1.4). Thus, it is generally inferred that global expansion has occurred in Ganymede's history, coincident with the formation of grooved terrain (Section 7.8).

7.3.1.4 Strike-Slip Faulting

Observations of regional and local-scale structures indicative of strike-slip processes suggest that a component of horizontal shear may have contributed to shaping grooved terrain and may also have played an important role in its evolution (Lucchitta, 1980; Murchie and Head, 1988; Pappalardo, 1994; Cameron et al., 2018; Rossi et al., 2020). In Galileo images, strong evidence is found for strike-slip displacement along some faults, as well as for a component of distributed shear in forming grooved terrain (Figure 7.6). Most apparently, *en echelon* fault segments are characteristic of strike-slip offsets and common within grooved terrain (Pappalardo et al., 1998; Pappalardo et al., 2004; Cameron et al., 2018).

Depending on the sense of strike-slip offset and the direction of the fault bend, regions of transtension (releasing bends), consisting of strike-slip plus extension, or transpression (restraining bends), consisting of strike-slip plus compression, will result (Davis et al., 2013). For example, at Galileo high resolution transtension is inferred within Uruk Sulcus (Figure 7.6a) (Pappalardo et al., 1998), while at the Voyager resolution Uruk Sulcus is framed in regional-scale transpressional kinematics (Rossi et al., 2018). Transpression is also implied by *en echelon* segments in Nun Sulci (Figure 7.6b).

High-resolution evidence for strike-slip motion lends credence to interpretations of apparent strike-slip offsets along major fault zones based on regional-scale images, as at: eastern Tiamat Sulcus (Murchie and Head, 1988; Patel et al., 1999;

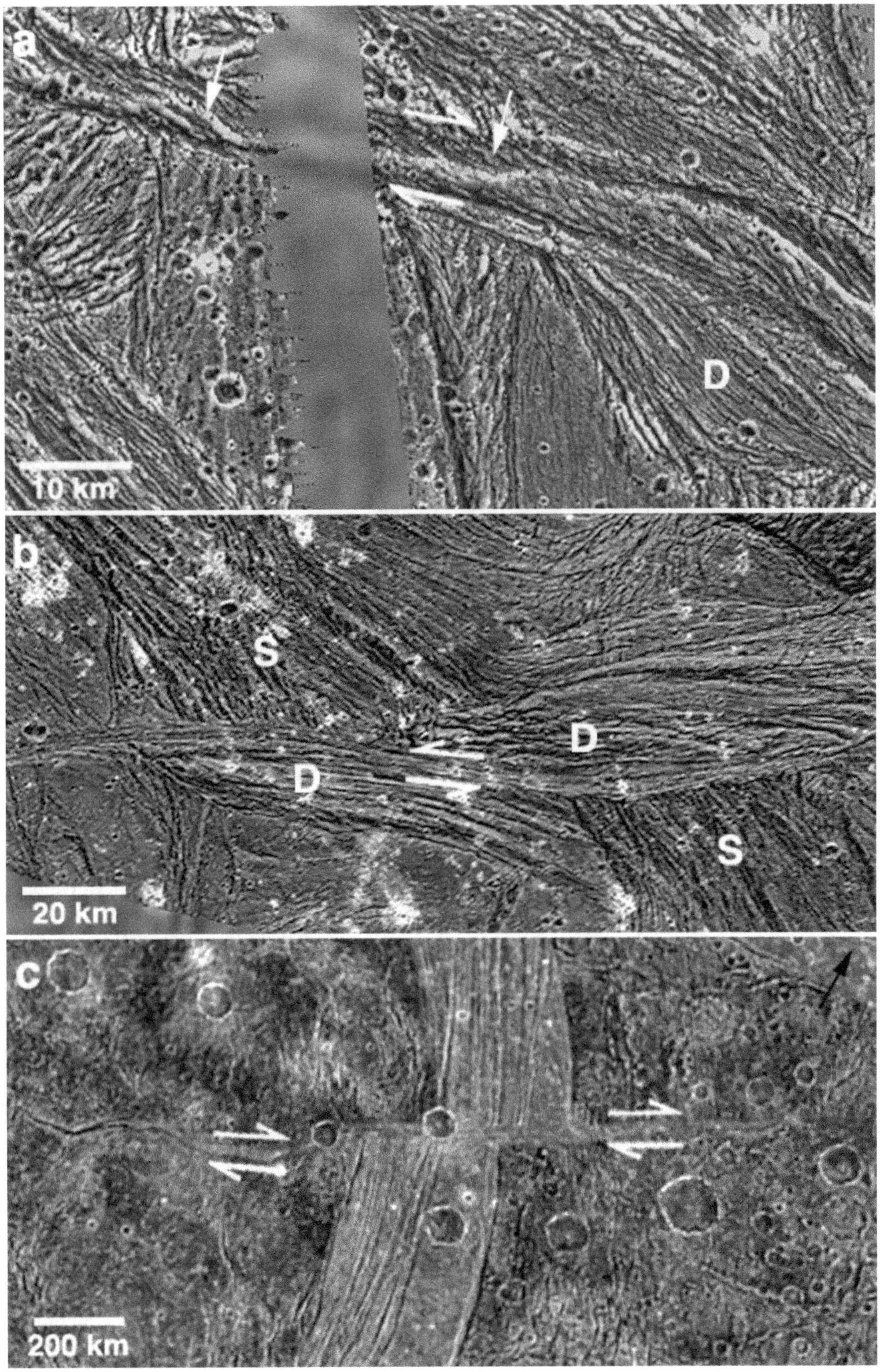

Figure 7.6 Examples of strike-slip tectonism on Ganymede. (a) In Uruk Sulcus, a right-lateral shear zone (half-arrows) ~7 to 10 km wide contains prominent *en echelon* structures (e.g., at arrows). A fault duplex D has formed in a releasing bend of this shear zone. (b) In Nun Sulci, an inferred left-lateral strike-slip fault (half-arrows) appears to have displaced preexisting structures at S (cf. Lucchitta, 1980), implying that fault duplexes D formed along a constraining bend and are transpressional. (c) Dardanus Sulcus has been displaced ~150 km along a right-lateral fault, which exhibits bright material where it cuts dark terrain (e.g., at half-arrows); black arrow indicates north. From Galileo observations G1GSURSUL01 (74 m/pixel) on Voyager 2 context (1 km/pixel, USGS [2020]); G7GSNUNSUL01 (164 m/pixel), and 28GSDARDAN01 (750 m/pixel), respectively.

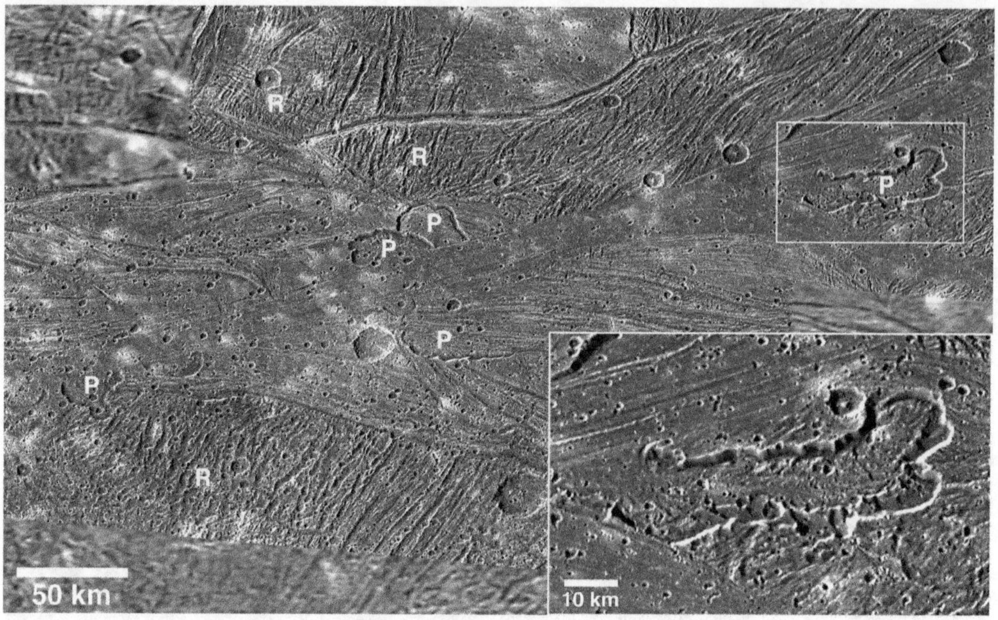

Figure 7.7 A portion of Sippar Sulcus (G8GSCALDRA01, 180 m/pixel) rich in scalloped depressions (paterae, marked P), as originally discussed by Lucchitta (1980). The most prominent patera (inset) displays surface texture that may be indicative of flow toward its open end, consistent with it being a source region for cryovolcanic material. Some tectonic structures in the region cross each other near-orthogonally, forming reticulate terrains (marked R); see Section 7.3.1.1.
Here, south is to the top.

Bedle and Jurdy, 2005; Cameron et al., 2018), Dardanus Sulcus (Figure 7.6c) (Smith, Soderblom, Johnson et al., 1979; Murchie and Head, 1988; Cameron et al., 2018), Arbela Sulcus (Head et al., 2002; Cameron et al., 2018), Nippur/Philus Sulcus, Byblus Sulcus, Anshar Sulcus (Cameron et al., 2018), and other grooved terrain sites (Pappalardo et al., 1998; Prockter et al., 2000). Analysis of strained craters also suggests that a component of distributed horizontal shear is common (Pappalardo and Collins, 2005).

Global mapping of the grooves (Rossi et al., 2020) shows that groove systems are arranged similarly to the strike-slip corridors on Earth. These are deformation zones delimited by long strike-slip structures that include secondary structures associated with their kinematics. Similarly, groove systems show longer and parallel grooves that bound areas with shorter grooves. The longer elements likely act as primary shear boundaries that confine an elongated region of deformation characterized by the presence of conjugate structures formed by the shear movement.

7.3.1.5 Rift-flank Uplift

Based on their characteristic shape, likely sites of flexural uplift alongside rift zones have been identified in stereo-derived topography (Nimmo et al., 2002). Flexural modeling suggests local heat fluxes $\sim$30–45 mW m^{-2} during their deformation (Nimmo et al., 2002; Nimmo and Pappalardo, 2004), consistent with estimates from extensional instability modeling (Dombard and McKinnon, 2001). Flexure of broad-scale topography seems to be a common characteristic of grooved terrains, while (as expected) small-scale ridge and trough topography has not been significantly affected by topographic relaxation.

7.3.2 Light Terrain Cryovolcanism

Though the extent of cryovolcanism in light terrain formation remains somewhat enigmatic, indirect evidence for cryovolcanic resurfacing of light terrain (Figure 7.3c) has been identified. Eruption of liquid water through an ice-rich crust may seem a difficult prospect because of the negative buoyancy of the melt compared to the solid. Possible means of overcoming this difficulty are by volatiles that exsolve with decreased pressure to decrease the density of a watery melt, by the presence of contaminants within the ice crust to increase its relative density, or by pressurization of a crustal melt reservoir, for instance by partial crystallization (Crawford and Stevenson, 1988; Fagents et al., 2000; Lesage et al., 2020). Alternatively, warm ice might rise through the colder and denser crust to erupt in the solid state (Kirk and Stevenson, 1987). Showman et al. (2004) suggest that topography could create subsurface pressure gradients that allow subsurface melt to move upward into topographic lows such as fault valleys while shutting down resurfacing before structural boundaries overflow.

At least 18, and probably more than 30, scalloped depressions ("paterae") are associated with light terrain on Ganymede and have been proposed to represent caldera-like source vents for icy volcanism (Lucchitta, 1980; Schenk and Moore, 1995; Head et al., 1998; Schenk et al., 2001; Chapter 9). The boundaries of Ganymede's scalloped depressions are typically concave-inward, with inward-pointed cusps, consistent with inward collapse along listric faults like at terrestrial calderas (Moore and Pappalardo, 2011). High-resolution Galileo images show that the largest patera of several within Sippar Sulcus (Figure 7.7) is associated with a ridged deposit, a possible icy flow (Head et al., 1998). Schenk et al. (2001) and Schenk et al. (Chapter 9, this volume) report that the smooth terrain in central Sippar

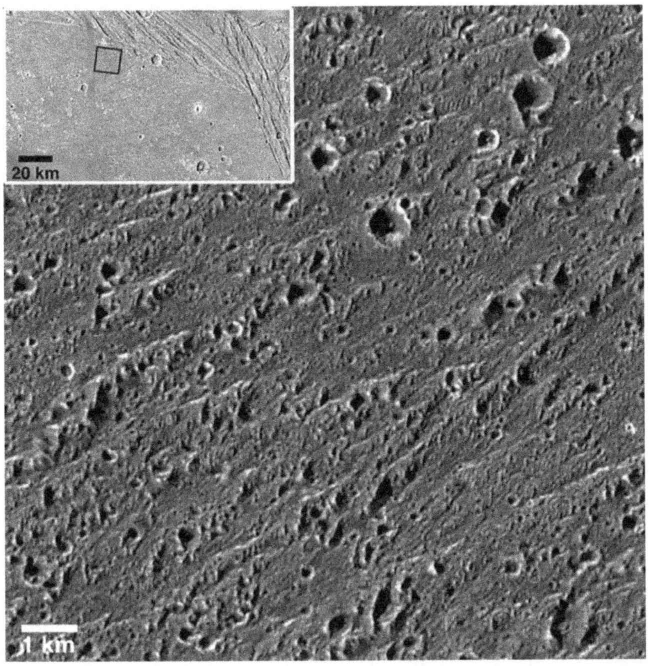

Figure 7.8 Light terrain within Harpagia Sulcus appears smooth in Galileo contextual images (inset; observation 28GSSMOOTH02, 119 m/pixel). Small-scale features with significant topographic relief are revealed at high resolution (28GSSMOOTH01, 15 m/pixel). This demonstrates that at least some of Ganymede's smoothest bright areas have undergone tectonic deformation at a fine scale. Relatively smooth material between these small ridges was plausibly emplaced by cryovolcanism or might result from seismic shaking causing mass wasting. North is toward the top in the inset contextual image.

Sulcus exhibits topographic characteristics consistent with icy volcanic resurfacing of structural troughs to an equipotential level by low-viscosity aqueous lavas to form the smooth terrains.

Virtually all smooth light terrains on Ganymede exhibit some degree of tectonic overprinting (Pappalardo et al., 2004). As one of the smoothest terrains observed in the Voyager images, a portion of Harpagia Sulcus (Figure 7.8) appears smooth in Galileo regional resolution images, but it is rough at the scale of hundreds of meters, containing many degraded linear ridge segments inferred to be tectonic in origin. Cryovolcanic embayment is a plausible origin for the relatively smooth material between individual ridges (Head et al., 2002), while an alternative hypothesis is that landform degradation through seismic shaking has smoothed this and other regions on a small scale (Mills et al., 2019). The search for pristine and unambiguous cryovolcanic landforms on Ganymede may be frustrated by near-simultaneous tectonism that would modify them.

7.3.3 Light Terrain Exogenic Processes

Mass wasting acts at a fine scale to smooth and level light terrain topography. High-resolution imaging, including stereo, indicates that dark streaks occur on steep slopes and are oriented downhill, representing mass movements (Pappalardo et al., 1998). Low albedo material within topographic lows further suggests downslope movement, with thermal segregation induced by sublimation probably contributing to the concentration of dark material in topographic lows (Spencer et al., 1998). Mills et al. (2019) estimate quake magnitudes from measured fault dimensions, indicating that mass wasting from tectonic shaking is a plausible mechanism to generate smooth mass-wasted material between ridges.

Several possible examples of mass movement and landform modification are recognizable in very high-resolution Galileo imagery (Moore et al., 1999). Almost nowhere is the surface found to be smooth at this high resolution. Slopes are usually less than about 10°, although in a few instances they approach 20°, suggesting that they are eroded massifs. The surface between hills is generally undulatory with a rough texture suggestive of erosional hummocks or block or boulder clusters. Some craters appear partially infilled by debris. Candidate talus slopes are recognized in light scattered from adjacent bright hills into shadows, and superimposed streaks, which appear to be controlled by the local topographic gradient, are interpreted as loose material that has moved downslope.

7.4 POLAR CAPS

Voyager observations showed a continuous diffuse, bright deposit without relief poleward of about ±40° latitude, affecting both dark and light terrains, known as the polar caps or hoods (Smith, Soderblom, Beebe et al., 1979). Based on Voyager data, it was suggested that water-frost deposits may simply remain longer in the colder environment of the high latitudes (Squyres, 1980b), or frosts may accumulate at high latitude by cold-trapping of water ice ablated by solar radiation (Purves and Pilcher, 1980; Spencer, 1987a). The caps might be a relict deposit formed during the emplacement of light terrain materials (Shaya and Pilcher, 1984) or may be a sink for ices that sublimate preferentially from the equatorial regions (Moore et al., 1996). Accumulation may be through cold-trapping of water ice bombarded by the Jovian radiation belts (Sieveka and Johnson, 1982; Johnson, 1985, Hillier et al., 1996).

Galileo magnetometer data revealed Ganymede's magnetic field. Kivelson and colleagues (Kivelson et al., 1998; Kivelson et al., 2004), and Khurana et al. (2007) showed that the polar cap margins coincide with the boundary between field lines that are open to polar plasma and field lines that are closed to plasma, suggesting a probable causal relationship. Johnson (1997) suggested that bombarding plasma damages ice grains and thus brightens the polar caps. Instead, or in addition, sputtering by charged particles may redistribute polar ice, with thermal segregation acting to create the patchiness and topographic muting observed in high-resolution Galileo images. The higher single scattering albedo and forward-scattering properties of the polar regions (Section 7.7.1) are consistent with sputter redistribution of ice there.

High-resolution Galileo observations show that the high-latitude surface (Pappalardo et al., 2004; Khurana et al., 2007) has a dominantly bright patchy mottled appearance, implying heterogeneity in the distribution and thickness of the polar frost deposits. Bright deposits preferentially occur on pole-facing slopes, and muting of some crater topography by bright deposits suggests that local thickness may be measured in terms of meters or more. The overall brightening of the polar hood is inferred to be the integrated effect of these bright mantling deposits.

7.5 IMPACT STRUCTURES

Ganymede's surface displays the greatest diversity of primary impact morphologies in the solar system (Figure 7.9), many of which are unique to Ganymede and Callisto. As reviewed by Kirchoff, Barr, Bland et al. (Chapter 8, this volume), these include bright ray, dark ray, and dark floor craters (Figure 7.9a–b); pedestal craters (Figure 7.9c); craters with central pits and domes (Figures 7.9d–e); the low-relief ancient impact scars called palimpsests (Figure 7.9f); and vast multi-ring structures expressed as sub-concentric furrows (Figure 7.1), interpreted to form by impact into a thin elastic lithosphere (McKinnon and Melosh, 1980). Differences in crater density can permit the age of various terrains to be constrained, as detailed in Chapter 8.

Large impact morphology is characterized by the formation of bright central domes (Figure 7.9e), presumed to be exposures of Ganymede's shallow interior (Moore and Malin, 1988; Schenk, 1993; Schenk et al., 2004; Senft and Stewart, 2011; Chapter 8, this volume). Palimpsests (up to 100s of kilometers in diameter) are recognized against dark terrain as low-relief bright patches (Figure 7.9f), and some (termed "penepalimpsests") show an associated basin-like ring structure (Shoemaker et al., 1982; Schenk et al., 2004). The two most likely palimpsest models suggest that the outer portions of the bright deposit represent either the entire continuous ejecta deposit of the impact structure or the edge of a pedestal-like ejecta deposit (analogous to the ejecta of pedestal craters; Figure 7.9c) (Jones et al., 2003; Schenk et al., 2004; Kirchoff et al., Chapter 8, this volume). Identifying the correct model is important for understanding the origins and geophysical implications of palimpsests and for determining the size distribution of these ancient impacts. Gilgamesh is Ganymede's largest basin at 590 km in diameter (Schenk et al., 2004); only impact-related materials have formed on Ganymede since, through the Gilgameshan period (Collins et al., 2013).

The subdued character of Ganymede's oldest impact structures, notably its palimpsests, implies a steep thermal gradient in Ganymede's early history, with more recent impact structures reflecting a thicker and more rigid elastic lithosphere (Passey and Shoemaker, 1982; Shoemaker et al., 1982; Schenk et al., 2004). Viscous relaxation has been suggested as the driving force for the evolution of crater morphology (Passey and Shoemaker, 1982; Singer et al., 2018; Bland et al., 2017); alternatively, some studies show that craters can retain their topography for realistic ice flow laws and reasonable thermal gradients (Dombard, 2000; Dombard and McKinnon, 2000).

Impact craters on Ganymede can have bright, presumably ice-rich impact ejecta (e.g., Figures 7.1 and 7.9a) or dark, presumably less icy ejecta (Figure 7.9b). Dark ray craters may represent impactor contamination, with colors generally most closely matching D-type asteroidal material (Schenk and McKinnon, 1991; Hibbitts, Hansen et al., 2003). Dark floor craters have the lowest albedos of any Ganymede materials (Helfenstein, 1986a, 1986b; Schenk and McKinnon, 1991; Helfenstein et al., 1997) (see Section 7.7.1), and they may represent concentrated deposits of dark impactor material (Helfenstein, 1986a, 1986b) or impact melt from which the volatile icy constituents were lost due to vaporization (Schenk and McKinnon, 1991).

7.6 SURFACE COMPOSITION

Ganymede's composition is summarized briefly here as pertinent to geology and in greater detail in Chapter 11 by Ahrens et al. (this volume). Early ground-based infrared reflectance spectra provided evidence of water ice as the major constituent of the surfaces of the three icy Galilean satellites (for reviews, see Calvin et al., 1995; McCord et al., 1998). Ice grains are likely finer (20–50 μm) on the leading side near the equator and coarser (>300 μm) in the trailing equatorial region, consistent with increased sputtering preferentially removing ice from the trailing side and redepositing as a frost elsewhere, especially at colder sites such as high latitudes (Section 7.4) and higher albedo areas. Modeling of Ganymede's ice spectra suggests that the surface in at least equatorial and mid-latitudes is made up of discrete patches of non-ice and ice material (McCord et al., 2001). The patches may be due to an uneven surface temperature distribution, resulting in uneven cold-trapping of H_2O molecules (Spencer, 1987a), and/or to downslope movement of dark material (see Section 7.2.2).

Analysis of the Galileo near-infrared mapping spectrometer (NIMS) data (Hansen et al., 1998; McCord et al., 2001) suggests considerable and widespread non-ice materials on and in Ganymede's surface. Ice abundance maps (Hansen, 2002) indicate many regions with >80% non-ice dark materials, consistent with photometric studies (Section 7.7.1). At least some regions of Ganymede's non-ice dark material are found to exhibit water absorptions indicative of hydrated minerals (McCord et al., 2001), similar to what is reported for Europa's surface.

Galileo NIMS data indicate CO_2 to be present in small (~100 molecule) clusters and widely distributed in complex patterns (McCord et al., 1997, 1998; Hibbitts et al., 2000; Hibbitts, Pappalardo et al., 2003). Hibbitts, Pappalardo et al. (2003) found that, in general, light terrain contains less CO_2 than dark terrain, there is little or no CO_2 observed at the poles, and dark rays of impact craters are commonly depleted in CO_2 relative to surrounding terrain. The CO_2 seems to be contained in non-ice materials, but its presence in large-grained ice would be difficult to detect due to the strong absorption of water ice in this spectral region. Areas with mostly fine-grained ice do not seem to contain CO_2, suggesting that CO_2 distribution is not related to magnetospheric effects but originated from Ganymede instead.

Other molecular species reported on Ganymede from analysis of Galileo infrared spectra include SO_2, $C \equiv N$ (perhaps as HCN), and CH (McCord et al., 1997; McCord et al., 1998a). The SO_2 has a complex spatial distribution, but the band weakness prevents generating detailed maps. The $C \equiv N$ and CH spectral features are consistent with those of organic-rich tholins, which could be the result of surface radiolysis or infall of asteroidal and cometary material.

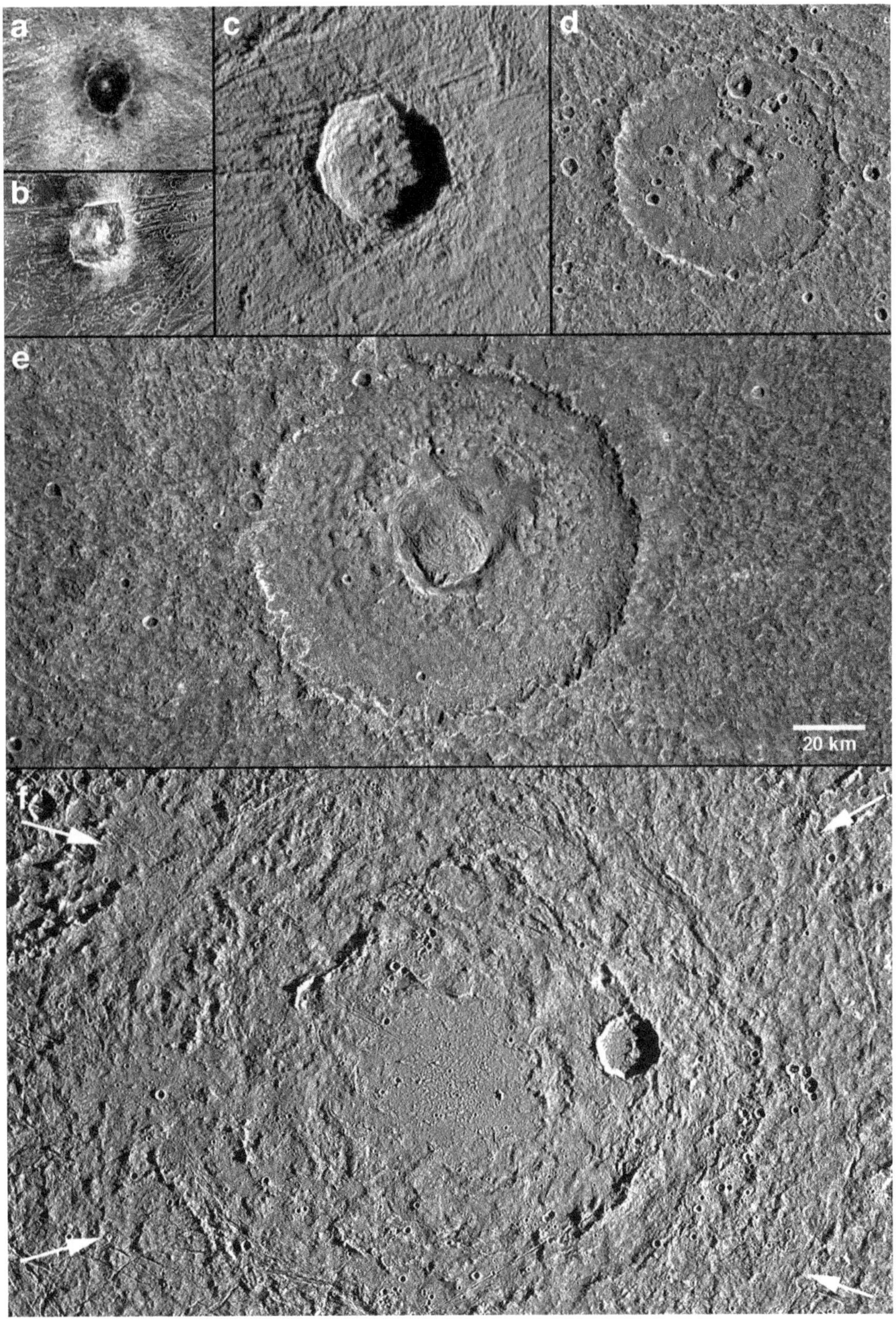

Figure 7.9 Examples of impact structures on Ganymede, shown to a common scale and with north to the top. (a) Dark floor crater Khensu, with surrounding bright ejecta (G2GSCRATER01, 219 m/pixel). (b) The rays of dark ray crater Kittu are subdued in this high-resolution view (G7GSKITTU_01, 145 m/pixel). (c) Pedestal crater Achelous shows characteristic outward-facing low ejecta scarp (G7GSACHELS01, 175 m/pixel). (d) Central pit crater Lugalmeslam, situated on the southwestern margin of Nippur Sulcus (G8GSPITCRA01, 146 m/pixel). (e) Central dome crater Melkart, on the boundary between dark terrain of Marius Regio (to the northeast) and light terrain (G8GSMELKRT01, 176 m/pixel). (f) Buto Facula within Marius Regio (12°N, 203°W) appears as an albedo feature with no obvious relief in low incidence angle Voyager images, while these ~ 80° incidence angle Galileo images (G8GSBUTOFC01, 180 m/pixel) show basin-like structure; the bright palimpsest margin corresponds to once-fluid ejecta (arrows), either a pedestal-like ejecta deposit (Schenk et al., 2004) or the outer edge of the more extensive continuous ejecta blanket (Jones et al., 2003).

7.7 REGOLITH PROPERTIES

The properties of Ganymede's regolith can be estimated using the tools of photometry, radiometry, and radar.

7.7.1 Photometric Properties

Photometric investigations of Ganymede fall into categories of whole-disk photometry that investigate average global behavior and studies of disk-resolved observations that explore the photometric properties of different terrain units. The most recent work on Ganymede's photometry has focused on application of the Hapke (e.g., 1981, 2021) photometric model as a means of estimating the physical properties of the uppermost regolith.

Earth-based rotational light curves for the icy Galilean satellites, which are obtained at small phase angles, have long shown that each body exhibits a global brightness asymmetry that approximately correlates with the leading and trailing hemispheres. The average particles on Ganymede's surface are intermediate in albedo between lunar regolith and Europa's high albedo grains (Verbiscer et al., 2013), and the macroscopic roughness is lunar-like (Buratti, 1991, 1995; Domingue and Verbiscer, 1997). However, uncertainties persist due to limited whole-disk phase coverage.

Hillier et al. (1996) used disk-resolved Voyager data for phase coverage from 4–128°, finding that bright-terrain regolith grains are significantly more transparent than those in dark terrains. Belgacem et al. (2021) find strong variations in regolith grain microstructures, ranging from nonpolar dark terrain, which exhibits moderately rough-shaped grains with high densities of internal scatterers, to brighter grains with less rugged shapes and fewer internal scatterers, and bright crater materials with grains that are much smoother and more transparent.

Only crude estimations of regolith porosity can be made based on photometry. Helfenstein (1986a) estimates porosities of 40–60% for nonpolar dark terrains, 84% for polar examples, and 48–89% for light terrain examples. For different crater materials and bright ejecta, the porosity estimates range from 35% to 97%.

The photometric characteristics of the segregated darkest and brightest materials visible at Galileo highest resolution have not yet been investigated in great detail. Normal albedos, r_n, for Ganymede terrains can be estimated from Voyager results (Helfenstein, 1986a) and yield average, nonpolar dark terrain materials with $r_n = 0.24 \pm 0.06$ and units varying from $0.16 \leq r_n \leq 0.34$. Average nonpolar light terrains have $r_n = 0.43 \pm 0.08$ with units having $0.39 \leq r_n \leq 0.82$. Studies of high-resolution Galileo images find that Ganymede's darkest materials have estimated normal albedos 0.12–0.34, with a mean of 0.25 ± 0.04 (Helfenstein et al., 1997). This is comparable to average values for dark materials on Callisto and dark deposits found in valley floors and maculae on Europa (Helfenstein et al., 1998) but about five times greater than D-type asteroidal materials which have been proposed as possible analogs for the dark contaminant at Ganymede (Section 7.5). Estimated normal reflectances of the brightest such materials from global albedo mapping exceed unity ($r_n \sim 1.14$), consistent with the contribution of the coherent-backscatter opposition effect (cf. Hapke, 2021) and comparable to Europa's bright icy materials (Helfenstein et al., 1998).

7.7.2 Thermal Characteristics

Ganymede's surface temperatures have been studied from Earth, from the Voyager InfraRed Interferometer Spectrometer (IRIS) instrument, and from the Galileo Photo Polarimeter-Radiometer (PPR). The observed day/night temperature night range of 150–90 K (Spencer, 1987b, Orton et al., 1996) can be fitted with a thermal inertia near 70 J m^{-2} K^{-1} s$^{-1/2}$, similar to the values for Callisto and Europa (Spencer, 1987b; Spencer et al., 1999). Eclipse cooling curves, like those of the other Galilean satellites, require an extremely uncompacted component with a low thermal inertia near 15 J m^{-2} K^{-1} s$^{-1/2}$. New disk-resolved millimeter-wave ALMA data (de Kleer et al., 2021), which probe beneath the surface, imply a rapid increase in compaction with depth, reaching low porosities of ~10% at tens of cm depth.

Light terrain and bright craters are respectively ~5 K and ~25 K colder during the day than dark terrain (Spencer, 1987b; Orton et al., 1996). At night, light terrain is ~3 K warmer than dark terrain, and bright craters are colder than either terrain type (de Kleer et al., 2021). This pattern is consistent with the surface consisting of two spatially segregated units, as originally suggested to explain the eclipse cooling curve (Spencer, 1987b): a dark, low thermal inertia unit, perhaps dust, and a bright, high thermal inertia unit, perhaps ice. The larger abundance of the ice component in the light terrain would explain its higher effective thermal inertia. However, bright ray craters are colder than other units during both day and night, suggesting that the ice in the fresh bright craters has low thermal inertia and is different in texture from the ice seen elsewhere on the surface. Global millimeter-wave daytime data confirm these patterns in some, but not all, surface areas and also reveal latitudinal and leading/trailing differences likely associated with exogenic processes (de Kleer et al., 2021).

Surface ice is likely colder than the mean surface temperature inferred from low-resolution remote radiometry, which is dominated by emission from low albedo materials if the surface is segregated. The daytime temperature of water ice on the leading side has been estimated as 126 ± 12 K by near-IR reflectance spectroscopy, using the temperature dependence of the water ice spectrum (Grundy et al., 1999).

7.7.3 Radar Characteristics

As with all of the Galilean satellites, Ganymede's radar signature is unusual. The radar albedo at both 13 and 3.5 cm is about 2.5 (Ostro et al., 1992), compared to 0.1 for typical terrestrial planets, and the circular polarization ratio is about 1.4 compared to 0.1 for the terrestrial planets. All three icy Galilean satellites scatter radar energy like rough surfaces, likely resulting from the near-surface material being only weakly absorbing, allowing for multiple scattering in the subsurface, whereas single scattering from the surface dominates the radar reflectivity of the Moon and other terrestrial planets. Coherent backscattering can then explain the circular polarization ratios (Peters, 1992). Black et al. (2001) were able to reproduce

Ganymede's scattering behavior with a model surface consisting of icy scatterers ranging up to at least 2 m in size, distributed in a porous ice matrix at least a few meters thick. The high albedos rule out silicates as the scatterers.

7.8 LINKS TO THE INTERIOR

Galileo gravity data suggest that Ganymede is highly differentiated, with an iron core and silicate mantle capped by an ocean and icy shell (Anderson et al., 1996; Hussmann et al., Chapter 6, this volume). Moreover, capture of Ganymede into the Laplace resonance with Io and Europa is predicted to have been chaotic (Malhotra, 1991; Showman and Malhotra, 1997), and increases in Ganymede's orbital eccentricity may have triggered enhanced tidal heating, internal melting, and geological activity, depending on the satellite's preexisting temperature structure (Showman et al., 1997; Bland et al., 2009). Ganymede may have differentiated prior to such a resonance passage through accretion (Showman et al., 1997) and subsequent impacts (Barr and Canup, 2010).

Changes in Ganymede's interior structure and orbital parameters would have changed the shape of the satellite, inducing lithospheric stresses. Moreover, comparison of the relative importance of stresses and timescales associated with tides and with freezing suggests that freezing processes may be more important in Ganymede's evolution, as with other icy moons (Nathan et al., 2021). Thus, it is tempting to relate Ganymede's history of differentiation, potential resonance passage, and associated heating and melting to the formation of Ganymede's light terrain. If core heating, internal melting, and bright terrain formation occurred during the same past tidal heating event, then Ganymede's current-day intrinsic magnetic field (Kivelson et al., 1996; Kivelson et al., 1998; Kivelson et al., 2004; Kivelson et al., Chapter 14, this volume) and implied hot iron core (Schubert et al., 2004) constrain the time of grooved terrain formation during the Harpagian period to $\leq$1 Gyr ago, the expected timescale of core cooling (Showman et al., 1997; Zahnle et al., 2003; Schubert et al., 2004).

Differentiation could have resulted in up to ~6% areal expansion of Ganymede (Mueller and McKinnon, 1988), while subsequent tidal resonance passage could have produced ~2% areal expansion (Bland et al., 2009). Bland and McKinnon (2015) reason that uneven distribution of strain across Ganymede's light terrains can help to reconcile its extremely high (~100%) local strains on the scale of individual faults or highly strained craters, its regional-scale strains (~10s of percent), and the (few percent) global strain that is possible from differentiation and/or heating and melting due to tidal resonance passage.

There is some indirect evidence for past nonsynchronous rotation of Ganymede, based on impact crater distributions (Zahnle et al., 2001). This period of rotation might have been triggered by an increase in orbital eccentricity and internal heating during entrance into the Laplace resonance (Malhotra, 1991; Showman and Malhotra, 1997; Showman et al., 1997; Bland et al., 2009). Cameron et al. (2019) infer that nonsynchronous rotation may have assisted the formation and evolution of strike-slip structures on Ganymede.

It has been proposed that light terrain formation may be related to convection within Ganymede's ice I outer shell (Shoemaker et al., 1982; Hammond and Barr, 2014), although convective stress is likely insufficient to fracture the lithosphere (McKinnon, 1998; Howell and Pappalardo, 2019). Bland and McKinnon (2018) consider that warming and global expansion associated with incipient mobile-lid convection (Hammond and Barr, 2014) may have resulted in nonuniform heat flux and localized strain to create regions of grooved light terrain and possibly regions of lithospheric separation. Implications of tidal evolution and convection for Ganymede's geology are considered in Steinbrügge et al. (Chapter 4, this volume) and Kalousová et al. (Chapter 12, this volume), respectively.

7.9 CONCLUDING REMARKS

The diversity and complexity of Ganymede's geology make it one of the most compelling solar system bodies, and its value in comparative planetology is not limited to interpreting the Galilean icy moons. To understand icy moons as a population of solar system bodies, it is important to constrain links between internal evolution and surface modifications, and how these connections preserve a record of the dominant global stresses important for past states of an icy moon's evolution. Ganymede's surface geology is an ideal laboratory for testing our understanding of icy moons, which have undergone past stress conditions that are different than those at present.

Galileo observations have greatly advanced understanding of Ganymede's geology, but many important questions remain, including:

- How does the epoch of dark terrain formation compare to Callisto's early geological history?
- Did entrance into the Laplace resonance affect internal evolution, and how is grooved terrain formation linked to internal processes (differentiation, phase changes, and/or convection)?
- What are the relative roles of geological processes (tectonism, cryovolcanism, mass wasting) that have shaped light terrain?
- What is the age of light terrain, and how extended was its formation period?
- Is any part of the surface still geologically active today?
- Have geological processes exposed Ganymede's oceanic material to the surface?

Future exploration of Ganymede from the JUICE mission (Grasset et al., 2013), along with opportunistic science by Europa Clipper (Howell and Pappalardo, 2020) will enable characterization of the global distribution, regional relationships, and detailed topography of geological features to better constrain their origin and unlock the secrets of this extraordinary world.

ACKNOWLEDGMENTS

Portions of this work by Robert T. Pappalardo, Marissa E. Cameron, and Samuel M. Howell were carried out at the Jet Propulsion Laboratory, California Institute of Technology, under a contract with the National Aeronautics and Space

Administration. Alice Lucchetti was supported for this activity by the Italian Space Agency (ASI) under the ASI-INAF contract 2018-25-HH.0.

REFERENCES

Allison, M. L. & Clifford, S. M. 1987, "Ice-covered water volcanism on Ganymede," *Journal of Geophysical Research: Solid Earth*, 92(B8), 7865–7876.

Anderson, J. D., Lau, E. L., Sjogren, W. L., Schubert, G., & Moore, W. B. 1996, "Gravitational constraints on the internal structure of Ganymede," *Nature*, 384(6609), 541–543.

Asphaug, E., Moore, J. M., Morrison, D. et al. 1998, "Crater-controlled fracture networks and the depth of ice lithospheres," *Lunar and Planetary Science Conference*, 29, 1587.

Barr, A. C. & Canup, R. M. 2010, "Origin of the Ganymede–Callisto dichotomy by impacts during the late heavy bombardment," *Nature Geoscience*, 3(3), 164–167.

Bedle, H. & Jurdy, D. M. 2005, "Ganymede's sulci on global and regional scales," *Lunar and Planetary Science Conference*, 36, 1161.

Belgacem, I., Schmidt, F., & Jonniaux, G. 2021, "Regional photometric study of Ganymede's surface," *Lunar and Planetary Science Conference*, 36, 1142.

Black, G. J., Campbell, D. B., & Nicholson, P. D. 2001, "Icy Galilean satellites: Modeling radar reflectivities as a coherent backscatter effect," *Icarus*, 151(2), 167–180.

Bland, M. T. & McKinnon, W. B. 2015, "Forming Ganymede's grooves at smaller strain: Toward a self-consistent local and global strain history for Ganymede," *Icarus*, 245, 247–262.

Bland, M. T. & McKinnon, W. B. 2018, "Viscous relaxation as a prerequisite for tectonic resurfacing on Ganymede: Insights from numerical models of lithospheric extension," *Icarus*, 306, 285–305.

Bland, M. T., McKinnon, W. B., Showman, A. P. 2010, "The effects of strain localization on the formation of Ganymede's grooved terrain," *Icarus*, 210(1), 396–410.

Bland, M. T. & Showman, A. P. 2007, "The formation of Ganymede's grooved terrain: Numerical modeling of extensional necking instabilities," *Icarus*, 189(2), 439–456.

Bland, M. T., Showman, A. P., & Tobie, G. 2009, "The orbital–thermal evolution and global expansion of Ganymede," *Icarus*, 200(1), 207–221.

Bland, M. T., Singer, K. N., McKinnon, W. B., & Schenk, P. M. 2017, "Viscous relaxation of Ganymede's impact craters: Constraints on heat flux," *Icarus*, 296, 275–288.

Buratti, B. J. 1991, "Ganymede and Callisto: Surface textural dichotomies and photometric analysis," *Icarus*, 92(2), 312–323.

Buratti, B. J. 1995, "Photometry and surface structure of the icy Galilean satellites," *Journal of Geophysical Research: Planets*, 100(E9), 19061–19066.

Calvin, W. M., Clark, R. N., Brown, R. H., & Spencer, J. R. 1995, "Spectra of the icy Galilean satellites from 0.2 to 5μm: A compilation, new observations, and a recent summary," *Journal of Geophysical Research: Planets*, 100(E9), 19041–19048.

Cameron, M. E., Smith-Konter, B. R., Burkhard, L. et al. 2018, "Morphological mapping of Ganymede: Investigating the role of strike-slip tectonics in the evolution of terrain types," *Icarus*, 315, 92–114.

Cameron, M. E., Smith-Konter, B. R., Collins, G. C., Patthoff, D. A., & Pappalardo, R. T. 2019, "Tidal stress modeling of Ganymede: strike-slip tectonism and Coulomb failure," *Icarus*, 319, 99–120.

Chuang, F. C. & Greeley, R. 2000, "Large mass movements on Callisto," *Journal of Geophysical Research: Planets*, 105(E8), 20227–20244.

Collins, G. C. 2006, "Global expansion of Ganymede derived from strain measurements in grooved terrain," *Lunar and Planetary Science Conference*, 37, 2077.

Collins, G. C., Head, J. W., & Pappalardo, R. T. 1998a, "Formation of Ganymede grooved terrain by sequential extensional episodes: Implications of Galileo observations for regional stratigraphy," *Icarus*, 135(1), 345–359.

Collins, G. C., Head, J. W., III, & Pappalardo, R. T. 1998b, "The role of extensional instability in creating Ganymede grooved terrain: Insights from Galileo high-resolution stereo imaging," *Geophysical Research Letters*, 25(3), 233–236.

Collins, G. C., Head, J. W., III, Pappalardo, R. T. et al. 2001, "The formation of Arbela Sulcus and other smooth linear features on Ganymede: Possible crustal spreading and shear," *Lunar and Planetary Science Conference*, 32, 1498.

Collins, G. C., Patterson, G. W., Head, J. W. et al. 2013, "Global geologic map of Ganymede," US Department of the Interior, US Geological Survey.

Cooper, J. F., Johnson, R. E., Mauk, B. H., Garrett, H. B., & Gehrels, N. 2001, "Energetic ion and electron irradiation of the icy Galilean satellites," *Icarus*, 149(1), 66–79.

Crawford, G. D. & Stevenson, D. J. 1998, "Gas-driven water volcanism and the resurfacing of Europa," *Icarus*, 73(1), 133–159.

Davis, J. R., Titus, S. J., & Horsman, E. 2013, "Non-steady homogeneous deformations: Computational techniques using Lie theory, and application to ellipsoidal markers in naturally deformed rocks," *Journal of Structural Geology*, 56, 142–155.

de Kleer, K., Butler, B., de Pater, I. et al. 2021, "Ganymede's surface properties from millimeter and infrared thermal emission," *Planetary Science Journal*, 2(1), 5.

Dombard, A. J. 2000, "Modeling of geodynamic processes on Ganymede and Callisto: Insight into thermal and tectonic histories," PhD Thesis, Washington University, St. Louis, MO.

Dombard, A. J. & McKinnon, W. B. 2000, "Long-term retention of impact crater topography on Ganymede," *Geophysical Research Letters*, 27(22), 3663–3666.

Dombard, A. J. & McKinnon, W. B. 2001, "Formation of grooved terrain on Ganymede: extensional instability mediated by cold, superplastic creep," *Icarus*, 154(2), 321–336.

Domingue, D. & Verbiscer, A. 1997, "Re-analysis of the solar phase curves of the icy Galilean satellites," *Icarus*, 128(1), 49–74.

Fagents, S. A., Greeley, R., Sullivan, R. J. et al. 2000, "Cryomagmatic mechanisms for the formation of Rhadamanthys Linea, triple band margins, and other low-albedo features on Europa," *Icarus*, 144(1), 54–88.

Fletcher, R. C. & Hallet, B. 1983, "Unstable extension of the lithosphere: A mechanical model for basin-and-range structure," *Journal of Geophysical Research Solid Earth*, 88(B9), 7457–7466.

Giese, B., Oberst, J., Roatsch, T. et al. 1998, "The local topography of Uruk Sulcus and Galileo Regio obtained from stereo images," *Icarus*, 135(1), 303–316.

Golombek, M. P. 1982, "Limits on the expansion of Ganymede and the thickness of the lithosphere," *Lunar and Planetary Science Conference*, 13, 269–270.

Golombek, M. P. & Allison, M. L. 1981, "Sequential development of grooved terrain and polygons on Ganymede," *Geophysical Research Letters*, 8(11), 1139–1142.

Grasset, O., Dougherty, M. K., Coustenis, A. et al. 2013, "JUpiter ICy moons Explorer (JUICE): An ESA mission to orbit Ganymede and to characterise the Jupiter system," *Planetary and Space Science*, 78, 1–21.

Grimm, R. E. & Squyres, S. W. 1985, "Spectral analysis of groove spacing on Ganymede," *Journal of Geophysical Research: Solid Earth*, 90(B2), 2013–2021.

Grundy, W. M., Buie, M. W., Stansberry, J. A., Spencer, J. R., & Schmitt, B. 1999, "Near-infrared spectra of icy outer solar system surfaces: Remote determination of H_2O ice temperatures," *Icarus*, 142(2), 536–549.

Hammond, N. P. & Barr, A. C. 2014, "Formation of Ganymede's grooved terrain by convection-driven resurfacing," *Icarus*, 227, 206–209.

Hansen, G. B. 2002, "A method of measuring surface ice abundance: Application to Ganymede," *Lunar and Planetary Science Conference*, 33, 1060.

Hansen, G. B., McCord, T. B., Smythe, W. D., & Carlson, R. W. 1998, "The state of water ice on the icy Galilean satellites from the Galileo NIMS investigation," paper presented at Nantes98 International Symposium: The Jovian System after Galileo, the Saturnian System before Cassini-Huygens. Nantes, France, May 11–15.

Hapke, B. 1981, "Bidirectional reflectance spectroscopy: 1. Theory," *Journal of Geophysical Research: Solid Earth*, 86(B4), 3039–3054.

Hapke, B. 2021, "Bidirectional reflectance spectroscopy: 8. The angular width of the opposition effect in regolith-like media," *Icarus*, 354, 114105.

Head, J. W., Pappalardo, R., Collins, G. et al. 2002, "Evidence for Europa-like tectonic resurfacing styles on Ganymede," *Geophysical Research Letters*, 29(24), 4-1–4-4.

Head, J. W., Pappalardo, R., Collins, G., Prockter, L., & Galileo Imaging Team 1997, "Nippur Sulcus region, Ganymede: Nature of high-latitude groove lanes and their relation to Marius Regio from Galileo SSI data," *Lunar and Planetary Science Conference*, 28, 537.

Head, J. W., Pappalardo, R. T., Kay, J. et al. 1998, "Cryovolcanism on Ganymede: Evidence in bright terrain from Galileo solid state imaging data," *Lunar and Planetary Science Conference*, 24, 1666.

Helfenstein, P. 1986a, "Derivation and analysis of geological constraints on the emplacement and evolution of terrains on Ganymede from applied differential photometry," PhD thesis, Brown University.

Helfenstein, P. 1986b, "Dark-floored craters on Ganymede: Evidence for devolatilized impact melt," *Lunar and Planetary Science Conference*, 17, 331–332.

Helfenstein, P., Currier, N., Clark, B. E. et al. 1998, "Galileo observations of Europa's opposition effect," *Icarus*, 135(1), 41–63.

Helfenstein, P., Veverka, J., Denk, T. et al. 1997, "Dark-floor craters: Galileo constraints on a Ganymede regolith component," *Lunar and Planetary Science Conference*, 28, 547.

Hibbitts, C. A., Hansen, G. B., McCord, T. B., & Stephan, K. 2003, "Impactor contamination of dark ray craters on Ganymede," *Lunar and Planetary Science Conference*, 34, 1925.

Hibbitts, C. A., McCord, T. B., & Hansen, G. B. 2000, "Distributions of CO_2 and SO_2 on the surface of Callisto," *Journal of Geophysical Research: Planets*, 105(E9), 22541–22557.

Hibbitts, C. A., Pappalardo, R. T., Hansen, G. B., & McCord, T. B. 2003, "Carbon dioxide on Ganymede," *Journal of Geophysical Research: Planets*, 108(E5).

Hillier, J., Helfenstein, P., & Veverka, J. 1996, "Latitude variations of the polar caps on Ganymede," *Icarus*, 124(1), 308–317.

Hirata, N., Suetsugu, R., & Ohtsuki, K. 2020, "A global system of furrows on Ganymede indicative of their creation in a single impact event," *Icarus*, 352, 113941.

Howell, S. M. & Pappalardo, R. T. 2018, "Band formation and ocean-surface interaction on Europa and Ganymede," *Geophysical Research Letters*, 45(10), 4701–4709.

Howell, S. M. & Pappalardo, R. T. 2019, "Can Earth-like plate tectonics occur in ocean world ice shells?," *Icarus*, 322, 69–79.

Howell, S. M. & Pappalardo, R. T. 2020, "NASA's Europa Clipper: A mission to a potentially habitable ocean world,' *Nature Communications*, 11(1), 1–4.

Johnson, R. E. 1985, "Polar frost formation on Ganymede," *Icarus*, 62(2), 344–347.

Johnson, R. E. 1997, "Polar 'caps' on Ganymede and Io revisited,' *Icarus*, 128(2), 469–471.

Johnson, R. E., Carlson, R. W., Cooper, J. F. et al. 2004, "Radiation effects on the surfaces of the Galilean satellites," in Bagenal, F., Dowling, T., & McKinnon, W. B. (eds.), *Jupiter: The Planet, Satellites and Magnetosphere*, Cambridge University Press, pp. 485–512.

Jones, K. B., Head, J. W., Pappalardo, R. T., & Moore, J. M. 2003, "Morphology and origin of palimpsests on Ganymede from Galileo observations," *Icarus*, 164(1), 197–212.

Kadel, S. D., Chuang, F. C., Greeley, R., & Moore, J. M. 2000, "Geological history of the Tyre region of Europa: A regional perspective on Europan surface features and ice thickness," *Journal of Geophysical Research: Planets*, 105(E9), 22657–22669.

Khurana, K. K., Pappalardo, R. T., Murphy, N., & Denk, T. 2007, "The origin of Ganymede's polar caps," *Icarus*, 191(1), 193–202.

Kirk, R. L. & Stevenson, D. J. 1987, "Thermal evolution of a differentiated Ganymede and implications for surface features," *Icarus*, 69(1), 91–134.

Kivelson, M. G., Bagenal, F., Kurth, W. S. et al. 2004, "Magnetospheric interactions with satellites," in Bagenal, F., Dowling, T., & McKinnon, W. B. (eds.), *Jupiter: The Planet, Satellites and Magnetosphere*, Cambridge: Cambridge University Press, pp. 513–536.

Kivelson, M. G., Khurana, K. K., Russell, C. T. et al. 1996, "Discovery of Ganymede's magnetic field by the Galileo spacecraft," *Nature*, 384(6609), 537–541.

Kivelson, M. G., Warnecke, J., Bennett, L. et al. 1998, "Ganymede's magnetosphere: Magnetometer overview," *Journal of Geophysical Research: Planets*, 103(E9), 19963–19972.

Lesage, E., Massol, H., & Schmidt, F. 2020, "Cryomagma ascent on Europa," *Icarus*, 335, 113369.

Lucchitta, B. K. 1980. 'Grooved terrain on Ganymede', *Icarus*, 44(2), 481–501.

Malhotra, R. 1991. "Tidal origin of the Laplace resonance and the resurfacing of Ganymede," *Icarus*, 94(2), 399–412.

Mandl, G. 1987, "Tectonic deformation by rotating parallel faults: The 'bookshelf' mechanism," *Tectonophysics*, 141(4), 277–316.

McCord, T. B., Carlson, R. W., Smythe, W. D. et al. 1997, "Organics and other molecules in the surfaces of Callisto and Ganymede," *Science*, 278(5336), 271–275.

McCord, T. B., Hansen, G. B., Clark, R. N. et al. 1998a, "Non-water-ice constituents in the surface material of the icy Galilean satellites from the Galileo near-infrared mapping spectrometer investigation," *Journal of Geophysical Research: Planets*, 103(E4), 8603–8626.

McCord, T. B., Hansen, G. B., & Hibbitts, C. A. 2001, "Hydrated salt minerals on Ganymede's surface: Evidence of an ocean below," *Science*, 292(5521), 1523–1525.

McKinnon, W. B. 1998, "Geodynamics of icy satellites," in de Bergh, C., Schmitt, B., & Festou, M. (eds.), *Solar System Ices*, Dordrecht: Kluwer Academic Publishers, pp. 525–550.

McKinnon, W. B. & Melosh, H. J. 1980, "Evolution of planetary lithospheres: Evidence from multiringed structures on Ganymede and Callisto," *Icarus*, 44(2), 454–471.

McKinnon, W. B. & Parmentier, E. M. 1986, "Ganymede and Callisto," in Burns, J. A., & Matthews, M. S. (eds.), *Satellites*, Tucson: University of Arizona Press, pp. 718–763.

Melosh, H. J. 1982, "A simple mechanical model of Valhalla basin, Callisto," *Journal of Geophysical Research: Solid Earth*, 87(B3), 1880–1890.

Mills, M., Pappalardo, R. T., & Panning, M. P. 2019, "Moonquake triggered mass wasting on icy worlds," *Lunar and Planetary Science Conference*, 51, 2013.

Moore, J. M., Asphaug, E., Belton, M. J. et al. 2001, "Impact features on Europa: Results of the Galileo Europa Mission (GEM),' *Icarus*, 151(1), 93–111.

Moore, J. M., Asphaug, E., Morrison, D. et al. 1999, "Mass movement and landform degradation on the icy Galilean satellites: Results of the Galileo nominal mission," *Icarus*, 140(2), 294–312.

Moore, J. M., Asphaug, E., Sullivan, R. J. et al. 1998, "Large impact features on Europa: Results of the Galileo nominal mission," *Icarus*, 135(1), 127–145.

Moore, J. M., Chapman, C. R., Bierhaus, E. B. et al. 2004, "Callisto," in Bagenal, F., Dowling, T., & McKinnon, W. B. (eds.), *Jupiter: The Planet, Satellites and Magnetosphere*, Cambridge: Cambridge University Press, pp. 397–426.

Moore, J. M. & Malin, M. C. 1988, "Dome craters on Ganymede," *Geophysical Research Letters*, 15(3), 225–228.

Moore, J. M., Mellon, M. T., & Zent, A. P. 1996, "Mass wasting and ground collapse in terrains of volatile-rich deposits as a solar system-wide geological process: The pre-Galileo view," *Icarus*, 122(1), 63–78.

Moore, J. M. & Pappalardo, R. T. 2011, "Titan: An exogenic world?," *Icarus*, 212(2), 790–806.

Mueller, S. & McKinnon, W. B. 1988, "Three-layered models of Ganymede and Callisto: Compositions, structures, and aspects of evolution," *Icarus*, 76(3), 437–464.

Murchie, S. L. & Head, J. W. 1988, "Possible breakup of dark terrain on Ganymede by large-scale shear faulting," *Journal of Geophysical Research: Solid Earth*, 93(B8), 8795–8824.

Murchie, S. L., Head, J. W., Helfenstein, P., & Plescia, J. B. 1986, "Terrain types and local-scale stratigraphy of grooved terrain on Ganymede," *Journal of Geophysical Research: Solid Earth*, 91(B13), E222–E238.

Murchie, S. L., Head, J. W., & Plescia, J. B. 1990, "Tectonic and volcanic evolution of dark terrain and its implications for the internal structure and evolution of Ganymede," *Journal of Geophysical Research: Solid Earth*, 95(B7), 10743–10768.

Nathan, E. M., Huber, C., & Head, J. W. 2021, "Understanding icy moons as a population of solar system bodies," *Lunar and Planetary Science Conference*, 52, 2548.

Neukum, G. 1997, "Bombardment history of the Jovian system," in Barbieri, C., Rahe, J. H., Johnson, T. V., & Sohus, A. M. (eds.), *The Three Galileos: The Man, the Spacecraft, the Telescope*, New York: Springer, pp. 201–212.

Nimmo, F. & Pappalardo, R. T. 2004, "Furrow flexure and ancient heat flux on Ganymede," *Geophysical Research Letters*, 31(19).

Nimmo, F., Pappalardo, R. T., & Giese, B. 2002, "Effective elastic thickness and heat flux estimates on Ganymede," *Geophysical Research Letters*, 29(7), 62-1–62-4.

Oberst, J., Schreiner, B., Giese, B. et al. 1999, "The distribution of bright and dark material on Ganymede in relationship to surface elevation and slopes," *Icarus*, 140(2), 283–293.

Orton, G. S., Spencer, J. R., Travis, L. D., Martin, T. Z., & Tamppari, L. K. 1996, "Galileo photopolarimeter-radiometer observations of Jupiter and the Galilean satellites," *Science*, 274(5286), 389–391.

Ostro, S. J., Campbell, D. B., Simpson, R. A. et al. 1992, "Europa, Ganymede, and Callisto: New radar results from Arecibo and Goldstone," *Journal of Geophysical Research: Planets*, 97(E11), 18,227–18,244.

Pappalardo, R. T. 1994, "The origin and evolution of ridge and trough terrain and the geological history of Miranda," PhD thesis, Arizona State University.

Pappalardo, R. T. & Collins, G. C. 2005, "Strained craters on Ganymede," *Journal of Structural Geology*, 27(5), 827–838.

Pappalardo, R. T., Collins, G. C., Head, J. W. et al. 2004, "Geology of Ganymede," in Bagenal, F., Dowling, T., & McKinnon, W. B. (eds.), *Jupiter: The Planet, Satellites and Magnetosphere*, Cambridge: Cambridge University Press, pp. 363–396.

Pappalardo, R. T., Head, J. W., Collins, G. C. et al. 1998, "Grooved terrain on Ganymede: First results from Galileo high-resolution imaging," *Icarus*, 135(1), 276–302.

Parmentier, E. M., Squyres, S. W., Head, J. W., & Allison, M. L. 1982, "The tectonics of Ganymede," *Nature*, 295(5847), 290–293.

Passey, Q. R. & Shoemaker, E. M. 1982, "Craters and basins on Ganymede and Callisto-Morphological indicators of crustal evolution," in Morrison, D. (ed.), *Satellites of Jupiter*, Tucson: University of Arizona Press, pp. 379–434.

Patel, J. G., Pappalardo, R. T., Head, J. W. et al. 1999, "Topographic wavelengths of Ganymede groove lanes from Fourier analysis of Galileo images," *Journal of Geophysical Research: Planets*, 104(E10), 24057–24074.

Patterson, G. W., Collins, G. C., Head, J. W. et al. 2010, "Global geological mapping of Ganymede," *Icarus*, 207(2), 845–867.

Peters, K. J. 1992, "Coherent-backscatter effect: A vector formulation accounting for polarization and absorption effects and small or large scatterers," *Physical Review B*, 46(2), 801.

Pizzi, A., Di Domenica, A., Komatsu, G. et al. 2017, "Spreading vs. rifting as modes of extensional tectonics on the globally expanded Ganymede," *Icarus*, 288, 148–159.

Prockter, L. M., Figueredo, P. H., Pappalardo, R. T., Head, J. W., & Collins, G. C. 2000, "Geology and mapping of dark terrain on Ganymede and implications for grooved terrain formation," *Journal of Geophysical Research: Planets*, 105(E9), 22,519–22,540.

Prockter, L. M., Head, J. W., Pappalardo, R. T. et al. 1998, "Dark terrain on Ganymede: Geological mapping and interpretation of Galileo Regio at high resolution," *Icarus*, 135(1), 317–344.

Prockter, L. M., Head, J. W., III, Pappalardo, R. T. et al. 2002, "Morphology of Europan bands at high resolution: A mid-ocean ridge-type rift mechanism," *Journal of Geophysical Research: Planets*, 107(E5), 4-1-4-26.

Purves, N. G. & Pilcher, C. B. 1980, "Thermal migration of water on the Galilean satellites," *Icarus*, 43(1), 51–55.

Rossi, C., Cianfarra, P., & Salvini, F. 2020, "Structural geology of Ganymede regional groove systems (60° N–60° S)," *Journal of Maps*, 16(2), 6–16.

Rossi, C., Cianfarra, P., Salvini, F., Mitri, G., & Massé, M. 2018, "Evidence of transpressional tectonics on the Uruk Sulcus region, Ganymede," *Tectonophysics*, 749, 72–87.

Schenk, P. M. 1993, "Central pit and dome craters: Exposing the interiors of Ganymede and Callisto," *Journal of Geophysical Research: Planets*, 98(E4), 7475–7498.

Schenk, P. M., Chapman, C. R., Zahnle, K., & Moore, J. M. 2004, "Ages and interiors: The cratering record of the Galilean satellites," in Bagenal, F., Dowling, T., & McKinnon, W. B. (eds.), *Jupiter: The Planet, Satellites and Magnetosphere*, Cambridge: Cambridge University Press, pp. 427–456.

Schenk, P. M. & McKinnon, W. B. 1989, "Fault offsets and lateral crustal movement on Europa: Evidence for a mobile ice shell," *Icarus*, 79(1), 75–100.

Schenk, P. M. & McKinnon, W. B. 1991, "Dark-ray and dark-floor craters on Ganymede, and the provenance of large impactors in the Jovian system," *Icarus*, 89(2), 318–346.

Schenk, P. M., McKinnon, W. B., Gwynn, D., & Moore, J. M. 2001, "Flooding of Ganymede's bright terrains by low-viscosity water-ice lavas," *Nature*, 410(6824), 57–60.

Schenk, P. M. & Moore, J. M. 1995, "Volcanic constructs on Ganymede and Enceladus: Topographic evidence from stereo images and photoclinometry," *Journal of Geophysical Research: Planets*, 100(E9), 19009–19022.

Schubert, G., Anderson, J. D., Spohn, T., & McKinnon, W. B. 2004, "Interior composition, structure and dynamics of the Galilean satellites," in Bagenal, F., Dowling, T., & McKinnon, W. B. (eds.), *Jupiter: The Planet, Satellites and Magnetosphere*, Cambridge: Cambridge University Press, pp. 281–306.

Senft, L. E. & Stewart, S. T. 2011, "Modeling the morphological diversity of impact craters on icy satellites," *Icarus*, 214(1), 67–81.

Shaya, E. J. & Pilcher, C. B. 1984, "Polar cap formation on Ganymede," *Icarus*, 58(1), 74–80.

Shoemaker, E. M., Lucchitta, B. K., Wilhelms, D. E., Plescia, J. B., & Squyres, S. W. 1982, "The geology of Ganymede," in Morrison, D. (ed.), *Satellites of Jupiter*, Tucson: University of Arizona Press, pp. 435–520.

Showman, A. P. & Malhotra, R. 1997, "Tidal evolution into the Laplace resonance and the resurfacing of Ganymede," *Icarus*, 127(1), 93–111.

Showman, A. P., Mosqueira, I., & Head J. W. 2004, "On the resurfacing of Ganymede by liquid-volcanism," *Icarus*, 172(2) 625–640.

Showman, A. P., Stevenson, D. J., & Malhotra, R. 1997, "Coupled orbital and thermal evolution of Ganymede," *Icarus*, 129(2), 367–383.

Sieveka, E. M. & Johnson, R. E. 1982, "Thermal- and plasma-induced molecular redistribution on the icy satellites," *Icarus*, 51(3), 528–548.

Singer, K. N., Bland, M. T., Schenk, P. M., & McKinnon, W. B. 2018, "Relaxed impact craters on Ganymede: Regional variation and high heat flows," *Icarus*, 306, 214-224.

Smith, B. A., Soderblom, L. A., Beebe, R. et al. 1979, "The Galilean satellites and Jupiter: Voyager 2 imaging science results," *Science*, 206(4421), 927–950.

Smith, B. A., Soderblom, L. A., Johnson, T. V. et al. 1979, "The Jupiter system through the eyes of Voyager 1," *Science*, 204(4396), 951–972.

Spencer, J. R. 1987a, "Thermal segregation of water ice on the Galilean satellites," *Icarus*, 69(2), 297–313.

Spencer, J. R. 1987b, "The surfaces of Europa, Ganymede, and Callisto: An investigation using Voyager IRIS thermal infrared spectra," PhD thesis, University of Arizona.

Spencer, J. R., Prockter, L., Pappalardo, R. et al. 1998, "Local volatile migration on Ganymede: Galileo SSI images, PPR radiometry, and theoretical considerations," *Lunar and Planetary Science Conference*, 29, 1149.

Spencer, J. R., Tamppari, L. K., Martin, T. Z., & Travis, L. D. 1999, "Temperatures on Europa from Galileo photopolarimeter-radiometer: Nighttime thermal anomalies," *Science*, 284(5419), 1514–1516.

Squyres, S. W. 1980a, "Volume changes in Ganymede and Callisto and the origin of grooved terrain," *Geophysical Research Letters*, 7(8), 593–596.

Squyres, S. W. 1980b, "Surface temperatures and retention of H_2O frost on Ganymede and Callisto," *Icarus*, 44(2), 502–510.

Squyres, S. W. 1982, "The evolution of tectonic features on Ganymede," *Icarus*, 52(3), 545–559.

USGS 2020, "Ganymede Voyager – Galileo SSI Global Mosaic 1km v1," https://astrogeology.usgs.gov/maps/ganymede-Voyager-Galileo-global-mosaics.

Vendeville, B., Cobbold, P. R., Davy, P., Choukroune, P., & Brun, J. P. 1987, "Physical models of extensional tectonics at various scales," *Geological Society, London, Special Publications*, 28(1), 95–107.

Verbiscer, A. J., Helfenstein, P., & Buratti, B. J. 2013, "Photometric properties of solar system ices," in Gudipati, M. S., & Castillo-Rogez, J. (eds.), *The Science of Solar System Ices*, New York: Springer, pp. 47–72.

Wernicke B. & Burchfiel, B. C. 1982, "Modes of extensional tectonics," *Journal of Structural Geology*, 4(2), 105–115.

Zahnle, K., Dones, L., & Levison, H. F. 1998, "Cratering rates on the Galilean satellites," *Icarus*, 136(2), 202–222.

Zahnle, K., Schenk, P., Levison, H., & Dones, L. 2003, "Cratering rates in the outer Solar System," *Icarus*, 163(2), 263–289.

Zahnle, K., Schenk, P., Sobieszczyk, S., Dones, L., & Levison, H. F. 2001, "Differential cratering of synchronously rotating satellites by ecliptic comets," *Icarus*, 153(1), 111–129.

8

The Cratering Record of Ganymede

Surface Ages, Impactor Populations, and Evolutional History

MICHELLE R. KIRCHOFF, AMY BARR, MICHAEL BLAND, VERONICA BRAY, EDGARD G. RIVERA-VALENTÍN AND PAUL SCHENK

8.1 INTRODUCTION

The impact cratering record of any planetary body is not only a chronicle of the population of bodies traversing that environment but also a window on the properties and history of that body. Ganymede, with its terrains of different ages, likely provides the best record of impactor populations that are drawn into the Jupiter system over time (Io and Europa have too few craters and Callisto's densely cratered surface is subjected to pervasive sublimation erosion and is less well observed than Ganymede). In particular, Ganymede's extensive resurfaced terrains provide a well-preserved record of Jupiter system cratering less encumbered by crater saturation effects. Examining major differences in the properties of Ganymede's icy outer shell relative to the 'water-free' Moon, which has comparable surface gravity, reveals the physical properties of that icy shell and its thermal history. Here, we review what is known about crater populations and morphologies on Ganymede, implications for bombardment histories and interior physical and thermal states, and anticipate results from future Ganymede observations.

8.2 CRATER MORPHOLOGIES

Planetary crater morphologies and transitions from one type to another generally scale inversely with planetary surface gravity (e.g., Pike, 1980). The large icy bodies Ganymede, Callisto, Europa, and Titan are at the high end of that scale. Craters on Ganymede follow a morphologic progression beginning with simple craters and developing into more complex morphologies as crater size increases, as summarized by Schenk et al. (2004). A change in relative crater depth is associated with, although not precisely, these transitions (Schenk, 2002), with three transitions in depth-diameter (d/D) ratio for Ganymede craters (Figure 8.1). The first occurs at craters of ~2 km in diameter, marking the simple-to-complex transition. The second at ~30 km is associated with a subtle decrease in crater depths and the transition from central peak morphology to central pit craters. Another, sharper reduction in crater depth occurs after ~60 km and marks the development of crater morphologies containing domes. Each morphologic type, and its alteration over time, is described in this section.

The excavation of material during impact crater formation, combined with the lower density of the fragmented sub-crater debris relative to the surrounding area, creates an isostatically uncompensated feature. The process by which crater topography gradually returns to the isostatic equilibrium of a level plain, viscous relaxation, is dependent on material strength, crater diameter, temperature, and time (see Section 8.6). Viscous relaxation is evident in the altered craters of Ganymede (Figure 8.2) and is particularly notable in the up-bowed floors and subdued rims of the oldest and largest Ganymede craters. These morphologies will be further discussed in Sections 8.2.5 and 8.2.7.

8.2.1 Simple Craters

Simple craters on Ganymede are <2 km in diameter and resemble those seen elsewhere: generally round, bowl-shaped, with a concentric raised rim, when fresh. The depth of these simple craters increases as crater diameter increases (Figure 8.1) and are statistically indistinguishable from lunar simple craters (Schenk, 2002 – the earlier Schenk (1993) Voyager data being unreliable for the smallest craters in that data set). Reliable topographic data for Ganymede simple craters are very limited, and detailed quantitative comparison is not possible at present. The few that are resolved sufficiently to make useful morphologic comparisons show slope streaks and floor debris indicative of the slope failure observed in craters on other planetary bodies.

8.2.2 The Simple-to-Complex (S-C) Transition

On most planetary bodies (e.g., Schenk et al., 2004; Schenk et al., 2021; Robbins and Hynek, 2012; Krüger, Hergarten and Kenkmann, 2018; Robbins et al., 2021), the transition from simple to complex crater morphology has been observed to occur first with rim failure and the formation of extended hummocky floor morphologies and then with the uplift of the crater floor at larger crater diameters to form peaks. If the formation of hummocky floors ahead of central peak formation occurs on Ganymede, then it has not been directly observed – due to inadequate resolutions of well-preserved small craters. The diameter at which craters on Ganymede begin to exhibit central peaks was first estimated at D = 3–5 km by Passey and Shoemaker (1982) and later refined by Schenk (1991) on the basis of Voyager data to D = 5 ± 1 km. This value was updated after more complete Galileo observations to reveal an even smaller S-C transition diameter of 1.9 ± 0.5 km (Schenk, 2002: fig. 2), due to improved resolution of crater interior morphologies.

Central peak craters on Ganymede structurally resemble central peak craters on the terrestrial bodies except for crater

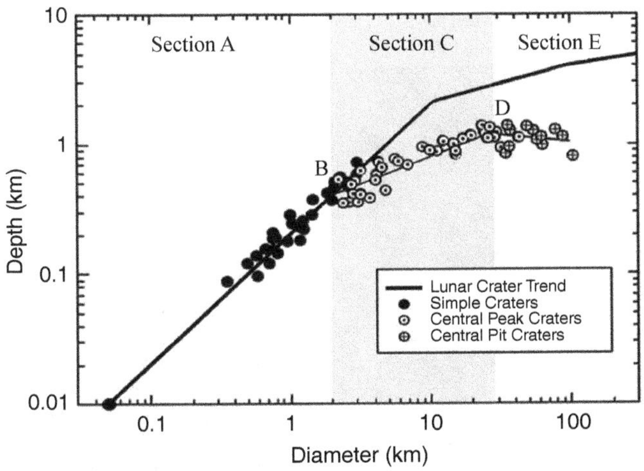

Figure 8.1 Crater rim-to-floor depth versus rim-to-rim diameter of relatively fresh unrelaxed impact craters on Ganymede. Section A demonstrates that simple craters on Ganymede have a similar depth and d/D as lunar craters. The simple to complex transition, and related depth-to-diameter transition are noted "B." The later transition from central peak to central pit morphology is noted "D." Sections C and E show diameter ranges over which central peak and central pit craters are the dominant crater form respectively.
Source: Updated from Schenk, P. M. (2002). Thickness constraints on the icy shells of the Galilean satellites from a comparison of crater shapes, *Nature*, 417, 419–421. Springer Nature.

depths that are 60–70% shallower than the same sized crater on the Moon and the paucity of observable rim terraces (Schenk, 1991, 2002). In craters with diameters smaller than 30 km, central peaks are mainly conical in shape (Figure 8.3A), albeit with asymmetries and off-center locations in some cases. Some of the larger central peaks on Ganymede have a two-tiered morphology with a peak slope that is considerably shallower at its base (Figure 8.3B; Bray et al., 2008). A similar progression in peak morphology with crater size has been noted on the Moon (Bray, Atwood-Stone and McEwen, 2012) and is associated with the development of peak-rings on the Moon (something not observed on Ganymede).

Peak diameters are generally ~1/3 the width of the crater, which is wider than peaks on terrestrial bodies (Moore, Horner and Greeley, 1985; Schenk, 1989), likely due to the more limited terrace development on icy satellites (less rim slumping produces less crater floor debris and preserves more of the central peak above the level of crater floor debris-fill). The height of the central peaks is extremely variable (cf Bray, Atwood-Stone and McEwen, 2012; Bray, Schenk et al., 2012) but generally remains below the pre-impact surface level and below the crater rim height, except for a few examples, or when the crater is viscously relaxed. In general, however, most craters were resolved at no better than 500 m/pixel and more detailed comparisons are not possible until a return to Jupiter.

8.2.3 Flat/Hummocky Floored and Central Peak Craters

Craters that have flat floors extending over >30% of the diameter and do not display a central peak or pit on Ganymede were originally designated 'smooth-floored' craters (Passey and Shoemaker, 1982), although they are not smooth at the higher Galileo resolutions. The terms 'hummocky' and 'flat-floored' have also been used to describe this crater type (Wood and Andersson, 1978). These craters are noted in the transition range between central peak to central pit morphology, and have diameters ranging from ~20 to 40 km (Figure 8.3C). The d/D ratios of flat-floored craters range from 0.05 to 0.17 in the bright terrain, and from 0.013 to 0.17 in dark terrain (Passey and Shoemaker, 1982). They comprise ~20% of the Ganymede craters between 20 and 30 km in diameter and 5% of the craters between 30 and 40 km (Passey and Shoemaker, 1982). These transitional craters often have shallow pitting in the central region, suggesting partial collapse or incipient central pits (see Section 8.2.4).

8.2.4 Central Pit Craters

Although craters on Ganymede exhibit the simple and central peak morphologies seen on the terrestrial planets, no clear peak-ring craters have been observed on Ganymede or other icy bodies (Chapman and McKinnon, 1986; McKinnon, Chapman and Housen, 1991; Croft, 1985). Instead, central pit craters replace the peak-ring morphology in the size-morphology progression. Pit craters are characterized by flattened or upbowed floors with a pit at or near the center (Passey and Shoemaker, 1982). "Summit pit" craters (Figure 8.3D) have the appearance of an incomplete or annular central peak with a depression at its center; larger "floor pit" craters (Figure 8.3E–F) have less pronounced, irregular pit-rims compared to their relatively wide pits. The largest floor pit craters display little or no raised rims around a vast central pit (Schenk et al., 2004). Generally, summit pit crater pits extend down within the summit, but pit depths do not exceed the summit height. In floor pit craters, pit depths extend below the base of the pit-summit (Figure 8.3E), and even below the level of the crater floor in some cases (Figure 8.3F).

Pits in smaller central pit craters tend to be conical in shape (Figure 8.3D–E). As crater size increases, central pits begin to exhibit a flat-floored appearance (Figure 8.3F), followed by a hummocky pit-floor and the emergence of interior central domes at D > 60 km (Schenk, 1993). Some central pit craters also possess an extra topographic high between the pit and crater rim (Figure 8.3E–F; Bray, Schenk et al., 2012). Many pit craters have floors that are domed rather than flat, although this is thought to be due to viscous relaxation after their formation, rather than being a morphological characteristic of the crater type (Schenk, 1993).

Central pit craters on Ganymede range in diameter from 5 to 120 km (Passey and Shoemaker, 1982), peaking in absolute frequency at D = 35–45 km (Alzate and Barlow, 2008). They are abundant on both bright and dark terrains, although slightly more common on dark terrains (Klaybor and Barlow, 2006). Summit and floor pits both increase in diameter (Dp) as crater size increases, but with differing trends: summit pit craters have smaller Dp/D ratios than floor pit craters (e.g. Passey and Shoemaker, 1982; Schenk, 1993; Barlow, 2006). Central pits increase in diameter as crater size increases, but not linearly. Pit diameters are ~20% of crater diameters at small diameters, increasing to ~40% at larger diameters (e.g., Schenk, 1993: fig. 1). The combined trend of pit diameters with crater size produces an exponential relationship, recorded as $Dp = 1.9 \times \exp[0.023D]$ by Passey and Shoemaker (1982) and supported by

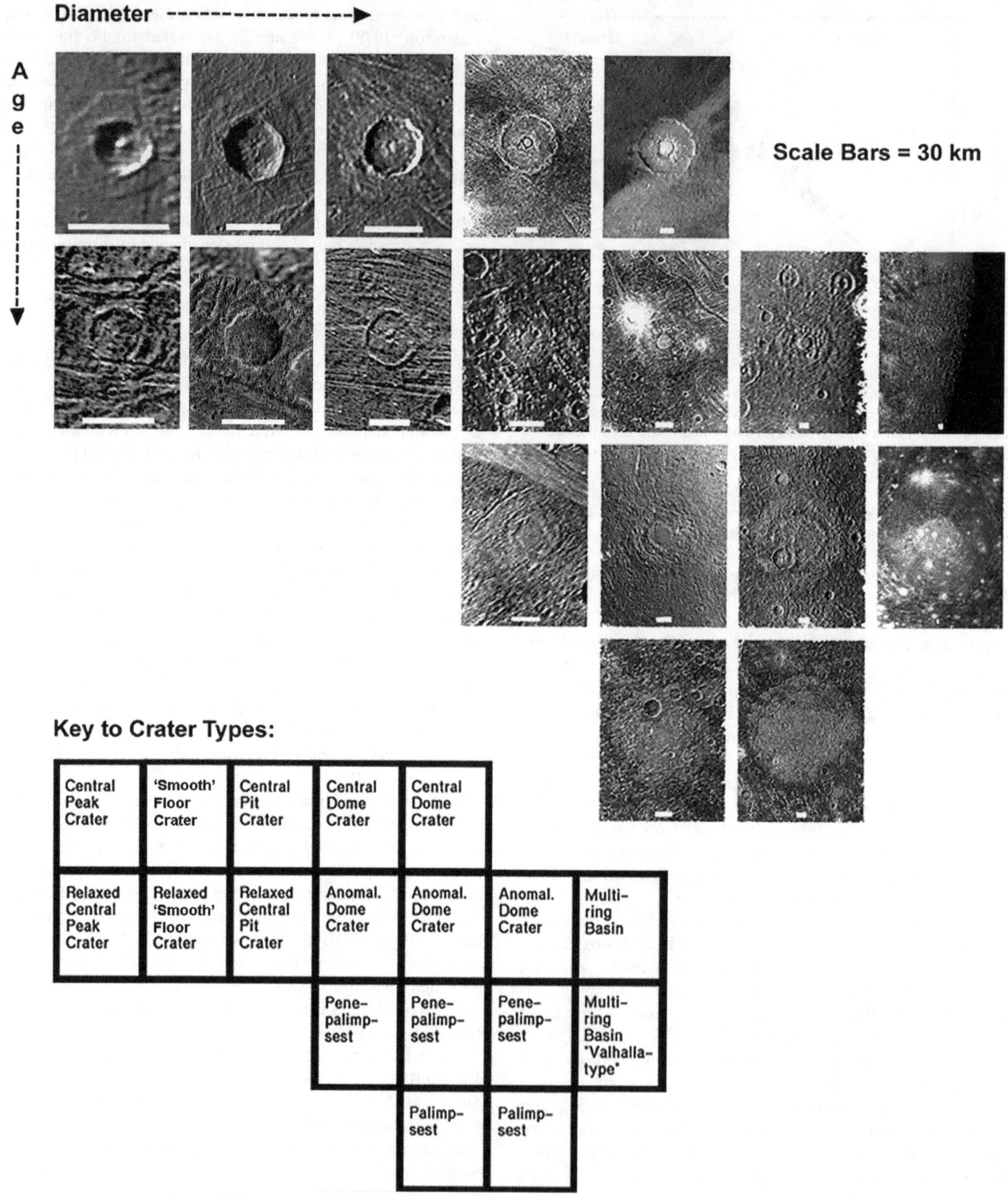

Figure 8.2 Mosaic showing the variety of complex crater morphologies observed on Ganymede. These image tiles are arranged according to likely relative age and crater size. Scale bars are shown on each image and are 30 km long. A companion image gives names for each crater type.
Source: Updated from Schenk (2002).

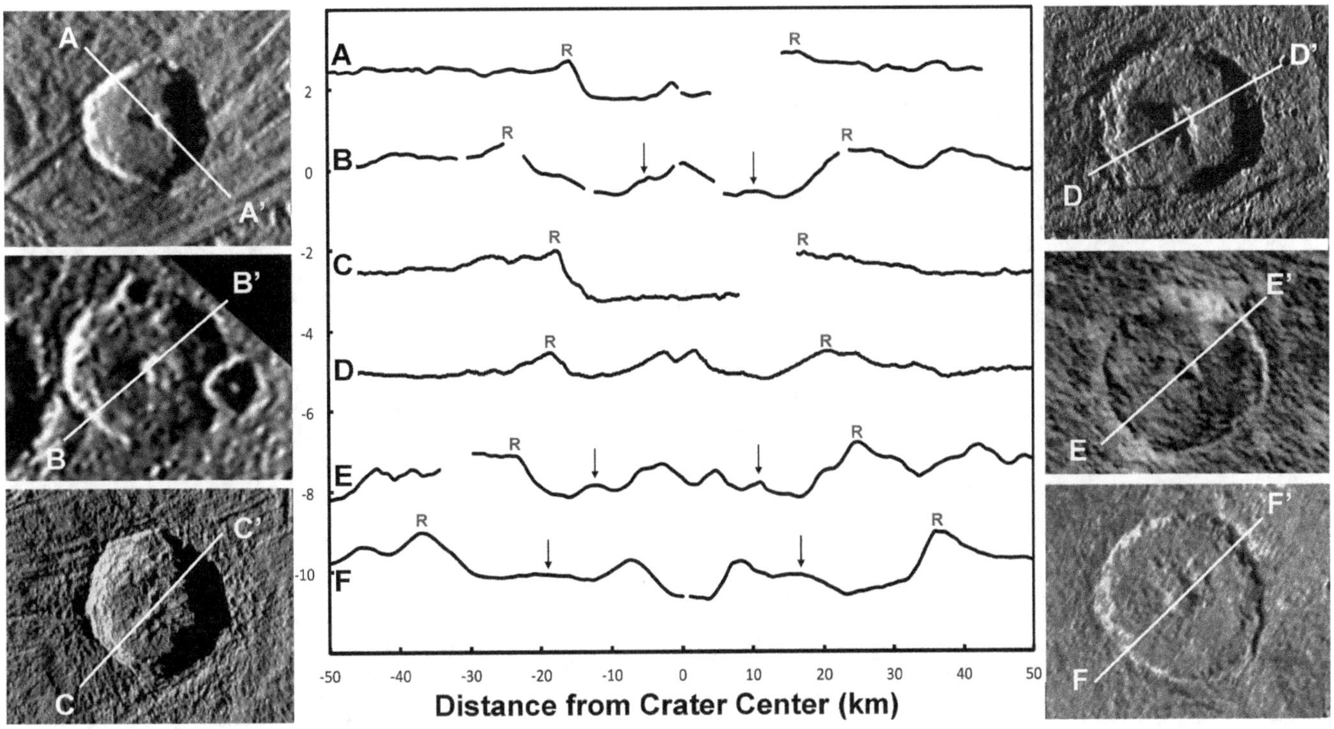

Figure 8.3 Topographic profiles across a selection of Ganymede craters. (A) Central peak crater with conical peak, (B) central peak crater with "tiered" peak, (C) Achelous, a flat/hummocky floors crater, (D) Gula, a summit-pit crater, (E) a central floor-pit crater with conical pit, and (F) Isis, a floor-pit crater with a flat-floored pit. Arrows indicate the presence of topographic highs between the crater center and crater rim. "R" marks the crater rim. Vertical axis shows relative elevation in kilometers.
Source: After Bray, V. J., Collins, G. S., Morgan, J. V. and Schenk, P. M. (2008). The effect of target properties on crater morphology: Comparison of central peak craters on the Moon and Ganymede. *Meteor. Planet. Sci.*, 43, 1979–1992. John Wiley and Sons.

Schenk (1991). Pit depth (dp) increases with increasing crater size, following a weak power law ($dp = 0.00002 \times D^{2.58}$), until crater diameters of ~70 km, at which point an apparent maximum pit depth of ~800 m is reached (Bray, Schenk et al., 2012). The diameter of the (putative) additional internal ring (Dr) around the central peak or pit (Figure 8.3A, E–F), when present, has a Dr/D ratio similar to peak-rings noted on rocky bodies: Dr/D = 0.53 (Bray, Atwood-Stone and McEwen, 2012).

An important constraint on pit origins is their near absence on the mid-sized icy satellites of Uranus and Saturn, which inverse gravity scaling from the Ganymede occurrences predicts at circa 70 km diameters. However, central pits are the dominant floor morphology in craters > 75 km on Ceres (Schenk et al., 2019; Schenk et al., 2021), consistent with inverse gravity scaling, although Ceres is a mixed ice+rock target in which 20–40% of the outer shell is phyllosilicates, clathrates, salts, and carbonates (Ermakov et al., 2017). The relative lack of central pits on the mid-sized icy moons is unexpected but could be due to lower internal temperatures on those bodies, which could be of importance if formation hypotheses that require full or partial melting of the central uplift are valid (see discussion in Schenk et al., 2019). Overall, the few central pit craters observed on the mid-sized icy satellites may imply localized processes.

8.2.5 Central Dome and Anomalous Dome Craters

In craters between D ~ 55 and 180 km on Ganymede, central pits are partially to extensively filled by subcircular domed deposits of slightly higher albedo material (e.g., Moore and Malin, 1988; Schenk, 1993). These domes can rise 500–1000 m above their base and scale with increasing crater diameter (Schenk, 1993). The highest resolution Galileo images show the upper surfaces of domes to be densely fractured compared to the surrounding crater floor, suggesting either subsurface inflation and doming, or flattening and widening by viscous creep, both producing fracturing in the cold brittle surface layers. Relatively older examples of this crater type are referred to as 'anomalous dome craters' as the central dome and pit becomes the predominant feature of the crater to remain topographically expressed but crater rims are poorly expressed or absent (Schenk, 1993; Schenk et al., 2004).

Similar to central pits, the lack of any central domes on mid-sized icy moons (or Mars) but formation of the 3-km-wide dome in Occator crater, Ceres (Schenk et al., 2019), also places constraints on dome origins. Although only one example is still preserved on Ceres, the Occator dome is similar in morphology to the Ganymede examples, including the network of surface fractures on the dome crest related to viscous or ductile deformation within the dome. In the Ceres case, it has been proposed that the dome formed from the freezing of a melt zone beneath the central uplift, but dome origins remain unresolved (e.g., Scully et al., 2019). Observations constraining the emplacement timing of these domes would be critical, and while Schenk (1993) and Schenk et al. (2004) argue that the formation of bright ray and floor material on both the floor and central dome of Osiris crater might suggest very rapid emplacement, *Voyager* imaging was limited to ~800 m/pixel.

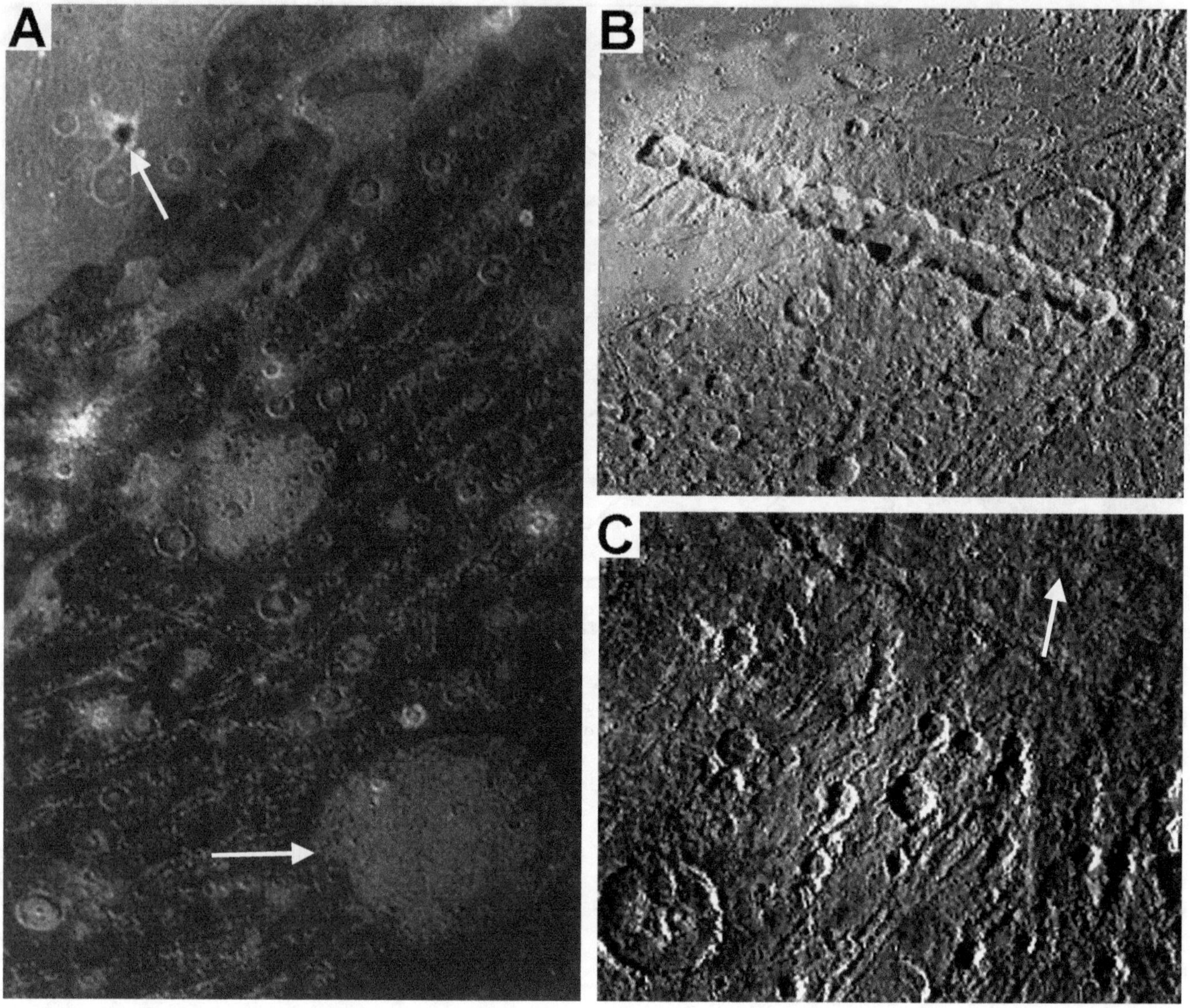

Figure 8.4 (A) Galileo Regio. The furrow system that might record the multi-ring structure of an ancient impact can be seen running from the bottom left to the upper right of this image, forming a wide arc of furrows around an original impact located far beyond the top left. The uppermost arrow points to Khensu, a fresh crater with both dark and bright ejecta deposits. The lower arrow points to Memphis Facula, a palimpsest. (B) Enki Catenae, one of the three crater chains on Ganymede, which crosses a boundary between bright and dark terrain. (C) Secondary crater chains at the edge of the Gilgamesh ejecta deposit. The arrow points towards the approximate crater center, which is not shown in this image.

8.2.6 Multi-Ring Basins

The next transition in the traditional size-morphology progression is the development of multi-ringed basins, which possess at least two asymmetric, inward-facing scarped rings, one of which may be the original crater rim (Hartmann and Wood, 1971; Spudis, Hawke and Lucey, 1984). The exact morphology of multi-ring structures varies across Ganymede, and on different bodies, suggesting that multi-ring basins are likely to be affected by differences in crustal structure and composition (Melosh and McKinnon, 1978). Gilgamesh (62°S, 124°W) is Ganymede's largest confirmed multi-ringed impact basin, with an estimated rim diameter of 585 km (Schenk et al., 2004). This rim location is marked by a prominent inward-facing scarp and is surrounded by a set of discontinuous but roughly concentric inward-facing scarps extending as far as 500 km from the crater center (Thomas, Forni and Masson, 1986). Radial scours outside of Gilgamesh mark the chains of secondary craters produced by the impact (Figure 8.4C), some as large as ~20 km across (Singer, McKinnon and Nowicki, 2013), and mark the extent of continuous ejecta deposited as a result of the Gilgamesh impact. A 144 km diameter depression at the center of Gilgamesh hosts a relatively smooth and high albedo surface. These central smooth areas are common in the largest Ganymede craters and are thought to record solidified impact melt, and/or the partial to full breach of the ice crust by the basin forming impact (e.g., Jones et al., 2003; Schenk et al., 2004).

Hemispheric-scale concentric furrow systems similar in appearance to the Valhalla rings on Callisto are present on the ancient dark terrains (Figure 8.4A). These could mark the remnants of huge multi-ring basins spanning hundreds and thousands of kilometers in diameter (e.g., Passey and

Shoemaker, 1982). The impact origin of these features has not been confirmed, as the ring center has been erased/overwritten by bright terrain formation (Zuber and Parmentier, 1984; Murchie, Head and Plescia, 1990). Hirata, Suetsugu and Ohtsuki (2020) have suggested that the furrow systems across multiple sections of dark terrain might be due to a single impact of a 300 km diameter impactor.

8.2.7 Palimpsests

The very oldest and largest of the preserved crater forms on Ganymede are the palimpsests; they are nearly featureless circular regions of high albedo with no discernable crater rim or rim-like features (Figure 8.4A; Passey and Shoemaker, 1982). Smooth, subcircular regions at the center of Ganymede's palimpsests are thought to be solidified impact melt water (Jones et al., 2003) and/or a subsurface water layer exposed during the impact. Palimpsests occur predominantly but not exclusively on the ancient dark terrains. Crater counts and cross-cutting relationships suggest that the formation of these features predated the emplacement of Ganymede's youngest geologic terrain (the bright terrain) (Shoemaker, 1982), making them approximately older than 1 Ga (Zahnle, Dones and Levison, 1998; Section 8.5). The original crater diameters associated with the palimpsests are estimated to be $\geq$36 km on the basis of secondary crater positions (Jones et al., 2003). As relatively younger craters of this diameter maintain topographically expressed central pit morphology, it suggests that the ancient Ganymede crust was thinner than in more recent times. A thinner crust is likely to be associated with higher heat flow and higher rates of viscous relaxation (see Section 8.6.2).

8.2.8 Ejecta Deposits and Secondary Craters

The material ejected during the impact process deposits around the crater, forming a 'blanket' of continuous ejecta from the rim until ~1 crater radii from the crater center (e.g., Schenk and Ridolfi, 2002). Pedestal, rampart, single layer and multi-layer ejecta morphologies have been noted around Ganymede craters (e.g., Horner and Greeley, 1985; Boyce et al., 2010). Secondary craters on Ganymede occur, as expected from other planetary bodies (e.g., Bierhaus et al., 2018), beyond the limit of the continuous ejecta deposit. Secondary craters can occur as isolated craters that exhibit standard simple bowl-shaped morphology and as clusters, chains, and radial scours (e.g., Figure 8.4C). The relatively low impact velocity of ejected debris may lead to shallower secondary craters (e.g., Shoemaker, 1965; Schultz and Gault, 1985; Bierhaus and Schenk, 2010; Bierhaus et al., 2018). Some of the ejected material does reach a high enough velocity to escape Ganymede's gravity and impact Ganymede or another jovian system body as a 'sesquinary' impact (Alvarellos et al., 2005, 2008; Alvarellos et al., 2017; Zahnle et al., 2008). On Ganymede, craters above ~2 km in diameter are unlikely to be sesquinaries because the largest known crater – Gilgamesh – does not appear to produce large enough spall fragments. Furthermore, it is not likely that large, high-velocity fragments remain intact while traveling (Singer, McKinnon and Nowicki, 2013).

The youngest impact craters on Ganymede display distinct rays radial to the crater. In most cases, this ejected material has a higher albedo than its surrounding terrain, presumably due to it being composed of freshly exposed water ice. The largest ray system on Ganymede emanates from Osiris crater (D ~ 107 km, 38°S, 166.31°W), and reaches at least 1 500 km from the crater center. Dark and mixed albedo ejecta deposits are also noted on Ganymede (e.g., Khensu, Figure 8.4A). The dark ejecta could be due to the excavation of dark material from depth, particularly in the cases of ejecta that displays both bright and dark layers, or the darkening of originally bright ejecta over time (Johnson et al., 2004).

8.2.9 Chains/Catanae

In addition to the scours and chains of impact craters formed by secondary impacts, to date three primary crater chains have been identified on Ganymede (Schenk et al., 1996; Figure 8.4B) and several on Callisto. The linear orientation of the individual craters that make up these chains suggests that these are not clusters of multiple impactors. Instead, it is likely that a comet was tidally disrupted during passage past Jupiter, as observed with Comet Shoemaker-Levy 9, and impacted the surface of Ganymede on the outbound trajectory (Melosh and Schenk, 1993).

8.3 CRATER DENSITIES

At present, measurements of terrain crater densities are the best resource to compare the relative formation of major geologic units on outer solar system bodies, as well as inter-satellite comparisons of relative surface ages. Typically, crater density is expressed as the cumulative number of craters above a given diameter per unit area and is represented as $N(> D)$, with uncertainty following Poisson statistics (CATWG, 1979). However, given advancements in computing techniques, bootstrap methods have been suggested to more robustly account for measurement uncertainties (Robbins et al., 2018). Given two terrains on the same body, and all impact bombardment being equal, the terrain with the higher cumulative crater density within uncertainty would represent an older surface. Secondary/sesquinary cratering (see Section 8.2.8), though, can potentially contaminate and mislead these interpretations. Furthermore, comparisons at specific diameter ranges provide information on relative crater erasure, such as by ejecta infill of small craters. Comparison of two satellites orbiting the same primary body is complicated due to differences in impact characteristics (e.g., velocity, target gravity, etc.). Nevertheless, inferences can be made by comparing the global crater densities between the two bodies.

For Ganymede, the first crater measurements were compiled from the *Voyager* 1 and 2 flybys at limited imaging resolutions of no better than ~2 km/pixel (Smith, Soderblom, Johnson et al., 1979; Smith, Soderblom and Beebe et al., 1979; Strom, Woronow and Gurnis, 1981; Neukum, 1982; Passey and Shoemaker, 1982; Woronow, Strom and Gurnis, 1982; Neukum and Pozio, 1984; Chapman and McKinnon, 1986; Melosh and Schenk, 1993; McKinnon and Schenk, 1995; Schenk

Table 8.1 *Summary of Ganymede crater dimension scaling trends.*

Dimension	Equation/value	Diameter Range	Reference	Notes
Depth, d	$d = 0.2D$ $d = 0.227\ D^{0.446}$	<1.9 (0.5) km 5–50 km	Schenk, 1991; Schenk, 2002; Bray et al., 2008	*Voyager, Galileo, Galileo*
Rim height, Hr	$Hr = 0.021\ D^{1.22}$ $Hr = 0.032\ D^{0.72}$	<9 km 9–51km	Bray, Schenk et al., 2012	*Galileo*
Wall Slope, S	$S = 173.41\ (d/D)^{0.84}$	4–50 km	Bray, Schenk et al., 2012	*Galileo*
Continuous ejecta radius, Er	$Er = 2.74\ D^{0.793}$		Croft, 1988	*Voyager*
Largest secondary crater diameter (Ds) about a primary (D)	$Ds = 0.12D^{0.805}$		Croft, 1988	*Voyager*
Central peak height, Hcp	$Hcp = 0.0004\ D^{2.557}$ $Hcp = 0.0018\ D^{1.54}$	<15 km 12–50 km	Schenk, 1991; Bray et al., 2008	*Voyager, Galileo*
Central pit diameter, Dp	$Dp = 0.005D$^0.89	4–150 km	Schenk, 1993	*Voyager*
Central pit depth dp	$dp = 0.0016D^{1.42}$	<100 km	Bray, Schenk et al., 2012	*Galileo*
Dome width, Dd	$Dd = 0.1D$ $Dd = 0.25D$	50–70 km >70 km	Moore and Malin, 1988	*Voyager*
Dome height, Hd	$Hd = 0.025(Dd)^{1.2}$	Dome widths of 5–30 km	Schenk, 1993	*Voyager*

et al., 1996). These early results indicated that Ganymede's dark terrain was relatively older than its bright terrain (Woronow, Strom and Gurnis, 1982; Casacchia and Strom 1984; Murchie, Head and Plescia, 1989). Additionally, although deeper within the jovian gravity well, *Voyager* results indicated that, overall, Ganymede was less cratered than Callisto (Strom, Woronow and Gurnis, 1981; Schenk et al., 2004). Based on these early crater densities, the dark terrain was interpreted to have been emplaced during a prolonged period of resurfacing early on in solar system history, while the bright terrain was later emplaced during subsequent activity (Murchie, Head and Plescia, 1989). Given that Ganymede's surface appears younger than Callisto's, the formation of dark terrain may have occurred at some 3.8–4.0 Gy (Murchie, Head and Plescia, 1989), while the age of the bright terrain may be at some 1–2 Gy (Neukum et al., 1998; Zahnle et al., 2003; Schenk et al., 2004; see Section 8.5).

Later, the *Galileo* mission provided considerable improvement over *Voyager* results, obtaining images with resolution down to ~70–200 m/pixel in small areas, but most of the data were at 0.8–3.6 km/pixel. These data further reinforced *Voyager*'s findings on the crater densities of the main units of Ganymede but also facilitated a detailed study of each of the major dark terrains – Galileo, Perrine, Nicholson, and Marius. In Table 8.2, we show example cumulative crater density results for these terrains for $N(D > 8\ \text{km})$ and $N(D > 30\ \text{km})$, that is, for craters larger than 8 km and 30 km, respectively (Seddio and Schenk, 2007; Rivera-Valentín, Kirchoff and Schenk, 2008). In their study, Rivera-Valentín, Kirchoff and Schenk (2008) also included counts from high-resolution images of Nicholson Regio, which found a crater density of $N(D > 1\ \text{km}) = 2.5(\pm 0.1) \times 10^{-2}\ \text{km}^{-2}$ while recent work by Martellato et al. (2021) in preparation for the JUpiter ICy moons Explorer (JUICE) mission by the European Space Agency (ESA), found $N(D > 1\ \text{km}) = 1(\pm 0.1) \times 10^{-2}\ \text{km}^{-2}$.

The measured cumulative crater densities on Ganymede's dark terrains indicate that they are ~75% less densely cratered than Callisto. In particular, Galileo Regio was found to be far less densely cratered than Callisto and less than half as densely cratered as other Ganymedian dark terrains. Considering craters >8 km in diameter, on average Galileo Regio is approximately one-third less cratered than other dark terrain. This may imply that Galileo Regio is a relatively young dark terrain unit, though still much older than the bright terrains. Generally, Galileo Regio may be some 2 to 50 times more cratered than the bright terrain (Table 8.2). The crater density for diameters larger than 30 km, though, is similar between dark terrains (Table 8.2), which may imply preferential erasure of smaller craters within Galileo Regio. Indeed, studies of the highest-resolution images over Galileo Regio revealed that the terrain has been modified by tectonic deformation, impact ejecta, and potentially cryovolcanism (Prockter et al., 1998).

Given its position on Ganymede (47°N, 129.6°W), though, the lower crater density within Galileo Regio may also be a result of the so-called apex–antapex asymmetry. This predicted cratering asymmetry results because the orbital velocity of a synchronously rotating satellite is large compared to impact velocities by a heliocentric population (Shoemaker and

Table 8.2 *Measured cumulative crater density on Ganymede. Measurements over the dark terrain are from Rivera-Valentín, Kirchoff and Schenk (2008) and from Seddio and Schenk (2007) for the bright terrain.*

Terrain	**N(D > 8 km) per 10^6 km^2**	**N(D > 30 km) per 10^6km^2**
Bright Terrain	62 ± 57	8 ± 7
Galileo Regio	246 ± 7	25 ± 2
Perrine Regio	764 ± 38	37 ± 7
Marius Regio	964 ± 84	28 ± 13
Nicholson Regio	512 ± 114	26 ± 26
Nicholson Regio *high resolution, N(D > 1 km)*	25000 ± 800 (Rivera-Valentín, Kirchoff and Schenk, 2008) 10000 ± 830 (Martellato et al., 2021)	

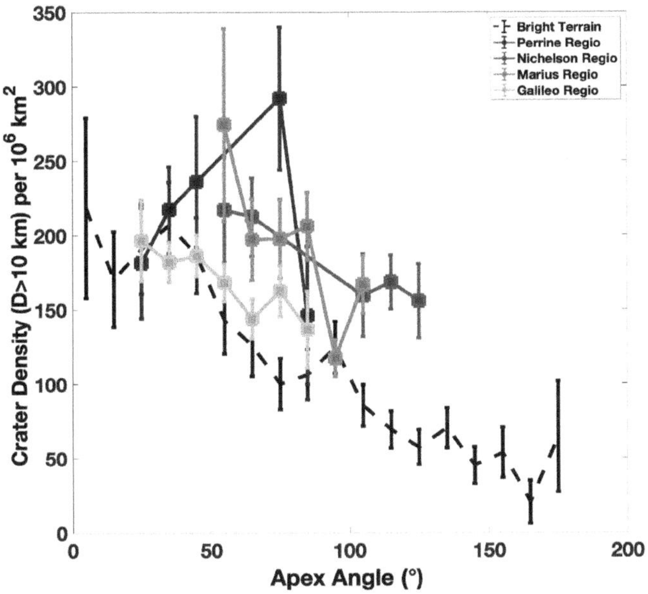

Figure 8.5 Crater densities for different terrains as a function of distance from the apex of motion (0°; leading hemisphere). No significant apex-antapex (180°) difference is observed for the dark terrain. A bigger difference is observed for the bright terrain, but not as large as predicted by theory.

Wolfe 1982; Horedt and Neukum 1984; Zahnle, Dones and Levison, 1998; Zahnle et al., 2001). This leads to a more heavily cratered leading than trailing hemisphere. Predictions based on Monte Carlo modeling indicated that the leading hemisphere of Ganymede should be 70 times more cratered than the trailing hemisphere (Zahnle et al., 2001). Although a cratering asymmetry was found, initial results showed that Ganymede's apex may only be ~4 times more heavily cratered than its antapex (Zahnle et al., 2001). Detailed crater counts on the dark terrains, though, found this asymmetry to be only of a factor of two (Figure 8.5; Rivera-Valentín, Kirchoff and Schenk, 2008; Yozzo, Kirchoff and Schenk, 2009), along with a factor of three for craters within the bright terrain (Figure 8.5; Xu, Hirata and Miyamoto, 2017). Xu, Hirata and Miyamoto (2017) further found that the youngest craters on Ganymede, the rayed craters, continue to exhibit an apex–antapex asymmetry, but only over the bright terrain. However, the authors point out that their null result over the dark terrain may be attributed to uncertainties due to lower counts of rayed craters over Ganymede's dark terrain. This discrepancy between predicted and observed crater density asymmetries may suggest that Ganymede's surface rotated nonsynchronously or its pole wandered in the past (Zahnle et al., 2003; Mohit, Greenhagen and McKinnon, 2012; McKinnon et al., 2014). Nonsynchronous rotation may also explain other features, such as Ganymede's grooved terrain (e.g., Nimmo and Pappalardo, 2004; Bland, Showman and Tobie, 2009). The observation of a large topographic dome at the current sub-Jovian region supports a potential polar wander episode on Ganymede (Schenk, McKinnon, Moore et al., Chapter 9, this volume). However, although apex–antapex crater density asymmetries have been observed on other satellites, they too have been much less than predicted (e.g., Leliwa-Kopystyński, Banaszek and Wlodarczyk, 2012).

An alternate hypothesis to explain the observed apex–antapex discrepancy is crater saturation, also known as crater equilibrium and crater saturation equilibrium. A surface is said to be saturated by craters when every newly formed crater erases a preexisting impact crater, and thus the overall surface crater density remains unchanged (Chapman, 1968; Gault, 1970; Woronow, 1977, 1985; Richardson, 2009; Kirchoff, 2018). The similarity in the crater density of the oldest planetary surfaces across the solar system may support the occurrence of crater saturation (Hartmann, 1984). Because smaller impact craters form more frequently than larger craters, saturation, or steady-state, may be reached more rapidly for smaller craters, while craters of diameters larger than ~10 km, which form less frequently, might not reach saturation (Strom, 1977; Woronow, 1977; Woronow, Strom and Gurnis, 1982; Strom et al., 2005; Marchi et al., 2012; Marchi et al., 2013). However, studies using spatial statistics argue that saturation can be reached even for kilometer-sized craters (Kirchoff, 2018; Lissauer et al., 1988; Squyers et al., 1997). Should a planetary surface reach saturation, the assumed association between measured crater densities and surface ages would no longer be valid. Additionally, apex–antapex asymmetries would get washed out. Robustly identifying a saturated surface, though, is complicated because the saturation process itself is dependent on the impactor population's size-frequency distribution and size dynamic range (Woronow, 1977; Hartmann 1984; Chapman and McKinnon, 1986; Hartmann and Gaskell, 1997; Richardson, 2009). Using crater counts of Ganymede along with detailed modeling of spatial statistics, Kirchoff (2018) showed large craters on the most heavily cratered terrains on Ganymede may be approaching saturation but are not yet saturated. As such, crater saturation in older terrains might not explain the apex–antapex discrepancy.

8.4 CRATER SIZE-FREQUENCY DISTRIBUTIONS

Initial crater size-frequency distributions (SFDs) were compiled from the *Voyager* 1 and 2 flybys (Smith, Soderblom, Johnson et al., 1979; Smith, Soderblom and Beebe et al., 1979; Strom, Woronow and Gurnis, 1981; Woronow, Strom and Gurnis, 1982; Neukum, 1982; Passey and Shoemaker, 1982; Neukum and Pozio, 1984; Chapman and McKinnon, 1986; Melosh and Schenk, 1993; McKinnon and Schenk, 1995; Schenk et al., 1996). As is their nature, flybys provide variable resolution and solar incidence angle coverage of the surfaces. The highest pixel scale imaging is at closest approach, with worsening pixel scale during the approach and retreat. Solar incidence angle is dependent on the position of the Sun, satellite, and spacecraft; and the sequence was planned to get solar incidence angles optimal for identification of topographic features (~55–80°) during the nearest approach. Thus, imaging of areas appropriate for identifying craters was available but limited. Fortunately, that area sampled both the relatively older dark terrain and the younger bright terrain on Ganymede, so the crater SFDs of both could be examined. However, the analysis was regional, so any potential global variability could not be constrained. A primary analytical tool to compare crater SFDs is the so-called R-plot, or relative size-frequency distribution plot (e.g., Figure 8.6). These are differential size-frequency distribution plots that show the distribution function relative to a slope of the form D^{-3}, where the slope of -3 is chosen because many crater populations follow a power law of this form (CATWG, 1979).

Several important features of Ganymede's crater SFDs came out of analyses based on *Voyager* 1 and 2 (e.g., Figure 8.6) (1) the SFDs on Ganymede (and Callisto) did not look like those from the inner solar system (e.g., Strom, Woronow and Gurnis, 1981; Woronow, Strom and Gurnis, 1982; Chapman and McKinnon, 1986), although a counterargument came from Neukum (1982) and Neukum and Pozio (1984) that "shifted" crater SFDs on Ganymede (and Callisto) were similar to the global lunar crater SFD; (2) at least the larger craters ($D \gtrsim 40$ km) are representative of the impactor production population; (3) the dark terrain (and Callisto) contains many large ($D \gtrsim 50$ km) degraded craters; (4) Ganymede has fewer relatively smaller craters ($D \approx 10$–40 km) in the dark terrain than Callisto; (5) clusters and chains of smaller secondary craters are identifiable on the bright terrain (e.g., Strom et al., 1981); and (6) chains of larger craters ($D \gtrsim 5$ km) are identifiable across the dark terrains (and on Callisto; Melosh and Schenk, 1993; Schenk

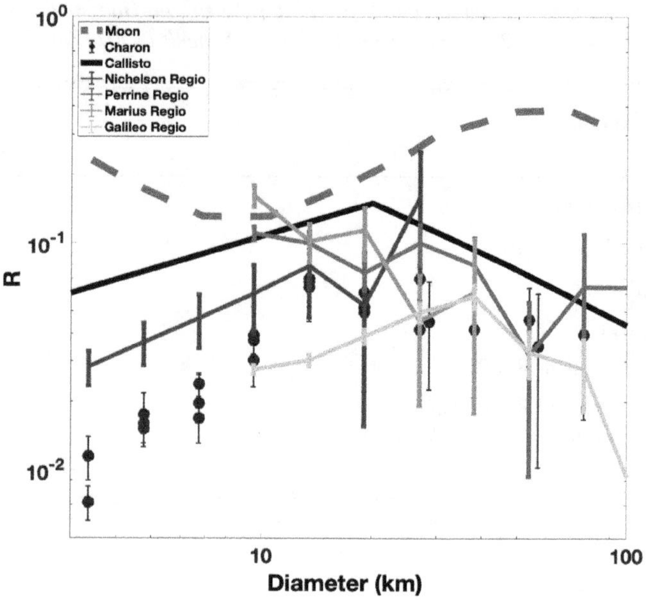

Figure 8.6 Crater size-frequency distributions for the jovian satellites, Charon (Singer et al., 2019), and the Moon (e.g., Neukum et al., 2001) in relative or R-plot format to compare the shapes and densities. Crater diameters have not been scaled and are original to the surface they formed on.

et al., 1996; see Section 8.2.9). Results (1) and (2) imply that there could be a different impactor population for the inner and outer solar systems (at least at Jupiter) and that this population has fewer "large" impactors ($D_i \sim 2$–4 km), but no consensus could ultimately be reached (e.g., Woronow et al., 1982; Neukum, 1982; Neukum and Pozio, 1984; Chapman and McKinnon, 1986). Results (3) and (4) imply that both satellites have been affected by a process(es) that degrades craters of all sizes, which leaves heavily degraded craters ≥50 km in diameter and has removed craters from 10 to 40 km in diameter on Ganymede, where the process(es) has likely lasted longer (e.g., Passey and Shoemaker, 1982; see Sections 8.2 and 8.6). Result (5) implies that young large craters on Ganymede have left a population of secondary craters that are readily identifiable on the bright terrain (e.g., Strom, Woronow and Gurnis, 1981). Finally, result (6) implies that the break up of larger comets by Jupiter (e.g., the Shoemaker-Levy 9 event) occurs and can leave traces on the surfaces of Jupiter's satellites (Melosh and Schenk, 1993; Schenk et al., 1996; see Section 8.2.9).

The return to the Jupiter system by the *Galileo* spacecraft provided global coverage of the satellites at improved resolutions and solar incidence angle coverage (see Section 8.3). These data finally allowed studies of global variations of the crater SFDs extended to smaller diameters (Figure 8.6) and further insight into the results gained from the *Voyager* studies. Early results indicated a dearth of small craters ($D < \sim 500$ m) compared to the larger craters and inner solar system bodies, excluding the continued identification of clusters and chains of secondary craters, which may have caused a steeper crater SFD slope for $D \sim 1$ km (Chapman and Merline, 1996; Chapman, 1997; Chapman et al., "Impactor populations," 1997; Chapman et al., "Populations," 1997; Chapman et al., 1998). From these data, however, it still was not clear if the dearth of small craters was a result of a reduced number of small impactors or geologic process(es) on Ganymede. Support for the former was described by Zahnle and colleagues (Zahnle et al., 1998; Zahnle et al., 2003) based upon other observations in the jovian system: population of small comets is deficient at Jupiter; Io has no observed impact craters, which is inconsistent with the existence of small impactors; and Callisto's surface would not have become so thermally weathered if bombardment were higher. Data from the Pluto system (Singer et al., 2019) have now further confirmed a relative paucity of small comets throughout the outer solar system.

Following on from Neukum and colleagues (Neukum, 1997; Neukum et al., 1997; Neukum et al., 1998), the counterargument is that the crater SFDs on Ganymede (and Callisto), when shifted horizontally to larger diameters (Fig. 8.6 and plotting Ganymede and Callisto data to the right on the R-plot), were similar to the global lunar crater SFD. Newer analyses first found differences in the shapes between the Ganymede and Callisto crater SFDs (Ivanov et al., 2002; Seddio and Schenk, 2007), potentially implying a changing impactor population and/or variable geological histories, and then later found they were similar, potentially implying similar impactor populations (Rivera-Valentín, Kirchoff and Schenk, 2008). Overall, however, consensus again could not be reached on the relevant impactor population(s) from the crater populations (it eventually took data returned by the *New Horizons* spacecraft at the Pluto system to better constrain the impactor population – see Section 8.5). The global data indicated that larger craters are heavily degraded across all of Ganymede's dark terrain and that generally all of the dark terrain has fewer craters than Callisto (e.g., Prockter et al., 1998; Ivanov et al., 2002; Wagner et al., 2003; Seddio and Schenk, 2007; Rivera-Valentín, Kirchoff and Schenk, 2008; Figure 8.6). Thus, the process(es) affecting at least the larger craters appears global.

8.5 ABSOLUTE MODEL AGES

To compute absolute ages of planetary terrains, one needs to know the primary impact crater density and details about the relevant heliocentric impactor population, such as its size-frequency distribution and flux. For any planetary body, these are not easy to obtain due to non-heliocentric impactor populations (i.e., secondary, sesquinary and planetocentric impactors; e.g., Wilhelms, 1987; Gladman et al., 1995; Zahnle et al., 2001; Schenk et al., 2004; Alvarellos et al., 2005, 2008; Alvarellos et al., 2017; Bierhaus, Chapman and Merline, 2005; McEwen and Bierhaus, 2006; Zahnle et al., 2008; Dones et al., 2009; Dobrovolskis et al., 2010; Bierhaus et al., 2012, 2018; Zanetti et al., 2017; Kirchoff et al., 2018; Schenk et al., 2020), a lack of observations of heliocentric populations, particularly at smaller sizes (e.g., Fraser, Brown and Schwamb, 2010; Mainzer et al., 2014; Harris and D'Abramo, 2015; Ryan et al., 2015; Shankman et al., 2016; Nesvorný et al., 2019; Yoshida et al., 2019), and a lack of constraints on the impact flux (e.g., Zahnle, Dones and Levison, 1998; Zahnle et al., 2003; Cohen, Swindle and Kring, 2005; Stöffler et al., 2006; Ćuk, 2012; Plescia, 2012; Robbins, 2014; Huang et al., 2018; Nesvorný and Roig, 2018). Sample return from known locations on the Moon (e.g., Stöffler et al., 2006) and extensive surveys of the near-Earth asteroid population (e.g., Mainzer et al., 2014) have helped build some constraints for the inner solar system, but the outer solar system has far fewer constraints (e.g., Zahnle et al., 1998, 2003; Nesvorný et al., 2019). The heliocentric population of the outer solar system likely originates from the Kuiper Belt (e.g., Duncan, Quinn and Tremaine, 1988; Levison and Duncan, 1997; Zahnle, Dones and Levison, 1998; Zahnle et al., 2003; Brunini, Di Sisto and Orellana, 2003; Gomes et al., 2005; Morbidelli et al., 2005; Di Sisto and Brunini, 2007, 2011; Di Sisto and Zanardi, 2013; Singer et al., 2019), which, given its distance from Earth (~30 AU), is difficult to directly observe. Additionally, a lack of correlated physical samples of terrains with measured crater densities and poor constraints on the impact flux – even current – make the absolute aging of icy moon terrains difficult. Owing to the large number of unknowns, models of the impactor population and flux have to be used, resulting in the computation of absolute *model* ages (AMAs). Given the difficulty in differentiating impacts features made by heliocentric objects from those of non-heliocentric objects (e.g., Wilhelms, 1987; McEwen and Bierhaus, 2006; Alvarellos et al., 2017; Zanetti et al., 2017; Bierhaus et al., 2018; Schenk et al., 2020), significant error remains in interpreting AMAs.

The first estimated AMAs for the outer solar system were computed by Shoemaker and Wolfe (1982). They assumed that active and extinct comet nuclei were the primary impactors and extrapolated from available observations of those objects (only one for extinct comet nuclei) and the cratering records of the jovian satellites. They used a dynamical calculation of the probability of impacts with each satellite and basic gravity-regime crater scaling laws developed for rock (for a more recent review, see Holsapple and Housen, 2007), assuming that impacts in ice would be proportional. This gave an estimate of the "current" cratering rate of 2.3×10^{-14} per km^2 per year for D $\geq$ 10 km. Then, assuming a similar evolution in the impactor flux as what had been suggested for the Earth–Moon system with a late heavy bombardment period and a recent increase in the flux, they assumed this "current" value was appropriate for 0–0.5 Ga, the rate for 0.5–3.0 Ga is half the "current" rate, and the rate exponentially increases for >3.0 Ga. As more observations of comets became available, they updated their current rate in Shoemaker, Wolfe and Shoemaker (1986) to 6.6×10^{-14}. In preparation for the *Galileo* mission, Nakamura and Yoshikawa (1995) and Shoemaker (1996) worked on improving the recent impact rates by incorporating new observations and understanding of the impactor dynamics and found values of 6×10^{-15} and 9.5×10^{-14} per km^2 per year, respectively. Note the Nakamura and Yoshikawa (1995) rate did not include extinct comets, thus explaining their generally lower value. Zahnle and colleagues (Zahnle, Dones and Levison, 1998; Zahnle et al., 2003) further improved on these analyses using modern observations of active and extinct comets and crater distributions, actual observed impacts like Shoemaker-Levy 9 and improved understanding of the dynamics, crater scaling and impact flux changes. Their estimated current impact rates for D $\geq$ 10 km craters on Ganymede were 6×10^{-14} per km^2 per year and 1.8×10^{-14} for the more recent update in 2003. They also assumed an exponential increase from this current flux for 3.0 Ga and older but did not include any variation in the recent flux, explicitly assuming that there is no tie to the bombardment history of the Earth–Moon system. Overall, these estimates indicate the recent flux is unknown to at least a factor of five (excluding Nakamura and Yoshikawa's rate due to their outlier difference in included impactor populations), and thus any AMAs computed using these rates should assume at least this uncertainty.

The rates for terrains older than 3 Ga are even less well constrained. The so-called Nice model and its predecessors (e.g., Levison et al., 2001; Gomes et al., 2005; Malhotra, 2007; Morbidelli et al., 2010; Levison et al., 2011), and other general ideas on the end of planetary accretion (Tera, Papanastassiou and Wasserburg, 1974; Neukum, 1977; Neukum, Ivanov and Hartmann, 2001; Kring and Cohen, 2002; Walsh et al., 2011; Agnor and Lin, 2012; Bottke et al., 2012; Morbidelli et al., 2012, 2018; Dones and Levison, 2013; Heller, Marleau and Pudritz, 2015; Bottke and Norman, 2017; Nesvorný et al., 2018), imply that the impact flux should have been higher early in solar system history. However, how much higher and with what decay rate are not well constrained. Lunar history might suggest that there was a fast decline in rates after accretion ended (i.e., accretion did not have a long-lived tail) with a relatively short increase or "spike" in impacts at ~3.9 Ga ago, and then another fast decline to current average rates, traditionally called the "lunar Late Heavy Bombardment or Cataclysm" (Tera, Papanastassiou and Wasserburg, 1974; Kring and Cohen, 2002; Bottke et al., 2007; Cohen, 2013; Marchi et al., 2013; Bottke and Norman, 2017). Nevertheless, counterarguments have been presented that (1) the lunar Late Heavy Bombardment did not occur and/or was not a true "spike" in flux exactly at 3.9 Ga (Neukum, 1977; Neukum, Ivanov and Hartmann, 2001; Hartmann, 2003, 2019; Chapman, Cohen and Grinspoon, 2007; Michael, Basilevsky and Neukum, 2018) and (2) that the Late Heavy Bombardment would not have looked the same in the outer solar system nor would it have occurred at the same time (Johnson et al., 2008; Charnoz et al., 2009; Dones and Levison, 2013), and thus could not be extrapolated from lunar conditions. While the Nice model provides the dynamical background for producing a relative increase in impacts in the inner and outer solar system, it does not preclude a "spike" of only a few hundred million years occurring at the same time in the inner and outer solar system (e.g., Levison et al., 2011). Furthermore, continuing investigations of lunar bombardment history have suggested that a spike may not have occurred (e.g., Hartmann, 2003, 2019; Chapman, Cohen and Grinspoon, 2007; Michael, Basilevsky and Neukum, 2018) or that the "spike" may be longer than previously assumed (e.g., Morbidelli et al., 2012; Morbidelli et al., 2018). The lunar sample ages used in support of this spike may all be influenced by the formation of Imbrium (e.g., Hartmann, 2003, 2019; Chapman, Cohen and Grinspoon, 2007; Michael, Basilevsky and Neukum, 2018). Dynamical work further suggests that if a spike did occur it potentially was several hundred million years, rather than a few hundred million years (e.g., Bottke et al., 2012; Morbidelli et al., 2012; Morbidelli et al., 2018). Even updates to the Nice model (to prevent destruction of smaller outer solar system satellites) have reduced the increase in flux (Morbidelli et al., 2018; Nesvorný and Roig, 2018; Nesvorný et al., 2019). However, recent work using the survival of binary Trojan asteroids (Nesvorný et al., 2018), as well as self-triggered instability models informed by exoplanet observations (Ribeiro et al., 2020), suggests that the giant planet migration occurred earlier than 100 Ma.

Nevertheless, there have been continued efforts to extrapolate the lunar chronology, because it is better constrained, to the outer solar system by assuming asteroids produce the craters on Jupiter's and other giant planet satellites (Horedt and Neukum, 1984; Neukum, 1997; Wagner et al., 1999; Wagner et al., 2006; Neukum et al., 2005; Schmedemann et al., 2009; Schmedemann and Neukum, 2011). In this scenario, asteroids from the Main Belt are flung into the outer solar system and captured by Jupiter (or other giant planets) into planetocentric orbits. Once there, their velocities are reduced by some unknown mechanism and thus can reproduce the "shifted" crater SFDs required to match the lunar crater production function (see Section 8.4). This mechanism also assumes that Gilgamesh was formed during the Late Heavy Bombardment, placing it as a marker with an age of 3.8 Ga. While the advantage is that the impactor production function and flux is better constrained, the dynamics and statistics of this scenario are difficult. The number of escaped asteroids from the Main Belt that would come into the Jupiter system is only ~1–2% and the probability of a ~1 km asteroid impacting Ganymede to create a 20 km crater is about 1 in 1 000 000 (Gladman et al., 1997; Zahnle, Dones and Levison, 1998). Furthermore, the dynamical mechanism for Jupiter to capture and reduce the impactor speed is not specified (Gladman et al., 1997; Zahnle, Dones and Levison, 1998). Finally, it is not clear if the Late Heavy Bombardment occurred and can be extrapolated to the outer solar system as discussed in the previous paragraph. Thus, we do not recommend using this extrapolation of the lunar chronology to the outer solar system.

Most recently, observations by *New Horizons* of the Pluto system have provided more robust constraints on the origination of the heliocentric impactor population in the outer solar system. The similarity in the shapes on the R-plot (Figure 8.6) of the crater SFDs from the jovian satellites to Charon imply these both are most likely produced by a heliocentric impactor population from the far outer solar system – the Kuiper Belt (Singer et al., 2019). This similarity is particularly unique for the shallow slope (increasing R-values for increasing diameter) observed for craters $D < 10$–20 km. If diameters are properly gravity scaled for the lunar data, assuming no velocity reduction, the lunar crater SFD shifts to the right aligning the steep slope (decreasing R-values for increasing diameter) with the shallow slope of the outer solar system (Figure 8.6). This is another nail in the coffin for inner solar system asteroids being the primary heliocentric impactor population in the outer solar system.

Absolute model ages of Ganymede's terrains have been estimated since *Voyager* using these various models. For completion, we include ages computed by others using the lunar extrapolation model, although as noted, we do not suggest its continued usage. Table 8.3 gives these ages and the models used by the given references to compute them. The outer solar system impactor models are identified by S&W82 (Shoemaker and Wolfe, 1982), Z98 (Zahnle, Dones and Levison, 1998), and Z03 (Zahnle et al., 2003), while the lunar extrapolation model is identified by N97 (Neukum et al., 1997). As noted, all of these ages should be considered to have at least a factor of five

Table 8.3 *Absolute model ages (AMAs) computed by the given references for various terrains on Ganymede.*

Terrain	AMA (Ga)	Model	Citation
Galileo Regio	3.8	S&W82	Passey and Shoemaker (1982)
	3.8–4.2	N97	Prockter et al. (1998)
	4.2	N97	Neukum (1997)
Nicholson Regio	3.9	S&W82	Passey and Shoemaker (1982)
	4.1	N97	Wagner et al. (2018)
	2.87–4.56	Z03	Wagner et al. (2018)
Gilgamesh	3.5	S&W82	Passey and Shoemaker (1982)
	0.7–2.0	Z98	Zahnle, Dones and Levison (1998)
	0.8	Z03	Zahnle et al. (2003)
	0.11–2.13	Z98	Wagner et al. (1999)
	0.55–3.14	Z03	Wagner et al. (2018)
Grooved Terrain Leading Hemisphere	3.4	S&W82	Shoemaker et al. (1982)
	0.5–2	Z98	Zahnle et al. (1998)
Grooved Terrain Trailing Hemisphere/Harpagian Sulcus	3.6	S&W82	Shoemaker (1982)
	0.5–3	Z98	Zahnle et al. (1998)
	3.85	N97	Wagner et al. (2018)
	0.93–3.97	Z03	Wagner et al. (2018)
Grooved Terrain South Pole	3.8	S&W82	Shoemaker et al. (1982)
Western Equatorial Basin	1–2	Z98	Zahnle, Dones and Levison (1998)
Uruk Sulcus	3.7	N97	Neukum (1997)
Osiris	0.26	N97	Wagner et al. (1999)
	0.003–0.01	Z98	Wagner et al. (1999)
	0.3–1.4	N97	Wagner et al. (2018)
	0.03–0.24	Z03	Wagner et al. (2018)

Outer solar system models are identified by S&W82, Z98 and Z03; lunar extrapolation model is identified by N97.

uncertainty, whether actual uncertainties are given or not, and to be more guidelines than actual ages.

8.6 CRATERING IMPLICATIONS FOR THE INTERIOR AND EVOLUTION

8.6.1 The Interior

Ganymede's moment of inertia C/MR^2 ~0.31 (Anderson et al., 1996; Sotin and Tobie 2004; Vance et al., 2014) and intrinsic dynamo (Schubert et al., 1996; Kimura, Nakagawa and Kurita, 2009) indicate that, much like Europa, Ganymede is a fully differentiated body (i.e., the satellite has experienced almost total separation of ice and rock and additionally rock from metal, see Hussmann, Schubert, Steinbrügge et al., Chapter 6, this volume). On the other hand, Callisto's axial moment of inertia of C/MR^2 ~0.35 and mean density of ~1800 kg m^{-3} suggest that it is only partially differentiated (Anderson et al., 2004), though both moons are similar in size – the radius of Ganymede is 2631 km and Callisto is 2410 km. These differences in interior state and surface geology have been difficult to reconcile with models of satellite formation (e.g., Canup and Ward, 2002). Impact-induced processes, though, may help to explain the observed dichotomy. Because Ganymede is deeper in the jovian gravity well, it would have received more impacts over the age of the solar system than Callisto (Zahnle et al., 2003), which may help explain the observed differences if these impacts led to differentiation on Ganymede and not Callisto (Barr and Canup, 2010).

The orbital migration of the planets almost certainly resulted in a shower of cometary material onto the moons of Jupiter and the other outer planets (e.g., Levison et al., 2001;

Gomes et al., 2005; Malhotra, 2007; Morbidelli et al., 2010; Levison et al., 2011). Rough estimates indicate that Ganymede may have been hit by $\sim 6 \times 10^{20}$ kg of total impacting mass during an outer solar system bombardment (Zahnle et al., 2003; Gomes et al., 2005; Barr and Canup 2010). At a typical impacting velocity of $\sim$20 km/s, this would deliver a total energy to Ganymede of 10^{29} Joule, five times higher than the amount of energy required to melt all of its ice (i.e., $E \sim (1 - m_r)L$, where $m_r \sim 0.52$ is the mass fraction of rock in Ganymede's interior and $L \sim 3 \times 10^5$ J/kg is the latent heat of ice) (Barr and Canup, 2010). However, Callisto would have only received twice the energy needed to fully melt its water ice (Barr and Canup, 2010). Because this calculation assumes an efficient process, where all of the impact energy is converted to heat that melts ice, it is an over-approximation. Thus, Callisto likely did not receive sufficient energy to fully differentiate.

Impacting energy would have been delivered to the outer layers of the young Ganymede by impactors with radii in the tens of kilometer range (Morbidelli et al., 2005). If Ganymede accreted slowly, from small satellitesimals in a cold disk environment (e.g., Canup and Ward 2002, 2006; Barr and Canup 2008; Johansen and Lambrechts, 2017; Shibaike et al., 2019), its outer layers would initially contain small fragments of rock. Each impact would create a pool of melted water, and any rock inside that pool would sink to the bottom of the melted area (Tonks and Melosh, 1993; Barr and Canup, 2010). Once consolidated, the now larger rock body at the base of the impact site would be denser than its surrounding ice/rock mantle and sink to the putative core by Stokes flow (Friedson and Stevenson, 1983). As the rock fragments sink, their gravitational potential energy would be liberated in Ganymede's interior as heat. If the amount of energy liberated is greater than the latent heat of ice, the partial differentiation would drive itself to completion, in a process called "runaway differentiation" (for discussion, see e.g., Friedson and Stevenson, 1983; Stevenson and Lunine, 1986).

The end result for Ganymede would be a warm interior that had experienced complete ice/rock separation following its impact history. This event would result in Ganymede having a global ocean, perched near the ice I/III phase boundary, at a depth of around 100 km (e.g., Kirk and Stevenson, 1987; Mueller and McKinnon, 1988). The size and extent of the early ocean would depend on the efficiency with which Ganymede could remove this energy via convection in its deep interior layers and conduction across its surface (Friedson and Stevenson, 1983).

This event would also have obliterated the surface of Ganymede created at the tail end of its accretion and any geological features or impact structures that dated from the time of its formation. It would also set the stage for Ganymede to experience significant tidal dissipation due to the Laplace resonance with Europa and Io. Already heated, Ganymede would deform significantly in response to tidal forces, with a relatively large k_2 Love number and low effective quality factor Q (see Steinbrügge, Hussmann, Tobie et al., Chapter 4, this volume). In this way, Ganymede's early bombardment history may have set the stage for its long-term evolution as a fully differentiated and active planetary body in contrast to its neighbor, Callisto.

8.6.2 Evolution of Ganymede's Craters and Lithospheric Thermal History

Crater morphologies on icy satellites are influenced by their icy composition both during impact (in the form of morphologies distinct from lunar craters) and in their subsequent modification, including by elevated internal heat flow discussed further in this section (e.g., Schenk et al., 2004, Bray et al., 2014; Schenk et al., 2021). The shallowness of unmodified complex craters on both Ganymede and Callisto compared to lunar craters is a primary effect related directly to the lower effective strength of water ice as a crustal material compared to basalt or anorthosite (e.g., Schenk, 1991; Schenk et al., 2004). The central pit and dome complex crater morphologies are dominant only on the larger icy bodies Ganymede and Callisto (Schenk et al., 2004; Alzate and Barlow, 2011; Barlow et al., 2017) or bodies that have enhanced crustal ice components such as Ceres or Mars (e.g., Barlow, 2006; Barlow et al., 2017; Schenk et al., 2019). The formation of these two morphologies is still poorly understood. Pits may form from vapor explosion (Wood et al., 1978; Williams et al., 2015), structural collapse of an uplift of relatively competent material (Greeley et al., 1982; Passey and Shoemaker, 1982) or the draining of impact-derived fluid (either melt or the molten core of a peak) through fractures in the crater floor (Bray et al., 2009; Bray, Schenk et al., 2012; Elder, Bray and Melosh, 2012). The above modeling work, as well as the similarity of summit pit dimensions to intact central peaks (see Section 8.2), favors the melted peak model; however, high-resolution observational constraints for unmodified pristine examples on Ganymede and Callisto are currently lacking, something ESA's JUICE and NASA's Europa Clipper missions could address.

Central dome and anomalous dome craters have no analogue in terrestrial cratering. They are hypothesized to form either from the breaching of the brittle ice crust and the exposure of a relatively ductile layer that rises buoyantly after impact (Moore and Malin, 1988) or dynamically during impact (Schenk, 1993) or from the refreezing of a subsurface quasi-conical zone of melted water ice directly under the central structure. The latter is presently favored for the central dome at Occator, Ceres (cf. Schenk et al., 2019), currently the only known central dome crater other than on Ganymede or Callisto. Certainly, the dense fracturing on dome surfaces (cf. Schenk et al., 2004) suggests inflation of the dome surface from inside, either by intrusion or by lateral viscous creep of the dome.

The change over time of primary/original morphologies from palimpsests to penepalimpsests and then to anomalous dome crater and finally to the large central dome craters that dominate the current large cratering record (Figure 8.2), as indicated by observed superposed crater densities, is not related to long-term thermal viscous relaxation (see this section) but is attributed to higher heat flow in ancient times leading to more 'unusual' landforms (cf. Schenk et al., 2004). Whether these changes are related to enhanced melt production, floor uplift or some other combination of processes is the subject of ongoing work and is a question that will require the improved observational constraints for future missions, particularly the abundance of impact melt.

Ganymede's impact craters also provide important clues to the subsequent thermal history of the moon. While the

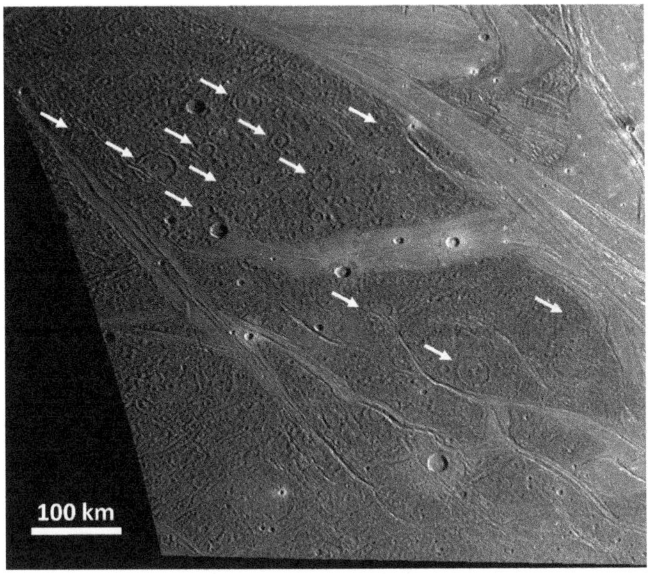

Figure 8.7 A portion of Marius Regio (just west of Nippur Sulcus) where numerous viscously relaxed impact craters (white arrows) are mixed with deep, pristine craters. The relaxed craters are shallow, with up-domed floors and intact rims. The image is 940 m/pixel.
Source: After Bland et al. (2017), reprinted from *Icarus*, 296. Bland, M. T., Singer, K. N., McKinnon, W. B. and Schenk, P. M. Viscous relaxation of Ganymede's impact craters: Constraints on heat flux 275–288, ©2017, with permission from Elsevier.

morphology of impact craters varies as a function of diameter (as described in Section 8.2), and the crustal properties at the time of impact as just described (e.g., Schenk et al., 2004; Bray et al., 2014; Schenk et al., 2021), crater morphology can also be modified over relatively long timescales (tens of millions to billions of years) by viscous flow in a process called viscous relaxation. Viscous relaxation is driven by differential stresses induced by the crater topography (see, e.g., Scott, 1967; Parmentier and Head, 1981; Dombard and McKinnon, 2006). The result of such relaxation is craters with shallow (or even flat) interiors, preserved rims and uplifted centers (Figure 8.7). The rate at which viscous relaxation occurs depends on both the wavelength of the topography (long wavelengths relax faster than short wavelengths) and the viscosity of the medium (ice on Ganymede). The viscosity of ice is strongly temperature-dependent (e.g., Durham, Stern and Kirby, 2001). Thus, the presence of viscously relaxed craters on an icy satellite may, in some cases, indicate elevated temperatures in the ice shell.

Voyager and *Galileo* images revealed that viscously relaxed impact craters are widespread on Ganymede and occur within both the dark and the bright terrains (e.g., Shoemaker et al., 1982; Singer et al., 2018). The degree to which craters are relaxed varies regionally. In some locations, few craters show evidence of viscous relaxation, while in others a continuous distribution of crater depths (for similar crater diameters) suggests a continuous process of relaxation after crater formation (Figure 8.8). In at least a few regions, such as within Marius Regio (Figure 8.7), there are two distinct populations of craters: a population of relatively deep, unrelaxed craters and a population of very shallow (nearly flat) craters (Singer et al., 2018; Figure 8.8). These two populations suggest an intense, short-lived heating event occurred, during which existing craters relaxed quickly, while craters that formed subsequently remain unrelaxed (Bland et al., 2017; Singer et al., 2018; Schenk, McKinnon, Moore et al., Chapter 9, this volume), implying that heat flow declined rapidly after bright terrain formation.

Observations of viscous relaxation can be combined with analytical (e.g., Parmentier and Head, 1981; Dombard and McKinnon, 2006) or numerical models (Dombard and McKinnon, 2000, 2006; Bland et al., 2017) to quantify the thermal state of Ganymede's lithosphere over time. Dombard and McKinnon (2006) focused on craters with $D \geq 25$ km and found that the range of relaxation states observed on Ganymede could be obtained by assuming heat fluxes in excess of 10 mW m^{-2}, roughly consistent with ancient radiogenic heat fluxes. Under such conditions, large craters (>100 km in diameter) relaxed in less than 100 000 years, but smaller craters (25 km in diameter) required ~1 Gyr. Bland et al. (2017) extended the investigation of crater relaxation on Ganymede to even smaller crater diameters and applied time-varying heat fluxes. They confirmed that radiogenic heating alone will result in substantial viscous relaxation of Ganymede's oldest and largest craters (i.e., craters larger than ~30 km and older than 2 Gyr will be ~50% relaxed simply due to radiogenic heating). Heat fluxes well in excess of radiogenic are required to explain the observed crater depths at small crater sizes. For craters with diameters between 5 and 20 km, the observed crater depths can result from heat fluxes of 40–50 mW m^{-2} acting over 1–2 Gyrs (Figure 8.9). However, such heat fluxes are more than twice Ganymede's maximum radiogenic heat flux and would imply an extended period of tidal heating, and they still fail to explain the population of small impact craters ≤ 5 km in diameter that are substantially relaxed (Singer et al., 2018). Bland et al. (2017) argue that the relaxation of these small craters likely requires a short-lived, high-intensity heat pulse of >100 mW m^{-2} acting over 10–100 Myr, perhaps during passage through a Laplace-like resonance (Showman, Stevenson and Malhotra, 1997; Bland, Showman and Tobie, 2009). Heat pulses can explain both the relaxation of small craters and the observation of populations of relaxed and unrelaxed craters in the same region, as described in this section: the short-lived pulse quickly relaxes any preexisting craters, while craters formed after the pulse subsides remain relatively pristine. Reality is potentially even more complex, with the signature of long-lived radiogenic heating, sustained elevated heat fluxes and short-lived, intense heat pulses all superimposed.

The heat fluxes implied by Ganymede's viscously relaxed impact craters add additional context to the substantial evidence of high heat flux on Ganymede at key points in its geologic history. The formation of the grooved terrain implies heat fluxes of 50–150 mW m^{-2} (e.g., Dombard and McKinnon, 2001; Bland, McKinnon and Showman, 2010), flexure along furrows (Nimmo and Pappalardo, 2004) and grooved terrain margins (Nimmo, Pappalardo and Giese, 2002) imply fluxes of 60–110 mW m^{-2} and viscously relaxed craters, as described here, suggest heat fluxes in excess of 50 mW $m^{-2}$and perhaps in excess of 100–150 mW m^{-2} (Bland et al., 2017). Thus, many of Ganymede's impact craters that formed prior to bright terrain have experienced heat fluxes similar to those inferred for Ganymede's grooved terrain. However, unlike the grooved terrain, which formed through substantial extensional strain (Pappalardo and Collins, 2005; Bland and McKinnon, 2015), Ganymede's relaxed craters formed in low-strain

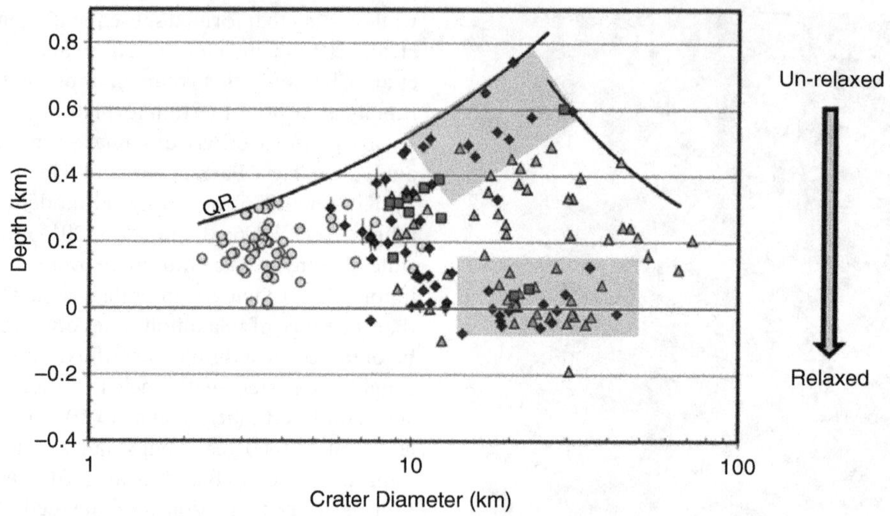

Figure 8.8 Measured apparent depths (depth relative to the surrounding terrain) for impact craters from four regions on Ganymede: Anshar Sulcus (yellow circles), the south pole (light blue triangles), Tiamat Sulcus (red squares), and Marius Regio (dark blue triangles). Gray boxes highlight the bimodality of crater depths observed in Marius Regio. Quantile regression fits to the upper envelope of the data (QR black curve) are used as a marker of initial crater depth. (A black and white version of this figure will appear in some formats. For the color version, please refer to the plate section.)
Source: Modified from Singer et al. (2018). Reprinted from *Icarus*, 306. Singer, K. N., Bland, M. T., Schenk, P. M. and McKinnon, W. B. Relaxed impact craters on Ganymede: Regional variation and high heat flows 214–224, ©2018, with permission from Elsevier.

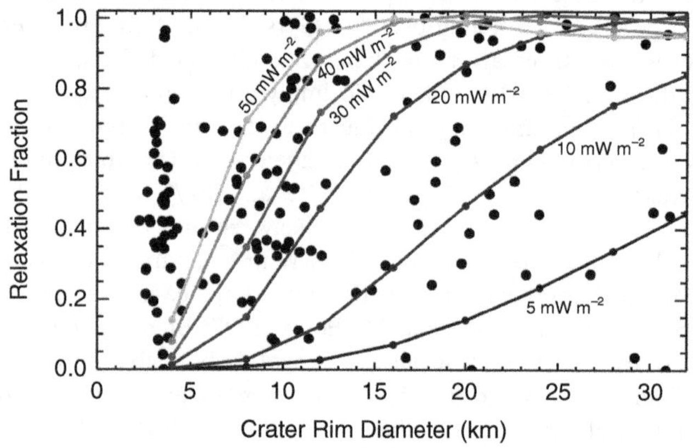

Figure 8.9 A comparison of observed crater depths (black points) and simulated crater depths after 2 Gyr for a range of heat fluxes (blue curves as labeled). Because crater depth varies with diameter, depths are parameterized as the final (present-day) depth relative to the assumed initial depth (i.e., the relaxation fraction, where 0.0 is pristine and 1.0 is fully relaxed). (A black and white version of this figure will appear in some formats. For the color version, please refer to the plate section.)
Source: Reprinted from *Icarus*, 296. Bland, M. T., Singer, K. N., McKinnon, W. B. and Schenk, P. M. Viscous relaxation of Ganymede's impact craters: Constraints on heat flux 275–288, ©2017, with permission from Elsevier.

conditions. Ganymede's impact craters therefore imply a regionally heterogeneous geologic evolution in which some regions remained relatively quiescent (the dark terrain, with relatively low heat fluxes and low strain), some experienced high heat fluxes and high strain (the bright, grooved terrain), some experienced low heat fluxes and high strain (the dark, lineated terrains) and some experienced high heat fluxes and low strain (viscously relaxed craters in dark terrain) (Bland and McKinnon, 2018). Throughout Ganymede's history, shear (Cameron et al., 2018) and cryovolcanic resurfacing (e.g., Schenk et al., 2001) complicate the geologic picture.

Although the fluxes implied by Ganymede's craters are consistent with other observations, as discussed in this section, viscous relaxation of impact craters is not unique to Ganymede and has been observed on Enceladus (Bland et al. 2012), Rhea (White, Schenk and Dombard, 2013), Dione and Tethys (White et al., 2017), Ariel (Schenk and Moore, 2020) and possibly Callisto (Passey and Shoemaker, 1982; Schenk, 1995). The prevalence of relaxation on these icy worlds, and the high heat fluxes they imply, could also suggest incomplete understanding of the material properties, especially the thermal conductivity, of the ice lithospheres of these bodies. Clathrates in particular have a much lower thermal conductivity (English and Tse, 2009) and could enable relaxation with less heat production (either radiogenic or tidal). We note, however, that the thermal conductivity of the ice cannot be too low, or the majority of craters on these bodies would be relaxed over short geologic timescales, which is not observed.

8.7 CONCLUSION

Ganymede's observed cratering on terrains of distinctly different ages records the flux of impactors in the Jupiter system over time in ways that either ancient Callisto (where degradation processes are more relevant and more intense) or the very young surfaces of Europa and Io do not. Although the observed record is incomplete, some general conclusions can be drawn. Crater SFDs from Ganymede's terrains have similar shapes to each other and Callisto's but occur at different relative densities, consistent with the recording of a persistent projectile population over time by terrains with different crater retention ages. With the exception of Galileo Regio, there are only a few minor differences in the relative densities of Ganymede's dark terrains, but they are all generally lower than that of Callisto, implying that even Ganymede's oldest terrains are younger than Callisto, probably <4 Ga in absolute age. Because of its lower crater density, particularly for small crater diameters, Galileo

Regio may have been the last dark terrain significantly altered by cryovolcanism. Bright terrains are younger than the dark terrains and average in AMA from 1 Ga to 4 Ga. The dark terrains do not indicate a strong apex–antapex difference in density, while the bright terrains show a factor of ~4 difference, much less than the predicted asymmetry. This discrepancy may be explained by changes in the impacting population, crater saturation effects in older terrains and resetting of the asymmetry by non-synchronous rotation and/or polar wonder. The shapes of Ganymede's (and Callisto's) crater SFDs are similar to those from the Pluto system but different from the Earth's Moon. This implies the jovian system heliocentric impactor population originated from the same place as for the plutonian system – the Kuiper Belt – and thus is not the same impactor population as for the Moon. Relaxation and other postimpact erasure mechanisms cannot account for this difference. Thus, the lunar chronology should not be used for the outer solar system.

Bombardment of Ganymede could also explain its interior evolution, particularly the fact that it appears differentiated while Callisto does not. Ganymede being closer to Jupiter, and thus deeper in its gravity well, experienced a higher impact bombardment than Callisto. Impacts result in heating of the crust and more impacts result in increased heating. Modeling has shown that the heating will likely be more intense on Ganymede, resulting in the melting of its crustal ice and the sinking of any rock contained in that ice. The descent of that rock then results in further melting due to the conversion of gravitational potential energy to heat. Overall, this produces "runaway differentiation" and a fully differentiated Ganymede. The modeling further indicates Callisto would not receive enough bombardment heating to start this process.

Ganymede's craters also record the thermal evolution of this geologically complex planetary body, particularly through their morphologies. Although there are significant gaps in mapping coverage and in resolution (with few craters observed at better than ~100 m/pixel), enough observational constraints have been documented to formulate hypotheses. Central pit craters are predominant at D > 35 km, and these are currently attributed to the melting of water ice in the uplifted central peak, leading directly to material loss and collapse, though other possibilities cannot be excluded. The primary unmodified crater morphologies at D > 60 km are even more complex, featuring large domes up to 1 km high and fractured rounded surfaces. These may be due to the extrusion of ductile mobile ices from depth due to either dynamic uplift during impact or subsequent freezing of melted subsurface ice, but again observational constraints are incomplete. There is also evidence that primary large crater landforms changed significantly over time. The causes are poorly understood, but these changes may be related to the heating associated with bright terrain formation.

Long-term modification of crater shapes due to viscous crustal relaxation in the ice shell (e.g., Dombard and McKinnon, 2006; Bland et al., 2017; Singer et al., 2018) provides important insights into Ganymede's thermal history, which suggests that enhanced heat flow in excess of 50 mW m^{-2} was prevalent regionally across the moon in the past. Mapping of the distribution of relaxed craters indicates that this period of enhanced heat flow was strongest around the time of bright terrain formation in some areas or heat flow may have been more prolonged in others areas.

A return to the Jupiter system by ESA's JUICE and NASA's Europa Clipper missions with improved global mapping (sufficient to map out D < 5 km cratering over most of the surface) and targeted high-resolution imaging of Ganymede (and Callisto) would help to further constrain many of the concepts discussed in this chapter. High-resolution targeted observations of major impact crater types, for morphology but also for crater retention ages, are required to resolve crater formation mechanisms. Similar imaging of the most pristine craters at a variety of sizes from simple craters D ~ 1–2 km to Osiris (D ~ 105 km) at better than 50 m/pixel would be required to investigate the prevalence of impact melt and its ice–rock content and the timing of central dome formation, as well as to map out structural deformation and other features. These will help place important constraints on the formation of Ganymede's unusual features and the nature of its shallow crustal structure or composition. The bright floor deposits of 105 km-diameter Osiris, for example, appear to cover the central dome, implying that the dome formed rapidly, an observation that can only be verified by imaging and spectral mapping at resolutions of <50 and <500 m/pixel, respectively, as well as comparable topographic mapping from stereo imaging. Spectral imaging in both visual and infrared will also provide compositional context to crater formation and perhaps projectile composition (e.g., Schenk and McKinnon, 1991) and ejecta crystallinity and ages (e.g., Dalle Ore et al., 2015; Kirchoff, Dalle Ore and Rivera-Valentín, 2018; Berdis et al., 2020; Rivera-Valentín et al., 2020). Global mapping of the topography would also constrain models for crater formation but also better map out the distribution of relaxed craters, when relaxation occurred and the associated high heat flow on this geologically complex icy ocean world. Finally, more detailed global mapping of the distributions and relative densities of relaxed and unrelaxed craters on both dark terrains and bright terrains is required to improve our understanding of the timing and duration of elevated heat flow, to further constrain the smaller impactor distribution and flux and to help discriminate secondary crater populations.

REFERENCES

Agnor, C. B. and Lin, D. N. C. (2012) "On the migration of Jupiter and Saturn: Constraints from linear models of secular resonant coupling with the terrestrial planets," *Astrophys. J.*, 745, 143, doi:10.1088/0004-637x/745/2/143.

Alvarellos, J. L., Dobrovolskis, A. R., Zahnle, K. J. et al. (2017) "Fates of satellite ejecta in the Saturn system, II," *Icarus*, 284, 70–89.

Alvarellos, J. L., Zahnle, K. J., Dobrovolskis, A. R. and Hamill, P. (2005) "Fates of satellite ejecta in the Saturn system," *Icarus*, 178, 104–123.

Alvarellos, J. L., Zahnle, K. J., Dobrovolskis, A. R. and Hamill, P. (2008) "Transfer of mass from Io to Europa and beyond due to cometary impacts," *Icarus,* 194, 636–646.

Alzate, N. and Barlow, N. G. (2011) "Central pit craters on Ganymede," *Icarus,* 211, 1274–1283.

Anderson, J. D., Lau, E. L., Sjogren, W. L., Schubert, G. and Moore, W. B. (1996) "Gravitational constraints on the internal structure of Ganymede," *Nature*, 384, 541–543.

Anderson, J. D., Schubert, G., Jacobson, R. A. et al. (2004) "Discovery of mass anomalies on Ganymede," *Science,* 305, 989–991.

Barlow, N. G. (2006) "Impact craters in the northern hemisphere of Mars: Layered ejecta and central pit characteristics," *Meteorit. Planet. Sci.*, 41, 1425–1436.

Barlow, N. G., Ferguson, S. N., Horstman, R. M. and Maine, A. (2017) "Comparison of central pit craters on Mars, Mercury, Ganymede, and the Saturnian satellites," *Meteorit. Planet. Sci.*, 52, 1371–1387.

Barr, A. C. and Canup, R. M. (2008) "Constraints on gas giant satellite formation from the interior states of partially differentiated satellites," *Icarus,* 198, 163–177.

Barr, A. C. and Canup, R. M. (2010) "Origin of the Ganymede–Callisto dichotomy by impacts during the late heavy bombardment," *Nature Geosci.*, 3, 164–167.

Berdis, J. R., Gudipati, M. S., Murphy, J. R. and Chanover, N. J. (2020) "Europa's surface water ice crystallinity: Discrepancy between observations and thermophysical and particle flux modeling," *Icarus,* 341, 113660, doi:10.1016/j.icarus.2020.113660.

Bierhaus, E. B., Chapman, C. R. and Merline, W. J. (2005) "Secondary craters on Europa and implications for cratered surfaces," *Nature,* 437, 1125–1127.

Bierhaus, E. B., Dones, L., Alvarellos, J. L. and Zahnle, K. (2012) "The role of ejecta in the small crater populations on the mid-sized saturnian satellites," *Icarus,* 218, 602–621.

Bierhaus, E. B., McEwen, A. S., Robbins, S. J. et al. (2018) "Secondary craters and ejecta across the solar system: Populations and effects on impact-crater based chronologies," *Meteorit. Planet. Sci.*, 53, 638–671.

Bierhaus, E. B. and Schenk, P. M. (2010) "Constraints on Europa's surface properties from primary and secondary crater morphology," *J. Geophys. Res.*, 115, E12004, doi:10.1029/2009JE003451.

Bland, M. T. and McKinnon, W. B. (2015) "Forming Ganymede's grooves at smaller strain: Toward a self-consistent local and global strain history for Ganymede," *Icarus*, 245, 247–262.

Bland, M. T. and McKinnon, W. B. (2018) "Viscous relaxation as a prerequisite for tectonic resurfacing on Ganymede: Insights from numerical models of lithospheric extension," *Icarus,* 306, 285–305.

Bland, M. T., Showman, A. P. and Tobie, G. (2009) "The orbital-thermal evolution and global expansion of Ganymede," *Icarus,* 200, 207–221.

Bland, M. T., McKinnon, W. B. and Showman, A. P. (2010) "The effects of strain localization on the formation of Ganymede's grooved terrain," *Icarus,* 210, 396–410.

Bland, M. T., Singer, K. N., McKinnon, W. B. and Schenk, P. M. (2012) "Enceladus' extreme heat flux as revealed by its relaxed craters," *Geophys. Res. Lett.*, 39, L17204, doi:10.1029/2012GL052736.

Bland, M. T., Singer, K. N., McKinnon, W. B. and Schenk, P. M. (2017) "Viscous relaxation of Ganymede's impact craters: Constraints on heat flux," *Icarus,* 296, 275–288.

Bottke, W. F., Levison, H. F., Nesvorný, D. and Dones, L. (2007) "Can planetesimals left over from terrestrial planet formation produce the lunar late heavy bombardment?," *Icarus,* 190, 203–223.

Bottke, W. F. and Norman, M. D. (2017) "The late heavy bombardment," *Annu. Rev. Earth Planet. Sci.*, 45, 619–647.

Bottke, W. F., Vokrouhlicky, D. et al. (2012) "An Archaean heavy bombardment from a destabilized extension of the asteroid belt," *Nature,* 485, 78–81.

Boyce, J., Barlow, N., Mouginis-Mark, P. and Stewart, S. (2010) "Rampart craters on Ganymede: Their implications for fluidized ejecta emplacement," *Meteorit. Planet. Sci.*, 45, 638–661.

Bray, V. J., Atwood-Stone, C. and McEwen, A. M. (2012) "Investigating the transition from central peak to peak-ring basins using central feature volume measurements from the Global Lunar DTM 100 m," *Geophys. Res. Lett.*, 39, L21201, doi:10.1029/2012GL053693.

Bray, V. J., Collins, G. S., Morgan, J. V., Melosh, H. J. and Schenk, P. M. (2014) "Hydrocode simulation of Ganymede and Europa cratering trends: How thick is Europa's crust?," *Icarus,* 231, 394–406.

Bray, V. J., Collins, G. S., Morgan, J. V. and Schenk, P. M. (2008) "The effect of target properties on crater morphology: Comparison of central peak craters on the Moon and Ganymede," *Meteor. Planet. Sci.*, 43, 1979–1992.

Bray, V. J., Schenk, P. M., Melosh, H. J., Collins, G. S. and Morgan, J. V. (2009) "Dimensions of Central Pits in Ganymede Craters," *40th Lunar Planet. Sci. Conf.*, Abst. #1350.

Bray, V. J., Schenk, P. M., Melosh, H. J., Morgan, J. V. and Collins, G. S. (2012) "Ganymede crater dimensions – Implications for central peak and central pit formation and development," *Icarus,* 217, 115–129.

Brunini, A., Di Sisto, R. P. and Orellana, R. B. (2003) "Cratering rate on the Jovian system: The contribution from Hilda asteroids," *Icarus,* 165, 371–378.

Cameron, M. E., Smith-Konter, B. R., Burkhard, L. et al. (2018) "Morphological mapping of Ganymede: Investigating the role of strike-slip tectonics in the evolution of terrain types," *Icarus,* 315, 92–114.

Canup, R. M. and Ward, W. R. (2002) "Satellite formation around gas giant planets," *AGU Fall Mtg.*, Abst. #P12C-03.

Canup, R. M. and Ward, W. R. (2006) "A common mass scaling for satellite systems of gaseous planets," *Nature,* 441, 834–839.

Casacchia, R. and Strom, R. G. (1984) "Geologic evolution of Galileo Regio, Ganymede," P*roc. 14th Lunar Planet. Sci. Conf. Part 2. J. Geophys. Res.*, B419–B428.

CATWG (Crater Analysis Techniques Working Group). (1979) "Standard techniques for presentation and analysis of crater size-frequency data," *Icarus,* 37, 467–474.

Chapman, C. R. (1968) "Interpretation of the diameter-frequency relation for lunar craters photographed by Rangers VII, VIII, and IX," *Icarus,* 8, 1–22.

Chapman, C. R. (1997) "Cratering on the Galilean satellites: Implications for the size distribution of cratering impacts in the solar system," *Met. Soc. Meet. #60,* pp. A27, Abst. #5261.

Chapman, C. R., Cohen, B. A. and Grinspoon, D. H. (2007) "What are the real constraints on the existence and magnitude of the late heavy bombardment?," *Icarus,* 189, 233–245.

Chapman, C. R. and McKinnon, W. B. (1986) "Cratering of planetary satellites," in J. A. Burns and M. S. Matthews (eds.) *Satellites*. Tucson: University of Arizona Press, pp. 492–580.

Chapman, C. R. and Merline, W. J. (1996) "Ganymede cratering: Preliminary Galileo G1 results," *Am. Astron. Soc., DPS Mtg. #28,* pp. 1139, Abst. #21.30.

Chapman, C. R., Merline, W. J., Bierhaus, B., Keller, J. and Brooks, S. (1997) "Impactor populations on the Galilean satellites," *Am. Astron. Soc., DPS Mtg. #29,* pp. 984, Abst. #12.10.

Chapman, C. R., Merline, W. J., Bierhaus, B. and Brooks, S. (1998) "Cratering in the jovian system: Intersatellite comparisons," *Lunar Planet. Sci. 29,* Abst. #1927.

Chapman, C. R., Merline, W. J., Bierhaus, B. et al. (1997) "Populations of small craters on Europa, Ganymede, and Callisto: Initial Galileo imaging results," *Lunar Planet. Sci. XXVIII,* Abst. #1221.

Charnoz, S., Morbidelli, A., Dones, L. and Salmon, J. (2009) "Did Saturn's rings form during the late heavy bombardment?," *Icarus,* 199, 413–428.

Cohen, B. A. (2013) "The Vestan cataclysm: Impact-melt clasts in howardites and the bombardment history of 4 Vesta," *Meteorit. Planet. Sci.*, 48, 771–785.

Cohen, B. A., Swindle, T. D. and Kring, D. A. (2005) "Geochemistry and 40Ar-39Ar geochronology of impact-melt clasts in feldspathic lunar meteorites: Implications for lunar bombardment history," *Meteorit. Planet. Sci.*, 40, 755–777.

Croft, S. K. (1985) "The scaling of complex craters," *Proc. 15th Lunar Planet. Sci. Conf., Part 2, J. Geophys. Res.: Solid Earth*, pp. C828–C842.

Croft, S. K. (1988) "Crater depth/diameter/morphology relationships on the icy satellites: Implications for ice rheology," *Lunar Planet. Sci. XIX*, pp. 219–220.

Ćuk, M. (2012) "Chronology and sources of lunar impact bombardment," *Icarus,* 218, 69–79.

Dalle Ore, C. M., Cruikshank, D. P., Mastrapa, R. M. E., Lewis, E. and White, O. L. (2015) "Impact craters: An ice study on Rhea," *Icarus,* 261, 80–90.

Di Sisto, R. P. and Brunini, A. (2007) "The origin and distribution of the Centaur population," *Icarus,* 190, 224–235.

Di Sisto, R. P. and Brunini, A. (2011) "Origin of craters on Phoebe: Comparison with Cassini's data," *Astron. Astrophy.*, 534, A68, doi:10.1051/0004-6361/201116988.

Di Sisto, R. P. and Zanardi, M. (2013) "The production of craters on the mid-sized Saturnian satellites by Centaur objects," *Astron. Astrophy.*, 553, A79, doi:10.1051/0004-6361/201220568.

Dobrovolskis, A. R., Alvarellos, J. L., Zahnle, K. J. and Lissauer, J. J. (2010) "Exchange of ejecta between Telesto and Calypso: Tadpoles, horseshoes, and passing orbits," *Icarus,* 210, 436–445.

Dombard, A. J., and McKinnon, W. B. (2000) "Long-term retention of impact crater topography on Ganymede," *Geophys. Res. Lett.*, 27, 3663–3666.

Dombard, A. J. and McKinnon, W. B. (2001) "Formation of grooved terrain on Ganymede: Extensional instability mediated by cold, superplastic creep," *Icarus,* 154, 321–336.

Dombard, A. J. and McKinnon, W. B. (2006) "Elastoviscoplastic relaxation of impact crater topography with application to Ganymede and Callisto," *J. Geophys. Res.*, 111, E01001, doi:10.1029/2005JE002445.

Dones, L., Chapman, C. R., McKinnon, W. B. et al. (2009) "Icy satellites of Saturn: Impact cratering and age determination," in M. K. Daugherty, L. W. Esposit and S. M. Krimigis (eds.) *Saturn from Cassini-Huygens*. Dordrecht: Springer, pp. 613–636.

Dones, L. and Levison, H. F. (2013) "The impact rate on giant planet satellites during the late heavy bombardment," *44th Lunar Planet. Sci. Conf.*, Abst. #2772.

Duncan, M., Quinn, T. and Tremaine, S. (1988) "The origin of short-period comets," *Astrophys. J.*, 328, L69–L73.

Durham, W. B., Stern, L. A. and Kirby, S. H. (2001) "Rheology of ice I at low stress and elevated confining pressure," *J. Geophys. Res.: Solid Earth,* 106, 11031–11042.

Elder, C. M., Bray, V. J. and Melosh H. J. (2012) "The theoretical plausibility of central pit crater formation via melt drainage," *Icarus*, 221, 831–843.

English, N. and Tse, J. S. (2009) "Mechanisms for thermal conduction in methane hydrate." *Phys. Rev. Lett.* 103. https://doi.org/10.1103/PhysRevLett.103.015901.

Ermakov, A. I., Fu, R. R., Castillo-Rogez, J. C. et al. (2017) "Constraints on Ceres' internal structure and evolution from its shape and gravity measured by the Dawn spacecraft," *J. Geophys. Res. Planets,* 122, 2267–2293.

Fraser, W. C., Brown, M. E. and Schwamb, M. E. (2010) "The luminosity function of the hot and cold Kuiper Belt populations," *Icarus,* 210, 944–955.

Friedson, A. J. and Stevenson, D. J. (1983) "Viscosity of rock-ice mixtures and applications to the evolution of icy satellites," *Icarus,* 56, 1–14.

Gault, D. E. (1970) "Saturation and equilibrium conditions for impact cratering on the lunar surface: Criteria and implications," *Radio Sci.*, 5, 273–291.

Gladman, B. J., Burns, J. A., Duncan, M. J. and Levison, H. F. (1995) "The dynamical evolution of lunar impact ejecta," *Icarus,* 118, 302–321.

Gladman, B. J., Migliorini, F., Morbidelli, A. et al. (1997) "Dynamical lifetimes of objects injected into Asteroid Belt resonances," *Science,* 277, 197–201.

Gomes, R., Levison, H. F., Tsiganis, K. and Morbidelli, A. (2005) "Origin of the cataclysmic late heavy bombardment period of the terrestrial planets," *Nature,* 435, 466–469.

Greeley, R., Fink, J. H., Gault, D. E. and Guest, J. E. (1982) "Experimental simulation of impact cratering on icy satellites," in D. Morrison (ed.) *Satellites of Jupiter.* Tucson: University of Arizona Press, pp. 340–378.

Harris, A. W. and D'Abramo, G. (2015) "The population of near-Earth asteroids," *Icarus,* 257, 302–312.

Hartmann, W. K. (1984) "Does crater 'saturation equilibrium' occur in the solar system?," *Icarus,* 60, 56–74.

Hartmann, W. K. (2003) "Megaregolith evolution and cratering cataclysm models: lunar cataclysm as a misconception (28 years later)," *Meteorit. Planet. Sci.*, 38, 579–593.

Hartmann, W. K. (2019) "History of the terminal cataclysm paradigm: Epistemology of a planetary bombardment that never (?) happened," *Geosciences,* 9, 285, doi: 10.3390/geosciences9070285.

Hartmann, W. K. and Gaskell, R. W. (1997) "Planetary cratering 2: Studies of saturation equilibrium," *Meteorit. Planet. Sci.*, 32, 109–121.

Hartmann, W. K. and Wood, C. A. (1971) "Moon: Origin and evolution of multi-ring basins" *Earth Moon Planet.*, 3, 3–78.

Heller, R., Marleau, G.-D. and Pudritz, R. E. (2015) "The formation of the Galilean moons and Titan in the Grand Tack scenario," *Astron. Astrophy.*, 579, L4, doi:101051/0004-6361/201526348.

Hirata, N., Suetsugu, R. and Ohtsuki, K. (2020) "A global system of furrows on Ganymede indicative of their creation in a single impact event," *Icarus* 352, 113941, doi: 10.1016/j.icarus.2020.113941.

Holsapple, K. A. and Housen, K. R. (2007) "A crater and its ejecta: An interpretation of Deep Impact," *Icarus,* 187, 345–356.

Horedt, G. P. and Neukum, G. (1984) "Planetocentric versus heliocentric impacts in the Jovian and Saturnian satellite system," *J. Geophys. Res.*, 89, 10405–10410.

Huang, Y.-H., Minton, D. A., Zellner, Z. E. B. et al. (2018) "No change in the recent lunar impact flux required based on modeling of impact glass spherule age distributions," *Geophys. Res. Lett.*, 45, 6805–6813.

Ivanov, M. A., & Basilevsky, A. T. (2002) "Morphology and size-frequency distribution of kilometer-scale impact craters on Callisto and Ganymede derived from Galileo data." *Solar System Research*, 36(6), 447–457.

Johansen, A. and Lambrechts, M. (2017) "Forming planets via pebble accretion," *Annu. Rev. Earth Planet. Sci.*, 45, 359–387.

Johnson, R. E., Carlson, R. W., Cooper, J. F. et al. (2004) "Radiation effects on the surfaces of the Galilean satellites," in F., Bagenal, T. Dowling and W. B. McKinnon (eds.) *Jupiter: The Planet, Satellites and Magnetosphere*. Cambridge: Cambridge University Press, pp. 485–512.

Johnson, T. V., Castillo-Rogez, J. C., Matson, D. L., Morbidelli, A. and Lunine, J. I. (2008) "Constraints on outer solar system early chronology," *Early Solar Sys. Imp. Bomb.*, Abst. #3018.

Jones, K. B., Head, J. W., III, Pappalardo, R. T. and Moore, J. M. (2003) "Morphology and origin of palimpsests on Ganymede based on Galileo observations," *Icarus,* 164, 197–212.

Kimura, J., Nakagawa, T. and Kurita, K. (2009) "Size and compositional constraints of Ganymede's metallic core for driving an active dynamo," *Icarus,* 202, 216–224.

Kirchoff, M. R. (2018) "Can spatial statistics help decipher impact crater saturation?," *Meteorit. Planet. Sci.*, 53, 874–890.

Kirchoff, M. R., Dalle Ore, C. M. and Rivera-Valentín, E. G. (2018) "Constraints on the recent Saturnian crater flux from Cassini VIMS and ISS: Crater ages on Rhea," *Cassini Science Symposium*, Abst. #IcySat3-4.

Kirchoff, M. R., Bierhaus, E. B., Dones, L. et al. (2018) "Cratering histories in the saturnian system," in Verbiscer, J. H. Waite, Jr., and R. N. Clark (eds.) *Enceladus and the Icy Moons of Saturn*. Tucson: University of Arizona Press, pp. 267–284.

Kirk, R. L. and Stevenson, D. J. (1987) "Thermal evolution of a differentiated Ganymede and implications for surface features," *Icarus,* 69, 91–134.

Klaybor, K. and Barlow, N. G. (2006) "Interior morphologies of impact craters on Ganymede," *37th Lunar Planet. Sci. Conf.*, Abst. #1360.

Kring, D. A. and Cohen, B. A. (2002) "Cataclysmic bombardment throughout the inner solar system 3.9–4.0 Ga," *J. Geophys. Res.* 107, doi:10.1029/2001JE001529.

Krüger, T., Hergarten, S. and Kenkmann, T. (2018) "Deriving morphometric parameters and the simple-to-complex transition diameter from a high-resolution, global database of fresh lunar impact craters (D $\geq\sim$ 3 km)," *J. Geophys. Res.: Planets,* 123, 2667–2690.

Leliwa-Kopystyński, J., Banaszek, M. and Wlodarczyk, I. (2012) "Longitudinal asymmetry of crater' density distributions on the icy satellites," *Planet. Space Sci.*, 60, 181–192.

Levison, H. F., Dones, L., Chapman, C. R. et al. (2001) "Could the lunar 'late heavy bombardment' have been triggered by the formation of Uranus and Neptune?," *Icarus,* 151, 286–306.

Levison, H. F. and Duncan, M. J. (1997) "From the Kuiper Belt to Jupiter-Family comets: The spatial distribution of ecliptic comets," *Icarus,* 127, 13–32.

Levison, H. F., Morbidelli, A., Tsiganis, K., Nesvorný, D. and Gomes, R. (2011) "Late orbital instabilities in the outer planets induced by interaction with a self-gravitating planetesimal disk," *Astron. J.*, 142, 152, doi:10.1088/0004-6256/142/5/152.

Lissauer, J. J., Squyres, S. W. and Hartmann, W. K. (1988) "Bombardment history of the Saturn system," *J. Geophys. Res.*, 93, 13776–13804.

Mainzer, A., Bauer, J., Grav, T. et al. (2014) "The population of tiny near-Earth objects observed by NEOWISE," *Astrophys. J.*, 784, 110, doi:10.1088/0004-637X/784/2/110.

Malhotra, R. (2007) "Dynamical cause of the late heavy bombardment," *Lunar Planet. Sci. Conf. XXXVIII*, Abst. #2373.

Marchi, S., Bottke, W. F., Cohen, B. A. et al. (2013) "High-velocity collisions from the lunar cataclysm recorded in asteroidal meteorites," *Nature Geosci.*, 6, 303–307.

Marchi, S., Bottke, W. F., Kring, D. A. and Morbidelli, A. (2012) "The onset of the lunar cataclysm as recorded in its ancient crater populations," *Earth Planet. Sci. Lett.*, 325–326, 27–38.

Martellato, E., Marchi, S., Galluzzi, V., Palumbo, P. and Rotundi, A. (2021) "Ganymede's cratering record," *52nd Lunar Planet. Sci. Conf.*, Abst. #2175.

McEwen, A. S. and Bierhaus, E. B. (2006) "The importance of secondary cratering to age constraints on planetary surfaces," *Annu. Rev. Earth Planet. Sci.*, 34, 535–567.

McKinnon, W. B., Chapman, C. R. and Housen, K. R. (1991) "Cratering of the Uranian satellites," in J. T. Bergstralh, E. D. Miner and M. S. Matthews (eds.) *Uranus*. Tucson: University of Arizona Press, pp. 629–692.

McKinnon, W. B., Mohit, P. S., Greenhagen, B. T. and Bland, M. T. (2014) "Polar wander on Ganymede and Callisto: A solution to the apex-antapex cratering conundrum," *45th Lunar Planet. Sci. Conf.*, Abst. #1869.

McKinnon, W. B. and Schenk, P. M. (1995) "Estimates of comet fragment masses from impact crater chains on Callisto and Ganymede," *Geophys. Res. Lett.*, 22, 1829–1832.

Melosh, H. J. and McKinnon, W. B. (1978) "The mechanics of ringed basin formation," *Geophys. Res. Lett.*, 5, 985–988.

Melosh, H. J. and Schenk, P. (1993) "Split comets and the origin of crater chains on Ganymede and Callisto," *Nature,* 365, 731–733.

Michael, G., Basilevsky, A. and Neukum, G. (2018) "On the history of the early meteoritic bombardment of the Moon: Was there a terminal lunar cataclysm?," *Icarus,* 302, 80–103.

Mohit, P. S., Greenhagen, B. T. and McKinnon, W. B (2012) "Polar wander on Ganymede: A possible solution to the apex-antapex cratering conundrum," *Early Solar Sys. Imp. Bomb. II*, Abst. #4043.

Moore, J. M. and Malin, M. C. (1988) "Dome craters on Ganymede" *Geophys. Res. Lett.*, 15, 225–228.

Moore, J. M., Horner, V. M. and Greeley, R. (1985) "The geomorphology of Rhea: Implications for geologic history and surface processes," *J. Geophys. Res.: Solid Earth,* 90, C785–C795.

Morbidelli, A., Brasser, R., Gomes, R., Levison, H. F. and Tsiganis, K. (2010) "Evidence from the asteroid belt for a violent past evolution of Jupiter's orbit," *Astron. J.*, 140, 1391–1401.

Morbidelli, A., Levison, H. F., Tsiganis, K. and Gomes, R. (2005) "Chaotic capture of Jupiter's Trojan asteroids in the early solar system," *Nature,* 435, 462–465.

Morbidelli, A., Marchi, S., Bottke, W. F. and Kring, D. A. (2012) "A sawtooth-like timeline for the first billion years of lunar bombardment," *Earth Planet. Sci. Lett.*, 355–356, 144–151.

Morbidelli, A., Nesvorny, D., Laurenz, V. et al. (2018) "The timeline of the lunar bombardment: Revisited," *Icarus,* 305, 262–276.

Mueller, S. and McKinnon, W. B. (1988) "Three-layered models of Ganymede and Callisto: Compositions, structures, and aspects of evolution," *Icarus,* 76, 437-464.

Murchie, S. L., Head, J. W. and Plescia, J. B. (1989) "Crater densities and crater ages of different terrain types on Ganymede," *Icarus,* 81, 271–297.

Murchie, S. L., Head, J. W. and Plescia, J. B. (1990) "Tectonic and volcanic evolution of dark terrain and its implications for the internal structure and evolution of Ganymede," *J. Geophys. Res.: Solid Earth,* 95, 10743–10768.

Nakamura, T. and Yoshikawa, M. (1995) "Close encounters and collisions of short-period comets with Jupiter and its satellites," *Icarus,* 116, 113–130.

Nesvorný, D. and Roig, F. (2018) "Dynamical origin and terrestrial impact flux of large near-Earth asteroids," *Astron. J.*, 155, 42, doi:10.3847/1538-3881/aa9a47.

Nesvorný, D., Vokrouhlický, D., Bottke, W. F. and Levison, H. F. (2018) "Evidence for very early migration of the solar system planets from the Patroclus–Menoetius binary Jupiter Trojan," *Nat. Astron.*, 2, 878–882.

Nesvorný, D., Vokrouhlický, D., Stern, A. S. et al. (2019) "OSSOS. XIX: Testing early solar system dynamical models using OSSOS Centaur detections," *Astron. J.*, 158, 132, doi:10.3847/1538-3881/ab3651.

Neukum, G. (1977) "Lunar cratering." Philosophical Transactions of the Royal Society of London, 285, 267–272.

Neukum, G. (1982) "Ancient cratering records of the terrestrial-type planets," *Lunar Planet. Sci. Conf. XIII*, 588–589.

Neukum, G. (1997) "Bombardment history of the jovian system," in C. Barbieri, J. H. Rahe, T. V. Johnson and A. M. Sohus (eds.) *The Three Galileos: The Man, the Spacecraft, the Telescope.* Dordrecht: Springer, pp. 201–212.

Neukum, G., Ivanov, B. A. and Hartmann, W. K. (2001) "Cratering records in the inner solar system in relation to the lunar reference system," *Space Sci. Rev.*, 96, 55–86.

Neukum, G. and Pozio, S. (1984) "The cratering record of Ganymede," *Lunar Planet. Sci. Conf. XV*, 601–602.

Neukum, G., Wagner, R., Denk, T., Porco, C. C. and Cassini ISS Team (2005) "The cratering record of the Saturnian satellites Phoebe, Tethys, Dione and Iapetus in comparison: First results from analysis of the Cassini ISS imaging data," *Lunar Planet. Sci. 36*, Abst. #2034.

Neukum, G., Wagner, R., Wolf, U. et al. (1997) "Bombardment history of the Galilean satellites and derived ages," *Am. Astron. Soc., DPS Mtg. #29*, pp. 984, Abst. #12.08.

Neukum, G., Wagner, R., Wolf, U. et al. (1998) "Cratering chronology in the jovian system and derivation of absolute ages," *Lunar Planet. Sci. 29*, Abst. #1742.

Nimmo, F. and Pappalardo, R. T. (2004) "Furrow flexure and ancient heat flux on Ganymede," *Geophys. Res. Lett.*, 31, L19701, doi:10.1029/2004GL020763.

Nimmo, F., Pappalardo, R. T. and Giese, B. (2002) "Effective elastic thickness and heat flux estimates on Ganymede," *Geophys. Res. Lett.*, 29, 62, doi:10.1029/2001GL013976.

Pappalardo, R. T. and Collins, G. C. (2005) "Strained craters on Ganymede," *J. Struct. Geol.*, 27, 827–838.

Parmentier, E. M. and Head, J. W. (1981) "Viscous relaxation of impact craters on icy planetary surfaces: Determination of viscosity variation with depth," *Icarus*, 47, 100–111.

Passey, Q. R. and Shoemaker, E. M. (1982) "Craters and basins on Ganymede and Callisto morphological indications of crustal evolution," in D. Morrison (ed.) *Satellites of Jupiter*. Tucson: University of Arizona Press, pp. 379–434.

Pike, R. J. (1980) "Control of crater morphology by gravity and target type: Mars, Earth, Moon," *Proc. Lunar Planet. Sci. Conf. 11th*, 2159–2189.

Plescia, J. B. (2012) "Uncertainties in the <3 Ga lunar impact cratering chronology," *43rd Lunar Planet. Sci. Conf.*, Abst. #1614.

Prockter, L. M., Head, J. W., Pappalardo, R. T. et al. (1998) "Dark terrain on Ganymede: Geological mapping and interpretation of Galileo Regio at high resolution," *Icarus*, 135, 317–344.

Ribeiro, R. de S., Morbidelli, A., Raymond, S. N. et al. (2020) "Dynamical evidence for an early giant planet instability," *Icarus*, 339, 113605, doi: 10.1016/j.icarus.2019.113605.

Richardson, J. E. (2009) "Cratering saturation and equilibrium: A new model looks at an old problem," *Icarus*, 204, 697–715.

Rivera-Valentín, E. G., Kirchoff, M. R. and Schenk, P. (2008) "An analysis of the geologic histories of Ganymede's dark terrain and Callisto through impact cratering distributions," *Lunar Planet. Sci. 39*, Abst. #2370.

Rivera-Valentín, E. G., López Oquendo, A. J., Kirchoff, M. R., Dalle Ore, C. M. and Scipioni, F. (2020) "Constraints on the recent cratering rate in the Saturn system from water ice crystallinity derived crater formation ages on Dione," *51st Lunar Planet. Sci. Conf.*, Abst #2839.

Robbins, S. J. (2014) "New crater calibrations for the lunar crater-age chronology," *Earth Planet. Sci. Lett.*, 403, 188–198.

Robbins, S. J. and Hynek, B. M. (2012) "A new global database of Mars impact craters ≥1 km: 2. Global crater properties and regional variations of the simple-to-complex transition diameter," *J. Geophys. Res.*, 117, E06001, doi:10.1029/2011JE003967.

Robbins, S. J., Riggs, J. D., Weaver, B. P. et al. (2018) "Revised recommended methods for analyzing crater size-frequency distributions," *Meteorit. Planet. Sci.*, 53, 891–931.

Robbins, S. J., Schenk, P. M., Riggs, J. D. et al. (2021) "Depths of Pluto's and Charon's craters, and their simple-to-complex transition," *Icarus*, 356, 113902, doi: 10.1016/j.icarus.2020.113902.

Ryan, E. L., Mizuno, D. R., Shenoy, S. S. et al. (2015) "The kilometer-sized Main Belt asteroid population revealed by Spitzer," *Astron. Astrophy.* 578, A42, doi:10.1051/0004-6361/201321375.

Schenk, P. M. (1989) "Crater formation and modification on the icy satellites of Uranus and Saturn: Depth/diameter and central peak occurrence," *J. Geophys. Res.*, 94, 3813–3832.

Schenk, P. M. (1991) "Ganymede and Callisto: Complex crater formation and planetary crusts," *J. Geophys. Res. Planets*, 96, 15635–15664.

Schenk, P. M. (1993) "Central pit and dome craters: Exposing the interiors of Ganymede and Callisto," *J. Geophys. Res.*, 98, 7475–7498.

Schenk, P. M. (1995) "The geology of Callisto," *J. Geophys. Res.*, 100, 19023–19040.

Schenk, P. M. (2002) "Thickness constraints on the icy shells of the Galilean satellites from a comparison of crater shapes," *Nature*, 417, 419–421.

Schenk, P. M., Asphaug, E., McKinnon, W. B., Melosh, H. J. and Weissman, P. R. (1996) "Cometary nuclei and tidal disruption: The geologic record of crater chains on Callisto and Ganymede," *Icarus*, 121, 249–274.

Schenk, P., Castillo-Rogez, J., Otto, K. A. et al. (2021) "Compositional control on impact crater formation on mid-sized planetary bodies: Dawn at Ceres and Vesta, Cassini at Saturn," *Icarus*, 359, 114343, doi:10.1016/j.icarus.2021.114343.

Schenk, P. M., Chapman, C. R., Zahnle, K. and Moore, J. M. (2004) "Ages and interiors: The cratering record of the Galilean satellites," in In F. Bagenal, T. Dowling and W. B. McKinnon (eds.) *Jupiter: The Planet, Satellites and Magnetosphere*. Cambridge: Cambridge University Press, pp. 427–456.

Schenk, P. M., Kirchoff, M. R. and Hoogenboom, T. (2020) "The anatomy of fresh complex craters on the mid-sized icy moons of Saturn and self-secondary cratering at the rayed crater Inktomi (Rhea)," *Meteor. Planet. Sci.*, 55, 2440–2460.

Schenk, P. M. and McKinnon, W. B. (1991) "Dark-ray and dark-floor craters on Ganymede, and the provenance of large impactors in the jovian system," *Icarus*, 89, 318–346.

Schenk, P. M., McKinnon, W. B., Gwynn, D. and Moore, J. M. (2001) "Flooding of Ganymede's bright terrains by low-viscosity water-ice lavas," *Nature*, 410, 57–60.

Schenk, P. M., & Moore, J. M. (2020) "Topography and geology of Uranian mid-sized icy satellites in comparison with Saturnian and Plutonian satellites." *Philosophical Transactions of the Royal Society A: Mathematical, Physical and Engineering Sciences*, *378*(2187), 20200102, doi:10.1098/rsta.2020.0102.

Schenk, P. M. and Ridolfi, F. J. (2002) "Morphology and scaling of ejecta deposits on icy satellites," *Geophys. Res. Lett.*, 29, 1590, doi:10.1029/2001GL013512.

Schenk, P., Sizemore, H., Schmidt, B., Castillo-Rogez, J., De Sanctis, M., Bowling, T., Scully, J., Buczkowski, D., Quick, L., Preusker, F., Park, R., Raymond, C. and Russell, C. (2019) "The central pit and dome at Cerealia Facula bright deposit and floor deposits in Occator crater, Ceres: Morphology, comparisons and formation," *Icarus*, 320, 159–187.

Schmedemann, N. and Neukum, G. (2011) "Impact crater size-frequency distribution (SFD) and surface ages on Mimas," *42nd Lunar Planet. Sci. Conf.*, Abst. #2772.

Schmedemann, N., Neukum, G., Denk, T. and Wagner, R. (2009) "Impact crater size-frequency distribution (SFD) on saturnian satellites and comparison with other solar-system bodies," *40th Lunar Planet. Sci. Conf.*, Abst. #1941.

Schubert, G., Zhang, K., Kivelson, M. G. and Anderson, J. D. (1996) "The magnetic field and internal structure of Ganymede," *Nature*, 384, 544-545.

Schultz, P. H. and Gault, D. E. (1985) "Clustered impacts: Experiments and implications," *J. Geophys. Res.: Solid Earth*, 90, 3701–3732.

Scott, R. F. (1967) "Viscous flow of craters," *Icarus*, 7, 139–148.

Scully, J. E. C., Bowling, T., Bu, C. et al. (2019) "Synthesis of the special issue: The formation and evolution of Ceres' Occator crater," *Icarus* 320, 213–225.

Seddio, S. and Schenk, P. (2007) "Crater densities, surface ages and evolution of Ganymede's Bright Terrain," *Lunar Planet. Sci. 38*, Abst. #2350.

Shankman, C., Kavelaars, JJ., Gladman, B. J. et al. (2016) "OSSOS. II: A sharp transition in the absolute magnitude distribution of the Kuiper belt's scattering population," *Astron. J.,* 151, 31, doi:10.3847/0004-6256/151/2/31.

Shibaike, Y., Ormel, C. W., Ida, S., Okuzumi, S. and Sasaki, T. (2019) "The Galilean satellites formed slowly from pebbles," *Astrophys. J.,* 885, 79, doi:10.3847/1538-4357/ab46a7.

Shoemaker, E. M. (1965) "Preliminary analysis of the fine structure of the lunar surface in Mare Cognitum Ranger 7, Part 2: Experimenters' analyses and interpretations," *JPL/NASA Technical Report 32-700*, 75–134.

Shoemaker, E. M., Lucchitta, B. K., Wilhelms, D. E., Plescia, J. B. and Squyres, S. W. (1982) "The geology of Ganymede," in D. Morrison (ed.) *Satellites of Jupiter*. Tucson: University of Arizona Press, pp. 435–520.

Shoemaker, E. M. and Wolfe, R. F. (1982) "Cratering time scales for the Galilean satellites," in D. Morrison (ed.) *Satellites of Jupiter*. Tucson: University of Arizona Press, pp. 277–339.

Shoemaker, E. M., Wolfe, R. and Shoemaker, C. (1986) "Extinct Jupiter-family comets and cratering rates on the Galilean satellites," *Lunar Planet. Sci. Conf. 17*, pp. 799–800.

Shoemaker, S. (1996) "The age of Europa's surface," in D. Matson and D. Nash (eds.) *Europa Ocean Conference*. San Juan Capistrano, CA, pp. 65–66.

Showman, A. P., Stevenson, D. J. and Malhotra, R. (1997) "Coupled orbital and thermal evolution of Ganymede," *Icarus,* 129, 367–383.

Singer, K. N., Bland, M. T., Schenk, P. M. and McKinnon, W. B. (2018) "Relaxed impact craters on Ganymede: Regional variation and high heat flows," *Icarus,* 306, 214–224.

Singer, K. N., McKinnon, W. B. and Nowicki, L. T. (2013) "Secondary craters from large impacts on Europa and Ganymede: Ejecta size–velocity distributions on icy worlds, and the scaling of ejected blocks," *Icarus,* 226, 865–884.

Singer, K. N., McKinnon, W. B., Gladman, B. et al. (2019) "Impact craters on Pluto and Charon indicate a deficit of small Kuiper Belt objects," *Science,* 363, 955.

Smith, B. A., Soderblom, L. A., Johnson, T. V. et al. (1979) "The Jupiter system through the eyes of Voyager 1," *Science,* 204, 951–972.

Smith, B. A., Soderblom, L. A., Beebe, R. et al. (1979) "The Galilean satellites and Jupiter: Voyager 2 imaging science results," *Science,* 206, 927–950.

Sotin, C. and Tobie, G. (2004) "Internal structure and dynamics of the large icy satellites," *C. R. Phys.,* 5, 769-780.

Spudis, P. D., Hawke, B. R. and Lucey, P. (1984) "Composition of orientale basin deposits and implications for the lunar basin-forming process," *J. Geophys. Res.: Solid Earth,* 89, C197–C210.

Squyres, S. W., Howell, C., Liu, M. C. and Lissauer, J. J. (1997) "Investigation of crater 'saturation' using spatial statistics," *Icarus,* 125, 67–82.

Stevenson, D. J. and Lunine, J. I. (1986) "Mobilization of cryogenic ice in outer solar system satellites," *Nature,* 323, 46–48.

Stöffler, D., Ryder, G., Ivanov, B. A. et al. (2006) "Cratering History and Lunar Chronology," *Rev. Mineral. Geochem.,* 60, 519–596.

Strom, R. G. (1977) "Origin and relative age of lunar and mercurian intercrater plains," *Phys. Earth Planet. Int.,* 15, 156–172.

Strom, R. G., Malhotra, R., Ito, T., Yoshida, F. and Kring, D. A. (2005) "The origin of planetary impactors in the inner solar system," *Science,* 309, 1847–1850.

Strom, R. G., Woronow, A. and Gurnis, M. (1981) "Crater populations on Ganymede and Callisto," *J. Geophys. Res.,* 86, 8659–8674.

Tera, F., Papanastassiou, D. A. and Wasserburg, G. J. (1974) "Isotopic evidence for a terminal lunar cataclysm," *Earth Planet. Sci. Lett.,* 22, 1–21.

Thomas, P. G., Forni, O. P. and Masson, P. L. (1986) "Geology of large impact craters on Ganymede: Implications on thermal and tectonic histories," *Earth Moon Planet.,* 34, 35–53.

Tonks, W. B. and Melosh, H. J. (1993) "Magma ocean formation due to giant impacts," *J. Geophys. Res.: Planets,* 98, 5319-5333.

Vance, S., Bouffard, M., Choukroun, M. and Sotin, C. (2014) "Ganymede's internal structure including thermodynamics of magnesium sulfate oceans in contact with ice," *Planet. Space Sci.,* 96, 62–70.

Wagner, R., Neukum, G., Giese, B. et al. (2006) "Geology, ages and topography of Saturn's satellite Dione observed by the Cassini ISS camera," *Lunar Planet. Sci. 37*, Abst. #1805.

Wagner, R., Neukum, G., Wolf, U., Greeley, R. and Klemaszewski, J. E. (2003) "Impact basins on Ganymede and Callisto and implications for the large-projectile size distribution," *EGS-AGU-EUG Joint Assem.*, Abst. #11475.

Wagner, J. R., Schmedemann, N., Werner, S. C. et al. (2018) "The large bright ray crater Osiris on Ganymede: Its age, role as a potential time-stratigraphic marker, and target for detailed imaging by the JUICE/JANUS Camera," *Euro. Planet. Sci. Cong.*, Abst. #EPSC2018-855.

Wagner, R., Wolf, U., Neukum, G. and Galileo SSI Team (1999) "Ages of individual craters on the Galilean satellites Ganymede and Callisto," *Lunar Planet. Sci. 30*, Abst. #1818.

Walsh, K. J., Morbidelli, A., Raymond, S. N., O'Brien, D. P. and Mandell, A. M. (2011) "A low mass for Mars from Jupiter's early gas-driven migration," *Nature,* 475, 206–209.

White, O. L., Schenk, P. M. and Dombard, A. J. (2013) "Impact basin relaxation on Rhea and Iapetus and relation to past heat flow," *Icarus,* 223, 699-709.

White, O. L., Schenk, P. M., Bellagamba, A. W. et al. (2017) "Impact crater relaxation on Dione and Tethys and relation to past heat flow," *Icarus* 288, 37–52.

Wilhelms, D. E. (1987) *The Geologic History of the Moon.* U. S. Geol. Surv. Prof. Paper 1348. Washington, DC: U. S. Geol. Surv.

Williams, N. R., Bell, J. F., III, Christensen, P. R. and Farmer, J. D (2015) "Evidence for an explosive origin of central pit craters on Mars," *Icarus,* 252, 175–185.

Wood, C. A. and Andersson, L. (1978) "New morphometric data for fresh lunar craters," *Proc. Lunar Planet. Sci. Conf. 9th*, 3669–3689.

Wood, C. A., Head, J. and Cintala, M. (1978) "Interior morphology of fresh Martian craters: The effects of target characteristics," *Proc. Lunar Planet. Sci. Conf. 9th.*, 3691–3709.

Woronow, A. (1977) "Crater saturation and equilibrium: A Monte Carlo simulation," *J. Geophys. Res.,* 82, 2447–2456.

Woronow, A. (1985) "A Monte Carlo study of parameters affecting computer simulations of crater saturation density," *Proc. 15th Lunar Planet. Sci. Conf. Part 2. J. Geophys. Res.*, C817–C824.

Woronow, A., Strom, R. G. and Gurnis, M. (1982) "Interpreting the cratering record: Mercury to Ganymede and Callisto," in D. Morrison (ed.) *Satellites of Jupiter*. Tucson: University of Arizona Press, pp. 237–276.

Xu, L., Hirata, N. and Miyamoto, H. (2017) "Ray craters on Ganymede: Implications for cratering apex-antapex asymmetry and surface modification processes," *Icarus,* 295, 140–148.

Yoshida, F., Terai, T., Ito, T. et al. (2019) "A comparative study of size frequency distributions of Jupiter Trojans, Hildas and Main Belt asteroids: A clue to planet migration history," *Planet. Space Sci.*, 169, 78–85.

Yozzo, J. E., Kirchoff, M. R. and Schenk, P. (2009) "Apex-antapex asymmetry of impact crater density on Ganymede's dark terrain," *Lunar Planet. Sci. Conf. 40*, Abst. #2214.

Zahnle, K., Alvarellos, J. L., Dobrovolskis, A. and Hamill, P. (2008) "Secondary and sesquinary craters on Europa," *Icarus*, 194, 660–674.

Zahnle, K., Dones, L. and Levison, H. F. (1998) "Cratering rates on the Galilean satellites," *Icarus*, 136, 202–222.

Zahnle, K., Schenk, P., Sobieszczyk, S., Dones, L. and Levison, H. F. (2001) "Differential cratering of synchronously rotating satellites by ecliptic comets," *Icarus*, 153, 111–129.

Zahnle, K., Schenk, P., Levison, H. and Dones, L. (2003) "Cratering rates in the outer solar system," *Icarus*, 163, 263–289.

Zanetti, M., Stadermann, A., Jolliff, B. et al. (2017) "Evidence for self-secondary cratering of Copernican-age continuous ejecta deposits on the Moon," *Icarus*, 298, 64–77.

Zuber, M. T. and Parmentier, E. M. (1984) "A geometric analysis of surface deformation: Implications for the tectonic evolution of Ganymede," *Icarus*, 60, 200–210.

9

The Topography of Ganymede

Geology, Global Characteristics, and Future Exploration

PAUL SCHENK, WILLIAM MCKINNON, JEFFREY MOORE AND FRANCIS NIMMO

9.1 INTRODUCTION

Strongly differentiated, possessing an active intrinsic dipole magnetic field and a deepwater ocean, larger than Mercury and with a complex geologic history including (cryo)volcanic and tectonic resurfacing, Ganymede stands unique among planetary satellites (see Chapters 5 and 7 of this volume).[1] As such it is a major target of scientific and exploration interest and is now the focus of a planned European Space Agency (ESA) orbital mission, Jupiter ICy moons Explorer (JUICE), (Grasset et al., 2013), and gravity-assisted encounters by NASA's Europa Clipper (Howell and Pappalardo, 2020). Determining the geologic and thermal history of Ganymede has importance for learning about the evolution of Ganymede and the entire Jupiter system in terms of when differentiation and tidal heating occurred (see Chapters 4 and 6 of this volume). Surface topography is a key tool for understanding the nature of Ganymede's ice shell and the geologic and geophysical processes that have altered its surface. Topography is also relevant for exploration as short-wavelength topography strongly influences how the shell can be probed with radar and other instrumentation (e.g., Berquin et al., 2013). Topography can also reveal or constrain variations in shell properties, including thickness and density, possibly related to phase changes in the interior. The objective of this report is to describe the available topography data for Ganymede, their nature and limitations, and their implications for geophysics and exploration.

Scientific objectives that can be addressed with topography on Ganymede relate to thermal history, resurfacing history and processes, and icy crust and mantle properties. Not all of these can be addressed here without explicit physical modeling beyond the scope of this work, but the data in hand provide possible solutions.

Ganymede can be divided into four basic common geomorphological units (Pappalardo et al., 1998; Pappalardo et al., 2004; Chapters 5 and 7 of this volume), the evolution and emplacement of which remain poorly understood. *Undisrupted dark terrains* are dominated by impact craters, both relaxed and unrelaxed, and by smaller-scale mass wasting of the surface. *Disrupted dark terrains* are swathes of dark terrain broken into elongate slabs by deep curvilinear troughs, some of which merge with grooved bright terrain. *Grooved bright terrains* (Fig. 9.15) have higher albedos and lower crater densities, indicating younger crater retention ages, but they are characterized by varying degrees of parallel ridge-and-trough morphology indicative of intense faulting. *Smooth bright terrains* are also brighter and lower in crater densities, but they are generally texturally smooth in kilometer-scale imaging, though they may exhibit subtly lineated or hummocky textures at higher resolutions. Topographic characteristics of each terrain will be a key tool in evaluating formation models. As tectonism is evident in both disrupted dark and smooth bright terrains, the relative importance of tectonic resurfacing (e.g., Head et al., 2002) and (cryo)volcanic resurfacing (e.g., Schenk et al., 2001) in these terrains remains unsettled. Ganymede's extensive cratering record (e.g., Schenk et al., 2004; Chapter 8 of this volume) also provides important constraints on its thermal history via the anomalous primary landforms created (such as central pit and central dome craters as well as multi-ringed basins, as reviewed by Schenk et al., 2004) and through viscous relaxation studies (e.g., Passey and Shoemaker, 1982; Dombard et al., 2006; Bland et al., 2017; Singer et al., 2018).

Some individual topographic data sets have been described previously. Giese and colleagues (1998) examined high-resolution topography of dark and bright terrains. Schenk (2002) and Schenk and colleagues (2004) have surveyed impact crater shapes and landforms while Bray et al. (2012) have used crater shapes to model crater formation. Berquin and colleagues (2013) used some of these data in evaluating the prospects and designs for radar sounding of the subsurface. Nimmo and colleagues (2002) and Nimmo and Pappalardo (2004) used stereo digital elevation models (DEMs) to identify flexural profiles and thus deduce heat fluxes for various Ganymede terrain types. Many studies of relaxation of Ganymede crater topography have been carried out, of which Singer and colleagues (2018) is a recent example. The topography of grooved terrain can be reproduced using models in which extension leads to local instabilities and boudinage (e.g., Dombard and McKinnon 2006). Schenk and colleagues (2001) examined several sites of smooth bright terrain formation. Here we describe the entirety of the available data and examine their geologic and global implications. We first reconsider our understanding of the elevation differences among these terrains as they relate to emplacement and resurfacing on Ganymede. We then examine the global topographic characteristics of Ganymede, of which these terrain differences are an essential element. Possible regional topographic variability on Ganymede is of increasing interest after the discovery of topographic undulations at >100 km scales on Enceladus (Schenk and McKinnon, 2009; 2023), another ocean world, the detection of an unusual broad dome at Ganymede's sub-Jovian point

[1] We note that whether Triton possesses an intrinsic magnetic field is not yet known, but Ganymede will always maintain its rank by size.

(Drake, 2015; Kay et al., 2018; Schenk et al., 2020), as well as indications of possible gravity anomalies in northern areas of Ganymede (Palguta et al., 2006; Palguta et al., 2009).

9.2 TOPOGRAPHIC DATA OF GANYMEDE

Except for six extant Galileo limb profile tracks covering <1% of the surface (Anderson et al., 2001; P. Thomas, personal comm.), most topographic data for Ganymede are derived from stereogrammetric (SG) and shape-from-shading (or photoclinometric [PC]) analyses of Galileo and Voyager images to produce DEMs. Shadow-length measurements (e.g., Schenk, 2002) are integrated into the PC toolkit and can be used to assess the reliability of PC mapping. As a direct result of the geometry of the two Voyager flybys and the Galileo main antenna deployment failure, these data are severely limited in scope (Fig. 9.13). The best Voyager encounter in the Jupiter system was reserved for Ganymede and Voyager 2 stereo images of part of the southern hemisphere were acquired with resolutions as high as 500 meters/pixel. Most Galileo imaging was restricted to individual targeted sites at resolutions of ~100–500 m/pixel supplemented by additional lower-resolution imaging. All possible stereo- and PC-suitable image combinations for Ganymede have been identified and processed by the lead author (PS). As for Europa (Schenk, 2014), most of the individual DEMs for Ganymede are not contiguous over large areas except near the south pole. In contrast to Cassini's global topographic mapping of the Saturnian satellites (Schenk and McKinnon, 2023), only limited regional medium resolution (hundreds of meters) imaging and stereo mapping were acquired on either Ganymede (Fig. 9.13) or Callisto and insights on these scales are limited.

Sufficient topographic data exist, however, to describe the slope and elevation characteristics of representative samples of each of the major terrain types. High-resolution stereo and PC mapping at better than 100 m/pixel were acquired in sufficient quantity and quality to characterize the major geologic terrains on Ganymede, but only over <35% of the surface. Stereogrammetric and photoclinometric processing of these images benefits from an updated control network (i.e., camera-pointing vectors) for Ganymede, including all medium- and high-resolution Voyager, Galileo, and New Horizons images (updated from Schenk, 2010). We note in revision that additional DEM coverage of the northern subJovian hemisphere was obtained by Juno (Ravine et al., 2022).

9.2.1 Topographic Methods

As all of our topographic data for Ganymede are derived from image analysis, it is critical that all users of these data be aware of the limitations of such data and how they might influence DEM construction and interpretation. These methods and issues associated with them are described in more detail in Section 6.1 but are summarized here along with their geographic distribution (Fig. 9.13).

Stereogrammetry uses stereo images to obtain feature parallax and height. As the imaging data from Voyager and Galileo are framing images, not line-scan images, the derived DEMs are undistorted across the image plane, but the DEMs are subject to blurring of factors of three to five relative to image resolution due to the use of scene recognition algorithms (see Section 6.1).

Useful stereo imaging and derived DEM data cover ~5 600 000 km^2, or ~6.5% of Ganymede's surface (Fig. 9.13), but DEM quality varies from site to site. Voyager acquired stereo coverage of Ganymede over significant areas, in the form of overlap between regional ~0.5 and 1 km/pixel mosaics from Voyager 2. These cover the area from the equator to the south pole along the terminator near meridian 240°E. In addition, three frames taken during an atmospheric search scan (c2063949.img, c2063949.img, c2063949.img) make useful stereo pairs with earlier regional mosaics. The highest-resolution Voyager stereo DEMs were of the south polar region south of ~60°S latitude, where imaging of ~500 meters was acquired. Galileo stereo imaging coverage of Ganymede is limited to a few targeted sites of geologic interest and serendipitous stereo with other imaging. Only two targeted stereo observations were acquired during the first four orbits (e.g., Giese et al., 1998), but most of orbit G28 was dedicated to several stereo observations in the eastern Nicholson Regio–Harpagia Sulcus region.

Stereo imaging at resolutions >1 km occurs elsewhere within Voyager 2 and Voyager 1 coverage areas from mosaic overlap but none of the areas has sufficient parallax to resolve any mappable geologic features except the deepest craters (~1–2 km depths; Schenk, 2002). Such products do reveal whether any major topographic deviations of >500 m are present (see Section 4. Some 0.8–1.0 km resolution Galileo imaging also can be combined with the images from Voyager. An example of this is the G29 ~800 m/pixel regional images, which make weak stereo with Voyager 1 imaging on the sub-Jovian hemisphere. The ~840 m pixel scale Galileo C9 Perrine–Galileo transect stereo DEM (using the C9 hemispheric mosaic at 2 km scales), which reveals the deepest bright terrain units and a prominent ridge (Fig. 9.14), is another example. While regional stereo coverage exists for Callisto, high resolution sampling is much more limited.

The PC-DEMs are constructed from low-sun monoscopic images. Pixel brightness values are used to estimate surface slope under certain assumptions (see Section 6.1). The PC-DEMs are valuable as indicators of the magnitudes of terrain and feature relief (assuming an ~25% uncertainty in the absence of more sophisticated PC modeling than can be done with these data), but they are not reliable as measures of relief across terrain boundaries or for assessing elevation differences across large areas (i.e., >100 pixels in the original images) due to the accumulation of uncertainties along profile lines. Long-distance components may be informative but have indeterminate reliability in the absence of additional constraints. Image interpretation is essential when using all PC-topographic data.

Topographic mapping from PC at 0–2 km/pixel resolutions covers ~26,042,000 km^2, or ~30% of Ganymede's surface (Fig. 9.13, bottom). The best PC-DEMs, at <500 m/pixel, cover only ~9% of the surface. The greatest extant PC mapping coverage is from the terminator imaging from both Voyager and Galileo mosaics along four longitude bands. These map swathes are at 0.8 to ~3.5 km/pixel resolution and are variable in quality, depending largely on the complexity of bright–dark terrain geography. The quality of these maps is inconsistent as they depend on the quality of the high-sun images to compensate for the brightness variations inherent over such large areas. Although the Voyager 1, Voyager 2, and Galileo G6 and Galileo C9 terminator PC-DEMs cannot be used for regional topographic studies of absolute elevations for the reasons described earlier in this chapter, useful topographic information can be obtained for specific craters and grooved terrain units at 3–0.6 km pixel scales along their respective longitudes (Fig. 9.13).

In those areas where we are fortunate to have both stereo and PC mapping coverage (Fig. 9.13), we can merge SG and PC DEMs. This procedure preserves the long-wavelength components of the stereo-DEM and the short-wavelength components of the PC-DEM, thereby suppressing the uncertain long-wavelength PC components while retaining their higher resolution fidelity. (Unfortunately, the multiple angles required in more advanced stereo-PC techniques are simply not available on any of the Galilean satellites.) When merged using filtering in the Fourier domain, these integrated SG-PC-DEMs provide both long-wavelength topographic fidelity and high-frequency topographic data at the original pixel scale of the input images.

For Ganymede, merged stereo-PC topographic mapping is more limited than on Europa (Schenk 2014) and was possible for only a handful of sites. The largest such areas are the Voyager 2 south polar mapping mosaic (Fig. 9.1), including Gilgamesh (Fig. 9.1). Smaller 3d-PC DEM sites are the Menhit crater (Fig. 9.2), Mumu Sulci, and Harpagia Sulcus. Only ~4% of Ganymede's surface is covered by merged stereo-PC coverage. The total useful global mapping coverage from all topographic sources is ~35%.

Absolute Elevations and Regional Slopes

Finally, none of the Galileo or Voyager DEMs produced by these imaging techniques are tied to any global topographic datum for the simple reason that, as on Europa and Callisto, none exist for Ganymede. Limb profiles could provide a global reference as on Io (Thomas et al., 1998; White et al., 2014) or Enceladus (Thomas, 2010), but are far too sparse on Ganymede to permit interpolation and registration of DEMs to them. In no case do we have reliable information regarding absolute elevations with respect to the center of Ganymede.

Each DEM has been normalized to a local mean of 0 (zero) meters by subtracting out a best-fit plane. This has the effect of reducing any residual biases in interpreting local small-scale features due to any (slight) misalignment of the original images. Negligible regional slopes are also consistent with the lack of significant shadows and shading in regional low-sun terminator images of Ganymede, the lack of lateral distortions of the terminator on Ganymede as seen on the images, and the lack of significant relief in the limb profiles greater than 1 km (P. Thomas, personal comm.). All these observations are consistent with a general lack of large-scale regional rises (with one important exception described in Section 9.4.1), depressions, or slope gradients on Ganymede. Lower-amplitude deviations from a global degree-2 shape could exist in areas where we do not have topography, however.

9.3 RELATIVE TOPOGRAPHY OF BRIGHT AND DARK TERRAINS

Dark Terrains

Dark terrain geology (e.g., Prockter et al., 1998; Prockter et al., 2000; Chapters 5 and 7 of this volume) and topography is dominated by an extensive population of diverse impact craters (Fig. 9.2–9.4) of increasing complexity with size (cf. Schenk et al., 2004; Bray et al., 2012) and in various stages of degradation and relaxation (e.g., Dombard and McKinnon, 2006; Bland et al., 2017; Singer et al., 2018). In addition to the concentric giant-impact-related rimmed troughs or furrows that cover many areas (Fig. 9.3a; Schenk and McKinnon, 1987; Hirata et al., 2020), some dark terrains have also been disrupted by extensional nonconcentric block faulting that can occur in intersecting bands (Fig. 9.5a–c).

The three dark terrain sites mapped at better than 250/m pixel scales were ~100 km-wide samples of Galileo Regio (Fig. 9.3), Eastern Nicholson Regio (Fig. 9.4), and the Nicholson-Arbela Sulcus region (Fig. 9.5a–c). First topographic and geophysical interpretations of the Galileo Regio site were reported by Nimmo and Pappalardo (2004) and those of the Nicholson-Arbela Sulcus site by Giese and colleagues (2001b) and Nimmo and colleagues (2002). The targeted SG-DEM Galileo and E. Nicholson sites have topographic ranges of ± 0.46 km and ± 0.196 km, respectively (all topographic ranges reported here are given at the 95% limits, with respect to their means). Each of these sites was targeted to avoid craters larger than ~15 km, though the Galileo Regio site includes several of the large concentric furrows that most likely formed during a giant impact (McKinnon and Melosh, 1980; Schenk and McKinnon, 1987; Hirata et al., 2020) explaining the greater topographic amplitude there. Individual furrows at this site have relief of up to 1.2 km (Giese et al., 1998, 2001a; this chapter). Similar relief of 0.5–1 km is observed for furrows at the Anshar Sulcus and Buto sites (Fig. 9.2). Galileo PC- and SG-DEMs of Callisto over one of the Valhalla outer furrows, which resemble the Galileo furrows, also indicate relief of 07–1.5 km (Fig. 9.3b). Voyager stereo DEMs of the inner concentric ridges of the Valhalla system, however, indicate relief of 2–3 km (Fig. 9.3c), suggesting higher degrees of strain in the inner parts of that system.

The Arbela–Nicholson serendipitous SG site (Fig. 9.5a–cb) combines G7 and G28 images of 135 and 180 m/pixel scales and is ~175 km across. The 95% topographic range of dark terrains at this site west of Arbela Sulcus, which includes craters as large as 17 km, is ± 0.285 km and is greater than at Eastern Nicholson due to bands of fractures disrupting dark terrains at this site (Fig. 9.5a–c). These fractures are up to 3 km wide, a few hundred km long, and ~600 m deep. Of course, relief could have been higher before the relaxation inferred to have occurred across dark terrains on Ganymede (e.g., Dombard and McKinnon, 2006).

The fractures associated with disrupted dark terrain in Nicholson Regio (Fig. 9.5a–c) and other areas are noteworthy in that most appear relatively pristine and preserve slopes of 35° or more, suggesting that they formed relatively late, perhaps as a prelude to or as part of the general process of dark terrain breakup during bright terrain resurfacing. Although these fractures often occur in parallel packets separated by undisrupted dark terrain, no recognizable volcanic resurfacing is associated with this fracturing and they do not form bands of bright terrain.

Bright Terrains

Bright terrain (see Chapters 5 and 7 of this volume.) replaced large sections of dark terrain in a variety of styles, divided into grooved to smooth bright terrains based on apparent relief at kilometer scales. Most bright terrain occurs in elongate bands that sharply cut across and truncate features within dark terrain

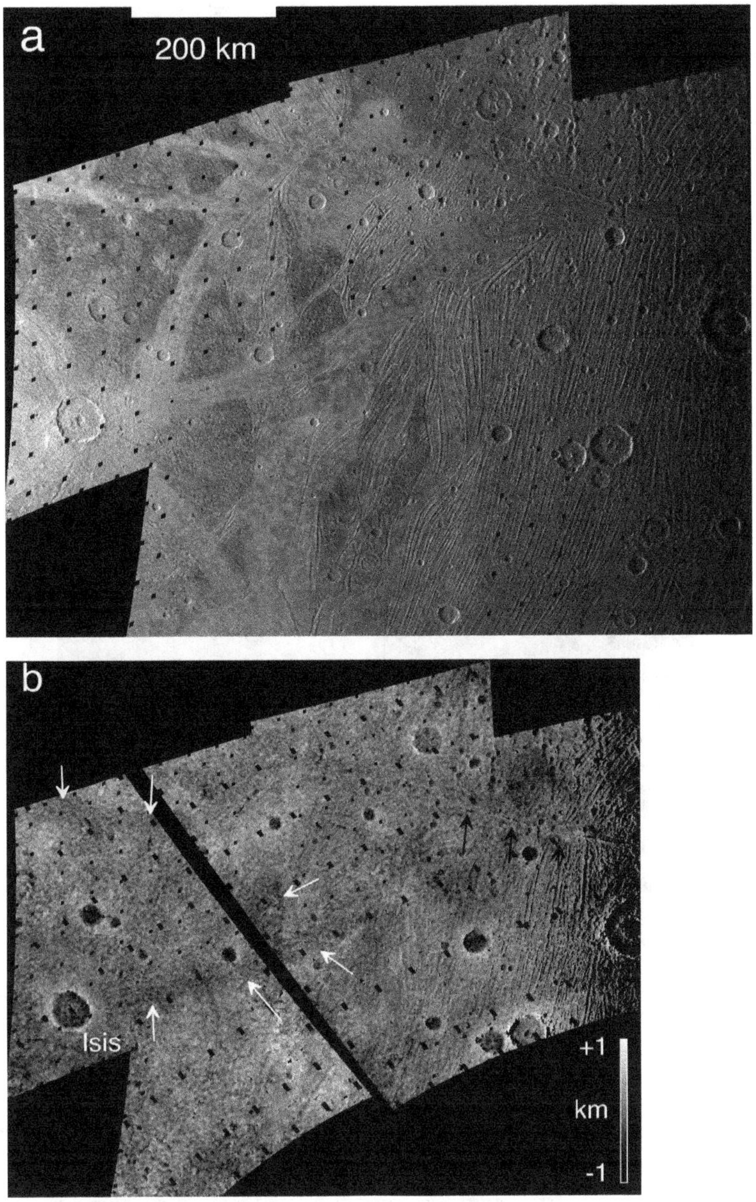

Figure 9.1 Voyager image (a) and SG-DEM (b) of the south polar region, including several exposures of bright and dark terrain and numerous craters, including 80 km central pit crater Isis and the large crater Anubis at far right. White arrows indicate low-standing lanes of bright terrain. Dark arrows indicate high-standing ridge similar to one evident in Fig. 9.14. Voyager images acquired at 500–600 m pixel scales; view centered at ~77° S, 180° E.
Source: NASA.

(e.g., Figs. 9.5a–c–9.7, Fig. 9.15) (e.g., Parmentier et al., 1982; Collins et al., 1998; Pappalardo et al., 1998. Hammond and Barr, 2014). Many bands are marked by closely spaced ridges and troughs seen at high resolution to consist of densely faulted terrains. The degree of faulting and ridge formation observed at high resolution is highly variable from site to site (e.g., Fig. 9.7), with some bands exhibiting faulting down to the limits of resolution and other bands either featureless or exhibiting subtle lineations or only poorly organized hummocky texture (Fig. 9.6).

Perhaps the most compelling unresolved geologic question on Ganymede involves the mechanism(s) of resurfacing, including dark terrain breakup and bright terrain emplacement. Galileo has provided key insights into these processes, including the partial fragmentation of dark terrain (e.g., Prockter et al., 2000), the complexity of groove formation (e.g., Patel et al., 1999; see Pappalardo et al., 2004), the heat flux at the presumed time of breakup (Dombard and McKinnon, 2001; Nimmo et al. 2002; Bland and McKinnon, 2015, 2018), and the combined depressed level topography and small-scale roughness of smooth bright terrain (Schenk et al., 2001).

The lack of preservation of preexisting topography opens the question of how bright terrain resurfacing is accomplished. Three mechanisms were considered during the Voyager era (Parmentier et al., 1982). The first is simple down dropping of lanes of dark terrain as structural graben and flooding by icy lava (e.g., Parmentier et al., 1982). This would require sufficient down dropping in all known exposures such that all

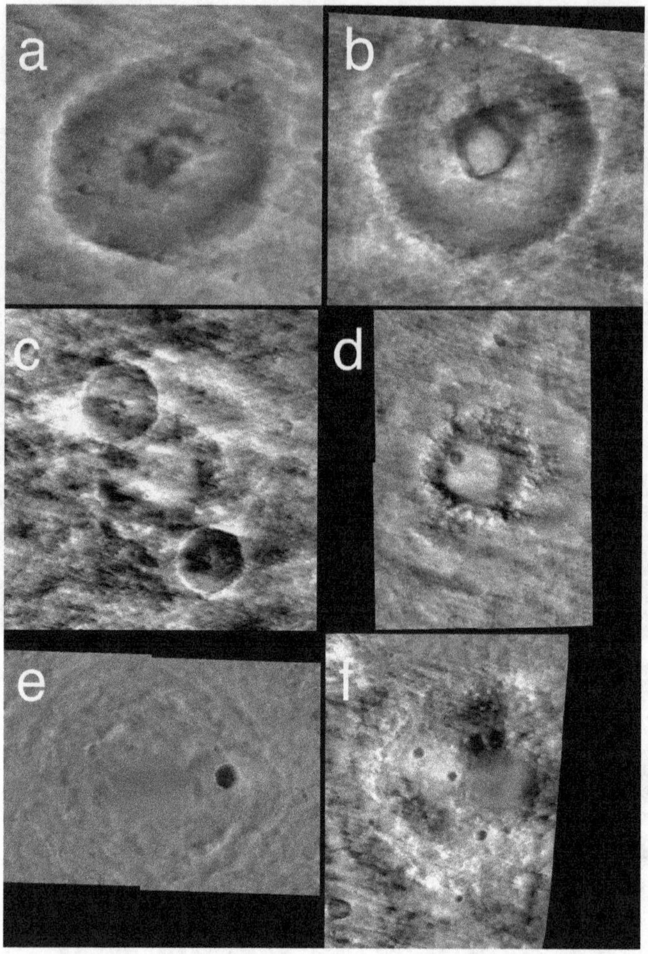

Figure 9.2 Topographic maps of major large-impact crater morphologic classes on Ganymede (see Schenk et al., 2004 and Chapter 8 of this volume for details), shown at same apparent size to facilitate comparison of features. From top left, in order of increasing size or age: (a) central pit crater Lugalmeslam (D = 64 km), (b) central dome crater Melkart (D = 107 km), (c) anomalous dome crater Menhit (D = 138 km), (d) anomalous dome crater Neith (D = 165 km), (e) penepalimpsest Buto (D = 157 km), (f) Gilgamesh basin (D ~ 570 km). All data are from photoclinometry except Gilgamesh, which is from stereo-PC.
Source: NASA.

Figure 9.3a Galileo image mosaic (left) and SG-DEM (right) of intersecting furrows in dark terrain in central Galileo Regio, which form the deep linear features at the upper right. Scene centered at 18°N, 212°E. Also evident are small unrelaxed craters and larger relaxed craters (bottom center).
Source: NASA.

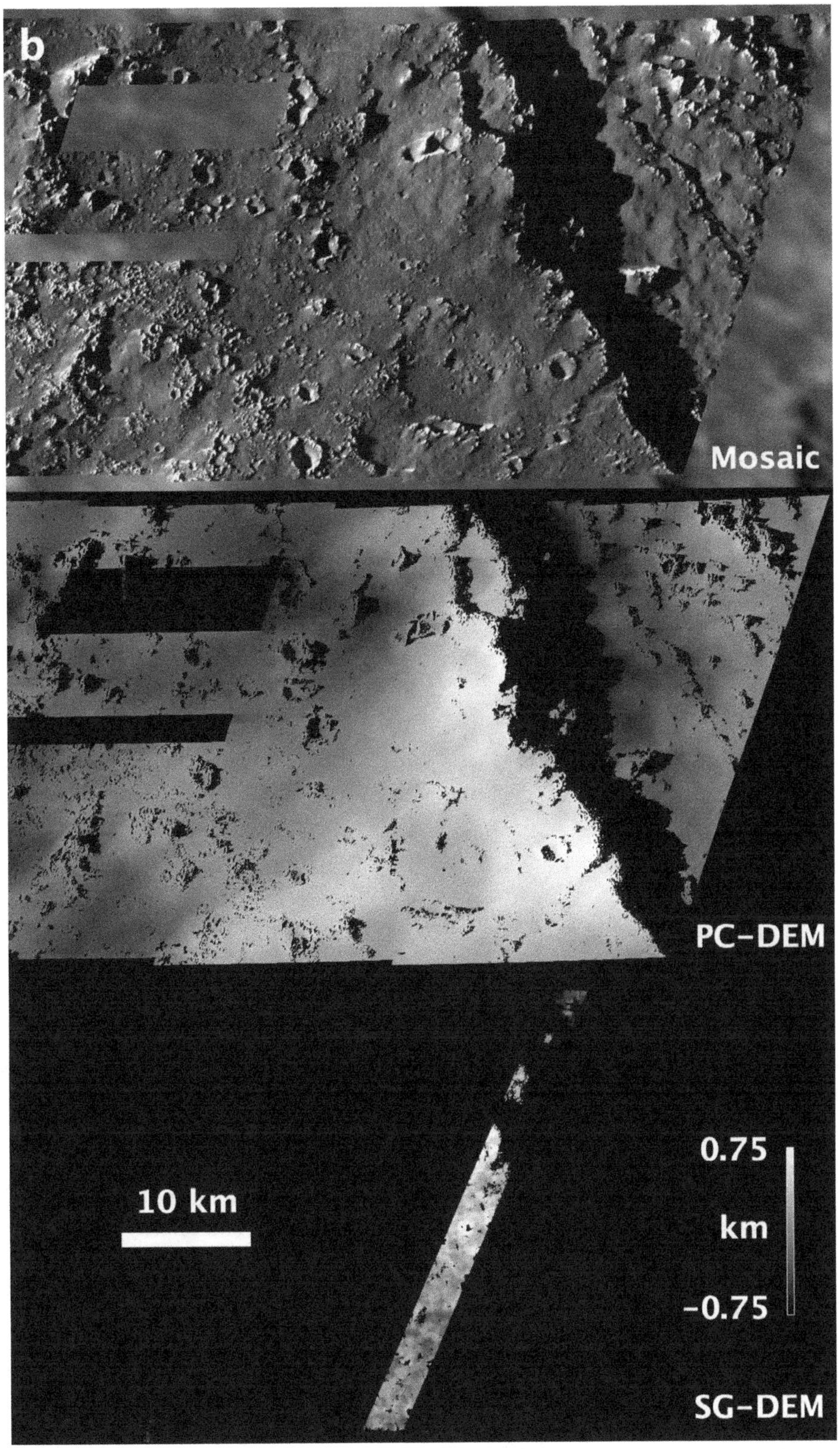

Figure 9.3b Galileo mosaic and DEMs of outer furrow in Valhalla multi-ring system. PC-DEM in center panel also calculates shadow heights. The SG-DEM at bottom is from stereo overlap in five-frame Galileo mosaic at top.
Source: NASA.

older surfaces are buried and none exposed. The second mechanism is similar to the first, except that graben formation involves pervasive deformation in some manner such that the relict topography of dark terrain is eliminated and flooding proceeds either in concert with the deformation or afterward to smooth the floor of the band. The third mechanism involves lateral expansion of the icy shell along vertical fractures and formation of new material extruded from below

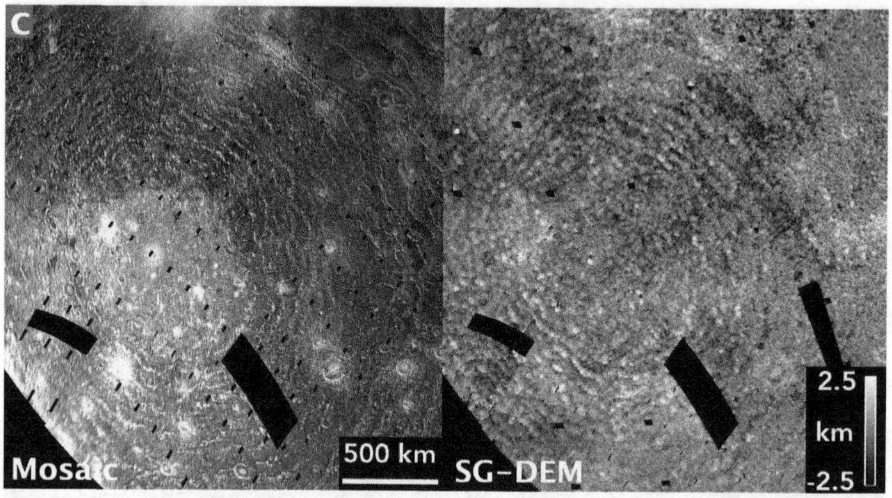

Figure 9.3c Voyager 1 Mosaic and SG-DEM of Valhalla multi-ring system, showing ridges in central portion. North is to the top.
Source: NASA.

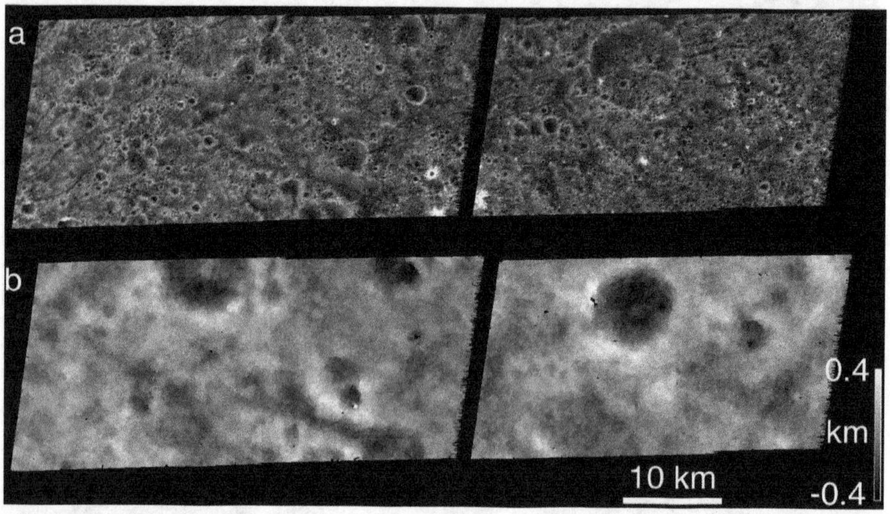

Figure 9.4 Galileo image mosaic (a) and SG-DEM (b) of dark terrains in Eastern Nicholson Regio. Although targeted as an area of undisturbed dark terrain, several degraded troughs at bottom left are evident. Centered at 15°S, 23°E.
Source: NASA.

(e.g., Head et al., 2002), similar to how pull-apart bands are formed on Europa (Prockter et al., 2002). Parmentier and colleagues (1982) argued from Voyager mapping that bright terrain contacts and cross-cutting relationships with dark terrain were most consistent with down dropping of graben-like blocks of dark terrain and burial with new material, rather than lateral divergence of crustal blocks along separating faults and replacement of the gap with completely new material. Although this latter concept has been revisited for some bands of smooth terrain (Head et al., 2002), with somewhat ambiguous results, basic questions remain on the actual emplacement mechanics of smooth terrain specifically and bright terrain generally.

Several sites were observed in stereo with sufficient resolution to unambiguously determine the relative elevations of bright and dark terrain units together, one of the most useful of which may be the Arbela Sulcus site (Fig. 9.5a–c). Here, the 25 km wide smooth band Arbela Sulcus, which exhibits very modest degrees of linear textures indicative of faulting, is depressed ~200 m relative to flanking dark terrains to the east and west (Fig. 9.5a–c), except for a 150 m high scarp bounded arch (or ridge) along its eastern margin. The 35 km wide band of grooved terrain immediately west of Arbela at this site is also elevated ~200 m above the mean elevation of Arbela Sulcus and has similar topographic amplitudes and mean elevation as nearby dark terrains. Whether the pervasive fracturing that formed this grooved band deformed dark or resurfaced bright terrains is unclear, but the lower albedo of the unit suggests that dark terrain deformation could have been involved.

In the Perrine-Galileo image transect for G9 (Fig. 9.14), a 260 km wide east–west composite band of bright material at 25°N, 302°E, consisting of both grooved and smooth terrains, is depressed ~500 m relative to adjacent dark terrain. A second 45 km wide north–south trending bright terrain band appears to be depressed ~200 m. Grooved terrain lanes near the Isis crater near the south pole (Figure 9.1) also appear to be depressed 100–350 meters relative to bright terrains. Both mapping sites also feature the only known occurrences of single ridges on Ganymede, each ~500 m high and 100–300 km long. Whether

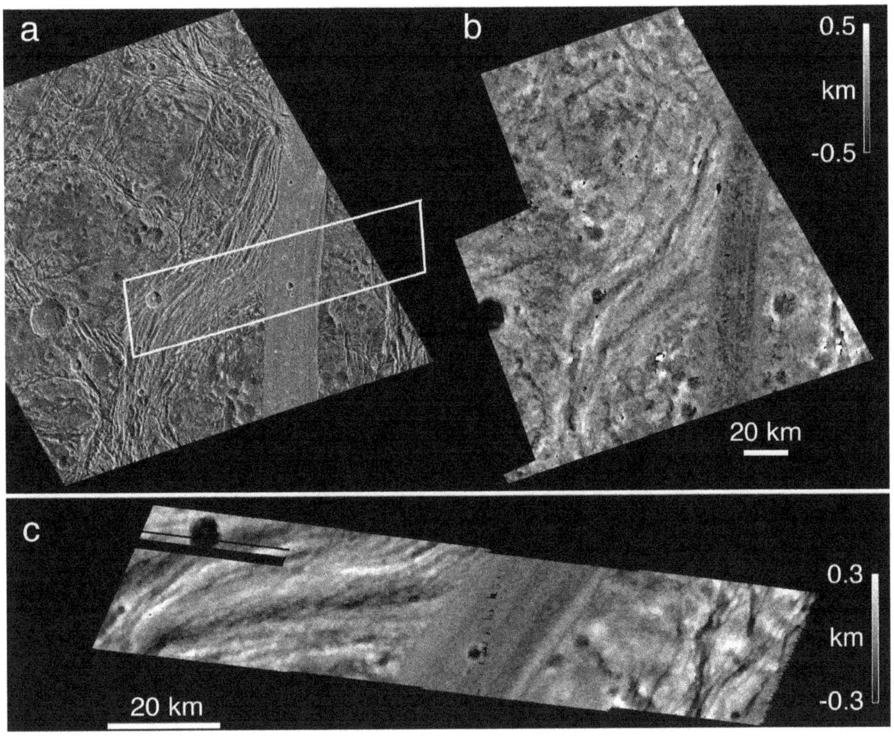

Figure 9.5a–c Galileo image (a) and DEM (b) of Arbela Sulcus and Nicholson Regio, centered near 15°S, 12°E. The SG-DEM is from Galileo G7 and G28 images. Inset box in (a) shows location of high-resolution DEM (c) from targeted G28 imaging.
Source: NASA.

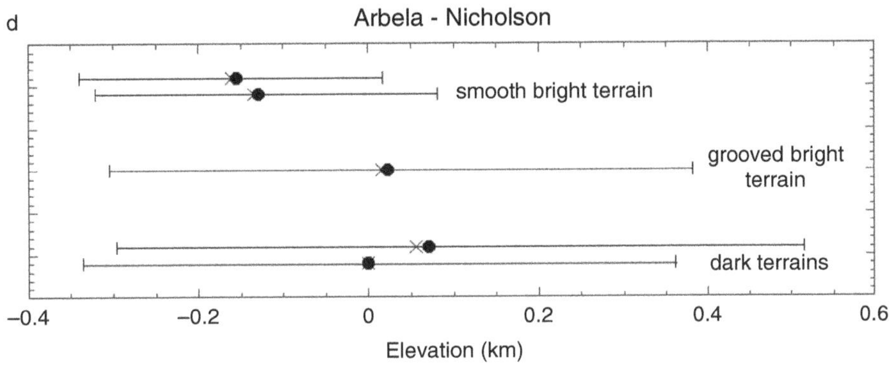

Figure 9.5d Histograms of elevation values of smooth and grooved bright terrains and dark terrains at Arbela Sulcus DEM site shown in Figure 9.5a–cb and c. Dot represents mean, "x" represents median, and horizontal lines show 95% of all elevations in each terrain sample. Dark and smooth bright terrains are sampled twice in these data.
Source: NASA.

compressional or constructional in origin is not evident from the 500–800 m pixel scale imaging.

The most extensive high-resolution mapping of a variety of bright terrain units is the G8 "calderas" site at 30°S, 176°E, and 180 m pixel scale (Fig. 9.6). Although the combination of Voyager and Galileo images at significantly different solar elevations and the application of PC at sun angles of 60–65° can produce locally spurious results, it is evident that the grooved terrain packets or blocks observed here occur at different elevations. The DEM suggests that the least deformed bands are the lowest and the most deformed highest in elevation, consistent with the grooved and smooth bright terrain bands at Arbela (Fig. 9.5a–c). The mean elevations of the different units at the G8 site span an ~1 km range. These limited results apparently confirm that smooth terrain bands are depressed relative to all other units, confirming the results of Schenk and colleagues (2001), but also that there could be a correlation between the degree of deformation and the mean elevation of the units. Whether these findings are valid globally for all bright terrains remains to be determined.

Two end-member mechanisms for bright terrain resurfacing have (arguably) received the most attention from Galileo analyses. Some grooved terrains may be formed by complete tectonic disruption of terrains in situ without new material emplacement (e.g., Pappalardo et al. 1998). Alternatively, some (or all) smooth bright terrains may have been emplaced by (cryo)volcanic resurfacing of down-dropped blocks of dark terrain by water or water–ice hybrid lavas (e.g., Schenk et al., 2001). Topographically, smooth (and to date grooved bright) terrains are characterized by a complete lack of

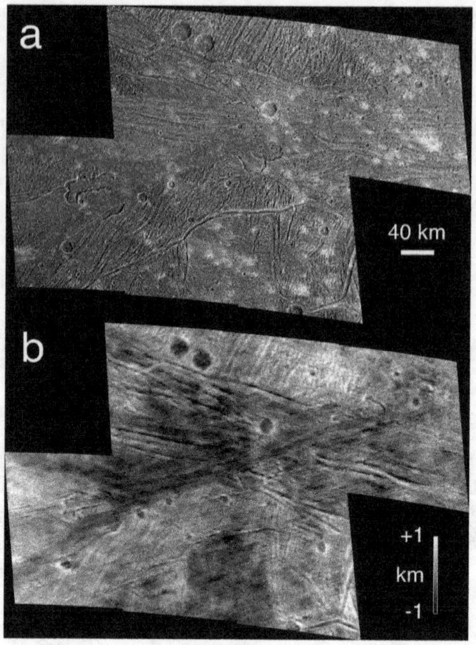

Figure 9.6 Galileo image mosaic (a) and Galileo–Voyager SG-PC-DEM of crosscutting bands of smooth and grooved bright terrain and associated caldera-like paterae in the Mummu and Sippar Sulci region of Ganymede.
Source: NASA.

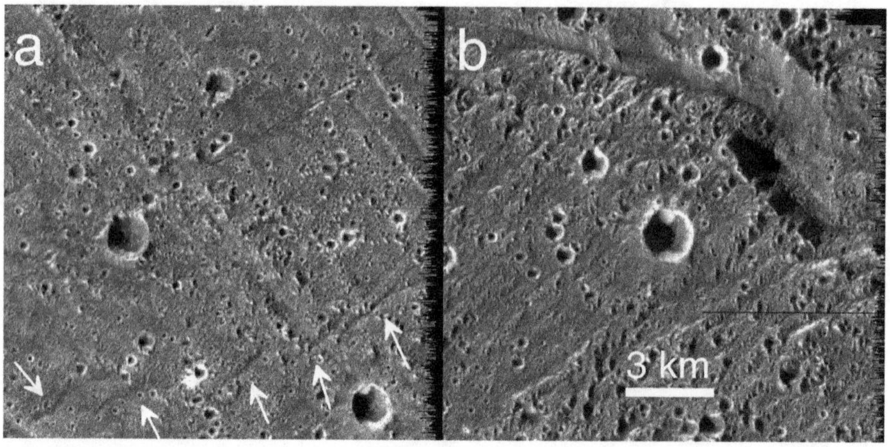

Figure 9.7 Highest-resolution views of smooth bright terrains in Harpagia Sulci on Ganymede. Galileo images s552443600 and s552443513, showing aligned hummocks or low hills with interspersed smoother materials. Note the darker sinuous lineations at the bottom of image (a, arrows), which could be flow textures. Image resolutions are ~15 to 22 m/pixel.
Source: NASA.

preservation of preexisting topography. The remnant crater depressions and furrows characteristic of dark terrain, which have magnitudes of 1 km or so, are not evident in the topographic signatures of bright terrains that have been mapped thus far (but might be observed in future data sets). Tectonic deformation would have to be pervasive enough to erase such signatures.

A global survey of such terrains is not possible here due to the paucity of topographic data. One key question is how frequently packets of grooves are due to pervasive faulting in dark terrain or to resurfacing of older terrains to form new bright terrain materials. Where a strong case can be made for tectonic resurfacing is the packet of curvilinear grooved terrain just west of Arbela Sulcus (Fig. 9.5a–c). This packet is cut by Arbela Sulcus and predates it. The contact between the groove packet and surrounding dark terrains is not as sharp as most bright terrain units, however, suggesting that structural deformation tended to "eat" into dark terrain, gradually decaying in intensity over a short distance. Dark material is also abundant on some groove flanks, suggesting involvement of dark terrain, but incomplete albedo evolution to be classified as bright terrain.

For the smooth bright terrains in Arbela and other locations (Figs. 9.5a–c, 9.7), a tectonic resurfacing mechanism may not be able to reduce topography of dark terrains by factors of two or more to match the smooth bright terrains. The lack of relict or unburied outcrops of dark terrain within smooth terrain and its general planar smoothness (despite undulations along trend) implies that burial by resurfacing obscures all such preexisting topography. For such flooding to occur within these narrow lanes, either the +450 m of relief observed at dark terrains sites such as Galileo Regio (Fig. 9.4) must be erased viscously (as proposed by Bland et al., 2018) or by other mechanisms, or the amount of down dropping must exceed the amount of fill to cover positive topographic features in dark terrain of this magnitude, plus the 200–500 m mean depressed relief relative to the other terrains, to explain the total topographic depression observed. The total calculated vertical displacement implied for older terrains would thus be on the order of ≥ 1 kilometers. Alternatively, disruption of dark terrain must be so pervasive, through small-scale closely spaced faulting and structural breakup, as to reduce topographic amplitude to low enough values prior to flooding to result in the combined total erasure of all relict topography within smooth terrain. Without flooding and burial, dark (i.e., low-albedo) materials themselves must be sufficiently superficial as to be able to slough off tilted surfaces and into troughs and other low spots (as proposed by Prockter et al. [2000]) and then be obscured by mass wasting.

The common occurrence of faint linear striations within smooth terrain (Fig. 9.7) suggests a continuum of degrees of deformation of grooved terrain, which can range from <10 to 800 meters amplitude. Within terrains designated "smooth," these striations have negligible (or unmeasurable) topographic relief (Fig. 9.7). Indeed, across the Achelous–Gula crater site, which appears quite smooth at ~ 1 km resolution Voyager imaging, grooves are pervasive but of very low amplitude, seldom exceeding 50 meters depth. Our comprehensive data suite expands the range over which groove amplitude and by inference strain magnitudes can be inferred by mechanical modeling (e.g., Bland and McKinnon, 2018; Howell and Pappalardo, 2018).

Several factors favor volcanic resurfacing for the smoothest bright terrain bands. First, these smoother bands tend to be lower than other terrains and at what appears to be a consistent level along their length, suggesting infilling to an equipotential surface. The second factor is the absence of significant relief across these bands. It is difficult to design deformational mechanisms that erase rather than enhance topographic relief, something that is relatively easy with (cryo)volcanic infilling. High-resolution imaging shows that these terrains are often deformed by parallel ridges, but these have relief on 10-meter scales at most and are not evidence of major deformation. Third, we have been unable to unambiguously realign terrains on either side of these bands to reconstruct a pre-spreading terrain at Arbela Sulcus, indicating that the pull-apart mechanism as proposed by Head and colleagues (2002) may be less likely than resurfacing by down dropping material, or that the pull-apart is not pure separation (see such evidence from bands on Europa by Culha et al. [2014]).

Smooth bright terrain bands are often associated with partially enclosed caldera-like depressions (Lucchitta, 1980; Schenk and Moore, 1995; Head et al., 1998; Kay and Head, 1999; McKinnon et al., 2000) of which there are ~ 35 mapped globally to date (see Fig. 9.6). These paterae are distinctly nonimpact landforms and at high resolution (125–180 m) paterae floors feature high, standing, mesa-like deposits inferred to be viscous flows (Head et al., 1998; Kay and Head, 1999; Giese et al., 2017). These features and their association with the low, smooth bright terrains strengthen the case for a (cryo)volcanic origin for both types of features. The thicker patera floor fill materials could represent a transition from low- to high-viscosity extrusions if volcanism is a key process in bright terrain emplacement.

Even if smooth bright terrains are volcanically emplaced, the mechanism(s) of resurfacing and the origins of the depressed topography are not yet resolved and may be uniquely Ganymedean rather than duplications of silicate volcanism processes. One puzzle is the absence of obvious protruding remnants of preexisting dark terrains within these bands. One proposed solution is pressure release melting under a down-dropped graben (Showman et al., 2004), in which melt is presumably extruded through multiple linear fracture-controlled vents. Liquid water has negligible viscosity and is unlikely to make flow fronts of much thickness. Flash freezing on eruption may assist in producing thick outer skins of flows, but experiments to verify this are lacking, as eruption mechanisms on Ganymede are unknown. Galileo never achieved resolutions of bright terrain at better than 100 meters except at four locations (Arbela Sulcus, central and western Harpagia Sulcus, and Uruk Sulcus), and, except for poorly defined sinuous dark tracks in western Harpagia Sulcus (Fig. 9.7), never unambiguously resolved flow fronts or related features at any of these sites. Inadequate resolution, post-emplacement modification, mass wasting and regolith formation, or a combination of these may have been responsible for this failure, and regolith formation may make detection challenging even for future spacecraft – but high-resolution imaging, laser altimetry, and especially subsurface radar sounding from the upcoming JUICE mission to Ganymede (Grasset et al., 2013) holds great promise.

The origin of the low hills within Harpagia Sulcus (Fig. 9.7; Galileo image s552443513) remains unclear. The hills are pervasive and are oriented in several distinct directions, suggesting structural control. They differ from striations and grooves in their short sub-kilometer scales and lack of linear continuity. Whether the hills formed prior to, during, or after resurfacing is also unclear. It is interesting to speculate that the hills are remnants of the lava extrusion process or represent what remains of an original preflood surface with smoother material between the hills being extruded (water) lavas. Alternatively, the hills could themselves be multiple low-volume extrusions that made their way to the surface along swarms of fractures, or the 3D expression of incipient tectonic topographic instabilities. Despite the pervasiveness of these features, there has been no pervasive large-scale block faulting in these terrains such as observed at Uruk Sulcus or west of Arbela Sulcus (Fig. 9.5a–c). It is also unclear whether the low hills of central and western Harpagia Sulcus (Fig. 9.7) are common to smooth terrain generally, as they are not evident in Arbela Sulcus (Fig. 9.5a–c) but are even more prominent in Xibalba Sulci (Pappalardo et al., 2004).

9.4 GLOBAL TOPOGRAPHIC CHARACTERISTICS

9.4.1 Regional Topographic Variability and Global Elevation Extremities

The global topographic characteristics of a planet record the thermal and geologic history of that body, and Ganymede is no different. Galileo gravity observations have also been interpreted as suggesting the existence of large-scale subsurface mass anomalies (Palguta et al., 2006; Palguta et al., 2009), for example, which might be manifest as topographic distortions of the surface. Although the limb profiles are continuous for >4000 km, they were extracted from low-resolution images that may have compression artifacts and lack meaningful geologic context. DEM elevations cannot be validly compared from site to site on "continental" scales because of the lack of a global reference datum, but valuable information comparing terrain elevations within a DEM site are possible in several locations, based on the inference that Ganymede does not possess significant deviations from sphericity >2 km (Anderson et al., 2001; Schubert et al., 2004). That is, analysis of (albeit limited) limb profiles yields a physical mean radius R of 2631.2 ± 1.7 km (Schubert et al., 2004), which can be compared with maximum triaxial radius variation expected, $(12J_2/5 + 2q)R$, where J_2 is the unnormalized degree-2 zonal gravity coefficient, and q is a dimensionless measure of the strength of the centrifugal potential due to Ganymede's rotation. For the parameter values in Anderson and colleagues (2004), this maximum variation is 1.8 km. Ganymede's true shape should be easily measurable with improved imagery and altimetry from JUICE.

Our analysis of bright terrain–dark terrains indicates mean elevation differences of 200–500 m. Dark terrains and the most tectonized bright terrains are apparently at similar elevations, but the least deformed smooth bright terrains are depressed several hundred meters, confirming the findings of Schenk and colleagues (2001). Our measurements of the range of elevations within terrain units varies, but is most commonly less than ±100 m to ±500 m. The differences between mean elevations of these units range from undetectable to ~750 meters. Though very limited, these observations would suggest that the topographic amplitude across all of Ganymede might also be relatively low, ±0.75 km, perhaps. As all our DEMs are referenced to a local zero mean, this supposition neglects any regional variations or possible higher-amplitude terrain sites (the G8 "calderas" site, Mummu Sulci, for example [Fig. 9.6]). No examples have been observed of long shadows being cast at bright-dark terrain boundaries, suggesting that these examples are typical.

The deepest known geologic features are the larger, younger impact craters (Fig. 9.2). Impact craters have been measured previously (Schenk, 2002; Bray et al., 2012) and our results confirm that the deepest craters are in the size range of ~20–100 km wide and ~1–1.25 km deep rim to floor. These depths apply to post-bright-terrain craters, formed when the heat flow on Ganymede had evidently diminished to levels whereby impact craters could retain their original shapes. That Ganymedean craters formed in a shallower initial state than comparably sized lunar craters has also been established, given the inherently weaker "impact strength" of ice as a crustal material (Schenk, 1991). The unusual primary crater landforms such as central domes and penepalimpsests are also recorders of the heat-flow history of Ganymede (e.g., Schenk et al., 2004), as is the subsequent longer-term relaxation of many older craters on dark terrains due to enhanced internal heat flow (Singer et al., 2018; Bland et al., 2017).

Data are limited when looking for regional topographic deviations, but some constraints are evident from 1 km vertical scale stereo imaging (e.g., Fig. 9.8). These include DEMs from Voyager 1 and Voyager 2 hemispheric 1–2 km pixel scale stereo imaging, which at best resolve only the deeper craters; higher-resolution Voyager 2 stereo of the southern regions at 500 m scales, and isolated but extended Galileo stereo pairs such as the G9 Galileo–Perrine transect mosaic. The largest of these DEMs from Voyager 1 mosaics and from the Galileo Perrine-Galileo four-frame transect show no topographic deviations of >1 km anywhere in the mapping areas, with one important exception described later in this chapter.

The issue of regional topographic variations from the mean radius comes to the fore in light of putative gravity anomalies inferred from the Galileo gravity data (Palguta et al., 2006; Palguta et al., 2009), and the observed long-wavelength 100–150 km wide, 1–1.5 km deep topographic depressions on Enceladus (Schenk and McKinnon, 2009). Unfortunately, none of the three anomalies reported by Palguta and colleagues (2009) occur within our DEM mapping areas, and the possibility of density or gravity anomaly–related topographic signatures remains unconstrained. The available limb profiles (Anderson et al., 2001) reveal variations of 1–2 km from the mean limb fit, but most occur in poorly imaged areas and correlation with geologic features is poor. The level of noise in these profiles from the compressed low-resolution images is also not well understood, and some of these extremes may be spurious. While much of Ganymede remains to be surveyed, broad topographic basins such as those seen on Enceladus (Schenk and McKinnon, 2023) are not yet evident.

The only verified large-scale topographic distortion on Ganymede is a 3 km high, 500-km wide dome (Fig. 9.8) discovered within 2° of the sub-Jovian point (1.6°N, 1.5°E; Schenk et al., 2020). This dome is observed in a Voyager 1 stereo pair with vertical precision (or sensitivity) of only ~450 m, and while it is not sensitive enough to detect geologic features, including the largest craters in the site, it does reveal major topographic deviations. The size and location of the dome, discussed in more detail later in this chapter, is fortuitously confirmed by one of the Galileo limb profiles (Fig. 9.8). The lack of similar doming on this scale in the other low-resolution stereo DEMs and its extreme elevation (for Ganymede) indicates that this is an unusual feature. (We note in proofs that Junocam stereo observations also capture the dome and show it to be an elongate feature ~750 by 500 km centered within 1.5° of the subJovian point (Ravine et al., 2022).)

9.4.2 Global Hypsogram

When we compile all valid SG-DEMs, including the Voyager 1 DEM revealing the sub-Jovian dome (Fig. 9.8), the resulting provisional global histogram of elevations (or hypsogram; Fig. 9.9) is somewhat broader than that suggested by the more limited topographic amplitudes of individual sites. This likely reflects higher amplitudes of ±1 km (at 95% levels) and very rugged km-scale relief across the Gilgamesh impact basin (Fig. 9.2). The 95% limits of this provisional Ganymede hypsogram are ±1.1 km, which is broader than the globally complete and controlled stereogrammetric hypsogram of Enceladus (Schenk and McKinnon, 2023;), which has 95% limits of −0.76 and +0.63 km. A second provisional hypsogram for Ganymede using the currently available limb profiles gives a 95% limits of ±1.6 km, slightly broader than but generally consistent

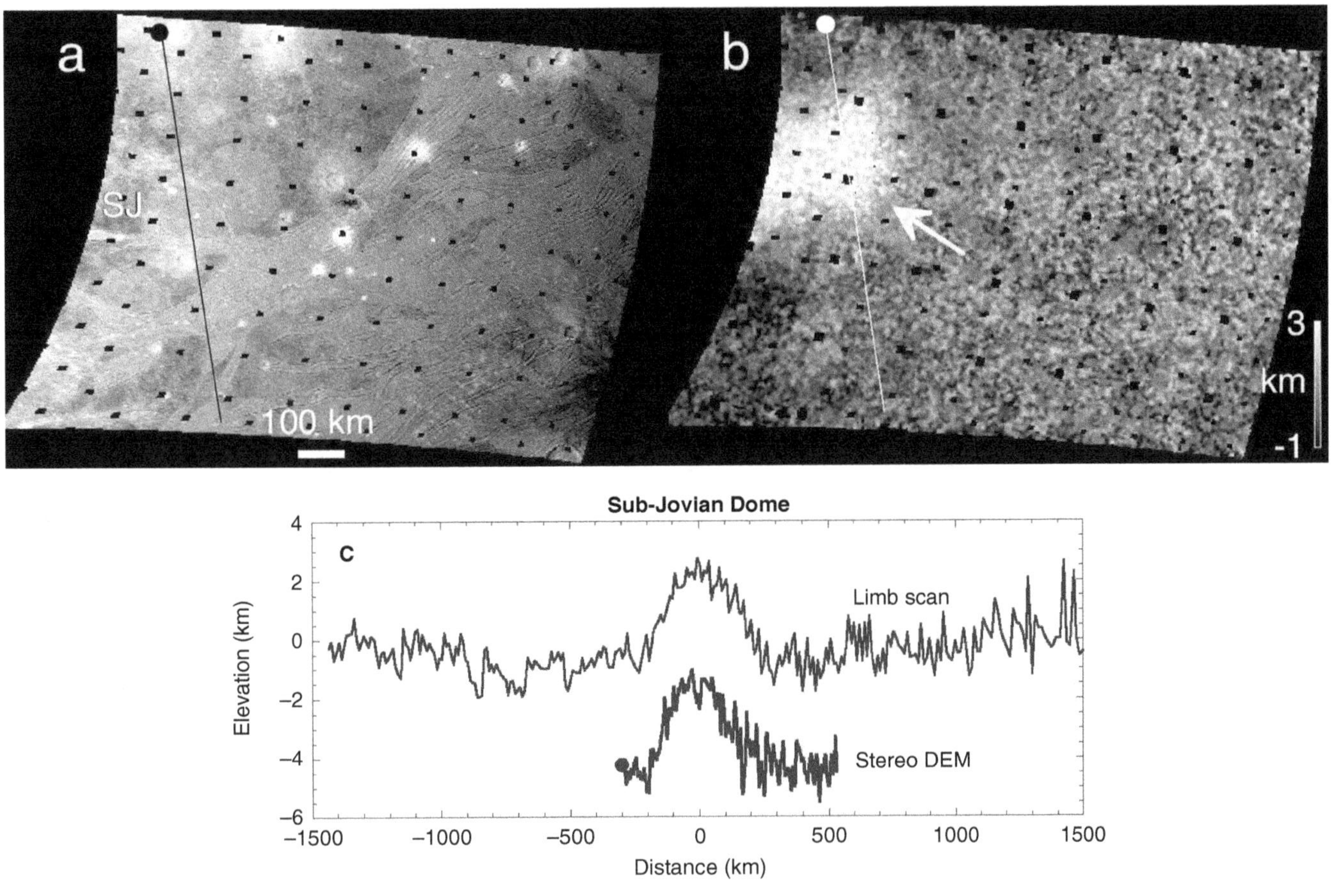

Figure 9.8 Voyager image and stereo DEM (a, b) of large dome feature at sub-Jovian point (SJ), with topographic profiles (c) from limb scan (P. Thomas, personal comm.) and stereo DEM.
Source: NASA.

with the stereogrammetric results (noting that these data are limited by extremely sparse coverage and noisy limb profiles related to image compression). We note that the hypsogram of the G9 site (Fig. 9.5a–c)is very similar to our provisional global SG estimate, with 95% limits of ~1.1 km. We thus provisionally consider the stereogrammetric data broadly representative of Ganymede's topography. Future mapping will permit a global shape model to be constructed from match-point radii estimates (in construction of a global image control network) combined with more comprehensive limb profiles. A provisional hypsogram of Europa, where we have even less regional stereo coverage, is narrower than Ganymede by a factor of ~3, but we caution again that the hypsograms of both Europa and Ganymede are untied to a global datum or shape model and thus are unlikely to stand the test of time.

9.4.3 Roughness Profiles

Slope and roughness characteristics of various terrain types can strongly affect interpretation of ground-penetrating radar data due to off-nadir scatter of the radar beam on contact with the surface (e.g., Berquin et al., 2013). Slopes are also important for future lander designs, although it is highly unlikely that a surface lander will be sent to Ganymede before improved topographic data are returned. Schenk (2009) in his studies of Europa surface slopes also examined available slopes on Ganymede. For Ganymede, we only have reliable slope estimates at length scales of $\gtrsim$100 meters and for grooved bright terrain and dark terrains, slopes are approx. 50–100% greater than at comparable length scales on Europa (Schenk, 2009). Whether this is difference is valid at smaller meter and decameter scales is unknown, pending higher-resolution DEMs

While not globally complete, a plot of root mean square (RMS) topographic variance (Fig. 9.9) suggests that the maximum relief on Ganymede exceeds that of both Europa and Enceladus. Another way of assessing topography is to calculate the RMS variance v as a function of step size Δx (Shepard et al., 2001):

$$v(\Delta x) = \left\{ \frac{1}{n} \sum_{i=1}^{n} [z(x_i) - z(x_i + \Delta x)]^2 \right\}^{1/2} \tag{9.1}$$

Typically, this variance increases with increasing step size and gives an indication of the roughness of terrain at different length scales. Figure 9.10 plots the RMS topographic variance for Ganymede, Europa, and Enceladus using limb profile data. At least over the length scales plotted, Ganymede is significantly rougher on average than both Europa and Enceladus, confirming the impression given in Figure 9.9. The contrast between Europa and Enceladus is more ambiguous and depends on wavelength, as discussed in Nimmo and colleagues (2011).

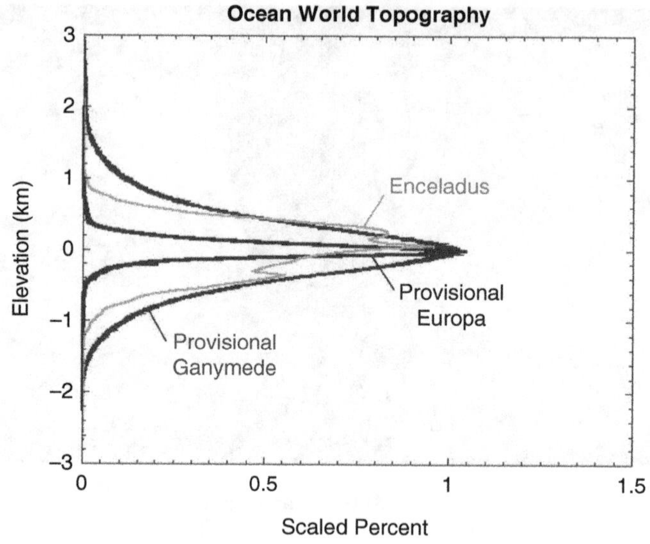

Figure 9.9 Provisional hypsogram of Ganymede, compared to Enceladus (gray line) and provisional data for Europa (narrowest hypsogram, data from PS).

Another way of assessing topography on Ganymede is to calculate the RMS slope (Shepard et al., 2001), which is defined as $v/\Delta x$ (cf. equation 9.1). Figure 9.10 shows the RMS slope as a function of step size based on limb profiles of Ganymede, Europa, and Enceladus. As expected from Figure 9.10 (top), Ganymede experiences the largest slopes at all wavelengths.

The local topography derived from DEMs suggests that Ganymede lacks extreme relief, yet the provisional hypsogram from the available data (Fig. 9.9), even with the arbitrary registration of each DEM to a hypothetical zero planum, indicates that Ganymede has more relief than either Europa or Enceladus. None of the DEMs and none of the four major N-S terminator observations from Galileo and the two Voyager spacecraft reveal any large plateaus or promontories on Ganymede. However, the preservation of higher relief across Gilgamesh (Figure 9.2) and the 3 km high sub-Jovian dome on Ganymede (discussed later in this chapter.) as well as the >2 km relief across many of the inner Valhalla ridges on Callisto (Figures 9.3, 9.4) imply that the icy shell of Ganymede is presently capable of supporting significant relief.

The limb profiles tell another part of the story. Figure 9.10 suggests that Ganymede is on average rougher than Europa or Enceladus. The DEMs contain 2D information and have higher spatial resolution, but they are geographically limited; the limb profiles have global coverage, but are lower resolution and may suffer from compression artifacts. The fact that Figure 9.9, which is based on DEMs, nonetheless indicates a rougher Ganymede suggests that both analysis techniques are correct. Ganymede's surface, though geologically old and more degraded, has not experienced high heat flow conditions for some time (or at least since most of the bright terrain craters formed, which are mostly unrelaxed) and supports a greater range of local and global topography than Europa, a body with similar surface gravity. Enceladus is locally the smoothest of all in terms of local slopes but has substantial regional variations that produce the global hypsogram (Fig. 9.9) and increase the RMS variation at long wavelengths (Fig. 9.10). These regional variations are up to 1.5 km deep and are plausibly due to subsurface density anomalies or shell thickness variations (Schenk and McKinnon, 2009), and thus likely isostatic. If strength is limited on higher-gravity bodies such as Europa and Ganymede, strength-limited anomalies and ice shell variations should be highly muted (reduced by more than an order of magnitude in topographic amplitude to on the order of 100–200 m assuming similar conditions).

Do the spatially limited topographic data described here represent a "random" and therefore reliable sampling of topographic variability on Ganymede? The Voyager stereo coverage was dictated solely by the Voyager trajectory. High-resolution Galileo observations were targeted at impact craters, cross-cutting terrain boundaries, and several "type" examples of the major terrains. These sites were not chosen with prior knowledge of topography, but in part due to Galileo trajectory and viewing, and with the benefit of available Voyager coverage (as the earlier passes were planned before Galileo's Jupiter arrival). Hence biases in the data sampling can be assumed to be due to non-geologic motives. No prominent undulations or promontories are betrayed by shading or shadows along the four longitudinal terminator strips that cover ~30% of the surface. Distortions in the Voyager stereo due to improperly calibrated reseaux marks are possible, but such distortions are easily correlated with specific reseaux marks and are unlikely to precisely "cancel" geologically defined topographic structures. We therefore tentatively conclude that, if Ganymede possesses any extreme (>2 km) deviations in topography, they are isolated and occur in only a few locations, such as the sub-Jovian dome, discussed next.

9.4.4 The Sub-Jovian Dome

The 3 km high sub-Jovian dome (Fig. 9.8) is the only large topographic distortion known on Ganymede (Schenk et al., 2020). The dome represents a previously unrecognized phenomenon and presents three questions: how to get (and preserve) such a dome, and why is it essentially on the sub-Jovian point? Formation of a broad dome of this scale could be due to

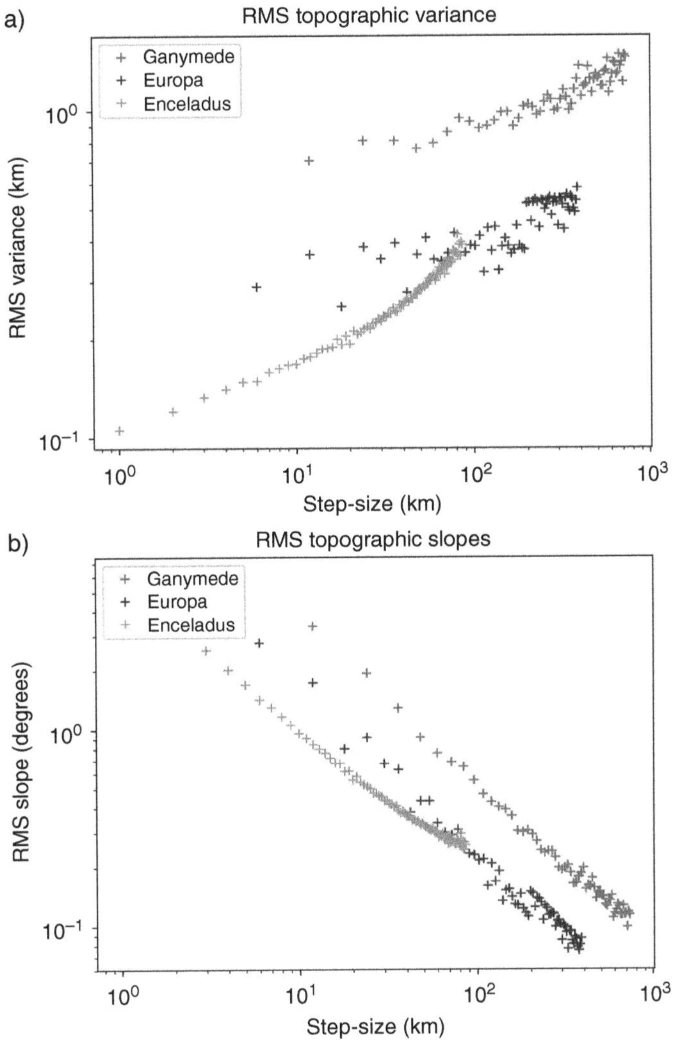

Figure 9.10 (top). RMS variance as a function of step size (equation 1) using limb profiles of Ganymede, Europa (Nimmo et al., 2007), and Enceladus (Thomas, 2010). Resolution is higher for Enceladus because of the smaller overall scale of this body. Variance is only calculated for length scales up to one-tenth of the total profile length (cf. Shepard et al., 2001). (bottom) RMS slope as a function of step size for Ganymede, Europa, and Enceladus, based on limb profiles. Data sources as in Figure 9.10 (top). Ganymede has the steepest slopes at all wavelengths.

(cryo)volcanism, ice shell thickening, subsurface intrusion of material inflating a dome (e.g., Squyres, 1980; Michaut and Manga, 2014), or crustal growth due to colder surface temperatures at that location (this latter proposed for Europa's poles by Ojakangas and Stevenson [1989]). Although imaged at low resolutions by both Voyagers, the sub-Jovian region, within Barnard Regio, is otherwise geologically unremarkable and betrays no surface expression of cryovolcano or a massive cryovolcanic deposit. Subsurface intrusion could be the result of tidal heating focused at a given location (e.g., Chapter 4 of this volume), though the sub- and anti-Jovian points are actually the regions of minimal tidal heating for a floating ice shell. Ganymede is not undergoing significant tidal heating presently, so this would most likely have occurred during or shortly after the tidal heating plausibly associated with bright terrain formation (Bland et al., 2009). The amount of intruded material required to raise a surface dome of this magnitude would depend on the densities of the materials involved and on what type of elastic support or isostasy is in play within the shell, both of which are unknown, but the extreme horizontal and vertical scales of the observed feature challenge present-day models of "laccolith-like" intrusions into ice shells (Manga and Michaut, 2017).

The alternative mechanism of thickening due to low surface temperatures (Ojakangas and Stevenson, 1989) is plausible only if the dome, or its precursor degree-2 shell thickness variation, formed at one of Ganymede's paleopoles and was afterwards displaced to the current equator via true polar wander (TPW) (McKinnon et al., 2014). Once relocated to the warmer equator, the longer wavelength topographic components above and below the surface might be expected to relax, leaving a residual flexurally supported dome observable today, which could explain how a large dome now of smaller size today could have been responsible for TPW in the past. The ~3 km elevation suggests ice shell thickening and possibly a crustal root of substantial size, depending on the degree of compensation and amount of flexural support. J. Kay et al. (2018) discuss how this dome can be preserved against basal flow of ice considering its likely substantial geologic age.

In the case of TPW of an ice shell thickened or domed at the poles, we would expect two such features antipodal to each other. Stereo imaging at the anti-Jove location is limited to wide-angle images that result in DEMs with best possible vertical precision of 1–2 km, and while no dome is evident in these data, we cannot yet be certain whether such a dome is present or absent. If polar shell thickening is a significant process, new polar domes could have formed after the present dome(s) (or their precursor topography) moved to their present location. Voyager 1 and Voyager 2 obtained partial stereo and DEM coverage of Ganymede's north and south polar regions (e.g., Figure 9.1) and no domes or rises are evident in current DEM coverage. As noted earlier, global limb profiles are not of sufficient quality and quantity to detect or constrain Galileo's degree-2 shape, so polar shell thickening could be lurking below detection limits. We also observe no "doming" in the Voyager stereo DEM of Callisto's north pole but note that the south pole and sub- and anti-Jovian regions of that body were not observed in stereo, and none of these four cardinal points were observed in stereo on Europa. Thus, the prevalence of this type of doming in the Jovian system is poorly constrained.

The apparent absence of polar shell thickening at Ganymede's current rotational poles might suggest that heat flow decreased rapidly after the TPW event, leading to elastic shell thickening globally sufficient to lock in present global topography and prevent further episodes of TPW. Lower heat flows and thicker shells could also trigger subsolidus convection at depth in the ice shell, whose effect is to reduce equator-to-pole shell thickness variations, and thus reduce or eliminate the drive for TPW (Nimmo et al. 2007).

As noted, the geologic record of the sub-Jovian dome site is indistinguishable from surrounding terrains, occurring within Barnard Regio dark terrains near several small bright terrain bands. Geologic mapping of this site is limited by the low image resolution of ~1 km/pixel, and high solar illumination precludes topographic shading. While no mappable deformation is evident at either the sub- or anti-Jovian locations, unresolved concentric or radial fractures associated with either uplift or subsequent partial collapse of the dome cannot be ruled out. Although we favor TPW of a polar dome by crustal thickening, we recognize that this is speculative, and final resolution of the origin of this feature awaits global mapping to search for additional such features at the predicted locations and more evolved modeling.

Regardless, secondary evidence for TPW on Ganymede may come from several sources. The dilution of the predicted apex-antapex of motion impact cratering asymmetry from factors of 20–30 to the observed factor of 4 could be explained by one or more TPW events (Zahnle et al., 2003; McKinnon et al., 2014; Chapter 8 of this volume). The distribution of late-forming bright terrain bands may also suggest TPW. The formation of bright terrain bands (or lanes) is known to be sequential, with narrow crosscutting lanes (often of smooth bright terrain) forming later (Collins et al., 1998). The mapped distribution of these lanes forms a broad global swathe oriented at large angles to Ganymede's equator, but when this pattern is rotated ~90° (as implied by TPW to the current dome location), a symmetric antipodal pattern results centered on the rotational paleo-equator (Fig. 9.11). This pattern is very similar to tidal or recessional stress patterns on tidally distorted bodies (e.g., Matsuyama and Nimmo, 2008) and suggests these late-stage bands formed under the influence of such stresses and then were rotated with the ice shell to their new location when the polar high topography moved to the current equator. Such a scenario is easily tested with new global data.

9.5 A PROVISIONAL GEOLOGICAL HISTORY OF GANYMEDE AND FUTURE PROSPECTS

The limited topographic data acquired by Voyager and Galileo help constrain a provisional evolutionary sequence for the development of geologic features on Ganymede. In the scenario, dark terrain prior to bright terrain formation has a long history of cratering, perhaps as long at 2 Gyr (see Chapter 8 of this this volume). The largest basin-forming events leave no obvious impact scars except for several large sets of concentric furrows formed by extensional faulting. Extensive topographic relaxation reduces crater depths to several hundred meters (from original depths up to 2 km; Fig. 9.12). Furrows are also reduced in height but retain 1–1.5 km of relief (Figs. 9.3; 9.15).

At some point, perhaps 2 Ga ago, faulting disrupts large swathes across dark terrain, truncating craters and furrows. Ice-shell volcanism resurfaces large areas, followed by pervasive tectonism in the form of grooves of varying intensity. In some areas, tectonic resurfacing, in the form of closely spaced parallel fault block formation, in the absence of volcanism, creates bright terrains in place with relief of up to 1 km in some areas. This style of deformation may have occurred throughout the period of bright terrain formation. The importance of volcanism versus tectonic resurfacing remains unresolved pending global high-resolution mapping but both processes are likely to be important.

The frequency of unrelaxed craters per unit area appears to be very similar on dark and bright terrains (Fig. 9.12), at least in the few sites for which we have reliable data (e.g., northern Marius Regio, Enki Catena, Perrine Regio, Tiamat Sulcus). As cratering rates were identical on the two terrains, a similar crater density would imply that the period of thermally induced relaxation of topography ceased late during bright terrain formation or immediately thereafter, as both terrains accumulated intact deep craters similarly after this time. Whether this observation is valid in all regions, at all latitudes, or on all scales will also require the global topographic mapping of Clipper and especially JUICE. Whether this episode of relaxation is related to a continuous, long-duration, high heat flow or to a shorter pulse of high heat associated with bright terrain formation is also unclear.

Future Prospects

Scenarios of this type can be added to the hypotheses to be tested from future mapping efforts with spacecraft observations from ESA and NASA in the 2030s, including high-resolution imaging, laser altimetry, and subsurface radar sounding from the upcoming Clipper and JUICE missions to Ganymede (e.g., Grasset et al., 2013). If gravity anomalies of the type suggested by Galileo (Palguta et al., 2006; Palguta et al., 2009) or other internal mass deviations occur within Ganymede, the discovery of (or lack of any) subtle topographic signatures will be

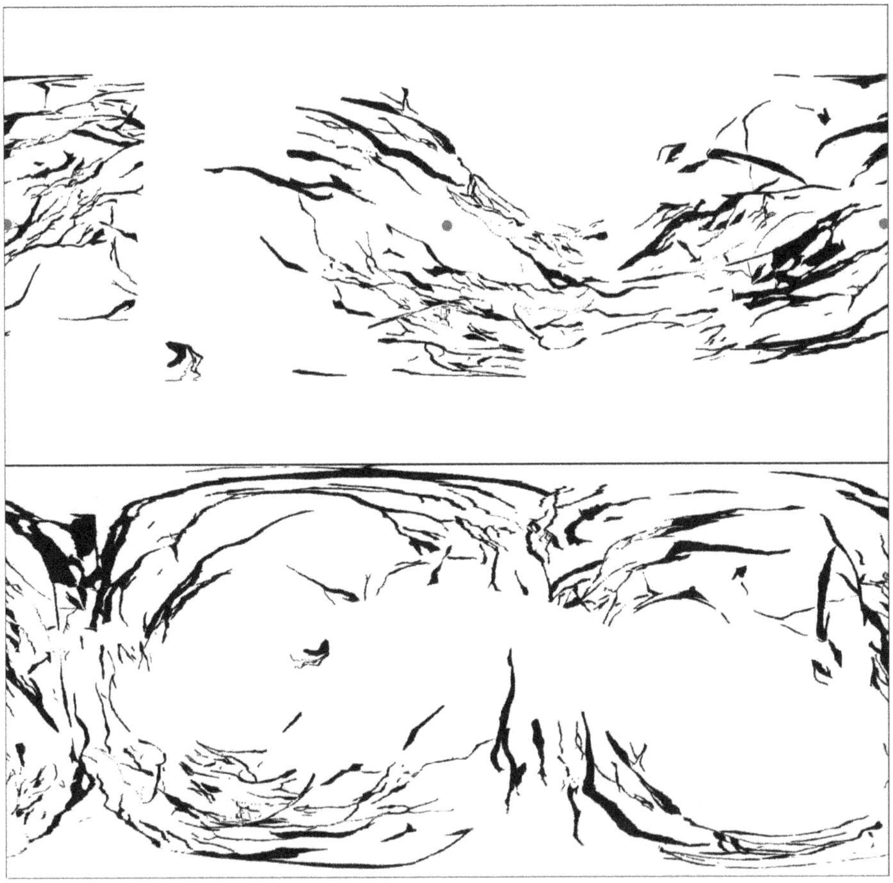

Figure 9.11 (top) Current global distribution of late-forming bright terrain bands (black lines and polygons; map courtesy of Geoff Collins). (bottom) Configuration of late-forming bright terrain bands after 88° of reversed true polar wander on Ganymede, highlighting antipodal axial symmetry of pattern suggestive of control by tidal stresses. Simple cylindrical map projection.

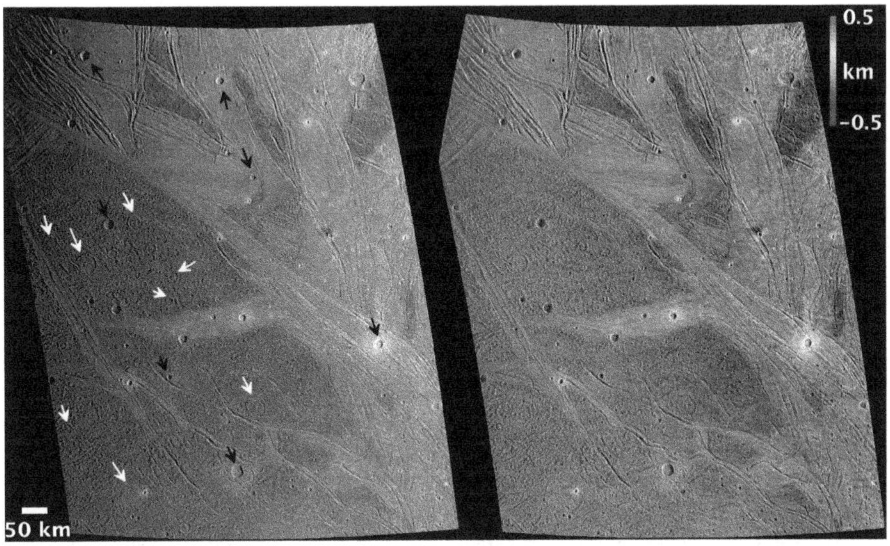

Figure 9.12 Galileo image and PC-DEM of northern Marius Regio area. Relaxed craters are white arrows, unrelaxed black arrows. Unrelaxed crater densities of N(>10 km) ~30 per 10^6 km2 are observed in both dark and bright terrains in this location. Image resolution is ~900 m/pixel; north is to top.

Source: NASA. (A black and white version of this figure will appear in some formats. For the color version, please refer to the plate section.)

important for constraining such deviations and the properties of the ice shell. The presence of a large antipodal 3 km high dome (Fig. 9.8) on the anti-Jovian hemisphere is unconfirmed, and the investigation of these and any possible remnant domes elsewhere must also await future observations. Global stereo mapping coupled with limb profiles and image match-point radius determinations with vertical precisions of < 50m will all be necessary to search for such signals.

Models of dark and bright terrain formation (including the tectonized and untectonized variations) rely on topographic and radar sounding constraints. Future missions to Ganymede will determine whether the <1 km elevation differences in the few examples of bright and dark terrain described here (Figures 9.1, 9.5a–c, 9.6), as well as the possible correlation of elevation with degree of tectonization (e.g., Figure 9.6), are representative of all such terrains across Ganymede.

The limited topographic data suggest Ganymede may be locally steeper on average than Europa(Schenk, 2009), with implications for lander designs, but again sampling was very limited and negligible at spatial scales of <100 m. Stereo observations at a range of resolutions, down to <10 m, supplemented by shape-from-shading mapping from low-sun imaging, will be important for geologic and slope investigations but will be complemented by global stereo mapping, extensive limb imaging, and occultation observations.

Prominent high-resolution and regional topographic targets for future flybys include: the large central dome crater Osiris and its very young mantling bright deposits (the most recently formed such crater), the anti-Jovian and current polar regions to search for putative domes, one or more proposed gravity anomaly sites (Palguta et al., 2009), Gilgamesh, and widely sampled bright terrains and their boundaries, including both grooved and smooth types.

9.6 CONCLUSIONS

Voyager- and Galileo-derived topographic mapping products for Ganymede include limb profiles and stereogrammetric and photoclinometric DEMs but are limited to <35% of the surface. These data reveal a topographically complex world but show that topographic variations are limited to +/−1 km or so. Surface slopes and roughness on Ganymede may be greater than both Europa and Enceladus despite the much greater surface ages of Ganymede compared to these active ocean worlds. The spatial density of relaxed craters appears to be similar on bright and dark terrains, suggesting that relaxation ceased shortly after bright terrain formation, and both terrains have been accumulating deep craters since then. Topographic variability within the major terrain types is more limited, rarely exceeding +/−500 m. Similarly, the mean elevations of older dark terrains and younger bright terrain are mostly no greater than ~500 m and sometimes negligible in current data. Smooth bright terrains are most often depressed with respect to older dark terrains, whereas the tectonized grooved bright terrains are more variable, with elevation possibly correlating with degree of deformation. These observations place useful constraints on bright terrain formation models.

The only larger-scale topographic feature is the 3 km high, 500 km wide dome at the current sub-Jovian point (Fig. 9.8). Although subsurface intrusion may be possible, we currently favor rotation and partial relaxation of a polar rise by true polar wander of the ice shell. The discovery of the sub-Jovian dome in otherwise marginally useful stereo data indicates the utility of stereo mapping and limb profiles at image resolutions between 0.2 and 2 km to identify gross deviations from sphericity, such as basins and swells or large impact basins or domes, provided it can resolve elevations of a hundred meters or less. Regional stereo mapping in the 50–200 m pixel scales successfully reveals terrain differences of <500 m in most cases but is very limited in quantity. Stereo imaging optimized to search for regional topographic variations (including regional rises or basins) and more fully mapping out terrain related topographic differences can be easily designed in upcoming missions to the Jupiter system. Regardless of technique, careful morphologic and geologic interpretation of the original stereo and non-stereo images, in concert with analysis of the DEMs, is key to understanding image-derived topographic data from all the icy Galileo satellites.

9.6.1 Methods

The full details of our primary topographic methods are described in earlier reports. Here we restrict ourselves to discussion of aspects unique to the sparsely distributed Ganymede data sets (Fig. 9.13) necessary to understand the uses and limitations of each data set. Our stereogrammetric DEMs were constructed using the methods described in Schenk and colleagues (2018), which produce identical results to similar techniques (e.g., Zubarev et al., 2017). The digital stereo mapping tool kit uses basic scene recognition software to map feature parallax displacements and calculate their heights. Because a discrete 3x3 pixel or larger patch must be used for scene recognition (5x5 is preferred if the stereo quality is not optimal), the surface ground scale of each topographic measurement footprint is at least a factor of three worse than the poorest image resolution of the stereo pair. The post spacing or distance between measurements is by default one pixel in our runs. The vertical resolution of each stereo pair, which is defined as the ability to detect surface elements of a given height, is unique to each site and is the fundamental measure of stereo quality. This value is calculated directly from the stereo convergence angle associated with the two images, height of the camera (effectively the surface resolution and IFOV), and the resolving power of the software (in this case one-fifth of a pixel; Schenk et al., 2018), using essentially the same equations as in standard aircraft photogrammetry (Lee and McGlone, 2013). The best DEM has a vertical resolution (or alternatively described as the sensitivity of the DEM in the vertical direction) of ~28 m; most of the extant Ganymede stereo DEMs have vertical resolutions of >50 m, with values of 250 m for the lower-resolution regional imaging (e.g., Fig. 9.14).

The DEMs derived from the Galileo framing camera are free from the lateral distortions in elevation values common to line-scan imager DEMs, as the camera distortion is well understood in the USGS Integrated Software for Imagers and Spectrometers processing package (e.g., Anderson et al., 2004). The most important aspect of SG on Ganymede is the camera distortions of the Voyager vidicon framing images, which are modelled using fiducial marks (reseaux) etched onto the vidicon image plate. Reseaux locations can shift location in each image by up to a pixel or so, depending on scene brightness and

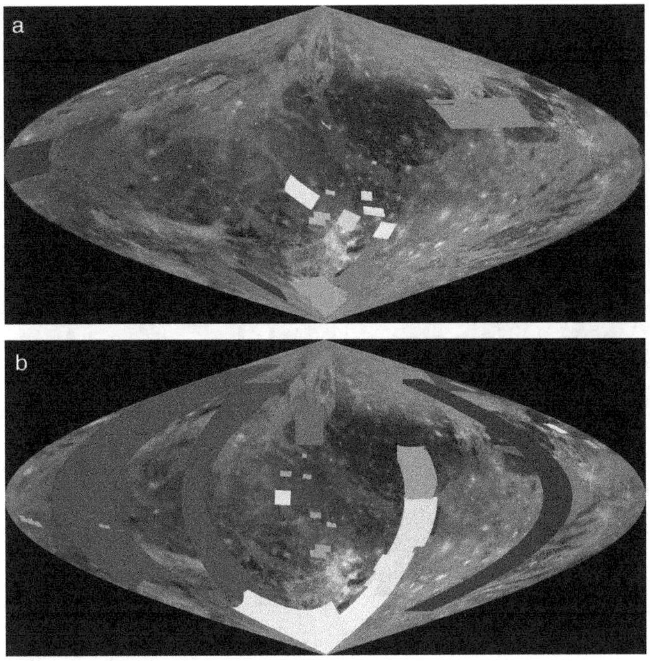

Figure 9.13 a, b Global sinusoidal image mosaic maps of Ganymede showing distribution of (a) stereogrammetric and (b) photoclinometric DEMs. DEM sites are in blank polygons: redder represents lower-resolution DEM products from each site. (a) includes DEM at sub-Jovian point (far left) described in main text showing large dome, although additional regional DEMs of this type in the Voyager 1 mapping area are not shown here. (A black and white version of this figure will appear in some formats. For the color version, please refer to the plate section.)

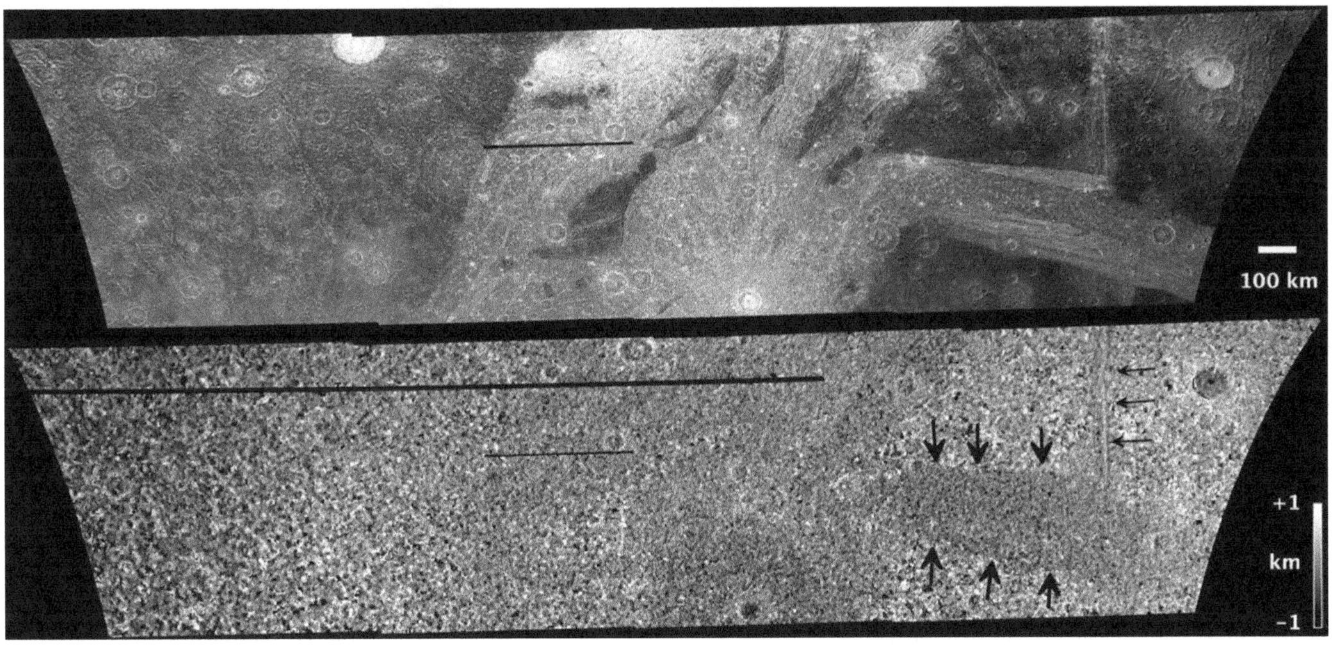

Figure 9.14 Regional SG-DEM of bright terrain units between Perrine and Galileo Regios. Stereo images from Galileo orbit C9 at pixel scales of 840 and 2000 meters. DEM vertical precision is ~500 m, not sufficient to resolve geologic structures, except the largest deep craters, the narrow ridge in the center of a bright terrain band (small arrows), and broad-scale terrain differences, such as the depressed wider bright terrain band (large arrows). *Source*: NASA.

camera operating conditions, and must be located in each image prior to removal. Any error in reseaux locating and the distortion removal process will result in image distortions interpreted as topography in the SG. Experience shows that this distortion removal is good to ~95% reliability in the central 80% or so of the image, but that those parts of derived DEMs derived from the outer 25–50 pixels or so of the original frames can sometimes exhibit topographic distortions of a few hundred meters. Normally, these areas are cropped unless there is no other imaging coverage in those areas. In rare occasions, a single

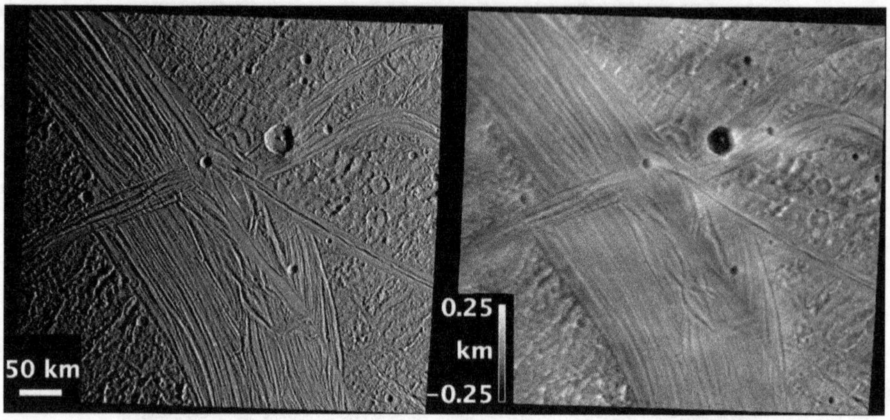

Figure 9.15 Example PC-DEM of Tiamat Sulcus (center) and portions of Marius Regio. Galileo image is 500 m/pixel. Groove and crater topographic information is mostly reliable, and, although suggestive, it is not possible to derive reliable relative elevations of bright and dark terrain in such DEMs due to the inherent long-wavelength ambiguities associated with this technique.
Source: NASA.

improperly registered reseau will result in a spurious large-amplitude feature, but these are easily identified by comparing the location with reseaux locations.

Radiation noise in the Ganymede images is almost negligible and easily removed with noise filters or manual editing. Galileo image compression artifacts are an issue in only a few cases, but map projection resampling and low-pass filters mitigate most of these, and we observe no significant degradation of DEM quality.

The photoclinometry (PC) or shape-from-shading technique as employed here (see Schenk, 2009; Bierhaus and Schenk, 2010) uses single low-Sun images and interprets brightness variations at each pixel as slopes (Fig. 9.15). The slope value is derived from a comparison of pixel brightness to a nominal photometric model based on the combined lunar–lambertian photometric equation described by McEwen (1991), in which an "L" term describes the relative contributions of the "lunar-like" and Lambertian additive components. The "L" term is estimated from a best-fit to the scene, when the area is large and after image smoothing to remove spikes. Our best estimate of the L-term phase angle (α) dependence for Ganymede (based on available data) is given by $L=0.879 - 0.0138\alpha + 0.00005\alpha^2$. Slopes are calculated and integrated along parallel lines across the selected scene. The estimated slope values are then converted into changes in height to produce a 2D DEM. Small errors within each line cause them to diverge, and the problem of integrating and adjusting the horizontal profiles was solved by adjusting each line after the map is produced, using stepped high- and low-pass filters, in a form of destriping.

Several assumptions are built into the PC approach, the most important of which is that brightness differences are due only to slope variations, which is mostly valid at low solar illumination angles. Photometric properties, including albedo and phase function, do vary across Ganymede, particularly around younger rayed craters or as frosts on steeper pole-facing slopes in terrains poleward of ~40°, and, within dark terrains, the brighter crests of topographic rises contrast with darker material mostly in the bottoms of local depressions. To mitigate the effects of albedo variations on the slope solutions, we use a photometrically corrected high-sun image of each area (at the highest resolution available) as a proxy for an albedo map. In most cases, only one suitable high-sun image is available for a given area due to the very limited imaging coverage generally, and they are almost always lower in resolution than the base mosaic by at least factors of two. The reliability of such maps is limited by the available data, but they provide insights into smaller-scale relief not otherwise available. Inadequate photometric models of the surface are the primary source of PC errors and manifest as warping of the DEM.

The PC technique as used here has proved useful (Fig. 9.15) in areas of Ganymede imaged with solar incidence angles of >70° and <85° (although sometimes the technique can successfully be pushed to areas where incidence angle is as low as 65° and albedo variations are small). Features must also be resolvable, limiting applicability to imaging with resolutions of better than 2 km/pixel, although specific features can be mapped at worse resolution. At resolutions better than ~750 m, the PC technique can also be applied between ~60 and ~70° incidence angles, except that results tend to be more sensitive to albedo variations and produce more artifacts. Success in mapping these areas using the described techniques is variable, with three best results in areas devoid of strong bright/dark material contrasts. Many small-scale features can be reliably mapped, but unless a local albedo model can be constructed or stereo control used, local topographic anomalies (artifacts) may occur.

One advantage of our PC-DEM technique (e.g., Schenk, 2009; Bierhaus and Schenk, 2010) is that it will threshold shadowed areas so that the software interpolates the topographic profiles along the "shadow surface" using the local incidence angle as the slope, until valid data are encountered again. This then becomes a 2D shadow-measuring tool that can also be used to evaluate the validity of the illuminated portions of the map, particularly for features that can be inferred to be symmetrical, such as large impact craters or double ridges. As always, any derived topographic data set should be interpreted with caution and in context with the coincident imaging.

REFERENCES

Anderson, J. A., Sides, S. C., Soltesz, D. L., Sucharski, T. L., and Becker, K. J. (2004). Modernization of the integrated software for imagers and spectrometers. Lunar Planet. Sci. Conf. 35, abstract no. 2039.

Anderson, J. D., Jacobson, R. A., Lau, E. L. et al. (2001). Shape, mean radius, gravity field and interior structure of Ganymede. *Bull. Am. Astron. Soc.* 33, 1101.

Berquin, Y., Kofman, W., Herique, A., Alberti, G., and Beck, P. (2013). A study on Ganymede's surface topography: Perspectives for radar sounding. *Planetary and Space Science* 77, 40–44.

Bierhaus, E. B., and Schenk, P. M. (2010). Constraints on Europa's surface properties from primary and secondary crater morphology. *J. Geophys. Res.* 115, E12004, 1–17.

Bland, M. T., and McKinnon, W. B. (2015). Forming Ganymede's grooves at smaller strain: Toward a self-consistent local and global strain history for Ganymede. *Icarus* 245, 247–262. https://doi.org/10.1016/j.icarus.2014.09.008.

Bland, M. T., and McKinnon, W. B. (2018). Viscous relaxation as a prerequisite for tectonic resurfacing on Ganymede: Insights from numerical models of lithospheric extension. *Icarus* 306, 214–224. https://doi.org/10.1016/j.icarus.2017.10.017.

Bland, M. T., Showman, A. P., and Tobie, G. (2009). The orbital-thermal evolution and global expansion of Ganymede. *Icarus* 200, 207–221. https://doi.org/10.1016/j.icarus.2008.11.016.

Bland, M. T., Singer, K. N., McKinnon, W. B., and Schenk, P. M. (2017). Viscous relaxation of Ganymede's impact craters: Constraints on heat flux. *Icarus* 296, 275–288. https://doi.org/10.1016/j.icarus.2017.06.012.

Bray, V., Schenk, P., Melosh, H. J., Morgan, J., and Collins, G. S. (2012). Ganymede crater dimensions: Implications for central peak and central pit formation and development. *Icarus* 217, 115–129.

Collins, G. C., Head, J. W., and Pappalardo, R. T. (1998). Formation of Ganymede grooved terrain by sequential extensional episodes: Implications of Galileo observations for regional stratigraphy. *Icarus* 135, 345–359.

Culha, H., Hayes, A. G., M., and Thomas, A. M. (2014). Double ridges on Europa accommodate some of the missing surface contraction. *J. Geophys. Res.* 119, 395–403.

Dombard, A. J., and McKinnon, W. B. (2001). Formation of grooved terrain on Ganymede: Extensional instability mediated by cold, superplastic creep. *Icarus* 154, 321–336.

Dombard, A. J., and McKinnon, W. B. (2006). Elastoviscoplastic relaxation of impact crater topography with application to Ganymede and Callisto. *J. Geophys. Res.* 111, E01001. 10.1029/2005JE002445.

Drake, N. (2015) Bizarre bulge found on Ganymede, Solar System's largest moon. *National Geographic* 227. http://bit.ly/3EH8B4Y.

Giese, B., Hauber, H., and Hussmann, H. (2017). Geological mapping of the Ganymede G8 calderas region: Evidence for cryovolcanism. *Lunar Planet. Sci. Conf.* XLVIII, abstract no. 2474.

Giese, B., Oberst, J., Roatsch, T. et al. (1998). The local topography of Uruk Sulcus and Galileo Regio obtained from stereo images. *Icarus* 135, 303–316.

Giese, B., Wagner, R., Neukum, G. et al. (2001a). The topography of bright/dark terrain on Ganymede. *Lunar Planet. Sci. Conf.* XXXI, abstract no. 1751.

Giese, B., Wagner, R., Neukum, G. et al. (2001b). The topography of Ganymede's Arbela Sulcus. *Lunar Planet. Sci. Conf.*, XXXII, abstract no. 1743.

Grasset, O., Dougherty, M. K., Coustenis, A. et al. (2013). Jupiter icy moons explorer (JUICE): An ESA mission to orbit Ganymede and to characterise the Jupiter system. *Planet. Space Sci.* 78, 1–21. https://doi.org/10.1016/j.pss.2012.12.002.

Hammond, N. P., and Barr, A. C. (2014). Formation of Ganymede's grooved terrain by convection-driven resurfacing. *Icarus* 227, 206–209.

Head, J. W., Pappalardo, R. T., Collins, G. C. et al. (2002). Evidence for Europa-like tectonic resurfacing styles on Ganymede. *Geophys. Res. Lett.* 29(24), 2151–2155.

Head, J. W., Pappalardo, R., Kay, J. et al. (1998). Cryovolcanism on Ganymede: Evidence in bright terrain from Galileo solid state imaging data. *Lunar Planet. Sci. Conf.* XXIX, abstract no. 1666.

Hirata, N., Suetsugu, R., and Ohtsuki, K. (2020). A global system of furrows on Ganymede indicative of their creation in a single impact event. *Icarus* 352, 113941.

Howell, S. M., and Pappalardo, R. T. (2018) Band formation and ocean-surface interaction on Europa and Ganymede. *Geophys. Res. Lett.* 45, 4701–4710.

Howell, S. M., and Pappalardo, R. T. (2020). NASA's Europa Clipper: A mission to a potentially habitable ocean world. *Nature Communications* 11, 1–4.

Kay, J. E., and Head, J. W. (1999). Geological mapping of the Ganymede G8 Calderas region: Evidence for cryovolcanism. *Lunar Planet. Sci. Conf.* XXX, abstract no. 1103.

Kay, J., Schenk, P., Dombard, A., and McKinnon, W. (2018). The 3-km-high megadome on Ganymede: Simulation of stability by Pratt isostasy. *Lunar Planet. Sci. Conf.* XLIX, abstract no. 2212.

Lee, G. Y. G., and McGlone, J. C. (2013). *Manual of Photogrammetry*. Baton Rouge, LA: American Society for Photogrammetry and Remote Sensing.

Lucchitta, B. K. (1980). Grooved terrain on Ganymede. *Icarus* 44, 481–501.

Manga, M., and Michaut, C. (2017). Formation of lenticulae on Europa by saucer-shaped sills. *Icarus* 286, 261–269. https://doi.org/10.1016/j.icarus.2016.10.009.

Matsuyama, I., and Nimmo, F. (2008). Tectonic patterns on reoriented and despun planetary bodies. *Icarus* 195, 459–473.

McEwen A. (1991). Photometric functions for photoclinometry and other applications. *Icarus* 92, 298–311.

McKinnon, W. B., and Melosh, H. J. (1980). Evolution of planetary lithospheres: Evidence from multi-ringed basins on Ganymede and Callisto. *Icarus* 44, 454–471.

McKinnon, W. B., Mohit, P. S., Greenhagen, B. T., and Bland, M. T. (2014). Polar wander on Ganymede and Callisto: A solution to the apex-antapex cratering conundrum. *45th Lunar Planet. Sci. Conf*, abstract no. 1869.

McKinnon, W. B., Schenk, P. M., and Moore, J. M. (2000). Bright terrain: Caldera associations on Ganymede. *AGU Fall Meeting Abstracts, EOS Trans. AGU* 81.

Michaut, C., and Manga, M. (2014). Domes, pits, and small chaos on Europa produced by water sills. *J. Geophys. Res. Planets* 119, 550–573.

Nimmo, F., Bills, B., and Thomas, P. (2011). Geophysical implications of the long-wavelength topography of the Saturnian satellites. *Icarus*, 116, 2011JE003835.

Nimmo, F., and Pappalardo, R. T. (2004). Furrow flexure and ancient heat flux on Ganymede. *Geophys. Res. Lett.* 31, https://doi.org/10.1029/2004GL020763.

Nimmo, F., Pappalardo, R. T., and Giese, B. (2002). Elastic thickness and heat flux estimates on Ganymede. *Geophys. Res. Lett.* 29(7), 10.1029/2001GL013976.

Nimmo, F., Thomas, P. C., Pappalardo, R. T., and Moore, W. B. (2007). The global shape of Europa: Constraints on lateral shell thickness variations. *Icarus* 191, 183–292.

Oberst, J., Schreiner, B., Giese, et al. (1999). The distribution of dark and bright surface material on Ganymede in relationship to surface elevations and slopes. *Icarus* 140, 283–293.

Ojakangas, G. W., and Stevenson, D. J. (1989). Thermal state of an ice shell on Europa. *Icarus* 81, 220–241. https://doi.org/10.1016/0019-1035(89)90052-3.

Palguta, J., Anderson, J. D., Schubert, G., and Moore, W. B. (2006). Mass anomalies on Ganymede. *Icarus* 180, 428–441.

Palguta, J., Schubert, G., Zhang, K., and Anderson, J. D. (2009). Constraints on the location, magnitudes, and dimensions of Ganymede's mass anomalies. *Icarus* 201, 615–625.

Pappalardo, R. T., Collins, G. C., Head, J. W. et al. (2004). Geology of Ganymede. In *Jupiter: The Planet, Satellites and Magnetosphere*, ed. F. Bagenal, T. E. Dowling, and W. B. McKinnon, pp. 363–396. Cambridge: Cambridge University Press.

Pappalardo, R. T., Head, J. W., Collins, G. C. et al. (1998). Grooved terrain on Ganymede: First results from Galileo high-resolution imaging. *Icarus* 135, 276–302.

Parmentier, E. M., Squyres, S. W., Head, J. W., and Allison, M. L. (1982). The tectonics of Ganymede. *Nature* 295, 290–293.

Patel, J. G., Pappalardo, R. T., Head, J. W. et al. (1999). Topographic wavelengths of Ganymede groove lanes from Fourier analysis of Galileo images. *J. Geophys. Res.*, 104, 24 057–24 074.

Patterson, G. W., Collins, G. C., Head, J. W. et al. (2010). Global geological mapping of Ganymede. *Icarus* 207(2), 845–867. https://doi.org/10.1016/j.icarus.2009.11.035.

Prockter, L. M., Figueredo, P. H., Pappalardo, R. T., Head, J. W., and Collins, G. C. (2000). Geology and mapping of dark terrain on Ganymede and implications for grooved terrain formation. *J. Geophys. Res.* 105, 22519–22540. https://doi.org/10.1029/1999JE001179.

Prockter, L. M., Head, J. W., Pappalardo, R. T. et al. (1998). Dark terrain on Ganymede: Geological mapping and interpretation of Galileo Regio at high resolution. *Icarus* 135(1), 317–344.

Prockter, L. M., Head, J. W., Pappalardo, R. T. et al. (2002). Morphology of Europan bands at high resolution: A mid-ocean ridge-type rift mechanism. *J. Geophys. Res. Planets* 107, 5028–5052. https://doi.org/10.1029/2000JE001458.

Ravine, M., C. Hansen, G. Collins, P. Schenk, M. Caplinger, L. Lipkaman, D. Krysak, R. Zimdar, J. Garvin, S. Bolton (2022). Ganymede Observations by JunoCam on Juno Perijove 34, *Geophysical. Res. Lett* 49, e2022GL099211.

Schenk, P. M. (1991). Ganymede and Callisto: Complex crater formation and planetary crusts. *J. Geophys. Res.* 96, 15635–15664.

Schenk, P. (2002). Thickness constraints on the icy shells of the Galilean satellites from a comparison of crater shapes. *Nature* 417, 419–421.

Schenk, P. (2009). Slope characteristics of Europa: Constraints for landers and radar sounding. *Geophys. Res. Lett.* 36, CiteID L15204.

Schenk, P. (2014). Mapping the topography of Europa: The Galileo-Clipper story, DPS meeting abstract #46, id.418.08.

Schenk, P. (2010). *Atlas of the Galilean Satellites*. Cambridge: Cambridge University Press.

Schenk, P. M., Chapman, C. R., Zahnle, K., and Moore, J. M. (2004). Ages and interiors: The cratering record of the Galilean satellites. In *Jupiter: The Planet, Satellites and Magnetosphere*, ed. F. Bagenal, T. E. Dowling, and W. B. McKinnon, pp. 427–456. Cambridge: Cambridge University Press.

Schenk, P., and McKinnon, W. B. (1987). Ring geometry on Ganymede and Callisto. *Icarus* 72, 209–234.

Schenk, P. M., and McKinnon, W. B. (1991). Dark ray and dark floor craters on Ganymede and the provenances of large impactors in the Jovian system. *Icarus* 89, 318–346.

Schenk, P., and McKinnon, W. (2009). One-hundred-km-scale basins on Enceladus: Evidence for an active ice shell. *Geophys. Res. Lett.* 36, CiteID L16202.

Schenk, P., and McKinnon, W. B. (2020). A new global topographic map of Enceladus: Polar signatures and large basin distributions, EOS, Fall Meeting, abstract no. P001-04.

Schenk, P., and McKinnon, W. B. (2023). New Global Topography of Enceladus: Hypsometry, Basins, Spherical Harmonics, Shell Thickness, and True Polar Wander Revisited, Icarus, in press.

Schenk, P. M., McKinnon, W., Gwynn, D., and Moore, J. (2001). Flooding of Ganymede's resurfaced terrains by low-viscosity aqueous lavas. *Nature* 410, 57–60.

Schenk, P., McKinnon, W. B. Moore, J., and Nimmo, F. (2020). The topography of Ganymede (and Callisto): Geology, global characteristics and future exploration. 52nd *Lunar Planet. Sci. Conf.* LII, abstract no. 2228.

Schenk, P., and Moore, J. M. (1995). Volcanic constructs on Ganymede and Enceladus: Topographic evidence from stereo images and photoclinometry. *J. Geophys. Res.* 100, 19009–19022.

Schenk, P., and 10 others (2018b). Basin, fractures and volcanoes: Global cartography and topography of Pluto from new horizons. *Icarus* 314, 400–433.

Schenk, P., White, O., Bryne, D., and Moore, J. (2018). Saturn's other icy moons: Geologically complex worlds in their own right. In *Enceladus and the Icy Moons of Saturn*, ed. P. Schenk, R. Clark, C. Howett, A. Verbiscer, and J. Waite. Tucson: University of Arizona Press, pp. 237–265.

Schenk, P., and 18 others (2021). Triton: Topography and geology of a probable ocean world with comparison to Pluto and Charon. *Remote Sensing*, 13, 3476–3533. https://doi.org/10.3390/rs13173476.

Schubert, G., Anderson, J. D., Spohn, T., and McKinnon, W. B. (2004). Interior composition, structure and dynamics of the Galilean satellites. In *Jupiter: The Planet, Satellites and Magnetosphere*, ed. F. Bagenal, T. E. Dowling, and W. B. McKinnon, pp. 281–306. Cambridge: Cambridge University Press.

Shepard, M., Campbell, B., Bulmer, M. et al. (2001). The roughness of natural terrain: A planetary and remote sensing perspective. *J. Geophys. Res.* 106, 32777–32795.

Showman, A. P., Mosqueria, I., Head, J. W. (2004). On the resurfacing of Ganymede by liquid-water volcanism. *Icarus* 172, 625–640.

Singer, K. N., Bland, M. T., Schenk, P. M., and McKinnon, W. B. (2018). Relaxed impact craters on Ganymede: Regional variation and high heat flows. *Icarus* 306, 214–224.

Squyres, S. W. (1980). Topographic domes on Ganymede: Ice vulcanism or isostatic upwarping. *Icarus* 44, 472–480.

Squyres, S. W. (1981). The topography of Ganymede's grooved terrain. *Icarus* 46, 156–168.

Thomas, P. C., and 9 others (1998). The shape of Io from Galileo limb measurements. *Icarus* 135, 175–180.

Thomas, P. (2010). Sizes, shapes, and derived properties of the saturnian satellites after the Cassini nominal mission. *Icarus* 208, 395–401.

White, O. L., Schenk, P. M., Nimmo, F., and Hoogenboom, T. (2014). A new stereo topographic map of Io: Implications for geology from global to local scales. *J. Geophys. Res.* 119, 1276–1301, https://doi.org/10.1002/2013JE004591.

Zahnle, K., Schenk, P., Levison, H., and Dones, L. (2003). Cratering rates in the outer solar system. *Icarus* 163, 263–289.

Zubarev, A., Nadezhdina, I., Brusnikin, E., Giese, B., and Oberst, J. (2017). A search for Ganymede stereo images and 3D mapping opportunities. *Planet. Space Sci.* 146, 40–54.

10

Ganymede's Surface Composition

KATRIN STEPHAN, CHARLES A. HIBBITTS, NICOLAS LIGIER, PHILIPPA M. MOLYNEUX, FRANCOIS POULET, LOUISE M. PROCKTER, AMANDA R. HENDRIX, GEOFFREY C. COLLINS, CAITLIN J. AHRENS, AND RALF JAUMANN

10.1 INTRODUCTION

Ganymede is an exceptional icy satellite with a complex geological history and a unique connection with the Jovian magnetosphere due to Ganymede's own magnetic field. The best possible knowledge about Ganymede's surface composition is essential to reconstruct past and present surface processes in Ganymede's geological evolution. Are Ganymede's so-called polar caps a result of the interaction of Ganymede with the Jovian magnetosphere? Could salts detected on Ganymede's surface originate from Ganymede's subsurface ocean, and/or could volatiles incorporated in the subsurface ice (clathrates) explain Ganymede's observed tectonic and potential cryovolcanic activity? Is the dark material primordial or of exogenic origin? In this chapter, we review the current knowledge about Ganymede's surface composition and discuss its implications for Ganymede's exogeneous and endogenous surface processes.

10.2 HISTORICAL BACKGROUND OF INVESTIGATING GANYMEDE'S SURFACE

The earliest ground-based telescopic observations in the visible (VIS) and near-infrared (NIR) wavelength range showed that up to 65% of Ganymede's surface is covered by H_2O ice with Ganymede's leading hemisphere exhibiting 20% more ice than on its trailing hemisphere (Kuiper, 1957; Moroz, 1965; Pilcher et al., 1972) While Caldwell (1975) presented the first broad-band measurements of the icy Galilean satellites in the ultraviolet (UV) wavelength range (~211 nm) from Earth orbit, the following UV observations from the International Ultraviolet Explorer (IUE) and the Hubble Space Telescope (HST) were the first to characterize the spectral properties of ice on Ganymede and to denote the presence of radiation products on its surface (Nelson and Lane, 1987; Spencer et al., 1995; Noll et al., 1996; Calvin and Spencer, 1997; Calvin et al., 1996) It was not, however, until the Galileo spacecraft performed its first Ganymede flyby in June 1996 that spectral surface properties could be studied with respect to their geological context. The Galileo Ultraviolet Spectrometer (UVS) covered the wavelength range from 115 to 430 nm and enabled the first disk-resolved observations of Ganymede (Hendrix et al., 1999; Hendrix et al., 2005). The Near Infrared Mapping Spectrometer (NIMS) (Carlson et al., 1992) acquired spectral images over the wavelength range of ~ 0.7 to 5.2 μm with spatial resolutions ranging from a couple of kilometers per pixel to about 100 km per pixel, which cover areas of a few 10s of kilometers to a nearly hemispherical coverage.

Currently, the Cosmic Origins Spectrograph of the Hubble Space Telescope (HST/COS) is available for observations of solar system objects in the wavelength range of 138 to 215 nm (Molyneux et al., 2020). Furthermore, Earth-based near-infrared data, such as from the SINgle Faint Object Near-IR Investigation (SINFONI) instrument of the Very Large Telescope (VLT) in Chile, investigated Ganymede's surface (Ligier et al., 2019) Although its wavelength range is much smaller than NIMS, covering only from 1.46 μm to 2.41 μm, its spectral resolution and signal-to-noise ratio (SNR) is 5 times higher, allowing it to perform convincing spectral modeling.

10.3 GENERAL RELATIONSHIPS BETWEEN COMPOSITION AND GEOLOGY

Ganymede's geology and geologic evolution are discussed in detail in Chapter 5 by Jaumann et al. and Chapter 7 by Pappalardo et al.; we here introduce the broader geological context of Ganymede's surface material. Ganymede exhibits a pronounced geological heterogeneity. The geologically oldest heavily cratered terrain, consisting of several *regiones* (singular *regio*) covers about 1/3 of Ganymede's surface (Pappalardo et al., 2004). Based on the measured densities of impact craters, this ancient terrain is predicted to be older than 4 Gyr (Neukum, 1997; Zahnle et al., 2003). The remaining 2/3 of the surface is dominated by cross-cutting lanes of tectonically resurfaced terrain (called *sulci,* singular *sulcus*) (Pappalardo et al., 2004), which is significantly less cratered than the ancient terrain and therefore younger, possibly emplaced between 500 Myr and 1 Gyr ago (Zahnle et al., 2003). Early during the process of the bright terrain formation, H_2O ice volcanism may have occurred, creating materials with a very smooth surface texture (Schenk et al., 2001). However, later geological history appears to be dominated by tectonism, in which lithospheric extension resulted in the significant deformation of the *sulci* by subparallel ridges and troughs (known as "grooves"), at a variety of scales (Collins et al., 1998; Pappalardo et al., 1998).

Both terrains are compositionally distinct, providing clues to their formation and evolution. The *regiones* are characterized by a high concentration of dark, non-ice materials (McCord et al., 1998; Prockter et al., 1998). The non-ice, rocky material is either thermally segregated primordial material and/or has been brought in through impact bombardment (asteroids, comets). Knowledge of the composition of the bright, ice-rich, tectonically resurfaced terrain could explain whether it originated from cryovolcanic eruptions, with possible connections to subsurface salty liquid reservoirs or an ocean, or by tectonic disruption of preexisting terrain.

In addition to the two major terrain types, impact features are ubiquitous on Ganymede and exhibit a wide range of morphologies, including dark-floor craters, pedestal craters, craters with a central peak, pit, or dome, and ancient impact features such as low-relief palimpsests, penepalimpsests, and large, multi-ring impact basins (Schenk et al., 2004). Many have unique compositions, and younger craters can have bright or dark rays. The youngest impact craters may have excavated relatively unweathered material from the subsurface, thus enabling a view into the composition of the uppermost crust in both dark and bright terrains, and/or that of the impactor itself (Schenk and McKinnon, 1991; Hibbitts et al., 2009).

Ganymede's bright polar caps, likely composed of H_2O ice (frost), extend toward the equator to about +/– 40° latitude. The margins of the polar deposits coincide with the boundary between the closed and open field lines of Ganymede's magnetic field, which suggests a possible formation mechanism resulting from the interaction of the Jovian magnetosphere with Ganymede's surface (Kivelson et al., 1996; Khurana et al., 2007). However, diurnal temperature variations are also postulated to be sufficient for the formation of the ice caps through thermal migration of H_2O molecules (Spencer, 1987; Stephan et al., 2020). Improved knowledge of the composition of the polar deposits, including the distribution of radiation products in the surface material, can help resolve the origin of the polar caps. Such measurements are also essential to determine the formational processes and history of the ancient dark and tectonically resurfaced bright terrain.

10.4 MAJOR SURFACE COMPOSITIONAL COMPONENTS

10.4.1 Ice

Like most of the outer planet satellites, Ganymede's surface has long been known to be dominated by H_2O ice (Pilcher et al., 1972; Pollack et al., 1978). Particularly in the NIR wavelength region, H_2O ice exhibits diagnostic absorptions centered at 0.9, 1.04, 1.25, 1.5, and 2 μm (Pollack et al., 1978; Clark, 1980) (Fig. 10.1). Depending on crystallinity and temperature, additional absorptions can occur at 1.31, 1.57, and 1.65 μm (Grundy and Schmitt, 1998; Mastrapa et al., 2013; Stephan et al., 2021). Absorptions near and beyond 3 μm are generally strongly saturated. Nevertheless, a reflectance peak at 3.6 μm appears in the spectra of small H_2O ice grains (Filacchione et al., 2016), and the Fresnel reflection of H_2O ice grain facets results in a reflectance peak near 3.1 μm (Grundy and Schmitt, 1998; Mastrapa et al., 2013). In the visible spectral region as well as in the near- and mid UV (280–400 nm) H_2O ice is very transparent, with a maximum transparency at 0.46 μm (Warren, 1984). Only in the far UV H_2O ice exhibits a prominent absorption edge near ~165 nm due to electronic transitions (Hendrix et al., 2013).

From the UV through the IR, the H_2O ice spectral signature is very sensitive to the H_2O ice abundance and various physical parameters such as grain size and shape, crystallinity, surface compactness, temperature, and thermal history of H_2O ice (Grundy and Schmitt, 1998). Although these effects make the interpretation of the H_2O ice spectra more challenging, they also offer valuable indicators for the chemical and physical conditions of H_2O ice on Ganymede's surface. Two major ways have been successfully used so far to obtain compositional information from reflectance spectra of icy surfaces: (a) radiative transfer modeling and (b) spectral indicators.

(a) Radiative Transfer Modeling

Radiative transfer models predict that the interaction of light with particulate surfaces can be linearized by converting reflectance data to single scattering albedo and by simulating intimate- or intra-mixtures (Hapke, 2002; Shkuratov et al., 2004). Another approach to simulate the behavior of light–matter interactions was developed with the objective to calculate bidirectional reflectance by a plane of parallel, absorbing, scattering, and slightly stratified medium (e.g., Quirico et al., 1999). Whatever the degree of physical realism of these models, each requires a knowledge of the intrinsic optical properties (*n* and *k*) of each of the compositional components. These properties vary with the temperature of the materials, crystallinity and hydration state, if relevant (Grundy and Schmitt, 1998; Dalton, 2010).

(b) Spectral indicators

Spectral indicators have proven very useful to characterize major differences in the abundance, grain size, crystallinity, and temperature of H_2O ice. Due to the high transparency of H_2O ice in the visible light, its visible albedo, for one, is a good indicator of its varying relative abundance on Ganymede's surface (Dozier, 1989). Although the NIR absorptions vary in band depth depending on both the abundance and grain size (Clark and Lucey, 1984; Clark and Lucey, 1984), the band depth ratios of different absorptions have been found largely indicative of the grain size (Stephan et al., 2020). In addition, the strength of the reflection peak at 3.6 μm has been used to map the grain size of H_2O ice (Filacchione et al., 2016).

Furthermore, spectral indicators enable us to distinguish between different physical states of H_2O ice. On Ganymede's surface, H_2O can exist in three distinct solid configurations (amorphous, cubic, hexagonal), depending on the condensation temperature, condensation rate, and temperature history (Jenniskens et al., 1998) (please see Chapter 11 for more details). In cases of both amorphous and crystalline H_2O ice present on an icy surface, a shift in the wavelength positions of the H_2O ice absorptions should be observed (Grundy and Schmitt, 1998; Mastrapa et al., 2013). In addition, amorphous H_2O ice is often indicated by a weak or absent absorption at 1.65 μm and a suppressed Fresnel reflection peak at 3.1 μm (Hansen and McCord, 2004). However, increasing temperatures and decreasing grain sizes have similar effects on spectra of crystalline H_2O ice, and differences in the band position of the strong absorption at 1.5 and 2 μm are only visible when these absorptions are not affected by saturation as observed for large H_2O ice grains (> ~500 μm) (Stephan et al., 2021) or by the unsuspected presence of hydrated non-ice material. Therefore, the comparative analysis of spectral parameters, including the geological context and the surface temperature, should be considered when investigating the crystallinity of H_2O ice on icy bodies (Mura et al., 2020).

Distribution of H_2O Ice and Variations in Its Physical Properties across Ganymede's Surface

The way H_2O ice is distributed across Ganymede's surface largely corresponds to the variations in the visible albedo as seen in the Voyager ISS and Galileo SSI images (Stephan et al., 2020). The H_2O ice is concentrated in the vicinity of Ganymede's numerous geologically young and visually bright impact craters, where its abundance can exceed 60% (Ligier et al., 2019) and thus dominates their spectral signature (Fig. 10.1a). In contrast, the *regiones* of the ancient dark terrain show the lowest amount of H_2O ice (<5%, Ligier et al., 2019) expressed by a low visible albedo and strongly reduced H_2O ice signature because of the presence of darker non-ice material (Section 10.4.2). The tectonically resurfaced bright terrain is characterized by an intermediate H_2O-ice abundance.

Superimposed onto the spectral signatures related to Ganymede's geology are latitudinal and hemispherical variations in the abundance, grain size, and crystallinity of H_2O ice (Figs. 10.1b and 10.2). The enrichment of H_2O ice at Ganymede's higher latitudes, first visible in the Voyager images (Smith et al., 1979), is often referred to as polar caps. The trend of an increasing H_2O ice abundance with increasing latitude is accompanied by a decreasing H_2O ice grain size with grains smaller than 10 μm prevalent in the polar regions, whereas grain sizes exceed diameters of 1 mm in the equatorial regions (Ligier et al., 2019; Stephan et al., 2020). Only at intermediate latitudes can variations in H_2O ice grain size still be recognized between Ganymede's major terrain types (Stephan et al., 2020). This dichotomy in grain size between the equator and poles is particularly seen in the spectra of Ganymede's fresh bright impact craters (Fig. 10.1b). Impact craters, such as Osiris and Harakhtes, located at the margin of the polar caps exhibit H_2O ice grain sizes of 10 to 50 μm (Ligier et al., 2019; Stephan et al., 2020), yet impact crater *Tammuz*, located closer to the equator, exhibits a distinctly steeper negative spectral slope between 0.7 and 2.7 μm characteristic of large-grained H_2O ice (200 to 1000 μm), respectively (Stephan et al., 2020).

Although Ganymede's surface is dominated by crystalline H_2O ice, amorphous H_2O ice appears to exist in Ganymede's polar regions (Hansen and McCord, 2004; Ligier et al., 2019) (Fig. 10.2). The polar ice is interpreted as predominantly crystalline by Mura and colleagues (2020) because of a lack of the shift expected in the amorphous H_2O ice absorption at 2 μm. However, this does not rule out small amounts (<20%) of amorphous H_2O ice. It is also important to note that the interpretation of amorphous ice could instead be explained by the presence of very small-grained crystalline ice, which also exhibits a weaker 3.1 μm-peak (Stephan et al., 2021).

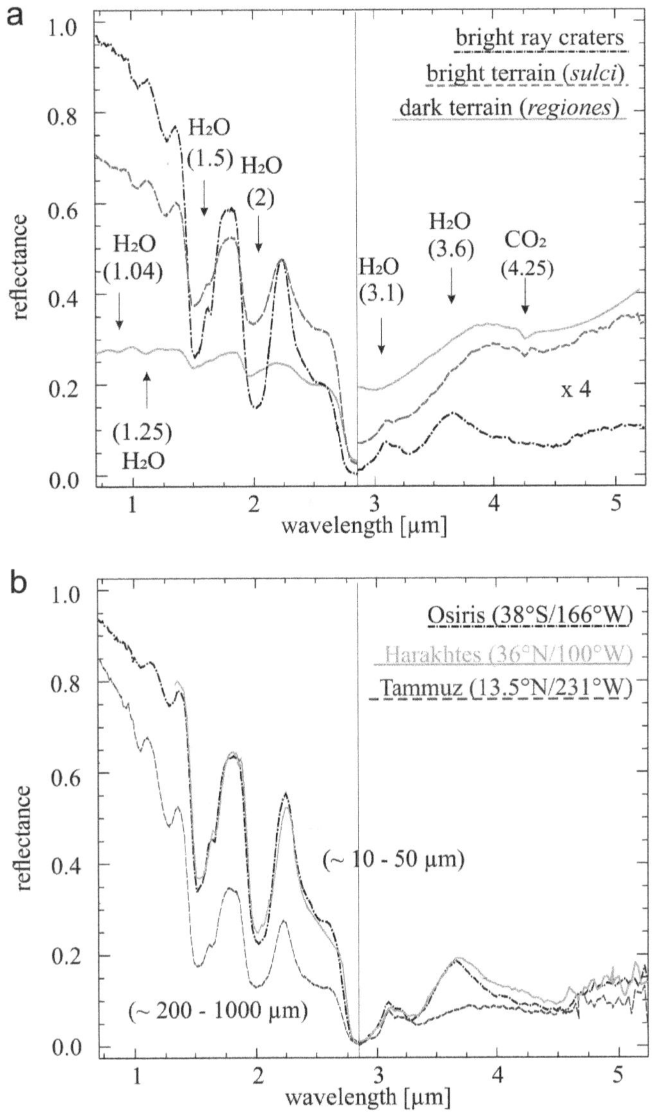

Figure 10.1 Example NIMS spectra of Ganymede: (a) three major surface terrains: bright ray impact craters, bright terrain (*sulci*) and ancient dark terrain (*regiones*) and (b) bright fresh impact craters located at different latitudes: *Osiris* (38°S/166°W), *Tammuz* (13.5°N/231°W) and *Harakhtes* (36°N/100°W).

Hemispherical differences in ice crystallinity have been described by Hansen and McCord (2004) and Ligier and colleagues (2019) with slightly more, partially amorphous, and larger-grained H_2O ice (>200 μm) in the equatorial region of the trailing hemisphere than on the leading hemisphere. This dichotomy has also been seen in the mid-UV (>170 nm) and the visible spectral range (Hendrix et al., 1999), with the trailing hemisphere of Ganymede darker than the leading hemisphere, similar to what has been observed for Europa (Nelson et al., 1987). Impact craters observed by NIMS on Ganymede at a local scale exhibit diameters of less than 10 km and up to more than 100 km and are located in both dark and bright terrain (Hibbitts et al., 2003; Stephan et al., 2020). So far, no significant variations in the H_2O ice properties of fresh crater material located in either ancient dark or tectonically resurfaced bright terrain have been recognized at the given spatial resolution (Stephan et al., 2008; Stephan et al., 2019). Moreover, ancient impact features such as penepalimpsests and palimpsests such as the 343 km large *Epigeus* (~23°N/~179°E) do not differ much with respect to their spectral properties, which are spectrally similar to those of the bright terrain (Fig. 10.3). Only very fresh small impact craters such as the one emplaced within *Epigeus* exhibit distinctly more, and, if young enough, finer-grained H_2O ice than their surroundings (Hibbitts

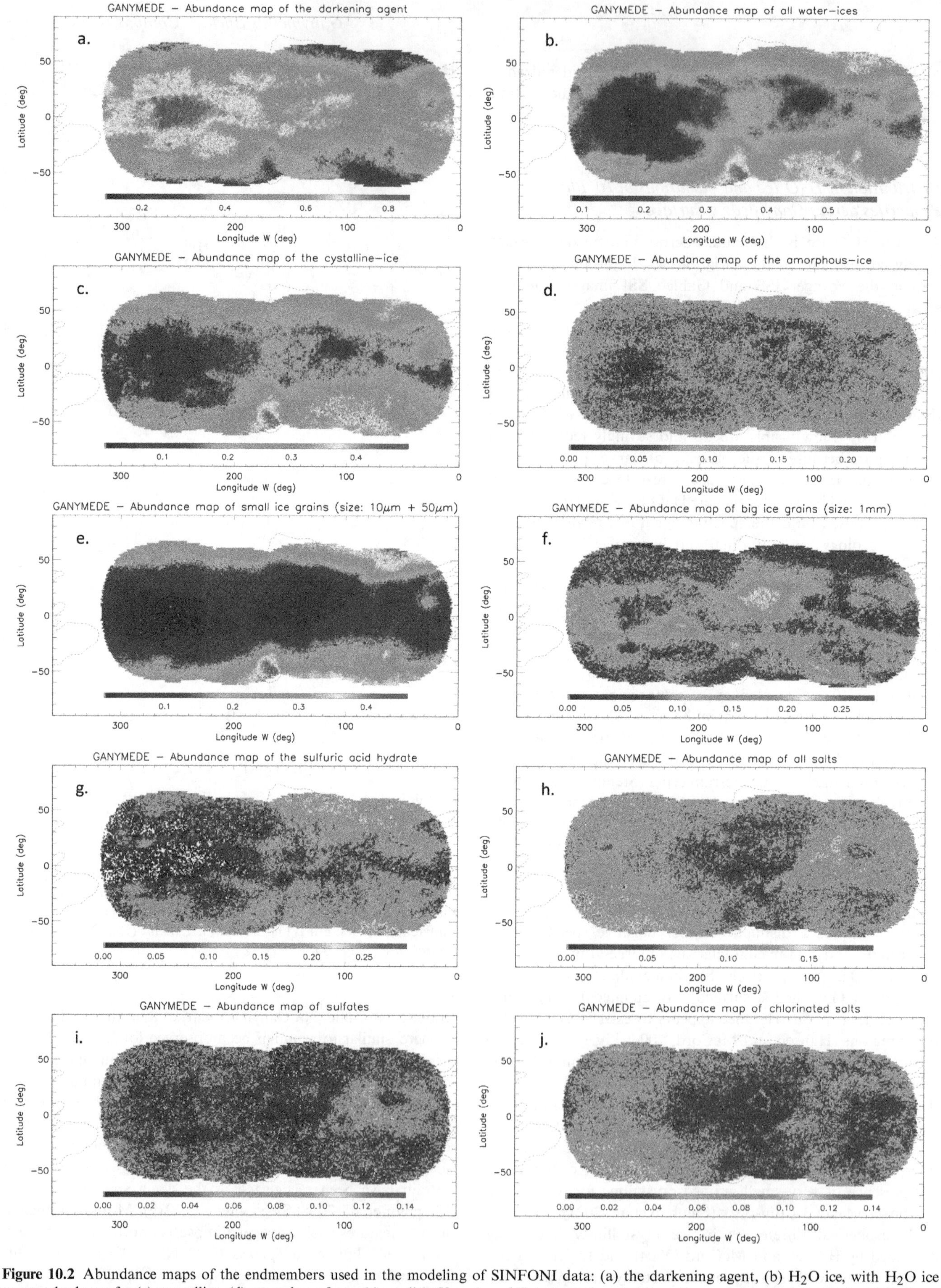

Figure 10.2 Abundance maps of the endmembers used in the modeling of SINFONI data: (a) the darkening agent, (b) H_2O ice, with H_2O ice separately shown for (c) crystalline, (d) amorphous form, (e) small ($\leq$50 μm), and (f) large grains, and the minor compounds (g) sulfuric acid hydrate and (h) salts in general, and separately shown for (i) sulfates and (j) chlorinate salts. Reprinted from *Icarus*, 333. N. Ligier, C. Paranicas, J. Carter, F. Poulet, W. M. Calvin, T. A. Nordheim, C. Snodgrass, L. Ferellec. Surface composition and properties of Ganymede: Updates from ground-based observations with the near-infrared imaging spectrometer, ©2019, with permission from Elsevier. (A black and white version of this figure will appear in some formats. For the color version, please refer to the plate section.)

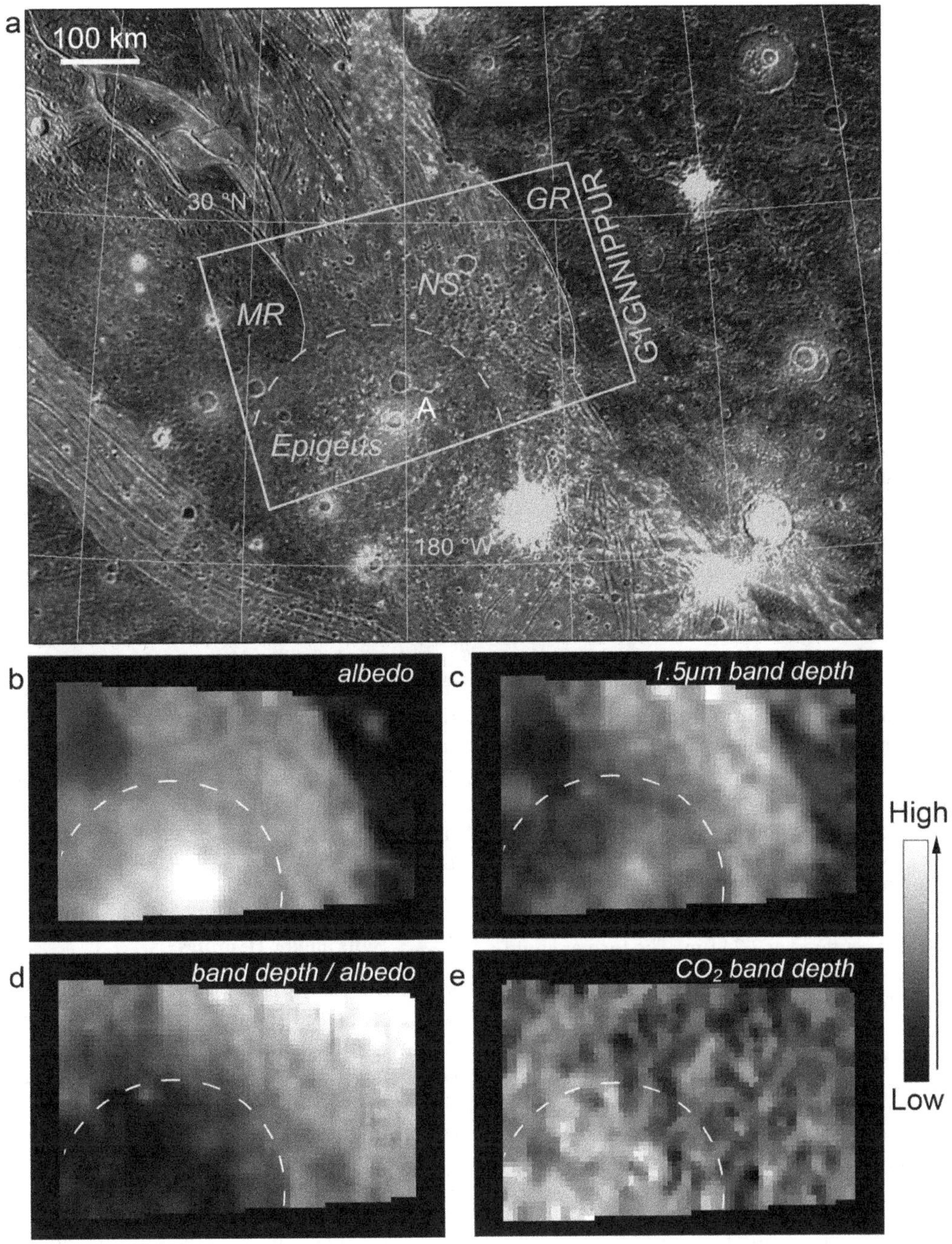

Figure 10.3 Penepalimpsest *Epigeus* (23°N/180.6°W): (a) geologic context of NIMS observation G1GNNIPPUR outlining *Epigeus*, a superimposed small fresh impact crater (*A*), and the boundary of *Nippur Sulcus* (*NS*) to *Marius Regio* (*MR*) and *Galileo Regio* (*GR*) in a Voyager/Galileo image mosaic and (b) the NIMS-derived maps showing (b) the visible albedo (scaled from 0.3 to 0.6), (c) the band depth of the absorption at 1.5 μm (scaled from 0.2 to 0.6) indicating the abundance of H_2O ice, (d) the ratio of the 1.5-μm band depth to albedo (scaled from 0.5 to 1) indicating larger (low values) and smaller (high values) H_2O ice grains, (e) CO_2 absorption band depth at 4.25 μm (scaled from 0 to 0.3) (after Hibbitts et al., 2003).

et al., 2003; Stephan et al., 2020). Possible processes causing the described pattern are discussed in Section 10.6.2.

10.4.2 Dark Non-ice Material(s)

In addition to H_2O ice, Ganymede's surface is composed of at least one darker material, first mentioned almost 50 years ago (Pilcher et al., 1972). Although, clearly apparent in the Voyager and then Galileo camera images because of its low albedo (Smith et al., 1979; Prockter et al., 1998), the mineralogy of the dark compound(s) remains unidentified. However, studies have been able to infer some compositional characteristics from color and spectral analysis/modeling in the ultraviolet through infrared wavelengths and from the nature of its spatial distribution across Ganymede's surface.

Spectral characteristics of Ganymede's Dark Non-ice Material

The most notable spectral feature observed in the far ultraviolet (FUV) is the masking of the prominent H_2O absorption edge at ~165 nm (Fig. 10.4a), which could be best explained by intraparticle models of H_2O grains containing inclusions of <1% UV-dark material. The composition of the UV-dark material that best matches the observed UV spectrum is a mixture of tholins – a mixture of organic residues formed by solar ultraviolet or cosmic ray irradiation of simple C-, H-, O-, and N-bearing compounds, including CO_2 and H_2O (Sagan and Khare, 1979; Khare et al., 1989), various hydrous or anhydrous silicate mineral assemblages, and H_2O_2 (Molyneux et al., 2020).

Spectral modeling of telescopic data highlights that the dark non-ice material on Ganymede is spectrally flat in the visible and near-infrared, which is consistent with carbon-dominated materials similar to the material dominating C- and D-type asteroids (Calvin et al., 1995; Ligier et al., 2019). Its reflectance ranging between 0.2 and 0.3, however, is not fully consistent with purely carbon-dominated materials (which would be 0.05–0.15), but more closely matches hydroxylated/hydrated minerals (Calvin and Clark, 1991). It is more hydrated than the rather anhydrous dark non-ice material identified on Callisto with the spectral differences remaining present when either the hydrated (Ganymede) or anhydrous (Callisto) non-ice endmember are mixed with H_2O ice, as shown in Figure 10.4c–e (Hibbitts et al., 2003). Analyses of spatially resolved NIMS data confirmed that Ganymede's dark material is mostly dominated by moderately hydroxylated/hydrated material showing a characteristic drop in the reflectance at 2.7 μm due to OH^- accompanied by bands of hydration at 1.4 and 1.9 μm – that is, slightly shifted to shorter wavelengths compared to pure H_2O ice (Hibbitts et al., 2003; Fig. 10.1). More detailed differences have been obtained through subsequent telescopic observations. Residuals between SINFONI telescopic spectra and modeled spectra (Ligier et al., 2019) show that two absorptions are not accounted for: a strong and sharp band at ~1.92 μm and a wider and shallower one centered at 2.24 μm. The former could correspond to a distorted H-O-H vibration and the latter to -OH or Si-OH bends from hydrated silicate minerals (Langer and Flörke, 1974; Goryniuk et al., 2004); both of which could be associated with the NIMS-derived dark non-ice material(s).

Spatial Distribution of Ganymede Dark Non-ice Material

Non-ice abundance has been most accurately modeled, globally, using telescopic observations. Modeling of SINFONI data showcases the prevalence and even dominance of dark non-ice material on Ganymede's surface (Fig. 10.2), ranging from 15% to almost 90%, with a mean value just above 50%. These results were previously observed in relatively high-spatial resolution data from the Galileo instruments, from which a dark material(s) abundance was estimated at around 80% in some areas (Prockter et al., 1998). At ultraviolet wavelengths, a clear hemispherical dichotomy exists, with the trailing hemisphere appearing darker than the leading hemisphere at wavelengths longer than ~170 nm (Molyneux et al., 2020)(Fig. 10.4a–b). At shorter wavelengths (<155 nm), the hemispheric asymmetry inverts and the trailing hemisphere appears brighter (Musacchio et al., 2017). The origin of the spectral inversion is not well understood, but may be a result of either compositional differences or a result of space weathering altering the optical properties of the surface material (Alday et al., 2017). The hemispherical dichotomy also persists into the visible and near-infrared as observed in both the Galileo NIMS and Earth-based telescopic data. To the first order, the dark non-ice material(s) is concentrated on Ganymede's trailing hemisphere. To the second order, however, the abundance of dark non-ice material(s) shows a clear latitudinal gradient with a sharp decrease in abundance at ~ 35–40° (both north and south), likely due to spectral masking by polar ice at the higher latitudes as discussed previously.

On a regional scale, the dark non-ice material dominates the geologically older *regiones* (Figs. 10.1 and 10.2). Variations in the spectral properties of the *regiones* observed by NIMS are mostly related to superimposed younger impact craters that tend to be relatively icy. Although the majority of fresh impact craters on Ganymede in both dark and bright terrain show an enhancement in H_2O ice abundance, not all do. NIMS observed some peculiar small fresh impact craters such as *Kittu* (Fig. 10.5) that seem to have ejected dark non-ice material(s), in lieu of H_2O ice, with the non-ice material spectrally distinct, less hydrated, and more red than the ubiquitous non-ice material (Hibbitts et al., 2003). These anomalously dark craters were previously observed in Voyager panchromatic and color images (e.g. Schenk and McKinnon, 1991). Whereas most of these dark craters exhibit a reddish visible near-infrared spectral slope, the rays of impact crater *Kittu* are characterized by a bluish slope, possibly implying a unique composition for the impactor responsible for that crater. Linear spectral unmixing models indicate that these dark ejecta deposits contain the highest concentrations of dark non-ice material on Ganymede. with several 10s of percent a less hydrated component spectrally similar to the dark non-ice material(s) on Callisto, and C-type or D-type asteroids.

10.4.3 Salts and/or Hydrates

While it is widely accepted that hydroxylated/hydrated minerals on Ganymede's surface could have been formed by aqueous alteration of endogenous or impact-emplaced materials (see Section 10.6.2), the existence and origin of salts, likely S-bearing, is still controversial (McCord et al., 2001; Ligier et al., 2019).

Spectral Characteristics of Salts

As with H_2O ice, most of the spectral features of the salty species appear in the near-infrared wavelength range (Dalton and Pitman, 2012; Hanley et al., 2014), although chlorinated salts like NaCl or KCl, especially after being irradiated, show some diagnostic absorptions in the ultraviolet and visible (Poston et al., 2017; Hibbitts et al., 2019). While the strong absorption bands between (i) 1.4 and 1.5 μm, (ii) 1.9 and 2.0 μm, and (iii) around 2.9 μm are observed for all hydrates, the shape, position, and depth of these spectral features depend on multiple parameters, such as the temperature and the degree of hydration (McCord et al., 2001; Dalton and Pitman, 2012; Hanley et al., 2014). However, it is sometimes the lee-obvious spectral features, which can be sharper or subtler, that can prove the most helpful for discriminating between the salty species. For example, mirabilite is the only hydrated sulfate salt showcasing a strong combination of two features at ~1.75 μm and ~2.20 μm, while nitrates and carbonates present distinct absorptions around 3 and 5 μm (Gaffey, 1987; Dalton et al., 2010; De Angelis et al., 2021).

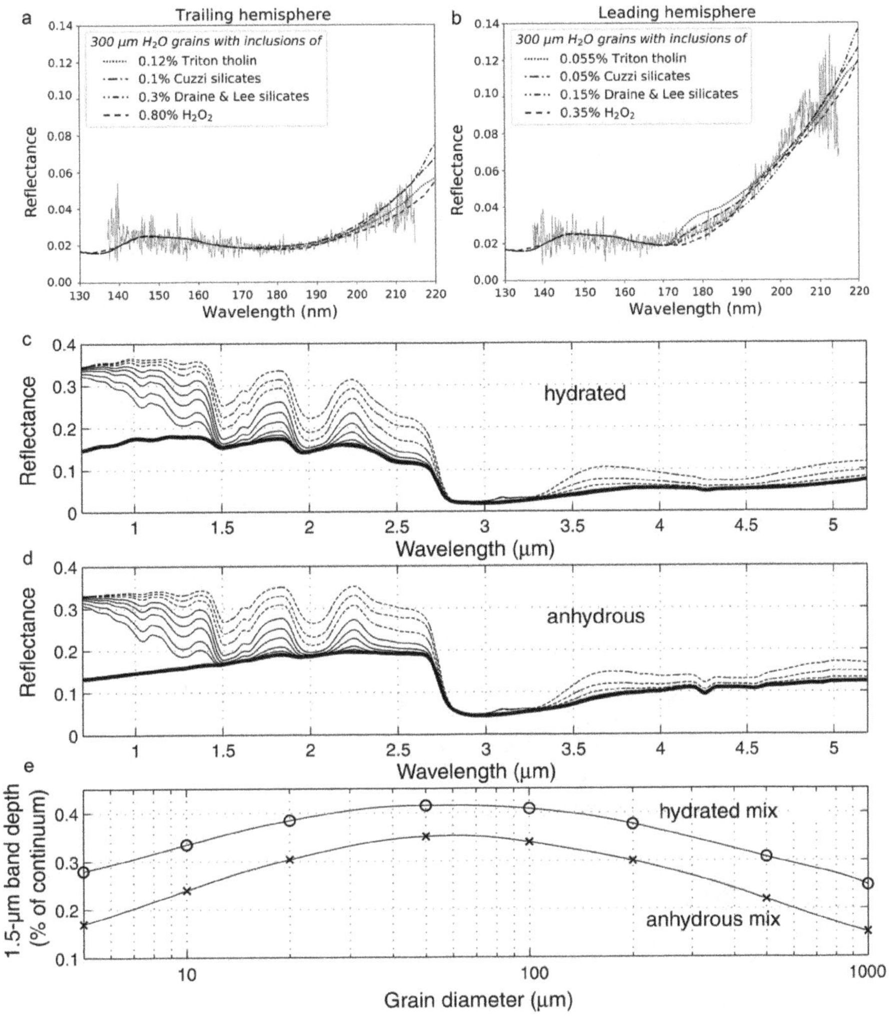

Figure 10.4 Ganymede's dark non-ice material candidates measured in the UV and NIR: (1) Intraparticle models of 300 μm H_2O ice grains with inclusions of rocky analogs and H_2O_2 in comparison with the FUV reflectance of Ganymede's (a) leading and (b) trailing hemisphere obtained by the HST/COS (adapted from Molyneux et al., 2019) and possible analog materials observed by Molyneux and colleagues (2020) using optical constants for silicates from Cuzzi and colleagues (2018) or Draine and Lee (1984). (2) The effect of varying ice abundance from 1% to 50% (areal mixture) on Galileo NIMS spectra of ice/non-ice mixtures for (c) Ganymede's hydrated non-ice material (o), (d) Callisto's anhydrous non-ice material x, and (e) the dependence of the band depth-to-albedo ratio on ice abundance in the mixtures. Ice grain size is 200 μm (solid line) and 20 μm (dashed line) (Hibbitts et al., 2003).

However, despite these diagnostic features, no salty species have ever been definitely identified on the surface of Ganymede. Recent observations, using either near-infrared high-spectral-resolution telescopic data (Ligier et al., 2019) (Fig. 10.2) or mid-infrared high-spatial-resolution data from the JIRAM instrument of the current Juno mission (Mura et al., 2020), only continue to confirm the likely existence of hydrated salts on the surface of Ganymede. Despite the lack of discriminating spectral absorptions in spectra of Ganymede, spectral analysis and modeling indeed highlight the likelihood of a few hydrates, specifically hydrated sulfuric acid, Mg-bearing sulfates, and chlorinated salts.

Spatial Distribution of Ganymede's Acid Hydrate and Salts

In contrast to Europa, Ganymede's sulfuric acid hydrate abundance is quite low (< 10%) and with a heterogeneous distribution and a hemispherical dichotomy and a latitudinal gradient, where the abundance increases with latitude, both suggesting an exogenic origin (Ligier et al., 2019). The compositions of the Mg-bearing sulfates and chlorinated salts are likely similar to those on the surface of Europa (McCord et al., 1999; Brown and Hand, 2013; Ligier et al., 2016), although at a much lower mean total abundance of about 8–20% between 65°S and 65°N latitude (Ligier et al., 2019).

As shown in Section 10.4.2, hydrated minerals on Ganymede were first detected by NIMS in the dark ancient terrain on the satellite's trailing hemisphere (McCord et al., 2001), and more recent observations show they might be related to the chlorinated species mapped using SINFONI data (Ligier et al., 2019). SINFONI data suggest that the highest abundances of hydrated salts largely coincide with extensively tectonically resurfaced portions of Ganymede's bright terrain. The southern portion of *Xibalba Sulcus* and *Babylon Sulci* in Ganymede's leading hemisphere appear to be the richest areas for hydrated salts (Ligier et al., 2019) (Fig. 10.2). This distribution could suggest, an endogenic origin for these species, most likely resulting from

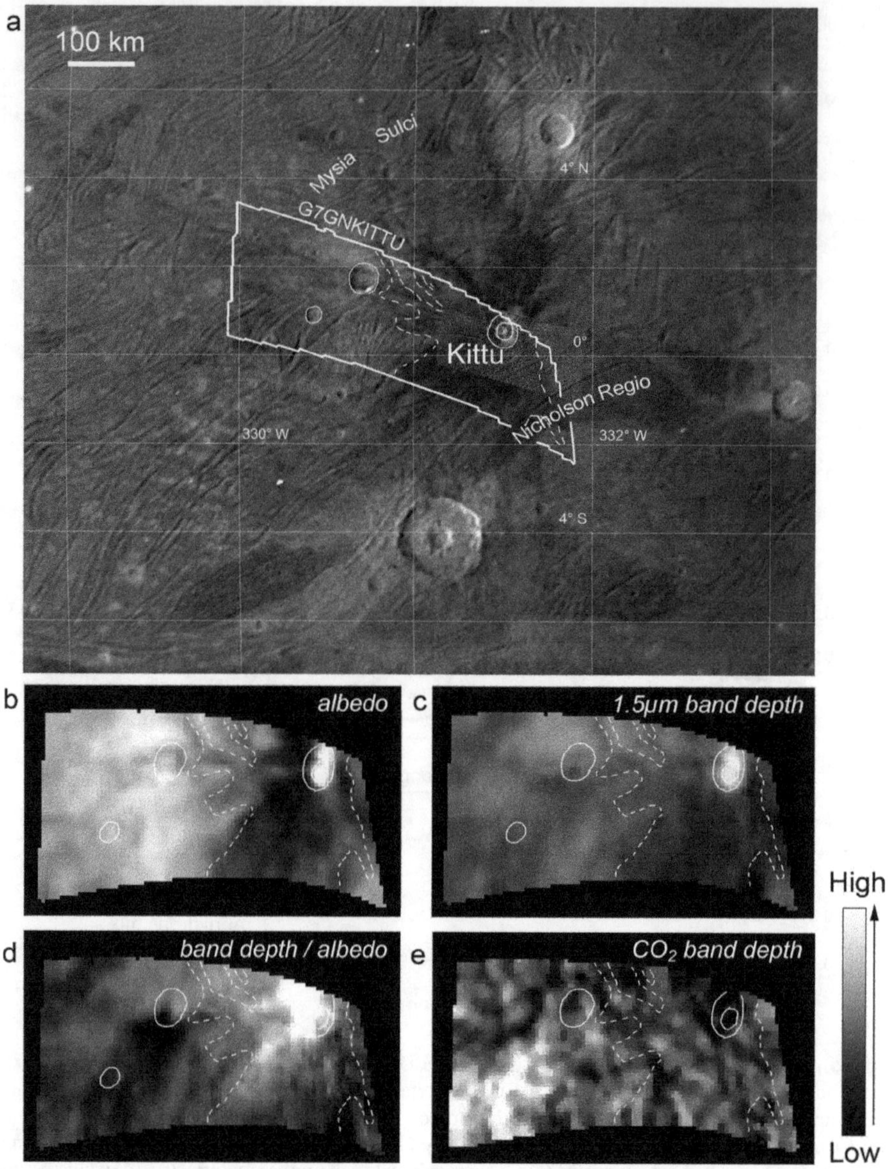

Figure 10.5 Dark ray impact crater *Kittu* (0.4°N/334.6°W) located at the border between *Nicholson Regio* and *Mysia Sulcus*: (a) geological context outlining Kittu, its dark ejecta, two adjacent older impact craters, and the corresponding NIMS observation (G7GNKITTU) on top of a Voyager/Galileo image mosaic in comparison with the NIMS-derived maps showing (b) the visible albedo (scaled from 0.11 to 0.28), (c) the band depth of the H_2O ice absorption at 1.5 μm (scaled from 0.15 to 0.6), indicating the abundance of H_2O, (d) the ratio of the 1.5-μm band depth to albedo (scaled from 1.2 to 3.2), indicating larger (low values) and smaller (high values) H_2O ice grains, and (e) CO_2 absorption band depth at 4.25 μm (scaled from 0 to 0.3) (after Hibbitts et al., 2003).

emplacement at the surface of an upwelling fluid enriched in Mg^{2+}, SO_4^{2-}, and Cl- ions, as discussed in Section 10.6.3. However, an exogenic origin of these compounds cannot be fully excluded.

10.5 TRACE CONSTITUENTS

In addition to the major surface compounds, traces of other molecular species have been identified on Ganymede's surface. Although they exist in small quantities, these trace species provide essential clues to the driving exogenic alteration processes influencing Ganymede's surface composition, and to the origin, formation, and evolution of Ganymede's ice and non-ice (including salty) material.

10.5.1 Oxygen (O_2), Ozone (O_3), and Hydrogen Peroxide (H_2O_2)

Ground-based visible reflectance spectroscopy led to the identification of two narrow absorptions at 0.5773 and 0.6275 μm associated with oxygen molecules (Spencer et al., 1995) (Fig. 10.6a). The asymmetry of the features observed on Ganymede suggests that O_2 resides in a high-density state with liquid, γ or β-O_2 providing the best fit (Calvin et al., 1996) Observations by the Hubble Space Telescope enabled the mapping of Ganymede's O_2 distribution using the narrow-band visible filters of the Wide Field Planetary Camera 2 (WFPC2) and showed that the absorption bands are strongest at low latitudes on the trailing hemisphere (Calvin and Spencer, 1997) (Fig. 10.6a and b) results that Spencer and Klesman (2001)

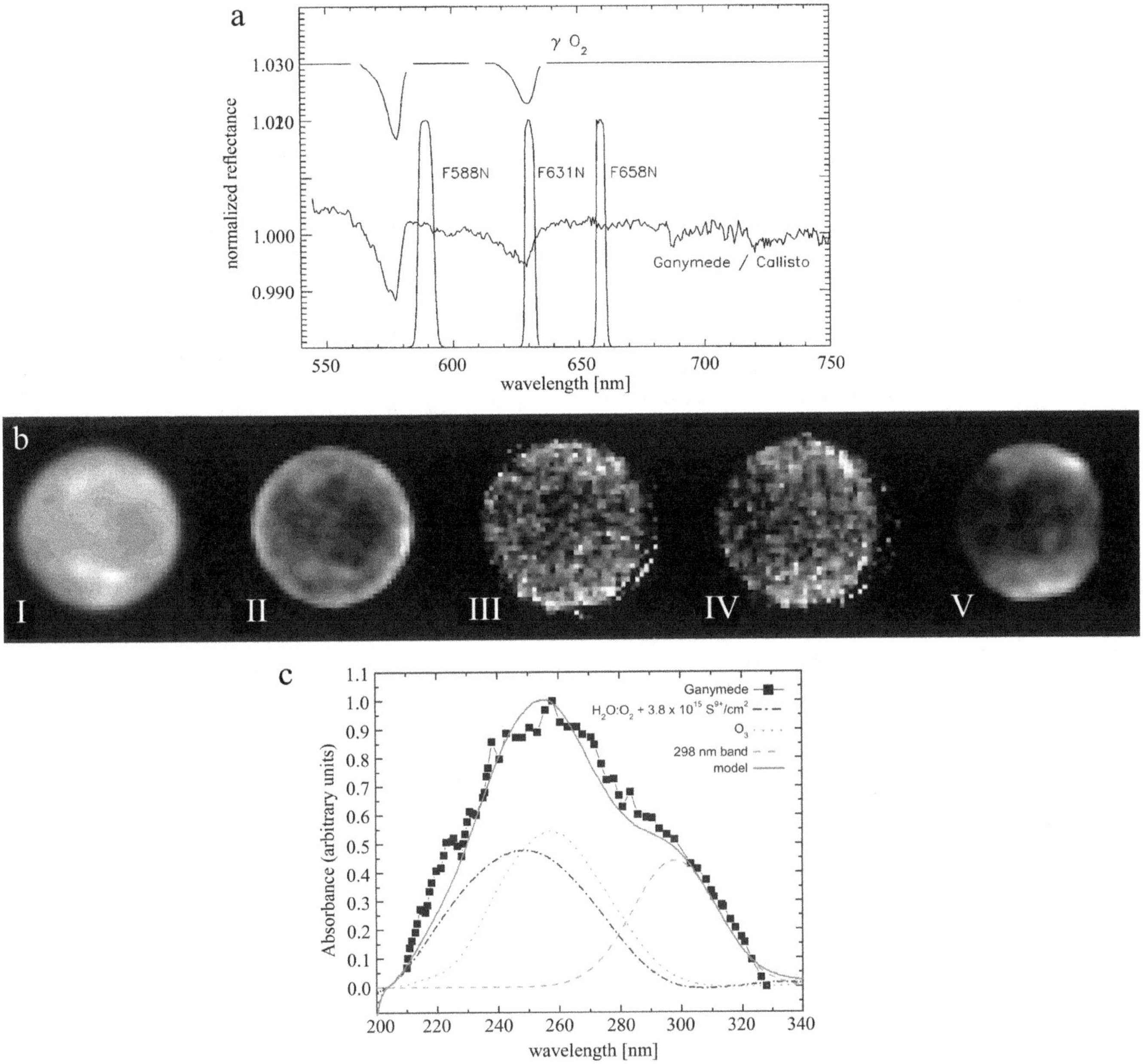

Figure 10.6 Trace compounds on Ganymede: (a) Visible reflectance spectrum of Ganymede containing the two absorptions of molecular oxygen (O_2) at 0.5773 and 0.6275 μm in comparison with the lab spectrum of solid (γ) O_2 (from Landau et al., 1962). Figure reprinted from *Icarus*, 130. Wendy M. Calvin, John R. Spencer. Latitudinal Distribution of O_2 on Ganymede: Observations with the Hubble Space Telescope, 505–516 ©1997, with permission from Elsevier. Also plotted are the wavelength positions of the three HST WFPC2 filters used for the acquisition of the images of Ganymede's trailing hemisphere shown in (b), which consist of (I) Continuum image based on the filters F588N (589.4 nm) and F658N (659 nm), (II) continuum ratio image of the 588/658 filters (scale from black to white 7%), (III) ratio of the filter F631N image (630.6 nm) to the continuum ratio image in frame II (first CCD position) (IV) second CCD position for mapping the O_2 absorption band strength, (V) average of frames III and IV smoothed by a Gaussian filter to reduce noise. The band strength of the 630 nm O_2 band extends from zero (white) to 0.014 (black). Reprinted from *Icarus*, 130. Wendy M. Calvin, John R. Spencer. Latitudinal Distribution of O2 on Ganymede: Observations with the Hubble Space Telescope, 505–516 ©1997, with permission from Elsevier. (c) UV spectrum of Ganymede shown in absorbance including the absorption of ozone (O_3) at 260 nm compared to a model spectrum including O_3, a sulfur-bearing species and an unknown absorber exhibiting an absorption at 298 nm. Reprinted from *Icarus*, 277. P. Boduch, R. Brunetto, J. J. Ding, A. Domaracka, Z. Kaňuchová, M. E. Palumbo, H. Rothard, G. Strazzulla. Ion processing of ices and the origin of SO_2 and O_3 on the icy surfaces of the icy jovian satellites. 424–432 ©2016, with permission from Elsevier.

extended to all latitudes on Ganymede. extended to all longitudes on Ganymede.

The observed O_2 distribution is different from the distribution of the ozone (O_3) feature seen at 260 nm in IUE and HST observations by Nelson and colleagues (1987) and Noll and colleagues (1996) (Fig. 10.6a and b). Subsequent disk-resolved measurements of this feature using Galileo UVS data showed that the O_3 is concentrated in Ganymede's polar ice as well as near sunrise and sunset at low latitudes (Hendrix et al., 1999). However, the 260nm-absorption appears broader than expected, with an apparent overlapping contribution from additional material that absorbs strongly near 300 nm. The identity of this extra absorber is currently unknown, despite efforts to reproduce the observed spectra experimentally by irradiating ice mixtures (Boduch et al., 2016).

In addition, H_2O_2 is thought be responsible for the reddening observed in the UV spectra (Hendrix et al., 2011) and has been shown to produce a good fit to HST FUV spectra (Molyneux et al., 2020) However, although, H_2O_2has been identified in NIR spectra of Europa via the detection of an absorption feature at 3.5 μm (Carlson et al., 1999a), no such feature has been identified on Ganymede to date.

10.5.2 Carbon Dioxide (CO_2)

Carbon dioxide is inferred to be present on the surface of Ganymede by the presence of an absorption near 4.26 μm in reflectance spectra in NIMS data (Fig. 10.7), which is negligibly different from the CO_2 absorption band detected in the nonice material on Callisto (Hibbitts et al., 2003). This is clearly distinct from that of solid CO_2 or CO_2 gas, implying the molecules are trapped or bound in the surface, affecting the energy of the asymmetric stretch vibration responsible for this band. This also provides a degree of thermal stability so that CO_2 is retained in these surfaces that are otherwise warmer than the ~80 K temperature at which CO_2 rapidly sublimes (e.g., McCord et al., 1998).

The CO_2 NIMS detected is in the non-ice material(s), not in the ice. The ice in all terrain types where CO_2 is detected is sufficiently large-grained so that its low reflectance prevents discrimination of the CO_2 absorption band from the continuum. Areas with relatively fine-grained ice that are sufficiently bright at 4 microns for CO_2 to be observed, such as the polar regions, do not contain detectable CO_2, unlike for Europa, for which very small amounts of CO_2 have been detected in the finer-grained ice on its leading hemisphere. Thus, in general, bright terrains (*sulci*) contain less CO_2 than dark terrains (*regiones*), little or no CO_2 is detected at the poles, and, unlike for Callisto, there does not appear to be any leading/trailing hemisphere asymmetry in the distribution of CO_2 (Fig. 10.7), nor do impact craters tend to be CO_2 rich. If anything, impact craters tend to be depleted in CO_2 relative to surrounding terrain, even for dark ray craters for which spectral masking by H_2O ice would not explain the absence of a spectral feature (Fig. 10.5)

10.5.3 S- and C-bearing Surface Compounds

Other molecules, specific bonds, or functional groups tentatively identified in observations of Ganymede by the NIMS instrument include SO_2, SH, C≡N (perhaps HCN) and C-H (McCord et al., 1997; McCord et al., 1998; Hibbitts et al., 2003). Unlike CO_2, however, these absorptions could only be inferred from average spectra of Ganymede and are too weak to be mapped across Ganymede's surface (Hibbitts et al., 2003)

The detected 4.05-micron NIMS absorption is very close to the strong absorptions of SO_2 ice at 4.07 μm and SO_2 gas, near 4.03 μm suggesting that SO_2 might be trapped or stabilized by bonding in Ganymede's nonice material (McCord et al., 1998; Hibbitts et al., 2003). The SO_2 has also been inferred from a time-variable feature at 0.32 μm in UV telescopic data, which may imply temporal changes of its existence on the surface on decadal timescales (Domingue et al., 1998). The absorption NIMS observed on Ganymede has also been observed on Callisto (McCord et al., 1998; Hibbitts et al., 2002). Recent investigations by Cartwright and colleagues (2020), however, demonstrate that Callisto's 4-micron absorption has a spectral signature more similar to S-bearing compounds such as thermally altered sulfur, as well as disulfanide (HS_2) Finally, sulfite ions have also been successfully used to model the broad absorption band shortward of 0.4 μm observed on Ganymede UV, which is superimposed by the UV ozone absorption (Boduch et al., 2016) (Fig. 10.6c). It is possible that Ganymede's sulfur compounds are related to the sulfuric acid hydrate observed at high latitudes and on the leading hemisphere (Ligier et al., 2019).

The absorption features at 4.57 μm and at 3.4 μm seen in NIMS data and interpreted as due to C≡N and C-H (McCord et al., 1997; McCord et al., 1998), respectively, are consistent with those seen in spectra of tholins, which have also been successfully used to model the UV signature of Ganymede's dark material (Fig. 10.5a), and may account for the long wavelength shoulder on the disk-averaged ozone absorption feature on Ganymede described in Boduch and colleagues (2016). Simple organic molecules could be the result of radiolysis on the surface or due to infall of material from asteroids and comets onto Ganymede's surface.

10.6 IMPLICATIONS FOR GANYMEDE'S SURFACE PROCESSES

While UV-to-NIR spectral observations deliver the compositional information "only" of the uppermost mm of Ganymede's surface, the cross-correlation with their location and the type of terrain, as well as with specific, local, surface features, including their stratigraphic position and geologic age, provide essential information to constrain Ganymede's geological evolution and its dominant surface processes. Therefore, in this section, we discuss the implications Ganymede's surface composition could have for the major exogenic and endogenic processes known to shape Ganymede's surface in the dark and bright terrain.

10.6.1 Impact Processes: Crustal Composition

Geologically young impact craters on planetary bodies commonly present a useful tool to examine the composition of the uppermost crust. Particularly, bright ray craters on icy satellites are known to represent younger geological units on their surfaces, where erosional processes have not yet completely removed the thin ice deposits (Stephan et al., 2019). On Ganymede bright, ice-rich crater rays may remain stable for up to ~ 1 Ga after the impact event (Wagner et al., 2019). Since excavation depths approximately vary with the size of impact events, broad stratigraphic profiles can be derived to reveal the nature of the subsurface material.

Morrison and Cruikshank (1973) inferred that variations in the intensity of emitted thermal infrared radiation during eclipses of Ganymede and Callisto are consistent with a thin surface layer of ice covering a denser mixture of rock and ice. Ganymede's fresh bright impact craters are composed predominantly of pure H_2O ice, and depending on the excavation depth of their ejecta material, support the inference of an ice-rich uppermost crust to a depth of at least about 9 km

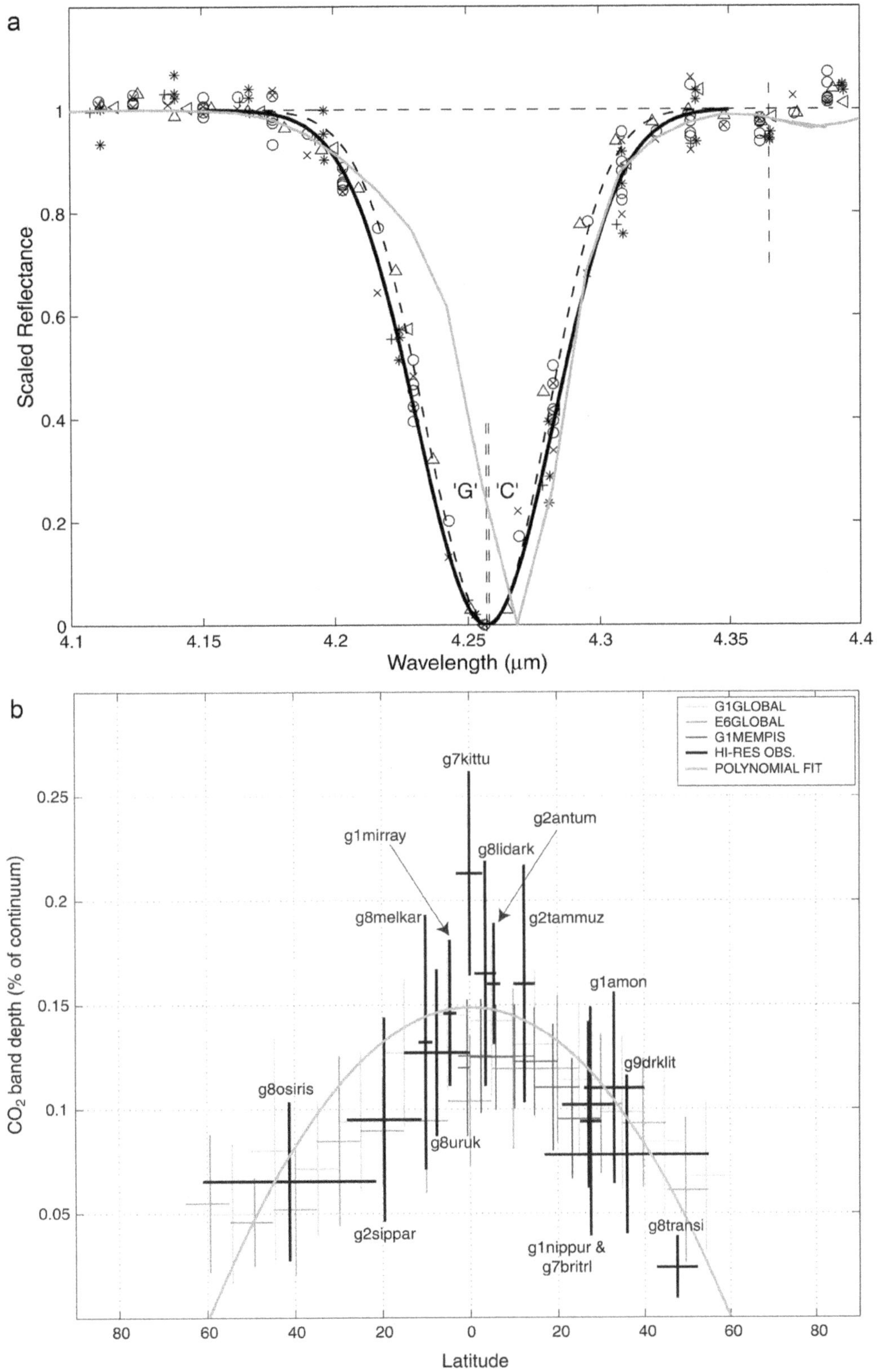

Figure 10.7 CO_2 on Ganymede: (a) the shape of the normalized CO_2 band on Ganymede (solid black) in comparison with Callisto (dashed) and crystalline CO_2 at 80 K (solid gray). Derived Ganymede band center at 4.257±0.004 μm (G) and Callisto-derived band center at 4.258±0.004 μm (C) are marked by dashed lines. The model of the CO_2 absorption on Ganymede is the best fit Gaussian curve to the normalized CO_2 absorption band (second-order polynomial continuum-removed) for 14 NIMS observations. The band-shaped model for the CO_2 absorption on Callisto was similarly derived (Hibbitts et al., 2002). Symbols group observations by orbit: orbit 1 (o); orbit 2 ('x'); orbit 6 ('<'); orbit 7 ('∧'); orbit 8 ('*'), orbit 9 ('+') (from Hibbitts et al., 2003). (b) The dependence of CO_2 band depth on latitude with the average CO_2 band depths from G1GLOBAL (light gray), E6GLOBAL (gray) and G1MEMPHIS (black thin line) plotted in latitude bins of 10, 10 and 5, respectively, along with other high-resolution data (thick black lines). Vertical lines represent a variation of one standard deviation in band depth. The thick, gray line is the expected variation in CO_2 band depths if they were to decrease according to the sine of the latitude. Reprinted with permission from John Wiley and Sons. C. A. Hibbitts, R. T. Pappalardo, G. Hansen, T. B. McCord. 2003. Carbon dioxide on Ganymede. *J. Geophys. Res.*, 108, 5036.

(Stephan et al., 2019). Fresh bright craters occur in both dark as well as bright terrain, implying that H_2O ice dominates Ganymede's uppermost crust just below the thin layer of dark material in Ganymede's ancient dark terrain. Furthermore, spectra from Ganymede's darkest areas still show a H_2O ice signature as strong as those for the iciest locations on Callisto (Hibbitts et al., 2003b), highlighting an uppermost Callistoean crust composed of less pure H_2O ice, which is consistent with an incomplete ice-rock differentiation of Callisto's uppermost crust in comparison to the fully differentiated interior of Ganymede (Sotin and Tobie, 2004). In contrast to Callisto, the fresh impact crater materials on Ganymede do not contain any significant amount of volatiles such as CO_2 (Hibbitts et al., 2003). Depletion of CO_2 in crater ejecta compared to the dark terrain may imply that the interior of Ganymede has over time become depleted in CO_2. This is also different on Callisto, where the nonice material associated with recent impact craters is often enriched in CO_2 compared to older surrounding terrain (Hibbitts et al., 2002). Carbon dioxide may have been lost from the interior of Ganymede as a consequence of Ganymede's more complete differentiation and continued tectonism. The association of CO_2 with the nonice material does not necessarily imply it is formed within it. It is very possible the CO_2 was originally endogenic to Ganymede, existing as a clathrate for solid solution with the ice, depending on the temperature/pressure conditions, and has either adsorbed onto the nonice material or become incorporated into the nonice material as inclusions (Hibbitts et al., 2003).

Care must be taken when comparing fresh material of impact craters located at different latitudes, due to the influence of polar ice covering the pristine crater material. Although the H_2O ice grain sizes of fresh crater material are usually smaller than those of their surroundings, comparing grain sizes from different fresh impact craters underlines a correlation with the geographic location and the associated surface temperature (Stephan et al., 2019). The influence of ice becomes particularly apparent for fresh impact craters at midlatitudes (Stephan et al., 2020). The topography of impact craters covered by polar H_2O ice deposits suggest a local thickness reaching several meters for the polar caps (Pappalardo et al., 2004), far more than required to mask the spectral signature of the underlying crater material.

10.6.2 Origin and Nature of Dark Material

The spectral properties of Ganymede's dark non-ice material(s) described in Section 10.4.3 provide information about its nature and origin together, especially when combined with the geological investigations of Ganymede based on Galileo SSI images. Detailed discussions of Ganymede's geology and its evolution are provided in Chapter 1.5 and Chapter 2.2.

Using Galileo SSI images, Prockter and colleagues (1998) investigated more than 16,000 km^2 of dark terrain in Galileo Regio at relatively high spatial resolution (<100 m/pixel), evaluating three end-member models to determine the origin and nature of the dark non-ice material and thus to better constrain the formation of Ganymede's dark terrains: (1) the crust of dark terrain is dark throughout, (2) the dark material is a low-albedo cryovolcanic layer, or (3) dark material is distributed in small quantities throughout the crust and geological processes have acted to concentrate the dark material on the surface. They found that the third model, that of a thin veneer of low-albedo material, concentrated on the surface as a sublimation lag, best fits the observations of Galileo Regio. In this area, the lowest albedo deposits are correlated with local topographic lows like crater floors and troughs, whereas brighter material is typically found on north facing slopes of ridges, craters, and other topographic highs, a finding further expanded upon by Oberst and colleagues (1999). This distribution was also found to be common within the dark terrain regions Nicholson Regio and Marius Regio, leading Prockter and colleagues (2000) to conclude that the processes that form dark terrain are similar across the surface of Ganymede. Explanations for the concentration of dark material on the surface of Ganymede include (i) micrometeoroid gardening, (ii) thermal segregation and migration of volatile molecules, and (iii) magnetospheric sputtering and subsequent redeposition of H_2O ice. Dark material is expected to be further concentrated by the actions of tectonism and mass wasting. These processes are discussed in the following subsections.

Micrometeorite Bombardment

Micrometeorite bombardment is expected to affect the abundance and distribution of dark material at smaller scales through the mixing of surface ice and excavated very near-surface material along with any refractory remnant of the stony component of the impactor (Hartmann, 1980). This theory is consistent with the presence of dust detected around Ganymede by the dust detector onboard the Galileo spacecraft and interpreted to originate from impacting micrometeorites (Krüger et al., 1999). Nearly all ejected refractory particles eventually fall back to the surface on ballistic trajectories and darken Ganymede's dark regolith.

There is also evidence for exogenous contribution of dark material by larger impacts. Ganymede's dark ray, floor, and halo craters (please see Chapter 1.6 and Chapter 2.2) possess infrared reflectance spectra that are consistent with the presence of remnants of the impactor material, suggesting an exogenous origin of at least some of Ganymede's dark non-ice material (Schenk and McKinnon, 1991; Hibbitts et al., 2009). Using Galileo images, Helfenstein and colleagues (1997) showed that the visible albedo of the dark terrain can be successfully modeled using a simple linear mixture of 55% dark material of dark floor craters found in both the dark and bright terrain and 45% average bright terrain material and therefore supported that meteoritic infall was important in the evolution of Ganymede's dark terrain and Ganymede's surface in general.

Consequently, impact craters with large amounts of dark non-ice material(s) apparent in their ejecta are hypothesized to result from a combination of preserved dark impactor material and preferential loss of ice in the ejecta and crater due to thermal segregation (e.g., Schenk and McKinnon, 1991). Initially, the excavation of a subsurface layer of dark non-ice material(s) was also considered. Although regions of dark terrain might have been obscured by later ice deposits and then re-excavated by younger impact events, this theory was dismissed as unlikely because of the association of dark craters with all terrain types and ages (e.g., Schenk and McKinnon, 1991). Infrared spectral unmixing models of NIMS observations of dark ray craters

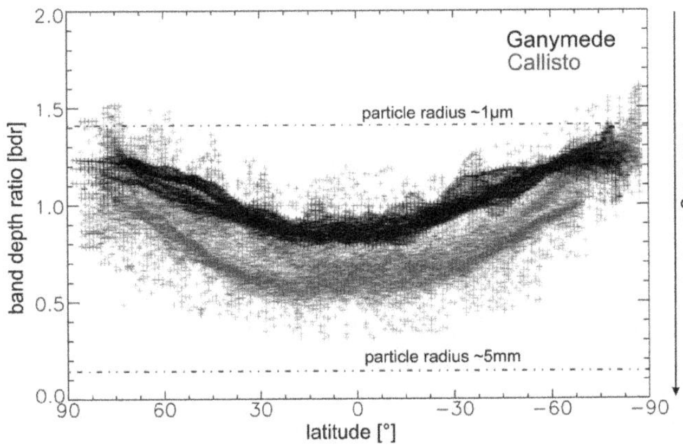

Figure 10.8 Comparison of grain size variations on Ganymede and Callisto based on measurements of the band depth ratio (BDR) of the two H_2O ice absorptions at 2 and 1.5 µm, respectively, which decreases with increasing H_2O ice grain size. Reprinted from *Icarus*, 337. Katrin Stephan, Charles A. Hibbitts, Ralf Jaumann. H_2O-ice particle size variations across Ganymede's and Callisto's surface. 113–440 ©2020, with permission from Elsevier.

show up to several 10s of percent of a less-hydrated component spectrally similar to the dark non-ice material(s) on Callisto, or C-type asteroids (Hibbitts et al., 2009). This suggests a slow accumulation of carbonaceous impactor and micrometeoroid material. Hence, in terms of mineralogy, two different families of species coming from carbonaceous impactors and micrometeoroids are commonly suggested as the best candidates for the dark non-ice material(s): (i) carbon-rich materials and (ii) hydroxylated or hydrated silicates (Pilcher et al., 1972; Pollack et al., 1978; Calvin et al., 1995; McCord et al., 1997) suggesting a slow accumulation of carbonaceous impactor and micrometeoroid material.

Thermal Segregation and Migration

Thermal segregation by sublimation and cold-trapping of volatiles is considered the primary process contributing to the distribution of dark non-ice material and H_2O ice in Ganymede's dark ancient terrain (Prockter et al., 1998). Whereas dark material is concentrated in topographic low regions and on sun-facing slopes forming a relatively thin layer over Ganymede's icy crust, H_2O ice is predominantly segregated from the dark material and generally resides exposed in topographic heights and polar-facing slopes.

These imaging observations are consistent with models predicting sublimation to be an important process for the redistribution of H_2O ice on Ganymede at small scales (Spencer, 1987), as well as being sufficient to account for the presence of Ganymede's polar caps. The caps might be formed through migration or "hopping" of H_2O ice from the equatorial region to higher latitudes by sublimation and subsequent redeposition (Spencer, 1987; Moore et al., 1999) with the H_2O ice molecules simply accumulating in the colder, more thermally stable environment of the high latitudes (Squyres, 1980). Stephan and colleagues (2020) also found a continuous trend of decreasing H_2O ice grain sizes with increasing latitudes on Ganymede as well as on Callisto (Fig. 10.8) implying more fine-grained ice (i.e., frost at high latitudes), with the same process active on both satellites. While the H_2O ice grain sizes in Callisto's equatorial region are larger than in the equatorial region on Ganymede, the grain sizes of H_2O ice on both satellites converge toward the poles. The differences in equatorial ice grain sizes are consistent with the differences in their surface temperatures. Whereas Ganymede exhibits a maximum temperature of 152 K on the equator (Orton et al., 1996), Callisto's equatorial region has a maximum of 165 +/− 5°K (Moore et al., 1999) and the Callistoan day is longer (~17 Earth days) than a day on Ganymede (~7 Earth days). Thus, ice is more stable on Ganymede for finer-grained ice to persist at its equator. In contrast, temperatures at either moon's poles do not exceed 80±5 K and ice will be thermally stable regardless of grain size (Hanel et al., 1979; Spencer, 1987). Local dependences in the transition from larger to smaller grains at midlatitudes appear to be mostly related to temperature, resulting in slightly larger grains in the warmer dark terrain and slightly smaller grains in the colder bright terrain or fresh ice-rich impact craters for a given latitude (Pappalardo et al., 2004; Stephan et al., 2020; de Kleer et al., 2021). No distinct thermal signature is attributable to the polar caps other than their relatively high albedo due to the accumulation of fine-grained ice (de Kleer et al., 2021).

Purves and Pilcher (1980) and Squyres (1980) concluded that both Ganymede and Callisto accumulate ice at their poles at the expenses of the equatorial region where exposed ice should entirely sublimate away. Possibly, not enough H_2O ice present on Callisto's surface anymore to produce a sufficiently thick layer of polar ice needed to mask the underlying dark terrain and to create polar caps as on Ganymede. Nevertheless, Ganymede is continually bombarded by plasma from Jupiter's magnetosphere, whereas Callisto, due to its larger distance from Jupiter, is not subjected to such a high flux (Bagenal et al., 2016; Belenkaya, 2020). This particle interaction, which can sputter H_2O ice, cannot be neglected when searching for the origin of Ganymede's polar caps, and will be discussed further in the next section.

Magnetospheric Sputtering and Redeposition

Detailed studies point out that because of extremely low temperatures in the polar regions, thermal migration alone is possibly incapable of transporting appreciable H_2O above a latitude of ~ +/−50° (Purves and Pilcher, 1980), requiring an additional process for polar cap formation. It is possible that a sputter-induced redistribution and subsequent cold trapping of H_2O molecules may have formed Ganymede's polar caps (Johnson, 1985). Johnson (1997) also suggests that bombarding plasma originating from the Jovian magnetosphere physically damages ice grains so that they scatter more sunlight and thus brightens the polar caps, mimicking the spectral signature of smaller-grained H_2O ice.

Ganymede orbits within Jupiter's corotating magnetospheric plasma, where charged particles together with solar and galactic ultraviolet radiation interact with the surface material causing physical and chemical alteration. Ganymede's strong internal magnetic field (Kivelson et al., 1996), however, partially blocks magnetospheric ions from bombarding the surface in the equatorial region yet funnels the particles onto the polar region, where it preferentially reaches the surface (see Chapter 3.2 for more details).

Intriguingly, Ganymede's polar cap margins coincide with the boundary between field lines that are open to polar ions and field lines that are closed to them (Khurana et al., 2007), supporting a possible relationship between the formation of the polar caps and Ganymede's interaction with impacting ions. A surface–ion interaction is also supported by the latitudinal and longitudinal dependent distribution of trace compounds trapped in Ganymede's material such as O_2 in the equatorial region, and O_3 and hydrated sulfuric acid predominantly at the poles (Calvin and Spencer, 1997; Hendrix et al., 1999; Ligier et al., 2019). O_3 is either readily photolyzed at the equator resulting in the dominant O_2 signature, or is enhanced via plasma processes at the poles. The H_2O_2 could be produced by the principal H_2O dissociation product OH in the absence of H that is lost from the system (e.g., Laffon et al., 2006). Also, the radiation-induced decomposition of hydrogen peroxide H_2O_2 dimers in hydrogen peroxide inclusions in H_2O ice could be a precursor to the formation of O_2 (Calvin et al., 1996; Cooper et al., 2001). Sulfuric acid and other S-bearing compounds could be the result of the interaction of Jovian magnetospheric S^+ implanted into the ice on the surface (Carlson et al., 1999b).

Therefore, the existence and the brightness of these caps may be attributed to both thermal migration and charged particle damage/sputtering. The Jovian charged particle bombardment redistributes H2O molecules via sputtering as well as physically altering H_2O ice grains at colder high latitudes (Johnson, 1997; Khurana et al., 2007; Plainaki et al., 2015) At lower latitudes, thermal sublimation and migration to colder higher latitudes dominates, a process also observed on Callisto, which contains no magnetic field (Stephan et al., 2020)

Ganymede's magnetic field can also explain the observed hemispherical differences between its leading and trailing hemispheres (Calvin and Spencer, 1997). Magnetospheric plasma corotates with Jupiter faster than the satellite orbits, thereby preferentially impinging on its trailing hemisphere (e.g., Kivelson et al., 2004). Charged particles with sufficient energy can penetrate the field lines to impact the surface, implanting oxygen ions and/or sputtering as well as inducing radiolysis in H_2O ice (Calvin et al., 1996).

10.6.3 Origin of Bright Terrain

The interpretation of McCord and colleagues (2001) that hydrated minerals identified on Ganymede's surface could be similar to those found on Europa, which is mostly frozen magnesium sulfate brines, has sparked the discussion about whether these materials could have been derived from Ganymede's liquid ocean reaching the surface at some point in its history. Such an exchange between the ocean and the surface could fundamentally affect whether Ganymede's subsurface is habitable. Subsurface oceans are expected to exist on Europa and Ganymede, as well as Callisto (Sohl et al., 2010). Despite some indications for Europa (Roth et al., 2014; Jia et al., 2018) and Triton (Smith et al., 1989), so far only the Saturnian satellite Enceladus has revealed direct evidence of cryovolcanic activity with a possible link to the interaction between surface and a liquid subsurface reservoir. High concentrations of sodium and sodium minerals (NaCl, $NaHCO_3$, and/or Na_2CO_3 as the main Na-bearing compounds) emitted from the geysers at Enceladus' south pole (Postberg et al., 2009) imply a slightly alkaline sodium salt H_2O reservoir enriched in carbonates (CO_3^{2-}) beneath the surface of Enceladus through warm cracks. The H_2O reservoir could even be linked with a subsurface ocean between the icy crust and a rocky core (Sohl et al., 2010). Furthermore, ice grains detected to be enriched in complex organics suggest the presence of a thin organic-rich film on top of the ocean (Postberg et al., 2018).

Ganymede's ocean, which is discussed in detail in Chapter 12, sandwiched between two layers of ice, at a depth of $\sim$150 km (Sohl et al., 2010), is potentially enriched in the ions of Na^+, Mg^{2+}, and SO_4^{2-}, because of the aqueous alteration of carbonaceous chondrites that represent the bulk material that presumably accreted to form the Galilean moons (Pilcher et al., 1972), but the composition may instead reflect current equilibria between solutes and solides (salts) above and/or below the ocean (Sohl et al., 2010 and references therein) and have a more evolved composition containing halides (Hand and Carlson, 2015). However, a clear link to a potential ocean–surface exchange has remained elusive so far.

An interaction between Ganymede's surface and subsurface ocean material may have occurred in regions where the surface shows sign of past disruption such as the bright terrain (Pappalardo et al., 2004 and references therein). The formation of Ganymede's bright terrain may be dominated by global expansion due to tidal heating and/or differentiation, possibly organized into coherent regional patterns by transient low-order convection or changes in the satellite's figure (Collins et al., 1998). During its formation, subsurface H_2O ice of Ganymede's upper crust became tectonically mixed with the (less icy) darker surface material, which is supported by the higher content of H_2O ice in these areas. Fresh impact craters emplaced in the bright terrain are dominated by H_2O ice, similar to their counterparts in the dark terrain (Stephan et al., 2008). In addition, ancient impacts emplaced across Ganymede's surface before bright terrain formation and penepalimpsests emplaced concurrent or after bright terrain formation (Collins et al., 2013; see Chapter 5 for more details) show similar compositional characteristics (Hibbitts et al., 2003). Therefore, a similar icy composition in the subsurface of both terrains is expected, at least in the areas where these features are situated, and supports icy tectonism rather than the introduction of salty material due to interaction with ocean material.

Nevertheless, partial fractional crystallization of upwelling fluids moving toward the surface during the bright terrain formation may explain the regional presence of both sulfates and chlorinated salts on the surface (Ligier et al., 2019). This result could have very important implications in the search for extraterrestrial life. An endogenous origin of sulfate minerals is consistent with the predictions of a subsurface ocean composed of aqueous magnesium sulfate (Vance et al., 2014). Numerical simulations show that, while a thick icy lithosphere such as of Ganymede may be tectonically disrupted after only a few kilometers of extension, the lithosphere is expected generally to be too thick to expose potential ocean material or facilitate ocean–surface interaction through the formation of bright terrain (Howell and Pappalardo, 2018). However, in these models, ocean material could be exposed after a total extension of at least 80 km, which is supposed to result in smooth terrain bounded by highly tectonically resurfaced terrains such as *Babylon Sulci*, which covers large portions of the leading hemisphere. In addition, magmatic activity might have generated local pockets of water melt or Ganymede could have as many as three ice layers with liquid ocean between them (Vance et al., 2014).

10.7 SUMMARY AND CONCLUSION

Our knowledge about Ganymede's surface composition has improved tremendously since the instruments onboard the Galileo spacecraft provided the first spatially resolved spectral data sets. Also, Earth-based telescopic observations with unprecedented spectral resolution revealed even more details of Ganymede's surface composition, confirming that the complexity and diversity of Ganymede's geology is met by the composition of its surface. The pattern of how the individual surface compounds and/or their physical parameters are distributed across Ganymede's surface provides an important key to understand and constrain their endogenous or exogenous formation processes, and brings essential input for theoretical simulations of Ganymede's internal evolution, formation of, and interaction with Ganymede's subsurface ocean.

As stated earlier in this chapter, however, the available spectral and imaging data are still limited with respect to coverage, spatial, and spectral resolution. More details are required to clearly identify the individual surface compounds and to better understand their association with the local geology, subsurface properties, and environment conditions, which could provide unequivocal evidence for an interaction between space environment, surface, and subsurface ocean interaction. Such measurements and many others are planned during the upcoming ESA-led JUICE mission (Grasset et al., 2013) with its capabilities for spectral imaging (UV to MIR), laser altimetry, subsurface radar sounding, and many other geophysical instruments (magnetometer, plasma detector) (Grasset et al., 2013). The synergy between these instruments will enable us to:

(1) evaluate the composition of Ganymede's upper crust
(2) characterize the detailed mineralogy of Ganymede's dark absorbing material in comparison to Callisto
(3) identify the nature and origin of organic constituents in the dark non-ice material
(4) verify and map the distribution of the different hydrated minerals that exist on Ganymede's surface and resolve their formation processes
(5) identify local potentially geologically active regions, which could be associated with the presence of salts in the bright terrain to provide unequivocal evidence for ocean–surface interaction
(6) monitor the competition between thermal segregation and plasma bombardment effects on a local scale
(7) resolve the origin of CO_2 and SO_2 in Ganymede's surface material

Nonetheless, since former laboratory measurements have shown that the spectral signatures of Ganymede's key surface compounds such as H_2O ice and salts or ice/non-ice mixtures can change dramatically within the surface temperatures (~70 to 150 K) expected on Ganymede's surface (Hanley et al., 2014; Stephan et al., 2021), additional laboratory spectra and optical constants from UV to IR spectral range are still required as input for all efforts of spectral analyses of Ganymede's surface.

REFERENCES

Alday, J., Roth, L., Ivchenko, N. et al., 2017. New constraints on Ganymede's hydrogen corona: Analysis of Lyman-α emissions observed by HST/STIS between 1998 and 2014. *PSS*, 148, 35–44.

Bagenal, F., Wilson, R. J., Siler, S., Paterson, W. R. & Kurth, W. S. 2016. Survey of Galileo plasma observations in Jupiter's plasma sheet. *JGR*, 121, 871–894.

Belenkaya, E. 2020. Callisto in the magnetosphere of Jupiter. *Solar System Research*, 54, 85–95.

Boduch, P., Brunetto, R., Ding, J. J., Domaracka, A., Kaòuchová, Z., Palumbo, M. E., Rothard, H. & Strazzulla, G. 2016. Ion processing of ices and the origin of SO_2 and O_3 on the icy surfaces of the icy Jovian satellites. *Icarus*, 277, 424–432.

Brown, M. E. & Hand, K. P. 2013. Salts and radiation products on the surface of Europa. *AJ*, 145, 110.

Caldwell, J. 1975. Ultraviolet observations of small bodies in the solar system by OAO-2. *Icarus*, 25, 384–396.

Calvin, W. M. & Clark, R. N. 1991. Modeling the reflectance spectrum of Callisto 0.25 to 4.1 μm. *Icarus*, 89, 305–317.

Calvin, W. M., Clark, R. N., Brown, R. H. & Spencer, J. R. 1995. Spectra of the icy Galilean satellites from 0.2 to 5 μm: A compilation, new observations, and a recent summary. *JGR* 100, 19041–19048.

Calvin, W. M., Johnson, R. E. & Spencer, J. R. 1996. O_2 on Ganymede: Spectral characteristics and plasma formation mechanisms. *GRL*, 23, 673–676.

Calvin, W. M. & Spencer, J. R. 1997. Latitudinal distribution of O_2 on Ganymede: Observations with the Hubble Space Telescope. *Icarus*, 130, 505–516.

Carlson, R. W., Anderson, M. S., Johnson, R. E. et al., 1999a. Hydrogen peroxide on the surface of Europa. *Science*, 283, 2062–2064.

Carlson, R. W., Johnson, R. E. & Anderson, M. S. 1999b. Sulfuric acid on Europa and the radiolytic sulfur cycle. *Science*, 286, 97–99.

Carlson, R. W., Weissman, P. R., Smythe, W. D. & Mahoney, J. C. 1992. Near-infrared mapping spectrometer experiment on Galileo. *SSR*, 60, 457–502.

Cartwright, R. J., Nordheim, T. A., Cruikshank, D. P. et al., 2020. Evidence for sulfur-bearing species on Callisto's leading hemisphere: Sourced from Jupiter's Irregular Satellites or Io? *Astrophysical Journal*, 902:L38, 1–11.

Clark, R. N. 1980. Ganymede, Europa, Callisto, and Saturn's rings: Compositional analysis from reflectance spectroscopy. *Icarus*, 44, 388–409.

Clark, R. N. & Lucey, P. G. 1984. Spectral properties of ice-particulate mixtures and implications for remote sensing. 1. Intimate mixtures. *JGR* 89, 6341–6348.

Clark, R. N. & Roush, T. L. 1984. Reflectance spectroscopy: Quantitative analysis techniques for remote sensing applications. *JGR: Solid Earth*, 89, 6329–6340.

Collins, G. C., Head, J. W. & Pappalardo, R. T. 1998. Formation of Ganymede grooved terrain by sequential extensional episodes: Implications of Galileo observations for regional stratigraphy. *Icarus*, 135, 345–359.

Collins, G. C., Patterson, G. W., Head, J. W. et al., 2013. Global geologic map of Ganymede. *U.S.G.S.*, 3227, https://doi.org/10.3133/sim3237.

Cooper, J. F., Johnson, R. E., Mauk, B. H., Garrett, H. B. & Gehrels, N. 2001. Energetic ion and electron irradiation of the icy Galilean satellites. *Icarus*, 149, 133–159.

Cuzzi, J. N., French, R. G., Hendrix, A. R. et al., 2018. HST-STIS spectra and the redness of Saturn's rings. *Icarus*, 309, 363.

Dalton, J. B. 2010. Spectroscopy of icy moon surface materials. *SSR*, 153, 219–247.

Dalton, J. B., Cruikshank, D. P., Stephan, K. et al., 2010. Chemical composition of icy satellite surfaces. *SSR*, 153, 113–154.

Dalton, J. B. & Pitman, K. M. 2012. Low temperature optical constants of some hydrated sulfates relevant to planetary surfaces. *JGR* 117 E09001, 1–15.

De Angelis, S., Tosi, F., Carli, C. et al., 2021. Temperature-dependent, VIS-NIR reflectance spectroscopy of sodium sulfates. *Icarus*, 357, 114165.

De Kleer, K., Butler, B., De Pater, I. et al., 2021. Ganymede's surface properties from millimeter and infrared thermal emission. *PSJ*, 2:5, 1–20, https://doi.org/10.3847/PSJ/abcbf4.

Domingue, D. L., Lane, A. L. & Beyer, R. A. 1998. IUE's detection of tenuous SO_2 frost on Ganymede and its rapid time variability. *GRL*, 25, 3117–3120.

Dozier, J. 1989. Remote sensing of snow in visible and near-infrared wavelengths. In ASRAR, G. (ed.) *Theory and applications of optical remote sensing.* John Wiley & Sons, 527–547.

Draine, B. T. & Lee, H. M. 1984. Optical properties of interstellar graphite and silicate grains. et al., *Astrophysical Journal*, 285, 89–108.

Filacchione, G., D'aversa, E., Capaccioni, F. et al., 2016. Saturn's icy satellites investigated by Cassini-VIMS. IV. Daytime temperature maps. *Icarus*, 271, 292–313.

Gaffey, S. 1987. Spectral reflectance of carbonate minerals in the visible and near infrared (0.35-2.55 um): Anhydrous carbonate minerals. *JGR: Solid Earth*, 92, 1429–1440.

Goryniuk, M. C., Rivard, B. A. & Jones, B. 2004. The reflectance spectra of opal-A (0.5-25 μm) from the Taupo Volcanic Zone: Spectra that may identify hydrothermal systems on planetary surfaces. *GRL*, 31, L24701, 1–4.

Grasset, O., Dougherty, M. K., Coustenis, A. et al., 2013. JUpiter ICy moons Explorer (JUICE): An ESA mission to orbit Ganymede and to characterise the Jupiter system. *PSS*, 78, 1–21.

Grundy, W. M. & Schmitt, B. 1998. The temperature-dependent near-infrared absorption spectrum of hexagonal H_2O ice. *JGR* 103, 25809–25822.

Hand, K. P. & Carlson, R. W. 2015. Europa's surface color suggests an ocean rich with sodium chloride. *GRL*, 42, 3174–3178.

Hanel, R., Pearl, J. C., Lowman, P., Kumar, S. & Horn, L. 1979. Preliminary results from Voyager 1 infrared observations of the Jovian satellites. *EOS* 60, 308.

Hanley, J., Dalton, J. B., Chevrier, V. F., Jamieson, C. S. & Barrows, R. S. 2014. Reflectance spectra of hydrated chlorine salts: The effect of temperature with implications for Europa. *JGR*, 119, 2370–2377.

Hansen, G. B. & McCord, T. B. 2004. Amorphous and crystalline ice on the Galilean satellites: A balance between thermal and radiolytic processes. *JGR* 109, E01012.

Hapke, B. 2002. Bidirectional reflectance spectroscopy. *Icarus*, 157, 523–534.

Hartmann, W. K. 1980. Surface evolution of two-component stone/ice bodies in the Jupiter region. *Icarus*, 44, 441–453.

Helfenstein, P., Veverka, J., Denk, T. et al., 1997. Dark-floor craters: Galileo constraints on a Ganymede regolith component. *LPSC XXVIII* 547.

Hendrix, A. R., Barth, C. A. & Hord, C. W. 1999. Ganymede's ozone-like absorber: Observations by the Galileo ultraviolet spectrometer. *JGR* 104, 14169–14178

Hendrix, A. R., Cassidy, T. A. & Johnson, R. E. 2011. Ganymede: The ultraviolet albedo and weathering effects. *AGU, Fall Meeting 2011* P11D–1613.

Hendrix, A. R., Domingue, D. L. & King, K. 2005. The icy Galilean satellites: ultraviolet phase curve analysis. *Icarus*, 173, 29–49.

Hendrix, A. R., Domingue, D. L. & Noll, K. S. 2013. Ultraviolet properties of planetary ices. *In: Gudipati M. & Castillo-Rogez J. (eds.), The Science of Solar System Ices*, 356, 73–105.

Hibbitts, C. A., Klemaszewski, J. E., McCord, T. B., Hansen, G. B. & Greeley, R. 2002. CO2-rich impact craters on Callisto. *JGR* 107, 5084, 14-1–14-12

Hibbitts, C. A., Pappalardo, R. T., Hansen, G. B. & McCord, T. B. 2003. Carbon dioxide on Ganymede. *JGR* 108, 5036.

Hibbitts, C. A., Stephan, K., Collins, G. & Hansen, G. B. 2009. Composition and distribution of nonice and trace materials on Ganymede as derived from Galileo observations. *EPSC 2009*, EPSC abstracts, 4, EPSC2009-632

Hibbitts, C. A., Stockstill-Cahill, K., Wing, B. & Paranicas, C. 2019. Color centers in salts: Evidence for the presence of sulfates on Europa. *Icarus*, 326, 37–47.

Hibbitts, C. A. 2023. Dark ray craters on Ganymede: Impactor or endogenous origin? Icarus, 394, 115400.

Howell, S. M. & Pappalardo, R. T. 2018. Band formation and ocean–surface interaction on Europa and Ganymede. *Geophysical Research Letters*, 45, 4701.

Jenniskens, P., Blake, D. F. & Kouchi, A. 1998. Amorphous water ice: A Solar System Material. *In: (Eds: Schmitt, B., De Bergh, C., Festou, M.) Solar System Ices*, 227, 139–155.

Jia, X., Kivelson, M. G., Khurana, K. K. & Kurth, W. S. 2018. Evidence of a plume on Europa from Galileo magnetic and plasma wave signatures. *Nature Astronomy*, 2, 459–464.

Johnson, R. E. 1985. Polar frost formation on Ganymede. *Icarus*, 62, 344–347.

Johnson, R. E. 1997. Polar "caps" on Ganymede and Io revisited. *Icarus*, 128 469–471.

Khare, B. N., Thompson, W. R., Chyba, C. F., Arakawa, E. T. & Sagan, C. 1989. Organic solids produced from simple C/H/O/N ices by charged particles: Applications to the outer solar system. *Advances in Space Research*, 9, 41–53.

Khurana, K. K., Pappalardo, R. T., Murphy, N. & Denk, T. 2007. The origin of Ganymede's polar caps. *Icarus*, 191, 193–202.

Kivelson, M. G., Bagenal, F., Kurth, W. S. et al., 2004. Magnetospheric interactions with satellites. In Fran Bagenal, T. & E. D., William B. McKinnon (Eds.) *Jupiter. The Planet, Satellites, Magnetosphere*. Cambridge University Press, 513–536.

Kivelson, M. G., Khurana, K. K., Russell, C. T. et al., 1996. Discovery of Ganymede's magnetic field by the Galileo spacecraft. *Nature*, 384, 537–541.

Krüger, H., Krivov, A. V., Hamilton, D. P. & Grün, E. 1999. Detection of an impact-generated dust cloud around Ganymede. *Nature*, 399, 558–560.

Kuiper, G. P. 1957. Infrared observations of planets and satellites. *AJ*, 62, 245.

Laffon, C., Lacombe, S., Bournel, F. & Parent, P. 2006. Radiation effects in water ice: A near-edge x-ray absorption fine structure study. *Journal of Chemical Physics*, 125, 204714-204714-8.

Landau, A., Allin, E. J. & Welsh, H. L. 1962. The absorption spectrum of solid oxygen in the wavelength region from 12,000 Å to 3300 Å. 18, 1–19.

Langer, K. & Flörke, O. W. 1974. Near infrared adsorption spectra (4000-9000 cm^{-1}) of opals and the role "water" in these SiO_2 nH2O minerals. *Fortschritte der Mineralogie*, 52, 17–51.

Ligier, N., Paranicas, C., Carter, J. et al., 2019. Surface composition and properties of Ganymede: Updates from ground-based observations with the near-infrared imaging spectrometer SINFONI/VLT/ESO. *Icarus*, 333, 496–515.

Ligier, N., Poulet, F., Carter, J., Brunetto, R. & Gourgeot, F. 2016. VLT/SINFONI Observations of Europa: New insights into the surface composition. *AJ*, 151:163, 1–16.

Mastrapa, R. M. E., Grundy, W. M. & Gudipati, M. S. 2013. Amorphous and crystalline H2O-Ice. In *Gudipati M. & Castillo-Rogez J. (eds.) The Science of Solar System Ices,Astrophysics and Space Science Library, Springer Science+Business Media*, 356, 371–408.

Mccord, T. B., Carlson, R., Smythe, W. et al., 1997. Organics and other molecules in the surfaces of Callisto and Ganymede. *Science*, 278, 271–275.

McCord, T. B., Hansen, G. B., Clark, R. N. et al., 1998. Non-water-ice constituents in the surface material of the icy Galilean satellites from the Galileo near-infrared mapping spectrometer investigation. *JGR* 103, 8603–8626.

McCord, T. B., Hansen, G. B. & Hibbitts, C. A. 2001. Hydrated salt minerals on Ganymede's surface: Evidence of an ocean below. *Science*, 292, 1523–1525.

McCord, T. B., Hansen, G. B., Matson, D. L. et al., 1999. Hydrated salt minerals on Europa's surface from the Galileo near-infrared mapping spectrometer (NIMS) investigation. *JGR* 104, 11827–11851.

Molyneux, P. M., Nichols, J. D., Becker, T. M., Raut, U. & Retherford, K. D. 2020. Ganymede's far-ultraviolet reflectance: Constraining impurities in the surface ice. *JGR*, 125, e2020JE006476, 1–18.

Moore, J. M., Asphaug, E., Morrison, D. et al., 1999. Mass movement and landform degradation on the icy Galilean satellites: Results of the Galileo Nominal Mission. *Icarus*, 140, 294–312.

Moroz, V. I. 1965. Infrared spectrophotometry of the moon and the Galilean satellites of Jupiter. *Astronomicheskii Zhurnal*, 42, 1287.

Morrison, D. & Cruikshank, D. P. 1973. Thermal properties of the Galilean satellites. *Icarus*, 18, 224–236.

Mura, A., Adriani, A., Sordini, R. et al., 2020. Infrared observations of Ganymede from the Jovian infrared auroral mapper on Juno. *JGR*, 125, e2020JE006508, 1–17.

Musacchio, F., Saur, J., Roth, L. et al., 2017. Morphology of Ganymede's FUV auroral ovals. *JGR Space Physics*, 122, 2855–2876.

Nelson, R. M. & Lane, A. L., Matson, D. L. 1987. Exploring the universe with the IUE satellite. In KONDO, A. W. W. (ed.) *Planetary Satellites*, 129. Springer, 67–99.

Nelson, R. M., Lane, A. L., Matson, D. L. et al., 1987. Spectral geometric albedos of the Galilean satellites from 0.24 to 0.34 micrometers: Observations with the international ultraviolet explorer. *Icarus*, 72, 358–380.

Neukum, G. 1997. Bombardment history of the Jovian system. *The Three Galileos: The Man, the Spacecraft, the Telescope, Proceedings of the conference held in Padova, Italy on January 7-10, 1997*. Kluwer Academic, 201

Noll, K. S., Johnson, R. E., Lane, A. L., Domingue, D. L. & Weaver, H. A. 1996. Detection of ozone on Ganymede. *Science*, 273, 341–343.

Oberst, J., Schreiner, B., Giese, B. et al., 1999. The distribution of bright and dark material on Ganymede in relationship to surface elevation and slopes. *Icarus*, 140, 283–293.

Orton, G. S., Spencer, J. R., Travis, L. D., Martin, T. Z. & Tamppari, L. K. 1996. Galileo photopolarimeter-radiometer observations of Jupiter and the Galilean satellites. *Science*, 274, 389–391.

Pappalardo, R. T., Collins, G. C., Head, J. W. et al., 2004. Geology of Ganymede. In Bagenal, F. (ed.) *Jupiter. The Planet, Satellites and Magnetosphere*. Cambridge University Press, 363–396.

Pappalardo, R. T., Head, J. W., Collins, G. C. et al., 1998. Grooved terrain on Ganymede: First results from Galileo high-resolution imaging. *Icarus*, 135, 276–302.

Pilcher, C. B., Ridgway, S. T. & Mccord, T. B. 1972. Galilean satellites: Identification of water frost. *Science*, 178, 1087–1089.

Plainaki, C., Milillo, A., Massetti, S. et al., 2015. The H_2O and O_2 exospheres of Ganymede: The result of a complex interaction between the jovian magnetospheric ions and the icy moon. *Icarus*, 245, 306–319.

Pollack, J. B., Witteborn, F. C., Erickson, E. F. et al., 1978. Near-infrared spectra of the Galilean satellites: Observations and compositional implications. *Icarus*, 36, 271–303.

Postberg, F., Kempf, S., Schmidt, J. et al., 2009. Sodium salts in E-ring ice grains from an ocean below the surface of Enceladus. *Nature*, 459, 1098–1101.

Postberg, F., Khawaja, N., Abel, B. et al., 2018. Macromolecular organic compound s from the depths of Enceladus. *Nature*, 558, 564–568.

Poston, M. J., Carlson, R. W. & Hand, K. P. 2017. Spectral behavior of irradiated sodium chloride crystals under Europa-like conditions. *JGR*, 122, 2644–2654.

Prockter, L. M., Figueredo, P. H., Pappalardo, R. T., Head, J. W. I. & Collins, G. C. 2000. Geology and mapping of dark terrain on Ganymede and implications for grooved terrain formation *JGR*, 105, 29315–29325.

Prockter, L. M., Head, J. W., Pappalardo, R. T. et al., 1998. Dark terrain on Ganymede: Geological mapping and interpretation of Galileo regio at high resolution. *Icarus*, 135, 317–344.

Purves, N. G. & Pilcher, C. B. 1980. Thermal migration of water on the Galilean satellites. *Icarus*, 43, 51–55.

Quirico, E., Douté, S., Schmitt, B. et al., 1999. Composition, physical state, and distribution of ices at the surface of Triton. *Icarus*, 139, 159–178.

Roth, L., Saur, J., Retherford, K. D. et al., 2014. Transient water vapor at Europa's south pole. *Science*, 343, 171–174.

Sagan, C. & Khare, B. N. 1979. Tholins: Organic chemistry of interstellar grains and gas. *Nature*, 277, 102–107.

Schenk, P. M., Chapman, C. R., Zahnle, K. & Moore, J. M. 2004. Ages and interiors: the cratering record of the Galilean satellites. In Bagenal, F., Dowling, T. E. & McKinnon, W. B. (eds.) *Jupiter: The Planet, Satellites and Magnetosphere*. Cambridge University Press, 427–456.

Schenk, P. M. & McKinnon, W. B. 1991. Dark-ray and dark-floor craters on Ganymede, and the provenance of large impactors in the Jovian system. *Icarus*, 89, 318–346.

Schenk, P. M., McKinnon, W. B., Gwynn, D. & Moore, J. M. 2001. Flooding of Ganymede's bright terrains by low-viscosity water-ice lavas. *Nature*, 410, 57–60.

Shkuratov, Y., Ovcharenko, A., Zubko, E. et al., 2004. The negative polarization of light scattered from particulate surfaces and of independently scattering particles. *Journal of Quantitative Spectroscopy Radiative Transfer*, 88, 267–284.

Smith, B. A. and 65 co-authors 1989. Voyager 2 at Neptune: Imaging science results. *Science*, 246, 1422–1449.

Smith, B. A., Soderblom, L. A., Beebe, R. et al., 1979. The Galilean satellites and Jupiter: Voyager 2 imaging science results. *Science*, 206, 927–950.

Sohl, F., Choukroun, M., Kargel, J. et al., 2010. Subsurface water oceans on icy satellites: Chemical composition and exchange processes. *SSR*, 153, 485–510.

Sotin, C. & Tobie, G. 2004. Internal structure and dynamics of the large icy satellites. *Comptes rendus - Physique*, 5, 769–780.

Spencer, J. R. 1987. Thermal segregation of water ice on the Galilean satellites. *Icarus*, 69, 297–313.

Spencer, J. R., Calvin, W. M. & Person, M. J. 1995. CCD Spectra of the Galilean satellites: Molecular oxygen on Ganymede. *JGR*, 100, 19049–19056.

Spencer, J. R. & Klesman, A. 2001. New observations of molecular oxygen on Europa and Ganymede. *Bulletin of the American Astronomical Society*, 33, 1125.

Squyres, S. W. 1980. Surface temperatures and retention of H2O frost on Ganymede and Callisto. *Icarus*, 44, 502–510.

Stephan, K., Ciarniello, M., Dalle Ore, C. et al., 2019. Spectral properties of H2O ice depending on particle size and surface temperature: Comparison between laboratory data and icy satellites observations. *EPSC-DPS Joint Meeting 2019*, 13, EPSC-DPS2019-1493.

Stephan, K., Ciarniello, M., Poch, O. et al., 2021. VIS-NIR/SWIR spectral properties of H2O ice depending on particle size and surface temperature. *Minerals*, 11, 1328, 1–27, https://doi.org/10.3390/min11121328.

Stephan, K., Hibbitts, C. A. & Jaumann, R. 2020. H2O-ice particle size variations across Ganymede's and Callisto's surface. *Icarus*, 337, 113440, 1–21.

Stephan, K., Wagner, R., Hibbitts, C. A., Hansen, G. B. & Jaumann, R. 2008. Ganymede's impact crater Melkart: An example for a combination of high-resolution spectral and geological analyses in the outer solar system. *The Science of Solar System Ices (ScSSI): A Cross-Disciplinary Workshop* Abstract No. #9060.

Vance, S., Bouffard, M., Choukroun, M. & Sotin, C. 2014. Ganymede's internal structure including thermodynamics of magnesium sulfate oceans in contact with ice. *PSS*, 96, 62–70.

Wagner, R. J., Schmedemann, N., Werner, S. C. et al., 2019. Ray craters on Ganymede as time-stratigraphic markers. *50th Lunar and Planetary Science Conference*, LPI Contribution 2132 1849.

Warren, S. G. 1984. Optical constants of ice from the ultraviolet to the microwave. *Applied Optics*, 23, 1206–1225.

Zahnle, K., Schenk, P., Levison, H. & Dones, L. 2003. Cratering rates in the outer solar system. *Icarus*, 163, 263–289.

11

Physical Chemistry and Thermal Properties of Ices at Ganymede

CAITLIN AHRENS, CHARLES HIBBITTS, KLÁRA KALOUSOVÁ, NICOLAS LIGIER, THOMAS MCCORD, ANEZINA SOLOMONIDOU AND KATRIN STEPHAN

11.1 INTRODUCTION

Icy bodies of the outer solar system, including satellites of the gas giants, harbor surface ices made of volatile molecules, clathrates, and complex molecules like hydrocarbons. Imaging of Ganymede's surface, as observed by the Voyager flyby and Galileo orbiter, reveals complex and mysterious geological structures composed of these materials and shaped by intricate physical processes. Understanding the nature of these geologic features on the Ganymedean surface and establishing their formational timescales and processes requires knowledge of their thermophysical and rheological properties. Rheological information currently rests on laboratory-derived empirical relationships, which are essential for landform evolution and thermal modeling. Major science questions at Ganymede (and other icy satellites) and respective laboratory-appropriate research offers a unique perspective for just how interdisciplinary thermophysical and rheological studies can be, as listed in Table 11.1.

The Pioneer and Voyager flyby missions paved the way for NASA's Jovian system spacecraft Galileo (Mura et al., 2020). This spacecraft acquired results by an onboard magnetometer (MAG), solid-state imaging camera (SSI), near-infrared mapping spectrometer (NIMS) operating in the range 0.7–5.2μm, and radio science experiment (Carlson et al., 1992). Further observations at infrared wavelengths of Ganymede were obtained by the flybys of the Cassini Visual and Infrared Mapping Spectrometer (VIMS) (Brown et al., 2004) and the New Horizons Linear Etalon Imaging Spectral Array (LEISA) (Grundy et al., 2007). From these instruments, the interior and surface processes of Ganymede have been estimated (see Chapter 2.1).

Ganymede's low normalized moment of inertia (Anderson et al., 1996) and dynamo-driven magnetic field (Kivelson et al., 2002) suggest that it experienced enough heating for differentiation of a liquid iron core from the silicate mantle (see Chapter 2.1). Vance and colleagues (2018) modeled that the thickness of the hydrosphere probably exceeds 800 km, with the rock–ice interface at $\sim$1.5 – 1.7 GPa. The internal ocean, sandwiched between the outer ice shell and deep high-pressure (HP) ice layer (e.g., Vance et al., 2014), is likely at a depth of $\sim$160 km (ice I–III liquid triple point), although its actual depth and thickness are not known (Barr and Pappalardo, 2005). Ice shell thickness depends mainly on the rheological parameters of ice I (Goldsby and Kohlstedt, 2001; Barr and Pappalardo, 2005; see also Chapter 2.7).

The salinity of Ganymede's ocean is closely linked with the structure of the HP ice layer and the stability of ice phases. Several studies have noted that briny fluids under pressure can have densities larger than those of HP ices (Journaux et al., 2013; Vance and Brown, 2013; Vance et al., 2014). While dense, salty fluids between HP ices are thermodynamically plausible, the stability of such fluids requires further geodynamic modeling like that performed for purer oceans (Kalousová et al., 2018). In addition, the exchange processes between the silicate mantle and the ocean may be affected by the presence of salts in the melt between the HP ice layer and the silicate crustal surface.

The surface of Ganymede (with a temperature range $\sim$90 – 160 K) is a mix of water ice phases and darker non-ice materials (Kieffer and Smythe, 1974; Clark and McCord, 1980; Mura et al., 2020). Reflectance spectra in the visible, infrared, and ultraviolet (namely from the NIMS aboard the Galileo spacecraft) indicate not only water ice but perhaps also hydrated salts like on Europa (McCord et al., 2001) or other hydrated materials (Carlson et al., 1999). Carbon dioxide (CO_2), sulfur (S = O bonds), oxygen (O_2), and ozone (O_3) molecules are also detected by visible and ultraviolet absorption features (Nelson et al., 1987; Spencer et al., 1995). While Hapke modeling is used as an estimation of the uppermost regolith physical properties (Hapke, 1993), this is still dependent on the grain size, albedo effects, and optical properties of the regolith, which are poorly studied in an experimental setting.

In the thermophysical and rheological sense, ice can be considered both a mineral and rock, as can the solar system's more volatile condensed species and hydrates. Past and present internal and surficial activity on the icy moons is driven by not only the frozen volatile components and respective phases, but also by internal and external energy sources (e.g., radiogenic, gravitational, tidal), and includes planetary surface-scale convection to localized relaxation of surface topography (Durham et al., 2010). Such landforms relax under gravitational forces controlled by the viscosity of the ice. Large impacts play several important roles for the rheological state of Ganymede's surface. An impact may trigger the merging of the ice I and HP ice layers and allow the warm ice to the near surface and fracture the crust, where the soluble impurities in the ocean would have otherwise made this spontaneous overturn impossible. Impacts can also puncture the lithosphere, allowing underlying warmer ice to flow onto the surface. This brings about an interesting relationship between exogenic and endogenic events of Ganymede (Murchie et al., 1990; Spencer et al., 1995).

Various ice phases and non-ice constituents have been estimated to fracture and flow, though understanding the material physics of the icy grains and rheological properties is difficult at low temperatures (see Section 11.5) (Parmentier, 1982; Jankowski and Squyres, 1988; Melosh and Janes, 1989). The purpose of this chapter is to present a current understanding of the temperature-dependent input thermophysical and rheological quantities and their implications for

Table 11.1 *General overview of science questions at Ganymede in relation to respective experimental thermophysical properties*

Ganymede Science Questions	Laboratory Measurements/Properties
What is the composition and structure of the subsurface material?	Dielectric constant, rheology
What is the composition and structure of the surface material?	Rheology, density, thermodynamic phase transitions
What power processes drive geological evolution?	Diffusivity, conductivity, rheology, phase transitions
What are the radiative processes and surface evolution over short timescales?	Heat capacity, sublimation pressures, phase stability, latent heat

geomorphological processes on the Ganymedean surface. In most experimental work, this information is more complete in the pure ice forms, but currently poorly known are the complex mixtures that these ices may form.

11.1.1 Rheology in Cryovolcanism and Tectonism

As described in other chapters throughout this book, Ganymede possesses several tectonic and possibly cryovolcanic terrains in which understanding the thermodynamic and rheologic nature of the ices is vital to understanding the lithospheric properties for the resurfacing and tectonic strains involved.

The surface of Ganymede is rife with various levels of resurfacing from smooth to grooved terrains (see more in-depth reviews of these features in Chapter 2.2). On icy bodies in general, cryovolcanism is either the end-member of effusive eruptive deposits or explosive (Smith et al., 1982, 1986; Squyres et al., 1983; Croft et al., 1995). The eruption of liquid constituents through an ice-rich crust may seem difficult given the buoyancy difference of the melt compared to the solid (including the possible presence of contaminants, such as hydrates, that would vary its relative density), but can be accommodated due to the presence of volatiles that exsolve to decrease the density of the melt, or by the pressurization of a crustal melt or partial crystallization (Fagents et al., 2000).

The bright terrains present on Ganymede's surface may be indirect evidence for volcanic resurfacing, even with the lack of source vents and flow fronts. Resurfacing of smooth terrain has been noted to be analogous to lunar mare with shallow flooding from low-viscosity lavas and a general lack of major edifice structures (Schenk et al., 2001), and may share similar qualities such as: (i) occupation at low-lying regions, (ii) embayed or partially buried craters and other rugged terrains, and (iii) large numbers of completely buried small craters. The subsurface sources of smooth floors at the centers of impact basins and caldera-like structures are also unknown (Schenk and Moore, 1995; Schenk et al., 2001).

Extensional tectonism is responsible for the down-dropping blocks of dark terrain, extending both dark and bright terrains and resurfacing via grooved textures (e.g., flat-floored graben and flat-topped horst ridges like those at Uruk Sulcus) (Collins et al., 1998; Schenk et al., 2001). Images from Galileo have aided in the observation of brittle faulting (and some domino-style faulting) at the surface with additional imbrication of these faults on a scale of greatest local strain (Golombek, 1982; Pappalardo et al., 1998; Patel et al., 1999), which also suggests that similar lithospheric properties are across Ganymede at the time of grooved terrain development (Grimm and Squyres, 1985). Flexural modeling suggests local heat fluxes $\geq 30 - 45$ mW m^{-2} during deformation, consistent with extensional instability models, though this still depends on the ice phases and localized extensional tectonism (Dombard and McKinnon, 2001; Nimmo et al., 2002; Pappalardo et al., 2003; Bland and Showman, 2007).

11.1.2 Landforms with Rheological Interest

It is with the variations of these Ganymedean morphologies that thermodynamical and rheological information must be required, which are discussed throughout this chapter. Here we discuss a brief overview of the different geologic structures that would have rheological aspects.

The fascinating bright furrowed terrain, called *sulcus* (pl. *sulci*), provides interesting insights into resurfacing and tectonic mechanisms (see Chapters 1.5 and 2.2 for more geologic information). Bright regions not necessarily associated with sulci are faculae, which also have implications of tectonism and resurfacing processes, typically labeled as palimpsests. The current implications of the emplacement of such brighter albedo formations include cryovolcanism or extensive heavily tectonized ice (Shoemaker et al., 1982; Collins, 2000; Schenk et al., 2001). Such bright terrain burying older features indicates smooth terrains to be formed by flooding in shallow troughs with low-viscosity water ice lavas (Schenk et al., 2001; see Section 11.5.3). Sippar Sulcus exhibits topographic characteristics of icy volcanic resurfacing, consistent with low-viscous flooding of structural troughs to form the smooth terrains (Schenk et al., 2001). Such tectonism and deformations at sulci and faculae depend on the density and stresses of the ices (see Sections 11.5.2 and 11.5.3, respectively).

The surface of Ganymede comprises a variety of crater formations, including dark-floor, pedestal, palimpsest, penepalimpsest, multi-ring, and central peak (or dome) craters (McKinnon and Melosh, 1980; Schenk and McKinnon, 1991; Schenk et al., 2004; see Chapter 2.3). Numerical models (temperature-dependent and a power law stress-dependent viscosity for ice) of viscous relaxation of craters were purported to have relatively short relaxation times (t_e) for some crater depth to become reduced from an initial impactor state (Dombard and McKinnon, 2006). Newtonian rheologies appear to essentially dominate crater relaxation activity, though this applies when the effective temperature is several tens of degrees warmer than the surface (e.g., insulating regolith) or where the temperature gradients are considerably small. This tends to be more so for crater diameters < 25 km (Thomas and Schubert, 1988). However, non-Newtonian rheological laws can occur at larger craters and have shorter relaxation times (e.g., $t_e \sim 10^7$ years for crater diameters < 100 km). From Dombard and McKinnon (2006), relaxed craters in the Marius Regio suggest ancient heat flows with $\gtrsim$ 10 mW m^2 (see Section 11.5.3). However, compositional constraints (e.g., silicious content) and dispersion hardening can extend these relaxation times. Modeling from Thomas and Schubert (1988) shows that a threshold crater diameter for the onset of non-Newtonian viscous behavior could estimate not only thermal gradients on a localized scale, but also localized stresses, although the variable resolution from Galileo plus the current lack of experimental viscous ice research make this problem unresolved.

Fossae (long, narrow depressions) and paterae (irregular, scalloped depressions) offer more interesting geomorphological structures relating to just how rheologically complex the surface of Ganymede is. Such scalloped textures could be considered caldera-like where these are sites of potential cryovolcanic vents (Lucchitta, 1980; Schenk and Moore, 1995). Fossae and their orientations also offer an interesting case for shear stresses (see Section 11.5.3) and brittleness of density-variable crustal material.

Dark terrain (*regiones*) comprises nearly 1/3 of the Ganymedean surface with a greater fraction of non-ice and rocky material and showing a variety of modification by processes such as sublimation, mass wasting, ejecta blanketing, and tectonism (Parmentier, 1982; Prockter et al., 2000). Based on impact crater counts, the dark terrain is considered ancient with an age > 4 Gy (see Chapter 2.3 for more details) (Zahnle et al., 2003). Also due to its low, heterogeneous albedo, this suggests thermally driven segregation of ice and non-ice constituents (Spencer, 1987). For example, as observed in Pappalardo et al. (2004), Nicholson Regio imaged by Galileo is heavily tectonized with furrows (e.g., Anshar Sulcus, Byblus Sulcus), implying preexisting weaknesses of the ice and non-ice materials.

11.2 WATER ICE AND OTHER VOLATILE PHASES

11.2.1 Water Ice Phases

The evolution of the outer solar system icy bodies has been significantly affected by water ice, which can exist in three different solid configurations (amorphous, cubic, hexagonal) at low temperatures and pressures (Jenniskens et al., 1998). These different phases present large differences in their nature and behavior and have a considerable impact on the physical properties of ices on the solar system (Mastrapa et al., 2013). Nevertheless, the presence of amorphous and/or crystalline (cubic, hexagonal) water ice on a body depends on many factors, such as condensation temperature/rate, and its thermal history connected to the plethora of forms of irradiation, impact heating, and thermal variations (e.g., Jenniskens et al., 1998).

Amorphous ice has also been observed in shallow (depths ofμm) layers of icy body surfaces. Different flavors of amorphous ices exist as well, including the high-density glassy water (HDG), which forms by vapor deposition at low temperatures (< 110 K) and irradiation, and could be a major phase present in many icy satellites (Palumbo, 2005; Giovambattista et al., 2013).

During heating to a temperature between 110 and 150 K, amorphous ice transforms itself irreversibly to a cubic crystal structure and above 190 K, another irreversible transformation takes place to the familiar hexagonal crystal structure (Jenniskens et al., 1998). Water ice frozen from liquid water ice is always expected to crystallize with the hexagonal structure and it maintains that structure, even when cooled to cryogenic temperatures (Schmitt et al., 1998). Hexagonal ice should be particularly abundant at the surfaces of solar system objects, freezing from standing water, cloud droplets, erupted water-rich melts, or impact melts (Schmitt et al., 1998).

However, even though one would expect energetic radiation to destroy the crystal structure and thus the remaining crystalline water ice on the Jovian, Saturnian, and Uranian icy satellites, it predominantly exists at the very shallow depths (mm) of their surfaces (Grundy et al., 1999; Mastrapa et al., 2013). The reason crystalline water ice can exist under such large doses of radiation and thermal variations is that at these locations, the temperature on the satellites reaches above 80 K, where amorphous ice can recrystallize (Hansen and McCord, 2004).

The presence of water ice also affects the surface dynamics and morphology depending on its phase, especially in the bright terrains of Ganymede (Allison and Clifford, 1987). Ganymede's total water abundance in the ice is ~30 – 50 wt.%, nearly up to ~90 wt.% in the shallow subsurface (in depths of meters; e.g., McCord et al., 1998). There are also climatological dynamics at the polar regions, the result of transportation from the warmer equator to higher latitudes (Clark, 1980; Calvin et al., 1995). The spectral signature (discussed in Chapter 2.5) can also tell us about grain sizes, ranging from a few hundred micrometers to a millimeter (Pollack et al., 1978; Calvin et al., 1995; Hendrix et al., 2013), where the equator and the trailing hemisphere of Ganymede have considerably larger grains of ice (Stephan et al., 2021). Note reflectance spectra and deconvolution techniques to determine ice abundances and phase species require some form of physical parameters obtained from experimental research, such as optical constants of different materials (Grundy and Schmitt, 1998; Dalton and Pitman, 2012), though these are temperature-dependent, and unfortunately lack experimental data at low temperatures (Umurhan et al., 2021).

The global pattern of ice phases present on Ganymede's surface indicates a balance between crystallization and disruption processes. On Ganymede, mainly crystalline water ice is present at the equator, where relatively high temperatures and lower radiation flux exist, whereas amorphous water ice can be predominantly found at Ganymede's poles, which experience relatively low temperatures and higher radiation flux (Hansen and McCord, 2004; Ligier et al., 2019). According to Jenniskens and colleagues (1998), as temperature varies between 105 K and 125 K, the crystallization of ice can vary by a factor of $\sim 10^5$. On Ganymede, the equatorial midday ice temperature varies by around 20 K, but from the equator to the poles, the temperature varies by approximately 10 K. This implies that a significant variation in crystallization rate as a function of temperature in the midlatitudes and polar regions is largely absent on Ganymede (Hansen and McCord, 2004). Hence, it seems possible that other factors are controlling the disruption and crystallization functions on Ganymede, and these are not simply controlled by the thermal crystallization rates (Hansen and McCord, 2004).

The triple point of H_2O is at 273.16 K and ~6.116 x 10^{-3} bar (Wagner and Pruss, 2002). Fray and Schmitt (2009) collected several experimental measurements ranging from 130 K to 273.15 K. Numerous experimental relations have been published to compute the vapor pressure of hexagonal water ice (I_h); however, the vapor pressure of cubic water ice (I_c) is ~10.5% higher than for I_h (Washburn, 1924; Yamamoto et al., 1983; Murphy and Koop, 2005; Huebner et al., 2006). Note that the I_c phase is a metastable version of I_h (Petrenko and Whitworth, 1999). All these works, and those included in the Fray and Schmitt (2009) summary, have also shown that amorphous ice strongly depends on condensation temperature and the rate of condensation. This dependence on condensation rate and temperature conditions basically applies to all forms of HP ices as well (Mastrapa et al., 2013). From the observations of Mastrapa and colleagues (2013), we know amorphous ice will convert to crystalline structures within 10^4 years, wherefore amorphous ices are only consistently stable at colder (<100 K) surfaces.

What is interesting for Ganymede, and for icy bodies in general, are their hydrospheres, where there are variable amounts of pressure (i.e., 0 – 1 600 MPa). At this pressure range exist several icy polymorphs, including HP ices (I_h, I_c, II, III, IV, V, VI, IX, XI, XII, XIII, XIV, and XV) (Schmitt et al., 1998; Nakamura and Ohtani, 2011), as well as a suite of clathrate hydrates (see Section 11.3). Ice phases relevant to Ganymedean conditions are shown in Figure 11.1. While ice VI is expected to occur for any ocean composition (pure H_2O or aqueous solution of $MgSO_4$ or other relevant salts) and thermal profile, the presence of ice V and III is predicted only for dense salty oceans and cold interiors (Vance et al., 2018). Ice III can form at the base of the ocean underlain by ice V and VI, if the ocean is cold and its salinity is ~ 10 wt.% $MgSO_4$(Vance et al., 2018). However, Ice III can be buoyant relative to modeled oceans, where dynamically short-lived ices may perhaps quicken the cooling of Ganymede and thicken the ice I crust, therefore accumulating underneath the ice I–ocean interface. In Figure 11.1, the gray area depicts the conditions expected for large ocean worlds. This area encompasses ice phases I, II, III, V, and VI, and the water liquid phase. This area also includes conditions for the existence of ammonia hydrate dissociation (in the monohydrate phase), and other various clathrates at temperatures ~280 – 300 K.

11.2.2 Other Volatile Species

Oxygen-type compounds have been found trapped in the surface ice within voids or bubbles, with O_2 predominantly existent in the equatorial region and O_3 dominating at the poles (Nelson et al., 1987; Spencer et al., 1995; Noll et al., 1996; Johnson and Jesser, 1997; Hendrix et al., 1999) and have been interpreted to form by a combination of solar UV radiation and impacting magnetospheric plasma. Solid O_2 has three different crystalline phases (Table 11.2); the temperatures of the phase transitions are 23.78 K (α-β transition) and 43.77 K (β-γ transition) (Mullins et al., 1963).

Carbon dioxide presumably trapped in Ganymede's dark material is widely distributed (± 60° from the equator; Hibbitts et al., 2003). Carbon dioxide has a triple point located at 216.58 K and 5.18 bar. Experimental and empirical data points have been collected from 69.69 K to 216.56 K (Table 11.2), listed in Fray and Schmitt (2009). However, its origin on Ganymede is not clear. See Chapters 2.5 and 3.2 for a detailed discussion of the identification, distribution, mixing, and formation processes of these compounds. From observations by Hibbitt and colleagues (2003), CO_2 distributions appear to be controlled by

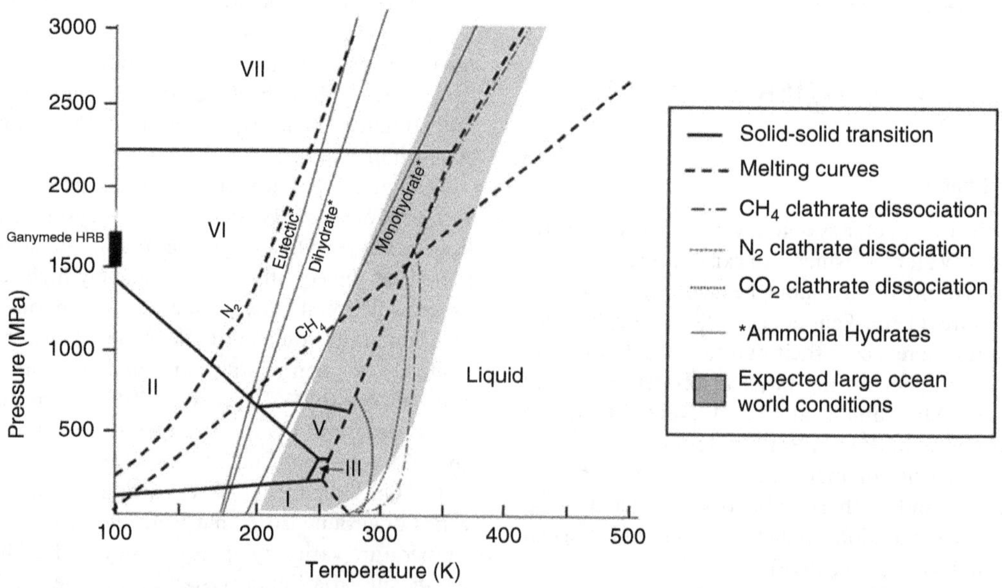

Figure 11.1 Temperature/pressure phase diagram of water ice polymorphs relevant to water-rich satellite interiors. Melting curves and solid-solid phase transitions are from Journaux and colleagues (2020). The shaded region corresponds to the estimated range of conditions expected in ocean worlds. Clathrate dissociation curves are from Durham and colleagues (1998). The hydrosphere–rock boundary (HRB) is adapted from Journaux and colleagues (2020). Baptiste Journaux et al. (2020). Large Ocean Worlds with High-Pressure Ices. *Space Science Reviews*. Reproduced with permission from Springer Nature.

Table 11.2 *Temperature and pressure ranges for non-H_2O volatile species observed on Ganymede. Values from Fray and Schmitt (2009)*

Species	Solid Phase	Temperature Range (K)	Pressure (bar)
O_2	α	20 – 23.78	$(1.49 \pm 0.03) \cdot 10^{-3}$
	β	23.78 – 43.77	
	γ	43.77 – 54.33	
O_3		< 79.6	$(5.6 \pm 0.6) \cdot 10^{-6}$
CO_2		40 – 194.7	5.185 ± 0.005
		194.7 – 216.58	
H_2S	I	126.2 – 187.57	0.229 ± 0.008
	II	103.6 – 126.2	
	III	< 103.6	
SO_2		15 – 197.63	0.0167 ± 0.0001

geological processes, with combination effects from the Jovian (and Ganymedean) magnetospheres, finding that: (i) the deepest CO_2 bands are found in darker, non-icy material regions; (ii) CO_2 is not detected in the icy polar regions; (iii) Ganymede's upper crust is depleted of CO_2 (the exception is the Mir crater); (iv) dark crater ejecta appear to be depleted in CO_2; (v) greater concentrations of CO_2 are correlated with areas of larger-grained ice; (vi) CO_2 detected by NIMS is mostly contained in the non-ice constituents; and (vii) CO_2 could exist in the large-grained ice near the equatorial and midlatitude regions, but this remains undetected by NIMS data due to the too-large grain sizes.

Sulfuric acid hydrate and other sulfur compounds, including sulfur dioxide (SO_2), might be endogenic or could be the product of the bombardment of the icy surface by exogenous sulfur, as well as sulfur molecules released by Io's very active volcanoes (Calvin et al., 1995; Strazzulla, 2011; Ligier et al., 2019). Radiolysis can also play a major role in the forming of $SO_2$and H_2SO_4 in either the ice or in water associated with Ganymede's hydrated non-ice material and the implantation of S-ions (Hibbitts et al., 2003; Ligier et al., 2019). The triple point of SO_2 is 197.64 K and 0.0167 bar. The heat capacity of the solid has been measured (15.20 – 197.64 K) in Giauque and Stephenson (1938). Hydrogen sulfide's (H_2S) triple point is at 187.57 K (0.229 bar). The H_2S exists in three different crystalline forms (Anderson et al., 1977), where phase III transforms to phase II at 103.6 K and to phase I at 126.2 K.

11.3 HYDRATED SALTS AND CLATHRATE CONSTITUENTS

Hydrates may have reached the surface on Ganymede at some point in the satellite's history. These briny deposits on the surface as observed from Galileo NIMS suggest movement from a conductive fluid layer, which is consistent with the interpretation of results from the Galileo MAG investigations (McCord et al., 2001; Pappalardo et al., 2004). This is also prevalent in the identification of observed surface disruptions, like those seen on Europa (McCord et al., 2001). Several studies have predicted such salts and hydrates to exist through exogenic and endogenic processes (Kargel et al., 2000; Fanale et al., 2001; Strazzulla et al., 2009). The thermodynamic modeling of these hydrates remains relatively unknown at higher pressures, where the solubility and activity of many planetary-relevant gases are poorly constrained, particularly at lower temperatures (Sloan and Koh, 2007; Choukroun et al., 2013).

Hydrated sulfates are anticipated to be a major component in rock-forming minerals in the interiors of the inner Galilean moons, probably a result of the leaching of chondritic sulfates from their cores during differentiation (Kargel et al., 1991). Ganymede's magnesium-bearing sulfates and salts are like those observed on Europa (McCord et al., 1998). Similarly with sulfuric acid hydrates, the abundance on Ganymede is lower than on Europa (< 36%; Dalton et al., 2013).

Nakamura and Ohtani (2011) identified several phases in the $MgSO_4$–H_2O system up to 4 GPa at 298 K through laboratory experiments: Ices VI and VII, $MgSO_4$–$7H_2O$ (magnesium heptahydrate), and a liquid phase. These phase relations suggest that the liquid layer may be at a depth of ~800 km in the interior of Ganymede. The $MgSO_4$ content of the bulk icy mantle composition is ~15 – 20 wt.% (as modeled by Kargel et al., 1991), which is higher than the $MgSO_4$ eutectic (12 – 14 wt.% at 2 GPa), so its liquidus phase is likely ($MgSO_4$–$7H_2O$, referred to as *MS7*) hydrate at the base of the mantle, where ($MgSO_4$–$11H_2O$, referred to as *MS11*) is at shallower depths (Nakamura and Ohtani, 2011). The eutectic between MS11 and ice I_h is at 269.25 K with 17.5 wt.% $MgSO_4$ (Fortes et al., 2010).

The low eutectic temperature indicates that the icy mantle may be molten, if not partially crystallized, and so the crystallization of the outer icy mantle is inevitable in the early stages of Ganymede's geologic evolution. A fractional crystallization of the icy mantle will produce a layered structure where the MS7 phase is at the bottom, an internal ocean close to the eutectic composition, covered by an icy crust (Fortes et al., 2010; Nakamura and Ohtani, 2011). Based on Zolotov and Shock (2001), such layering could lead to Cl-salts at the top. However, the thickness of the icy crust and that of the bottom layer depend on the temperature profile of Ganymede, silicate grains and other possible hydrate mixtures involved, and heat flux from the inner silicate layer (where the internal ocean would be deeper) (Hussmann et al., 2006; Bland et al., 2009).

Clathrate hydrates are essentially skeleton-like cages of hydrogen-bonded H_2O molecules that trap neutral guest molecules (e.g., CH_4, N_2, CO_2, etc.). No bond occurs between the host and guest species, but a van der Waals interaction stabilizes the structure. The cage structure depends on the pressure and temperature conditions and the nature of the guest species, where several species may occur at the same void space. Carbon dioxide presents an interesting case as a flavor of clathrate in that a single CO_2 hydrate structure destabilizes into a filled, solid ice structure <1 GPa (Amos et al., 2017; Massani et al., 2017). At conditions >1 GPa, H_2O and CO_2 form a solid mixture (Hirai et al., 2010). However, at >3 GPa, a change in the CO_2 speciation leads to greater solubility of C and the appearance of crystalline, hydrated carbonic acid (Abramson et al., 2018).

11.4 THE "DARKENING" COMPONENT

Before the arrival of Galileo, a great number of studies proposed that the non-ice (or frost-free) areas seem to be covered with a fine dust of carbonaceous chondritic material derived from ejecta from the outer irregular satellites, hence ancient meteoritic material (Hartmann, 1980; Squyres, 1980; Schenk and McKinnon, 1991). Later, Galileo high-resolution images suggested that the dark material is composed of a relatively thin dark lag deposit of clays and organic materials overlying brighter icy material (McCord et al., 1997; Patterson et al., 2010). Roughly 34 – 35% of Ganymede's surface is ancient dark terrain that consists of a greater fraction of non-ice and rocky material (Collins, 2000). Murchie and colleagues (1990) suggest that ammonia-rich fluids may also play a role in the dark deposits.

In a geophysical sense, dark terrain is suggested to be the product of a relatively thin lag deposit that punched through a thinner, more brittle lithosphere into a more mobile material (usually observed at the furrowed regions on Ganymede) and affected by a multitude of processes, such as sublimation, mass wasting, and tectonism (McKinnon and Melosh, 1980; Prockter et al., 2000). As water vapor is removed, a refractory material forms a sublimation lag and thus a darker surface. Over time, this low-refractory lag deposit may thicken for it to slough downslope. Based on the retention of small craters in dark terrains, the depth of the lag deposits is probably a few meters on slopes, but thicker in the topographic lows at furrowed regions (Pappalardo et al., 2004).

These dark clays and organic components are also suggested to be endogenic at regions with increased fracturing or reactivation of preexisting structures (Shoemaker et al., 1982; Murchie et al., 1986; Prockter et al., 2000). As mentioned earlier, the non-ice compounds could also originate from cryovolcanism, with the embayment of large diameter craters and smooth areas around furrows (Croft and Strom, 1985; Murchie et al., 1989, 1990; Lucchitta et al., 1992). Understanding that this may hinder low-albedo lag deposits or thicken from endogenic processes presents its own complexities as to the role this dark material would play in the overall rheology.

11.5 RHEOLOGICAL CONSIDERATIONS

11.5.1 Microstructure, Composition, and Polymorphism

As we have reviewed, the surface of Ganymede presents a variety of ice polymorphs and non-ice constituents. The rheology of ice is mainly controlled by temperature, grain size, porosity, and stress with the corresponding material parameters highly influenced by impurities (Kubo et al., 2009). Although the observed impurities exist only in small quantities, they could play a role in the deformation of Ganymede's crust, especially if they are transported deeper into the icy crust. Even more macroscopic mass transport of trapped species by tectonic processes as supposed on Europa (Prockter and Pappalardo, 2000) could also carry impurities to Ganymede's ocean (see Chapter 2.7).

In the case of water ice (both crystalline and amorphous), it is accepted that the interatomic cohesive forces are generally strong in H_2O ice crystals owing to the strength and complexity of hydrogen bonds, and that the presence of dislocation or surface heterogeneities is the main cause of relative weaknesses (Macmillan, 1972). The polymorphism of water ice, as described in this chapter, results in the formation of concentric ice shells in the outermost layers of a differentiated icy satellite consisting of a variety of phases (and thus densities) of ice (Fortes and Choukroun, 2010). Journaux and colleagues (2020) have made considerable pressure-volume-temperature computations for ices III, V, and VI, to determine the values of the main thermodynamic properties (density, thermal expansivity, heat capacity, etc.) for 250 – 300 K temperatures using the computational SeaFreeze package that uses the Gibbs energy representation to derive all thermodynamic properties at proper P-T conditions. This approach constrains material parameters in a wide range of conditions and prevents their extrapolation outside the range of laboratory experiments, which is not fully reliable.

Chemistry also influences HP ices. For example, ice VI is the only HP polymorph with ice VII to have been investigated for salt incorporation. Journaux and colleagues (2017) found that even small amounts of salts incorporated can have a significant effect on lowering its density. The extent to which other planetary-relevant salts (e.g., $MgSO_4$, $MgCl_2$, Na_2SO_4, etc.) incorporate into HP ices remains unconstrained.

11.5.2 Density

The chemical differentiation and thermal evolution of the HP ice layers, and the silicate core (and any subsequent impurities), depend greatly on the relative densities of important phases and whether they tend to sink or float and (if large enough) become isolated from the residual liquids. The low viscosity and large density differentials of phases in saltwater brines should result in effective fractional crystallization of brine flows, plus water ice would float to the top while the salt hydrates would sink. Other works on density estimates can be found for water ice polymorphs (Bridgman, 1912; Journaux et al., 2020), ammonia-water systems (Croft et al., 1988), and saline solutions (Kargel et al., 1991; Hogenboom et al., 1995).

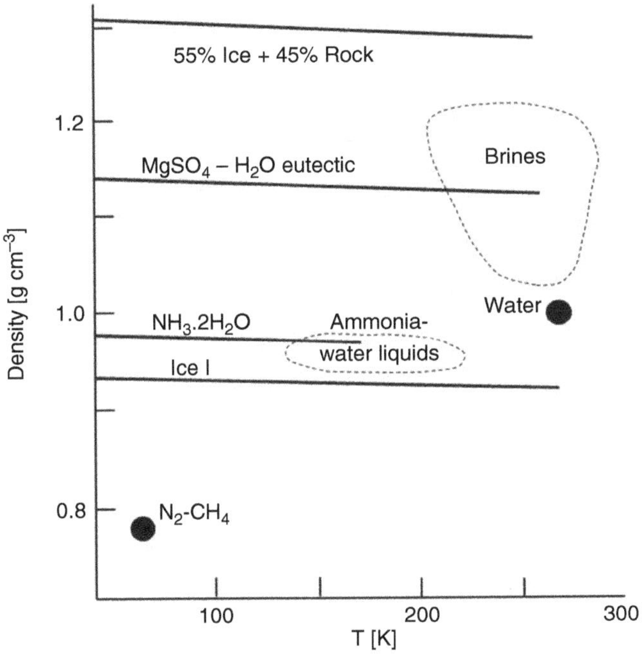

Figure 11.2 Ice I and cryofluid material densities across temperatures relevant to Ganymede. Also included is the $MgSO_4$-H_2O eutectic. Adapted from Kargel, J. S. (1998). Physical chemistry of ices in the outer solar system. *Astrophysics and Space Science Library*. Reproduced with permission from Springer Nature.

For water ice polymorphs, Jenniskens and colleagues (1998) documented comparative density values between crystalline and amorphous water ice phases. For hexagonal ice (I_h), the density is ~0.93 g cm^{-3}, comparably the same as cubic ice (I_c), both at temperatures 120 – 150 K. Main phases of amorphous ice are based on density: I_ah (high-density amorphous) = 1.1 g/cm^3 (Narten et al., 1976) and low-density amorphous (I_al) = 0.94 g/cm^3 (Jenniskens and Blake, 1994). However, these experiments did not take porosity into account. At 90 K, the bulk density of water ice (from vapor deposition) is 0.82 g/cm^3, while at 35 K the density is 0.74 g/cm^3 (I_h) or 0.68 g/cm^3 (I_al with higher deposition rates) (Narten et al., 1976; Berland et al., 1996; Brown et al.). At Ganymede, the upper HP ice layers (ice III) are modeled to be ~1.166 g/cm^3, while the lower, deeper ice VI is ~1.396 g/cm^3 (Journaux et al., 2020).

In the pure H_2O setting, all HP phases are denser than liquid water. However, the inclusion of salts (e.g., $MgSO_4$) would significantly alter the liquid density, which may lead to density inversions and formation of layers (Vance et al., 2014, 2018). For non-ice materials, Kargel (1998) has summarized several densities for salts and saline solutions pertinent for briny volcanism and large ocean worlds. Figure 11.2 depicts several potential hydrate and salt solutions that can be present at Ganymedean surface conditions.

11.5.3 Ductile, Brittle, and Elastic Behavior

From a thermodynamic perspective, if the ice flows readily internally, the satellite is cooling. If the ice resists, the satellite will overheat, or essentially melt internally and increase its differentiation (Durham et al., 2010). Such warmer ice flowing in the subsurface against a colder crust may be sufficient to overcome the strength of the crust, which will cause crustal failure. Ice flow owes its origins to glaciologic laboratory studies to create the Glen law for the creep behavior of polycrystalline ices, including the applications of strain rate and shear stresses.

Table 11.3 *Deformation behavior coefficient values by grain boundary mechanism type. *Does not depend on grain size*

Grain Boundary Mechanism	*n*	*p*
Dislocation creep	4	—*
Grain-boundary sliding	1.8	1.4
Basal slip	2.4	0
Diffusion creep	1	2

Creep is a general term in which volume-conserving deformation mechanisms involve the motion of grain-boundary or crystal-lattice defects (Poirier, 1985). The deformation mechanism depends on the local stress and temperature conditions, as well as on the ice grain size. Based on laboratory experiments, three distinct mechanisms were described (e.g., Durham et al., 2001; Goldsby and Kohlstedt, 2001; Umurhan et al., 2021): grain boundary sliding, basal slip, and dislocation creep. All these mechanisms are nonlinear – that is, strain-rate $\dot{\varepsilon}$ and stress σ are not linealy related and are scalar measures of tensor quantities:

$$\dot{\varepsilon} = \frac{A}{d^p}\sigma^n \exp\left(\frac{-Q}{RT}\right) \tag{11.1}$$

Here A is a pre-exponential constant, d is the grain size with p its exponent, n is the stress exponent, Q is activation energy, R is the universal gas constant, and T is the temperature. Viscosity is proportional to the ratio between stress and strain rate, and is thus stress-dependent. Suggested values of n and p are listed in Table 11.3.

For very high stresses (>1MPa) and large grains, deformation is accommodated by the movement of dislocations in the crystal lattice. Dislocation creep (disl) is highly nonlinear and is characterized by the stress exponent of $n=4$. However, viscosity does not depend on the grain size. Enhanced deformation is observed if temperatures are within 15 K of the melting curve. For lower stresses (between ~0.01 and 1 MPa), grains deform along the crystal planes by basal slip (bs). In crystals, whose planes are not well oriented for basal slip, another process occurs to accommodate deformation, the grain boundary sliding (gbs). These two mechanisms are not independent but act together and are combined as follows (Goldsby and Kohlstedt, 2001):

$$\dot{\varepsilon}_{gss} = \left(\frac{1}{\dot{\varepsilon}_{gbs}} + \frac{1}{\dot{\varepsilon}_{bs}}\right)^{-1} \tag{11.2}$$

into so-called grain-size sensitive creep (gss). However, only grain boundary sliding viscosity depends on the grain size ($p=1.4$) while basal slip viscosity is grain size independent ($p=0$). Both mechanisms have stress exponents close to 2 ($n=1.8$ for gbs and 2.4 for bs). Finally, for even lower stresses, relatively high temperatures ($\gtrsim$150 K) and small grain sizes (< 1 mm), ice deforms by the diffusion of vacancies through the crystal lattice.

Diffusion creep (diff) is a linear mechanism (*n=1*), so its viscosity does not depend on stress. It depends on the square of the grain size (*p=2*).

Altogether, the four mechanisms combine as:

$$\dot{\varepsilon} = \dot{\varepsilon}_{diff} + \dot{\varepsilon}_{gss} + \dot{\varepsilon}_{disl} = \dot{\varepsilon}_{diff} + \left(\frac{1}{\dot{\varepsilon}_{gbs}} + \frac{1}{\dot{\varepsilon}_{bs}}\right)^{-1} + \dot{\varepsilon}_{disl} \quad (11.3)$$

(Goldsby and Kohlstedt, 2001). This flow law fits the laboratory data. For a more detailed review and the values of all parameters, see Durham and Stern (2001) or Qi and colleagues (2018). Depending on the grain size, which can range between <1 mm and a few cm, viscosity at the melting temperature (boundary with the ocean) is between 10^{13} and 10^{16} Pa s. At the surface, the experimentally derived flow law predicts too-large viscosity values (>10^{20} Pa s), indicating that another mechanism (elasticity or brittle failure) accommodates the stresses. In some numerical models, an Arrhenius-type viscosity law is often used that takes into account the most important effect of temperature but neglects the stress dependence (e.g., Han and Showman, 2008).

The viscosity of HP ices has been measured in the experimental setting (Sotin et al., 1985; Durham et al., 1996). The deformation of ice VI has been measured at large values of shear stress (10^8 Pa; Sotin et al., 1985) and found to have deformation rates of ~10^{-2} s^{-1} and a viscosity of 10^{10} Pa s. Extrapolation of these results to typical convective shear stresses (10^3 – 10^5 Pa) gives viscosities on the order of 10^{12} – 10^{14} Pa s, similar to the ice I viscosities at the melting point, as observed in field measurements by Hudleston (2015). Durham and colleagues (1996) measured the viscosity of ice VI at smaller pressures and found stronger stress dependence (e.g., larger values of the stress exponent *n*) and larger melting-point viscosity values. The discrepancy between viscosities predicted by these two studies may be due to different experimental conditions. Alternatively, a change in deformation mechanisms observed for ice I_h (e.g., Goldsby and Kohlstedt, 2001) may be occurring in the HP phases as suggested by Kalousová and colleagues (2018), although no laboratory experiments were so far performed to confirm this hypothesis. Both studies (Sotin et al., 1985; Durham et al., 1996) suggest that viscosity is nearly constant along the melting curve, justifying the use of Arrhenius's law in numerical simulations (Kalousová et al., 2018). Due to the extrapolation from the high-stress laboratory experiments to lower-stress conditions and large uncertainty in grain size, numerical simulations usually consider a wide range of melting-point viscosity values between 10^{13} and 10^{17} Pa s (Choblet et al., 2017; Kalousová et al., 2018; Kalousová and Sotin, 2018). Ice V viscosity, compared to ice VI, can be determined at the transition point between the two phases. Sotin and Poirier (1987) found that ice V is harder to deform than ice VI with their viscosity ratio close to two orders of magnitude at the experimental conditions, while Durham and colleagues (1996) found that ice VI and ice V deform in a similar manner. Finally, ice III is the weakest in deformation (Durham et al., 1997).

Ice viscosity may get as low as 10^{10} Pa s if other cryogenic liquids are present, such as methane or NH_3 hydrates, suggesting that solid-state pure water ice is less likely to explain observed flow features or overall smoothing (Stevenson and Lunine, 1986; Schenk and Moore, 1995). Ammonia-water

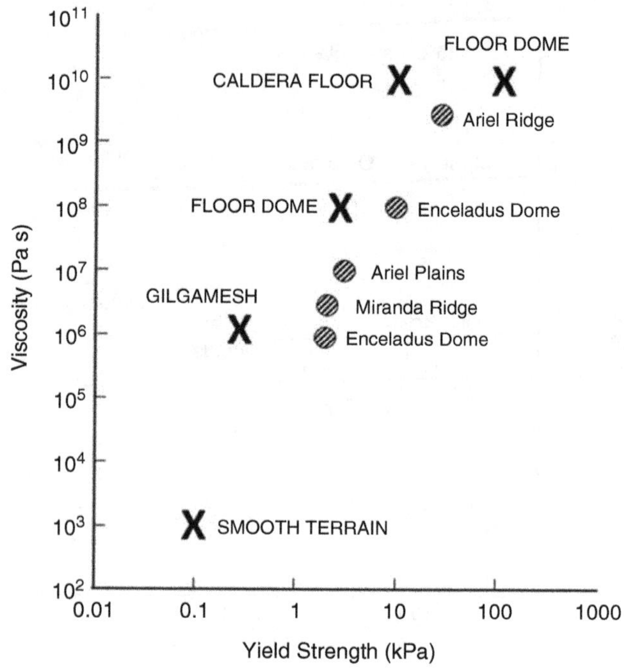

Figure 11.3 Rheologic strength versus viscosity of Ganymedean surface units compared to features from Enceladus, Miranda, and Ariel. Adapted from Schenk and Moore (1995) with permission of John Wiley and Sons, from 'Volcanic constructs on Ganymede and Enceladus: Topographic evidence from stereo images and photoclinometry', AGU, Journal of Geophysical Research 100, 1995, permission conveyed through Copyright Clearance Center, Inc.

systems may also be important in investigating viscous flows (Lunine and Stevenson, 1982). Kargel and colleagues (1991) observed that the behavior of ammonia-water eutectic and peritectic liquids closely resembles basaltic behavior. Brines also have water-like viscosities (Kargel et al., 1991; Hogenboom et al., 1995).

Several viscosity-related processes, such as erosion and impact crater relaxation, may have acted on Ganymede's surface (Parmentier et al., 1982; Dombard and McKinnon, 2006). For example, Gilgamesh floor deposits with thickness of ~150 m and ~10 km across (as upper limits) have an estimated viscosity of ~10^6 Pa s, similar to ridges at Miranda and domes on Enceladus (Figure 11.3). Smooth surface materials have viscosities of ~10^6 Pa s with yield strengths of 0.2 kPa (Schenk and Moore, 1995). These are in range of materials to behave (and deform) plastically with these viscosities and yield strengths. Alternatively, a partially crystalline melt could also be extruded, though achieving significant topography (>150 m) may not be possible due to its extreme low viscosity. Assuming that the caldera-like floor materials (cryolava) were continuous extrusions and had a thickness of ~300 m, Schenk and Moore (1995) estimated their viscosities to be 10^{10} Pa s with yield strengths of ~10 kPa. Comparably, domes on Europa have 10^6 – 10^8 Pa s viscosity estimates, with yield strengths of ~1 – 10 kPa (Schenk and Moore, 1995), somewhat weaker than on Ganymede.

Ductile flows result in the deformation (strain ε) at a strain rate ($\dot{\varepsilon}$) that depends on several factors, including temperature, pressure, and effective grain size of the ice. Ice has a strength at failure nearly independent of temperature, but confining on pressure, namely in the range $0 < P < 50$ MPa. Brittle

Table 11.4 *List of ice compounds and relevant Young's (E) or bulk moduli (B). Table adapted from Umurhan and colleagues (2021) and references therein. Ice III and VI modeled from Vance and colleagues 2018*

Ice Species	Modulus Type	Temperature (K)	Modulus (GPa)
Ice I_h	E	<273	0.009 – 0.0112
	E	90	0.1 – 0.5
	E	100	0.2 – 0.9
CO_2	E	80	13.12
CH_4 clathrate SI	E	<273	8.4
	B	<273	8.76
CH_4 clathrate SII	E	<273	8.2
	B	<273	8.48
NH_3-H_2O	E	90	0.85
	E	100	0.1
Ice III	B	250	9.761
Ice VI	B	300	20.27

behavior is regulated by frictional properties, although these are less scale-dependent and therefore rudimentary to extrapolate from laboratory data, at least for the uppermost layers of an icy moon (Schulson, 1979; Beeman et al., 1988; Petrenko and Whitworth, 1999; Durham and Stern, 2001). Frictional strength strongly depends on pressure, not so much on temperature, and so increases with depth. Ductile strength, however, depends on temperature, thus decreasing with depth.

Strain is a measure of how much an ice substance of a certain size stretches or deforms. Stress is a measure of the internal forces and is defined in terms of a force per unit area (i.e., pressure). Generally, a block of polycrystalline ice subjected to some amount of applied stress (e.g., compressive, tensile, or shear) will respond with either ductile (or plastic) flow or will experience brittle fracture failure when the applied stress exceeds the material's yield stress (or elastic motions if the applied stress is less than the yield stress).

To describe elasto-viscoplastic rheology (see Dombard and McKinnon, 2006 and references therein), the total strain is a linear summation of the elastic, viscous, and plastic strains (a Maxwell viscoelastic solid extended to include plasticity). Such an extended Maxwell solid captures the relative behavior on different geologic timescales, thus it behaves as a solid (elastic) material on shorter timescales, as a viscous (fluid) material on longer timescales, and exhibits brittle (plastic) failure should the yield strength be exceeded (Durham and McKinnon, 2006). Linear, isotropic elasticity can be characterized by two main parameters: *Young's modulus* (E) and *Poisson's ratio* (ν). While these parameters are inherently compressible, as E approaches infinity, the elastic response trends to incompressibility. For relaxation modeling on Ganymede, some use the values by Gammon and colleagues (1983) of water ice $E = 9.33$ GPa and $\nu = 0.325$.

The bulk elastic properties of an ice are the ratio of the pressure change to the fractional volume compression. The ratio of the stress to the strain, regarding compressional forces, defines the Young's modulus. Both variables give us insight into the planar forces that can be supported at icy satellites and the pressure-temperature conditions that an ice sample can endure. Table 11.4 shows Ganymede-relevant ices and their respective Young's or bulk moduli.

11.5.4 Thermal Conductivity

Thermal conductivity (k) is defined as a bulk property of condensed matter and a main element of the magnitude (and resultant type) of geologic activity on planets and satellites. Thermal conductivity models tend to rely on pure ice I_h parameters, but this variable relies on the composition, porosity, crystallinity, and temperature/pressure conditions, all of which are highly variable on icy satellites (Ross and Kargel, 1998; Umurhan et al., 2021). The thermal conductivity is a prime element in determining the physical processes related to cryovolcanism, mantle convection, viscous relaxation, and differentiation – each playing a role at the Ganymedean surface in some form or other.

Ice I_h to a degree is considered disordered in its crystallinity in that it exhibits nonzero entropy as temperature decreases, therefore not necessarily conforming to the third law of thermodynamics. For an overview of the disorder of crystalline materials and their respective thermal conductivities, see Ross and colleagues (1981). Both ice I_h and HP ices show a variation in thermal conductivity that is nearly proportional to the inverse of temperature (Ross and Kargel, 1998; Andersson and Inaba, 2005). Over the relevant temperatures, the ice I_h thermal conductivity (Figure 11.4) ranges from $\sim$ 2 W m^{-1} K^{-1} at the ocean interface to $\sim$6 W m^{-1} K^{-1} at the icy moons' surfaces. The lower thermal conductivities of ices II and III suggests the possibility of layers within an icy mantle that could exceed their melting point, producing cryovolcanic fluids (Fortes and Choukroun, 2010).

The existence of a maximum thermal conductivity at low temperatures is characteristic of crystalline substances (Klinger, 1973). The other extreme for thermal conductivity

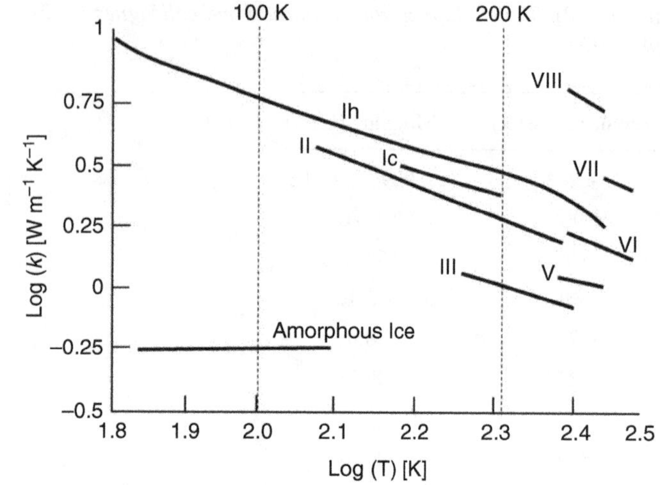

Figure 11.4 Thermal conductivities of water ice polymorphs in Ganymede-relevant temperatures. Note the lower thermal conductivity of amorphous ice compared to the crystalline water phases. Adapted from Ross and Kargel (1998). Ross, R. G., Kargel, J. S. (1998). Thermal conductivity of solar system ices, with special reference to Martian polarcaps. *Astrophysics and Space Science Library*. Reproduced with permission from Springer Nature.

arises for amorphous materials where thermal conductivity is relatively small (< 1 W m^{-1} K^{-1}) and exhibits a monotonic and very minuscule increase with T. Figure 11.4 illustrates the extreme forms of thermal conductivity over temperature, comparing crystalline water ice phases and amorphous ices. However, there is disagreement over the thermal conductivity of amorphous ice as determined from laboratory work, which has also been suggested, for instance, to be 10^{-5} W m^{-1} K^{-1} by Kouchi and colleagues (1992).

11.5.5 Activation Energies

Activation energy is defined as the energy an ice must possess to go through a physical or chemical reaction. Models relating the activation energy of the ice flow law and the critical Rayleigh number were derived by Solomatov (1995) and have been used to determine the conditions required to start convection in an ice shell (Pappalardo et al., 1998; McKinnon, 1999). The activation energy of the ice sample indicates the material's Arrhenius temperature dependence, reflecting the interior grain susceptibility to deformation (Umurhan et al., 2021). That is, the activation energy is usually the sum of a free energy activation energy plus an activation energy volume that considers the pressure dependencies.

Activation energies pertinent to ductility of ice have been reported in Leonteva and colleagues (1970) and Yamashita and colleagues (2010). Nucleation activation energy of amorphous (solid) water is 158 kJ/mol while the activation energy of crystal growth is ~47 kJ/mol (Safarik and Mullins, 2004). Sugisaki and colleagues (1969) observed that the activation energies of cubic ice (at 160 – 200 K) is ~21.3 kJ/mol where the hexagonal phase at those same temperatures have an activation energy of ~44.7 kJ/mol.

Sublimation energy is the heat required to change an icy constituent from the solid phase to the gas phase. From current experimental results, we see CO_2 and NH_3 have higher sublimation energies than CH_4 and CO by a factor of ≥ 2 (see Luna et al., 2014). Carbon dioxide is 80 – 90 K; the sublimation energy is 22.37 kJ/mol, then increases to 29.3 kJ/mol at 91.5 – 92.5 K (Luna et al., 2014).

These energies, though often experimentally with pure ices, are mostly inferred from theoretical predictions based on observed timescales of stress relaxation (Leonteva et al., 1970). This thermally dependent mechanism for activation (and sublimation) energy has only been explored for pure ice samples in an experimental setting, so more work certainly needs to be accomplished regarding mixtures and hydrate-type components relevant for Ganymede.

11.5.6 Rheological Properties of Clathrates

Clathrates are very peculiar thermodynamically compared to water ice. The H_2O skeleton strongly resembles the ice I_h structure on an atomical level from Raman and infrared studies (Nakano et al., 1998; Dartois and Deboffle, 2008; Dartois and Schmitt, 2009). However, the presence of guest gases dramatically influences the molecular interactions and subsequently their thermophysical properties. Thermodynamically, some clathrate hydrates are similar to that of water ice. For example, the thermal expansion, heat capacity, and elasticity of the CS-I and CS-II (clathrate structures) are very similar to ice I_h (Choukroun et al., 2013). While clathrates are usually more stable than water ices at MPa – GPa pressures, the stability is dependent on the abundance of the guest species in the environment (Sohl et al., 2010; Choukroun et al., 2013). Clathrate hydrates with volatiles like N_2, CH_4, or CO_2 are generally less dense than HP ices, but can achieve buoyancy or sink in aqueous solutions (Journaux et al., 2020).

Thermal conductivity has been the subject of numerous studies (e.g., Ross et al., 1981; Cook and Leaist, 1983; Tse and White, 1988), some of which noted a glass-like behavior that is temperature-dependent. At temperatures below 100 K, the conductivity increases, as expected from crystalline materials (Tse and White, 1988; English and Tse, 2010). On average, the thermal conductivity of clathrate hydrates is 0.4 – 0.7 W m^{-1} K^{-1} at pressures lower than 100 MPa and temperatures 100 – 270 K (Choukroun et al. 2013). The methane DO-clathrate (a 1.3-dioxolane clathrate type) has been measured to have a higher thermal conductivity than solid CH_4 uniformly with temperature by a factor of > 10 (Ahmad and Phillips, 1987). Such a large difference could indicate how porosity and the cage-like structure of a clathrate can affect thermal conductivity. Lorenz and Shandera (2001) recorded an NH_3 ice sample (with 16% molar ratio of NH_3: H_2O) being ~48% lower in thermal conductivity than pure water ice. Sumarokov and colleagues (2003) noted that CO_2 reaches higher thermal conductivities at low temperatures up to ~700 W m^{-1} K^{-1}.

Despite the elastic properties being somewhat similar, the flow properties of clathrates have been subject to recent studies (e.g., Stern et al., 1996; Durham et al., 2005). Mechanical experiments conducted under high differential stress with a sustained gas pressure within the samples have reported that: (i) the initial results were likely due to significant dissociation (~30%); (ii) gas hydrates exhibit a much greater strength than water ice in that the strain rate requires applying a stress 10 – 30 times greater for clathrates than for water ice (Choukroun et al., 2013). Note that density values in Table 11.4 have been

Table 11.5 *Physical property comparison between water ice I_h phase and clathrate hydrates based on cage structures (CS) – I and II. Values from Choukroun and colleagues (2013)*

Physical Property	Ice I_h	Clathrate CS I	Clathrate CS II
Dielectric constant (at 273 K)	94	58	58
Young's modulus (at 268 K) [GPa]	8 – 9.5	8.5	8.5
Poisson's ratio	0.331	0.31403	0.31119
Thermal Expansion (at 200 K) [10^{-6} K^{-1}]	56	77	52
Thermal Conductivity (at 263 K) [W m^{-1} K^{-1}]	2.18	0.51	0.50

omitted due to their extreme variability and preliminary measurements dependent on the temperature, pressure, and guest gas molecule (Sloan and Koh, 2007; Choukroun et al., 2013).

The rheological properties of clathrate hydrates have serious implications for the structures and processes of ices and icy worlds (see Table 11.5). With variations in density, structure, occupancy, and guest species, these parameters can create such complex mechanisms, though unfortunately still poorly known experimentally. Combined with larger viscosity, the clathrate hydrate may form mechanically stiffer thermally insulating layers, which are particularly interesting for ice–ocean interfaces of icy bodies (Kamata et al., 2019).

11.6 SYNTHESIS AND RECOMMENDATIONS

Accounting for important surface features and processes across Ganymede's surface is necessary and possible for experimental characterization and modeling. Tools and facilities to monitor rock and ice at low temperatures have developed in the past 50 years and can provide a basis to solve material science processes at such extreme conditions that are not otherwise natural here on Earth. These engineering approaches for such experimental work can be designed within logistical, technical, and economic constraints. We briefly identify the recommendations for research and development as they are intended to guide a framework toward understanding ices at Ganymede.

Recommendation 1. Take an interdisciplinary approach and use geologic, material science, and geomechanical, geophysical, hydrologic, and biogeochemical information to conceptualize fracturing, grain boundary mechanisms and sliding, and porosities (especially with possible hydrate contamination and fluids).

Recommendation 2. Characterize the thermal, chemical, mechanical, and hydraulic processes, their interactions, and the conditions that can lead to their coupling to better understand the dynamic surface–interior interactions of Ganymede.

Recommendation 3. Improve characterization and monitoring of such mechanisms through new and expanded technologies, such as thermal conductivity and expansivity, stress-strain measurements, seismo-acoustic properties, and expansion of viscosity experiments of ices to include hydrates and organics.

11.7 SUMMARY

As we have observed through this chapter, the thermodynamic and rheologic properties of ice phases and hydrates remain mostly theoretical, and we note that more experimental measurements are needed. Our knowledge of Ganymede's surface composition has greatly improved our modeling and understanding of spectral behaviors, but it is the complexity and diversity of Ganymede's surface that we must peer through to understand the thermal properties of these variable ice phases. Their physical parameters, such as grain size and density, are key to understanding the endogenous and exogeneous processes, especially in modeling the HP ice layer and geologic formations. We hope this chapter is used as a compendium for thermodynamic parameters of ice phases and hydrates relevant to Ganymede.

ACKNOWLEDGMENTS

The authors would like to thank our reviewers for their insight. K. Kalousova was supported by the Czech Science Foundation through project No. 19-10809S and by Charles University Research Program No. UNCE/SCI/023.

REFERENCES

Abramson, E., Bollengier, O., Brown, J. et al. (2018). Carbonic acid monohydrate. *American Mineralogist: Journal of Earth and Planetary Materials*, 103(9), 1468–1472.

Ahmad, N., and Phillips, W. (1987). Thermal conductivity of ice and ice clathrate. *Solid State Communications*, 63(2), 167–171.

Allison, M., and Clifford, S. (1987). Ice-covered water volcanism on Ganymede. *J. Geophys. Res.*, 9(2), 7865–7876.

Amos, D., Donnelly, M., Teeratchanan, P. et al. (2017). A chiral gas–hydrate structure common to the carbon dioxide–water and hydrogen–water systems. *Journal of Physical Chemistry Letters*, 8(17), 4295–4299.

Anderson, A., Binbrek, O., and Tang, H. (1977). Raman and infrared study of the low temperature phase of solid H_2S and D_2S. *Journal of Raman Spectroscopy*, 6(5), 213–220.

Anderson, J., Lau, E., Sjogren, W., Schubert, G., and Moore, W. (1996). Gravitational constraints on the internal structure of Ganymede. *Nature*, 384(6609), 541–543.

Andersson, O., and Inaba, A. (2005). Thermal conductivity of crystalline and amorphous ices and its implications on amorphization and glassy water. *Physical Chemistry Chemical Physics*, 7(7), 1441–1449.

Barr, A., and Pappalardo, R. (2005). Onset of convection in the icy Galilean satellites: Influence of rheology. *Journal of Geophysical Research: Planets*, 110(E12).

Barr, A., and Showman, A. (2009). Heat transfer in Europa's icy shell. In: R. T. Pallardo, W. B. McKinnon, and K. Khurana (eds.), *Europa* (pp.405–430). University of Arizona Press.

Beeman, M., Durham, W., and Kirby, S. (1988). Friction of ice. *Journal of Geophysical Research: Solid Earth*, 93(B7), 7625–7633.

Berland, B., Brown, D., Tolbert, M., and George, S. (1995). Refractive index and density of vapor-deposited ice. *Geophysical Research Letters*, 22(24), 3493–3496.

Blake, S. (1990). Viscoplastic models of lava domes. In koilk, L, J. Fink (ed.), *Lava Flows and Domes: Emplacement mechanisms and hazard implications* (pp. 25-46). Springer.

Bland, M., and Showman, A. (2007). The formation of Ganymede's grooved terrain: Numerical modeling of extensional necking instabilities. *Icarus*, 189(2), 439–456.

Bland, M., Showman, A., and Tobie, G. (2009). The orbital–thermal evolution and global expansion of Ganymede. *Icarus*, 200(1), 207–221.

Bridgman, P. (1912). Water, in the liquid and five solid forms, under pressure. In *Proceedings of the American Academy of Arts and Sciences* (Vol. 47, No. 13, pp. 441–558). American Academy of Arts & Sciences.

Brown, D., George, S., Huang, C. et al. (1996). H_2O condensation coefficient and refractive index for vapor-deposited ice from molecular beam and optical interference measurements. *Journal of Physical Chemistry*, 100(12), 4988–4995.

Brown, R., Baines, K., Bellucci, G. et al. (2004). The Cassini visual and infrared mapping spectrometer (VIMS) investigation. *Space Science Reviews*, 115(1–4), 111–168.

Calvin, W., Clark, R., Brown, R., and Spencer, J. (1995). Spectra of the icy Galilean satellites from 0.2 to 5μm: A compilation, new observations, and a recent summary. *Journal of Geophysical Research: Planets*, 100(E9), 19041–19048

Carlson, R., Anderson, M., Johnson, R. et al. (1999). Hydrogen peroxide on the surface of Europa. *Science*, 283(5410), 2062–2064.

Carlson, R., Weissman, P., Smythe, W., and Mahoney, J. (1992). Near-infrared mapping spectrometer experiment on Galileo. *Space Science Reviews*, 60(1–4), 457–502.

Choblet, G., Tobie, G., Sotin, C., Kalousova, K., and Grasset, O. (2017). Heat transport in the high-pressure ice mantle of large icy moons. *Icarus*, 285, 252–262.

Choukroun, M., Kieffer, S., Lu, X., and Tobie, G. (2013). Clathrate hydrates: Implications for exchange processes in the outer solar system. In M. Gudipati and J. Castillo-Rogez (eds.), *The science of solar system ices* (pp. 409-454). Springer.

Clark, R. (1980). Ganymede, Europa, Callisto, and Saturn's rings: Compositional analysis from reflectance spectroscopy. *Icarus*, 44(2), 388–409.

Clark, R., and McCord, T. (1980). The Galilean satellites: New near-infrared spectral reflectance measurements (0.65 – 2.5 mm) and a 0.325– 5μm summary. *Icarus*, 41, 323–339.

Collins, G. (2000). Driving mechanisms for grooved terrain tectonics on Ganymede and chaotic terrain formation on Europa: Constraints from Galileo data (Doctoral dissertation, Brown University).

Collins, G., Head, J., and Pappalardo, R. (1998). Formation of Ganymede grooved terrain by sequential extensional episodes: Implications of Galileo observations for regional stratigraphy. *Icarus*, 135(1), 345–359.

Cook, J., and Leaist, D. (1983). An exploratory study of the thermal conductivity of methane hydrate. *Geophysical Research Letters*, 10(5), 397–399.

Croft, S., and Strom, R. (1985). Ganymede's crust: Structural indicators in the Tiamat Sulcis quadrangle. In *Lunar and Planetary Science Conference* (pp. 156–157).

Croft, S., Lunine, J., and Kargel, J. (1988). Equation of state of ammonia-water liquid: Derivation and planetological applications. *Icarus*, 73(2), 279–293.

Croft, S., Kargel, J., Kirk, R. et al. (1995). The geology of Triton. In: D. Cruikshank (ed.), *Neptune and Triton*. University of Arizona Press.

Dalton III, J., Cassidy, T., Paranicas, C. et al. (2013). Exogenic controls on sulfuric acid hydrate production at the surface of Europa. *Planetary and Space Science*, 77, 45–63.

Dalton III, J., and Pitman, K. (2012). Low temperature optical constants of some hydrated sulfates relevant to planetary surfaces. *Journal of Geophysical Research: Planets*, 117(E9).

Dartois, E., and Deboffle, D. (2008). Methane clathrate hydrate FTIR spectrum: Implications for its cometary and planetary detection. *Astronomy & Astrophysics*, 490(3), L19–L22.

Dartois, E., and Schmitt, B. (2009). Carbon dioxide clathrate hydrate FTIR spectrum: Near infrared combination modes for astrophysical remote detection. *Astronomy & Astrophysics*, 504(3), 869–873.

Dombard, A., and McKinnon, W. (2001). Formation of grooved terrain on Ganymede: Extensional instability mediated by cold, superplastic creep. *Icarus*, 154(2), 321–336.

Dombard, A., and McKinnon, W. (2006). Elastoviscoplastic relaxation of impact crater topography with application to Ganymede and Callisto. *Journal of Geophysical Research: Planets*, 111(E1).

Durham, W., Kirby, S., and Stern, L. (1997). Creep of water ices at planetary conditions: A compilation. *Journal of Geophysical Research: Planets*, 102(E7), 16293–16302.

Durham, W., Kirby, S., and Stern, L. (1998). Rheology of planetary ices. In B. Schmitt, C. de Bergh, and M. Festou (eds.), *Solar system ices* (pp. 63–78). Springer.

Durham, W., Prieto-Ballesteros, O., Goldsby, D., and Kargel, J. (2010). Rheological and thermal properties of icy materials. *Space Science Reviews*, 153(1), 273–298.

Durham, W., and Stern, L. (2001). Rheological properties of water ice: Applications to satellites of the outer planets. *Annual Review of Earth and Planetary Sciences*, 29, 295–330.

Durham, W., Stern, L., and Kirby, S. (1996). Rheology of water ices V and VI. *Journal of Geophysical Research: Solid Earth*, 101(B2), 2989–3001.

Durham, W., Stern, L., and Kirby, S. (2001), Rheology of ice I at low stress and elevated confining pressure, *Journal of Geophysical Research*, 106(6), 11031–11042.

Durham, W., Stern, L., Kirby, S., and Circone, S. (2005). Rheological comparisons and structural imaging of sI and sII endmember gas hydrates and hydrate/sediment aggregates. In *Proceedings of the 5th International Conference on Gas Hydrates* (pp. 607–614). Tapir Academic.

English, N., and Tse, J. (2010). Mechanisms of thermal conduction in clathrate hydrates. In *12th International Conference on the Physics and Chemistry of Ice*, Abstract #6-A3-1 (Saporro).

Fagents, S., Greeley, R., Sullivan, R. et al. (2000). Cryomagmatic mechanisms for the formation of Rhadamanthys Linea, triple band margins, and other low-albedo features on Europa. *Icarus*, 144(1), 54–88.

Fanale, F., Li, Y., De Carlo, E. et al. (2001). An experimental estimate of Europa's "ocean" composition independent of Galileo orbital remote sensing. *Journal of Geophysical Research: Planets*, 106(E7), 14595–14600.

Fortes, A., and Choukroun, M. (2010). Phase behaviour of ices and hydrates. *Space Science Reviews*, 153(1–4), 185–218.

Fortes, A., Wood, I., Sclater, G., and Tucker, M. (2010). Recovery and characterization of high-pressure phases in the systems H_2O–NH_3, H_2O–H_2SO_4, and H_2O–$MgSO_4$. ISIS Experimental Report, Rutherford Appleton Laboratory, 1010009.

Fray, N., and Schmitt, B. (2009). Sublimation of ices of astrophysical interest: A bibliographic review. *Planetary and Space Science*, 57(14–15), 2053–2080.

Gammon, P., Kiefte, H., and Clouter, M. (1983). Elastic constants of ice samples by Brillouin spectroscopy. *Journal of Physical Chemistry*, 87(21), 4025–4029.

Giauque, W., and Stephenson, C. (1938). Sulfur dioxide. The heat capacity of solid and liquid. Vapor pressure. Heat of vaporization. The entropy values from thermal and molecular data. *Journal of the American Chemical Society*, 60(6), 1389–1394.

Giovambattista, N., Amann-Winkel, K., and Loerting, T. (2013). Amorphous ices. *Liquid Polymorphism*, 152, 139–173.

Goldsby, D., and Kohlstedt, D. (2001). Superplastic deformation of ice: Experimental observations. *Journal of Geophysical Research: Solid Earth*, 106(B6), 11017–11030.

Golombek, M. (1982). Constraints on the expansion of Ganymede and the thickness of the lithosphere. *Journal of Geophysical Research: Solid Earth*, 87(S01), A77–A83.

Grimm, R., and Squyres, S. (1985). Spectral analysis of groove spacing on Ganymede. *Journal of Geophysical Research: Solid Earth*, 90(B2), 2013–2021.

Grundy, W., Buie, M., Stansberry, J., Spencer, J., and Schmitt, B. (1999). Near-infrared spectra of icy outer solar system surfaces: Remote determination of H_2O ice temperatures. *Icarus*, 142(2), 536–549.

Grundy, W., Buratti, B., Cheng, A. et al. (2007). New Horizons mapping of Europa and Ganymede. *Science*, 318, 234.

Grundy, W., and Schmitt, B. (1998). The temperature-dependent near-infrared absorption spectrum of hexagonal H_2O ice. *Journal of Geophysical Research: Planets*, 103(E11), 25809–25822.

Han, L., and Showman, A. (2008). Implications of shear heating and fracture zones for ridge formation on Europa. *Geophysical Research Letters*, 35(3).

Hansen, G., and McCord, T. (2004). Amorphous and crystalline ice on the Galilean satellites: A balance between thermal and radiolytic processes. *Journal of Geophysical Research*, 109, E01012.

Hanson, D., and Mauersberger, K. (1986). The vapor pressures of solid and liquid ozone. *Journal of Chemical Physics*, 85(8), 4669–4672.

Hapke, B. (1993). *Theory of reflectance and emittance spectroscopy*. Cambridge University Press.

Hartmann, W. (1980). Surface evolution of two-component stone/ice bodies in the Jupiter region. *Icarus*, 44(2), 441–453.

Hendrix, A., Barth, C., Stewart, A., Hard, C., and Lane, A. (1999). Hydrogen peroxide on the icy Galilean satellites. In *Lunar and Planetary Science Conference*, Abstracts p. 2043.

Hendrix, A., Domingue, D., and Noll, K. (2013). Ultraviolet properties of planetary ices. In: M. Gudipati and J. Castillo-Rogez (eds.), *The science of solar system ices* (pp. 73-105). Springer.

Hibbitts, C., Pappalardo, R., Hansen, G., and McCord, T. (2003). Carbon dioxide on Ganymede. *Journal of Geophysical Research*, 108(5), 5036.

Hirai, H., Komatsu, K., Honda, M. et al. (2010). Phase changes of CO_2 hydrate under high pressure and low temperature. *Journal of chemical physics*, 133(12), 124511.

Hogenboom, D., Kargel, J., Ganasan, J., and Lee, L. (1995). Magnesium sulfate-water to 400 MPa using a novel piezometer: Densities, phase equilibria, and planetological implications. *Icarus*, 115(2), 258–277.

Hudleston, P. (2015). Structures and fabrics in glacial ice: A review. *Journal of Structural Geology*, 81, 1–27.

Huebner, W., Benkhoff, J., Capria, M. et al. (2006). *Heat and gas diffusion in comet nuclei* (Vol. 133). International Space Science Institute.

Huppert, H., Shepherd, J., Sigurdsson, R., and Sparks, S. (1982). On lava dome growth, with application to the 1979 lava extrusion of the Soufriere of St. Vincent. Journal of Volcanology and Geothermal Research, 14(3–4), 199-222.

Hussmann, H., Sohl, F., and Spohn, T. (2006). Subsurface oceans and deep interiors of medium-sized outer planet satellites and large trans-neptunian objects. *Icarus*, 185(1), 258–273.

Jankowski, D., and Squyres, S. (1988). Solid-state ice volcanism on the satellites of Uranus. *Science*, 241(4871), 1322–1325.

Jenniskens, P., and Blake, D. (1994). Structural transitions in amorphous water ice and astrophysical implications. *Science*, 265(5173), 753–756.

Jenniskens, P., Blake, D., and Kouchi, A. (1998). Amorphous water ice. In B. Schmitt, C. de Bergh, M. Festou (eds.), *Solar system ices* (pp. 139–155). Springer.

Johnson, R., and Jesser, W. (1997). O2/O3 microatmospheres in the surface of Ganymede. *Astrophysical Journal Letters*, 480(1), L79.

Journaux, B., Daniel, I., Caracas, R., Montagnac, G., and Cardon, H. (2013). Influence of NaCl on ice VI and ice VII melting curves up to 6 GPa, implications for large icy moons. *Icarus*, 226(1), 355–363.

Journaux, B., Daniel, I., Petitgirard, S. et al. (2017). Salt partitioning between water and high-pressure ices. Implication for the dynamics and habitability of icy moons and water-rich planetary bodies. *Earth and Planetary Science Letters*, 463, 36–47.

Journaux, B., Kalousová, K., Sotin, C. et al. (2020). Large ocean worlds with high-pressure ices. *Space Science Reviews*, 216(1), 1–36.

Kalousová, K., and Sotin, C. (2018). Melting in high-pressure ice layers of large ocean worlds: Implications for volatiles transport. *Geophysical Research Letters*, 45(16), 8096–8103.

Kalousová, K., Sotin, C., Choblet, G., Tobie, G., and Grasset, O. (2018). Two-phase convection in Ganymede's high-pressure ice layer: Implications for its geological evolution. *Icarus*, 299, 133–147.

Kamata, S., Nimmo, F., Sekine, Y. et al. (2019). Pluto's ocean is capped and insulated by gas hydrates. *Nature Geoscience*, 12(6), 407–410.

Kargel, J. (1998). Physical chemistry of ices in the outer solar system. In: B. Schmitt, C. de Bergh, and M. Festou (eds.), *Solar system ices* (pp. 3–32). Springer.

Kargel, J., Croft, S., Lunine, J., and Lewis, J. (1991). Rheological properties of ammonia-water liquids and crystal-liquid slurries: Planetological applications. *Icarus*, 89(1), 93–112.

Kargel, J., Kaye, J., Head III, J. et al. (2000). Europa's crust and ocean: Origin, composition, and the prospects for life. *Icarus*, 148(1), 226–265.

Kieffer, H., and Smythe, W. (1974). Frost spectra: Comparison with Jupiter's satellites. *Icarus*, 21, 506–512.

Kivelson, M., Khurana, K., and Volwerk, M. (2002). The permanent and inductive magnetic moments of Ganymede. *Icarus*, 157(2), 507–522.

Klinger, J., Whalley, E., Jones, S., and Gold, L. (1973). Thermal conductivity of ice single crystals at low temperatures. In *Physics and chemistry of ice* (pp. 114–116). Royal Society of Canada.

Kouchi, A., Greenberg, J., Yamamoto, T., and Mukai, T. (1992). Extremely low thermal conductivity of amorphous ice: Relevance to comet evolution. *Astrophysical Journal*, 388, L73–L76.

Kubo, T., Nakata, H., and Kato, T. (2009). Effects of insoluble particles on grain growth in polycrystalline ice: Implications for rheology of ice shells of icy satellites. *Journal of Mineralogical and Petrological Sciences*, 104(5), 301–306.

Leonteva, A., Stroilov, Y., Lakin, E., and Bolshutkin, D. (1970). Zero point oscillation energy effect on plastic deformation in solidified gases. *Physica status solidi (b)*, 42(2), 543–549.

Ligier, N., Paranicas, C., Carter, J. et al. (2019). Surface composition and properties of Ganymede: Updates from ground-based observations with the near-infrared imaging spectrometer SINFONI/VLT/ESO. *Icarus*, 333, 496–515.

Lorenz, R., and Shandera, S. (2001). Physical properties of ammonia-rich ice: Application to Titan. *Geophysical Research Letters*, 28(2), 215–218.

Lucchita, B. (1980). Grooved terrain on Ganymede. *Icarus*, 44(2), 481–501.

Lucchitta, B., Barnes, C., and Glotfelty, M. (1992). Geological map of the Memphis Facula quadrangle (Jg-7) of Ganymede, US Geological Survey Map. 1-2289, Jg-9.

Luna, R., Satorre, M., Santonja, C., and Domingo, M. (2014). New experimental sublimation energy measurements for some relevant astrophysical ices. *Astronomy & Astrophysics*, 566, A27.

Lunine, J., and Stevenson, D. (1982). Formation of the Galilean satellites in a gaseous nebula. *Icarus*, 52(1), 14–39.

Macmillan, N. (1972). The theoretical strength of solids. *Journal of Materials Science*, 7(2), 239–254.

Massani, B., Mitterdorfer, C., and Loerting, T. (2017). Formation and decomposition of CO_2-filled ice. *Journal of Chemical Physics*, 147(13), 134503.

Mastrapa, R., Grundy, W., and Gudipati, M. (2013). Amorphous and crystalline H_2O-Ice. In M. Gudipati and J. Castillo-Rogez (eds.), *The science of solar system ices* (pp. 371-408). Springer.

McCord, T., Carlson, R., Smythe, W. et al. (1997). Organics and other molecules in the surfaces of Callisto and Ganymede. *Science*, 278(5336), 271–275.

McCord, T., Hansen, G., Clark, R. et al. (1998). Non-water ice constituents in the surface material of the icy Galilean satellites from the Galileo near-infrared mapping spectrometer investigation. *Journal of Geophysical Research*, 103(E4), 8603–8626.

McCord, T., Hansen, G., and Hibbitts, C. (2001). Hydrated salt minerals on Ganymede's surface: Evidence of an ocean below. *Science*, 292(5521), 1523–1525. https://doi.org/10.1126/science.1059916.

McKinnon, W. (1999). Convective instability in Europa's floating ice shell. *Geophysical Research Letters*, 26(7), 951–954.

McKinnon, W., and Melosh, H. (1980). Evolution of planetary lithospheres: Evidence from multiringed structures on Ganymede and Callisto. *Icarus*, 44(2), 454–471.

Melosh, H., and Janes, D. (1989). Ice volcanism on Ariel. *Science*, 245(4914), 195–196.

Mullins, J., Ziegler, W., and Kirk, B. (1963). The thermodynamic properties of oxygen from 20 to 100 K. In: K. Timmerhaus (ed.), *Advances in cryogenic engineering* (pp. 126-134). Springer.

Mura, A., Adriani, A., Sordini, R. et al. (2020). Infrared observations of Ganymede from the Jovian infrared auroral mapper on Juno. *Journal of Geophysical Research: Planets*, 125(12), e2020JE006508.

Murchie, S., Head, J., Helfenstein, P., and Plescia, J. (1986). Terrain types and local-scale stratigraphy of grooved terrain on Ganymede. *Journal of Geophysical Research: Solid Earth*, 91(B13), E222–E238.

Murchie, S., Head, J., and Plescia, J. (1989). Crater densities and crater ages of different terrain types on Ganymede. *Icarus*, 81(2), 271–297.

Murchie, S., Head, J., and Plescia, J. (1990). Tectonic and volcanic evolution of dark terrain and its implications for the internal structure and evolution of Ganymede. *Journal of Geophysical Research: Solid Earth*, 95(B7), 10743–10768.

Murphy, D., and Koop, T. (2005). Review of the vapour pressures of ice and supercooled water for atmospheric applications. *Quarterly Journal of the Royal Meteorological Society: A Journal of the Atmospheric Sciences, Applied Meteorology and Physical Oceanography*, 131(608), 1539–1565.

Nakamura, R., and Ohtani, E. (2011). The high-pressure phase relation of the $MgSO_4$–H_2O system and its implication for the internal structure of Ganymede. *Icarus*, 211(1), 648–654.

Nakano, S., Moritoki, M., and Ohgaki, K. (1998). High-pressure phase equilibrium and Raman microprobe spectroscopic studies on the CO_2 hydrate system. *Journal of Chemical & Engineering Data*, 43(5), 807–810.

Narten, A., Venkatesh, C., and Rice, S. (1976). Diffraction pattern and structure of amorphous solid water at 10 and 77 K. *Journal of Chemical Physics*, 64(3), 1106–1121.

Nelson, R., Lane, A., Matson, D. et al. (1987). Spectral geometric albedos of the Galilean satellites from 0.24 to 0.34 micrometers: Observations with the International Ultraviolet Explorer. *Icarus*, 72(2), 358–380.

Nimmo, F., Pappalardo, R., and Giese, B. (2002). Effective elastic thickness and heat flux estimates on Ganymede. *Geophysical Research Letters*, 29(7), 1–4.

Noll, K., Johnson, R., Lane, A., Domingue, D., and Weaver, H. (1996). Detection of ozone on Ganymede. *Science*, 273(5273), 341–343.

Palumbo, M. (2005). The morphology of interstellar water ice. *Journal of Physics: Conference Series*, 6(1), 025.

Pappalardo, R., Collins, G., Head, J. et al. (2004). Geology of Ganymede. In: F. Bagenal, T. Dowling, and W. McKinnon (eds.), *Jupiter* (pp. 363–396). Cambridge University Press.

Pappalardo, R., Head, J., Collins, G. et al. (1998). Grooved terrain on Ganymede: First results from Galileo high-resolution imaging. *Icarus*, 135(1), 276–302.

Pappalardo, R., Nimmo, F., Giese, B. et al. (2003). Furrow topography and the elastic thickness of Ganymede's dark terrain lithosphere. In: *Lunar and Planetary Science Conference* (p. 1511). The Woodlands, TX.

Parmentier, E., Squyres, S., Head, J., and Allison, M. (1982). The tectonics of Ganymede. *Nature*, 295(5847), 290–293.

Patel, J., Pappalardo, R., Head, J. et al. (1999). Topographic wavelengths of Ganymede groove lanes from Fourier analysis of Galileo images. *Journal of Geophysical Research: Planets*, 104(E10), 24057–24074.

Patterson, G., Collins, G., Head, J. et al. (2010). Global geological mapping of Ganymede. *Icarus*, 207(2), 845–867.

Petrenko, V., and Whitworth, R. (1999). *Physics of ice*. Oxford University Press.

Poirier, J. (1985). *Creep of crystals: High-temperature deformation processes in metals, ceramics and minerals.* Cambridge University Press.

Pollack, J., Witteborn, F., Erickson, E. et al. (1978). Near-infrared spectra of the Galilean satellites: Observations and compositional implications. *Icarus*, 36, 271–303.

Prockter, L., Figueredo, P., Pappalardo, R., Head, J., and Collins, G. (2000). Geology and mapping of dark terrain on Ganymede and implications for grooved terrain formation. *Journal of Geophysical Research: Planets*, 105(E9), 22519–22540.

Prockter, L., and Pappalardo, R. (2000). Folds on Europa: Implications for crustal cycling and accommodation of extension. *Science*, 289(5481), 941–943.

Qi, C., Stern, L., Pathare, A., Durham, W., and Goldsby, D. (2018). Inhibition of grain boundary sliding in fine-grained ice by intergranular particles: Implications for planetary ice masses. *Geophysical Research Letters*, 45(23), 12757–12765.

Ross, R., Andersson, P., and Bäckström, G. (1981). Unusual PT dependence of thermal conductivity for a clathrate hydrate. *Nature*, 290(5804), 322–323.

Ross, R., and Kargel, J. (1998). Thermal conductivity of solar system ices, with special reference to Martian polar caps. In B. Schmitt, C. de Bergh, and M. Festou (eds.), *Solar system ices* (pp. 33–62). Springer.

Safarik, D., and Mullins, C. (2004). The nucleation rate of crystalline ice in amorphous solid water. *Journal of Chemical Physics*, 121(12), 6003–6010.

Schenk, P., Chapman, C., Zahnle, K., and Moore, J. (2004). Ages and interiors: The cratering record of the Galilean satellites. *Jupiter: The Planet, Satellites and Magnetosphere*, 2, 427.

Schenk, P., and McKinnon, W. (1991). Dark-ray and dark-floor craters on Ganymede, and the provenance of large impactors in the Jovian system. *Icarus*, 89(2), 318–346.

Schenk, P., and Moore, J. (1995). Volcanic constructs on Ganymede and Enceladus: Topographic evidence from stereo images and photoclinometry. *Journal of Geophysical Research*, 100 (E9), 19009–19022.

Schenk, P., McKinnon, W., Gwynn, D., and Moore, J. (2001). Flooding of Ganymede's bright terrains by low-viscosity water-ice lavas. *Nature*, 410(6824), 57–60.

Schmitt, B., Quirico, E., Trotta, F., and Grundy, W. (1998). Optical properties of ices from UV to infrared. In B. Schmitt, C. de Bergh, M. Festou (eds.), *Solar System Ices* (pp. 199–240). Springer.

Schulson, E. (1979). An analysis of the brittle to ductile transition in polycrystalline ice under tension. *Cold Regions Science and Technology*, 1(2), 87–91.

Shoemaker, E., Lucchita, B., Wilhelms, D., Plescia, J., and Squyres, S. (1982). The geology of Ganymede. In D. Morrison (ed.), *Satellites of Jupiter* (pp. 435–520). University of Arizona Press.

Sloan, E., and Koh, C. (2007) *Clathrate hydrates of natural gases*. 3rd ed. CRC Press.

Smith, B., Soderblom, L., Batson, R. et al. (1982). A new look at the Saturn system: The Voyager 2 images. *Science*, 215(4532), 504–537.

Smith, B., Soderblom, L., Beebe, R. et al. (1986). Voyager 2 in the Uranian system: Imaging science results. *Science*, 233(4759), 43–64.

Sohl, F., Choukroun, M., Kargel, J. et al. (2010). Subsurface water oceans on icy satellites: Chemical composition and exchange processes. *Space Science Reviews*, 153(1), 485–510.

Solomatov, V. (1995). Scaling of temperature-and stress-dependent viscosity convection. *Physics of Fluids*, 7(2), 266–274.

Sotin, C., Gillet, P., and Poirier, J. (1985). Creep of high-pressure ice VI. In J. Klinger, D. Benest, A. Dollfus, and R. Smoluchowski (eds.), *Ices in the solar system* (pp. 109–118), Springer Netherlands.

Sotin, C., and Poirier, J. (1987). Viscosity of ice V. *Le Journal de Physique Colloques* 48(C1), 233–238.

Spencer, J. (1987). Icy Galilean satellite reflectance spectra: Less ice on Ganymede and Callisto? *Icarus*, 70(1), 99–110.

Spencer, J., Calvin, W., and Person, M. (1995). CCD spectra of the Galilean satellites: Molecular oxygen on Ganymede (Paper 95JE01503). *Journal of Geophysical Research-Part E-Planets-Printed Edition*, 100(9), 19049–19056.

Squyres, S. (1980). Topographic domes on Ganymede: Ice vulcanism or isostatic upwarping? *Icarus*, 44(2), 472–480.

Squyres, S., Reynolds, R., Cassen, P., and Peale, S. (1983). Liquid water and active resurfacing on Europa. *Nature*, 301(5897), 225–226.

Stephan, K., Ciarniello, M., Poch et al. (2021). Spectral properties of H_2O ice depending on particle sizes and temperatures expected on Ganymede and Callisto. In *Lunar and Planetary Science Conference* (No. 2548, p. 1341). The Woodlands, TX.

Stern, L., Kirby, S., and Durham, W. (1996). Peculiarities of methane clathrate hydrate formation and solid-state deformation, including possible superheating of water ice. *Science*, 273(5283), 1843–1848.

Stevenson, D., and Lunine, J. (1986). Mobilization of cryogenic ice in outer solar system satellites. *Nature*, 323(6083), 46–48.

Strazzulla, G. (2011). Cosmic ion bombardment of the icy moons of Jupiter. *Nuclear Instruments and Methods in Physics Research Section B: Beam Interactions with Materials and Atoms*, 269(9), 842–851.

Strazzulla, G., Garozzo, M., Gomis, O. (2009). The origin of sulfur-bearing species on the surfaces of icy satellites. *Advances in Space Research*, 43(9), 1442–1445.

Sugisaki, M., Suga, H., and Seki, S. (1969). Calorimetric study of glass transition of the amorphous ice and of the phase transformation between the cubic and hexagonal ices. In Physics of ice (pp. 329–343). Platinum Press.

Sumarokov, V., Stachowiak, P., and Jeżowski, A. (2003). Low-temperature thermal conductivity of solid carbon dioxide. *Low Temperature Physics*, 29(5), 449–450

Thomas, P., and Schubert, G. (1988). Power law rheology of ice and the relaxation style and retention of craters on Ganymede. *Journal of Geophysical Research: Solid Earth*, 93(B11), 13755–13762.

Tse, J., and White, M. (1988). Origin of glassy crystalline behavior in the thermal properties of clathrate hydrates: A thermal conductivity study of tetrahydrofuran hydrate. *Journal of Physical Chemistry*, 92(17), 5006–5011.

Umurhan, O., Ahrens, C., and Chevrier, V. (2021). Rheological and thermophysical properties and some processes involving common volatile materials found on Pluto's surface. In: S. Stern, R. Binzel, W. Grundy, J. Moore, and L. Young (eds.), *The Pluto system after New Horizons*. University of Arizona Press.

Vance, S., and Brown, J. (2013). Thermodynamic properties of aqueous $MgSO_4$ to 800 MPa at temperatures from- 20 to 100 C and concentrations to 2.5 mol kg- 1 from sound speeds, with applications to icy world oceans. *Geochimica et Cosmochimica Acta*, 110, 176–189.

Vance, S., Bouffard, M., Choukroun, M., and Sotin, C. (2014). Ganymede's internal structure including thermodynamics of magnesium sulfate oceans in contact with ice. *Planetary and Space Science* 96, 62–70.

Vance, S., Panning, M., Sthler, S. et al. (2018). Geophysical investigations of habitability in ice-covered ocean worlds. *Journal of Geophysical Research: Planets*, 123(1), 180–205.

Wagner, W., and Pruss, A. (2002). The IAPWS formulation 1995 for the thermodynamic properties of ordinary water substance for general and scientific use. *Journal of Physical and Chemical Reference Data*, 31(2), 387–535.

Washburn, E. (1924). The vapor pressure of ice and of water below the freezing point. *Monthly Weather Review*, 52(10), 488–490.

Yamamoto, T., Nakagawa, N., and Fukui, Y. (1983). The chemical composition and thermal history of the ice of a cometary nucleus. *Astronomy and Astrophysics*, 122, 171–176.

Yamashita, Y., Kato, M., and Arakawa, M. (2010). Experimental study on the rheological properties of polycrystalline solid nitrogen and methane: Implications for tectonic processes on Triton. *Icarus*, 207(2), 972–977.

Zahnle, K., Schenk, P., Levison, H., and Dones, L. (2003). Cratering rates in the outer solar system. *Icarus*, 163(2), 263–289.

Zolotov, M., and Shock, E. (2001). Composition and stability of salts on the surface of Europa and their oceanic origin. *Journal of Geophysical Research: Planets*, 106(E12), 32815–32827.

12

Structure and Evolution of Ganymede's Hydrosphere

KLÁRA KALOUSOVÁ, KRISTA M. SODERLUND, ANEZINA SOLOMONIDOU, AND CHRISTOPHE SOTIN

12.1 INTRODUCTION

Ganymede belongs to the family of ocean worlds (Nimmo and Pappalardo, 2016; Hendrix et al., 2019). Its mean density of 1 942 kg m^{-3} indicates that it contains roughly 30 wt% of ice/water (Hussmann et al., 2015). The low value of the moment of inertia factor (MoI) of 0.3115 ± 0.0028 (Schubert et al., 2004) and the presence of a dynamo (Schubert et al., 1996) suggest that Ganymede is fully differentiated into an iron-rich liquid core, a rocky mantle, and a hydrosphere (Figure 12.1; for more details, see Chapter 6), which is estimated to be roughly 800–900 km thick (Anderson et al., 1996; Kuskov and Kronrod, 2001; Sohl et al., 2002; Vance et al., 2014, 2018). Due to the anomalous melting behavior of ice I and the high pressures predicted at the top of the rocky mantle, the hydrosphere likely has a three-layered structure of a deep high-pressure (HP) ice layer, a liquid ocean, and an outer ice I shell (Figure 12.1; e.g., Kivelson et al., 2002; Spohn and Schubert, 2003; Sotin and Tobie, 2004; Vance et al., 2014, 2018).

At least two lines of evidence argue for the presence of a deep liquid ocean. First, Galileo's magnetometer data at Ganymede can be explained by the sum of two dipoles – one internally generated (Kivelson et al., 1996) and one induced by Jupiter's time-variable magnetic field (Kivelson et al., 2002). Second, Saur and colleagues [2015] showed that the magnitude of Ganymede's auroral ovals oscillations measured by the Hubble Space Telescope is also best explained by the presence of an electrically conductive ocean.

In this chapter, we describe the structure and composition of Ganymede's hydrosphere, the dynamical processes expected in each layer (Sections 12.2–12.4), and the thermal evolution of the whole hydrosphere (Section 12.5). We also discuss surface morphologies possibly related to exchange processes through the ice shell, and the distribution of surface temperatures (Section 12.4). We discuss open questions and future exploration in Section 12.6 before concluding this chapter in Section 12.7.

12.2 HIGH-PRESSURE ICE LAYER

Ganymede's HP ice layer is between 300 and 700 km thick (Vance et al., 2018) and separates the rocky mantle from the ocean, which is often regarded as an impediment for Ganymede's habitability (Hand et al., 2020). The presence of the HP ice layer distinguishes Ganymede (along with Titan) from smaller moons such as Europa or Enceladus where the ocean is in direct contact with the rock.

12.2.1 Structure

High pressures at the rocky interface (1.5–1.7 GPa) lead to crystallization of a layer of tetragonal ice VI that is likely more than 400 km thick (Vance et al., 2018). The presence of other HP ice phases depends on the hydrosphere's composition and thermal structure (Figure 12.2). For the warmest case (light gray), the outer ice I shell is thin (D_I<30 km) and only ice VI crystallizes below the thick ocean (D_o~500 km). The corresponding high surface heat fluxes $\gtrsim$20 mW/m^2 are probably not realistic in the present (Section 12.4.1) but may be representative of younger Ganymede. For the intermediate profile (dark gray), both the HP ice VI layer and the ice I shell are thicker, resulting in a thinner ocean (D_o~300 km). For even colder interiors (black), other HP ice phases crystallize. In the case of a pure H_2O ocean (solid black), a layer of monoclinic ice V appears on top of ice VI and the ocean thickness is reduced to ~100 km. However, as indicated by the Galileo measurements, the ocean contains dissolved ions. Their composition is not well constrained, but magnesium sulfate is often considered (McCord et al., 2001) by analogy with Europa's ocean for which $MgSO_4$ was suggested to be dominant (Kargel, 1991; Hogenboom et al., 1995; Kargel et al., 2000; Fanale et al., 2001; McKinnon and Zolensky, 2003). If 10 wt% of $MgSO_4$ is present in Ganymede's ocean, ice V and tetragonal ice III can crystallize atop ice VI, thus reducing the ocean thickness to a few tens of kilometers (Figure 12.2, dashed black line). Moreover, ice III becomes buoyant with respect to the ocean water (Figure 12.2b), which may lead to the formation of a dense fluid layer between ices V and III (Vance and Brown, 2013). Similarly, ice VI may become buoyant with respect to highly saline water at the rocky interface (Journaux et al., 2013; Vance and Brown, 2013), resulting in the formation of a liquid layer in contact with rock. Such a setting could create habitable niches suitable for life (Vance et al., 2018). Note, however, that the model with one or more liquid layers sandwiched between the HP ice layers (Vance and Brown, 2013; Vance et al., 2014, 2018) satisfies thermodynamic relations at equilibrium but does not include a study of thermal evolution and building of such layers. Moreover, its stability against large-scale flow predicted in the HP ice layer (next section) remains to be investigated.

The presence of dissolved ions has important effects on the hydrosphere structure and, consequently, on the dynamic processes therein. So far, the ions were assumed to incorporate only into liquid, thus leaving pure solid phases. Although this assumption is supported by the small value of the partition coefficient of NaCl in ice I (~3×10^{-3}; Gross and Svec, 1997), recent studies show that small amounts of salt can be trapped in the pores of the freezing ice I at the ice–ocean interface, which

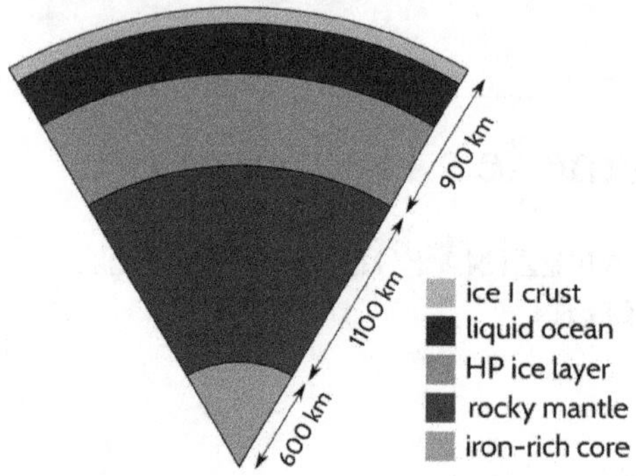

Figure 12.1 Possible internal structure of Ganymede suggested by Vance and colleagues [2018]

may eventually lead to the formation of larger brine pockets (Vance et al., 2019; Buffo et al., 2020; Steinbrügge et al., 2020). Even small briny inclusions may also attenuate radar signals (Pettinelli et al., 2015; Kalousová et al., 2017; Culha et al., 2020), which has strong implications for future radar sounding of Ganymede's ice shell by the JUICE mission. Measurements by Journaux and colleagues [2017] indicate that ice VI behaves similarly to ice I and rejects salts from its structure with the partition coefficient of NaCl analogues $\sim 5\times 10^{-3}$ close to that of ice I. However, as shown by Journaux and colleagues [2017], even small amounts of salts incorporated into ice VI substantially affect its density. Measurements for other relevant HP ices (V, III) and salt species ($MgSO_4$, $MgCl_2$, Na_2SO_4) are key for further constraining the complex processes occurring in the hydrosphere of Ganymede as well as other ocean worlds.

12.2.2 Dynamics

The HP ice layer is underlain by the rocky mantle that supplies it with heat from radioactive decay (Chapter 6), which drives the layer dynamics. Viscosity of ice VI, which mainly controls the dynamics, was measured by Poirier and colleagues [1981], Sotin and colleagues [1985], and Durham and colleagues [1996] under high stresses ($>10^6$ Pa). Extrapolation of their results to convective stresses ($\sim 10^3$–10^4 Pa) results in a wide range ($\sim 10^{13}$–10^{20} Pa s) of melting-point viscosity values, which may be caused by the change in deformation mechanism (Kalousová et al., 2018) as also reported for ice I (Goldsby and Kohlstedt, 2001). Results of both groups indicate that viscosity is almost constant along the melting curve, justifying the use of an Arrhenius law in simulations (Choblet et al., 2017a; Kalousová et al., 2018). Sotin and Poirier (1987) found that ice V is harder to deform than ice VI, with their viscosity ratio larger than 10, while Durham and colleagues [1996] found their viscosities comparable. However, the temperature and stress dependencies of the two phases are different, and Durham and colleagues [1996] suggested that under the same experimental conditions, they may deform by distinct mechanisms – ice V by dislocation creep and ice VI by grain-size sensitive creep. Ice III, possibly present in the case of a salty cold ocean (Section 12.2.1), is the weakest under laboratory conditions (Durham et al., 1997). For further details, see Journaux and colleagues [2020]. Let us note that despite the number of measurements, the HP ice viscosity remains poorly constrained and new measurements are required to better describe the corresponding dynamics.

Even though the occurrence of convection in HP ices was suggested several decades ago (Poirier et al., 1981), few numerical models have been developed due to the expected melt production. With the top boundary at the pressure-melting point and the temperature contrast across the layer a few tens of Kelvins (Figure 12.2a), the HP ice layer temperature is always close to the melting curve. Choblet and colleagues [2017a] modeled thermal convection in an ice VI layer and observed melt production mostly in the upwellings and at the top boundary. In their model, melt was instantaneously extracted into the ocean, which allows researchers to estimate its amount but neglects its effect on convection – since H_2O melt (water) is less dense than HP ice, it migrates upward through the ice, thus adding buoyancy that enhances the thermally driven flow. This is similar to silicate melts that are buoyant with respect to their solids but different from an ice I shell, where melt generation leads to the collapse of any warm upwelling (Tobie et al., 2003; Kalousová et al., 2016).

In magma dynamics, two-phase models are used to treat the melting process and the subsequent melt evolution (e.g., McKenzie, 1984). Figure 12.3 shows a snapshot from Kalousová and colleagues [2018], who applied a two-phase model to Ganymede's HP ice layer. Since the melting temperature T_m decreases with decreasing depth (pressure, full black line in panel a), a temperate (partially molten) layer where $T = T_m$ forms just below the top boundary (dashed line in panel a, black area in panel b). A few percent of melt is present in the top temperate layer (panel c), which is extracted into the overlying ocean. Note that this amount of melt depends on the percolation threshold (porosity above which liquid can percolate through solid), which is not well constrained but typically considered on the order of a few percent. The larger the threshold value, the more melt can accumulate in the temperate layer before being extracted into the ocean (Kalousová et al., 2018). The presence of a top temperate layer eliminates the cold thermal boundary layer (TBL) and convection is thus only driven by the hot TBL – the convection pattern is characterized by the upwelling plumes and passive background downwelling (Kalousová and Sotin, 2018). Apart from the top temperate layer, some melt may be present in the upwelling plumes as well as at the bottom boundary (Choblet et al., 2017a; Kalousová et al., 2018). The occurrence of bottom melting depends on the rocky interface temperature (T_r; Figure 12.2a): if $T_r = T_m$, melt is produced that can dissolve volatiles present at the interface and transfer them to the ocean via warm upwelling plumes, thus ensuring the exchange of material between the rock and the ocean.

The temperature at the rocky interface is determined by the efficiency of heat transfer through the HP ice layer by thermal convection, which can be described by the Rayleigh number:

$$Ra = \frac{\alpha \rho^2 c_p g \Delta T H^3}{k\eta}, \tag{12.1}$$

with thermal expansivity α, density ρ, heat capacity at constant pressure c_p, gravity g, the temperature contrast ΔT across the layer of thickness H, thermal conductivity k, and viscosity η. This nondimensional number compares the gravitationally induced thermal buoyancy (i.e., the mechanism driving the

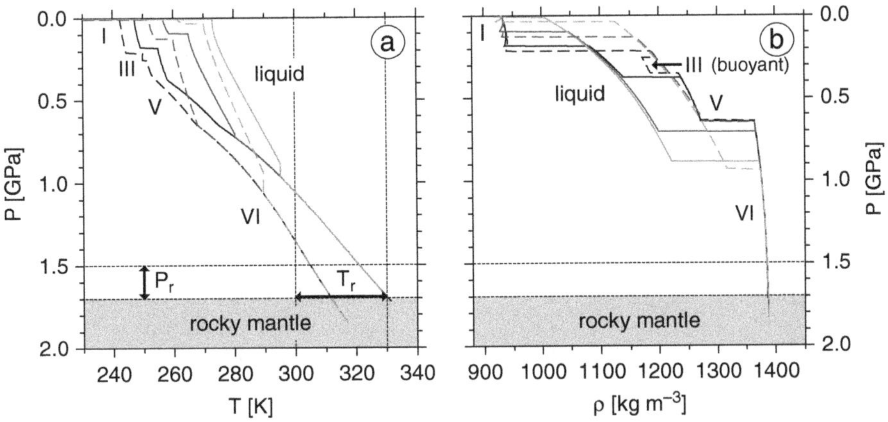

Figure 12.2 Temperature (a) and density (b) versus pressure for H_2O oceans (solid lines) and oceans with 10 wt% of $MgSO_4$ (dashed lines) and three thermal profiles: warm (D_I<30 km, light gray), intermediate (dark gray), and cold (D_I>130 km, black). Data from Vance (2017). Roman numerals indicate the particular ice phase. Top of rocky mantle (gray rectangles) and the pressure at the rocky interface (P_r~1.5–1.7 GPa, horizontal dashed lines) are indicated in both panels. Vertical dashed lines (panel a) mark the expected range of rocky interface temperatures (T_r~300–330 K, Kalousová and Sotin, 2018).

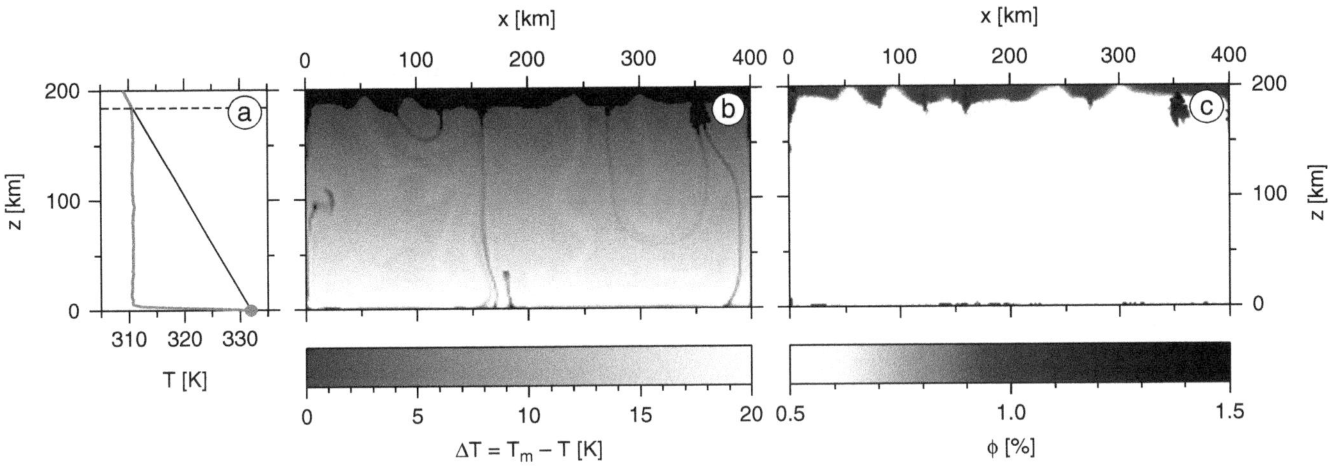

Figure 12.3 (a) Horizontally averaged temperature (solid gray), melting curve (solid black), and top temperate layer (dashed black). (b) Temperature difference from melting temperature ($\Delta T = T_m - T$). Black color marks temperate ice ($T = T_m$). (c) Water content (porosity). Modified from Kalousová and colleagues [2018]. Reprinted from *Icarus*, 299, Klára Kalousová, Christophe Sotin, Gaël Choblet, Gabriel Tobie, Olivier Grasset, Two-phase convection in Ganymede's high-pressure ice layer — Implications for its geological evolution, 133–147, ©2018, with permission from Elsevier.

flow) with two mechanisms that tend to resist the flow – diffusion of heat (described by heat diffusivity, $\kappa = \frac{k}{\rho c_p}$) and momentum (described by viscosity). Convection occurs if $Ra \gtrsim 10^3$, which is satisfied in the HP ice layer. For large values of Ra, convection is very efficient and $T_r < T_m$, while for smaller values of Ra, convection is less efficient and melting can occur at the rocky interface. The critical value of Ra for bottom melting depends on the rocky mantle heat flux q_r (Kalousová and Sotin, 2018) and can be inverted to provide the scaling for critical heat flux q_r^c necessary for bottom melting (Kalousová and Sotin, 2020a). Figure 12.4 shows q_r^c as a function of HP ice layer thickness H and reference viscosity η_0. If the mantle heat flux, q_r, is larger than the critical value, q_r^c, bottom melting will occur. On the other hand, if $q_r < q_r^c$, no melt is generated. Currently, q_r is expected to be between 9 and 11 mW/m^2 (Chapter 6, Figure 12.4, gray rectangle). Assuming that $H > 400$ km (Vance et al., 2018), no melt is predicted on Ganymede's rocky interface unless the reference ice viscosity $\eta_0 \gtrsim 10^{16}$ Pa s. However, melting may have occurred earlier during Ganymede's history when the HP ice layer was thinner and the rocky heat flux larger (Kalousová et al., 2018; Kalousová and Sotin, 2018).

The presented results were obtained for an ice VI layer, but other HP ices may be present (Figure 12.2). To investigate the effect of phase transitions, Bercovici and colleagues [1986] performed a linear stability analysis of a two-layered model assuming that both layers have the same viscosity and thickness but different density and latent heat. They found that exothermic phase transitions (ices III–V, V–VI) can enhance or diminish whole-layer convection, depending on ice viscosity, while endothermic phase transitions (ices I–III) always led to two-layer convection. Sotin and Parmentier (1989) considered two layers with different thickness and viscosity and predicted one-layer convection for a large range of phase transitions. Recently, Kalousová and Sotin (2020a) investigated the effect of the ice VI–V phase change numerically, including different densities, latent heats, and several ice VI/V viscosity ratios. They found that even with a viscosity contrast of 100, convection still

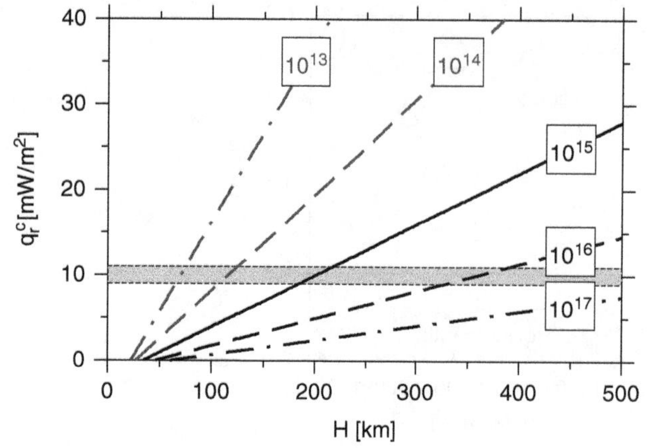

Figure 12.4 Critical heat flux q_r^c for bottom melting as a function of HP ice layer thickness H for different ice VI reference viscosities $\eta_0=10^{13}-10^{17}$ Pa s (line styles). The range of possible present-day values of q_r (Chapter 6) is marked by the gray rectangle.

occurs in one layer, in agreement with Sotin and Parmentier (1989).

The predicted water outflow into the ocean raises a question of whether it can influence the ocean circulation. The models of oceanic thermal convection (Section 12.3) give ocean flow velocities in the range of meters per second (Figure 12.5), several orders of magnitude larger than the outflow velocity from Ganymede's HP ice layer (~1 m/yr; Kalousová and Sotin, 2018). Therefore, the laterally varying water outflow from the HP ice layer does not directly influence the oceanic flow, but it may provide an uneven heating pattern that can modulate heat transfer through Ganymede's ocean. This should be investigated in future studies.

12.3 LIQUID OCEAN

12.3.1 Structure

The ocean can reach thicknesses of approximately 500 km for the warmest pure water and 10 wt% $MgSO_4$ scenarios described by Vance and colleagues [2018]. The saline ocean is about 100 kg/m^3 denser than the fresh ocean, and the density contrasts across the ocean are $\rho_i - \rho_o \sim 200$ kg/m^3 for both compositions (Figure 12.2b). These radial increases represent a fraction of a density scale height $N_\rho = \ln(\rho_o/\rho_i) \lesssim 0.2$. The equation of state for water can be approximated linearly as

$$\rho = \rho_0 + \rho' = \rho_0[1 - \alpha(T - T_0) + \beta(S - S_0)], \tag{12.2}$$

when the thermal- and salinity-derived density anomalies (ρ') are small relative to the reference state (ρ_0, T_0, S_0). The fractional changes in density of a material as its temperature and salinity changes are the thermal expansion and haline compaction coefficients, respectively:

$$\alpha = -\frac{1}{\rho}\frac{\partial\rho}{\partial T}, \quad \beta = \frac{1}{\rho}\frac{\partial\rho}{\partial S}. \tag{12.3}$$

While most materials expand with increased temperature ($\alpha > 0$), water at icy ocean world conditions can have negative expansivity values (e.g., Melosh et al., 2004). At Ganymede, this is unusual but may occur in a ~100 m thick layer near the outermost ice–ocean interface as indicated by the 10 wt% $MgSO_4$ ocean scenario considered by Vance and colleagues [2014]. Thus, warm fluid parcels are expected to rise when $\alpha > 0$ and sink in the less common scenarios where $\alpha < 0$. In contrast, $\beta > 0$ is expected for all ocean scenarios such that salty fluids will nominally sink and fresher water will rise. More complex behaviors can occur when density anomalies due to thermal and compositional perturbations are comparable in amplitude but opposite in direction, as described in Section 12.3.2.

Ocean temperatures are generally assumed to follow the convective adiabat, which depends on density, thermal expansivity, and specific heat capacity: $\partial T/\partial P = \alpha T/(\rho C_p)$. Thermal expansivity can vary by an order of magnitude across the ocean, while C_p varies much less significantly with depth (Vance et al., 2014). Temperatures are found to increase by 4 K every 100 km in a pure water ocean by Vance and colleagues [2018]; the adiabatic gradient is smaller for a 10 wt% $MgSO_4$ ocean (Figure 12.2a).

The bulk ocean salinity is not well known, but it is thought to be at least 0.9 grams $MgSO_4$ per kilogram of ocean water in order to be consistent with measurements of Ganymede's induced magnetic field and auroral oval oscillations (Saur et al., 2015). The vertical distribution of salts is essentially unconstrained.

12.3.2 Dynamics

There are several potential mechanisms to drive flows within Ganymede's ocean: mechanical stirring through tides, libration, and/or precession; electromagnetic pumping due to coupling with Jupiter's magnetic field; and thermo-compositional convection associated with temperature and salinity anomalies within the ocean (e.g., Soderlund et al., 2020).

Tides are one possible source of fluid motions, including both eccentricity and obliquity contributions. The simplest tidal models solve the Tidal Laplace Equations that assume a constant ocean depth that is much larger than tidal displacements, depth-independent (i.e., barotropic) flows where horizontal velocities greatly exceed those in the radial direction, and hydrostatic equilibrium. These studies show that oceans respond through gravity waves and Rossby–Haurwitz waves that are most prominent for eccentricity and obliquity forcings, respectively (e.g., Hay and Matsuyama, 2019). Eccentricity tides focus tidal heating at low latitudes and the leading/trailing hemispheres, while obliquity tides lead to peak dissipation at the poles with an equatorial minimum (e.g., Chen et al., 2014; Matsuyama, 2014; Matsuyama et al., 2018). At Ganymede, ocean dissipation from obliquity tides is predicted to significantly exceed that from eccentricity tides (on the order of 10^9 W compared to 10^3 W) (Hay and Matsuyama, 2019).

Additional fluid motions are allowed when this shallow water assumption is relaxed, such as internal inertial waves where energy can be focused along shear layers within the bulk fluid and lead to enhanced tidal dissipation near the poles. These flows have been considered for Europa and Enceladus, but not in the context of Ganymede explicitly (e.g., Rekier et al., 2019; Rovira-Navarro et al., 2019).

Libration, or periodic variations in the axial rate of rotation, may also drive motions (e.g., Le Bars et al., 2015). In a spherical

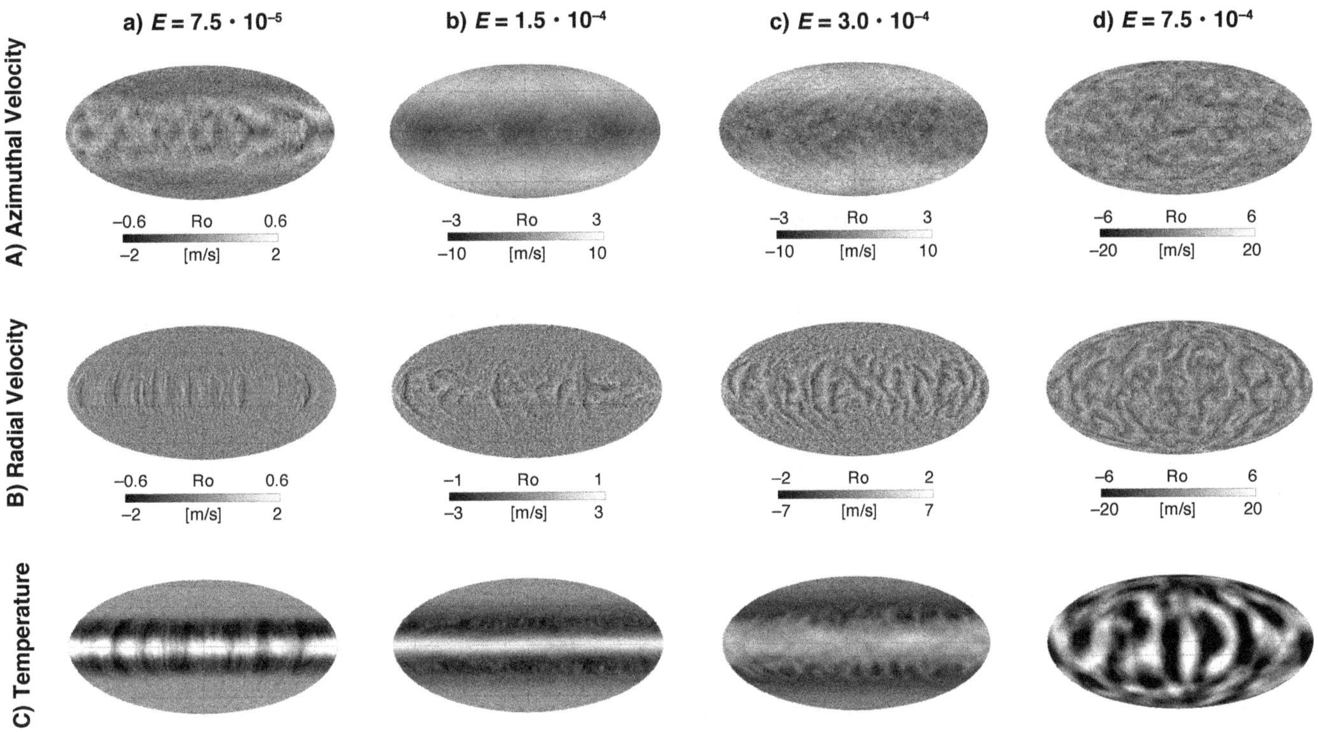

Figure 12.5 A) Azimuthal velocity, B) radial velocity, and C) temperature fields at mid-ocean depths for the four possible ocean convection regimes identified by Soderlund (2019), where the influence of rotation decreases from a) to d). The velocity fields are shown at an instantaneous snapshot in time, while the temperature field has been time-averaged. Black (white) indicates westward (eastward) zonal winds in A, black (white) indicates downwelling (upwelling) radial flows in B, and black (white) indicates cool (warm) fluid in C. Speeds are given in both dimensionless Rossby number units $Ro = U/\Omega D$ and dimensionalized units assuming rotation rate $\Omega = 1.0 \times 10^{-5}\ \mathrm{s}^{-1}$ and an ocean thickness of $D = 340$ km. (A black and white version of this figure will appear in some formats. For the color version, please refer to the plate section.)

shell, three regimes have been identified (Noir et al., 2009): inertial modes, longitudinal rolls beneath the outer boundary, and boundary layer turbulence. Longitudinal rolls were predicted to develop in Ganymede's core and application of their arguments to the subsurface ocean suggests rolls would form here as well (cf. Zhang et al., 2013). If an elliptical geometry is instead considered, the system may become unstable to libration-driven elliptical instability (LDEI) where two inertial modes interact with the basic flow response to the harmonic forcing. Lemasquerier and colleagues [2017] conducted a stability analysis to assess which planetary bodies may develop elliptical instabilities, and predicted Ganymede's ocean to be stable to LDEI.

Precession is a third possible mechanism to mechanically stir the ocean. However, the associated changes in the orientation of the rotation axis are poorly constrained and not considered further.

Ganymede's traverse through the Jovian magnetosphere may drive a second type of fluid motions through electromagnetic pumping (Gissinger and Petitdemange, 2019). The rotation of Jupiter's magnetic field and its interaction with the satellite's induced field will generate a Lorentz force that drives a retrograde equatorial jet with speeds reaching a few mm/s and an associated, weaker overturning circulation at low latitudes (Gissinger and Petitdemange, 2019). An additional consequence is ohmic dissipation within the ocean, which would be concentrated in a thin region below the outer boundary at high latitudes; at Europa, up to 10^8 W of power may be anticipated from this mechanism, and lower values would be expected at Ganymede.

Thermo-compositional convection is the third potential driver of fluid motions in Ganymede's ocean. Thermal buoyancy will result from heat transferred from the underlying HP ice layer as well as latent heat associated with phase changes, where melting absorbs heat and freezing releases it. Similarly, compositional buoyancy may result from the influx of salts along the seafloor, the rejection of salts as water freezes, and the release of fresh water as ice melts. If temperature decreases and salinity increases with radius, thermal and compositional buoyancy sources may reinforce each other to increase the vigor of convection. When the temperature and salinity gradients oppose each other, however, double-diffusive convection (DDC) may occur since heat and salt diffuse at different rates (e.g., Rosenblum et al., 2011; Radko, 2013).

When salinity increases with radius, overturning convection can exist even in the presence of a weak stable temperature contrast, and fingering DDC, where salty fluid descends into the underlying fresher water in characteristic "finger" plumes, can develop for stronger contrasts. Conversely, when salinity decreases with radius, the system is stable for a larger range of temperature gradients. As temperature instability is increased, a "semi-convective" regime where transport is controlled by diffusive processes develops and overturning convection is still possible with large contrasts.

In addition to the relative contributions of thermal and salinity gradients, the effect of rotation must also be considered (e.g.,

Soderlund et al., 2014; Soderlund, 2019; Amit et al., 2020). Near the onset of convection, the fluid motions are strongly constrained by the Coriolis force and tend to form columns that are aligned with the rotation axis. For a fixed rotation rate, increased convective driving weakens this vertical stiffness until they become fully three-dimensional and ultimately are no longer impacted by rotation. Determining what controls the transitions between these regimes at extreme conditions relevant to planetary interiors is an active area of research (e.g., Cheng et al., 2018). Given the uncertainties in scaling behaviors and limited constraints on Ganymede's internal properties, predictions for oceanic convection span from quasi-two-dimensional (quasi-2D) to quasi-isotropic turbulence (Soderlund, 2019). These regimes also correspond to different velocity and heat transfer patterns (Figure 12.5). At mid-ocean depths, the quasi-2D case has prograde (eastward) flow at low latitudes and retrograde (westward) flow at higher latitudes that can reach speeds on the order of m/s, instantaneous radial flows with similar speeds that are most prominent near the equator, and warmest/coolest mean ocean temperatures at the equator/mid latitudes (Figure 12.5a). The quasi-isotropic case has no clear zonal or radial orientations with peak speeds of tens of m/s and no preferred heat transfer orientations (Figure 12.5d). The intermediate cases have strong (several m/s) zonal flows organized into three jets with low-latitude radial flows that are only slightly weaker and an ocean temperature distribution that is qualitatively similar to the quasi-2D case (Figure 12.5b–c). This similarity in heat transfer occurs because all three of these cases have warm upwelling fluid near the equator with adjacent downwellings of cooler water when averaged in time (with flow speeds about 10–100 times weaker) (Soderlund, 2019). It is also worth noting that the velocity and heat flow patterns may be dependent on the boundary conditions assumed in the models; Amit and colleagues [2020] instead find enhanced heat flux at low latitudes when rotational effects are strong and enhanced polar flux when rotation plays a lesser role in their models that employ no-slip boundary conditions rather than the stress-free boundaries assumed in Soderlund (2019).

Models with regions of stable stratification have not yet been investigated in the context of Ganymede (cf. Lobo et al., 2021; Zeng and Jansen, 2021, for Enceladus), nor has the influence of mid-ocean HP ice formation or mantle outflow variations as described in Section 12.2, which are directions for future work. However, given the large flow speeds described earlier, especially for convection, it might be expected that any stable layers would be transient and mid-ocean ice layers would be difficult to maintain.

12.3.3 Implications

Ocean dynamics may have detectable consequences for the overlying ice shell and magnetic induction signal that can be used to test hypotheses for ocean flows and ice–ocean interactions. Ice shell thickness is controlled by its viscosity and thermal properties, heat production in the ice shell and deep interior, and heat transport characteristics of the ocean. While spatial variations in heating rates may promote thickness variations, viscous relaxation and ice pumping due to pressure-induced variations in the melting point act to reduce basal topography (e.g., Lewis and Perkin, 1986; Kamata and Nimmo, 2017). Viscous relaxation is most efficient at smaller wavelengths, but can also occur over global length scales and serve as a mechanism for horizontal ice transport (Ashkenazy et al., 2018). The ice pump mechanism, where melting (freezing) occurs where ice is relatively thick (thin) can similarly operate over local to global scales (e.g., Vance and Goodman, 2009; Soderlund et al., 2014). As ice freezes, it effectively rejects impurities such as salts that are present in the ocean, which might further establish spatial variations in ice shell composition and thermo-mechanical properties (Buffo et al., 2020). Ice shell thickness measurements may, conversely, be used to infer where melting and freezing are occurring at the base of the ice shell and the heat flux coming from the ocean, as has been done for Titan and Enceladus (Kvorka et al., 2018; Čadek et al., 2019).

Ganymede's ocean is embedded within two magnetic fields, that of its core dynamo and the Jovian magnetosphere. While ocean detection relied on interactions with the Jovian magnetosphere (Kivelson et al., 2002) and the diffusive induction response to this field may reveal compositional stratification in the ocean via sensitivity to depth-dependent electrical conductivity (Vance et al., 2021), motional induction is sensitive to the satellite's intrinsic magnetic field. Here, salty ocean water flowing within this background magnetic field generates a second type of induced magnetic field (Tyler, 2011; Vance et al., 2021). Vance and colleagues [2021] used a simple scaling relationship to estimate an upper bound for such a field at the satellite's surface:

$$b_{surface} \sim \mu_0 \sigma D U B_0 \left(\frac{r_{ocean}}{r_{satellite}} \right)^{\ell+2}, \tag{12.4}$$

where μ_0 is the permeability of free space, B_0 is the background magnetic field strength, and ℓ is spherical harmonic degree. Interior models were used to self-consistently estimate electrical conductivity σ, ocean thickness D, radius of the ocean r_{ocean}, and radius of the satellite $r_{satellite}$ for a range of potential ocean compositions, salinities, and thicknesses. Flow speed U was taken from global convection models (Soderlund, 2019). Induced surface field strengths up to ~300 nT for the dipole $\ell = 1$ component were found, which is nearly half of the intrinsic field strength at the equator. Because of this potentially large signal, it may be diagnostic of ocean flows if it can be distinguished from other potential origins. If the signal is present but not accounted for, it may bias inversions of the ocean's electrical conductivity and the core's intrinsic field. Further investigation of motional induction at Ganymede is, therefore, an important direction for future work.

12.4 OUTER ICE I SHELL

12.4.1 Thickness and Thermal State

Ganymede's ice shell thickness is not well constrained, but estimates have been obtained using different techniques and data sets. Based on a systematic study of impact crater morphology, Schenk (2002) inferred that Ganymede's ocean must currently be at least 80 km below the surface. Another piece of information is obtained from fitting Galileo magnetometer data, which favors an ice shell thickness ~150 km (Kivelson et al., 2002). The maximum shell thickness, given by the triple point

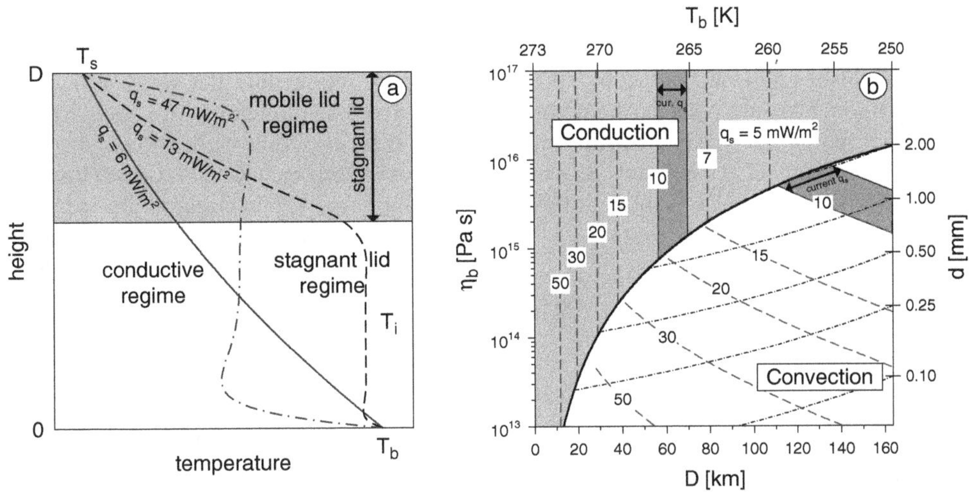

Figure 12.6 (a) Temperature profiles for different regimes: conduction (gray solid line), stagnant lid convection (black dashed line), and mobile lid convection (gray dash-dotted line) with corresponding values of surface heat flux q_S. Stagnant lid (gray region), and basal (T_b), surface (T_S), and convective (T_i) temperatures are indicated. (b) Basal viscosity η_b critical for the onset of convection (white area, assuming volume diffusion creep) as a function of shell thickness D. The corresponding basal temperature T_b is shown on the top axis. Constant grain size d evaluated from T_b and η_b (eq. 12.5) is marked by black dash-dotted lines with the values given on the right axis. The surface heat flux q_S in both regimes (conduction, convection), is shown by gray dashed lines with the values indicated in the figure. The currently estimated surface heat flux q_S~8–10 mW/m^2 is highlighted by dark gray areas (see text for details).

of ice I–ice III–liquid, is ~160 km in the case of pure H_2O hydrosphere, but is likely altered by the presence of other dissolved compounds (Section 12.2.1). Fitting the mass, radius, and MoI while using equations of state of pure water and aqueous $MgSO_4$, Vance and colleagues (2014) and Vance and colleagues (2018) obtained a large range of ice shell thicknesses between a few tens of kilometers up to more than 130 km (Figure 12.2). However, thin shell models result in heat fluxes incompatible with the current estimate (Figure 12.6b). It therefore seems likely that Ganymede's ice I shell is between 80 and 160 km thick.

Depending mainly on the shell thickness and viscosity, Ganymede's internal heat is transferred by conduction or convection (Figure 12.6a). The corresponding temperature profiles differ substantially with the conductive shell being colder (gray solid line) and the convective shell warmer. Furthermore, convection can occur in different regimes (e.g., Moresi and Solomatov, 1995). For a strongly temperature-dependent rheology with the viscosity contrast across the shell $>10^6$, convection occurs in the stagnant lid regime (black dashed line) with a thick conductive lid (gray area) present above a quasi-isoviscous convecting layer. In the mobile lid regime, viscosity is limited by the material strength, the lid is much thinner, and plate tectonics can occur. To obtain an approximate thermal profile, the viscosity contrast was limited to 10^3 (gray dash-dotted line). While in the stagnant lid regime, heat transfer is slowed by the conductive lid (surface heat flux q_S~13 mW/m^2; Figure 12.6a) and the exchange of materials is limited, in the mobile lid regime, the transfer of material and heat is more efficient (q_S~47 mW/m^2; Figure 12.6a).

The conductive heat transfer is slow and would allow the long-term stability of a deep ocean without the presence of antifreezes or additional heating (e.g., Consolmagno and Lewis, 1976). Thermal convection is more efficient (compare conductive q_S~6 mW/m^2 with convective q_S~13 mW/m^2, Figure 12.6a), and it was shown that Ganymede's H_2O ocean would freeze in geologically short times if the shell convects vigorously and is not tidally heated (e.g., Reynolds and Cassen, 1979; Spohn and Schubert, 2003). The critical thickness for convection to occur in Ganymede's outer shell depends on the ice I viscosity – a nonlinear function of temperature, stress, and grain size. Based on experiments, Goldsby and Kohlstedt (2001) described three creep regimes – dislocation, grain boundary sliding, and basal slip – and a fourth creep, diffusion, was predicted to occur at low stresses (Raj and Ashby, 1971). A constitutive equation combining these four creeps was proposed by Goldsby and Kohlstedt (2001) (see also Chapter 11). While the pressure/stress and temperature conditions within the icy shells are well constrained, the grain size can vary over several orders of magnitude. Barr and McKinnon (2007) developed models of grain size evolution in a convective ice shell and found two end-member states: (i) for pure H_2O ice, grains may grow due to dynamic recrystallization up to several centimeters, which is hard to reconcile with a convective interior – even if convection started with smaller grains, it would probably cease when the grain size reached a few centimeters (see also Barr and Pappalardo, 2005); (ii) in the presence of soluble ions or silicate microparticles, the grains are kept much smaller (1–5 mm), which permits convection in shells as thin as ~30 km. Tidal stresses may also limit the grain growth and keep its size close to 1 mm (McKinnon, 1999), while tidal heating may provide a sufficiently large temperature perturbation to initiate convection in case of larger grains / thinner shells (Barr et al., 2004; Barr and Pappalardo, 2005).

For diffusion creep, the grain size d can be linked to ice viscosity η and temperature T as follows (Goldsby and Kohlstedt, 2001):

$$d = \left[\frac{84 V_m \eta D_0}{RT} \exp\left(-\frac{Q}{RT}\right)\right]^{1/2}, \qquad (12.5)$$

with R the universal gas constant, and V_m, D_0, and Q the molar volume, preexponential, and activation energy of volume diffusion creep (Table 12.1; Goldsby and Kohlstedt, 2001). If the Rayleigh number (eq. 12.1), which mainly depends on the effective layer thickness H and viscosity η, exceeds a critical value, convection occurs. In the stagnant lid regime and assuming linear rheology (such as diffusion creep), the critical Rayleigh number is given by:

$$Ra_{\mathrm{cr}} = 20.9\theta^4 \tag{12.6}$$

(Solomatov, 1995). Here $\theta = Q\Delta T/RT_{\mathrm{i}}^2$, $\Delta T = T_{\mathrm{b}} - T_{\mathrm{s}}$ is the temperature contrast across the shell, T_{s} is the surface temperature and T_{i} is the convective temperature (Figure 12.6a), which can be evaluated from basal temperature T_{b} (see Deschamps and Sotin, 2000, for more details). We follow the approach of McKinnon (2006) to derive a critical basal viscosity η_{b} necessary for the onset of convection as a function of basal temperature T_{b}, which itself depends on the shell thickness D. For the sake of simplicity, we only assume deformation by volume diffusion creep (the possible effect of other creeps is discussed later). The basal temperature T_{b} and the shell thickness D are linked through the melting curve

$$D = \frac{1}{\rho g}\left\{P_0 + a\left[\left(\frac{T_{\mathrm{b}}}{T_0}\right)^b - 1\right]\right\}, \tag{12.7}$$

with the experimentally derived parameters P_0, T_0, a, and b (Table 12.1). We assume temperature-dependent thermal conductivity

$$k(T) = \frac{k_0}{T} \tag{12.8}$$

(Petrenko and Whitworth, 2002) and rescale the effective layer thickness H (evaluated for k=*const.*) that appears in Rayleigh number (eq. 12.1) to obtain the true ice shell thickness D that takes into account eq. (12.8):

$$D = H\frac{k_0}{k_{\mathrm{i}}\Delta T}\ln\left(\frac{T_{\mathrm{b}}}{T_{\mathrm{s}}}\right) \tag{12.9}$$

(see, e.g., Barr and Showman, 2009). Here k_{i} is the thermal conductivity at T_{i}. The critical basal viscosity can be obtained by inverting eq. (12.1) and plugging in the critical Rayleigh number Ra_{cr}:

$$\eta_{\mathrm{b}} = \frac{\alpha_{\mathrm{i}}\rho^2 c_p^{\mathrm{i}} g\Delta T H^3}{k_{\mathrm{i}} Ra_{\mathrm{cr}}}, \tag{12.10}$$

with α_{i} and c_p^{i} evaluated at T_{i} from

$$\alpha(T) = \alpha_0\left(\frac{T}{T_{\mathrm{ref}}}\right), \quad c_p(T) = c_p^0\left(\frac{T}{T_{\mathrm{ref}}}\right) \tag{12.11}$$

(Kirk and Stevenson, 1987). For the sake of simplicity, we assume constant ρ and g (Table 12.1).

Figure 12.6b shows a critical basal viscosity η_{b} (eq. 12.10) for convection to occur as a function of Ganymede's ice shell thickness D (eq. 12.9). In the gray region above the critical curve (black solid line), viscosity is too large and heat is transferred by conduction. Below the critical curve (white region), viscosity is small enough (for given D) to allow convection. Note that, for simplicity, we do not distinguish between different convection regimes and assume only steady-state stagnant lid convection. Grain size d, evaluated from T_{b} (top axis) and η_{b} (left axis) using eq. (12.5), is depicted by black dash-dotted lines. We found that convection is possible for $d \lesssim 2$ mm (right axis), roughly in agreement with the Callisto-focused study by McKinnon (2006).

Conductive surface heat flux depends on the true ice shell thickness D but is independent of viscosity:

$$q_{\mathrm{s}}^{\mathrm{cd}} = \frac{k_0}{D}\ln\left(\frac{T_{\mathrm{b}}}{T_{\mathrm{s}}}\right)\left(1 - \frac{D}{R_{\mathrm{s}}}\right), \tag{12.12}$$

with R_{s} the surface radius (McKinnon, 2006). Convective heat flux can be evaluated from the definition of the Nusselt number (ratio of convective to conductive heat flux):

$$q_{\mathrm{s}}^{\mathrm{cv}} = q_{\mathrm{s}}^{\mathrm{cd}} Nu. \tag{12.13}$$

For steady-state stagnant lid convection, Nu can be estimated from Ra and θ:

$$Nu = 2.0\,\theta^{-1} Ra_{\mathrm{i}}^{1/5} \tag{12.14}$$

(e.g., Dumoulin et al., 1999), with Ra_{i} evaluated at T_{i}. In Figure 12.6b, the obtained heat fluxes are indicated by gray dashed lines. Note the jump in heat flux for a fixed shell thickness D at the transition from conduction to convection, showing that convection is more efficient in transferring heat (see Mitri and Showman, 2008, for more details). Using different scaling for Nu or assuming other creeps would change the values of q_{s} – this is left for another study.

The surface heat flux is a sum of radiogenic heat q_r from the mantle (Chapter 6), heat due to hydrosphere cooling q_c, and tidal heat q_t (Chapter 4). Currently, tidal heating is not expected to significantly contribute to Ganymede's heat budget [e.g., Cassen et al., 1980; Kirk and Stevenson, 1987; Chapter 4]. Taking the radiogenic heat fluxes at the mantle–hydrosphere interface between 9 and 11 mW/m^2 (Chapter 6), the surface radiogenic heat flux q_{r} is between ~4 and 6 mW/m^2. The thermal evolution models (Section 12.5, Figure 12.8b; Bland et al., 2009) suggest an ice shell crystallization rate (assuming pure H_2O hydrosphere) of ~100 km/Gyr, which corresponds to surface heat flux q_c~4 mW/m^2. We thus arrive at the total surface heat flux q_{s}~8–10 mW/m^2 (dark gray regions in Figure 12.6b). Both conductive and convective shell can explain these values.

So far, we assumed only diffusion creep. However, Barr and Pappalardo (2005) showed that grain boundary sliding controls the plume growth for grain sizes between ~0.4 mm and 1 cm. In cases of modest temperature perturbations (~15 K) and large shell thicknesses (~130 km), grain sizes up to ~5 mm are compatible with convection. Even larger grain sizes require larger temperature perturbations, which might have been provided by tidal heating during the periods of increased eccentricity

Table 12.1 *List of parameters*

Symbol	Physical property	Value	Unit
R	universal gas constant	8.314	J/mol/K
V_m	molar volume[a]	1.97×10^{-5}	m^3/mol
D_0	volume diffusion creep preexponential[a]	9.1×10^{-4}	m^2/s
Q	volume diffusion creep activation energy[a]	59.4	kJ/mol
k_0	thermal conductivity constant (eq. 12.8)[b]	651	W/m
α_0	thermal expansion constant (eq. 12.11)[c]	1.56×10^{-4}	1/K
c_p^0	heat capacity constant (eq. 12.11)[c]	1925	J/kg/K
T_{ref}	reference temperature[c]	250	K
ρ	ice density	930	kg/m^3
g	gravity acceleration	1.428	m/s^2
T_s	surface temperature[d]	110	K
R_s	surface radius[e]	2631	km
P_0	reference pressure[f]	0	Pa
T_0	reference melting temperature[f]	273.15	K
a	melting curve parameter[f]	−395.2	MPa
b	melting curve parameter[f]	9	–

[a]Goldsby and Kohlstedt (2001), [b]Petrenko and Whitworth (2002), [c]Kirk and Stevenson (1987), [d]Squyres (1980a), [e]Hussmann et al. (2015), [f]Chizhov (1993)

(Section 12.5; Malhotra, 1991; Showman and Malhotra, 1997). For $d>2$ cm, the plume growth could be triggered by large temperature perturbations and accommodated by highly non-linear dislocation creep (Barr and Pappalardo, 2005), but the necessary shell thickness would be larger than ~160 km (Ganymede's maximum shell thickness), which excludes dislocation as a convection-triggering creep in Ganymede's shell. A creep enhancement close to the melting temperature observed by Goldsby and Kohlstedt (2001) may substantially reduce basal viscosity and thus allow convection for larger grains or thinner shells. A strong, localized heat source such as frictional heating on a fault can also trigger an upwelling (e.g., Han and Showman, 2008; Kalousová et al., 2016).

The presence of non-ice compounds such as ammonia, (hydrated) salts, or silicate microparticles can impede the grain growth (Barr and McKinnon, 2007). Furthermore, the rheological parameters of the shell material may be quite different from pure H_2O ice and thus the described results may be significantly altered. However, no study to date has investigated these effects in detail.

12.4.2 Surface Morphologies Related to Exchange Processes

The observations by Voyager and Galileo spacecraft showed that Ganymede's surface has a variety of geological units, formed at different times and under different conditions (Chapter 7). Smith and colleagues [1979] described four terrain types: (i) dark, densely cratered terrain, (ii) grooved terrain of diverse patterns and ages, (iii) young rugged impact basin and ejecta blanket, and (iv) smooth terrain, which occurs in patches that locally obscure older terrains. Grooves are formed during extension periods that are connected with the interior evolution. Relaxation of impact craters provides information on the brittle/ductile transition depth and therefore the heat flux and temperature profiles at the time of the impact. Smooth terrain patches may be related to resurfacing.

The bright, ancient, and heavily tectonized grooved terrain (Figure 12.7a) covers about two-thirds of Ganymede's surface. Smith and colleagues [1979] suggested that it formed during the era of global surface expansion that resulted from satellite differentiation (Squyres, 1980b). Although the stresses in Ganymede's present-day ice shell seem too low for fracture initiation (e.g., Cameron et al., 2020), the tidal stresses could have been up to an order of magnitude higher during the period of increased eccentricity [Showman et al., 1997; see also Section 12.5], thus allowing the initiation of tectonic processes (Showman and Malhotra, 1997; Cameron et al., 2020). The resulting cracks and graben faults are imprinted on the grooved terrains' surfaces (Prockter et al., 2010). Different groups modeled the grooves' formation and inferred the brittle/ductile transition depth consistent with the observed topography (e.g., Herrick and Stevenson, 1990; Dombard and McKinnon, 2001; Bland et al., 2010). A major difficulty in understanding these processes comes from the large variations of creep deformation at the brittle/ductile transition. Dombard and McKinnon (2001) proposed thermal gradients of 20–40 K/km, which are equivalent to heat fluxes >100 mW/m^2. Bland and colleagues [2010] found smaller (about half) values, but still much higher than radiogenic heat flux from the mantle. To explain these high heat fluxes, they suggested transient processes such as changes in tidal dissipation related to variations in eccentricity (Section 12.5). However, the recent models of grooved terrain

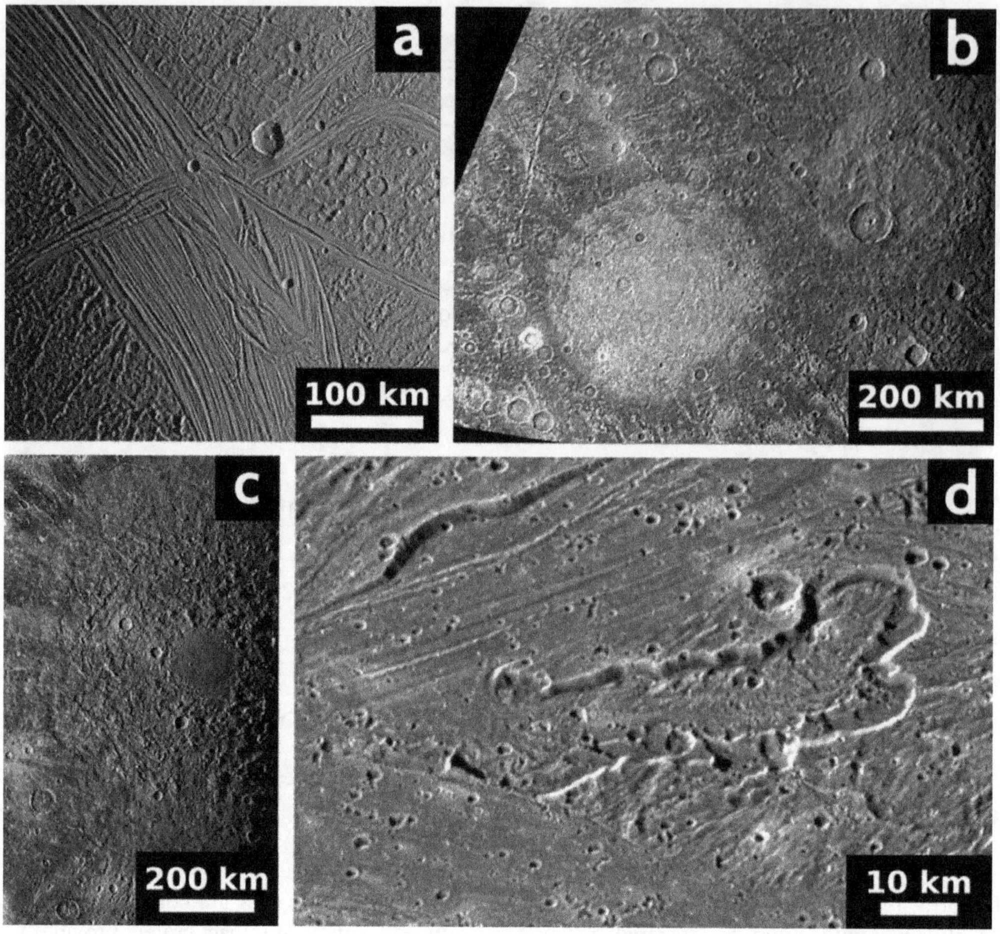

Figure 12.7 Examples of surface morphologies that may be related to surface–ocean exchange: (a) grooved terrain (Tiamat Sulcus), (b) palimpsest (Memphis Facula), (c) multi-ringed crater (Gilgamesh), (d) cryovolcanic area (in Sippar Sulcus). Credits: NASA/JPL/Brown University (a, d), Paul M. Schenk, Lunar and Planetary Institute (b, c)

formation by Howell and Pappalardo (2018) do not support surface exposition of oceanic material.

Ganymede experienced a heavy cratering epoch 3.5–4 Gyr ago, and impact craters can be found on both the bright and dark terrains. The craters are relatively flat, which is attributed to the weak nature of the icy shell. Palimpsests (Figure 12.7b), which preferentially occur in the densely cratered, old, dark terrains, are impact craters with a complete absence of a relief. Dombard and McKinnon (2006) suggested that palimpsests may have formed after preexisting impact craters were entirely erased by viscous surface relaxation that occurred several hundreds of million years ago. Alternatively, they may have formed over a thin, brittle lithosphere, indicating a direct and/or indirect ice–ocean exchange at the time of formation (Shoemaker et al., 1982). The large-scale impact excavation suggested by the size of the multi-ringed impact crater basins (Figure 12.7c) represents another possibility of direct ice–ocean exchange, although the ice shell thickness at the time of their formation is unknown. It is thus not clear whether the ice shell would have been ruptured even by these large impactors (Soderlund et al., 2020).

Smooth terrains could be expressions of resurfacing events. The Voyager images suggested that cryovolcanic activity is responsible for the formation of these terrains (e.g. Smith et al., 1979; Murchie et al., 1990), before the Galileo camera brought new evidence that weakened the previous claims (e.g., Belton et al., 1996). However, the small, isolated, scalloped depressions called paterae have been interpreted as potential caldera-like cryovolcanic source vents (Schenk and Moore, 1995; Kay and Head, 1999; Spaun et al., 2001), and Head and colleagues [1998] interpreted a ridged deposit associated with the largest patera within Sippar Sulcus (Figure 12.7d) as an icy flow. Pappalardo and colleagues [2004] suggested that volcanism and extensional tectonism operate together in particularly smooth regions of Ganymede's bright terrain where the paterae have been observed. In the absence of high-resolution imaging data, cryovolcanism is assumed to have acted locally, which is also supported by the relatively small size of the paterae. The most favorable hypothesis for their formation is diapirism, which could have brought ocean material to the surface or near the surface (Barr and Showman, 2009). Apart from water ice, Ganymede's surface displays an unknown darkening component for which the measured spectra suggest either hydrated materials or some oxygen-bearing molecules (Chapter 11). However, the correlation between the terrains suggestive of material exchange and the distribution of non-ice material is not as high as on Europa (Chapter 10). The high-resolution observations by the JUICE mission that will target these potentially cryovolcanic areas should bring more insight on the possible ice–ocean exchange (Stephan et al., 2021).

12.4.3 Surface Temperatures

Ganymede's surface temperature varies significantly between the equator and the poles, which has important implications for the ice shell dynamics. Surface temperatures were measured by the photopolarimeter-radiometer (PPR) instrument onboard the Galileo spacecraft (Orton et al., 1996) and more recently by the Atacama Large Millimeter/Submillimeter Array (ALMA) observatory (de Kleer et al., 2021). These observations show that the diurnal variations are large with brightness temperature peaks around 150 K at the equatorial subsolar point. The time-averaged subsurface temperature, where diurnal variations are negligible, shows latitudinal variations between 120 and 60 K (Squyres, 1980a). This boundary condition plays a critical role in determining the amount and spatial distribution of heat that can be transferred by the ice shell. Assuming only conductive heat transfer and taking into account the temperature-dependent thermal conductivity (eq. 12.8), the heat flux at the pole would be almost twice as large as at the equator assuming that the ice shell thickness and the temperature at the ice/ocean interface are the same (eq. 12.12). Conversely, models of ice shell convection suggest that pole-to-equator insolation variations can lead to higher heat flow with extensional deformation near the equator and lower heat flow with compressional deformation at higher latitudes (Weller et al., 2019). Such large latitudinal variations should be taken into account when building thermal evolution models.

12.5 HYDROSPHERE EVOLUTION

Although several lines of evidence support the current presence of a deep ocean in Ganymede's interior (Section 12.1), it is not clear what process is responsible for keeping it liquid on geological timescales. Tidal heating likely sustains the oceans of Europa and Enceladus (e.g., Tobie et al., 2005; Choblet et al., 2017b), and the presence of ammonia or an insulating layer of methane clathrates has been proposed for Titan (Tobie et al., 2006; Kalousová and Sotin, 2020b). However, in the case of Ganymede, no significant tidal heating is expected at present [e.g., Cassen et al., 1980; Kirk and Stevenson, 1987; Chapter 4] and there is no argument in favor of a strongly insulating surface layer such as methane clathrates on Titan.

The presence of oceans in the icy moons' interiors was first suggested by Lewis (1971), who predicted that satellites with radius larger than ~500 km should have a molten interior. The long-term thermal evolution of Ganymede was then investigated by several studies spanning the past few decades. While Consolmagno and Lewis (1976) assumed only conductive heat transfer through the ice shell that preserved the liquid ocean till today, Reynolds and Cassen (1979) predicted the occurrence of thermal convection and showed that it would efficiently freeze the ocean on geologically short timescales. Kirk and Stevenson (1987) developed a thermal evolution model that took into account convection in both the ice shell and the HP ice layer. They showed that, depending on the ice viscosity, the ocean could either stay liquid (large viscosity) or solidify (low viscosity). For the latter case, they predicted a heat pulse at the time of merging of the two layers (ice I and HP ice) and suggested its connection with the resurfacing and formation of Ganymede's grooved terrain.

Currently, tidal heating is probably negligible due to Ganymede's small eccentricity (0.0013) and large distance from Jupiter (Chapter 4), but it might have been more important throughout Ganymede's history. A number of papers have indicated that Ganymede (as well as Europa) might have had dramatically larger eccentricities before the Laplace resonance was established between the three inner Galilean moons. While Tittemore (1990) suggested that the two moons might have undergone a period of chaotic motion, Malhotra (1991) proposed that Io, Europa, and Ganymede might have gone through another three-body resonance with different orbital frequencies. Both authors argued that Ganymede's significantly higher past eccentricity is responsible for its resurfacing and thus the striking difference from its twin, Callisto. Showman and Malhotra (1997) investigated a larger range of parameters and initial conditions than Malhotra (1991) and found a capture into a different orbital resonance that might have pumped Ganymede's eccentricity to as high as 0.07, resulting in tidal heating several hundred times larger than at present. Showman and colleagues [1997] investigated conditions necessary for the occurrence of global thermal runaway due to passage through an eccentricity-pumping resonance using a thermal-orbital evolution model. They found that a global process was not likely, but they did not exclude the occurrence of smaller events that might have led to resurfacing.

Apart from strong tidal heating earlier in Ganymede's history, compounds such as magnesium sulfate (e.g., Kargel, 1991) or ammonia (e.g., Kargel, 1992) have been proposed to help preserve the ocean liquid. Spohn and Schubert (2003) used equilibrium models of heat transfer to investigate the effect of ammonia and found that with a pure H_2O model, Ganymede would not have an ocean at present, but the addition of 5 wt% of ammonia would enable an ocean as thick as 200–350 km. However, ammonia was never identified in the spectra recorded by the Galileo spacecraft although it was found in the spectra of some Saturnian moons (e.g., Clark et al., 2014), and its absence is in agreement with the higher accretion temperatures expected in the Jovian system when compared with the Saturnian system (Lunine and Stevenson, 1982). Spohn and Schubert (2003) pointed out the important role of a stagnant lid (Section 12.4.1) in reducing the amount of transferred heat, which was also reported by Freeman (2006).

Bland and colleagues [2009] investigated the long-term stability of Ganymede's ocean using a thermal-orbital evolution model that took into account grain-size sensitive ice rheology. They started with a frozen hydrosphere and compared two scenarios. First, they investigated thermal evolution without tidal heating and found that an ocean up to 600 km thick could have formed if the ice grain size (d) was $\gtrsim$1 mm (Figure 12.8a). For $d = 1$ mm, the ocean would freeze after ~4 Gyr due to efficient heat removal by ice shell convection (full black line), but for $d = 10$ mm (gray line), they found that an ocean was stable until today due to conduction in the shell. In the second scenario, Bland and colleagues [2009] included the passage through a Laplace-like resonance that increased Ganymede's eccentricity up to 0.015, leading to tidal heating as strong as

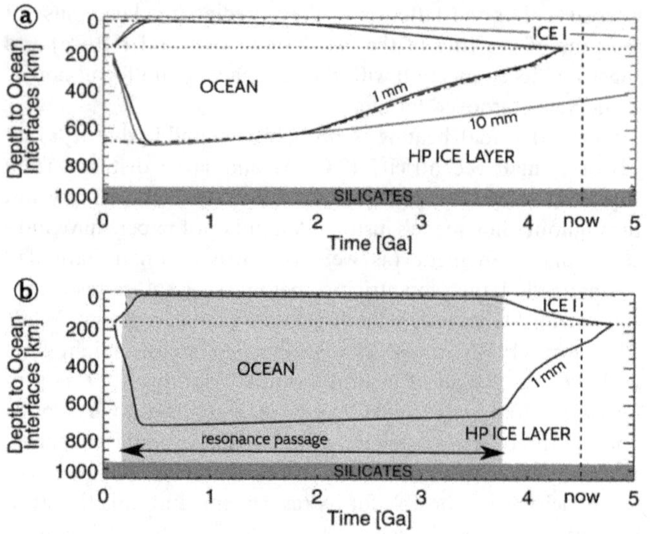

Figure 12.8 Thermal evolution of Ganymede's hydrosphere – depth to the ice I/ocean and ocean/HP ice interfaces as a function of time: (a) without tidal dissipation, (b) with resonance passage (light gray region). The top of rocky mantle (dark gray region), the depth of ice I/HP ice transition (horizontal dotted line), and present time (vertical dashed line) are marked in both panels. Two grain sizes (1 and 10 mm) are considered in panel a; only $d = 1$ mm is assumed in panel b. Modified from Bland and colleagues [2009]. Reprinted from *Icarus*, 200(1), Michael T. Bland, Adam P. Showman, Gabriel Tobie, The orbital–thermal evolution and global expansion of Ganymede, 207–221, ©2009, with permission from Elsevier.

$\sim$1 TW (Figure 12.8b). During the resonance passage (light gray area), the ocean thickness decreases only slowly due to very high tidal heating. Upon exiting the resonance, the ocean starts to thin rapidly, but it takes $\sim$1 Gyr before it freezes completely. Depending on the time of the resonance escape, such a scenario could explain the current presence of Ganymede's ocean. Importantly, the surface heat flux is also increased during the resonance passage with the maximum reaching almost 40 mW/m^2 – a value roughly consistent with the lower-end estimates of Ganymede's ancient heat flux from flexural models (Nimmo et al., 2002; Nimmo and Pappalardo, 2004), as well as from models of grooved terrain formation [Dombard and McKinnon, 2001; Bland and Showman, 2007; Section 12.4.2]. A modeled surface heat flux at $t\sim4.5$ Gyr is $\sim$8 mW/m^2, in good agreement with the estimate inferred in Section 12.4.1 (Figure 12.6b).

12.6 DISCUSSION

12.6.1 Dynamics with Non-ice Components

Various compounds (NaCl, $MgSO_4$, …) may be present in Ganymede's hydrosphere (Section 12.2.1). In general, ice (both low- and high-pressure polymorphs) does not incorporate chemical impurities very well (e.g., Gross and Svec, 1997; Journaux et al., 2017), so these compounds tend to stay in the ocean. However, gradual freezing of the ocean as well as other processes (such as large impacts or freezing of fractures) may lead to the capture of oceanic solutes within the solid ice with their amount, distribution, and composition depending on the rate of ice formation (e.g., Zolotov and Kargel, 2009). Such a process would generate compositional variations within the ice layers (e.g., Buffo et al., 2020), which could have a profound effect on the layer dynamics in three regards. First, since these compounds can depress the melting temperature by up to few tens of degrees (e.g., Hogenboom et al., 1995; Journaux et al., 2013), their presence would result in an earlier onset of melting and production of more melt. Second, the density contrast between the solid and liquid phase would be altered, possibly leading to density inversion and creation of local brine pockets or layers (e.g., Vance et al., 2014). Finally, non-ice components can impede the ice grain growth (Barr and McKinnon, 2007), thus keeping the ice viscosity relatively low. While the effect of salts on solid-state convection was investigated for example, for Europa (Han and Showman, 2005), no study so far has investigated the effect of salts on the process of melt generation in the ice layers and its subsequent transport. Future work should be devoted to this very important topic.

12.6.2 Future Exploration

Ganymede is the main target of the European Space Agency (ESA) JUpiter ICy moons Explorer (JUICE) mission, which launched on April 14, 2023. JUICE will spend more than a year in orbit around Ganymede (Grasset et al., 2013), thus becoming the first spacecraft orbiting an ocean world. Several of the 11 onboard instruments will make observations in an effort to determine the hydrosphere structure and dynamics. Measurements of the magnetic field should narrow down the range of ocean conductivity, which in turn provides constraints on the ocean composition, depth, and thickness (Dougherty and the J-MAG Team, 2013). Gravity measurements will improve our understanding of mass distribution in Ganymede's interior (the MoI factor) and, via the tidal Love number k_2, update the constraints on the ocean density and thus its composition. Determining the phase lag should also provide an important argument in the question of the presence of Ganymede's ocean (Kamata et al., 2016). The Ganymede laser altimeter (GALA) that will study the moon's tidal deformation as well as its surface characteristics (topography, morphology) should provide more arguments for or against the ocean presence via measurements of the tidal response, the libration, and the determination of h_2 that describes the radial surface displacement (Moore and Schubert, 2003; Baland and Van Hoolst, 2010; Steinbrügge et al., 2015). These observations should also improve the ice shell thickness estimates with an error of less than $\sim$20 km (Baland and Van Hoolst, 2010; Steinbrügge et al., 2015). Combining topographic maps and gravity field will allow the detection of subsurface mass anomalies (mascons) that were suggested by Palguta and colleagues [2009] based on the Galileo gravity data analysis. Finally, the ice-penetrating radar (RIME) will characterize compositional, thermal, and/or structural variations in the outer icy shell to test hypotheses about tectonic, convective, and impact processes (Bruzzone et al., 2015). The RIME will also search for present or past reservoirs of liquid water by detecting strong reflections from large bodies as well as scatter from smaller reservoirs (Bruzzone et al., 2015). Similarly, any shallow compositional interfaces should be detectable (Di Paolo et al., 2017; Kalousová et al., 2017; Sbalchiero et al., 2019). The sounding of Ganymede's ice shell may also constrain its thickness by either directly detecting the ocean (if the shell is only a few kilometers thick) or, more likely, by detecting the brittle/ductile transition and deeper structures within a thick convective shell (Bruzzone et al., 2015; Di Paolo et al., 2017; Kalousová et al., 2017).

12.7 CONCLUSIONS

Ganymede's hydrosphere consists of three layers: a deep HP ice layer, a liquid water ocean, and an outer ice I shell. Vigorous convection and melting are predicted to occur in Ganymede's HP ice layer that, earlier in its history, might have led to volatile transport from the rocky core into the ocean. In recent years, much progress has been made in modeling the dynamics of the deep ocean by taking into account the relative contributions of thermal and compositional density anomalies as well as the effect of rotation. The intense tectonics affecting Ganymede's surface points to a convective outer ice I shell in the past with potential local cryovolcanism. The preservation of a deep ocean can be explained by either a late cooling of the hydrosphere due to strong tidal dissipation episodes in the past or a conductive ice shell. Future modeling efforts should couple the models of each of the three layers to allow the use of self-consistent boundary conditions – the presence of a temperate layer at the HP ice–ocean interface should affect the ocean circulation models as would the ice pump process at the ocean–ice shell interface. Also, laboratory experiments investigating the properties of ices and fluids under high pressures and different composition would be critical. Such work could be performed before ESA's JUICE mission arrives at Ganymede and provides the wealth of observations that will help us understand the complex dynamical processes within this magnificent ocean world.

ACKNOWLEDGMENTS

We thank an anonymous reviewer for useful comments that helped to improve the manuscript. K. K. was supported by the Czech Science Foundation through project No. 19-10809S and by Charles University Research Program No. UNCE/SCI/023. K. M. S. was partially supported by NASA Grant NNX14AR28G. C. S. was partially supported by NASA award 80NSSC19K1427 "Exploring Ocean Worlds: Ocean System Science to support the search for life."

REFERENCES

Amit, H., Choblet, G., Tobie, G. et al. (2020). Cooling patterns in rotating thin spherical shells: Application to Titan's subsurface ocean. *Icarus* **338**, 113509.

Anderson, J. D., Lau, E. L., Sjogren, W. L., Schubert, G., and Moore, W. B. (1996). Gravitational constraints on the internal structure of Ganymede. *Nature* **384**(6609), 541–543.

Ashkenazy, Y., Sayag, R., and Tziperman, E. (2018). Dynamics of the global meridional ice flow of Europa's icy shell. *Nature Astronomy* **2**(1), 43–49.

Baland, R.-M., and Van Hoolst, T. (2010). Librations of the Galilean satellites: The influence of global internal liquid layers. *Icarus* **209**(2), 651–664.

Barr, A. C., and McKinnon, W. B. (2007). Convection in ice I shells and mantles with self-consistent grain size. *Journal of Geophysical Research: Planets* **112**(E02012).

Barr, A. C., and Pappalardo, R. T. (2005). Onset of convection in the icy Galilean satellites: Influence of rheology. *Journal of Geophysical Research: Planets* **110**(E12).

Barr, A. C., Pappalardo, R. T., and Zhong, S. (2004). Convective instability in ice I with non-Newtonian rheology: Application to the icy Galilean satellites. *Journal of Geophysical Research: Planets* **109**(E12).

Barr, A. C. and Showman, A. P. (2009). Heat transfer in Europa's icy shell. In R. T. Pallardo, W. B. McKinnon, and K. Khurana, eds., *Europa*, University of Arizona Press, pp. 405–430.

Belton, M. J. S., Head, J. W., Ingersoll, A. P. et al. (1996). Galileo's first images of Jupiter and the Galilean satellites. *Science* **274**(5286), 377–385.

Bercovici, D., Schubert, G., and Reynolds, R. T. (1986). Phase transitions and convection in icy satellites. *Geophysical Research Letters* **13**(5), 448–451.

Bland, M. T., McKinnon, W. B., and Showman, A. P. (2010). The effects of strain localization on the formation of Ganymede's grooved terrain. *Icarus* **210**(1), 396–410.

Bland, M. T., and Showman, A. P. (2007). The formation of Ganymede's grooved terrain: Numerical modeling of extensional necking instabilities. *Icarus* **189**(2), 439–456.

Bland, M. T., Showman, A. P., and Tobie, G. (2009). The orbital-thermal evolution and global expansion of Ganymede. *Icarus* **200**(1), 207–221.

Bruzzone, L., Plaut, J. J., Alberti, G. et al. (2015). JUpiter ICy moon Explorer (JUICE): Advances in the design of the radar for icy moons (RIME), *in* 2015 IEEE International Geoscience and Remote Sensing Symposium (IGARSS). pp. 1257–1260.

Buffo, J. J., Schmidt, B. E., Huber, C., and Walker, C. C. (2020). Entrainment and dynamics of ocean-derived impurities within Europa's ice shell. *Journal of Geophysical Research: Planets* **125**(10), e2020JE006394.

Čadek, O., Souček, O., Běhounková, M. et al. (2019). Long-term stability of Enceladus uneven ice shell. *Icarus* **319**, 476–484.

Cameron, M. E., Smith-Konter, B. R., Collins, G. C., Patthoff, D. A., and Pappalardo, R. T. (2020). Ganymede, then and now: How past eccentricity may have altered tidally driven coulomb failure. *Journal of Geophysical Research: Planets* **125**(7), e2019JE005995.

Cassen, P., Peale, S. J., and Reynolds, R. T. (1980). On the comparative evolution of Ganymede and Callisto. *Icarus* **41**(2), 232–239.

Chen, E., Nimmo, F., and Glatzmaier, G. (2014). Tidal heating in icy satellite oceans. *Icarus* **229**, 11–30.

Cheng, J. S., Aurnou, J. M., Julien, K., and Kunnen, R. P. J. (2018). A heuristic framework for next-generation models of geostrophic convective turbulence. *Geophysical & Astrophysical Fluid Dynamics* **112**(4), 277–300.

Chizhov, V. E. (1993). Thermodynamic properties and thermal equations of the state of high-pressure ice phases. *Journal of Applied Mechanics and Technical Physics* **34**, 253–262.

Choblet, G., Tobie, G., Sotin, C., Kalousová, K., and Grasset, O. (2017a). Heat transport in the high-pressure ice mantle of large icy moons. *Icarus* **285**, 252–262.

Choblet, G., Tobie, G., Sotin, C. et al. (2017b). Powering prolonged hydrothermal activity inside Enceladus. *Nature Astronomy* **1**, 841–847.

Clark, R. N., Swayze, G. A., Carlson, R., Grundy, W., and Noll, K. (2014). Spectroscopy from Space. *Reviews in Mineralogy and Geochemistry* **78**(1), 399–446.

Consolmagno, G. J., and Lewis, J. S. (1976). Structural and thermal models of icy Galilean satellites, *in* T. A. Gehrels, ed., Jupiter. University of Arizona Press, pp. 1035–1051.

Culha, C., Schroeder, D. M., Jordan, T. M., and Haynes, M. S. (2020). Assessing the detectability of Europa's eutectic zone using radar sounding. *Icarus* **339**, 113578.

de Kleer, K., Butler, B., de Pater, I. et al. (2021). Ganymede's surface properties from millimeter and infrared thermal emission. *Planetary Science Journal* **2**, 5.

Deschamps, F., and Sotin, C. (2000). Inversion of two-dimensional numerical convection experiments for a fluid with a strongly temperature-dependent viscosity. *Geophysical Journal International* **143**(1), 204–218.

Di Paolo, F., Lauro, S. E., Castelletti, D. et al. (2017). Radar signal penetration and horizons detection on Europa through numerical simulations. *IEEE Journal of Selected Topics in Applied Earth Observations and Remote Sensing* **10**(1), 118–129.

Dombard, A. J., and McKinnon, W. B. (2001). Formation of grooved terrain on Ganymede: Extensional instability mediated by cold, superplastic creep. *Icarus* **154**(2), 321–336.

Dombard, A. J., and McKinnon, W. B. (2006). Elastoviscoplastic relaxation of impact crater topography with application to Ganymede and Callisto. *Journal of Geophysical Research: Planets* **111**(E1).

Dougherty, M., and the J-MAG Team (2013). J-MAG: Magnetometer science on the JUICE mission, *in* European Planetary Science Congress 2013. Vol. 8, pp. EPSC2013-470.

Dumoulin, C., Doin, M.-P., and Fleitout, L. (1999). Heat transport in stagnant lid convection with temperature- and pressure-dependent Newtonian or non-Newtonian rheology. *Journal of Geophysical Research: Solid Earth* **104**(B6), 12759–12777.

Durham, W. B., Kirby, S. H., and Stern, L. A. (1997). Creep of water ices at planetary conditions: A compilation. *Journal of Geophysical Research: Planets* **102**(E7), 16293–16302.

Durham, W. B., Stern, L. A., and Kirby, S. H. (1996). Rheology of water ices V and VI. *Journal of Geophysical Research: Solid Earth* **101**(B2), 2989–3001.

Fanale, F. P., Li, Y.-H., De Carlo, E. et al. (2001). An experimental estimate of Europa's 'ocean' composition independent of Galileo orbital remote sensing. *Journal of Geophysical Research: Planets* **106**(E7), 14595–14600.

Freeman, J. (2006). Non-Newtonian stagnant lid convection and the thermal evolution of Ganymede and Callisto. *Planetary and Space Science* **54**(1), 2–14.

Gissinger, C., and Petitdemange, L. (2019). A magnetically driven equatorial jet in Europa's ocean. *Nature Astronomy* **3**, 401–407.

Goldsby, D. L., and Kohlstedt, D. L. (2001). Superplastic deformation of ice: Experimental observations. *Journal of Geophysical Research: Solid Earth* **106**(B6), 11017–11030.

Grasset, O., Dougherty, M., Coustenis, A. et al. (2013). JUpiter ICy moons Explorer (JUICE): An ESA mission to orbit Ganymede and to characterise the Jupiter system. *Planetary and Space Science* **78**, 1–21.

Gross, G. W., and Svec, R. K. (1997). Effect of ammonium on anion uptake and dielectric relaxation in laboratory-grown ice columns. *Journal of Physical Chemistry B* **101**, 6282–6284.

Han, L., and Showman, A. P. (2005). Thermo-compositional convection in Europa's icy shell with salinity. *Geophysical Research Letters* **32**(20).

Han, L., and Showman, A. P. (2008). Implications of shear heating and fracture zones for ridge formation on Europa. *Geophysical Research Letters* **35**(3).

Hand, K. P., Sotin, C., Hayes, A., and Coustenis, A. (2020). On the habitability and future exploration of ocean worlds. *Space Science Reviews* **216**, 95.

Hay, H. C., and Matsuyama, I. (2019). Nonlinear tidal dissipation in the subsurface oceans of Enceladus and other icy satellites. *Icarus* **319**, 68–85.

Head, J. W., Pappalardo, R. T., Kay, J. et al. (1998). Cryovolcanism on Ganymede: Evidence in bright terrain from Galileo solid state imaging data, *in* 'Lunar and Planetary Science Conference. Vol. 29, p. 1666.

Hendrix, A. R., Hurford, T. A., Barge, L. M. et al. (2019). The NASA roadmap to ocean worlds. *Astrobiology* **19**(1), 1–27.

Herrick, D. L., and Stevenson, D. J. (1990). Extensional and compressional instabilities in icy satellite lithospheres. *Icarus* **85**(1), 191–204.

Hogenboom, D., Kargel, J., Ganasan, J., and Lee, L. (1995). Magnesium sulfate-water to 400 MPa using a novel piezometer: Densities, phase equilibria, and planetological implications. *Icarus* **115**(2), 258–277.

Howell, S. M., and Pappalardo, R. T. (2018). Band formation and ocean–surface interaction on Europa and Ganymede. *Geophysical Research Letters* **45**(10), 4701–4709.

Hussmann, H., Sotin, C., and Lunine, J. (2015). 10.18: Interiors and evolution of icy satellites, *in* G. Schubert, ed., Treatise on geophysics (Second Edition). Elsevier, pp. 605–635.

Journaux, B., Daniel, I., Caracas, R., Montagnac, G., and Cardon, H. (2013). Influence of NaCl on ice VI and ice VII melting curves up to 6GPa, implications for large icy moons. *Icarus* **226**(1), 355–363.

Journaux, B., Daniel, I., Petitgirard, S. et al. (2017). Salt partitioning between water and high-pressure ices: Implication for the dynamics and habitability of icy moons and water-rich planetary bodies. *Earth and Planetary Science Letters* **463**, 36–47.

Journaux, B., Kalousová, K., Sotin, C. et al. (2020). Large ocean worlds with high-pressure ices. *Space Science Reviews* **216**, 7.

Kalousová, K., Schroeder, D. M., and Soderlund, K. M. (2017). Radar attenuation in Europa's ice shell: Obstacles and opportunities for constraining the shell thickness and its thermal structure. *Journal of Geophysical Research: Planets* **122**(3), 524–545.

Kalousová, K., and Sotin, C. (2018). Melting in high-pressure ice layers of large ocean worlds: Implications for volatiles transport. *Geophysical Research Letters* **45**(16), 8096–8103.

Kalousová, K., and Sotin, C. (2020a). Dynamics of Titan's high-pressure ice layer. *Earth and Planetary Science Letters* **545**, 116416.

Kalousová, K., and Sotin, C. (2020b). The insulating effect of methane clathrate crust on Titan's thermal evolution. *Geophysical Research Letters* **47**(13), e2020GL087481.

Kalousová, K., Sotin, C., Choblet, G., Tobie, G., and Grasset, O. (2018). Two-phase convection in Ganymede's high-pressure ice layer: Implications for its geological evolution. *Icarus* **299**, 133–147.

Kalousová, K., Souček, O., Tobie, G., Choblet, G., and Čadek, O. (2016). Water generation and transport below Europa's strike-slip faults. *Journal of Geophysical Research: Planets* **121**, 2444–2462.

Kamata, S., Kimura, J., Matsumoto, K. et al. (2016). Tidal deformation of Ganymede: Sensitivity of Love numbers on the interior structure. *Journal of Geophysical Research: Planets* **121**(7), 1362–1375.

Kamata, S., and Nimmo, F. (2017). Interior thermal state of Enceladus inferred from the viscoelastic state of the ice shell. *Icarus* **284**, 387–393.

Kargel, J. S. (1991). Brine volcanism and the interior structures of asteroids and icy satellites. *Icarus* **94**(2), 368–390.

Kargel, J. S. (1992). Ammonia-water volcanism on icy satellites: Phase relations at 1 atmosphere. *Icarus* **100**(2), 556–574.

Kargel, J. S., Kaye, J. Z., Head, J. W. et al. (2000). Europa's crust and ocean: Origin, composition, and the prospects for life. *Icarus* **148**(1), 226–265.

Kay, J. E., and Head, J. W. (1999). Geologic mapping of the Ganymede g8 calderas region: Evidence for cryovolcanism, *in* Lunar and Planetary Science Conference. Vol. 30, p. 1103.

Kirk, R., and Stevenson, D. (1987). Thermal evolution of a differentiated Ganymede and implications for surface features. *Icarus* **69**(1), 91–134.

Kivelson, M. G., Khurana, K. K., Russell, C. T. et al. (1996). Discovery of Ganymede's magnetic field by the Galileo spacecraft. *Nature* **384**, 537–541.

Kivelson, M., Khurana, K., and Volwerk, M. (2002). The permanent and inductive magnetic moments of Ganymede. *Icarus* **157**(2), 507–522.

Kuskov, O. L., and Kronrod, V. A. (2001). Core sizes and internal structure of Earth's and Jupiter's satellites. *Icarus* **151**(2), 204–227.

Kvorka, J., Čadek, O., Tobie, G., and Choblet, G. (2018). Does Titan's long-wavelength topography contain information about subsurface ocean dynamics? *Icarus* **310**, 149–164.

Lemasquerier, D., Grannan, A. M., Vidal, J. et al. (2017). Libration-driven flows in ellipsoidal shells. *Journal of Geophysical Research: Planets* **122**(9), 1926–1950.

Lewis, E. L., and Perkin, R. G. (1986). Ice pumps and their rates. *Journal of Geophysical Research: Oceans* **91**(C10), 11756–11762.

Lewis, J. S. (1971). Satellites of the outer planets: Their physical and chemical nature. *Icarus* **15**(2), 174–185.

Lobo, A. H., Thompson, A. F., Vance, S. D., and Tharimena, S. (2021). A pole-to-equator ocean overturning circulation on Enceladus. *Nature Geoscience* **14**(4), 185–189.

Lunine, J. I., and Stevenson, D. J. (1982). Formation of the Galilean satellites in a gaseous nebula. *Icarus* **52**(1), 14–39.

Malhotra, R. (1991). Tidal origin of the Laplace resonance and the resurfacing of Ganymede. *Icarus* **94**(2), 399–412.

Matsuyama, I. (2014). Tidal dissipation in the oceans of icy satellites. *Icarus* **242**, 11–18.

Matsuyama, I., Beuthe, M., Hay, H. C., Nimmo, F., and Kamata, S. (2018). Ocean tidal heating in icy satellites with solid shells. *Icarus* **312**, 208–230.

McCord, T. B., Hansen, G. B., and Hibbitts, C. A. (2001). Hydrated salt minerals on Ganymede's surface: Evidence of an ocean below. *Science* **292**(5521), 1523–1525.

McKenzie, D. (1984). The generation and compaction of partially molten rock. *Journal of Petrology* **25**(3), 713–765.

McKinnon, W. B. (1999). Convective instability in Europa's floating ice shell. *Geophysical Research Letters* **26**(7), 951–954.

McKinnon, W. B. (2006). On convection in ice I shells of outer solar system bodies, with detailed application to Callisto. *Icarus* **183**(2), 435–450.

McKinnon, W. B., and Zolensky, M. E. (2003). Sulfate content of Europa's ocean and shell: Evolutionary considerations and some geological and astrobiological implications. *Astrobiology* **3**, 879–897.

Melosh, H., Ekholm, A., Showman, A., and Lorenz, R. (2004). The temperature of Europa's subsurface water ocean. *Icarus* **168**(2), 498–502.

Mitri, G., and Showman, A. P. (2008). Thermal convection in ice-I shells of Titan and Enceladus. *Icarus* **193**(2), 387–396.

Moore, W. B., and Schubert, G. (2003). The tidal response of Ganymede and Callisto with and without liquid water oceans. *Icarus* **166**(1), 223–226.

Moresi, L., and Solomatov, V. S. (1995). Numerical investigation of 2D convection with extremely large viscosity variations. *Physics of Fluids* **7**(9), 2154–2162.

Murchie, S. L., Head, J. W., and Plescia, J. B. (1990). Tectonic and volcanic evolution of dark terrain and its implications for the internal structure and evolution of Ganymede. *Journal of Geophysical Research: Solid Earth* **95**(B7), 10743–10768.

Nimmo, F., and Pappalardo, R. T. (2004). Furrow flexure and ancient heat flux on Ganymede. *Geophysical Research Letters* **31**(19).

Nimmo, F., and Pappalardo, R. T. (2016). Ocean worlds in the outer solar system. *Journal of Geophysical Research: Planets* **121**(8), 1378–1399.

Nimmo, F., Pappalardo, R. T., and Giese, B. (2002). Effective elastic thickness and heat flux estimates on Ganymede. *Geophysical Research Letters* **29**(7), 62–1–62–4.

Noir, J., Hemmerlin, F., Wicht, J., Baca, S., and Aurnou, J. (2009). An experimental and numerical study of librationally driven flow in planetary cores and subsurface oceans. *Physics of the Earth and Planetary Interiors* **173**(1), 141–152.

Orton, G. S., Spencer, J. R., Travis, L. D., Martin, T. Z., and Tamppari, L. K. (1996). Galileo photopolarimeter-radiometer observations of Jupiter and the Galilean satellites. *Science* **274**(5286), 389–391.

Palguta, J., Schubert, G., Zhang, K., and Anderson, J. D. (2009). Constraints on the location, magnitude, and dimensions of Ganymede's mass anomalies. *Icarus* **201**(2), 615–625.

Pappalardo, R. T., Collins, G. C., Head, J. W. et al. (2004). *Geology of Ganymede*, Vol. 1, Cambridge University Press.

Petrenko, V. F., and Whitworth, R. W. (2002). *Physics of ice*, Oxford University Press.

Pettinelli, E., Cosciotti, B., Di Paolo, F. et al. (2015). Dielectric properties of Jovian satellite ice analogs for subsurface radar exploration: A review. *Reviews of Geophysics* **53**(3), 593–641.

Poirier, J. P., Sotin, C., and Peyronneau, J. (1981). Viscosity of high-pressure ice VI and evolution and dynamics of Ganymede. *Nature* **292**, 225–227.

Prockter, L. M., Lopes, R. M. C., Giese, B. et al. (2010). Characteristics of icy surfaces. *Space Science Reviews* **153**, 63–111.

Radko, T. (2013). *Double-diffusive convection*, Cambridge University Press.

Raj, R., and Ashby, M. F. (1971). Grain boundary sliding and diffusional creep. *Metallurgical Transactions* **2**, 1113–1127.

Rekier, J., Trinh, A., Triana, S. A., and Dehant, V. (2019). Internal energy dissipation in Enceladus's subsurface ocean from tides and libration and the role of inertial waves. *Journal of Geophysical Research: Planets* **124**(8), 2198–2212.

Reynolds, R. T., and Cassen, P. M. (1979). On the internal structure of the major satellites of the outer planets. *Geophysical Research Letters* **6**(2), 121–124.

Rosenblum, E., Garaud, P., Traxler, A., and Stellmach, S. (2011). Turbulent mixing and layer formation in double-diffusive convection: Three-dimensional numerical simulations and theory. *Astrophysical Journal* **731**(1), 66.

Rovira-Navarro, M., Rieutord, M., Gerkema, T. et al. (2019). Do tidally-generated inertial waves heat the subsurface oceans of Europa and Enceladus?. *Icarus* **321**, 126–140.

Saur, J., Duling, S., Roth, L. et al. (2015). The search for a subsurface ocean in Ganymede with Hubble Space Telescope observations of its auroral ovals. *Journal of Geophysical Research: Space Physics* **120**(3), 1715–1737.

Sbalchiero, E., Thakur, S., and Bruzzone, L. (2019). 3d radar sounder simulations of geological targets on Ganymede Jovian moon, *in* Image and signal processing for remote sensing XXV. Vol. 11155, International Society for Optics and Photonics, p. 111551J.

Schenk, P. M. (2002). Thickness constraints on the icy shells of the Galilean satellites from a comparison of crater shapes. *Nature* **417**, 419–421.

Schenk, P. M., and Moore, J. M. (1995). Volcanic constructs on Ganymede and Enceladus: Topographic evidence from stereo images and photoclinometry. *Journal of Geophysical Research: Planets* **100**(E9), 19009–19022.

Schubert, G., Anderson, J., Spohn, T., and McKinnon, W. (2004). Interior composition, structure, and dynamics of the Galilean satellites, *in* F. Bagenal, T. Dowling, and W. McKinnon, eds., Jupiter: The planet, satellites, and magnetosphere. Cambridge University Press, pp. 281–306.

Schubert, G., Zhang, K., Kivelson, M. G., and Anderson, J. D. (1996). The magnetic field and internal structure of Ganymede. *Nature* **384**, 544–545.

Shoemaker, E., Lucchitta, B., Wilhelms, D., Plescia, J., and Squyres, S. (1982). The geology of Ganymede, *in* D. Morrison, ed., Satellites of Jupiter. University of Arizona Press, pp. 435–520.

Showman, A. P., and Malhotra, R. (1997). Tidal evolution into the Laplace resonance and the resurfacing of Ganymede. *Icarus* **127**(1), 93–111.

Showman, A. P., Stevenson, D. J., and Malhotra, R. (1997). Coupled orbital and thermal evolution of Ganymede. *Icarus* **129**(2), 367–383.

Smith, B. A., Soderblom, L. A., Beebe, R. et al. (1979). The Galilean satellites and Jupiter: Voyager 2 imaging science results. *Science* **206**(4421), 927–950.

Soderlund, K. M. (2019). Ocean dynamics of outer solar system satellites. *Geophysical Research Letters* **46**(15), 8700–8710.

Soderlund, K. M., Kalousová, K., Buffo, J. J. et al. (2020). Ice–ocean exchange processes in the Jovian and Saturnian satellites. *Space Science Reviews* **216**, 80.

Soderlund, K. M., Schmidt, B. E., Wicht, J., and Blankenship, D. D. (2014). Ocean-driven heating of Europa's icy shell at low latitudes. *Nature Geoscience* **7**, 16–19.

Sohl, F., Spohn, T., Breuer, D., and Nagel, K. (2002). Implications from Galileo observations on the interior structure and chemistry of the Galilean satellites. *Icarus* **157**(1), 104–119.

Solomatov, V. S. (1995). Scaling of temperature- and stress-dependent viscosity convection. *Physics of Fluids* **7**(2), 266–274.

Sotin, C., Gillet, P., and Poirier, J. P. (1985). *Creep of high-pressure ice VI*, Springer Netherlands, pp. 109–118.

Sotin, C., and Parmentier, E. (1989). On the stability of a fluid layer containing a univariant phase transition: Application to planetary interiors. *Physics of the Earth and Planetary Interiors* **55**(1), 10–25.

Sotin, C., and Poirier, J. (1987). Viscosity of ice V. *Le Journal de Physique Colloques* **48**(C1), 233–238.

Sotin, C., and Tobie, G. (2004). Internal structure and dynamics of the large icy satellites. *Comptes Rendus Physique* **5**(7), 769–780.

Spaun, N. A., Head, J. W., Pappalardo, R. T., and the Galileo SSI Team (2001). Scalloped depressions on Ganymede from Galileo (g28) very high resolution imaging, *in* Lunar and Planetary Science Conference. Vol. 32, p. 1448.

Spohn, T., and Schubert, G. (2003). Oceans in the icy Galilean satellites of Jupiter? *Icarus* **161**(2), 456–467.

Squyres, S. W. (1980a). Surface temperatures and retention of H_2O frost on Ganymede and Callisto. *Icarus* **44**(2), 502–510.

Squyres, S. W. (1980b). Volume changes in Ganymede and Callisto and the origin of grooved terrain. *Geophysical Research Letters* **7**(8), 593–596.

Steinbrügge, G., Stark, A., Hussmann, H., Sohl, F., and Oberst, J. (2015). Measuring tidal deformations by laser altimetry: A performance model for the Ganymede Laser Altimeter. *Planetary and Space Science* **117**, 184–191.

Steinbrügge, G., Voigt, J. R., Wolfenbarger, N. S. et al. and Schroeder, D. M. (2020). Brine migration and impact-induced cryovolcanism on Europa. *Geophysical Research Letters* **47**(21), e2020GL090797.

Stephan, K., Roatsch, T., Tosi, F. et al. (2021). Regions of interest on Ganymede's and Callisto's surfaces as potential targets for ESA's JUICE mission. *Planetary and Space Science*, 208, 105324.

Tittemore, W. C. (1990). Chaotic motion of Europa and Ganymede and the Ganymede–Callisto dichotomy. *Science* **250**(4978), 263–267.

Tobie, G., Choblet, G., and Sotin, C. (2003). Tidally heated convection: Constraints on Europa's ice shell thickness. *Journal of Geophysical Research* **108**, 5124.

Tobie, G., Lunine, J. I., and Sotin, C. (2006). Episodic outgassing as the origin of atmospheric methane on Titan. *Nature* **440**, 61–64.

Tobie, G., Mocquet, A., and Sotin, C. (2005). Tidal dissipation within large icy satellites: Applications to Europa and Titan. *Icarus* **177**(2), 534–549.

Tyler, R. H. (2011). Magnetic remote sensing of Europa's ocean tides. *Icarus* **211**(1), 906–908.

Vance, S. (2017). vancesteven/PlanetProfile: Release for use in reproducing results submitted to Journal of Geophysical Research – Planets. https://doi.org/10.5281/zenodo.844131.

Vance, S., Barge, L. M., Cardoso, S. S., and Cartwright, J. H. (2019). Self-assembling ice membranes on Europa: Brinicle properties, field examples, and possible energetic systems in icy ocean worlds. *Astrobiology* **19**(5), 685–695.

Vance, S., Bouffard, M., Choukroun, M., and Sotin, C. (2014). Ganymede's internal structure including thermodynamics of magnesium sulfate oceans in contact with ice. *Planetary and Space Science* **96**, 62–70.

Vance, S., and Brown, J. M. (2013). Thermodynamic properties of aqueous $MgSO_4$ to 800MPa at temperatures from -20 to 100°C and concentrations to 2.5molkg^{-1} from sound speeds, with applications to icy world oceans. *Geochimica et Cosmochimica Acta* **110**, 176–189.

Vance, S. D., Panning, M. P., Stähler, S. et al. (2018). Geophysical investigations of habitability in ice-covered ocean worlds. *Journal of Geophysical Research: Planets* **123**(1), 180–205.

Vance, S. D., Styczinski, M., Bills, B. et al. (2021). Magnetic induction responses of Jupiter's ocean moons including effects from adiabatic convection. *Journal of Geophysical Research: Planets* **126**(2), e2020JE006418.

Vance, S., and Goodman, J. (2009). Oceanography of an ice-covered moon. *Europa, in* R. T. Pallardo, W. B. McKinnon, and K. Khurana, eds., Europa. University of Arizona Press,pp. 459–482.

Weller, M. B., Fuchs, L., Becker, T. W., and Soderlund, K. M. (2019). Convection in thin shells of icy satellites: Effects of latitudinal surface temperature variations. *Journal of Geophysical Research: Planets* **124**(8), 2029–2053.

Zeng, Y., and Jansen, M. F. (2021). Ocean circulation on Enceladus with a high- versus low-salinity ocean. *Planetary Science Journal* **2**(4), 151. https://doi.org/10.3847/psj/ac1114.

Zhang, K., Chan, K. H., Liao, X., and Aurnou, J. M. (2013). The non-resonant response of fluid in a rapidly rotating sphere undergoing longitudinal libration. *Journal of Fluid Mechanics* **720**, 212–235.

Zolotov, M. Y., and Kargel, J. (2009). On the chemical composition of Europa's icy shell, ocean, and underlying rocks, *in* R. T. Pallardo, W. B. McKinnon, and K. Khurana, eds., Europa. University of Arizona Press, pp. 431–458.

13

The Origin of Ganymede's Internal Magnetic Field

ULRICH CHRISTENSEN, TINA RÜCKRIEMEN-BEZ AND GERALD SCHUBERT

13.1 INTRODUCTION

In terms of understanding planetary magnetism, Ganymede is a unique object. It is the only known satellite in the solar system with an internal magnetic field that is most likely generated by an active dynamo (Kivelson et al., 1996, 2002; Schubert et al., 1996). Neubauer (1978) was the first to suggest that Ganymede (and other Galilean moons as well as Titan) may host an active dynamo, and estimated that the surface magnetic field at the equator could have a strength of 220 nT. The dynamo origin for Ganymede's internal field is supported by its main properties. It is of global scale and to a good approximation described by a planetocentric dipole aligned with the rotation axis. This contrasts, for example, with the highly irregular small-scale magnetic field of Mars (Acũna et al., 1999), which results from remanent crustal magnetization acquired in the field of a now extinct dynamo. Most of Ganymede's internal field is invariable on timescales shorter than a few years – that is, the same field model fits the magnetometer data obtained during Galileo flybys that occurred up to four years apart. At Europa and Callisto, internal field components have been detected that change with Jupiter's rotation (Saur et al., 2010). They result from electromagnetic induction in conducting interior regions, supposedly salty water oceans. In these satellites, which are embedded in Jupiter's magnetospheric field, the time variation is mainly driven by Jupiter's tilted and rotating dipole (Khurana et al., 1998). At Ganymede, an induced component is likely to also contribute (Kivelson et al., 2002; Saur et al., 2015). However, the internal field at Ganymede's surface has a strength of order 1,000 nT compared to 100 nT for Jupiter's time-variable field at Ganymede, which is an upper bound for the strength of the induced field. This shows that the induced component, if present, would be secondary.

Like the dynamos of other solid planets, the dynamo in Ganymede is thought to operate in a liquid core made predominantly of iron. The existence of a likely dynamo field is strong evidence that Ganymede has indeed formed a metallic core and that the core is at least partially liquid. Given that Ganymede is slightly larger than Mercury, which exhibits a dynamo-generated magnetic field (Ness, 1979), it may not be entirely surprising that Ganymede hosts a dynamo. However, the internal structure of the two objects is very different. While Mercury's iron core has a radius of slightly more than 2,000 km (Hauck et al., 2013; Wardinski et al., 2019), a probable upper limit for Ganymede's core is 1,000 km. Hence Ganymede hosts the smallest (in size) presently active dynamo that we know of.

Another unique property of Ganymede's dynamo is that it operates in a sizable external magnetic field, imposed by Jupiter. This has led to the suggestion that the dynamo may not be self-sustained. Rather, the internal field could result from "magnetoconvection," which strongly amplifies the imposed field, but would not create a field on its own (Sarson et al., 1997).

In this chapter, we discuss theory and models for the origin of Ganymede's internal magnetic field. We start by summarizing the field properties derived from observations. We continue by reviewing our knowledge of the internal structure of Ganymede, more fully treated in Chapter 6, as far as it is essential for the understanding of the origin of the magnetic field. In the following section, we discuss the basic requirements for convection-driven dynamos in planetary cores. We also touch on alternative suggestions for the origin of the internal magnetic field that do not rely on a dynamo operating in a metallic core. Unlike the geodynamo, which at present is driven mainly due to the freezing of iron at the top of the growing solid inner core, a viable model for convection in Ganymede's core relies on top-down freezing of an iron–sulfur FeS alloy, first suggested by McKinnon (1996) and later elaborated upon by Hauck and colleagues (2006) and Rückriemen and colleagues (2015, 2018). It may lead either to an iron-snow regime or to freezing of a FeS layer at the top of the core. In Section 13.5, we first review the relevant properties of FeS alloys, such as their phase diagram. We then explain the essential points of this freezing regime and the general principles of modeling the thermo-chemical evolution of a coupled core–mantle system. Section 13.6 deals with scaling laws that allow us to deduce the flow velocity in the core and the strength of the magnetic field from the thermo-compositional fluxes predicted by thermal evolution models. We then proceed in Section 13.7 with reviewing the results of such evolution models, which, in some parameter ranges, can successfully predict the existence of a contemporary magnetic field of the observed strength. Section 13.8 summarizes the results of numerical dynamo simulations. They suggest that in the iron-snow regime, a strongly dipolar magnetic field with only weak higher multipole contributions can be obtained, as is observed at Ganymede. The concluding section summarizes our current understanding of Ganymede's dynamo, puts it into perspective with respect to other planetary dynamos in the solar system, and gives an outlook to what we might learn from magnetometer observations of the JUpiter ICy moons Explorer (JUICE) mission.

13.2 INTERNAL MAGNETIC FIELD PROPERTIES

In this section, we briefly summarize the knowledge of Ganymede's internal magnetic field.

Somewhat unexpectedly, magnetometer measurements taken during the first two close passes of the Galileo spacecraft at Ganymede in 1996 revealed the existence of a relatively strong magnetic field of internal origin (Kivelson et al., 1996). It is dominantly that of a dipole whose axis is nearly aligned with the rotation axis and that remains constant on long timescales. This has been confirmed by additional data from later flybys. A small magnetosphere is formed around Ganymede. At low latitudes, the field lines close inside the satellite, whereas at high latitudes, they link up with Jupiter's field.

Kivelson and colleagues (2002) presented several models for Ganymede's magnetic field obtained by fitting the magnetometer data of all flybys with a combination of a permanent internal field expanded to a maximum harmonic degree of two, a time-varying external component resulting from Jupiter's field and magnetospheric currents, and, in some cases, a time-varying internal field created by electromagnetic induction. All models give similar values for the permanent internal dipole moment, with a preferred value of 719 nT$\times R^3$, where R is Ganymede's radius. This puts the mean surface field strength at about 1,000 nT. The dipole axis deviates by 4° from the rotation axis. More precisely, its tilt angle is 176° – that is, it is nearly antiparallel to the rotation axis and also opposite to the orientation of Jupiter's dipole. While the combination of an internal dipole and the external field already gives a good fit to the data, Kivelson and colleagues (2002) found that it could be somewhat improved by adding either a quadrupole component or an induced internal field. In the first case, the required quadrupole terms are relatively minor and their power at Ganymede's surface amounts to only 0.25% of the power in the dipole field. The magnetometer data can be fitted equally well without the need for an internal quadrupole by adding to the permanent dipole an induced dipole field, modeled for a perfect conductor located not far below the surface of Ganymede. Kivelson and colleagues (2002) prefer the latter model because it requires fewer fitting parameters. Also, the observation of induced fields at Europa and Callisto suggest that Ganymede should have this field component. The case for the presence of an induced field has been reinforced by observations of oscillations in the location of auroral emissions near Ganymede, which occur close to the boundary between open and closed field lines. Without an induced field, larger oscillation amplitudes would be expected than have been observed (Saur et al., 2015).

A point of concern in analyzing the magnetometer data is the separation of internal and external field components . Kivelson and colleagues (2002) assumed a simple uniform external field at each of the Galileo flybys. Jia and colleagues (2008) developed a more realistic model for the field generated by magnetospheric currents and suggested that the internal dipole moment could be slightly higher than the previously inferred value. However, uncertainties excluded a reliable revision of its value.

In summary, Ganymede's permanent internal magnetic field is well described by a dipole antiparallel to the rotation axis, with a surface field strength of 1,000 nT. Given the trade-off with a likely induced magnetic field, higher multipole components could not be identified unambiguously, but must be small. In view of the increasing evidence for an induced signal, the ratio of quadrupole strength to dipole strength obtained in the model without induction might be considered an upper bound.

13.3 INTERIOR STRUCTURE

In this section, we describe the features of Ganymede's interior that are relevant to understanding the origin of its magnetic field. A detailed discussion of Ganymede's internal structure is given in Chapter 6.

Ganymede is differentiated into a metallic core and an ice-silicate mantle. The core is liquid or sufficiently liquid for a dynamo to operate within it and create a magnetic field. Two main observations support this inference. One line of evidence is the average density and moment of inertia of the satellite, and the other is the existence of the magnetic field itself. Data from the Pioneer and Voyager spacecrafts determined that the average density of Ganymede was 1,940 kg m^{-3} (Null, 1976; Campbell & Synott, 1985). The Galileo spacecraft essentially confirmed this value, finding that the mean density was 1,936 $\pm$ 22 kg m^{-3} (Anderson et al., 1996a). Galileo also measured the gravitational coefficients J_2 and C_{22}, from which the satellite's axial moment of inertia C could be determined. Using equilibrium theory, it was found that $C/MR^2 = 0.3105\pm0.0028$, where M is Ganymede's mass and $R = 2634$ km its radius (Anderson et al., 1996a).

For a uniform density body, $C/MR^2 = 0.4$. The smaller the value of C/MR^2, the more concentrated is the body's mass toward its center. The value of C/MR^2 for Ganymede is among the smallest of any planet or satellite in the solar system. Io's value of C/MR^2 is 0.378; Earth's is 0.334. Only the giant outer planets have C/MR^2 values smaller than Ganymede's. Accordingly, it is immediately clear that Ganymede is strongly differentiated with a large concentration of mass toward its center. More quantitative models of Ganymede's interior based on its average density and moment of inertia can be constructed, but the models are not unique and assumptions about the values of other structural properties such as the densities of the mantle and core are required. The simplest models of planetary interiors are concentric spherical shells. While two-layer models of Ganymede can be constructed, three-layer models are more relevant because Ganymede's density is sufficiently low that it must contain a substantial amount of ice. Ganymede is, after all, one of the icy Galilean moons together with Europa and Callisto and has abundant ice on its surface (Carlson et al., 1996). This suggests that upon differentiation, Ganymede formed a metallic core surrounded by a silicate mantle that in turn is surrounded by an ice shell. Anderson and colleagues (1996a) explored such three-layer models and concluded that Ganymede has a metallic core of radius 400–1,300 km surrounded by a silicate mantle, which is in turn enclosed by an ice shell ~800 km thick. Depending on whether the core is pure iron or an alloy of iron and iron sulfide, it could account for as little as 1.4% or as much as one-third of Ganymede's total mass. If the ice were stripped away, Ganymede could look much like Io (Anderson et al., 1996b) in terms of its size and internal mass distribution. Similar conclusions on the size range of a metallic core have been reached in later investigations by Kuskov and Kronrod (2001) and Sohl and colleagues (2002).

One of the major uncertainties in estimating the size of Ganymede's core is its composition. The core could contain a light element. The nature of this element and the amount of it in the core are unknown. The light element is likely sulfur, given cosmochemical abundances and the preponderance of sulfur on the surface of Io. The core size increases for larger amounts of light elements in the core. Kuskov and Kronrod (2001) and Sohl and colleagues (2002) also considered a more restricted set of three-layer models in which they set the densities of the ice and silicate shells to plausible values and varied the core composition between that of pure iron and FeS. They find core radii ranging from 580 km to 900 km and 660 km to 880 km, respectively – that is, roughly between one-fourth and one-third of Ganymede's radius. This may represent the most likely range, although a somewhat larger or smaller core cannot be excluded.

Differentiation of Ganymede into a metallic core, silicate mantle, and outer ice shell is energetically plausible (Schubert et al., 1986). Accretional heating is probably insufficient for early formation of a metallic core, though it may be adequate to separate the ice from the rock and metal. Radiogenic heating during the course of Ganymede's evolution could provide enough heating to separate an Fe–FeS core if the eutectic melting temperature of about 1,250 K is all that is required. Although tidal heating is unimportant in Ganymede at present, tidal dissipation during passage through a dynamical resonance at some earlier time in its evolution may have played a role in the formation of Ganymede's core. For the magnetic field of Ganymede to originate in its metallic core it is necessary that the core is presently molten, or at least partly liquid. This constrains the temperature at the iron-core–silicate boundary at or above the melting temperature of the core material. For a core of pure iron, the melting temperature at the pressure of the core–silicate boundary is about 1,980 K; for an eutectic Fe–FeS core, the temperature is about 1,325 K. These temperatures place important constraints on the thermal structure of the satellite and provide end points for its thermal history.

It is likely that Ganymede's outer ice shell is not completely frozen but harbors an ocean that could be sandwiched between an outer layer consisting of the low-pressure modification (Ih) of water ice and a deep solid layer made of high-pressure forms of ice. The presence of dissolved salts would make the ocean reasonably electrically conducting (in contrast to solid ice, which is a poor conductor). In the context of Ganymede's magnetic field, the presence of an ocean is relevant because electrical currents induced by the time-variable magnetospheric field of Jupiter give rise to an additional internal field component. In fact, the likely presence of an induced field component (Kivelson et al., 2002; Saur et al., 2015) is the strongest indication for the existence of an ocean. Thermal evolution models suggest that an internal ocean could well have survived until the present day (Spohn & Schubert, 2003). The amplitude of the inductive response depends (aside from the frequency) on the depth, thickness, and conductivity of the ocean, which are as yet poorly constrained.

In summary, based on the mean density and moment of inertia of Ganymede, and the existence of a magnetic field, the differentiation of the satellite into a metallic, core, silicate mantle and outer ice shell seems highly probable. The compositions and sizes of these regions are uncertain, though some constraints on these properties are provided by the density and moment of inertia models. Given that the internal magnetic field originates in the metallic core, it must be partly or completely liquid.

13.4 SOURCES FOR A MAGNETIC FIELD

An active dynamo in the core is the most probable explanation for the observed strong field (Schubert et al., 1996). In this section, we will first discuss some fundamental requirements for a dynamo to operate and then consider other possible sources of the internal field.

13.4.1 Requirements for a Dynamo

A planetary dynamo requires the existence of a region inside the planet that is (i) sufficiently electrically conducting, (ii) of sufficient size, and (iii) consisting of parts that are in sufficiently rapid (differential) motion, which can occur only if it is liquid. What "sufficient" means is quantified in what follows. These basic requirements of a dynamo, together with the observation of a strong internal magnetic field, suggest that Ganymede has indeed formed a metallic core that is at least partially liquid and is stirred by some process. In addition to requirement (iii), the fluid motion must have a certain complexity. For example, a simple differential rotation of different parts of the core would be insufficient for generating a magnetic field. Helical motion, where a fluid particle moves on a corkscrew-type trajectory, is particularly suitable for magnetic field generation (e.g., Moffatt & Dormy, 2019).

Mechanical excitation of flow in planetary cores – for example by tides, libration, or nutation – has been discussed (e.g., Le Bars, 2016); however, its suitability for driving a dynamo has not been explored in great detail. While mechanical driving remains a distinct possibility, thermally or compositionally driven convection is a plausible mechanism for motion in Ganymede's core and has been studied extensively. In a rotating core, Coriolis forces are expected to play a dominant role. The ratio of inertial to Coriolis forces is measured by the Rossby number $Ro = U/(D\Omega)$, where U is a typical flow velocity, D a characteristic length scale such as the core radius, and Ω the rotation frequency. In the Earth's core Ro is of order 10^{-6}. Ganymede rotates seven times slower and D may be smaller by a factor of a few compared to Earth, but this still means $Ro < 10^{-4}$. Hence also in Ganymede, convection should be rotationally dominated, which typically leads to helical flow pattern.

If Ganymede's dynamo is thermally driven, then the temperature in the core must lie along an adiabat and heat transport in the core must exceed the heat flow carried by conduction along the adiabat. Although it is not known when Ganymede first acquired its magnetic field, it could be energetically difficult to maintain core thermal convection over a long period of evolution. The feasibility of present-day thermal convection in Ganymede's core is further discussed in Section 13.5. Compositionally driven core convection occurs when solidification processes increase the concentration of light elements in the outer layers of the core while increasing the concentration of heavy elements (Fe) in the inner region of the core. This is occurring in the Earth through the freezing out of its relatively iron-rich inner core. On Ganymede, the freezing of its core could occur differently – for example, through the formation and sinking of iron snow near the top of its core. This possibility was first suggested by McKinnon (1996) and later studied in detail by Hauck and colleagues (2006) and Rückriemen and colleagues (2015, 2018). It will be elaborated on further later in this chapter.

A key parameter for the operation of a dynamo is the magnetic Reynolds number

$$Rm = UD/\eta, \tag{13.1}$$

where U is a typical fluid velocity in the core, D is a characteristic length describing the geometry of the dynamo region such as the radius of the metallic core or the thickness of its fluid shell, and η is the magnetic diffusivity in the core. In a self-sustained dynamo, Rm must exceed a critical value, which was found to be around 50 in geodynamo simulations (Christensen & Aubert, 2006). When the condition is satisfied, the fluid velocities in the core are fast enough that magnetic induction overcomes magnetic diffusion. This allows estimates of the necessary convective velocities in the core given assumptions of the values for D and η. Scaling laws that relate flow velocity to the driving thermal or compositional flux can then be used to derive constraints on the thermal evolution of Ganymede that must be satisfied to have sufficiently fast flow in its core. Similar scaling laws for the magnetic field strength can be used to test if a core dynamo can plausibly explain the observed dipole moment. Such scaling laws are introduced in Section 13.6.

13.4.2 Other Potential Sources for the Internal Magnetic Field

Processes other than a core dynamo have been proposed for the origin of Ganymede's magnetic field. While they cannot be strictly rejected, they are much less plausible.

Conceivably, remnant magnetization of Ganymede's rocky shell could be the source for the present-day magnetic field. However, in order to explain the observed field strength, it requires a very high concentration of magnetite of order 10% and in addition a strong magnetizing field generated by a paleodynamo (Crary & Bagenal, 1998). Furthermore, while remanent magnetization can typically cause local magnetic field structures of short wavelength such as those found at the surface of Mars, there is no known example where it leads to a global dipolar field of significant strength.

A dynamo might operate in a salty subsurface ocean in Ganymede. However, because its electrical conductivity would be at least four orders of magnitude lower than that of a metallic core, the lower conductivity has to be balanced by much faster flow in order for the magnetic Reynolds number to exceed the critical value. Schubert and colleagues (1996) estimated that a velocity in excess of 1 m/s would be needed, which was deemed improbable for interior fluid layers in planetary bodies that are Earth-sized or smaller. For example, the typical flow velocity in the Earth's core is at least three orders of magnitude lower.

Ganymede is immersed in Jupiter's magnetic field. If convection in Ganymede's core was too sluggish to sustain a dynamo, the flow may still amplify the imposed magnetic field. This process is referred to as magnetoconvection (Weiss & Proctor, 2014). However, in simplified mean-field models, Sarson and colleagues (1997) found that in the case of Ganymede, magnetoconvection cannot generate a magnetic field of the observed strength, and they concluded that most likely a dynamo process in its own right operates in the core. They speculated that the antialignment of Ganymede's dipole with that of Jupiter could be due to the seed field provided by Jupiter when Ganymede's dynamo started.

13.5 CORE PROPERTIES AND PRINCIPLES OF EVOLUTION MODELS

Ganymede's core is likely not made of pure iron but of a mixture of iron and light alloying elements. As discussed in Section 13.3, assuming sulfur as the light alloying element is reasonable but remains an educated guess given the current inability to directly access core properties. A mixture of iron and sulfur forms a simple binary, eutectic alloy for the pressure prevailing in Ganymede's core (6–10 GPa) (Figure 13.1). Eutectic systems feature a eutectic point that divides the Fe–S phase diagram into alloys that are either iron- or sulfur-rich compared to the eutectic composition. Upon cooling below the liquidus, iron-rich Fe–FeS alloys crystallize iron, whereas sulfur-rich alloys form solid iron sulfide.

Convection in Ganymede's core as the driving force for the dynamo may be caused by temperature and/or concentration differences (see Section 13.4).

An essential condition for thermal convection is that the core heat flow, which is controlled by the mantle, exceeds the conductive heat flow along an adiabatic temperature gradient. The latter depends on the steepness of the adiabat and the thermal conductivity. For pure iron, the thermal conductivity at (P,T)-conditions of Ganymede's core is rather high in the range 40–60 $Wm^{-1}K^{-1}$ (de Koker et al., 2012; Ezenwa & Yoshino, 2021), which makes it unlikely that the adiabatic heat flow can be exceeded in the present stage of evolution. Recent experimental studies for S-bearing iron alloys and pure FeS (Pommier & Davies, 2018; Manthilake et al., 2019; Littleton et al., 2021) arrive at somewhat discrepant results, but agree on a significant decrease of conductivity with sulfur content,

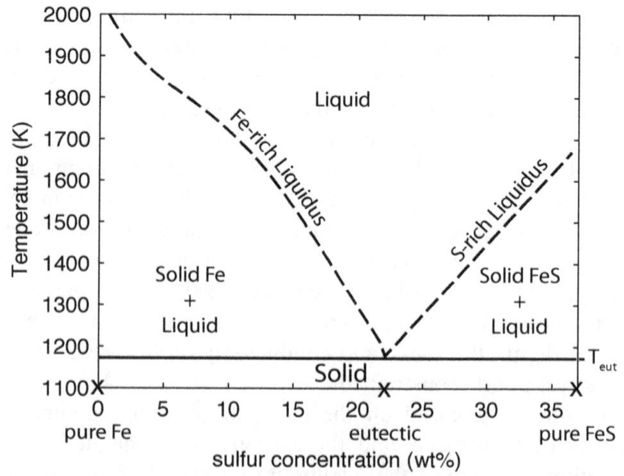

Figure 13.1 Phase diagram of the binary eutectic Fe–FeS alloy at 6 GPa (the approximate pressure at Ganymede's core mantle boundary). Iron- and sulfur-rich liquidi are calculated according to Rückriemen and colleagues (2018). Below the eutectic temperature, the alloy is completely solid, and above the liquidus, the alloy is completely liquid. In between liquidus and eutectic temperatures, solid and liquid coexist.

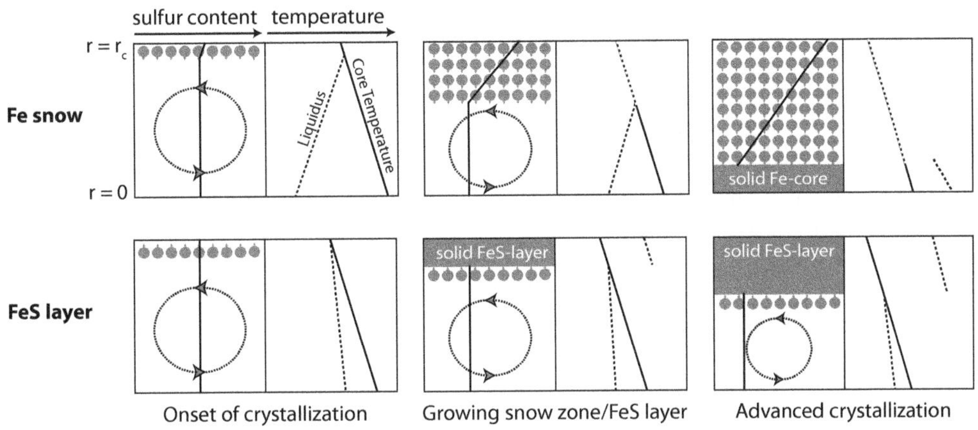

Figure 13.2 Qualitative thermal and structural evolution in the iron-snow (upper panel) and FeS-layer regime (lower panel). The gray dots and areas mark solid material, which is either iron (upper panel) or iron sulfide (lower panel). Arrows indicate compositional convection. The evolution as cooling progresses is shown from left to right (from Rückriemen et al. 2018).

to values that fall into the range 4–10 $\mathrm{Wm^{-1}K^{-1}}$ for FeS. For a sulfur-rich core, thermal convection could be possible (Littleton et al., 2021), not only in the early evolution but sustained until today.

Compositional convection is linked to freezing of the core alloy and is a consequence of gravitationally unstable situations caused by phase changes. The pressure at Ganymede's center is around 10 GPa (Sohl et al., 2002). Experimental work on Fe–FeS alloys at pressures below 14 GPa (e.g., Fei et al. 1997; Fei et al. 2000; Chudinovskikh & Boehler 2007; Morard et al. 2007; Stewart et al. 2007; Chen et al. 2008; Morard et al. 2008; Buono & Walker 2011) suggests that core differentiation of small planetary bodies like Ganymede is peculiar and unlike freezing in Earth's core. The location of incipient core freezing is dictated by the relative slopes of the core temperature and the liquidus of the core alloy. If $\partial T_l/\partial P < \partial T_c/\partial P$ in the entire core, where T_l is the liquidus temperature of the core alloy and T_c is the core temperature, the core freezes from the bottom to the top, forming a solid inner iron core, a scenario typically envisioned for Earth's core. However, experiments (Chen et al., 2008; Buono & Walker, 2011) show that at pressures below 14 GPa, an increasing amount of sulfur reduces not only the melting temperature but also the melting temperature slope $\partial T_l/\partial P$. For more than $\sim$15 wt% sulfur $\partial T_l/\partial P$ may even be negative. If the iron core of a moderate-sized planetary body such as Ganymede contains more than a few percent of sulfur, it crystallizes from the top to the bottom (Hauck et al., 2006; Stewart et al., 2007; Chen et al., 2008; Williams, 2009).

When the core temperature drops below the liquidus, iron-rich Fe–FeS alloys may form free iron crystals initially close to or at the core–mantle boundary (Figure 13.2). Formation of crystals in the free fluid (homogeneous nucleation) is impeded by a high energy barrier and may be difficult to achieve (Huguet et al., 2018). Heterogeneous nucleation on a pre-existing solid surface such as the core–mantle boundary would be easier to achieve. Here the question arises of to what extent the iron crystals would stick to the mantle despite their excess density relative to the fluid or if they would easily detach. The question of nucleation remains an open problem for the iron snow scenario.

The formation of iron crystals releases latent heat and increases the local sulfur concentration in the remaining liquid. In thermodynamic equilibrium, the sulfur concentration increases until the liquidus temperature is lowered to the local temperature. As a result, a stable chemical stratification evolves across the snow zone, provided the crystals settle and do not remain entrained in the fluid (e.g., Hauck et al., 2006; Dumberry & Rivoldini, 2015; Rückriemen et al., 2015; Davies & Pommier, 2018). The iron crystals, being denser than the liquid, "snow" down onto the deeper, entirely liquid core. Here they melt, because the local core temperature is higher than the liquidus temperature. The bottom of the snow zone is defined by melting of iron crystals. During melting, latent heat is consumed, and due to the addition of iron, the local sulfur concentration is reduced. The latter gives rise to a gravitationally unstable setting, which facilitates compositional convection in the deeper core (e.g., Hauck et al., 2006; Rückriemen et al., 2015; Olson et al., 2017; Davies & Pommier, 2018).

The sulfur concentration in the deeper core decreases due to melting and efficient mixing. Progressive cooling of the core leads to a downward growth of the snow zone. Eventually, the snow zone covers the entire core and crystals accumulate in the core center, forming a solid inner core. Rückriemen and colleagues (2015) hypothesized that the conversion of gravitational energy to magnetic energy due to stirring by the remelted iron crystals powers the dynamo in the deeper liquid core. Dynamo models by Christensen (2015a) support this hypothesis (see Section 13.8). Given that the flux of iron from the snow layer into the deep core provides enough gravitational power for running a dynamo, its maximum lifetime in the iron-snow regime is, however, limited by the time that it takes for the snow zone to grow across the entire core.

If the Fe–FeS alloy in Ganymede's core is more sulfur-rich than the eutectic composition (which depends on pressure, is subject to experimental uncertainties, and falls into the range of 18–26 wt% S), crystallization produces solid iron sulfide (e.g., Usselman, 1975; Fei et al., 1997). If the density of iron sulfide crystals is lower than the density of the surrounding more iron-rich liquid, as argued for by Hauck and colleagues (2013), freezing at the top of the core forms a layer of solid iron sulfide likely attached to the rocky mantle. The liquid at the solidification front becomes enriched in iron. Compositional convection mixes the enriched liquid with the deeper core, thereby increasing the iron concentration of the deeper core. As long as the core cools, the FeS layer grows downward. Once the eutectic concentration is reached in the deeper liquid core, the remaining liquid freezes as a eutectic solid. Similarly to the iron-snow regime, compositional convection in the deep core may drive the dynamo. In contrast to the iron-snow regime, the dynamo lifetime is limited only by the complete solidification of the core upon reaching the eutectic temperature (if there is enough gravitational power to run the dynamo).

The question of whether the iron-snow regime and/or the FeS-layer regime can explain Ganymede's present-day magnetic field can be answered with the help of thermal evolution models, whose results will be discussed in Section 13.7. Due to their fast computation, one-dimensional models are used to calculate the global thermo-chemical evolution of planetary bodies. Usually, the evolution models assume a fully differentiated Ganymede as the initial state, which is reasonable given the short timescales of core formation (e.g., Yin et al., 2002; Rubie et al., 2007; Kleine et al., 2009; Dauphas & Pourmand, 2011) compared to the subsequent long episode of cooling. It should be noted, however, that some studies argue for a late formation of the metal core (McKinnon, 1996; Spohn & Breuer, 1998). Such scenarios require cold accretion (initial temperature profile between 100 and 400 K). A fully differentiated Ganymede shortly after accretion would involve a metallic core at the center, a silicate mantle in the middle, and an ice shell as the outermost part (see Section 13.3). Heat transport through the outer layers dictates how efficiently the deep interior can cool (e.g., Schubert et al., 1979; Stevenson et al., 1983; Breuer & Spohn, 2003; Gubbins, 2003; Nimmo, 2015). Because the much lower viscosity of ice compared to that of silicates eases convective heat transport through the ice shell, the rocky mantle is the bottleneck for heat loss from the core. Different factors such as the mechanism of heat transport or the presence of heat sources affect the efficiency of mantle cooling. If the relevant conditions are met, silicate mantles cool through convection. One-dimensional simulations describe the heat transport in a convecting system by scaling laws, which are obtained, for example, from two- or three-dimensional simulations or from laboratory experiments (e.g., Schubert & Zhang, 2001).

The strong temperature dependence of the viscosity of silicate minerals inevitably leads to stagnant lid convection (Davaille & Jaupart, 1993; Solomatov, 1995; Grasset & Parmentier, 1998; Deschamps & Sotin, 2000; Solomatov & Moresi, 2000; Hüttig & Breuer, 2011), where the upper colder parts of the mantle do not convect, but cool conductively instead. The radioactive decay of long-lived unstable isotopes such as ^{238}U, ^{235}U, ^{232}Th, and ^{40}K is an important heat source in the silicate mantle, which slows down the degree of cooling depending on the absolute abundance of those isotopes (e.g., Lodders & Fegley, 1998).

If the silicate mantle is cooling, heat is continuously extracted from the core, which eventually leads to freezing of core material. As stated, freezing proceeds in the iron-snow or FeS-layer regime, depending on the core being either iron- or sulfur-rich relative to the eutectic composition. The associated thermo-chemical evolution of the core is based on some important assumptions. Iron-snow models presume (1) the system to be in thermodynamic equilibrium, (2) that crystallization in the snow zone produces a slurry (Loper & Roberts, 1977), (3) that iron crystals form, sink, and remelt rapidly, (4) that the snow zone is stably stratified and cools conductively, and (5) that remelting of iron crystals triggers compositional convection, which maintains the deeper liquid core at a well-mixed isentropic state. These assumptions have been discussed in detail in Rückriemen and colleagues (2015). Models of a growing FeS layer also assume equilibrium, conductive heat transport across the solid FeS layer, as well as a well-mixed isentropic state of the deeper liquid core caused by compositional convection. In addition, iron sulfide crystals are assumed to form directly attached to the overlying silicate mantle, with a solidification front proceeding downward concentrically. Finally, the models assume that an adiabatic temperature structure is maintained in the liquid part of the core at all times. While this is appropriate for fully developed convection, it does not apply during the stage of subadiabatic heat flow and a stagnant liquid core before crystallization starts. This is a potential weakness of the evolution models, whose consequences still need to be assessed. Details of the modeling assumptions for iron-snow and FeS-layer thermo-chemical models are given in Rückriemen and colleagues (2018).

One important output of the evolution models is the rate of gravitational energy release associated with the thermo-compositional convection. As outlined in the following section, it controls the flow velocity, for which an estimate is needed in order to determine if the magnetic Reynolds number is sufficient to allow for a dynamo. Furthermore, when a dynamo exists, the energy flux controls the magnetic field strength in a rapidly rotating convection-driven dynamo. While the energy release associated with the compositional component is always positive, the thermal part will in general be negative, except for low thermal conductivities that are likely for very sulfur-rich alloys. In Ganymede's core, the heat flux is likely less than the conductive flux along the adiabatic gradient and part of the compositional energy is needed to advect heat downward against the temperature gradient in order to maintain an isentropic state.

A major advantage of one-dimensional evolution models is their potential to scan a large parameter space due to the low computing effort. Many input parameters are indeed not well constrained and a thorough analysis therefore requires parameter variations within their reasonable ranges. While many parameters have an influence on thermal evolution, the most relevant ones for the freezing history of Ganymede's core are the reference viscosity of the silicate mantle (largely controlling the heat output of the core), the abundance of radioactive heat sources (influencing mantle cooling and, indirectly, core cooling), and the sulfur concentration of the core (influencing the liquidus temperature and hence the timing of core crystallization).

The viscosity of silicates is a strong function of temperature and pressure. The reference viscosity – that is, the viscosity at fixed values of temperature and pressure that are representative of Ganymede's rocky mantle – is important because it controls the efficiency of mantle convection, which in turn has a strong influence on how much heat is extracted form the core. Mantle rheology is a complex subject and is, aside from its dependence on temperature and pressure, influenced by many other factors such as the presence of water, the grain size, and the mineral phases or creep mechanisms. In the models discussed in Section 13.7, a range between 10^{19} Pa s and 10^{22} Pa s is considered for the reference viscosity.

The sulfur concentration of Ganymede's core depends on the primordial composition and on sulfur partitioning during metal-silicate differentiation. Both topics are subject to ongoing debate and allow for any concentration between the two end-members iron and iron sulfide (Kuskov and Kronrod, 2001; Scott et al., 2002).

The abundance of radiogenic heat sources is, like the sulfur concentration, dependent on the primordial composition. Different chondritic compositions are plausible (Schubert et al., 2004). Hauck and colleagues (2006) suggested varying the heat source abundance based on a CI chondritic composition. In the models discussed in Section 13.7, the effect of fewer and more heat sources is assessed by varying the abundance between half and double of that in CI chondrites.

13.6 SCALING LAWS FOR CONVECTION-DRIVEN DYNAMOS

In order to estimate the strength of a dynamo-generated magnetic field from the results of thermal evolution models, scaling laws are needed. Widely accepted scaling laws for convection-driven dynamos in a regime of sufficiently rapid rotation relate the average magnetic field strength B_{rms} and average velocity U_{rms} to the convective power per unit volume P (Christensen & Aubert, 2006; Aubert et al., 2009; Christensen, 2010). Knowledge of the velocity, together with information on the size of the convecting region and the electrical conductivity, is needed to calculate the magnetic Reynolds number Rm, which must exceed a critical value for a self-sustained dynamo to operate. Thermal evolution models can provide the mass anomaly flux Q_B, associated with density anomalies of thermal and/or compositional origin in the moving fluid. The convective power is then obtained by $P = Q_B \Delta\psi$, where $\Delta\psi$ is the average difference in the gravitational potential between the source and sink regions of the mass anomaly flux (Buffett et al., 1996).

The scaling laws have been based on numerical dynamo simulations, originally with a setup that is relevant for the present geodynamo, which is mostly driven by thermo-compositional fluxes originating at the inner-core boundary. Aubert and colleagues (2009) extended the analysis to cases without inner core, driven from above, and found basically the same scaling, which can be written as:

$$\frac{B_{rms}}{\sqrt{\rho\mu_o}\Omega D} = c_B \left(\frac{P}{\rho\Omega^3 D^2}\right)^{\beta}, \tag{13.2}$$

where ρ is density, μ_o is vacuum permeability, Ω is rotation frequency, and D is a characteristic length scale such as the core radius. The exponent β was found to be close to one-third and the numerical constant c_B is of order one. Remarkably, for $\beta = 1/3$, the dependence on Ω drops out in Equation (13.2) and the field strength is independent of the rotation rate:

$$B_{rms} = c_B \rho^{1/6} \mu_o^{1/2} D^{1/3} P^{1/3}. \tag{13.3}$$

The scaling law for the *rms* flow velocity U_{rms} is

$$\frac{U_{rms}}{\Omega D} = c_U \left(\frac{P}{\rho\Omega^3 D}\right)^{\gamma}, \tag{13.4}$$

with numerical values for γ in the range 0.41–0.43. Davidson (2013) gave a theoretical underpinning for the scaling laws and showed that, based on the assumption that the magnetic field strength is independent of rotation and controlled by the convective power, the exponents must be $\beta = 1/3$ and $\gamma = 4/9$. The reason for the slightly smaller numerical values for γ could be that the asymptotic regime has not been fully reached in the numerical simulations.

Equation (13.2) makes a prediction for the mean field strength inside the dynamo region. To compare with observation, the ratio b_{dip} between B_{rms} and the mean dipole field strength at the outer radius of the dynamo region r_{dyn} must be estimated. Assuming that the field is a potential field outside the core, the dipole field strength at the planetary surface at r_p is then obtained by multiplying with a factor $(r_{dyn}/r_p)^3$. In numerical simulations, the factor b_{dip} has been found to be of order 10, although its value is not well constrained, being smaller for bottom-driven dynamos and larger for top-driven dynamos (Aubert et al., 2009). Furthermore, values of order 10 apply only when the dynamo field is dominated by the axial dipole. Although conditions for this dipole dominance have been proposed (e.g., Christensen & Aubert, 2006; Aubert et al., 2009), their applicability is rather uncertain.

For Earth and Jupiter, the predictions using Equation (13.2), together with the procedure of relating the internal field to the surface field outlined earlier in this chapter, agree well with the observed field strength. For other solar system planets, this is more difficult to test because of greater uncertainties about the convective power or the dynamo region geometry. The finding that the observed magnetic fields of rapidly rotating stars are also well matched by the same scaling rules (Christensen et al., 2009) lends additional support to their validity. However, because of the remaining uncertainties in the scaling laws, in particular for dynamos that differ significantly from the geodynamo, for example, by including a snow zone, one should not expect more than order-of-magnitude estimates when applying them to Ganymede.

13.7 RESULTS FROM EVOLUTION MODELS

Several core evolution models that investigate the magnetic history of Ganymede have been presented. Most of these models focus on compositionally driven convection in the core, because of the low efficiency of a thermally driven dynamo (cf. Section 13.4). Kimura and colleagues (2009) suggested a present-day thermal dynamo if Ganymede was accreted in a cold state and subsequently heated by the decay of radioactive elements. However, the required minimum heat flux they assumed for obtaining thermal convection was rather low ($\approx$ 1 mW/m^2). Other studies found much higher values (Hauck et al., 2006; Bland et al., 2008; Rückriemen et al., 2018). Evolution models by Hauck and colleagues (2006) and Bland and colleagues (2008) suggested dynamo generation triggered directly by the sinking of iron crystals in the snow zone. Particulate-driven flow, such as by iron snow fall, has not been studied in dynamo models. Its ability to generate a dynamo remains unclear. Hauck and colleagues (2006) and Bland and colleagues (2008) assessed the feasibility for a dynamo by purely energetic considerations – that is, the power available to drive a dynamo. Hauck and colleagues (2006) found that magnetic fields caused by iron snow fall are feasible and on the order of 600–1600 nT, which is, given the uncertainties, in good agreement with the observations. However, to calculate the field strength, they used scaling laws developed for the geodynamo and their applicability to particulate flow is questionable. Bland and colleagues (2008) argued that the latent heat of snow formation is released close to the core–mantle boundary and is largely unavailable in the dynamo power budget. In contrast to Hauck and colleagues (2006), he arrived at the conclusion that iron snow fall cannot generate enough power to drive a dynamo.

Rückriemen and colleagues (2015, 2018) and Christensen (2015a) suggested that the dynamo is not produced in the snow zone but in the region below, due to compositional convection driven by remelting of iron crystals. Such a scenario is supported by the analog experiments by Olson and colleagues (2017), in which a rain of droplets precipitated through an upper layer and then dissolved in a lower layer, whose density was slightly lower than that of the droplets. They observed a coherent large-scale circulation driven by compositional convection in the lower layer, but no such flow in the precipitation layer. The possibility of a dynamo operating directly in the snow-layer remains an open question.

In addition to iron snow, Rückriemen and colleagues (2018) studied the FeS-layer regime, which occurs in sulfur-rich Fe–FeS cores. Reasonable input parameters were varied by conducting a Monte Carlo–like simulation consisting of 10,000 calculations. For each calculation, a pseudorandom combination of input values was set for parameters such as the initial temperatures of the mantle and the core, the reference viscosity of the silicate mantle, the initial core sulfur concentration, and the abundance of radiogenic heat sources in the mantle. For each case, the magnetic Reynolds number Rm, based on the flow velocity obtained by the energy-flux scaling (Eqn. 13.4) has been calculated. To assess if a dynamo is possible, a conservative value of 100 has been chosen for the critical value of Rm. If this is exceeded, the magnetic field strength inside the core has been calculated via Equation (13.3) and related to the observable dipole field strength above the core by assuming a ratio $b_{dip}=12$ between the two. Twenty-three percent of the models freezing in the iron-snow regime were successful by exhibiting a present-day magnetic field. The fraction of successful models in the FeS-layer regime was 93%. None of the entirely liquid models showed a present-day magnetic field, supporting the argument that thermally driven dynamos are unlikely at present, at least for the assumed high thermal conductivity that is appropriate for alloys with a rather low sulfur content.

Figure 13.3 illustrates the evolution of two representative models that predict a presently active dynamo, in the iron-snow and the FeS regimes, respectively. In these examples, the assumed initial condition with a rather strong temperature contrast between core and mantle resulted in an early thermally driven dynamo. However, if such a dynamo existed in Ganymede, it would have been short-lived because for most of the evolution history, the core heat flux is substantially less than the conductive adiabatic heat flow (Fig. 13.3b). Only when crystallization sets in, leading to a growing snow layer or solid FeS layer (Fig. 13.3c), respectively, and facilitating compositional convection, can the dynamo (re)start and generate a magnetic field with a strength comparable to the observed one (Fig. 13.3d).

The majority of iron-snow models was shown to yield surface magnetic field strengths between 400 and 700 nT, which is slightly below the observed (equatorial) value of 719 nT (Fig. 13.4). The FeS layer models were found to produce magnetic fields with a strength at the surface clustering between ~1000 and 2000 nT. Overall, FeS-layer models tend to feature higher field strengths than iron-snow models. This is because sulfur-rich models tend to have larger cores, which lead to dynamos that are closer to the surface and therefore less affected by the radial drop-off of the dipole field with r^{-3}. If the FeS layers are thick, they also allow for lower field strengths ($<$ 719 nT) because the dynamo region in those cases is thin and deep-seated. While thin dynamo regions invoke less available power, fields of deep-seated dynamos experience more attenuation before reaching the surface. Given the uncertainties in parameters that enter into the scaling, such as b_{dip}, the ranges of field strengths obtained in both the iron-snow and the FeS regimes compare very favorably with the observed strength of Ganymede's field.

Comparing the ages of dynamos in the iron-snow and FeS-layer regimse shows some differences (Fig. 13.5). All iron-snow models were found to feature dynamos that started only within the past 1000 Myr. Higher initial core sulfur concentrations allow for slightly older dynamos compared to lower sulfur concentrations. Nevertheless, iron-snow dynamos must be very recent. Their total lifetime cannot exceed ~1 Gyr and in most successful models they are younger than 500 Myr, implying that we would presently be in a special era in Ganymede's evolution. FeS-layer models feature much more variation in dynamo lifetime. A dependence on initial core sulfur concentration can clearly be seen. Higher sulfur concentrations can lead to young as well as quite old dynamos, lasting almost during Ganymede's entire history. Lower sulfur concentrations (26–29 wt%) allow for dynamos being at most 1000 Myr old. This behavior can be explained by the higher liquidus temperatures for higher sulfur concentrations on the sulfur-rich side of the eutectic. Higher liquidus temperatures can lead to an earlier start of core solidification and thus dynamo action. Such old dynamos require a low reference viscosity in the mantle (10^{19}–10^{20} Pa s) in order to allow for efficient cooling that facilitates an early onset of dynamo action. Young dynamos are possible for high initial sulfur concentrations only if the reference viscosity is above $\sim 10^{20}$ Pa s.

For both successful iron-snow and FeS-layer models, a trade-off between the initial core sulfur concentration and the reference mantle viscosity was found (Fig. 13.6). For iron-snow models, the reference mantle viscosity must fall into a narrow range for a given sulfur concentration and mantle heat sources. The narrow range of admissible reference viscosities for each sulfur concentration is an expression of the narrow temperature range, allowing for a growing snow zone. The narrow range also explains why iron-snow dynamos have a rather limited lifetime.

Most of the successful iron-snow (46 %) and FeS-layer models (60 %) have a low abundance of heat sources in the mantle (1/2 CI). Almost all models with a low number of heat sources

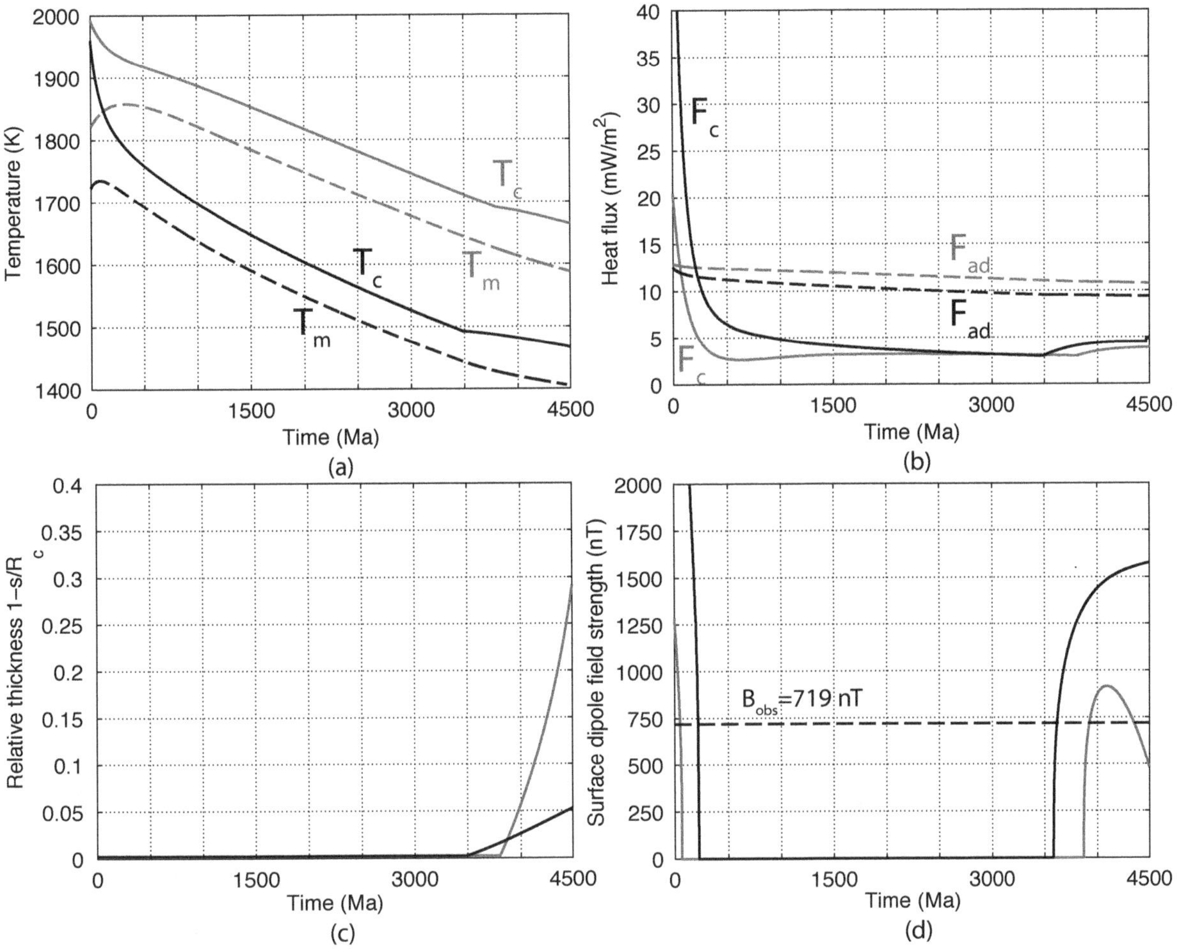

Figure 13.3 Representative examples for thermal evolution models in the iron-snow regime with 10.9% initial sulfur concentration (gray), and in the FeS regime with 31.6% initial S (black). (a) Temperature of the core T_c and of the mantle T_m, (b) heat flux at the top of the core F_c and adiabatic conductive heat flux F_{ad}, (c) thickness of the snow layer or the solid FeS layer, respectively, relative to the core radius, (d) predicted dipole field strength at Ganymede's surface (from Rückriemen et al. 2018)

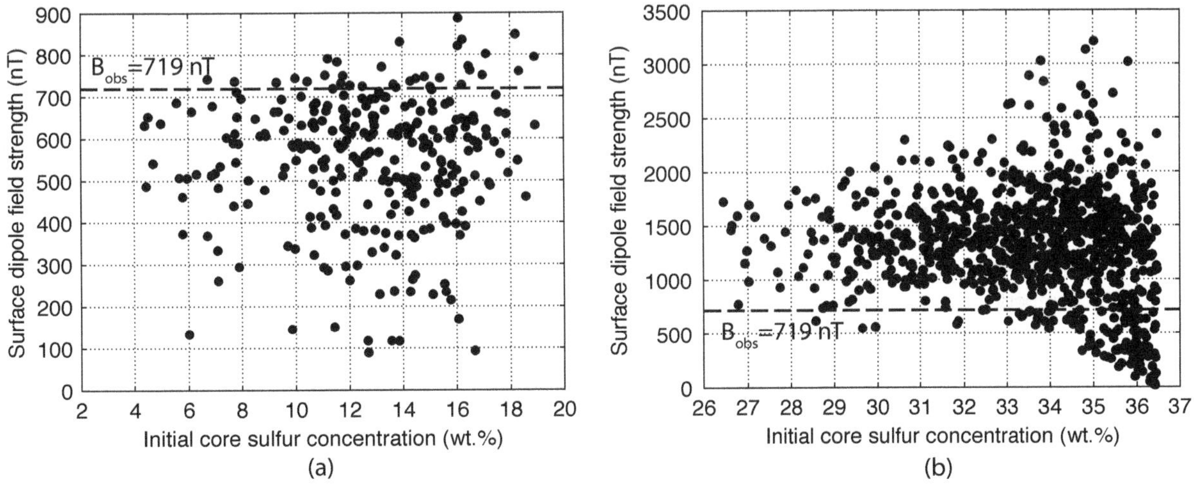

Figure 13.4 Present-day surface magnetic dipole field strength for successful iron-snow models (a) and FeS-layer models (b) as a function of sulfur concentration. The dashed line indicates the observed field strength (Kivelson et al., 2002) (from Rückriemen et al. 2018).

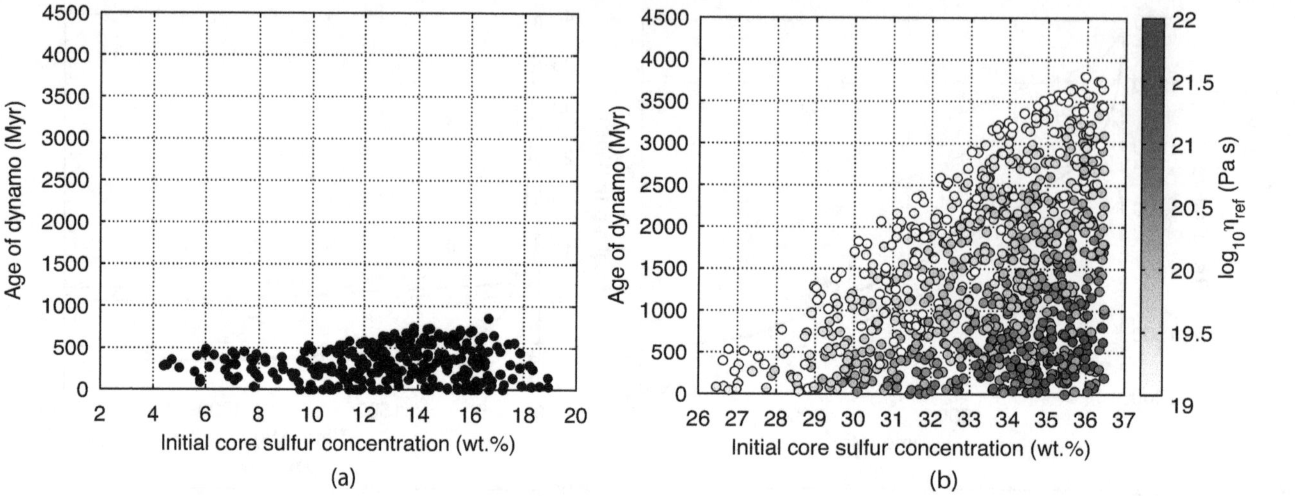

Figure 13.5 Age of the dynamo for successful iron-snow models (a) and FeS-layer models (b) as a function of sulfur concentration. Gray shading of circles in (b) corresponds to the reference viscosity in the mantle (from Rückriemen et al. 2018).

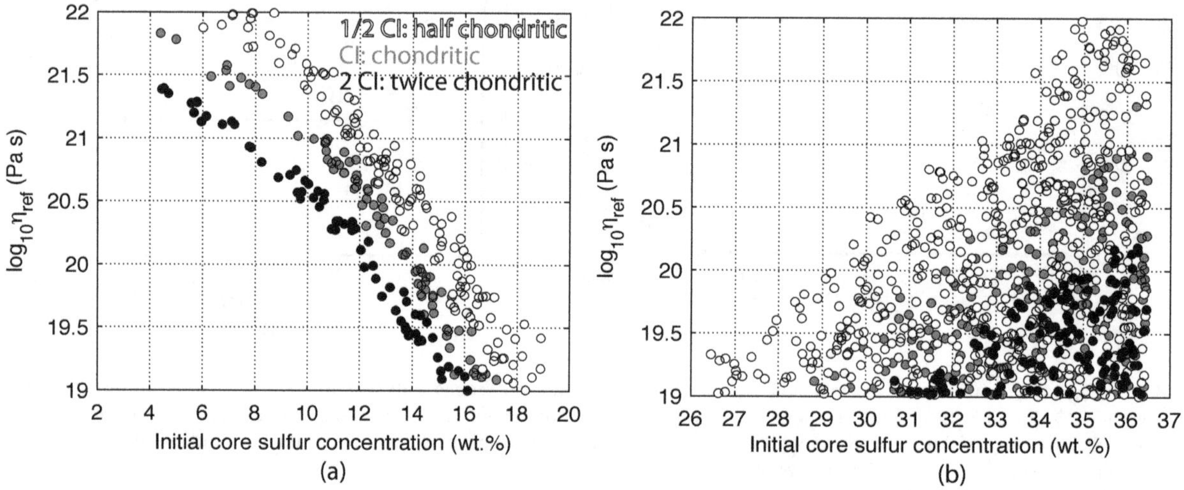

Figure 13.6 The reference mantle viscosity for successful iron-snow models (a) and FeS layer models (b) as a function of sulfur concentration. The abundance of chondritic heat sources is marked by the shading of symbols: half the chondritic abundance (white), chondritic abundance (gray), twice the chondritic abundance (black) (from Rückriemen et al. 2018).

lead to a present-day temperature at the core–mantle boundary that is in the range of liquidus temperatures of iron- or sulfur-rich Fe–FeS alloys. A considerable fraction of models with a higher abundance of heat sources in the mantle imply present-day temperatures at the core–mantle boundary above the liquidus temperatures, which precludes compositional convection.

In summary, reasonable thermal evolution models can be found that predict dynamo action at the present time, powered by the crystallization either of iron or of FeS near the top of Ganymede's core. The predicted magnetic field strength is of the right order of magnitude. In the iron-snow regime, the dynamo must have started rather recently and requires a rather special combination of parameters that control the thermal evolution. In the regime of FeS crystallization, successful dynamos can be found in a wider range of parameters, and they can also be longer-lived.

In this section, we concentrated on dynamos driven by crystallization in Ganymede's core, which have been studied in more detail than thermally driven dynamos. We note that according to new experimental results for the thermal conductivity of sulfur-rich alloys (e.g., Littleton et al., 2021), the adiabatic heat flow shown in Figure 13.3b may be overestimated by a factor of four in the case of the high sulfur concentration. In that case, thermal convection could be a viable driving mechanism. However, because thermal and electrical conductivity are linked through the Wiedemann–Franz law, the electrical conductivity should also be lower by a factor of four. In the evolution models presented here, the magnetic Reynolds number typically peaks at values of 400–700 with the assumption of high conductivity values (Rückriemen et al., 2015). When the electrical conductivity is reduced significantly, the magnetic Reynolds number may still be sufficient for a dynamo, although it would be somewhat marginal.

13.8 NUMERICAL DYNAMO MODELS

Direct numerical simulations of planetary dynamos typically all follow the same scheme, which was developed originally for the geodynamo (see Christensen & Wicht, 2015, for a review). In short, for a rapidly rotating spherical shell filled with an electrically conducting fluid, the coupled equations for fluid flow driven by thermal or compositional buoyancy and affected by electromagnetic forces, the magnetic induction equation, and a transport equation for temperature or composition are solved. Models can differ, for example, in the distribution of sources and sinks of buoyancy, the radius of a solid inner core, and the choice of boundary conditions. They also differ in the values of basic control parameters, where it must be kept in mind that for some of them, the adopted values are far from planetary values in all models. This reflects mainly that for numerical reasons, a much larger viscosity must be chosen than would be realistic.

The first numerical dynamo simulations dedicated to Ganymede were presented by Zhan and Schubert (2012). They compared two sets of simulations. In one that they identify with the iron-snow the mass anomaly flux decreases from the outer boundary to zero at the boundary of an inner core. In the second regime, the flux decreases from the inner core boundary to zero on the outer boundary. This is associated either with a compositional flux arising from inner core growth as in case of the present Earth, or with the flotation of FeS that crystallizes at depth. In these simulations, dipole-dominated magnetic fields were found only in the second regime, whereas in the first regime, weak multipolar magnetic fields have been obtained, in contrast to the observed structure of Ganymede's field. The authors therefore concluded that iron-snow-driven convection does not play a significant role in Ganymede's core.

Christensen (2015a) studied dynamo models for the iron-snow regime in more detail (see also the correction in Christensen 2015b). On the outer boundary of the active dynamo region, the addition of melting iron snow was modeled as a spatially uniform mass anomaly flux from above. The overlying snow-forming region was modeled as a fluid and electrically conducting shell that is coupled viscously and magnetically to the dynamo region. For this layer, a very strongly stabilizing stratification was assumed, which would allow horizontal flow but no radial motion. In these models, dipole-dominated magnetic fields were found in a rather broad range of control parameters. This holds also in the limiting case without an outer stable layer, which is basically the same setup as in the iron-snow models of Zhan and Schubert (2012). The reason for the discrepancy in the obtained field structure is not entirely clear. However, aside from differences in control parameters and the size of the (passive) inner core, the compositional boundary condition at the top of the dynamo region also differed. The condition of homogeneous flux used by Christensen (2015a) appears more favorable for a dipolar field than the condition of fixed composition (or temperature) used by Zhan and Schubert (2012), as was shown in a comparison study by Sakuraba and Roberts (2009).

In comparison to other planets, Ganymede seems to have an unusually weak quadrupole field relative to the dipole field. Denoting by R_n the magnetic field power in spherical harmonic degree n at the planet's surface, the ratio R_2/R_1 is $\leq 2.5 \times 10^{-3}$ for Ganymede, when we assume that the field model without induced component by Kivelson and colleagues (2002) provides an upper bound for the quadrupole power. For the Earth, this ratio is 4.6×10^{-2} (in 2020). Some part of the difference can be attributed to the differences in the core radii relative to the planetary radii, since above the core the ratio R_2/R_1 must drop with radius as r^{-2}. However, even if we assume for Ganymede a rather small core with one quarter of the planetary radius, the ratio R_2/R_1 at the top of Ganymede's core is still only 0.04 compared to 0.15 at the top of Earth's core.

Figure 13.7 compares snapshots of the radial magnetic field for two dynamo models from Christensen (2015a). In panel (a), there is no stable snow layer above the active dynamo region. This may represent the case where crystallization of iron or of FeS has just started and the snow layer or solid FeS layer, respectively, is still very thin. In panel (b), the snow layer comprises the upper 28% in radius of the core. In (b), the magnetic field is shown at the core–mantle boundary. Because in case (a) the active dynamo reaches up to the core–mantle boundary, the magnetic field at this point contains much more small-scale structure than in case (b). To better facilitate a comparison, the field shown in (a) is upward continued to a nominal radius r_1, such that the ratio between r_1 and the core radius is the same as the ratio between the core radius and the radius of the active dynamo region in case (b). The magnetic field is dominated by the axial dipole in both cases. However, without a snow layer of significant thickness (case a), higher multipole components contribute significantly. In particular, there is a significant quadrupole component of degree two and order one present at this instant in time. In contrast, in case (b) with a significant snow-forming layer, the multipole contributions are substantially weaker.

The presence of a stably stratified and electrically conducting snow-forming layer above the dynamo can explain the relative weakness of the multipole components and in particular the low quadrupole-to-dipole ratio found at Ganymede. In the dynamo models by Christensen (2015a), the values R_2/R_1 dropped systematically with increasing thickness of the stable region. This is illustrated by the comparison of time-averaged relative power spectra for cases with and without stable layer in Figure 13.8. For an active dynamo that extends to the top of Ganymede's core, the power in the quadrupole field would probably exceed the upper limit from observation or would be at best marginally compatible. The presence of a stable layer lowers the relative quadrupole power well below the upper limit. The lowering of R_2/R_1 at Ganymede's surface is to some degree a simple geometric effect, as the outer boundary of the dynamo is located at greater depth when there is a stable layer. More important is an electromagnetic skin effect of the conducting stable layer, which attenuates time-varying magnetic field components that must pass through the layer to the outside. As found in dynamo models for Mercury and Saturn (Christensen, 2006; Christensen & Wicht, 2008), the axial dipole is varying much more slowly in time and is therefore less attenuated by the skin effect in comparison to other field components, in particular the non-axisymmetric ones, which show much more rapid time variations.

The skin effect of a stable layer may also explain the rather low tilt angle of the dipole axis with respect to the rotation axis since it attenuates the equatorial dipole components more strongly than the axial component. Furthermore, the skin effect

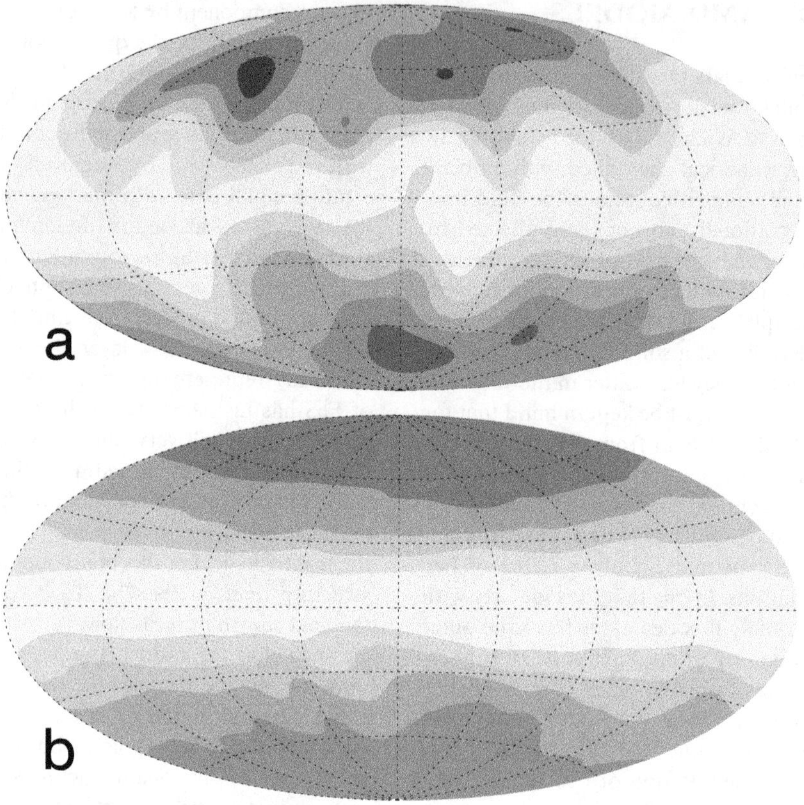

Figure 13.7 Snapshots of the radial magnetic field component for two dynamo models. (a) Case 14.0 without a stable top layer from Christensen (2015a), (b) case 14.4 with a stable layer (corrected as described in Christensen, 2015b). Red for positive B_r, blue for negative B_r. Gray-scale indicates value of $|B|$. See the text for more explanations. (A black and white version of this figure will appear in some formats. For the color version, please refer to the plate section.)

would prolong the characteristic timescale of secular variation. Based on his models, Christensen (2015a) estimated that the characteristic timescale for the change of the direction and tilt of the dipole axis would be prolonged from $\approx$ 60 yr without a stable layer to $\approx$ 200 yr in the presence of a stable layer. The difference in the rate of secular change may provide one handle to discriminate among dynamo models for Ganymede.

Gómez-Pérez and Wicht (2010) and Christensen (2015a) also addressed the question of whether Ganymede's dynamo is influenced by the field the Jupiter imposes on the core. The former authors found that an imposed field as weak as 2% of the dynamo-generated field can still have an effect. Christensen (2015a) found no significant influence for an external field strength of 1%. Because at the top of Ganymede's core, Jupiter's field has 0.5% or less of the strength of the internal dipole field, it may be concluded that Jupiter's field has at most a minor effect on Ganymede's present-day dynamo.

13.9 SUMMARY AND OUTLOOK

Ganymede's internal magnetic field is most likely generated by a dynamo in a metallic core that is at least partially liquid. Alternative explanations, such as a dynamo in a salty water ocean or remanent magnetisation in the rocky shell, can most probably be discarded. The highly probable existence of a core dynamo is strong evidence for Ganymede being a fully differentiated body including metal-silicate separation.

Driving of the dynamo by mechanical forcing, such as libration or tides, is a possibility. However, these mechanisms have not been explored much and it remains an open question if they could work in the case of Ganymede. The standard concept for driving a dynamo by thermo-compositional buoyancy forces is capable of explaining Ganymede's internal magnetic field, provided certain conditions for the composition of the core and for the thermal evolution of the planet are fulfilled. It is likely that the buoyant driving of flow in Ganymede's core differs from the mechanism operating presently in the Earth's core, where it is bottom-up – that is, associated with the release of light components and latent heat at the boundary of the growing inner core. Simple thermal convection in a completely liquid core could be possible for a sulfur-rich alloy, for which experimental data suggest a comparatively low thermal conductivity. If the sulfur content in Ganymede's core exceeds a few percent, top-down freezing is expected. If the sulfur concentration is less than eutectic, iron crystals form near the top of the core, sink, and remelt at greater depth. The top layer becomes enriched in sulfur and stably stratified. Remelting enriches the region just below the snow-forming layer in iron and drives compositional convection. If the sulfur concentration in the core is larger than the eutectic concentration, solid FeS would form at the top of the core, probably float, and form a growing FeS

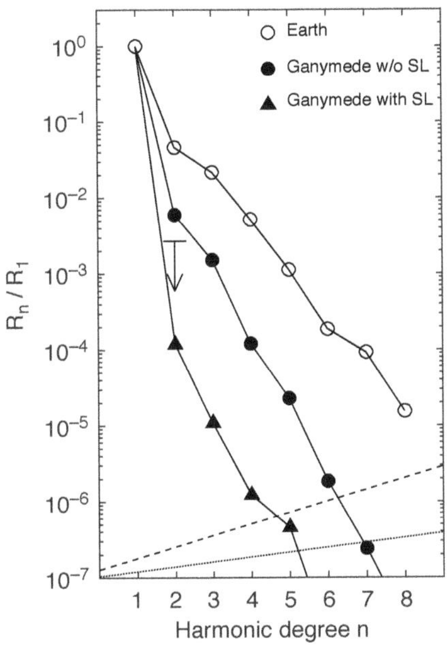

Figure 13.8 Ratio of time-average multipole power R_n to dipole power R_1 at Ganymede's surface from numerical dynamo models, assuming a core radius of one-third of Ganymede's radius. Filled circles for case 14.0 in Christensen (2015a) without an outer stable snow-forming layer (SL) and triangles for case 14.4 (corrected as described in Christensen 2015b) with a stable layer (SL) in the outer 29% of the core. Open circles show for comparison the ratio at Earth's surface in 2020. The probable upper limit for Ganymede's quadrupole-to-dipole ratio is marked by the downward-pointing arrow. Magnetometer observations by the JUICE mission are expected to resolve field components that lie above the broken and dotted lines, from the planned circular orbits at 500 km and 200 km altitude above Ganymede's surface, respectively.

layer sandwiched between the liquid core and the rocky mantle. The sulfur depletion in the remaining liquid likewise drives compositional convection.

Thermal evolution models, which include a parameterization of heat transport in the rocky mantle, show that both the iron-snow scenario and the FeS freezing scenario could operate at present, provided certain conditions for the viscosity and the radiogenic heat production in the silicate shell are fulfilled. Both are capable of driving a dynamo in the liquid central region below the snow-forming or the FeS layer, respectively. The flow velocities can be fast enough to support a dynamo and scaling laws predict a magnetic field strength of the right order. The iron-snow regime ceases to support a dynamo once the snow-forming layer has grown downward to the center of the planet, when iron does not remelt but collects in a growing inner core. In this regime, the lifetime of the dynamo is significantly less than the age of Ganymede, 1 Gyr at most. In the FeS-layer regime, the dynamo lifetime could be much longer if the sulfur concentration exceeded the eutectic composition by a substantial amount.

Dedicated numerical dynamo simulations for top-down driven convection in Ganymede suggest that the magnetic field could be dipole dominant. They also show that filtering of magnetic field components by a skin effect in the snow-forming layer can explain the rather low tilt angle of the dipole and the low ratio of quadrupole power to dipole power. A similar skin effect would be associated with a solid FeS layer, provided that its electrical conductivity is large enough and that it has grown to a sufficient thickness.

Our understanding of planetary magnetism is enhanced by the existence of pairs of similarly structured bodies with different magnetic properties. Our solar system has two such pairs. Ganymede stripped of its ice covering and Io form one pair; Earth and Venus comprise the other. Ganymede has an active dynamo; Io does not. Earth has an active dynamo; Venus does not. Each of these bodies has a liquid metallic core (although its size or even existence is less well constrained than in case of the Earth), potentially able to support dynamo action and create a magnetic field. Therefore, the existence of such a core (or highly electrically conducting fluid region) is a necessary but not sufficient requirement for a dynamo-driven magnetic field. The thermo-chemical evolution of these bodies and the special circumstances in which they evolve have the final say in whether the planetary body has a magnetic field. In the Io–Ganymede case, the intense tidal heating of Io has likely precluded its evolution into a core state similar to that described for Ganymede. These examples from our own solar system should be at the forefront of our thinking when we speculate about the occurrence of magnetic fields on extra solar planets.

All solar system dynamos seem unique in one way or another. Ganymede may be the only example of a dynamo that is predominantly driven from above by compositional convection, although also in Mercury, an iron-snow regime could operate (e.g., Chen et al., 2008). Convection in the Earth's core could have been driven mainly by the exsolution of mantle components such as MgO or SiO_2 in the past (e.g., O'Rourke & Stevenson, 2016), although after inner-core nucleation its contribution would probably be minor. Despite differences in the conducting material and the ways of driving convection, Ganymede may share with Mercury and Saturn the existence of an electrically conducting but stable layer above the dynamo. This would explain some common traits in the magnetic morphology of the three objects. The skin effect makes the magnetic field, as seen from the outside, more dipolar and more axisymmetric, with the dipole axis more aligned with the rotation axis, than in cases where the dynamo reaches up to the outer boundary of the conducting region. In Saturn, whose magnetic field is extremely axisymmetric (Cao et al., 2020), this effect would be most pronounced, somewhat weaker in Mercury, and probably weakest but still significant in Ganymede.

The planned circular orbits of the JUICE mission at 500 km and 200 km above Ganymede's surface will provide an unprecedented opportunity to improve on the internal magnetic field, using data from the J-MAG magnetometer. The data will allow the unambiguous separation of the time-dependent induced signal from the permanent internal magnetic field. Observation of an induced component at different periods, such as Jupiter's synodic rotation period, Ganymede's orbital period, and possibly longer stochastic perturbations of Jupiter's magnetosphere, can put constraints on the depth, conductivity, and thickness of a salty water ocean. The multipole components of Ganymede's dynamo field can probably be resolved up to degree six if the spatial power spectrum at the top of Ganymede's core is similar to that at Earth's core. This holds when assuming that the residual error from imperfect separation of induced internal and external field components from the permanent internal field will be 0.3 nT. If higher multipoles are more strongly suppressed due to the presence of a stable conducting layer at the top of Ganymede's core, the limit may be reached at degree four or five (see Fig. 13.8). The steepness of the spectral decay will also depend on the radius of Ganymede's core and might be used to infer the core size, although there will be trade-offs with the influence of a stably stratified layer.

After reevaluating the Galileo magnetometer measurements, with the knowledge of the induced internal field and an improved characterization of the external magnetic field based on JUICE data, it may be possible to constrain the secular variation of the field. Assuming an error of 5 nT for the reevaluated internal field, a field change from the Galileo to the JUICE epoch can be detected if it is at least 1% for the dipole field and $\gtrsim$ 10% for the quadrupole. The shape of the magnetic power spectrum, the degree of axisymmetry of the dipole and multipole field, and constraints on the rate of secular change (even if only upper limits can be derived) will be very valuable to discriminate between different dynamo models – for example, if the presence of a stably stratified layer above the dynamo is warranted.

REFERENCES

Acũna, M. H., Connerney, J. E. P., Ness, N. F. et al. 1999. Global distribution of crustal magnetization discovered by the Mars Global Surveyor MAG/ER experiment. *Science* 284, 790–793.

Anderson, J. D., Lau, E. L., Sjogren, W. L., Schubert, G. & Moore, W. B. 1996a. Gravitational constraints on the internal structure of Ganymede. *Nature* 384, 541–543.

Anderson, J. D., Sjogren, W. L., & Schubert, G. 1996b. Galileo gravity results and the internal structure of Io. *Science* 272, 709–712.

Aubert, J., Labrosse, S., & Poitou, C. 2009. Modelling the paleo-evolution of the geodynamo. *Geophysical Journal International* 179, 1414–1429.

Bland, M. T., Showman, A. P., & Tobie, G. 2008. The production of Ganymede's magnetic field. *Icarus* 198, 384–399.

Breuer, D. & Spohn, T. 2003. Early plate tectonics versus single-plate tectonics on Mars: Evidence from magnetic field history and crust evolution. *Journal of Geophysical Research* 108(E7), 5072.

Buffett, B. A., Huppert, H. E., Lister, J. R., & Woods, A. W. 1996. On the thermal evolution of the Earth's core. *Journal of Geophysical Research* 101, 7989–8006.

Buono, A. S., & Walker, D. 2011. The Fe-rich liquidus in the Fe-FeS system from 1 bar to 10 GPa. *Geochimica et Cosmochimica Acta* 75, 2072–2082.

Campbell, J. K., & Synott, S. P. 1985. Gravity field of the Jovian system from Pioneer and Voyager tracking data. *Astronomical Journal* 90, 364–372.

Cao, H., Dougherty, M., Hunt, G. J. et al. 2020. The landscape of Saturn's internal magnetic field from the Cassini Grand Finale. *Icarus* 344, 113541.

Carlson, R., Smythe, W., Baines, K. et al. 1996. Near-infrared spectroscopy and spectral mapping of Jupiter and the Galilean satellites: Results from Galileo's initial orbit. *Science* 274, 385–388.

Chen, B., Li, J., & Hauck, S. A. 2008. Non-ideal liquidus curve in the Fe-S system and Mercury's snowing core. *Geophysical Research Letters* 35, L07201.

Christensen, U. R. 2006. A deep dynamo generating Mercury's magnetic field. *Nature* 444, 1056–1058.

Christensen, U. R. 2010. Dynamo scaling laws: Applications to the planets. *Space Science Reviews* 152, 565–590.

Christensen, U. R. 2015b. Corrigendum to "Iron snow dynamo models for Ganymede." *Icarus* 256, 63–65.

Christensen, U. R. 2015a. Iron snow dynamo models for Ganymede. *Icarus* 247, 248–259.

Christensen, U. R., & Aubert, J. 2006. Scaling properties of convection-driven dynamos in rotating spherical shells and applications to planetary magnetic fields. *Geophysical Journal International* 166, 97–114.

Christensen, U. R., Holzwarth, V., & Reiners, A. 2009. Energy flux determines magnetic field strength of planets and stars. *Nature* 457, 167–169.

Christensen, U. R., & Wicht, J. 2008. Models of magnetic field generation in partly stable planetary cores: Applications to Mercury and Saturn. *Icarus* 196, 16–34.

Christensen, U. R., & Wicht, J. 2015. Numerical dynamo simulations. In Schubert, G. (ed), *Treatise on Geophysics*, 2nd ed., Volume 8: Core Dynamics. Amsterdam: Elsevier, pp. 245–277.

Chudinovskikh, L., & Boehler, R. 2007. Eutectic melting in the system Fe-S to 44 GPa. *Earth and Planetary Science Letters* 257(1), 97–103.

Crary, F. J., & Bagenal, F. 1998. Remanent ferromagnetism and the interior structure of Ganymede. *Journal of Geophysical Research* 103, 25 757–25 773.

Dauphas, N., & Pourmand, A. 2011. Hf-W-Th evidence for rapid growth of Mars and its status as a planetary embryo. *Nature* 473(7348), 489–492.

Davaille, A., & Jaupart, C. 1993. Transient high-Rayleigh-number thermal convection with large viscosity variations. *Journal of Fluid Mechanics* 253, 141–166.

Davidson, P. A. 2013. Scaling laws for planetary dynamos. *Geophysical Journal International* 195, 67–74.

Davies, C.J., & Pommier, A. 2018. Iron snow in the Martian core? *Earth and Planetary Science Letters* 481, 189–200.

de Koker, N., Steinle-Neumann, G., & Vlček, V. 2012. Electrical resistivity and thermal conductivity of liquid Fe alloys at high P and T, and heat flux in the Earth's core. *Proceedings of the National Academy of Sciences* 109, 4070–4073.

Deschamps, F., & Sotin, C. 2000. Inversion of two-dimensional numerical convection experiments for a fluid with a strongly temperature-dependent viscosity. *Geophysical Journal International* 143(1), 204–218.

Dumberry, M., & Rivoldini, A. 2015. Mercury's inner core size and core crystallization regime. *Icarus* 248, 254–268.

Ezenwa, I. C., & Yoshino, T. 2021. Martian core heat flux: Electrical resistivity and thermal conductivity of liquid Fe at Martian P,T-conditions. *Icarus* 360, 114367.

Fei, Y., Bertka, C. M., & Finger, L. 1997. High-pressure iron-sulfur compound, Fe_3S_2, and melting relations in the Fe-FeS system. *Science* 275(5306), 1621–1623.

Fei, Y., Li, J., Bertka, C. M., & Prewitt, C. T. 2000. Structure type and bulk modulus of Fe_3S, a new iron-sulfur compound. *American Mineralogist* 85(11–12), 1830–1833.

Gómez-Pérez, N., & Wicht, J. 2010. Behavior of planetary dynamos under the influence of external magnetic fields: Application to Mercury and Ganymede. *Icarus* 209, 53–62.

Grasset, O., & Parmentier, E. M. 1998. Thermal convection in a volumetrically heated, infinite Prandtl number fluid with strongly temperature-dependent viscosity: Implications for planetary thermal evolution. *Journal of Geophysical Research* 103(B8), 18171–18181.

Gubbins, D. 2003. Thermal Core–Mantle Interactions: Theory and Observations. In Earth's Core: Dynamics, Structure, Rotation. Wiley Online Library, pp. 163–179.

Hauck, S. A., Aurnou, J. M., & Dombard, A. J. 2006. Sulfur's impact on core evolution and magnetic field generation on Ganymede. *Journal of Geophysical Research* 111, E09008.

Hauck, S. A., Margot, J. L., Solomon, S. C. et al. 2013. The curious case of Mercury's internal structure. *Journal of Geophysical Research* 118, 1204–1220.

Huguet, L., Van Orman, J. A., Hauck, S. A., & Willard, M. E. 2018. Earth's inner core nucleation paradox. *Earth and Planetary Science Letters* 487, 9–20.

Hüttig, C., & Breuer, D. 2011. Regime classification and planform scaling for internally heated mantle convection. *Physics of the Earth and Planetary Interiors* 186(3), 111–124.

Jia, X., Walker, R. J., Kivelson, M. G., Khurana, K. K., & Linker, J. A. 2008. Three-dimensional MHD simulations of Ganymede's magnetosphere. *Journal of Geophysical Research* 113, A06212.

Khurana, K. K., Kivelson, M. G., Stevenson, D. J. et al. 1998. Induced magnetic fields as evidence for subsurface oceans in Europa and Callisto. *Nature* 395, 777–780.

Kimura, J., Nakagawa, T., & Kurita, K. 2009. Size and compositional constraints of Ganymede's metallic core for driving an active dynamo. *Icarus* 202, 216–224.

Kivelson, M.G., Khurana, K.K., Russell, C. et al. 1996. The discovery of Ganymede's magnetic field by the Galileo spacecraft. *Nature* 384, 537–541.

Kivelson, M.G., Khurana, K.K., & Volwerk, M. 2002. The permanent and induced magnetic moments of Ganymede. *Icarus* 157, 507–522.

Kleine, T., Touboul, M., Bourdon, B. et al. 2009. Hf–W chronology of the accretion and early evolution of asteroids and terrestrial planets. *Geochimica et Cosmochimica Acta* 73(17), 5150–5188.

Kuskov, O. L., & Kronrod, V. A. 2001. Core size and internal structure of Earth's and Jupiter's satellites. *Icarus* 151, 204–227.

Le Bars, M. 2016. Flows driven by libration, precession and tides in planetary cores. *Physical Review Fluids* 1, 060505.

Littleton, J. A., Secco, R. A., & Yong, W. 2021. Electrical resistivity of FeS at high pressures and temperatures: Implications for thermal transport in the core of Ganymede. *Journal of Geophysical Research* 126, e2020JE006793.

Lodders, K., & Fegley, B. 1998. *The Planetary Scientist's Companion*. Oxford: Oxford University Press.

Loper, D. E., & Roberts, P. H. 1977. On the motion of an iron-alloy core containing a slurry: I. General theory. *Geophysical and Astrophysical Fluid Dynamics* 9(1), 289–321.

Manthilake, G., Chantel, J., Monteux, J. et al. 2019. Thermal conductivity of FeS and its implications for Mercury's long-sustaining magnetic field. *Journal of Geophysical Research* 123, 2359–2368.

McKinnon, W. B. 1996. Core evolution in the icy Galilean satellites, and the prospects for dynamo-generated magnetic fields. *Bulletin of the American Astronomical Society* 28, 1076.

Moffatt, K., & Dormy, E. 2019. *Self-exciting Fluid Dynamos*. Cambridge: Cambridge University Press.

Morard, G., Andrault, D., Guignot, N. et al. 2008. In situ determination of Fe–Fe_3S phase diagram and liquid structural properties up to 65 GPa. *Earth and Planetary Science Letters* 272(3), 620–626.

Morard, G., C., Sanloup., Fiquet, G., Mezouar, M. et al. 2007. Structure of eutectic Fe–FeS melts to pressures up to 17 GPa: Implications for planetary cores. *Earth and Planetary Science Letters* 263(1), 128–139.

Ness, N. F. 1979. The magnetic field of Mercury. *Physics of the Earth and Planetary Interiors* 20, 209–217.

Neubauer, F. M. 1978. Possible strengths of dynamo magnetic fields of the Galilean satellites and of Titan. *Geophysical Research Letters* 5, 905–908.

Nimmo, F. 2015. Energetics of the core. In Schubert, G. (ed), *Treatise on Geophysics*, 2nd ed., Volume 8: Core Dynamics. Amsterdam: Elsevier, pp. 27–55.

Null, G. W. 1976. The gravity field of Jupiter and its satellites from Pioneer-10 and Pioneer-11 tracking data, *Astronomical Journal* 81, 1153–1161.

Olson, P., Landeau, M., & Hirsh, B. H. 2017. Laboratory experiments on rain-driven convection: Implications for planetary dynamos. *Earth and Planetary Science Letters* 457, 403–411.

O'Rourke, J. G., & Stevenson, D. J. 2016. Powering the Earth's dynamo with magnesium precipitation from the core. *Nature* 529, 387–389.

Pommier, A., & Davies, C. 2018. Influence of sulfur on the electrical resistivity of a crystallizing core in small terrestrial bodies. *Earth and Planetary Science Letters* 496, 37–46.

Rubie, D. C., Nimmo, F., & Melosh, H. J. 2007. Formation of Earth's Core. In Schubert, G. (ed), *Treatise on Geophysics*. Amsterdam: Elsevier, pp. 51–90.

Rückriemen, T., Breuer, D., & Spohn, T. 2015. The Fe snow regime in Ganymede's core: A deep-seated dynamo below a stable snow zone. *Journal of Geophysical Research* 1095–1118, 120.

Rückriemen, T., Breuer, D., & Spohn, T. 2018. Top-down freezing in a Fe-FeS core and Ganymede's present-day magnetic field. *Icarus* 307, 172–196.

Sakuraba, A., & Roberts, P. H. 2009. Generation of a strong magnetic field using uniform heat flux at the surface of the core. *Nature Geosciences* 2, 802–805.

Sarson, G. R., Jones, C. A., Zhang, K., & Schubert, G. 1997. Magnetoconvcetion dynamos and the magnetic fields of Ganymede and Io. *Science* 276, 1106–1108.

Saur, J., Duling, S., Roth, L. et al. 2015. A subsurface ocean in Ganymede from observations of its auroral oval. *Journal of Geophysical Research* 120, 1715–1737.

Saur, J., Neubauer, F., & Glassmeier, K.-H. 2010. Induced magnetic fields in solar system bodies. *Space Science Reviews* 152, 391–421.

Schubert, G., Cassen, P., & Young, R. E. 1979. Subsolidus convective cooling histories of terrestrial planets. *Icarus* 38(2), 192–211.

Schubert, G., Chan, K. H., Liao, X., & Zhang, K. 2004. Planetary dynamos: Effects of electrically conducting flows overlying turbulent regions of magnetic field generation. *Icarus* 172, 305–315.

Schubert, G., Spohn, T., & Reynolds, R. T. 1986. Thermal histories, compositions and internal structures of the moons in the solar system. In: Burns, J. A., & Matthews, M. S. (eds.), *Satellites*. Tucson: University of Arizona Press, pp. 224–293.

Schubert, G., & Zhang, K. 2001. Effects of an electrically conducting inner core on the planetary and stellar dynamos. *Astrophysical Journal* 557, 930–942.

Schubert, G., Zhang, K., Kivelson, M., & Anderson, J. 1996. The magnetic field and internal structure of Ganymede. *Nature* 384, 544–545.

Scott, H. P., Williams, Q., & Ryerson, F. J. 2002. Experimental constraints on the chemical evolution of large icy satellites. *Earth and Planetary Science Letters* 203(1), 399–412.

Sohl, F., Spohn, T., Breuer, D., & Nagel, K. 2002. Implications from Galileo observations of the interior structure and chemistry of the Galilean satellites. *Icarus* 157, 104–119.

Solomatov, V. 1995. Scaling of temperature-and stress-dependent viscosity convection. *Physics of Fluids* 7(2), 266–274.

Solomatov, V. S. & Moresi, L. N. 2000. Scaling of time-dependent stagnant lid convection: Application to small-scale convection on Earth and other terrestrial planets. *Journal of Geophysical Research* 105(B9), 21795–21817.

Spohn, T., & Breuer, D. 1998. Planetary Interiors. In *Implications from Galileo Observations on the Interior Structure and Evolution of the Galilean Satellites*, vol. 1. Editions Frontiers, pp. 135–145.

Spohn, T., & Schubert, G. 2003. Oceans in the icy Galilean satellites of Jupiter. *Icarus* 161, 456–467.

Stevenson, D. J., Spohn, T., & Schubert, G. 1983. Magnetism and thermal evolution of terrestrial planets. *Icarus* 54, 466–489.

Stewart, A.J., Schmidt, M.W., van Westrenen, W., & Liebske, C. 2007. Mars: A new core crystallization regime. *Science* 316(5829), 1323–1325.

Usselman, T. M. 1975. Experimental approach to the state of the core: Part II, Composition and thermal regime. *American Journal of Science* 275(3), 291–303.

Wardinski, I., Langlais, B., & Thébault, E. 2019. Correlated time-varying magnetic fields and the core size of Mercury. *Journal of Geophysical Research* 124, 2178–2197.

Weiss, N. O., & Proctor, M. R. E. 2014. *Magnetoconvection*. Cambridge: Cambridge University Press.

Williams, Q. 2009. Bottom-up versus top-down solidification of the cores of small solar system bodies: constraints on paradoxical cores. *Earth and Planetary Science Letters* 284, 564–569.

Yin, Q., Jacobsen, S. B., Yamashita, K. et al. 2002. A short timescale for terrestrial planet formation from Hf–W chronometry of meteorites. *Nature* 418(6901), 949–952.

Zhan, X., & Schubert, G. 2012. Powering Ganymede's dynamo. *Journal of Geophysical Research* 117, E08011.

Part III

ATMOSPHERE, IONOSPHERE, MAGNETOSPHERE

14

Ganymede: Its Magnetosphere and Its Interaction with the Jovian Magnetosphere

MARGARET G. KIVELSON, FRAN BAGENAL, XIANZHE JIA, KRISHAN K. KHURANA, MARTIN VOLWERK, AND PHILIPPE ZARKA

14.1 INTRODUCTION

Ganymede is the only moon in the solar system known to harbor a dynamo field, one strong enough to carve a magnetosphere out of the Jovian plasma that flows around it. The discovery of Ganymede's large internal field came as a surprise to most scientists, who imagined that small bodies in regions far from the sun would lose their fluid cores as a result of cooling over geological time. However, almost two decades before this discovery, Neubauer (1978) had considered it possible that even in the present era, the large moons of Jupiter might be capable of dynamo action and had calculated the magnitudes of the internal fields that could be anticipated. He assumed that all the Galilean moons contained fluid cores of iron–ironsulfide (Fe–FeS) eutectic mixtures surrounded by rocky mantles covered (except for Io) by a layer of ice. Using Busse's (1976) dynamo model, he estimated field magnitudes (at the equatorial surface) of 2200 nT at Io, 810 nT at Europa, 200 nT at Ganymede, and 63 nT at Callisto (see Chapter 2.8).

Kivelson and colleagues (1979), noting that the field magnitudes Neubauer suggested were large enough to stand off the flowing Jovian magnetospheric plasma in which they are embedded, proposed that some of the Galilean moons, most probably Io, might form magnetospheres. The nature of the plasma interaction with outer planet moons was developed by Neubauer (1980) for unmagnetized moons and by Southwood and colleagues (1980) for both magnetized and unmagnetized moons. In a neutral gas, energy and momentum are carried by sound waves. In a magnetized plasma, energy and momentum are carried by magnetohydrodynamic (MHD) waves. The MHD wave speeds are combinations of the sound speed ($c_S = (\gamma kT/m_i)^{1/2}$) and the Alfvén speed ($v_A = B/\sqrt{\mu_o \rho}$). Here γ is the polytropic index, k is the Boltzmann constant, T is the temperature, m_i is the mass of a plasma ion, B is the magnetic field strength, μ_o is the permeability of free space, and ρ is the plasma mass density. The structure of the interaction region surrounding a magnetized body takes very different forms depending on whether these phase speeds are larger or smaller than the flow speed of the ambient plasma relative to the body. The field and plasma perturbations imposed by flow onto a moon propagate into the surrounding plasma and modify its magnetic configuration. If the relative flow is sub-magnetosonic (as is the case for Ganymede), the moon's magnetosphere is largely confined within an approximately cylindrical region roughly aligned with the external field direction (see Chapter 3.6). By contrast, a magnetized body such as a planet embedded in the solar wind, where the relative flow speed dominates wave speeds, carves out a bullet-shaped cavity approximately aligned with the direction of the upstream flow.

Near Ganymede, the ambient plasma approximately corotates with Jupiter at a speed (relative to Ganymede) that is typically smaller than the phase speeds of MHD waves. In this sub-Alfvénic case, the theory proposes that disturbances arising from a plasma–moon interaction with the flow nearly perpendicular to the field direction radiate primarily along the background field. (Neubauer [1980] describes the responses for other relative orientations of field and flow.) The perturbations create a front tilted relative to the background field at an angle whose tangent is the ratio of the flow speed of the background plasma to the Alfvén speed. Consequently, the magnetospheric cavity that a magnetized moon carves out of the incident flow is a bent cylinder folded slightly around the moon and stretching north and south along the background magnetic field, as illustrated in Figure 14.1c. (Unmagnetized, electrically conducting moons perturb their magnetic environment in a similar way, bending the field above and below into what are referred to as Alfvén wings [Kivelson et al., 2004; Saur, 2021].) At Ganymede, currents that flow along the bent field lines link the moon's environment to Jupiter's polar ionosphere.

Before 1996, our understanding of the interaction was purely theoretical, and none of the first five spacecraft to probe the Jovian system (Pioneer 10 and 11, Voyager 1 and 2, Ulysses) came close enough to the moons to obtain the data needed to test conjectures about either the magnetic fields of the moons or the nature of the interaction. However, on its inbound pass, as Pioneer 11 crossed Jovian magnetic shells that map magnetically to equatorial regions near Ganymede's orbit and approximately 190 R_G (R_G is the radius of Ganymede, 2 632 km) downstream of the instantaneous position of the moon, it recorded the signature of an intense field-aligned current system (see figure 4 of Kivelson & Winge, 1976). Although the scale of the disturbed region was large compared with the 2 R_G cross-section of the moon, it seemed plausible that the current was somehow linked to its presence, suggesting that this was an early detection of the Alfvénic interaction of flowing plasma with Ganymede.

In December 1995, the Galileo spacecraft arrived at Jupiter and was placed into a capture orbit. It began a tour skillfully designed to pass close to a moon near each periapsis. No clear signatures of permanent fields were found at Io, Europa, or Callisto (Khurana, 1997; Neubauer, 1998; Jia, 2009 and references therein). Ganymede, however, turned out to have a very substantial internal field, more than 700 nT at the surface equator

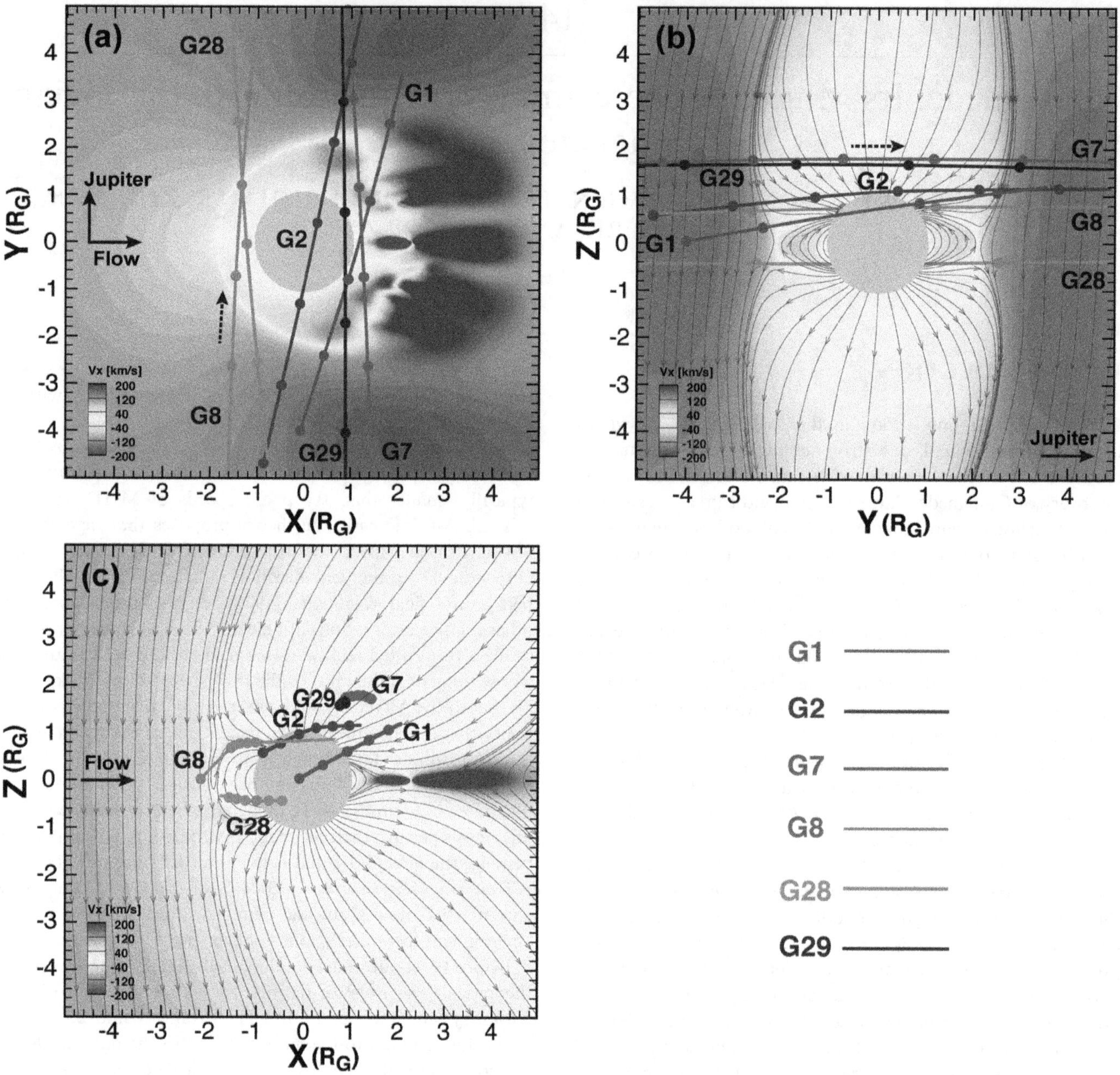

Figure 14.1 Trajectories of Galileo's six passes through Ganymede's magnetosphere plotted for the ~1 hour during which the highest-resolution data were acquired. Dots are separated by 10 minutes. The passes are projected into the (a) XY, (b) YZ, and (c) XZ planes of the GphiO coordinate system (X along the direction of corotating plasma flow at closest approach, Z parallel to Jupiter's spin axis, and Y pointing inward toward Jupiter. Panel (b) looks into the incident flow. The background is from the MHD simulation of Jia and colleagues (2009) for the Galileo G8 flyby with color representing the X-component of the flow velocity. Gray lines show the magnetic field lines, significantly distorted by the presence of the moon. The dashed arrows in panels (a) and (b) indicate the direction of the spacecraft motion, which is from –Y to +Y for all close encounters. (A black and white version of this figure will appear in some formats. For the color version, please refer to the plate section.)

(Gurnett et al., 1992; Kivelson et al., 1992; Gurnett et al., 1996; Kivelson et al., 1996). An internal field not only existed but was far larger than predicted. In a background field of order 110 nT, the field magnitude measured on the first Galileo pass near Ganymede peaked at 480 nT at an altitude of 840 km, and varied as the inverse cube of the distance from the center of the moon. The region in which the field displayed dipolar characteristics was bounded by a sharp field rotation (Kivelson et al., 1997) concurrent with abrupt changes of energetic particle anisotropies (Williams et al., 1992; Williams et al., 1997a; Williams et al., 1997b) at a radial distance consistent with crossing the magnetopause of a magnetosphere not unlike the one Southwood and colleagues (1980) had sketched for a magnetized body. sketched for a magnetized body. Although different in shape, the size of Ganymede's magnetosphere is comparable with that of Mercury's magnetosphere (Williams et al., 1998; Kivelson & Bagenal, 2014).

Galileo's six close encounters with Ganymede (trajectories are illustrated in Figure 14.1 with quantitative details provided in Table 14.1) revealed field and particle properties of

Table 14.1 *Information regarding Galileo's encounters with Ganymede*

Pass*	Date UT	Distance from Jupiter (RJ)	Local time (hours)	Jupiter latitude	SIII east long	Altitude at C/A	GN latitude	GN longitude	Region relative to Ganymede
G1	06/27/96 06:29:07	14.9–15.2	11.2–11.3	0.13	185.1	838.0 km	30.4	246.7	Central Wake
G2	09/06/96 18:59:34	14.8–15.2	10.7–10.9	0.19	202.0	264.4 km	79.3	236.4	Polar
G7	04/05/97 07:09:58	14.8–15.2	19.7–19.8	0.23	340.2	3104.9 km	55.8	270.4	Central Wake
G8	05/07/97 15:56:10	14.8–15.1	8.0–8.1	0.07	73.3	1606.2 km	28.3	84.8	Upstream
G28	05/20/00 10:10:10	14.7–15.3	0.7–0.9	−0.10	8.22	808.7 km	−19.0	92.4	Upstream
G29	12/28/00 08:25:27	14.7–15.3	23.9–30.0	0.23	139.5	2337.5 km	62.2	269.0	Polar
C9G	06/26/97 16:00–19:00	15.16	8.04	−0.16	300.1	30.3 R_G 79,750 km	−0.0	261.2	Ganymede Distant Wake
E12G	12/15/97 08:02–12:00	15.01	6.75	−0.12	349.8	5.47 R_G 14,300 km	−5.8	266.1	Ganymede Distant Wake

*The Galileo mission labeled each orbit by the first letter of the name of the moon encountered near periapsis followed by the orbit number, starting with I0. On two orbits targeted at other moons (C9 and E12), Galileo passed close enough to Ganymede to observe some disturbances arising from its interaction with the corotating plasma. R_G is the radius of Ganymede, 2632 km. Entries apply to closest approach unless a range is given.

the mini-magnetosphere, the first encountered in a sub-Alfvénic flow. The background of the figure shows the field lines and the plasma flow in the direction of the unperturbed upstream flow extracted from an MHD simulation for the Galileo G8 flyby (Jia et al., 2009). The flow speed decreases abruptly across the magnetopause boundary so, to a good approximation, the light-colored regions in the figures can be regarded as within Ganymede's magnetosphere. The images are cuts through the primary planes of the GphiO coordinate system (X along the direction of corotating plasma flow at closest approach, Z parallel to Jupiter's spin axis, and Y pointing inward toward Jupiter). As evident from the figure, Galileo's multiple passes probed many different parts of the low-altitude magnetosphere, allowing spacecraft instruments to obtain quantitative information about its large-scale structure.

As we have noted, the structure and dynamics of any magnetosphere are governed by the dimensionless parameters of the plasma that flows onto it as well as the properties of the field of the central body, so we start by summarizing what is known about the field and plasma conditions near Ganymede's orbit. We turn then to the intrinsic properties of the moon, including its magnetic field, which, as we shall see, consists of a permanent and a time-varying component. We then discuss the form and temporal variability of the magnetosphere itself, including disturbances on its boundary, its aurora, and the auroral signatures it imposes on Jupiter. We describe the properties of MHD waves and plasma waves observed near the moon. Much of our understanding of the system has been provided by numerical models, which we next discuss. We end by speculating on what we will learn from spacecraft expected to visit Ganymede in the near future.

14.2 OVERVIEW OF FIELD AND PLASMA CONDITIONS NEAR GANYMEDE'S ORBIT

Ganymede orbits Jupiter at 14.97 ± 0.017 R_J from the planet's center close to the jovigraphic equator (orbit inclination = 0.2°). Because Jupiter is gaseous and not spherical, its radius, R_J, is not uniquely defined; the Galileo project adopted the value 71,492 km. Jupiter's −10° dipole tilt imposes a variation of the plasma and field environment of Ganymede at the 10.53-hour synodic period of Jupiter. The range of values of Jupiter's magnetic field experienced by Ganymede reflects the ± 9.5° variation of Ganymede's magnetic latitude with System III longitude between where the jovigraphic and magnetic equators align (Wlon = 111°, 291°, Elon = 249°, 69°) and where there is a maximum dipole tilt angle (Wlon = 21°, 201°, Elon = 339°, 159°). The location of Jupiter's dipole equator relative to Ganymede shifts by ± 2.5 R_J parallel to Jupiter's spin axis on each synodic period. The range in plasma properties reflects the variation along the field line of plasma density, which is maximum at the farthest point from Jupiter's spin axis (the centrifugal equator, the location where the radial component of the field vanishes). At Ganymede's orbital distance, the centrifugal equator of a dipole field is tilted by ± 8.3° from the jovigraphic equator (Phipps & Bagenal, 2021), with plasma off the equator tending to have low density and high temperature compared with denser, colder plasma found near the equator, which passes over Ganymede twice per synodic period. The configuration of Ganymede's magnetosphere correspondingly varies with the synodic period. Because the plasma/field properties of Ganymede's environment vary systematically, Table 14.2 provides a range of values that bound the changes typically encountered over a synodic period.

The data in Table 14.2 have been updated from the equivalent table in Kivelson and colleagues (2004), combining data recently acquired by the Juno spacecraft with Voyager and Galileo measurements made near Ganymede's orbit (but >10 R_G from the moon). The range of values reflects the systematic variations in displacement from the magnetic and centrifugal equators as well as temporal variability arising from magnetospheric dynamics. Note that the plasma instruments carried by the three missions covered different energy ranges with different resolutions, which complicates the measurement of plasma temperature. We provide temperature estimates for the hot and cold components for electrons, protons, and heavy ions (comprising O^+, O^{++}, S^+, S^{++}, S^{+++}). The energy spectra of electrons and heavy ions are not purely Maxwellian. Often the Maxwellian distributions extend into supra thermal tails; a good fit can be obtained with two Maxwellian distributions, described here as "cold" and "hot." The temperatures are defined in the usual way from the fitted Maxwellians.

14.3 GANYMEDE'S MAGNETIC FIELD (OBSERVATIONS AND ANALYSIS)

14.3.1 Permanent Field

The vector magnetic field observed in the vicinity of Ganymede ($\mathbf{B}_{obs}$) is the sum of contributions from several different sources:

$$\mathbf{B}_{obs} = \mathbf{B}_{Jup} + \mathbf{B}_{Jmsph} + \mathbf{B}_G + \mathbf{B}_{Gmsph} + \mathbf{B}_{Ind}. \quad (14.1)$$

Here $\mathbf{B}_{Jup}$ is the field from Jupiter's interior, $\mathbf{B}_{Jmsph}$ is the field from the current sources in Jupiter's magnetosphere, $\mathbf{B}_G$ is the permanent field from Ganymede's interior (see Chapter 2.8), $\mathbf{B}_{Gmsph}$ is the field arising from currents in Ganymede's magnetosphere, and $\mathbf{B}_{Ind}$ is the inductive field from a putative ocean within Ganymede. In order to constrain the permanent magnetic field ($\mathbf{B}_G$) from the interior of Ganymede, all other sources of magnetic field must be evaluated and subtracted from the observations.

In the vicinity of the moon, magnetic fields from Jupiter and its magnetosphere do not vary significantly over length scales typical of Ganymede's magnetosphere; thus they can be regarded as constant and uniform during any one flyby. As a result, they are the easiest to evaluate. The internal field of Jupiter is Laplacian outside of the source region (i.e., at distances >1 R_J) and can be represented as the gradient of a scalar potential composed of multipoles (Connerney et al., 1998). The field of the Jovian magnetosphere near Ganymede is mainly dominated by current flowing in Jupiter's current sheet and is obtained either from a vector potential (Connerney et al., 1981) or two Euler potentials (Khurana, 1997). The combined contribution to the local field from these Jovian sources is of order 100 nT and can be treated as an external field of order one (i.e., a uniform field). This external field penetrates Ganymede's magnetosphere largely unimpeded because there is continual reconnection between the field of Ganymede and the external field of Jovian origin (Kaweeyanu et al., 2020; Zhou et al., 2020). Analysis of magnetic field observations from multiple passes and modeling confirm that, even in the closed regions of Ganymede's magnetosphere, the external Jovian field is present at full strength (Jia et al., 2008, 2009, 2010a). Because Jupiter's internal field is axially asymmetric, Ganymede senses it as a field varying at several different periods dominated by the synodic period of Jupiter's rotation (~10.5 hours). The field magnitude (65–110 nT) and direction change cyclically as the field of the current sheet flaps up and down (Connerney et al., 2018; Connerney et al., 2020) in sync with Jupiter's internal field. The combined field points almost southward when Ganymede is located at the center of Jupiter's current sheet and is inclined at ~ 45° to Ganymede's spin axis (toward or away from Jupiter) when Ganymede is located farthest from the center of the current sheet. In the remainder of this section, we shall assume that the uniform field (UF) ($\mathbf{B}_{uniform} = \mathbf{B}_{jup} + \mathbf{B}_{Jmsph}$) at the time of the flyby has been subtracted from the observations and will discuss the inversion that extracts the components of the residual field:

$$\mathbf{B}_{residual} = \mathbf{B}_{obs} - \mathbf{B}_{uniform} = \mathbf{B}_G + \mathbf{B}_{Gmsph} + \mathbf{B}_{Ind} \quad (14.2)$$

Using Galileo observations from the first two flybys of Ganymede, Kivelson and colleagues (1996) showed that the residual field is consistent with a Ganymede-centered magnetic dipole with an equatorial surface dipole strength of ~ 750 nT and a dipole tilt of ~10° relative to Ganymede's spin axis. In deriving these values, Kivelson and colleagues obtained the UF from the Jovian magnetospheric model of Khurana (1997) and did not allow for field contributions from the magnetosphere of Ganymede or from induction in the interior of Ganymede. Figure 14.2a, adapted from Kivelson and colleagues (1997), shows that, despite the simplifications, this vacuum superposition model reproduces the observations from the G2 flyby extremely well even though it does not capture the stepwise changes at the magnetopause.

As will be discussed in Sections 14.5 and 14.8, the principal contribution to the field within the magnetosphere of Ganymede, in addition to that of the internal dipole, is imposed by the magnetopause current system, which is analogous to the Alfvén wing current system that develops around a conducting object as a flowing plasma is diverted around it. Kivelson and colleagues (1997) showed that the observed field rotation at the magnetopause (~ 200 nT) is consistent with the strength of the

Table 14.2 *Plasma and field properties near Ganymede*

Property Symbol (units)	Typical & Range of Values			Ref
Magnetic Field Magnitude B_o (nT)		95 ± 25		1
Magnetic Pressure $B_o^2/2\mu_o$ (nPa)		4 ± 2		
Vertical Field B_θ (nT)		80 ± 8		1
Radial Field B_r (nT)		0 ± 90		1
Azimuthal Field B_ϕ (nT)		0 ± 15		1
Corotation Speed, V_{co} (km/s)		187		
Azimuthal Flow/V_{co}		80% ± 10%		2
Azimuthal Flow Speed (km/s)		150 ± 20		2
Ganymede's Orbital Speed, V_{orbit} (km/s)		11		
Azimuthal Flow – V_{orbit}		140 ± 20		2
	Highest magnetic latitude	**Intermediate magnetic latitude**	**Near magnetic equator**	
Magnetic Field Magnitude B_o (nT)	120	95	70	
Total Electron Density N_e (cm^{-3})	3	6	10	3
Cold Electron Temperature T_e (eV)	60	40	30	3
Hot Electron Density N_{eh}	10%	10%	10%	3
Hot Electron Temperature T_{eh} (eV)	200	200	200	3
Proton Density N_{proton} (cm^{-3})	0.5	0.8	1.2	4, 5
Proton Temperature T_p (eV)	300	200	100	4, 5
Heavy Ion Density N_{heavy} (cm^{-3})	1.7	3.5	6	4, 5
Heavy Ion Temperature T_{heavy} (eV)	100	75	50	4, 5
Hot Heavy Ion Density N_{heavyh}	20%	15%	10%	5
Hot Heavy Ion Temperature T_{heavyh} (eV)	3000	2000	1500	5
Heavy Ion Mass/Charge A/Z (amu/q)	19/1.2	24/1.7	25/1.6	6
Ion Mass Density (amu cm^{-3})	48	85	160	
Total Ion Thermal Pressure $P_{ion}=P_{proton}+P_{heavy}=P_{heavyh}=\Sigma nkT$(nPa)	0.25	0.25	0.20	
Energetic Ion Pressure $P_{energetic}$(nPa)	2.5	2.5	2.5	7
Electron Pressure P_{total}(nPa)	0.04	0.05	0.08	
Total Plasma Pressure P_{total}(nPa)	2.8	2.8	2.8	
Plasma $\beta=P_{total}/(B^2/2\mu_o)$	0.5	0.7	1.4	
Alfvén Speed V_A (km/s)	380	220	120	
Sound Speed C_s (km/s)	110	95	85	8
Alfvén Mach Number (M_A)	0.4	0.6	1	
Magnetosonic Mach Number (M_{ms})	0.4	0.6	1	8
Ram Pressure (nPa)	0.6	2	4	

[1] From Connerney and colleagues (2018) JRM09 internal field model, Connerney and colleagues (2020) current sheet model, plus Galileo and Juno data.
[2] Voyager (Dougherty et al., 2017) and Galileo (Bagenal et al., 2016) data.
[3] Voyager data (Sittler & Strobel, 1987).
[4] Voyager (Bagenal et al., 2017; Bodisch et al., 2017; Dougherty et al., 2017).
[5] Voyager (Dougherty et al., 2017), Galileo (Bagenal & Delamere, 2011; Bagenal et al., 2016), and Juno (Huscher et al., 2021).
[6] Based on physical chemistry models of the torus (Delamere et al., 2005) plus Voyager (Dougherty et al., 2017) and Juno (Kim et al., 2019; Kim et al., 2020).
[7] Mauk et al. (2004).
[8] Evaluated using total plasma pressure and ion mass density from the same column.

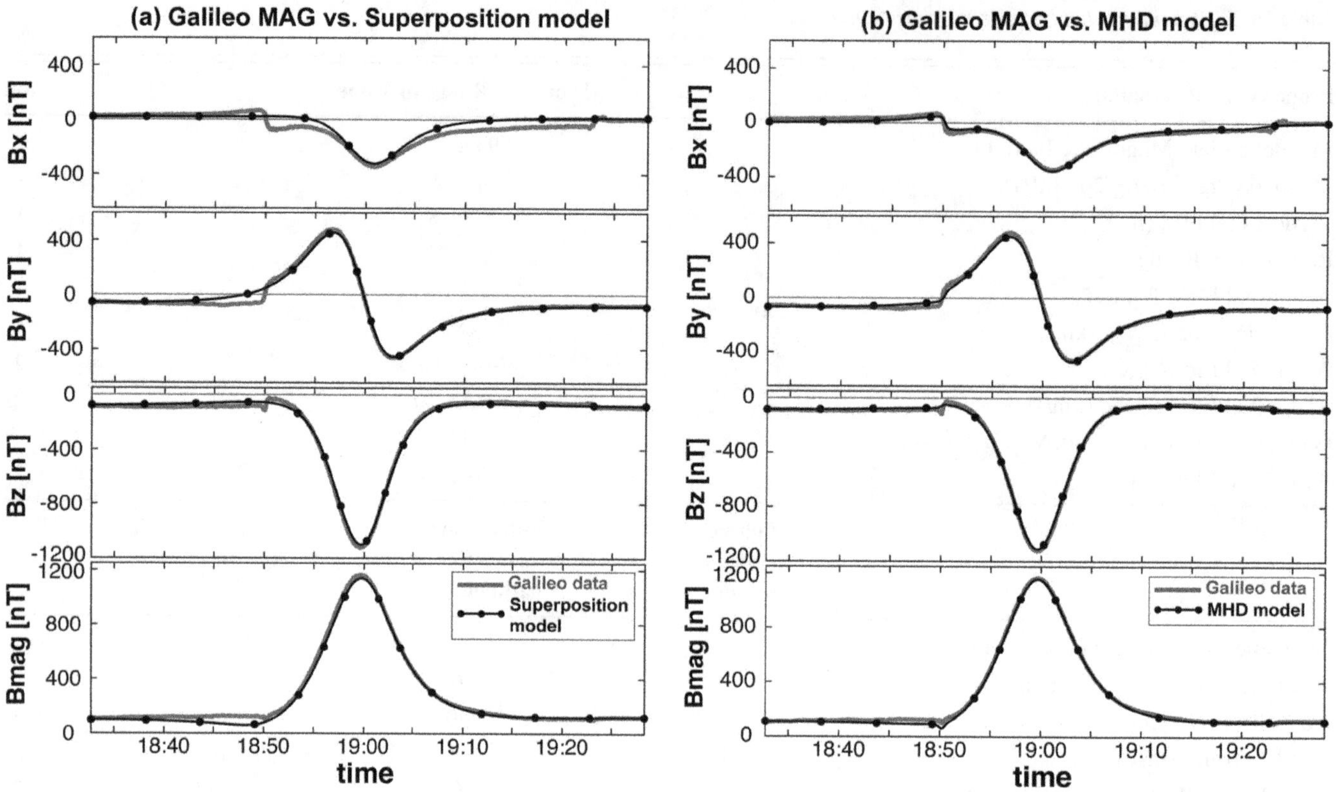

Figure 14.2 (a) The three components of the magnetic field and the field magnitude observed by Galileo during the G2 flyby (thick gray lines). The thin black lines with dots depict the simple vacuum superposition model discussed in the text (Kivelson et al., 1997). (b) The Galileo observations (thick gray lines) and an MHD model fit (thin black lines with dots) to these observations again from the G2 pass of Ganymede. Figure adapted from Jia and colleagues (2009a). All of the data and model plots are in GphiO coordinates.

field-aligned currents flowing on an Alfvén wing, if one assumes that the obstacle is not Ganymede but its entire magnetosphere with a cross-field scale size of 4–5 R_G. The field-aligned components of the magnetopause current system contribute a negative perturbation to B_X at positive latitudes and a positive perturbation to B_X at negative latitudes. The field from the closure currents that flow through Ganymede or its ionosphere produces a southward-oriented perturbation field upstream and a northward-oriented perturbation field downstream. Based on global MHD simulations conducted for various Galileo close flybys, the perturbation field strength of the external current systems is estimated to be of order of a few tens of nT inside the magnetosphere (Jia, 2009; Olsen et al., 2010), which is comparable to the strength of the induced field in Ganymede's interior (Kivelson et al., 2002).

The data from four additional flybys were used to refine models of the field arising from Ganymede's magnetospheric currents and the permanent field from Ganymede's interior (Kivelson et al., 2002). To lowest order, the contribution of currents flowing in Ganymede's magnetosphere to the field observed in regions near the moon on an individual flyby can be approximated as a UF. However, just as the UF from Jupiter and its magnetosphere differs from one flyby to another, the UF generated by Ganymede's magnetosphere differs from one flyby to another, depending on upstream field and plasma conditions and the location of the flyby (above or below Ganymede, or upstream or downstream of Ganymede, etc.). First removing the approximately UFs of external sources, Kivelson and colleagues (2002) modeled the residual field using Galileo data for the six passes whose closest approach distance was less than 2.5 R_G from Ganymede's center. This improved the estimate of the best-fit dipole moment, yielding an equatorial field magnitude of 719 nT and a tilt of 4° from the southern rotational pole. The southern magnetic pole was rotated by 24° from the Jupiter-facing meridian plane toward the trailing hemisphere.

Because the contribution to the field of higher-order moments decreases rapidly with distance, an estimate of the internal field through quadrupolar order requires data taken within 1.4 R_G of the center of the moon. Beyond that distance, perturbations arising from currents flowing in the local plasma dominate over plausible contributions from quadrupole fields. Only three of the flyby trajectories came within 1.4 R_G from Ganymede's center. Using data from these trajectories, fits to the residual field were performed in terms of fixed dipole and quadrupole moments with the external UF allowed to vary from pass to pass (see Table 14.3). The inverted dipole moment remained relatively unchanged and the power in the inferred quadrupole moments was extremely weak (~850 nT^2). The ratio of power in the quadrupole moments relative to the dipole moment is 0.0016, which is notably small compared to all other known planetary dynamos (Glassmeier & Heyner, 2021). The best-fit model had an RMS deviation of 13.5 nT. The implications of this extremely low quadrupole-to-dipole moment ratio for models of dynamo field generation are explored in Chapter 2.8.

Table 14.3 *Fitted parameters for Ganymede's internal field*

Fitted dipole plus quadrupole moments (in nT)							
g_1^0	g_1^1	h_1^1	g_2^0	g_2^1	g_2^2	h_2^1	h_2^2
−711.0±4.2	46.8±1.0	22.3±1.9	0.9±2.6	27.0±1.2	−0.4±1.7	1.8±2.2	−11.0±0.8
Fitted dipole moment (in nT) and induction efficiency							
g_1^0	g_1^1	h_1^1			α		
−716.8±1.6	49.3±0.7	22.2±1.1			0.84 ± 0.018		

14.3.2 Evidence for Subsurface Ocean: Inductive Response

Based on its surface geology, density, and moment of inertia, Ganymede appears to be intermediate to Europa and Callisto in surface age (Pappalardo et al. 2004; Schubert et al., 2004) and may be expected to share many features of their internal structures. Evidence of subsurface oceans had been found in both Europa and Callisto based on their strong inductive response to the time-varying field of Jupiter (Khurana et al., 1998; Kivelson et al., 1999; Zimmer et al., 2000). Thus it was natural to consider that Ganymede might also manifest an inductive response. As shown by Parkinson (1983), a spherical conducting shell exposed to a uniform time-varying field (order 1 in external spherical harmonics) creates a dipolar response (order 1 in internal spherical harmonics). In order to assess whether Ganymede responds inductively to the time-varying field of Jupiter, Kivelson and colleagues (2002) fitted the residual field from each of the three close Galileo flybys to the sum of the field of a permanent internal dipole, a UF representing the Jovian field at Ganymede's location at the time of each pass, the field resulting from Ganymede's magnetospheric currents (also different for each flyby), and the field of a time-varying internal dipole, **M**, whose strength (polar field at the surface of Ganymede) is determined by the time-varying part of the primary external field, $\mathbf{B}(t)_{primary}$

$$\mathbf{M} = -\alpha e^{i\phi}\mathbf{B}(t)_{primary} \tag{14.3}$$

where α is the response factor (also called induction efficiency) and ϕ is the phase delay of the response. Based on findings from Europa and Callisto (Zimmer et al., 2000), Kivelson and colleagues (2002) assumed that the phase delay of the induced field is small owing to the high conductivity of the putative ocean and that any reduction of the response from unity results from the burial depth of the ocean under Ganymede's ice layer (assuming that the thickness of the ocean layer is larger than the effective skin depth of the inducing signal). The best-fit model (RMS difference = 11.5 nT) yielded a response factor of 0.84, implying that a nonconducting layer with a thickness of $(1 - (0.84)^{1/3})R_G \approx 150$ km overlays the conducting layer. In order to carry the current required to produce the inductive response inferred from the data, the current-carrying layer must be global in extent and highly conducting. These requirements can be met only if the current-carrying layer is mostly or fully melted and has a significant electrolyte component. The dipole moment (g_1^1 and h_1^1) differed only slightly (see Table 14.3) from the value obtained from the dipole plus

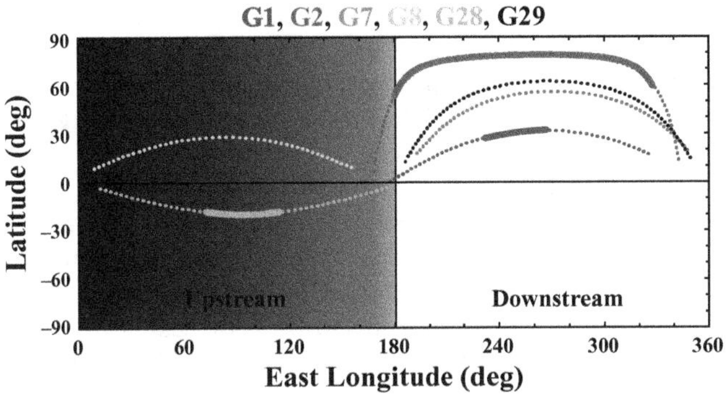

Figure 14.3 The latitude/longitude coverage of Ganymede by Galileo. The solid parts of the segments show the intervals when Galileo was within 1.4 RG from the center of Ganymede. Figure adapted by permission from Springer Nature: *Space Science Reviews*. Xianzhe Jia et al., Magnetic Fields of the Satellites of Jupiter and Saturn. ©2009. (A black and white version of this figure will appear in some formats. For the color version, please refer to the plate section.)

quadrupole fit shown in the table. Noting that the inductive response model requires only four internal parameters to fit the data compared to the eight required for the quadrupole fit, the improvement in the RMS deviation for the induction model (11.5 nT vs. 13.5 nT) is significant. The RMS errors for both models are quite small and may be underestimated because the simple model ignores contributions from higher-order multipole terms, fluctuations produced by local plasma currents, and temporal variations in the magnetosphere over time scales of minutes (passing waves or instabilities). Therefore, Kivelson and colleagues (2002) asserted that the error analysis is equally consistent with the dipole plus induced field model or the dipole plus quadrupole model.

An analysis of Galileo's trajectories (see Figure 14.3) near Ganymede reveals why it has been difficult to distinguish between the induction and quadrupole models for Ganymede. In order to define all four non-axial terms of the quadrupole moment unambiguously, trajectories must provide data at relatively low altitude from all four quadrants of Ganymede's surface. However, data are available from only three of Ganymede's quadrants, and the three flybys (G1, G2, and G28) that came within 0.4 R_G of the surface of Ganymede probed only two of the quadrants. This lack of global coverage is the root cause of the nonuniqueness problem. If activities now well underway are successful, the ambiguity will be resolved in the 2030s when the European Space Agency's (ESA) JUpiter ICy moons Explorer (JUICE) spacecraft reaches Ganymede and returns magnetometer measurements from planned orbits that will cover all sectors.

The model field of Table 14.3b was used by Jia and colleagues (2008, 2009, 2010a) in three-dimensional global MHD simulations of the interaction of Ganymede and its dipolar field with the corotating Jovian plasma for all six flybys of Ganymede. These self-consistent models of Ganymede's magnetosphere reproduce the observations much better than did the simple vacuum superposition model (compare Figures 14.2a and 14.2b). However, the nonuniqueness related to fitting the internal field as a permanent dipole plus an inductive response or as permanent dipole plus quadrupole fields remains. The results from simulations are discussed further in Section 14.8.

Saur and colleagues (2015) presented an independent approach to the search for a subsurface ocean in Ganymede by observing the location of auroral emissions on Ganymede in Hubble Space Telescope images. The auroral ovals mark the open/closed boundary of field lines in Ganymede's magnetosphere (Khurana et al., 2007; McGrath et al., 2013), where emissions are stimulated by the precipitation of Jovian magnetospheric electrons that have gained access to the magnetosphere of Ganymede through magnetic reconnection. The auroral ovals oscillate in position in response to the rocking of Jupiter's time-varying magnetospheric field. However, if an electrically conducting ocean is present within Ganymede, the amplitude of the time-varying part of the external field in the vicinity of Ganymede decreases because of the inductive response, thereby reducing the amplitude of the latitudinal excursions of auroral boundaries. An oscillation of $5.8° \pm 1.3°$ is predicted from a model without an ocean. The observed oscillation of $2.0° \pm 1.3°$ is close to the oscillation of $2.2° \pm 1.3°$ predicted by Saur and colleagues' model if a conducting ocean lies beneath the surface. The analysis of Saur and colleagues (2015) also supports the assertion that Ganymede's internal field has an extremely low quadrupole-to-dipole moment ratio.

Vance and colleagues (2021) have refined the two layer ice/water model of the inductive signal by considering potential effects of depth-dependent variations of the electrical conductivity of the ocean and effects of convective motion within the ocean. These effects cannot be identified in the limited data available from Galileo, but for some of the parameters investigated, the upper bound of the signals are predicted to be large enough to be detectable on future missions.

14.4 GANYMEDE'S MINI-MAGNETOSPHERE–PLASMA INTERACTION

The flowing plasma of Jupiter slows and diverts as it approaches any one of the Galilean moons, but the field and flow structures that develop are unique at Ganymede because of its internal magnetic field. The pressure exerted by the internal field stands off the flow well upstream of the moon's surface and creates a mini-magnetosphere, a large region surrounding the moon from which the upstream plasma is largely excluded and within which the field is significantly perturbed. The size and structure of the interaction region are determined by the magnitude of the internal field and the properties of the local Jovian plasma. It has been noted that near Ganymede's orbit, the Alfvén Mach number (M_A, the square root of the ratio of the dynamic pressure of the flowing plasma to twice the magnetic pressure in the upstream field) and the fast magnetosonic Mach number (M_{ms}) are both less than one (see Table 14.2). (It should be noted that the extreme values of entries in Table 14.2 can be combined to produce Mach numbers larger than one, but such combinations have not been encountered in existing spacecraft measurements.) The small value of M_A implies that the upstream magnetic pressure dominates the plasma pressure in establishing the configuration of Ganymede's magnetosphere. The small value of M_{ms} implies that, unlike the situation in planetary magnetospheres, no bow shock forms upstream of Ganymede. In this section, we describe in some detail the unusual magnetosphere of Ganymede.

The properties of Ganymede's magnetosphere have been established principally from measurements made in situ by the fields and particles instruments on Galileo: the Magnetometer (MAG: Kivelson et al., 1992), the Energetic Particle Detector (EPD: Williams et al., 1992), the Plasma Wave Subsystem (PWS: Gurnett et al., 1992), and the Plasma Analyzers (PLS: Frank et al., 1992). We show the full set of measurements for three critical encounters, a low-altitude pass across the polar cap (G2 in Figure 14.4) and two passes across the upstream region, one with closest approach at ~30° Ganymede latitude (G8 in Figure 14.5), and one closer to Ganymede and at less than 20° latitude (G28 in Figure 14.6). These representative passes reveal unambiguously that Ganymede's magnetosphere is a distinct structure embedded in the background plasma/field environment of Jupiter's magnetosphere.

Figure 14.1 shows that the G2 pass went over Ganymede's northern polar region at low altitude. On approach, the field rotated abruptly from radially outward from Ganymede's spin axis to radially inward (green curve in Figure 14.4) at the inbound magnetopause–indicated by a dashed line–and the field magnitude (black curve in the top panel) began to change systematically from its background value near 100 nT to well over 1000 nT near closest approach, revealing unambiguously that Ganymede has a large internal magnetic field. Entry into and exit from the magnetosphere (dashed lines) are evident in data from all instruments: a field rotation in MAG, a change of the noise level in EPD, an abrupt discontinuity of the ion energy spectrum in PLS (not clear on the outbound crossing), and a sharp change of the wave spectrum in PWS.

When energetic particle fluxes are not isotropic, the EPD data fluctuate both at the spacecraft spin period (20 s) and at longer periods because the detector steps through changes of look direction relative to the direction of the spacecraft spin axis. This complicates the interpretation of changing fluxes, but several features are clear. The sharp drops of both electron and ion count rates evident in the second panel of Figure 14.4 occur when the detector is measuring particles moving along the field toward the spacecraft from the direction of Ganymede. For a pass above Ganymede, such as G2, the deep drop in count rate occurs for particles with pitch angles near 180°. Such particles are absent because their bounce motion has led to impact on Ganymede (see Figure 14.1 and Chapter 3.6). On field lines linked to the polar cap, electrons returning from the north (pitch angles near 0°) are absorbed by Ganymede only after passing the spacecraft, but those returning from the south never reach the detector because they are absorbed by Ganymede.

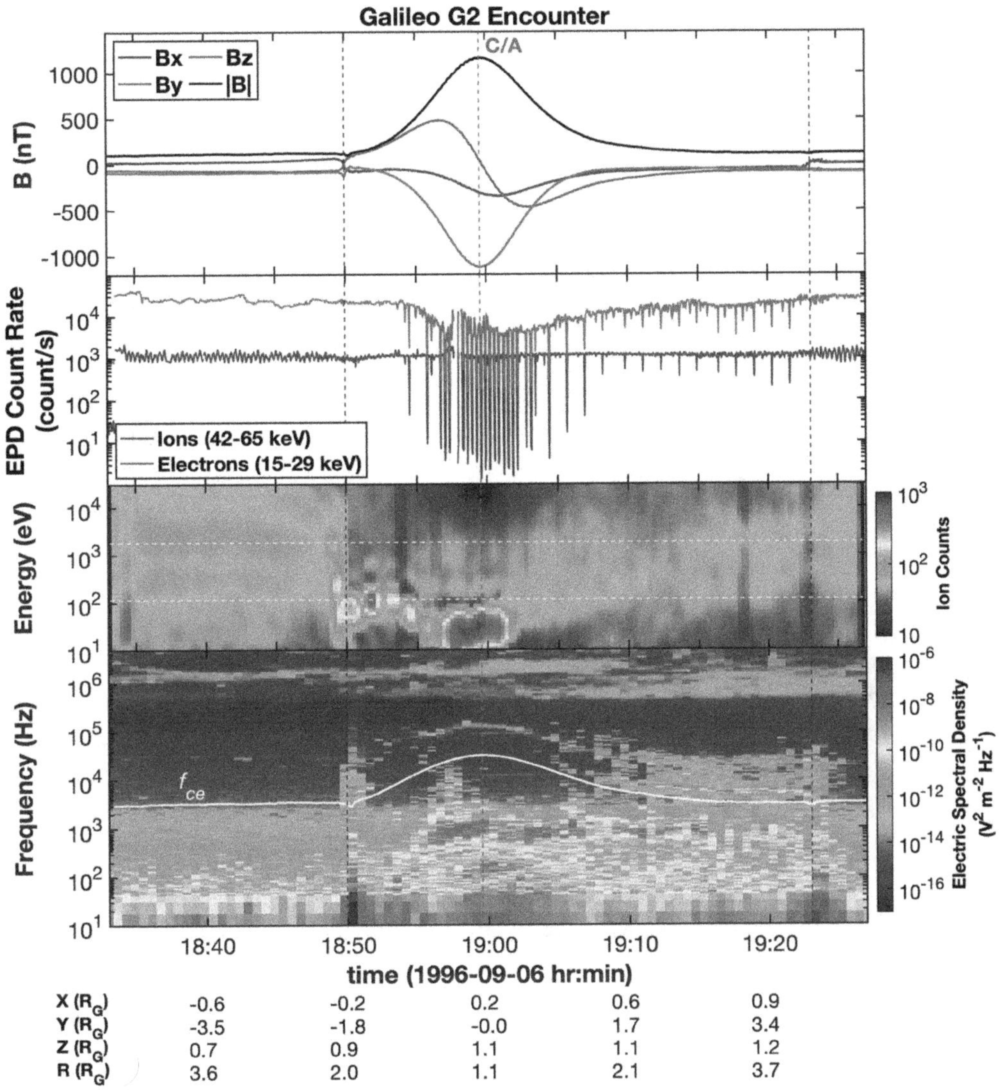

Figure 14.4 Combination of the MAG, EPD, PLS, and PWS data for the G2 encounter. The top panel shows the magnetic vector components and field strength in GphiO coordinates. The second panel shows the EPD counting rates for the ions (blue trace) from energy channel A1 (42–65 keV) and the electrons (red trace) from the energy channel E0 (15–29 keV), respectively. The third panel shows the ion energy spectrogram measured by the PLS. The color-coded ion count rate represents the maximum response from all available sensors at a given E/Q during one spacecraft spin period. For guidance, two white horizontal dashed lines at energies corresponding to the flow energy of ions with a convective flow speed of 150 km/s are plotted for $M/Q = 16$ and $M/Q = 1$, respectively. The bottom panel contains the PWS electric field spectrum with the electron gyrofrequency superimposed as white trace. The magenta vertical line marks closest approach (C/A), and the two gray vertical lines indicate the inbound and outbound magnetopause crossing identified based on the magnetic field data. (A black and white version of this figure will appear in some formats. For the color version, please refer to the plate section.)

Diffusion can refill the loss cone, leading over multiple bounces to decreases of counts at all pitch angles. This effect accounts for the decreasing maximum value of the electron count rate within a spacecraft spin period as the spacecraft moves deeper into the magnetosphere. Close to Ganymede's pole, the peak count rate in each cycle becomes roughly an order of magnitude smaller than in the surroundings, whereas the maximum ion count rate changes little from its value outside the magnetosphere. This difference reflects the large difference in the bounce periods of the two types of particles. Electrons of tens of keV can travel from Ganymede to their mirror point near Jupiter and return in a fraction of the time required for the flowing plasma to cross the magnetosphere, but O^+ and S^{++} bounce times are not short compared with the time for the plasma to flow across the polar cap.

In the third panel of Figure 14.4, before Galileo enters Ganymede's magnetosphere, the thermal ion counts (third panel) peak at two energies, one near 2000 eV and one near 200 eV. In this low energy range, the flow kinetic energy dominates the thermal energy so the most intense signals measure flow kinetic energy, which is a function of the ion mass, and the spread about the mean of those signals arises from thermal motion. The ratio of flow energies in the two bands at the start of the interval plotted indicates that ions of different mass per unit charge, likely H^+, and O^+/S^{++}, are present in the upstream flow. On entry into the magnetosphere, the flow slows abruptly across the magnetopause, identified by rotation of the magnetic field. The return to background does not show any abrupt boundary. Cold ($<\sim$ 100 eV) and dense plasma outflowing from Ganymede's ionosphere was detected by the PLS in the polar cap (see the

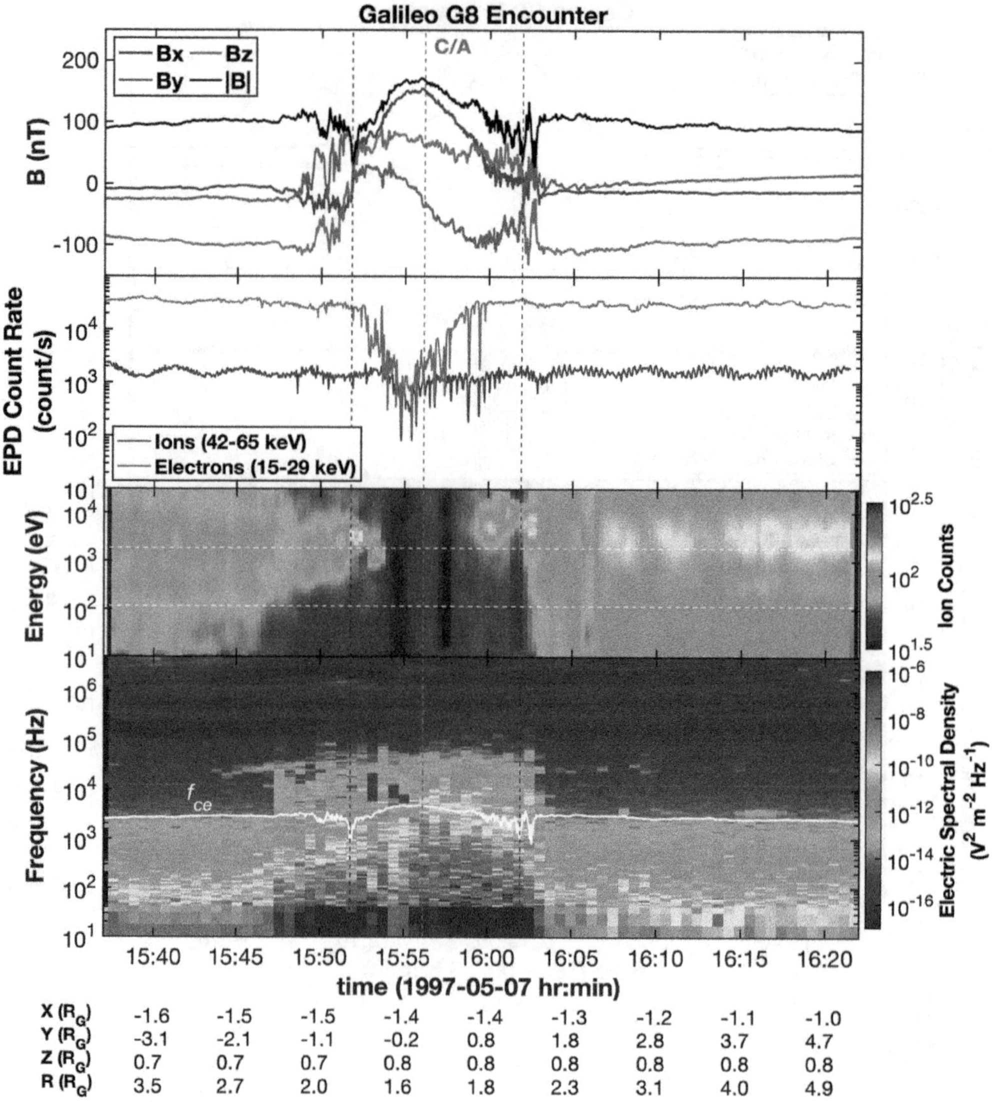

Figure 14.5 Same as Figure 14.4 but for the G8 encounter. (A black and white version of this figure will appear in some formats. For the color version, please refer to the plate section.)

third row plot near 19:00 UT). The ionospheric ions were initially assumed to be protons with a peak density of ~ 75 cm^{-3} at closest approach and an average convection speed of ~ 70 km/s (Frank et al., 1997). Assuming that the ions are protons, the derived plasma moments differ by a factor of a few from the total electron density inferred from the PWS data and the upper limit of flow speed inferred from the anisotropies of EPD ion counts. The assumption that the ions were dominantly protons was questioned by Vasyliunas and Eviatar (2001); they reinterpreted the PLS data, suggesting that the ions were oxygen. This suggestion brings measured plasma parameters from different instruments into close agreement. The conclusion is also supported by global magnetosphere simulations (Jia et al., 2009; Zhou et al., 2019). The lowest panel of Figure 14.4 shows the plasma wave spectrum in Ganymede's vicinity and is discussed in Section 14.7.

Figure 14.5 shows data from the same set of instruments for the G8 flyby, an upstream pass at relatively low Ganymede latitude but possibly close to the upstream cusp (see Figure 14.1). Again the interface between the background plasma and field is apparent in all panels, but the interface region shows large fluctuations and the boundary crossings in the different plots do not align as well as for G2. Near the boundaries, both inbound and outbound, there are large fluctuations in the magnetic field. Within the magnetopause, the magnetic field variations are dominated by the properties of Ganymede's internal dipole magnetic field. The flux of energetic electrons moving toward Ganymede decreases with decreasing distance from the surface (Cooper et al., 2001) where the strong magnetic field excludes their entry (see also Liuzzo et al., 2020). The large gyroradii of the energetic ions allows them to penetrate quite deeply and they develop strong loss cone signatures.

The thermal plasma data in the third panel of Figure 14.5 shows high count rates in the background plasma centered close to 2000 eV, the kinetic energy of O^+ or S^{++}ions moving at close to the velocity of corotation at Ganymede's orbit. On entry into the magnetopause, the flow continues for a few minutes but near closest approach (magenta line), the signal disappears. Near the

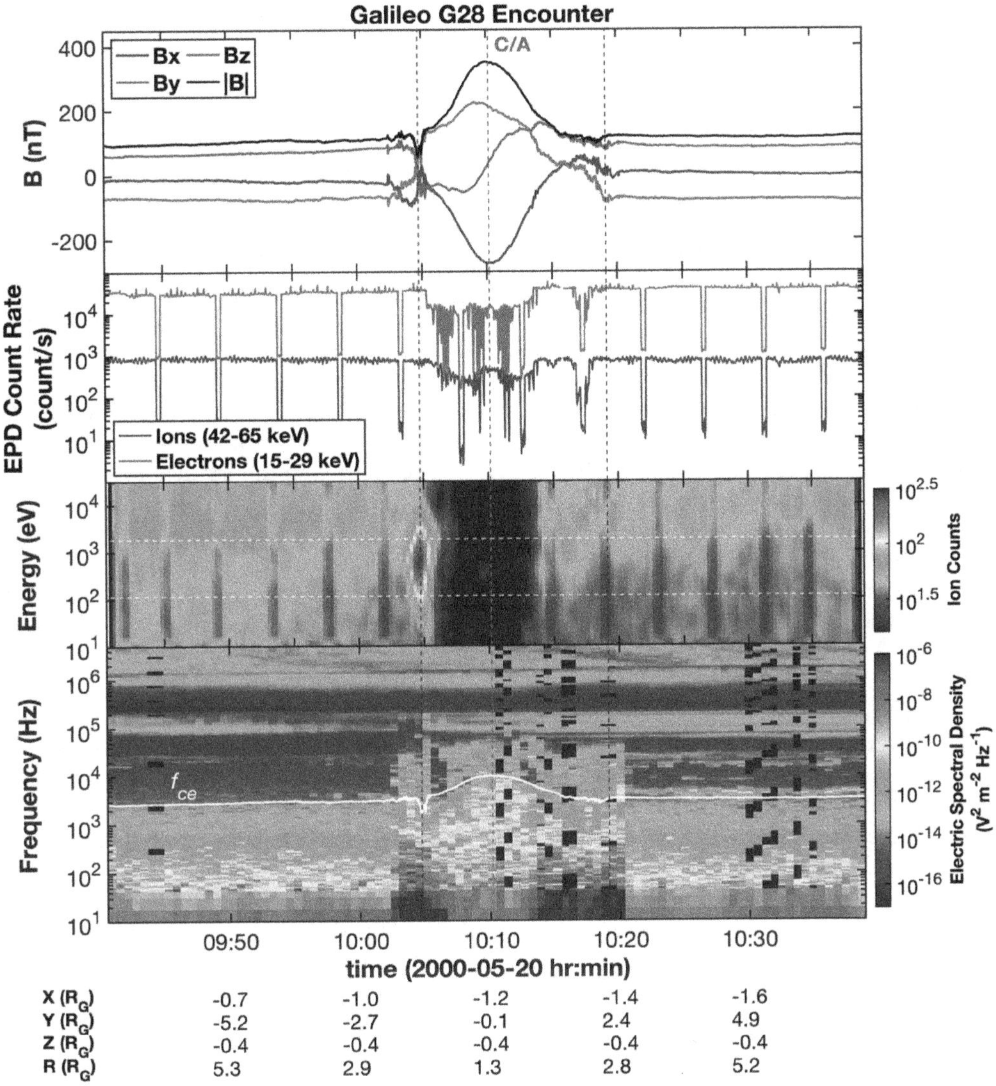

Figure 14.6 Same as Figure 14.4 but for the G28 encounter. (A black and white version of this figure will appear in some formats. For the color version, please refer to the plate section.)

outbound magnetopause crossing (black dashed line), the signature of the ambient plasma returns at slightly higher energy, implying acceleration of the sort expected for flow around an obstacle. The plasma wave spectrum shows strong discontinuities at both magnetopause crossings. Further discussion of the implications of this plot is provided in Section 14.7.

Figure 14.6 shows data from the same set of instruments for the G28 flyby, an upstream pass at southern latitudes and closer to Ganymede's equator than the G8 pass. Once again there are fluctuations of the magnetic field near the magnetopause, somewhat smaller in amplitude. Closest approach on this pass was at half the altitude of the G8 closest approach and the dropout of thermal plasma (third panel) in the interior of the magnetosphere is striking. Enhancements of energetic electrons fluxes and changes of counts of the thermal plasma just inside Ganymede's magnetopause both inbound and outbound on G8 (Figure 14.5) and outbound on G28 (see Figure 14.6) coincide with times during which Galileo was in the close vicinity of the cusp (see the trajectories in Figure 14.1).

14.5 INTERACTION WITH THE PLASMA OF JUPITER'S MAGNETOSPHERE AND GANYMEDE'S AURORA

Particle and field instruments often record rapid fluctuations in the vicinity of planetary magnetopause crossings. As Galileo entered Ganymede's magnetosphere, the field fluctuated, with the fluctuations particularly clear in data from G8 and G28, acquired at low latitudes (Figures 14.5 and 14.6). On G8, during the interval from 15:48 to 15:52 UT, the imminent inbound crossing of the magnetopause is heralded by fluctuating disturbances of the magnetic field with variations of order 10 nT (top panel) concurrent with abrupt changes in the power spectrum (bottom panel) measured by the PWS instrument. Within minutes, the field begins to rotate, the dominant field direction changing from southward to northward, followed by abrupt rotations (at the vertical black dashed line that marks the magnetopause crossing) of the other two components of the field. At almost the same time, the flux of energetic electrons (red curve in the second panel) begins to decrease. Galileo's outbound crossing of the magnetopause (again black dashed line) is again accompanied by a rotation of the field, this time from dominantly northward to dominantly southward, an abrupt decrease of the PWS power, and an increase of energetic electron flux. Again, large amplitude fluctuations of the field are present at the boundary. Boundary crossings are not evident in the ion flux, probably because the gyroradii of ions of energy of order 20 keV kinetic energy in a 100 nT field (~200 km for protons, 800 km for O^+ and 570 km for S^{++}) are comparable with or greater than the width of the magnetopause, but in the interior of the magnetosphere, the ion flux decreases relative to its level in the ambient plasma.

The field fluctuations at the upstream magnetopause were initially attributed to the Kelvin–Helmholtz instability (KHI) producing waves that propagate along the boundary (Kivelson et al., 1998). Analogous fluctuations on the dayside magnetopause of Mercury's magnetopause have tentatively been attributed to the KHI (Sundberg et al., 2010). However, results from a magnetohydrodynamic simulation carried out for the ambient conditions of the G8 pass (Jia et al., 2010a) supported a different interpretation. The fluctuations in the simulation arose from intermittent reconnection in a steady background flow with a typical recurrence time of tens of seconds. Assuming that the boundary fluctuations do arise from short duration reconnection, it is interesting to compare their recurrence time with that reported for flux transfer events (FTEs) on the magnetopause of Mercury (Slavin et al., 2012), a body of size similar to that of Ganymede and with a magnetic field only a factor of two smaller. Flux transfer events are signatures of intermittent reconnection between the fields inside and outside of the magnetopause. Sun and colleagues (2020) find that showers of FTEs at Mercury are more prevalent when the shear angle between the fields on the two sides of the boundary is large and the upstream β (ratio of thermal to magnetic pressure) is small compared with one, conditions satisfied for Galileo's encounters with Ganymede. Flux transfer events at Mercury recur at typical separations of five seconds, interestingly of the order of the fluctuation period at Ganymede's magnetopause. On Earth's magnetopause, the recurrence typical of FTEs is eight minutes. Jia and colleagues (2010a) propose that the "FTE recurrence times decrease with the scale length of the system." Two papers by Zhou and colleagues (2019) and Zhou and colleagues (2020) also address this question.

Ganymede's magnetosphere is in a quasi-steady state, perhaps similar to the state of "steady magnetospheric reconnection" in Earth's magnetosphere (see, for example, Sergeev et al., 1996), which develops when the interplanetary magnetic field remains steadily southward for extended intervals. The process may not be steady on timescales short compared with the time for the plasma to flow from the dayside to the night side, but, if averaged over times comparable with that reference time (a few minutes for Ganymede), convection is steady.

Reconnection at Ganymede's magnetopause is expected to accelerate electrons, some of which precipitate into Ganymede's ionosphere, contributing to its aurora (see Chapters 3.4 and 3.5), and some of which precipitate into Jupiter's ionosphere, contributing to the highly localized emission at the Jovian footprint of the flux tubes in which Ganymede is embedded (Clarke et al., 2002; Grodent et al., 2009; also see Chapter 3.6).

Reconnection on the upstream magnetopause of Ganymede's magnetosphere continually delivers flux tubes carrying plasma and energetic particles across the boundary. In simulations, they then convect across the polar regions and, following a second reconnection, return to the upstream boundary at low latitudes as emptied flux tubes (e.g., Jia et al., 2009). The validity of this picture for Ganymede is supported by analysis of energetic electrons in the polar cap (Williams et al., 1997b). Newly reconnected flux tubes near the magnetopause remain filled with energetic electrons over all measureable pitch angles, but near the center of the polar cap, when the detector looks toward Ganymede, the electron flux is substantially diminished (note the sharp decreases of electron flux within the regions bounded by magnetopause encounters in Figures 14.4, 14.5 and 14.6). The flux of electrons with pitch angles near 90° changes little from its upstream value, but the range of pitch angles near 90° for which electron flux decreases becomes larger as the distance from the upstream boundary increases. The missing electrons have either precipitated into Jupiter's ionosphere, where they must also contribute to the aurora at the footprint of the Ganymede flux tube (Clarke et al., 2002; Grodent et al., 2009; Bonfond et al., 2013; Bonfond et al., 2017a; Bonfond et al., 2017b), or have precipitated into Ganymede's ionosphere, again creating an aurora (Feldman et al., 2000; Eviatar et al., 2001; McGrath et al., 2013).

Collinson and colleagues (2018) analyzed Galileo plasma (Frank et al., 1992) and field data from the G1 pass, the only one that encountered the downstream sector of Ganymede's magnetosphere (see Figure 14.1), identifying localized variations that are consistent with reconnection across the central plasma sheet in the downstream equatorial magnetosphere. However, the data analyzed were acquired very near the magnetopause boundary and it is not clear whether the observed reconnection is part of the flux return process or arises from reconnection on the flank magnetopause.

The energetic particles that impact Ganymede in its polar regions have been shown to leave their mark on the surface. Khurana and colleagues (2007) argue that sputtering by energetic particles on field lines that link to Jupiter (open field lines) is responsible for the bright terrain seen at high latitudes in global color images of Ganymede. The evidence is quite persuasive; contours of the open/closed field line boundary inferred from the magnetic geometry follow quite closely the boundary between bright and dark regions on Ganymede's surface. Juno data lend support to this interpretation of the surface characteristics (Paranicas et al., 2021).

In a planetary magnetosphere (other than at Mercury), the plasma close to the central body approximately corotates with the planet, the rotation period being short compared with the orbital period. Ganymede's magnetosphere is different because the moon rotates about an axis parallel to Jupiter's spin axis and is phase-locked to Jupiter; its rotation period is identical to its orbital period. Consequently, the upstream plasma always flows toward the same surface longitude, the return flow in the low latitude magnetosphere is approximately antiparallel to the upstream flow, and, as contrasted with planetary magnetospheres, there is no corotation electric field. The absence of rotation relative to the direction of the external flow, combined with the fact that much of the volume of Ganymede's magnetosphere is occupied by the moon itself, greatly reduces the density of thermal plasma within Ganymede's near equatorial magnetosphere. On the upstream side, reconnected flux tubes convect over the poles and do not modify the plasma in the equatorial regions. Following downstream reconnection, a large fraction of streamlines returning plasma back upstream intersect Ganymede and are lost (Jia et al., 2009). For the upstream passes G8 and G28, the plasma data of Figures 14.5 and 14.6 (see also figure 5 of Paty et al. [2008] [G8] and figure 9 of Volwerk et al. [2013]) show regions in which the thermal plasma density becomes extremely low, dropping below the threshold of the low energy plasma instrument (Frank et al., 1992). The loss of plasma over an extended region near closest approach is especially clear on G28. Energetic particles are subject to gradient-curvature drift, and some can drift fast enough to avoid being absorbed by the solid surface (Eviatar et al., 2000), but it is uncertain whether they form stable radiation belts (Williams et al., 1997b; Williams, 2001).

14.6 ULTRALOW-FREQUENCY WAVES

Ultralow frequency (ULF) waves, defined as waves in the frequency band ranging from ~0.5 mHz to ~5000 mHz, are frequently observed in planetary magnetospheres. They are often resonant oscillations of closed flux tubes – that is, ones with one end embedded in the northern ionosphere and the other end embedded in the southern ionosphere. This type of wave is referred to as a field line resonance (FLR). Such waves have been identified in Ganymede's magnetosphere (Volwerk et al., 1999; Volwerk et al., 2013).

Disturbances of the plasma and field propagate along the field line at the Alfvén speed and can set up steady-state harmonic standing modes. The frequencies are imposed by the length of the field line, the strength of the magnetic field, and the distribution of plasma density. In a dipole field, if one knows the L-shell on which a spacecraft is located, then the equatorial density can be estimated by using, for example, the analysis of Cummings and colleagues (1969).

On two Galileo flybys, G8 and G28, Galileo entered deep into the closed field line region of the moon, and remained for several minutes on each of two fixed L-shells: 2.1 and 1.5. In both cases a harmonic spectrum of ULF waves, a superposition of five harmonics at frequencies between 59 and 709 mHz and 39 and 401 mHz, respectively, was present (Volwerk et al., 1999; Volwerk et al., 2013). Assuming that in this region of the magnetosphere a dipole field model is a good approximation, the solutions provided by Cummings and colleagues (1969) imply plasma densities of 3 AMU cm^{-3} and 90 AMU cm^{-3}, respectively. If the main ion species was oxygen, the observed frequencies imply that the ion number density at the equator, n_i, was $\approx 0.16\ cm^{-3}$ at 2.1 R_G (G8) and 5.5 cm^{-3} at 1.5 R_G (G28). The estimated density for G28 is in reasonable agreement with the electron density of $n_e \approx 9.5\ cm^{-3}$ determined from the upper-hybrid frequency by PWS (Kurth et al., 2000) for a charge-per-ion of 1.5.

Ultralow-frequency waves of a different type can be created by the ionization of neutrals and subsequent pick-up by the magnetic field. The ring-beam distribution of pick-up ions is unstable to the generation of ion cyclotron (IC) waves (e.g., Gary, 1991). Volwerk and Khurana (2010) identified IC waves near Ganymede, where emissions at frequencies consistent with generation by water-group ions (H_2O^+ or O^+) and O_2^+ were detected on open magnetic field lines (Williams, 2001). The assumption of a correlation between the wave power and the pickup rate applied to comets by Huddleston and Johnstone (1992) implies a pickup rate of $\dot{N} \approx 5 \times 10^{23}\ s^{-1}$ (equivalent to 13-27 g s^{-1}). This value is about two orders of magnitude below the ionospheric outflow rate (Frank et al., 1997) and indicates that Ganymede is a minor source of plasma in the Jovian magnetosphere (Bolton et al., 2015).

As discussed, the magnetopause separates magnetic flux tubes connected to Ganymede from those with both ends on Jupiter. Such boundaries can be unstable to oscillations (e.g., Plaschke et al., 2009, for the Earth's magnetopause). Oscillations can arise either through intermittent reconnection or through the Kelvin–Helmholtz instability driven by shears in the flow. Analysis of the magnetic field data (Volwerk et al., 2013) showed that for most magnetopause crossings, the data seem consistent with periodicity arising from intermittent reconnection because kinked flux tubes of the form created by reconnection were observed, but the waves observed during the G8 inbound crossing could have been Kelvin–Helmholtz-generated. However, the amplitude of the waves was much larger for G8 than for G28, ~50 nT versus ~20 nT, and the period of the fluctuations, which varied between 20 and 40 s for G8, remained near 90 s for G28. Numerical modeling by Jia and colleagues (2010a) does not show significant difference in the occurrence rate of bursts of reconnection between G8 and G28 conditions (near the center of the plasma sheet for G8 and at high negative magnetic latitude for G28). Other factors need to be considered to explain the differences in amplitude and period. Further investigations based on observations acquired from a spacecraft in orbit around Ganymede (such as those anticipated from the upcoming JUICE mission) should shed light on magnetopause dynamics at Ganymede.

14.7 GANYMEDE PLASMA WAVES

In various regions of planetary magnetospheres and near their boundaries, heating or acceleration processes create anisotropic plasma distributions, favoring the generation of electromagnetic waves with frequencies larger than 0.5 Hz, the highest frequency associated with the magnetohydrodynamic field line resonances described in the previous section. Such waves are referred to as plasma waves; waves in the frequency band from ~10 Hz to a few kHz are regarded as low-frequency plasma waves. They have been well documented at all giant planets visited by spacecraft (Kurth and Gurnett, 1991). Many wave signatures were observed on Galileo passes near Ganymede, leading to the suggestion that this moon has a magnetosphere and hence an internal field (Gurnett et al., 1996), as confirmed by magnetic field measurements (Kivelson et al., 1996).

Plasma waves with frequencies below ~100 kHz in fluctuating electric field measurements and below ~10 kHz in fluctuating magnetic field measurements were detected by the PWS during the Ganymede flybys G1 (Figures 14.S1 and 14.S4, both in Supplementary Information online) on June 27, 1996, and G2 (Figures 14.4 and 14.S4 in Supplementary Information) on September 6, 1996. It is useful to distinguish electrostatic waves in which there is no oscillating magnetic field and the electric field fluctuations are purely longitudinal from electromagnetic plasma waves in which an oscillating magnetic field is present and the electric field fluctuations have a finite transverse component. The distinction is clear in Figure 14.S5, which provides an overview of the plasma waves (including both magnetic and electric field fluctuations) detected in the vicinity of Ganymede on passes G1 and G2.

The most intense waves, detected below ~6 kHz, have a magnetic field contribution and are thus clearly electromagnetic. By analogy with chorus and hiss observed at high latitudes in planetary magnetospheres, they have been identified as whistler-mode waves. These waves exist only below f_{pe} [the plasma frequency $= \frac{1}{2\pi}(N_e e^2/\varepsilon_o m_e)^{1/2}$] and f_{ce} (the electron cyclotron frequency $= eB/2\pi m_e$) where N_e is the electron number density, e is the electron charge, m_e is the mass of an electron, and B is the magnitude of the background magnetic field; these waves generally occur below $f_{ce}/2$. Thus the emissions observed imply $f_{ce} \geq 12$ kHz and hence $B > 400$ nT, significantly larger than the Jovian ambient field of ~100 nT. Indeed, Gurnett and colleagues (1996) inferred correctly that Ganymede has an internal field from the high frequency of the whistler mode and electron cyclotron emissions observed on the G1 flyby.

Whistler-mode waves are usually produced by high energy (10–100 keV) electrons with a velocity distribution possessing a loss cone. It has not been unambiguously established whether Ganymede has stable radiation belts, but at least some of the energetic electrons present within Ganymede's magnetosphere must originate in Jupiter's magnetosphere. On the field lines connecting Ganymede to Jupiter, where whistler-mode waves are most often observed, these electrons encounter magnetic mirrors at both ends; consequently, most electrons moving nearly parallel or antiparallel to the field are absorbed by the planet or the moon, giving birth to distributions with loss cones in both directions (see Williams et al., 1997b) that are unstable to wave generation. Figure 14.4 shows that the anticipated energetic electron loss cones were observed in the portion of the G2 pass where whistler-mode radiation was observed.

At frequencies above 12 kHz, narrow bands of electrostatic plasma waves appear inside the magnetosphere of Ganymede. Comparison with the electron cyclotron frequency, f_{ce}, calculated for the field measured locally by the magnetometer, identifies them as electron cyclotron harmonics (ECH, also called Bernstein waves) centered at $(n + \frac{1}{2})f_{ce}$ with n an integer. These waves are generally observed with maximum intensity at low magnetospheric latitudes, and are produced by anisotropic low-energy (10–100 eV) electron distributions.

The narrow electrostatic emission band labeled UHR in Figure 14.S4 is the upper hybrid emission band at $f_{UH} = (f_{pe}^2 + f_{ce}^2)^{1/2}$. The frequencies of these emissions reached almost 100 kHz near 0630 SCET on G1 and exceeded that frequency at about 1900 SCET on G2. Here $f_{UH} \sim f_{pe}$ because f_{pe} was much larger than f_{ce}. The f_{UH} emissions, almost symmetrical in displacement from closest approach, appeared where Galileo encountered Ganymede's ionosphere with a peak electron density ~100 cm^{-3} and a scale height of ~1000 km. (This density greatly exceeds the ambient Jovian magnetospheric plasma density which is of order 3–10 cm^{-3} as indicated in Table 14.2.)

At the edges of the region where the aforementioned waves were detected (from 0640 to 0700 SCET on G1 and from 1910 to 1925 SCET on G2 in Figure 14.S4), the Galileo PWS observed broadband bursts of electrostatic noise up to tens of kHz, characteristic of magnetospheric boundaries, and here associated with Ganymede's magnetopause.

Finally, a weak emission structured in broader bands than ECH was detected in the range of a few tens of kHz outside of Ganymede's magnetosphere, up to a distance of about 8 R_G. Some of the emissions in this high-frequency band are connected to the most intense patches of upper hybrid emission. The emission is weak and at too high frequencies to show up in magnetic field measurements, but Kurth and colleagues (1997) compared its frequency to the ambient characteristic frequencies; they assumed that power in the signal varies with distance from Ganymede and concluded that it is, indeed, electromagnetic. It has been identified as a radio emission analogous to the nonthermal continuum (NTC) radiation that is ubiquitous in planetary magnetospheres (partly trapped in the magnetospheric plasma cavity and partly escaping it), and that, in spite of its name, often displays a banded structure (Gurnett et al., 1983). This radiation is thought to result from the conversion of electrostatic waves at f_{UH} into ordinary mode electromagnetic radio waves via a linear (Jones, 1988) or nonlinear (Rönnmark, 1992) mechanism. In both processes, the presence of a plasma density gradient nearly perpendicular to the local magnetic field vector is a necessary condition; hence this type of radiation is produced at a plasma boundary (such as the plasmapause, the Io torus, or the magnetopause – as in Ganymede's case). The emitted power was estimated to be ~100 W, comparable to that of Earth's NTC but only a fraction (10^{-6}) of Jupiter's (Kurth et al., 1997; Kurth et al., 2000). Figure 14.7 summarizes schematically the distribution of plasma waves detected around Ganymede. Intense electromagnetic waves were detected within ~4 R_G of the moon, electrostatic waves were detected up to ~8 R_G, and weak radio emissions were detected at distances as large as 15–20 R_G.

Based on wave power levels, Kurth and colleagues (2000) compared the strength of the interaction with Jupiter's magnetic field at the four Galilean satellites. At Ganymede, as at Europa, this strength is categorized as *intermediate* – that is, below the Io–Jupiter interaction strength. However, only at Ganymede (among all Jovian and even all solar system moons) is a full zoo of plasma waves – as described earlier – present. A statistical analysis of all eight Galileo–Ganymede encounters between 1996 and 2000 (see Table 14.1) showed that these waves and the interactions they reveal is a persistent – albeit variable – feature (Shprits et al., 2018), and that the level of whistler-mode waves around Ganymede is up to 10^6 times stronger than the ambient level of these waves in Jupiter's magnetosphere at Ganymede's orbit (see Figure 14.S4).

The high-frequency waves present within a few R_G of Ganymede, especially whistler-mode and electrostatic waves, can cause electron pitch-angle scattering, acceleration, and precipitation (into either Jupiter or Ganymede) These effects deeply modify the electron velocity distributions, favoring new instabilities and wave growth (Tripathi et al., 2014a, 2014b; Singhal et al., 2016; Shprits et al., 2018). The subsequent electron precipitation toward Ganymede produces ultraviolet (UV) auroras

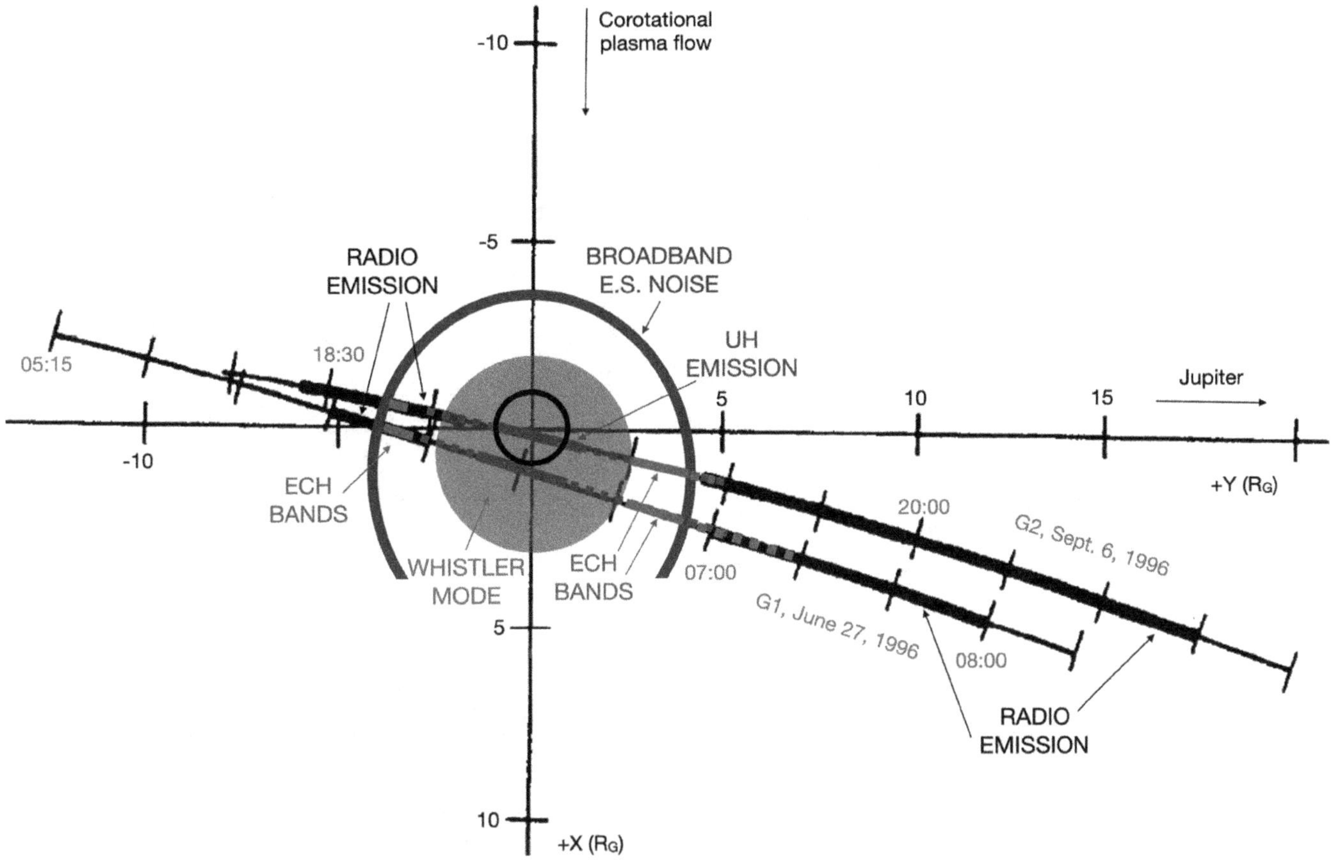

Figure 14.7 Schematic of the distribution of different types of plasma waves and radio emission in the region around Ganymede, deduced from observations during the first two Ganymede flybys. The black circle at the center represents the moon, Ganymede. (A black and white version of this figure will appear in some formats. For the color version, please refer to the plate section.)

in oxygen lines (Feldman et al., 2000; Eviatar et al., 2001) as previously noted. Electron precipitation toward Jupiter produces UV auroral spots in its atmosphere and induces decameter-wave Cyclotron Maser radio emission at higher altitudes (up to a few tenths of an R_J) above Jupiter's ionosphere (cf. Zarka et al., 2001; Zarka et al., 2018 and Chapter 3.6).

Radio emissions resulting from the Cyclotron Maser instability are absent in Ganymede's magnetosphere because the ratio f_{pe}/f_{ce} is everywhere too high ($\geq$0.4) to allow the instability to develop (Zarka, 1998; Treumann, 2006).

14.8 MODELS OF GANYMEDE'S MAGNETOSPHERE AND ITS VARIABILITY

The discovery of Ganymede's intrinsic field and its magnetosphere spurred the development of a number of global simulations of the plasma–magnetic field interaction that creates Ganymede's magnetosphere and accounts for its response to variations in the external and internal conditions. Those simulations not only establish a global context valuable for interpreting existing observations discussed in previous sections but also provide predictions relevant to future exploration of Ganymede's space environment. This section reviews the current state of modeling work on Ganymede's magnetosphere, including both global simulations based on MHD and those that incorporate kinetic physics.

14.8.1 Magnetohydrodynamic Descriptions

For the plasma and magnetic field conditions shown in Table 14.2, the typical gyroradii of ambient Jovian thermal plasma ions near Ganymede are of the order of 100 km, which is much smaller (<1%) than the radius of Ganymede's magnetosphere. Consequently, an MHD approach in which ions and electrons are treated as a single fluid is suitable for modeling the large-scale behavior of the magnetosphere. As discussed earlier in this chapter, Ganymede is embedded in a plasma whose sub-Alfvénic and subsonic flow produces a magnetosphere whose global configuration differs significantly from those of planetary magnetospheres; the differences are well captured by global MHD simulations, and some of these differences could not have been obtained from limited spacecraft observations. As an example, Figure 14.8a shows the magnetic configuration and plasma convection pattern simulated by a single-fluid MHD model (Jia et al., 2008, 2009). The simulated magnetosphere takes the form of a folded cylinder with a pair of Alfvén wings that extend nearly vertically in the north and south directions, consistent with theoretical predictions based on Galileo observations (e.g., Kivelson et al., 1998; Neubauer, 1998). The nearly anti-parallel orientation of Jupiter's magnetospheric field and Ganymede's internal field at the upstream magnetopause facilitates reconnection and the development of open magnetic flux tubes that introduce Jupiter's magnetospheric plasma into the mini-magnetosphere. The subsequent convection of those open flux tubes over the polar caps brings the external plasma

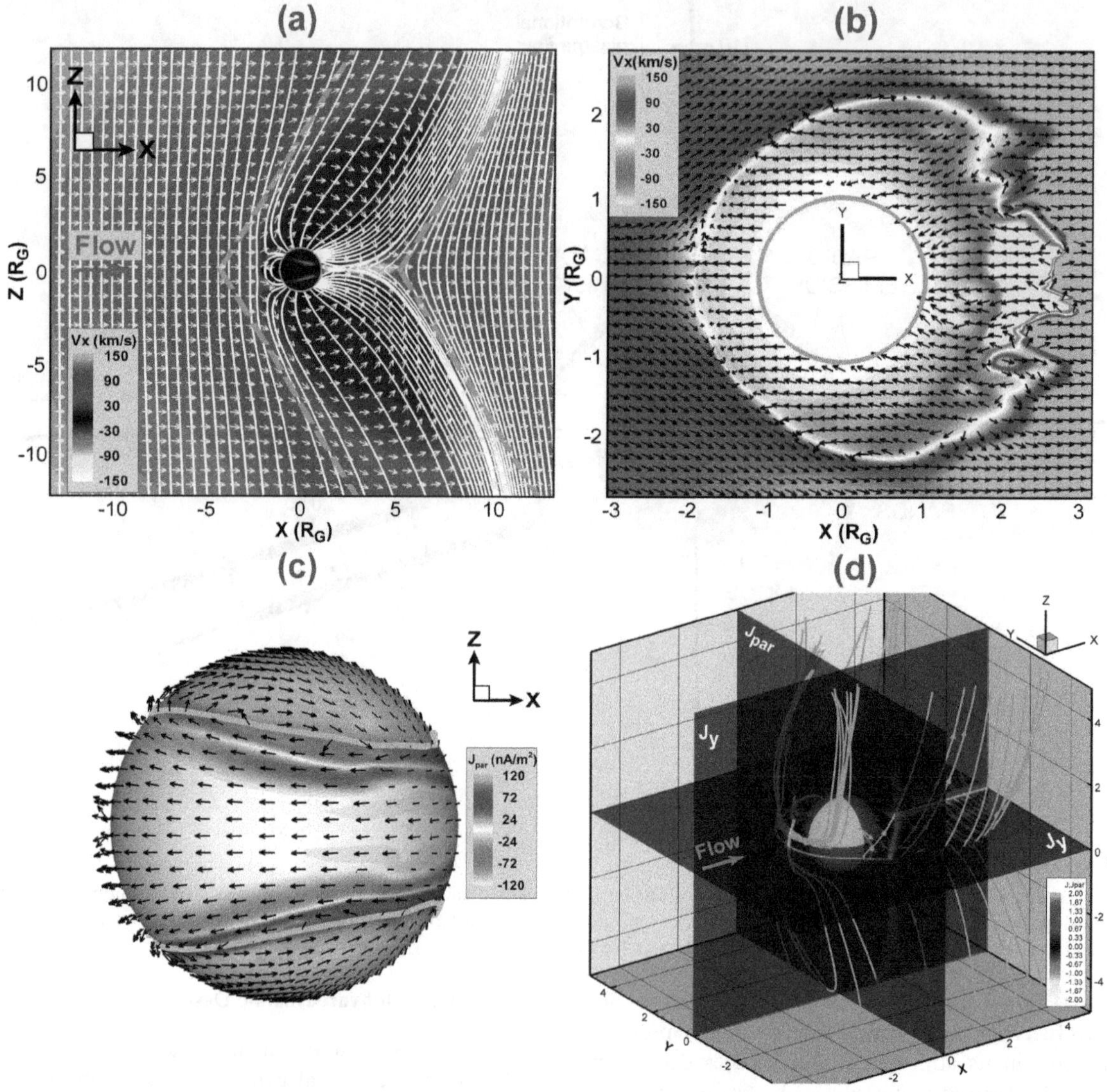

Figure 14.8 Global magnetospheric configuration and current systems extracted from MHD simulations. (a) Global plasma convection (colors: Vx contours; arrows: unit flow vectors) and magnetic field topology (white lines) in the XZ plane in GphiO coordinates. Orange dashed lines show the Alfvén characteristics for reference. (b) Convection pattern in the XY plane. Colors show the Vx contours whereas arrows show unit flow vectors. (c) Field-aligned current density along with unit flow vectors shown on a sphere of radius r= 1.08 R_G. The green curves show the flow separatrix between flows moving tailward and those moving toward the upstream. (d) 3D view of the current flowlines from the upstream flank side. Panels (a–c) reprinted by permission from John Wiley and Sons: *Journal of Geophysical Research: Space Physics*. Xianzhe Jia et al., Properties of Ganymede's magnetosphere inferred from improved three-dimensional MHD simulations. ©2009, Panel (d) reprinted by permission from John Wiley and Sons: *Journal of Geophysical Research: Space Physics*. Xianzhe Jia et al., Three-dimensional MHD simulations of Ganymede's magnetosphere. ©2008. (A black and white version of this figure will appear in some formats. For the color version, please refer to the plate section.)

to the downstream side, where magnetotail reconnection and subsequent convection back upstream complete the cycle of global convection.

To complement Figure 14.8a, Figure 14.8b illustrates the convection pattern in the equatorial (XY) plane. The presence of magnetotail reconnection is revealed by the diverging fast plasma flows 2–3 R_G down tail. The absence of a corotation electric field in Ganymede's rest frame implies that convection of plasma returning upstream from the tail reconnection zone depends largely on the properties of Ganymede's ionosphere, especially its conductance. Ganymede's ionosphere is expected to be only weakly conducting with an estimated Pedersen conductance of a few Siemens (Kivelson et al., 2004), which was assumed in the simulation shown in Figure 14.8b. As a result, the plasma flowing towards Ganymede from the downstream reconnection site does not divert around the moon, but instead impinges directly on the surface, although, as previously noted, high-energy particles, not modeled in the MHD simulation, may be able to avoid impact due to gradient-curvature drifts. Taken together, the global configuration and plasma convection from the MHD simulations are generally consistent with properties inferred from the Galileo observations discussed in Section 14.4.1.

Global simulations that go beyond single-fluid MHD treatment have shed new light on how nonideal MHD effects may affect Ganymede's magnetosphere. For example, Paty and Winglee (2006) and Paty and colleagues (2008) applied a multi-fluid MHD model in which ions originating from different sources are treated as separate fluids. By tracking the plasmas from the ambient Jovian environment and from Ganymede's ionosphere, the multi-fluid model reveals more clearly source, loss, and transport processes within Ganymede's magnetosphere. Dorelli and colleagues (2015) adapted the BATSRUS Hall-MHD code to Ganymede to investigate the effects of the Hall physics on the global structure of its magnetosphere. Their Hall–MHD simulation shows that significant asymmetries of the convection pattern develop between the sub- and anti-Jovian sides as a result of the separate bulk motion of ions and electrons. Similar asymmetries are seen in the 10-moment MHD simulation of Wang and colleagues (2018) and the Hall MHD with Embedded Particle-in-Cell (MHD-EPIC) simulation of Toth and colleagues (2016). These simulation results confirm the importance of taking into account nonideal MHD physics in studying Ganymede's magnetosphere. However, as shown by the simulation studies of Jia and colleagues (2009) and Duling and colleagues (2014), magnetospheric convection is sensitive not only to processes absent in an MHD treatment but is also to how the coupling between the magnetosphere and the moon's ionosphere and interior is modeled. Through comparisons with the Galileo data, these studies have revealed the sensitivity of simulation results to the boundary conditions at both upstream and inner boundaries and have shown that appropriate boundary conditions must be adopted in order to obtain a realistic representation of the magnetosphere. Simulations that go beyond ideal MHD predict convection patterns that are asymmetric about the upstream flow direction, but it has not yet been established how plasma convection and related processes are influenced by the boundary conditions adopted; future observations are needed to establish the magnitude of the asymmetry.

As discussed in Section 14.3, Ganymede's magnetospheric interaction generates large-scale current systems, which, in turn, contribute magnetic perturbations to the measured field at the moon. Figure 14.8d illustrates the 3D structure of the current systems extracted from the simulation of Jia and colleagues (2008). At high latitudes, the currents primarily flow at small angles relative to $\pm\mathbf{B}$, forming the previously mentioned Alfvén wing current system that produces the bend-back of the field lines that link to Jupiter. The bend-back was observed by Galileo and is well represented in Figure 14.8a. These nearly field-aligned currents flow toward Ganymede on the sub-Jovian side and away from the moon on the anti-Jovian side. In the vicinity of Ganymede, the Alfvén wing currents close on themselves not only through the moon and its ionosphere, in analogy to the interaction of Io with the Jovian plasma (e.g., Neubauer, 1980; Southwood et al., 1980), but also through the magnetopause and tail current sheets, a pattern familiar from the interaction between the solar wind and a planetary magnetosphere.

As Jupiter's plasma sheet flaps over Ganymede and the ambient plasma and field properties change at the synodic period of $\sim$10.5 hours, the global configuration of the moon's magnetosphere also varies The external conditions change slowly enough that runs for different steady state external conditions can be used to represent different phases of the periodically varying configuration of Ganymede's magnetosphere (e.g., Ip & Kopp, 2002; Kopp & Ip, 2002). It is found that the total field-aligned currents flowing within each of the Alfvén wings can vary by more than a factor of two in response to changes in upstream plasma densities and velocities, with a maximum total current produced when Ganymede is located close to Jupiter's central plasma sheet (Jia et al., 2008, 2009). As mentioned, the field-aligned current system arising from Ganymede's magnetospheric interaction produces an auroral footprint in Jupiter's ionosphere, which has been imaged by the Hubble Space Telescope (HST) (see Chapter 3.6). The magnitude of the emitted power of Ganymede's footprint appears to vary with a nearly five-hour periodicity, consistent with the variation of the Alfvén wing currents associated with the flapping of Jupiter's plasma sheet over Ganymede predicted by the MHD simulations (Grodent et al., 2009).

The portion of the Alfvén wing current system that closes through Ganymede's ionosphere is closely related to the auroral emissions observed in HST images (see Section 14.3.2 and Chapter 3.5). As shown in Figure 14.8(c), the most intense field-aligned currents colocate with the open-closed field line boundary where strong flow shears are present between the open field lines convecting at high latitudes over the polar cap from upstream to downstream and closed field lines at mid-to-low latitudes that contain plasma returning from the tail to the upstream side. This situation is analogous to the well-known Region-1 current system found in the terrestrial magnetosphere. McGrath and colleagues (2013) analyzed the morphology of Ganymede's aurora imaged under different external conditions and found a close correspondence between the location of the observed auroral emissions and the open-closed field line boundary predicted by the MHD model of Jia and colleagues (2009), supporting the idea that Ganymede's aurora is likely produced by parallel acceleration associated with strong field-aligned currents near the polar cap boundary (Eviatar et al., 2001). Other simulations that incorporate physics beyond ideal MHD, such as Hall physics (Dorelli et al., 2015) and high-order moment MHD (Wang et al., 2018), reinforce this conclusion and reveal additional features of the large-scale current systems. In particular, a Hall current system may develop along the magnetic separatrices associated with the upstream magnetopause and tail reconnection, such that additional field-aligned currents (in addition to the Region-1-like currents) that extend down to the moon's atmosphere/ionosphere can also excite auroras.

14.8.2 Beyond Magnetohydrodynamics

While global MHD simulations provide a reasonably good description of the large-scale structure of Ganymede's magnetosphere, as shown in the previous section, they do not model kinetic effects such as energy-dependent drifts and temperature anisotropy that develop in the magnetospheric plasma. Simulations based on a kinetic approach provide insight into such effects, which are important for understanding particle transport and energization. For example, Fatemi and colleagues (2016) conducted the first hybrid simulation for Ganymede. In this model, ions are treated as kinetic particles whereas electrons are represented as a massless, charge-neutralizing fluid. The hybrid simulation shows that the distribution of precipitating ion flux onto the moon's surface correlates well with the observed variations of surface brightness, supporting the idea that space weathering plays an important role in altering the moon's surface properties (Khurana et al., 2007; Ligier et al., 2019; Liuzzo et al., 2020).

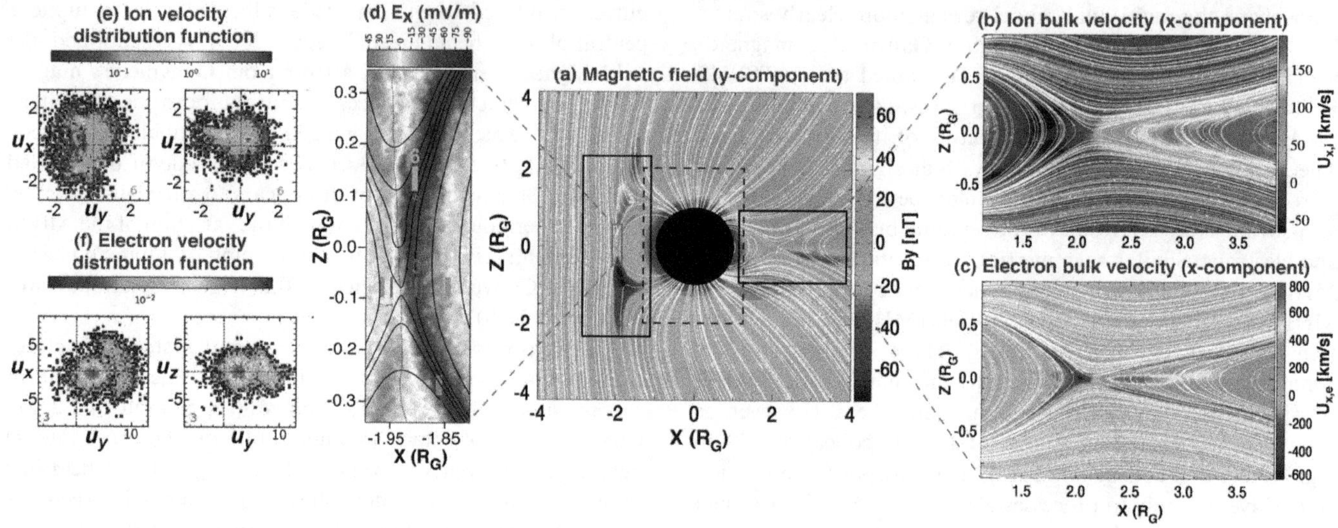

Figure 14.9 Reconnection-driven dynamics in Ganymede's magnetosphere extracted from MHD-EPIC simulations. (a) Contours of By overlaid with magnetic field lines (white lines) in the XZ plane in GphiO coordinates. The black solid and dashed boxes indicate the PIC regions embedded within the global MHD model. (b) and (c) Close-up views of the ion (b) and electron (c) flows (x-component of the bulk flow velocity) associated with reconnection in Ganymede's magnetotail as simulated by MHD-EPIC. (d) Close-up view of the upstream magnetopause reconnection, corresponding to the magenta box in (a). Colors indicate the x-component of the electric field, and black lines show magnetic field lines. (e) and (f) Normalized phase-space distribution functions in two cuts for ions and electrons, respectively. The velocity values shown along the axes are normalized to the ambient Alfvén speed. The ion distribution function shown here was extracted from location "6" labeled in (d), whereas the electron result shown was extracted from location "3" in (d). Panels (a–c) reprinted by permission from John Wiley and Sons: *Journal of Geophysical Research: Space Physics*. Gábor Tóth, et al., Extended magnetohydrodynamics with embedded particle-in-cell simulation of Ganymede's magnetosphere. ©2016, Panels (d–f) reprinted by permission from John Wiley and Sons: *Journal of Geophysical Research: Space Physics*. Hongyang Zhou, et al., Reconnection-Driven Dynamics at Ganymede's Upstream Magnetosphere: 3-D Global Hall MHD and MHD-EPIC Simulations. ©2020. (A black and white version of this figure will appear in some formats. For the color version, please refer to the plate section.)

As discussed, the ambient plasma and field conditions imply that Ganymede's magnetosphere is conducive to magnetic reconnection (e.g., Kaweeyanun et al., 2020). Toth and colleagues (2016) adapted the MHD-EPIC model to Ganymede to study reconnection-driven dynamics using a physics-based approach that does not rely on ad hoc or numerical resistivity to allow reconnection to occur, as is required for MHD models. In MHD-EPIC simulations, regions where kinetic physics is expected to be important, such as at the magnetopause and in the tail current sheet, are modeled by the PIC code that treats both ions and electrons as kinetic particles. The example shown in Figure 14.9 is from an MHD-EPIC simulation for the Galileo G8 flyby, during which the background field is approximately aligned with the -Z-axis with a very small B_y component. In the MHD-EPIC simulated magnetosphere, significant magnetic perturbations in the out-of-plane component (B_y) develop in the current sheets (both at the magnetopause and in the tail) as a result of Hall currents that arise from independent motions of ions and electrons (Fig. 14.9a). Figures 14.9b and 14.9c provide detailed views of the ion and electron flows associated with tail reconnection as modeled by the PIC code. As expected, both electron and ion jets develop in the reconnection outflow region. However, the acceleration regions corresponding to the two particle species are well separated in space because of the different inertia of the different charge carriers. The clear separation between the ion and electron outflows gives an indication of the different characteristic scales of the ion and electron diffusion regions in the tail. The PIC code also provides information about particle phase-space distributions, which offer additional insight into the kinetic aspects of the system (Zhou et al., 2019; Zhou et al., 2020). As an example, Figures 14.9e and 14.9f show the velocity distribution functions for ions and electrons sampled in the vicinity of a reconnection site at the upstream magnetopause (Zhou et al., 2020). The sampling region for the electron distribution is located very close to the X-point (within $\sim$ 1 electron inertial length) on the magnetospheric side of the neutral line (Fig. 14.9d). Here the distribution function exhibits a crescent shape in the plane perpendicular to the background field (left panel of Fig. 14.9f), characteristic of the electron population found within the electron diffusion region in an asymmetric reconnection. Crescent-shaped distributions are also found for ions in the ion diffusion region (within $\sim$ 1 ion inertial length from the X-point). The ion distribution function shown in Figure 14.9e was extracted from a location along the magnetic separatrix in the reconnection outflow region, where ions are being accelerated northward to approximately the ambient Alfvén speed.

Because the ambient Jovian plasma and field conditions vary at the 10.5-hour synodic period, much longer than the time required for the ambient plasma to flow across the magnetosphere ($\sim$ minutes), Ganymede's magnetosphere provides a unique laboratory for studying the nature of reconnection in a relatively steady external environment. As discussed in Section 14.4.1, large-amplitude magnetic fluctuations, initially interpreted as potential signatures of Kelvin–Helmholtz waves (Kivelson et al., 1998), were observed during the magnetopause crossings on the G8 flyby that traversed the upstream magnetosphere. However, global simulations suggest that the boundary fluctuations are more likely to arise from reconnection at the magnetopause (e.g., Jia et al., 2010a; Toth et al., 2016). In simulating the G8 flyby, Jia and colleagues (2010a) used a resistive MHD simulation with an anomalous resistivity model to capture fast reconnection. They found that magnetopause reconnection occurs in a nonsteady manner, producing structures resembling FTEs that recurred every 20–50 seconds even though the upstream conditions remained fixed throughout the simulation. Subsequent simulations using Hall MHD and MHD-EPIC for the same flyby show very similar behavior (Toth et al., 2016; Zhou et al., 2020). Although the three simulations differ in the details of the physics that triggers reconnection, all of them show that Ganymede's magnetopause reconnection is nonsteady under steady upstream conditions, and the characteristic timescale for FTE formation is on the order of tens of seconds. The consistency of these results based on different analysis tools suggests that the intrinsic timescale of Ganymede's magnetospheric reconnection is governed by the characteristic scale length of the magnetosphere and the upstream conditions.

It is worth pointing out that the existing simulations of Ganymede's magnetosphere, whether fluid or kinetic, do not directly model energetic particles, which are important for understanding the radiation environment at Ganymede. Nonetheless, the electromagnetic fields extracted from physics-based models can be used to drive test particle simulations that can provide insight into the behavior of energetic particles in the near-Ganymede environment. A number of test particle simulations have been performed to interpret the properties of the energetic particles measured by the Galileo EPD and to quantify the access of energetic particles to Ganymede's surface, including their precipitation rates and energy fluxes. The energetic particles that have access to Ganymede's near-surface regions contribute to generating the moon's atmosphere (e.g., Plainaki et al., 2015; Plainaki et al., 2020) and ionosphere (e.g., Carnielli et al., 2019; Carnielli et al., 2020) as well as contributing to surface weathering (e.g., Khurana et al., 2007; Poppe et al., 2018, Ligier et al., 2019; Liuzzo et al., 2020). Details of these modeling works are covered in Chapters 3.2 (plasma–surface interactions), 3.3 (atmosphere) and 3.4 (ionosphere).

14.9 EXPECTATIONS FOR THE FUTURE

Ganymede's next terrestrial visitor will provide a different view of the magnetosphere. On its PJ34 orbit around Jupiter on June 7, 2021, the Juno spacecraft flew by Ganymede at low altitude, allowing its fields and particles detectors as well as remote-sensing instruments to acquire further data on the unique magnetosphere. In the early 2030s, NASA's Europa Clipper spacecraft will encounter Ganymede multiple times as its orbit is modified to allow it to drop periapsis to Europa's orbit. In the same decade, ESA's JUICE mission will reach Jupiter, where it will spend 2.5 years learning about the magnetospheric environment, making flybys of the moons Europa, Callisto, and Ganymede and ultimately going into orbit around Ganymede where it will dwell for more than half a year. The continuous data set acquired in orbit will fully characterize the moon's magnetosphere and establish how it influences the moon itself through surface weathering. By providing high-resolution measurements in space and time, JUICE will determine internal magnetic moments beyond dipolar order and its measurements of the inductive response at multiple frequencies will greatly augment our knowledge of the moon's internal structure.

ACKNOWLEDGMENTS:

K. K. Khurana and X. Jia acknowledge support from NASA contract #NNN13AA07T through the Applied Physics Laboratory at Johns Hopkins University. The authors wish to acknowledge the contributions to the field by our dear friend, Aharon Eviatar (02/20/1931–06/27/2021) – known to all as Arkee. He would have read this chapter with interest, peppered us with questions, and then immediately offered his explanations. We miss him and remember his friendship and his many scientific contributions.

REFERENCES

Bagenal, F. & Delamere, P. A. (2011). Flow of mass and energy in the magnetospheres of Jupiter and Saturn. *Journal of Geophysical Research* 116, A15, A05209.

Bagenal, F., Dougherty, L. P., Bodisch, K. M., Richardson, J. D., & Belcher, J. M. (2017). Survey of Voyager plasma science ions at Jupiter: 1. Analysis method. *Journal of Geophysical Research* 122, 8241–8256. https://doi.org/10.1002/2016JA023797.

Bagenal, F., Wilson, R. J., Siler, S., Paterson, W. & Kurth, W. (2016). Survey of Galileo plasma observations in Jupite's plasma sheet. *Journal of Geophysical Research* 121.

Bodisch, K. M., Dougherty, L. P. & Bagenal, F. (2017). Survey of Voyager plasma science ions at Jupiter: III protons and minor ions. *Journal of Geophysical Research* 122, 8277–8294. https://doi.org/10.1002/2017JA024148.

Bolton, S. J., Bagenal, F., Blanc, M. et al. (2015). Jupiter's magnetosphere: Plasma sources and transport. *Space Science Reviews* 192, 209–236. https://doi.org/10.1007/s11214-015-0184-5.

Bonfond, B., Grodent, D., Badman, S. V. et al. (2017a). Similarity of the Jovian satellite footprints: Spots multiplicity and dynamics. *Icarus* 292, 208–217.

Bonfond, B., Hess, S., Bagenal, F. et al. (2013). The multiple spots of the Ganymede auroral footprint. *Geophysical Research Letters* 464, 40(19), 4977–4981.

Bonfond, B., Saur, J., Grodent, D. et al. (2017b). The tails of the satellite auroral footprints at Jupiter. *Journal of Geophysical Research* 122(8), 798–7996.

Busse, F. H. (1976). Generation of planetary magnetism by convection. *Phys. of the Earth and Plan. Int.* 12, 350–358.

Carnielli, G., Galand, M., Leblanc, F. et al. (2019). First 3D test particle model of Ganymede's ionosphere. *Icarus* 330, 42–59. https://doi.org/10.1016/j.icarus.2019.04.016.

Carnielli, G., Galand, M., Leblanc, F. et al. (2020). Simulations of ion sputtering at Ganymede. *Icarus*. https://doi.org/10.1016/j.icarus.2020.113918.

Clarke, J. T., Ajello, J., Ballester, G. et al. (2002). Ultraviolet auroral emissions from the magnetic footprints of Io, Ganymede, and Europa on Jupiter. *Nature* 415*(6875)*, 997–1000.

Collinson, G., Paterson, W. R., Bard, C. et al. (2018). New results from Galileo's first flyby of Ganymede: Reconnection-driven flows at the low-latitude magnetopause boundary, crossing the cusp, and icy ionospheric escape. *Geophysical Research Letters* 45, 3382–3392. https://doi.org/10.1002/2017GL075487.

Connerney, J. E. P., Acuna, M. H. & Ness, N. F. (1981). Modeling the Jovian current sheet and inner magnetosphere. *Journal of Geophysical Research* 86, 8370

Connerney, J. E. P., Acuna, M. H., Ness, N. F. & Satoh, T. (1998). New models of Jupiter's magnetic field constrained by the Io flux tube footprint. *Journal of Geophysical Research* 103, 11929.

Connerney, J. E. P., Kotsiaros, S., Oliversen, R. J. et al. (2018). A new model of Jupiter's magnetic field from Juno's first nine orbits. *Geophysical Research Letters* 45, 2590–2596. https://doi.org/10.1002/2018GL077312.

Connerney, J. E. P., Timmins, S., Herceg, M. & Joergensen, J. L. (2020). A jovian magnetodisc model for the Juno era. *Journal of Geophysical Research* 125. https://doi.org/10.1029/2020JA028138.

Cooper, J. F., Johnson, R. E., Mauk, B. H., Garrett, H. B. & Gehrels, N. (2001). Energetic ion and electron irradiation of the icy Galilean satellites. *Icarus* 149, 133–159. https://doi.org/10.1006/icar.2000.6498.

Cummings, W. D., O'Sullivan, R. J. & Coleman, P. J. (1969). Standing Alfvén waves in the magnetosphere. *Journal of Geophysical Research* 74, 778–793. https://doi.org/10.1029/JA074i003p00778.

Delamere, P. A., Bagenal, F. & Steffl, A. (2005). Radial variations in the Io plasma torus during the Cassini era. *Journal of Geophysical Research* 110, A12223.

Dorelli, J. C., Glocer, A., Collinson, G., & Toth, G. (2015). The role of the Hall effect in the global structure and dynamics of planetary magnetospheres: Ganymede as a case study. *Journal of Geophysical Research*. https://doi.org/10.1002/2014JA020951.

Dougherty, L. P., Bodisch, K. M. & Bagenal, F. (2017) Survey of Voyager plasma science ions at Jupiter: II Heavy ions. *Journal of Geophysical Research* 122, 8257–8276. https://doi.org/10.1002/2017JA024053.

Duling, S., Saur, J. & Wicht, J. (2014). Consistent boundary conditions at nonconducting surfaces of planetary bodies: Applications in a new Ganymede MHD model. *Journal of Geophysical Research* 119, 4412–4440, https://doi.org/10.1002/2013JA019554.

Eviatar, A., Strobel, D. F., Wolven, B. C. et al. (2001). Excitation of the Ganymede ultraviolet aurora. *Astrophys. J.* 555, 1 013–1 019.

Eviatar, A., Williams, D. J., Paranicas, C., McEntire, R. W. & Mauk, B. H. (2000). Trapped energetic electrons in the magnetosphere of Ganymede. *Journal of Geophysical Research* 105, 5547–5553.

Fatemi, S., Poppe, A. R., Khurana, K. K., Holmstrom, M. & Delory, G. T. (2016). On the formation of Ganymede's surface brightness asymmetries: Kinetic simulations of Ganymede's magnetosphere. *Geophysical Research Letters*. https://doi.org/10.1002/2016GL068363.

Feldman, P. D., McGrath, M. A., Strobel, D. F. et al. (2000). HST/STIS ultraviolet imaging of polar aurora on Ganymede. *Astrophys. J.* 535, 1085–1090.

Frank, L. A., Ackerson, K. L., Lee, J. A., English, M. R. & Pickett, G. L. (1992). The plasma instrumentation for the Galileo mission. *Space Science Reviews* 60, 283–307.

Frank, L. A., Paterson, W. R., Ackerson, K. L., Bolton, S. J. (1997). Outflow of hydrogen ions from Ganymede. *Geophysical Research Letters* 24, 2 151–2 154.

Gary, S. P. (1991). Electromagnetic ion/ion instabilities and their consequences in space plasmas: A review. *Space Science Reviews* 56, 37–415. https://doi.org/10.1007/BF0019663.

Glassmeier, K.-H. & Heyner, D. (2021). Planetary magnetic fields,ch24 in*Space Physics and Aeronomy Collection Volume 2: Magnetospheres in the Solar System*, Geophysical Monograph 259, Edited by Romain Maggiolo, Nicolas André, Hiroshi Hasegawa & Daniel T. Welling. American Geophysical Union. https://doi.org/10.1002/9781119815624.

Grodent, D., Bonfond, B., Radioti, A. et al. (2009). Auroral footprint of Ganymede. *Journal of Geophysical Research* 114, A07212. https://doi.org/10.1029/2009JA014289.

Gurnett, D. A., Kurth, W. S. & Scarf, F. L. (1983), Narrowband electromagnetic emissions from Jupiter's magnetosphere. *Nature*, 302 (5907), 385–388.

Gurnett, D. A., Kurth, W. S., Shaw, R. R. et al. (1992). The Galileo plasma wave investigation. *Space Science Reviews* 60, 341–355.

Gurnett, D. A., Kurth, W. S., Roux, A., Bolton, S. J. & Kennel, C. F. (1996). Evidence for a magnetosphere at Ganymede from plasma-wave observations by the Galileo spacecraft. *Nature* 384, 535–537.

Huddleston, D. E. & Johnstone, A. D. (1992). Relationship between wave energy and free energy from pickup ions in the comet Halley environment. *Journal of Geophysical Research* 97, 12 217–12 230.

Huscher, E., Bagenal, F., Wilson, R. J. et al. (2021). Survey of Juno observations in Jupite's plasma disk: Density. *Journal of Geophysical Research* (in review).

Ip, W. H., & Kopp, A. (2002). Resistive MHD simulations of Ganymede's magnetosphere: 2. Birkeland currents and particle energetics. *Journal of Geophysical Research*. https://doi.org/10.1029/2001JA005072.

Jia, X. (2009). Ganymede's magnetosphere: Observations and modeling. *PhD dissertation*. University of California at Los Angeles.

Jia, X., Kivelson, M. G., Khurana, K. K. & Walker, R. J. (2009). Magnetic fields of the satellites of Jupiter and Saturn. *Space Science Reviews*. https://doi.org/10.1007/s11214-009-9507-8.

Jia, X., Walker, R. J., Kivelson, M. G., Khurana, K. K. & Linker, J. A. (2008). Three dimensional MHD simulation of Ganymed's magnetosphere. *Journal of Geophysical Research* 13. https://doi.org/10.1029/2007JA01254.

Jia, X., Walker, R. J., Kivelson, M. G., Khurana, K. K. & Linker, J. A. (2009). Properties of Ganymed's magnetosphere inferred from improved three-dimensional MHD simulations. *Journal of Geophysical Research* 114, A09209. https://doi.org/10.1029/2009JA014375.

Jia, X., Walker, R. J., Kivelson, M. G., Khurana, K. K. & Linker, J. A. (2010a). Dynamics of Ganymede's magnetopause: Intermittent reconnection under steady external conditions. *Journal of Geophysical Research*. https://doi.org/10.1029/2010JA015771.

Jones, D. (1988). Planetary radio emissions from low magnetic latitudes: Observations and theories. In *Planetary radio emissions II*, edited by H. O. Rucker, S. J. Bauer & B.M. Pedersen. Austrian Academy of Sciences Press, 245–281.

Kaweeyanun, N., Masters, A. & Jia, X. (2020). Favorable conditions for magnetic reconnection at Ganymede's upstream magnetopause. *Geophysical Research Letters*. https://doi.org/10.1029/2019GL086228.

Khurana, K. K. (1997). Euler potential models of Jupiter's magnetospheric field. *Journal of Geophysical Research* 102, 11 295.

Khurana, K. K., Kivelson, M. G. Russell, C. T. Walker, R. J. & Southwood, D.J. (1997). Absence of an internal magnetic field at Callisto. *Nature* 387, 262.

Khurana, K. K., Kivelson, M. G., Stevenson, D. J. et al. (1998). Induced magnetic fields as evidence for subsurface oceans in Europa and Callisto. *Nature* 395, 749.

Khurana, K. K., Pappalardo, R. T., Murphy, N. & Denk, T. (2007). The origin of Ganymede's polar caps. *Icarus* 191, 193–202. https://doi.org/10.1026/j.icarus,2007.04.022.

Kim, T. K.-H., Ebert, R. W., Valek, P. W. et al. (2020). Survey of ion properties in Jupiter's plasma sheet: Juno JADE-I observations. *Journal of Geophysical Research* 125, e2019JA027696. https://doi.org/10.1029/2019JA02769.

Kim, T. K., Ebert, R. W., Valek, P. W. et al. (2019). Properties of Jupiter's plasma sheet ions observed by Juno JADE including estimates of in-situ abundance for O^+ and S^{2+}, *Journal of Geophysical Research*. https://doi.org/10.1029/2018ja026169.

Kivelson, M. G. & Bagenal, F. (2014). in *Planetary Magnetospheres, Encyclopedia of the Solar System, Third Edition* by D. Breuer, T. Johnson, T. Spohn.

Kivelson, M. G., Bagenal, F., Kurth, W. S. et al. (2004). Magnetospheric interactions with satellites. In *Jupiter: The planet, satellites and magnetosphere*, edited by F. Bagenal, T. E. Dowling & W. B. McKinnon. Cambridge University Press, 513–536.

Kivelson, M. G. & Winge, C. R. (1976). Field-aligned currents in the Jovian magnetosphere: Pioneer 10 and 11. *Journal of Geophysical Research* 81, 5853.

Kivelson, M. G., Slavin, J. A. & Southwood, D. J. (1979). Magnetospheres of the Galilean satellites. *Science* 205, 491.

Kivelson, M. G., Khurana, K. K., Means, J. D., Russell, C. T. & Snare, R. C. (1992). The Galileo magnetic field investigation. *Space Science Reviews* 60, 357.

Kivelson, M. G., Khurana, K. K., Russell, C. T. et al. (1996). Discovery of Ganymed's magnetic field by the Galileo spacecraft. *Nature* 384, 53–541.

Kivelson, M. G., Khurana, K. K., Coroniti, F. V. et al. (1997). The magnetic field and magnetosphere of Ganymede. *Geophysical Research Letters* 24, 2155.

Kivelson, M. G., Warnecke, J., Bennett, L. et al. (1998). Ganymede's magnetosphere: Magnetometer overview. *Journal of Geophysical Research* 103, 19 963–19 972.

Kivelson, M. G., Khurana, K. K., Stevenson, D. J. et al. (1999). Europa and Callisto: Induced or intrinsic fields in a periodically varying plasma environment. *J. Geophys. Res. Space Phys.* 104(A3), 4609–4625.

Kivelson, M. G., Khurana, K. K. & Volwerk, M. (2002). The permanent and inductive magnetic moments of Ganymede. *Icarus*, 157, 507–522.

Kopp, A. & Ip, W. H. (2002). Resistive MHD simulations of Ganymed's magnetosphere: 1. Time variabilities of the magnetic field topology. *Journal of Geophysical Research* 107(A12), 1490. https://doi.org/10.1029/2001JA005071.

Kurth, W. S., Gurnett, D. A. & Menietti, J. D. (2000). The influence of the Galilean satellites on radio emissions from the Jovian system. *Radio astronomy at long wavelengths*. Geophysical monograph series, 119. Edited by R. G. Stone, K. W. Weiler, M. L. Goldstein & J.-L. Bougeret. American Geophysical Union, 213–225.

Kurth, W. S. & Gurnett, D. A. (1991). Plasma wave in planetary magnetospheres. *Journal of Geophysical Research* 96, 18977–18991.

Kurth, W. S., Gurnett, D. A., Roux, A. & Bolton, S. J. (1997). Ganymede: A new radio source. *Geophysical Research Letters* 24, 2167–2170.

Ligier, N., Paranicas, C., Carter, J. et al. (2019). Surface composition and properties of Ganymede: Updates from ground-based observations with the near-infrared imaging spectrometer SINFONI/VLT/ESO. *Icarus*, 333(June), 496–515. https://doi.org/10.1016/j.icarus.2019.06.013.

Liuzzo, L., Poppe, A. R., Paranicas, C. et al. (2020). Variability in the energetic electron bombardment of Ganymede. *Journal of Geophysical Research* 125, e2020JA028347. https://doi.org/10.1029/2020JA028347.

Mauk, B. H., Mitchell, D. G., McEntire, R. W. et al. (2004). Energetic ion characteristics and neutral gas interactions in Jupite's magnetosphere. *Journal of Geophysical Research* 109, A09S12. https://doi.org/10.1029/2003JA010270.

McGrath, M. A., Jia, X., Retherford, K. D. et al. (2013). Aurora on Ganymede. *Journal of Geophysical Research*. https://doi.org/10.1002/jgra.50122.

Neubauer, F. M. (1978). Possible strengths of dynamo magnetic fields of the Galilean satellites and of Titan. *Geophysical Research Letters* 5, 905.

Neubauer, F. M. (1980). Non-linear standing Alfvén wave current system at Io: Theory. *Journal of Geophysical Research* 85, 1 17–1 178.

Neubauer, F. M. (1998). The sub-Alfvénic interaction of the Galilean satellites with the Jovian magnetospheres. *Journal of Geophysical Research* 103, 19 84–19 866.

Olsen, N., Glassmeier, K.-H. & Jia, X. (2010). Separation of the magnetic field into external and internal parts. *Space Science Reviews*, 152(1). https://doi.org/10.1007/s11214-009-9563-0.

Pappalardo, R. T., Collins, G. C., Head, J. W. et al. (2004). Geology of Ganymede. In *Jupiter: The planet, satellites & magnetosphere*, edited by F. Bagenal, T. E. Dowling & W. B. McKinnon.Cambridge University Press, 363–396.

Paranicas, C., Szalay, J. R., Mauk, B. H. et al. (2021). Energy spectra near Ganymede from Juno data. *Geophysical Research Letters* 48, e2021GL093021. https://doi.org/10.1029/2021GL093021.

Parkinson, W. D. (1983). *Introduction to geomagnetism*. Scottish Academic Press.

Paty, C., Paterson, W. & Winglee, R. (2008). Ion energization in Ganymede's magnetosphere: Using multifluid simulations to interpret ion energy spectrograms. *Journal of Geophysical Research*. https://doi.org/10.1029/2007JA012848.

Paty, C. & Winglee, R. (2006). The role of ion cyclotron motion at Ganymede: Magnetic morphology and magnetospheric dynamics. *Geophysical Research Letters*. https://doi.org/10.1029/2005GL025273.

Phipps, P., & Bagenal, F. (2021). Centrifugal equator in Jupiter's plasma sheet. *Journal of Geophysical Research* 126, e2020JA028713, https://doi.org/10.1029/2020JA028713.

Plainaki, C., Massetti, S., Jia, X. et al.(2020). Kinetic simulations of the Jovian energetic ion circulation around Ganymede. *Astrophys. J.* https://doi.org/10.3847/1538-4357/aba94c.

Plainaki, C., Milillo, A., Massetti, S. et al. (2015). The H_2O and O_2 exospheres of Ganymede: The result of a complex interaction between the Jovian magnetospheric ions and the icy moon. *Icarus*. https://doi.org/10.1016/j.icarus.2014.09.018.

Plaschke, F., Glassmeier, K.-H., Auster, H. U. et al. (2009). Standing Alfvén waves at the magnetopause. *Geophysical Research Letters* 36, L02104. https://doi.org/10.1029/2008GL036411.

Poppe, A. R., Fatemi, S. & Khurana, K. K. (2018). Thermal and energetic ion dynamics in Ganymede's magnetosphere. *Journal of Geophysical Research*. https://doi.org/10.1029/2018JA025312.

Rönnmark, K. (1992). Conversion of upper hybrid waves into magnetospheric radiation. In *Planetary radio emissions III*, edited by H. O. Rucker, S. J. Bauer & M. L. Kaiser. Austrian Academy of Sciences Press, 405–417.

Saur, J., Duling, S., Roth, L. et al. (2015). The search for a subsurface ocean in Ganymede with Hubble Space Telescope observations of its auroral ovals. *Journal of Geophysical Research* 120. https://doi.org/10.1002/2014JA020778.

Saur, J. (2021). Overview of moon–magnetosphere interactions. In *Space Physics and Aeronomy Collection Vol. 2: Magnetospheres in the Solar System*. Geophysical Monograph 259. Edited by R. Maggiolo, N. André, H. Hasegawa & D. T. Welling, https://doi.org/10.1002/ 9781119815624.ch36.

Schubert, G., Anderson, J. D., Spohn, T. & McKinnon, W. B. (2004). Interior composition, structure and dynamics of the Galilean satellites. In: *Jupiter: The planet, satellites and magnetosphere*, edited by F. Bagenal, T. E. Dowling & W. B. McKinnon. Cambridge University Press, pp. 281–306.

Sergeev, V. A., Pellinen, R. J. & Pulkkinen, T. I. (1996). Steady magnetospheric convection: A review of recent results. *Space Science Reviews* 75, 551–604. https://doi.org/10.1007/BF00833344.

Shprits, Y. Y., Menietti, J. D., Drozdov, A. Y. et al. (2018). Strong whistler mode waves observed in the vicinity of Jupiter's moons. *Nature Communications* 9, id. 3131, 1–6.

Singhal, R. P., Tripathi, A. K., Halder, S. & Singh, O. N., II (2016). Diffuse aurora on Ganymede driven by electrostatic waves. *Astrophys. J.* 832, id. 172, 1–9.

Sittler, E. C. & Strobel, D. F. (1987). Io plasma torus electrons: Voyager 1, *Journal of Geophysical Research* 92, 5741.

Slavin, J. A. & 16 co-authors (2012). MESSENGER observations of a flux-transfer event shower at Mercury. *Journal of Geophysical Research* 117, A00M06. https://doi.org/10.1029/2012JA017926, 2012.

Southwood, D. J., Kivelson, M. G. Walker, R. J. & Slavin J. A. (1980). Io and its plasma environment. *Journal of Geophysical Research* 85, 5959.

Sun, W. J., Slavin, J. A., Smith, A. W. et al. (2020). Flux transfer event showers at Mercury: Dependence on plasma β and magnetic shear and their contribution to the Dungey cycle. *Geophysical Research Letters* 47, e2020GL089784. https://doi.org/10.1029/2020GL08978.

Sundberg, T., Boardsen, S. A., Slavin, J. A., Blomberg, L. G. & Korth, H. (2010). The Kelvin-Helmholtz instability at Mercury: An assessment. *Planet. Space Sci.* 58, 1434–1441. https://doi.org/10.1016/j.pss.2010.06.008.

Toth, G., Jia, X., Markidis, S. et al. (2016). Extended magnetohydrodynamics with embedded particle-in-cell simulation of Ganymede's magnetosphere. *Journal of Geophysical Research*. https://doi.org/10.1002/2015JA021997.

Treumann, R. A. (2006). The electron-cyclotron maser for astrophysical application. *Astron. Astrophys. Rev.* 13, 229–315.

Tripathi, A. K., Singhal, R. P., Singh, K. P. & Singh II, O. N. (2014a). Whistler mode instability and pitch-angle diffusion near Ganymede. *Planet. Space Sci.* 92, 2014, 150–156. https://doi.org/10.1016/j.pss.2014.01.005.

Tripathi, A. K., Singhal, R. P., Singh, K. P., and Singh II, O. N. (2014b). Electrostatic electron cyclotron harmonic instability near Ganymede. *Astrophys. Space Sci.* 352, 421–427. https://doi.org/10.1007/s10509-014-1957-0.

Vance, S. D., Styczinski, M. J., Bills, B. G. et al. (2021). Magnetic induction responses of Jupiter's ocean moons including effects from adiabatic convection. *Journal of Geophysical Research Planets* 126, e2020JE006418. https://doi.org/10.1029/2020JE006418.

Vasyliunas, V. M. & Eviatar, A. (2000). Outflow of ions from Ganymede: A reinterpretation. *Geophysical Research Letters*, 27(9), 1347–1349.

Volwerk, M., Jia, X., Paranicas, C. et al. (2013). ULF waves in Ganymede's upstream magnetosphere. *Ann. Geophys.* 31, 45–59. https://doi.org/10.5194/NFWO-31-45-2013.

Volwerk, M. & Khurana, K. K. (2010). Ion pick-up near the icy Galilean satellites. *AIP Conference Proceedings* 1302, 263–269. https://doi.org/10.1063/1.3529982.

Volwerk, M., Kivelson, M. G., Khurana, K. K. & McPherron, R. L. (1999). Probing Ganymede's magnetosphere with field line resonances. *Journal of Geophysical Research* 104, 14729–14738. https://doi.org/10.1029/1999JA900161.

Wang, L., Germaschewski, K., Hakim, A. et al. (2018). Electron physics in 3-D two-fluid 10-moment modeling of Ganymede's magnetosphere. *Journal of Geophysical Research*. https://doi.org/10.1002/2017JA024761.

Williams, D. J. (2001). Ganymede's ionic radiation belts. *Geophysical Research Letters* 28, 3793–3796. https://doi.org/10.1029/2001GL01335.

Williams, D. J., McEntire, R. W., Jaskulek, S., & Wilken, B. (1992), The Galileo Energetic Particles Detector. *Space Science Reviews* 60, 385–412.

Williams, D. J., Mauk, B. & McEntire, R. W. (1998). Properties of Ganymed's magnetosphere as revealed by energetic particle observations. *Journal of Geophysical Research* 103, 17 523–17 534.

Williams, D. J., Mauk, B. H., McEntire, R. W. et al. (1997a). Energetic particle signatures at Ganymede: Implications for Ganymed's magnetic field. *Geophysical Research Letters* 24, 2 163–2 166.

Williams, D. J., Mauk, B. & McEntire, R. W. (1997b). Trapped electrons in Ganymede's magnetic field. *Geophysical Research Letters* 24, 2953–2956.

Zarka, P. (1998). Auroral radio emissions at the outer planets: Observations and theories. *Journal of Geophysical Research* 103, 2015–20194. https://doi.org/10.1029/98JE0132.

Zarka, P., Queinnec, J. & Crary, F. (2001). Low-frequency limit of Jovian radio emissions and implications on source locations and Io plasma wake. *Planet. Space Sci.* 49, 1137–1149.

Zarka, P., Marques, M. S., Louis, C. et al. (2018). Jupiter radio emission induced by Ganymede and consequences for the radio detection of exoplanets. *Astron. Astrophys.* 618, A84.

Zhou, H., Toth, G. Jia, X. & Chen, Y. (2020). Reconnection-driven dynamics at Ganymede's upstream magnetosphere: 3D global Hall MHD and MHD-EPIC simulations. *Journal of Geophysical Research*. https://doi.org/10.1029/2020JA028162.

Zhou, H., Toth, G., Jia, X., Chen, Y. & Markidis, S. (2019). Embedded kinetic simulation of Ganymede's magnetosphere: Improvements and inferences. *Journal of Geophysical Research*. https://doi.org/10.1029/2019JA026643.

Zimmer, C., Khurana, K. K. & Kivelson, M. G. (2000). Subsurface oceans on Europa and Callisto: Constraints from Galileo magnetometer observations. *Icarus* 147, 329–347.

Figure 1.3 Ganymede throwing a shadow on Jupiter. Image credit and copyright: Damian Peach, Chilescope.

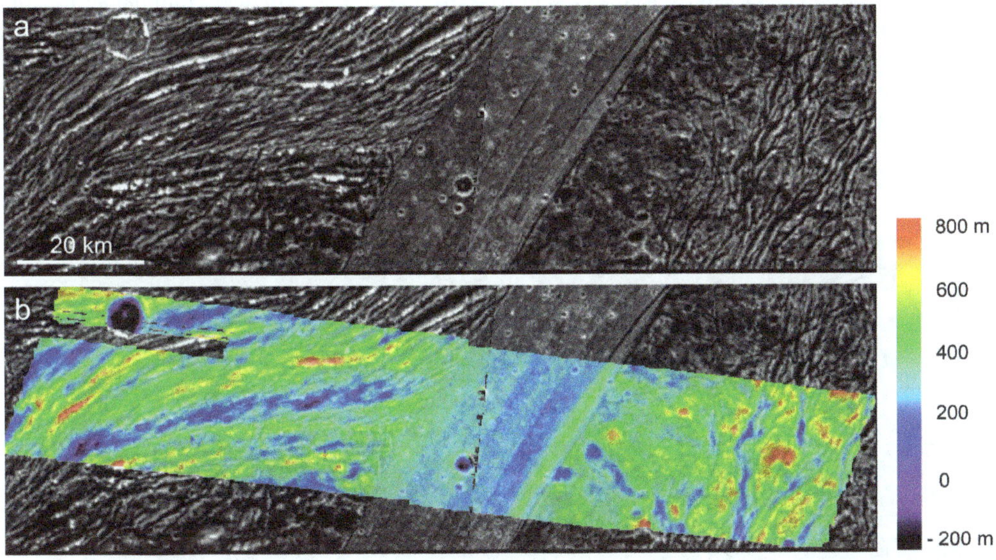

Figure 5.7 (a) High resolution image (34 m/px) and (b) topographic data of Arbela Sulcus (15° S/347° E) derived from stereo images. Arbela Sulcus exhibits small-scale lineations and straight margins indicative of tectonic deformation.
Note that the eastern margin of Arbela is elevated relative to local depressions in the adjacent dark-appearing terrain, arguing against a primary origin by cryovolcanic flooding. The topographic model has a horizontal resolution of 600 m and a vertical resolution of 15–25 m. North is to the top.
Source: adapted from Giese et al. (2001).

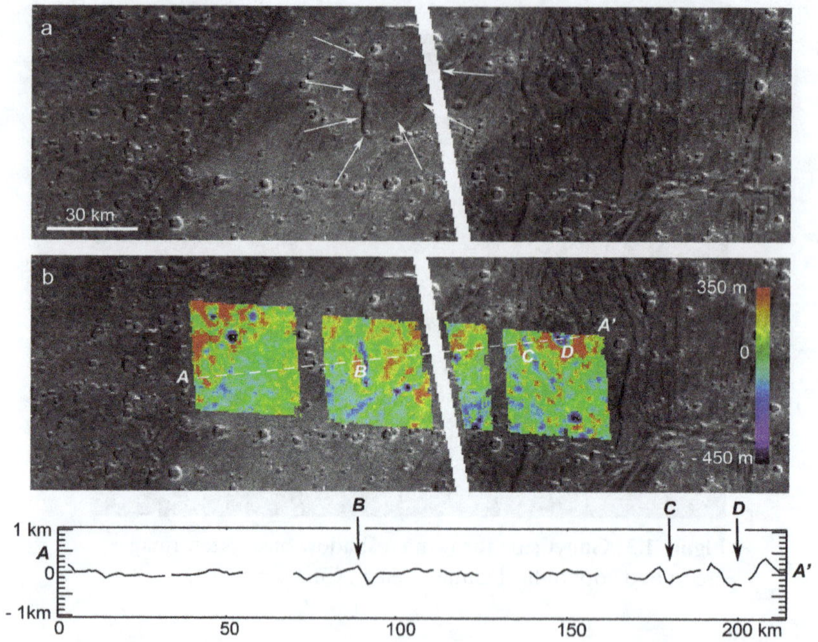

Figure 5.8 (a) Galileo observation G28GSCALDRA01 (180 m/px) covering 20 × 40 km near Hammamat Patera (ID #1.1) at 24°S/42°E (arrows) surrounded by bright-appearing terrain with (b) the DTM-derived topographic information (Giese et al., 2017). The DTM derived from stereo image analysis exhibits a spatial resolution of 600 m with a horizontal resolution of about 1 km and vertical precision of 20–40 m with (c) topographic profiles. The profiles are marked as dashed lines A to A'. Additional surface features such as prominent grooves (B, C, and D) and impact craters can be well recognized in the DTM and profile.

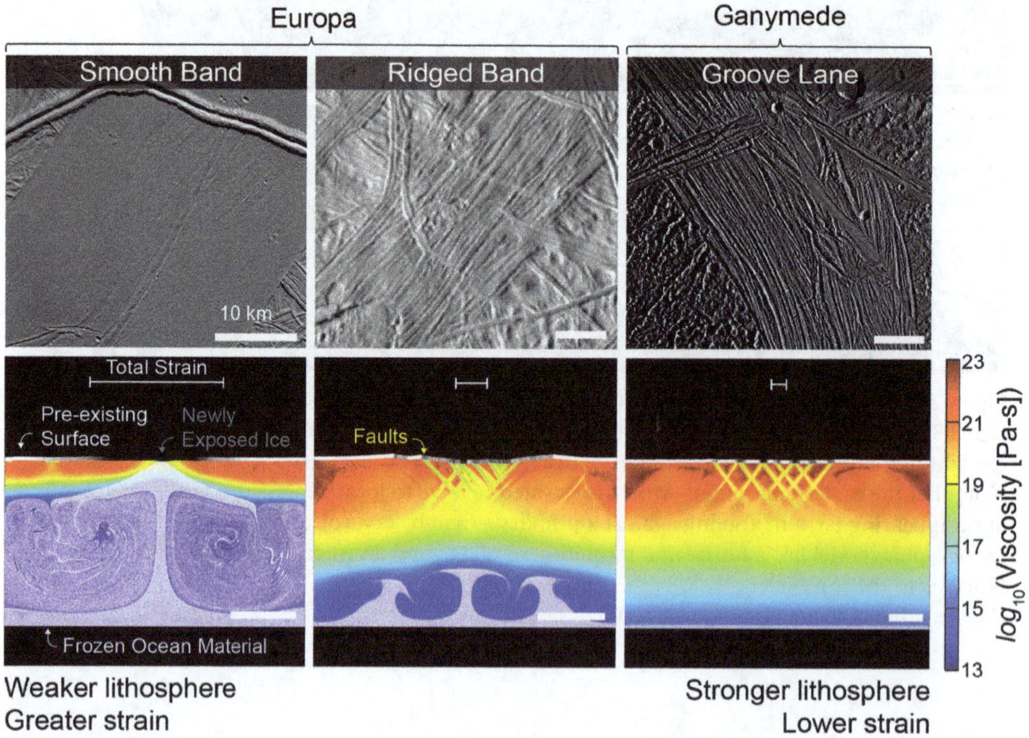

Figure 7.5 Galileo spacecraft images of Europa compared to Ganymede, showing extensional landform morphology in map view (top row) and associated numerical models in cross section (bottom row), modified from Howell and Pappalardo (2018). From left to right, the images are of Europa smooth band Thynia Linea (Galileo observation E17ESTHYLIN01, 40 m/pixel), a ridged band in Europa's northern leading hemisphere (E15ESREGMAP02, 234 m/pixel), and Ganymede groove lane Tiamat Sulcus (G8GSTIAMAT01, 494 m/pixel). From left to right, models increase in icy lithosphere strength and decrease in total strain. For the Europa smooth band, selected models exhibit symmetrical axial yielding with central axis for high regional strain, whereas the Europa ridged band model exhibits widespread asymmetrical faulting. Ganymede's Tiamat Sulcus is the most tectonized, exhibiting large-scale rifting at lower modeled strain than the Europa examples. Model outputs are colored by viscosity and show ice shell material derived in part from ocean material (white dots within ductile ice). The model surfaces are colored white where the original surface has not been significantly disrupted and black where the model surface strain is the greatest and fresh interior ice is exposed.

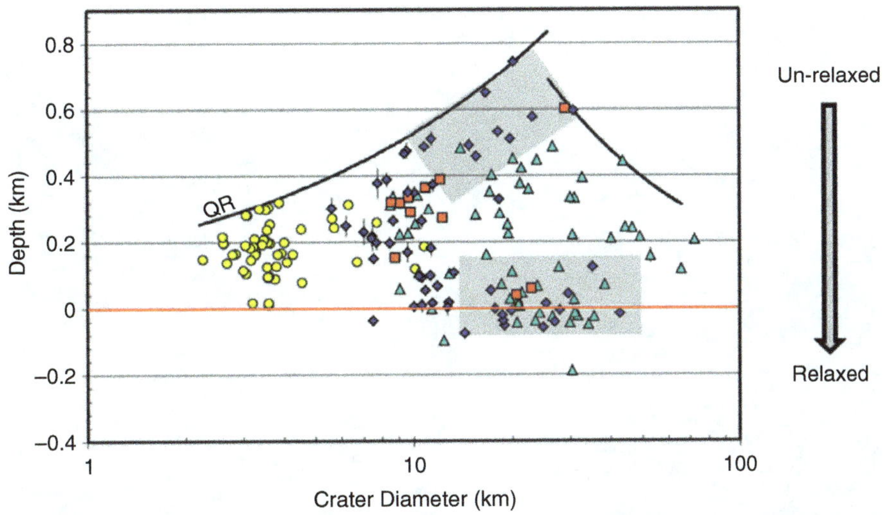

Figure 8.8 Measured apparent depths (depth relative to the surrounding terrain) for impact craters from four regions on Ganymede: Anshar Sulcus (yellow circles), the south pole (light blue triangles), Tiamat Sulcus (red squares), and Marius Regio (dark blue triangles). Gray boxes highlight the bimodality of crater depths observed in Marius Regio. Quantile regression fits to the upper envelope of the data (QR black curve) are used as a marker of initial crater depth.
Source: Modified from Singer et al. (2018). Reprinted from *Icarus*, 306. Singer, K. N., Bland, M. T., Schenk, P. M. and McKinnon, W. B. Relaxed impact craters on Ganymede: Regional variation and high heat flows 214–224, © 2018, with permission from Elsevier.

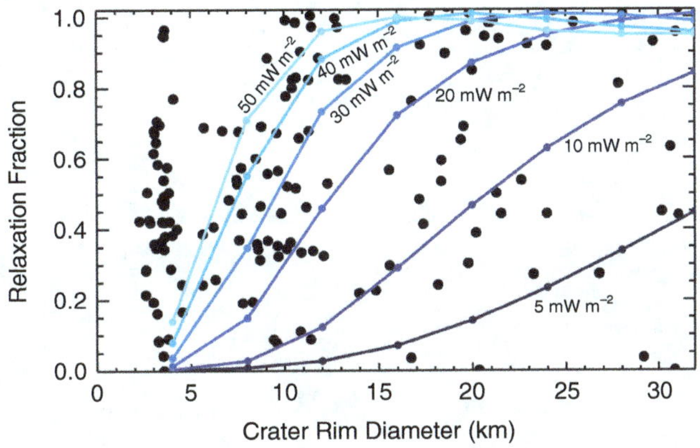

Figure 8.9 A comparison of observed crater depths (black points) and simulated crater depths after 2 Gyr for a range of heat fluxes (blue curves as labeled). Because crater depth varies with diameter, depths are parameterized as the final (present-day) depth relative to the assumed initial depth (i.e., the relaxation fraction, where 0.0 is pristine and 1.0 is fully relaxed).
Source: Reprinted from *Icarus*, 296. Bland, M. T., Singer, K. N., McKinnon, W. B. and Schenk, P. M. Viscous relaxation of Ganymede's impact craters: Constraints on heat flux 275–288, © 2017, with permission from Elsevier.

Figure 9.12 Galileo image and PC-DEM of northern Marius Regio area. Relaxed craters are white arrows, unrelaxed black arrows. Unrelaxed crater densities of N(>10 km) ~30 per 10^6 km2 are observed in both dark and bright terrains in this location. Image resolution is ~900 m/pixel; north is to top.
Source: NASA.

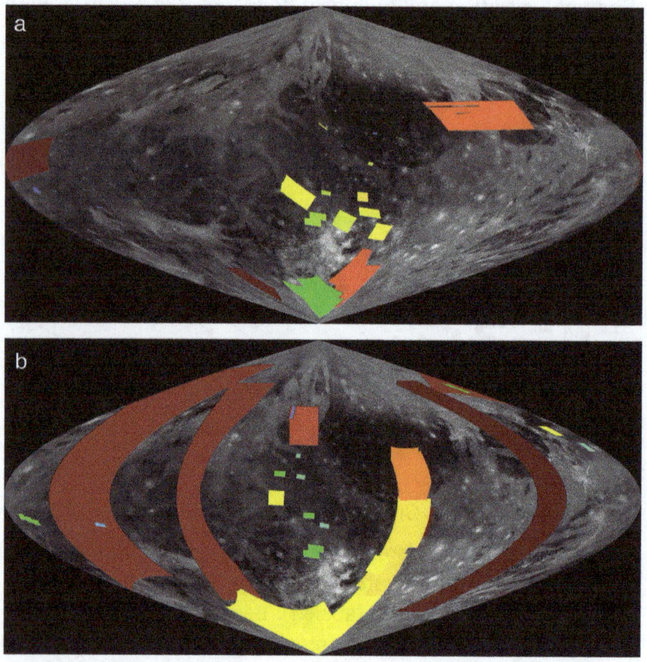

Figure 9.13 a, b Global sinusoidal image mosaic maps of Ganymede showing distribution of (a) stereogrammetric and (b) photoclinometric DEMs. DEM sites are in blank polygons: redder represents lower-resolution DEM products from each site. (a) includes DEM at sub-Jovian point (far left) described in main text showing large dome, although additional regional DEMs of this type in the Voyager 1 mapping area are not shown here.

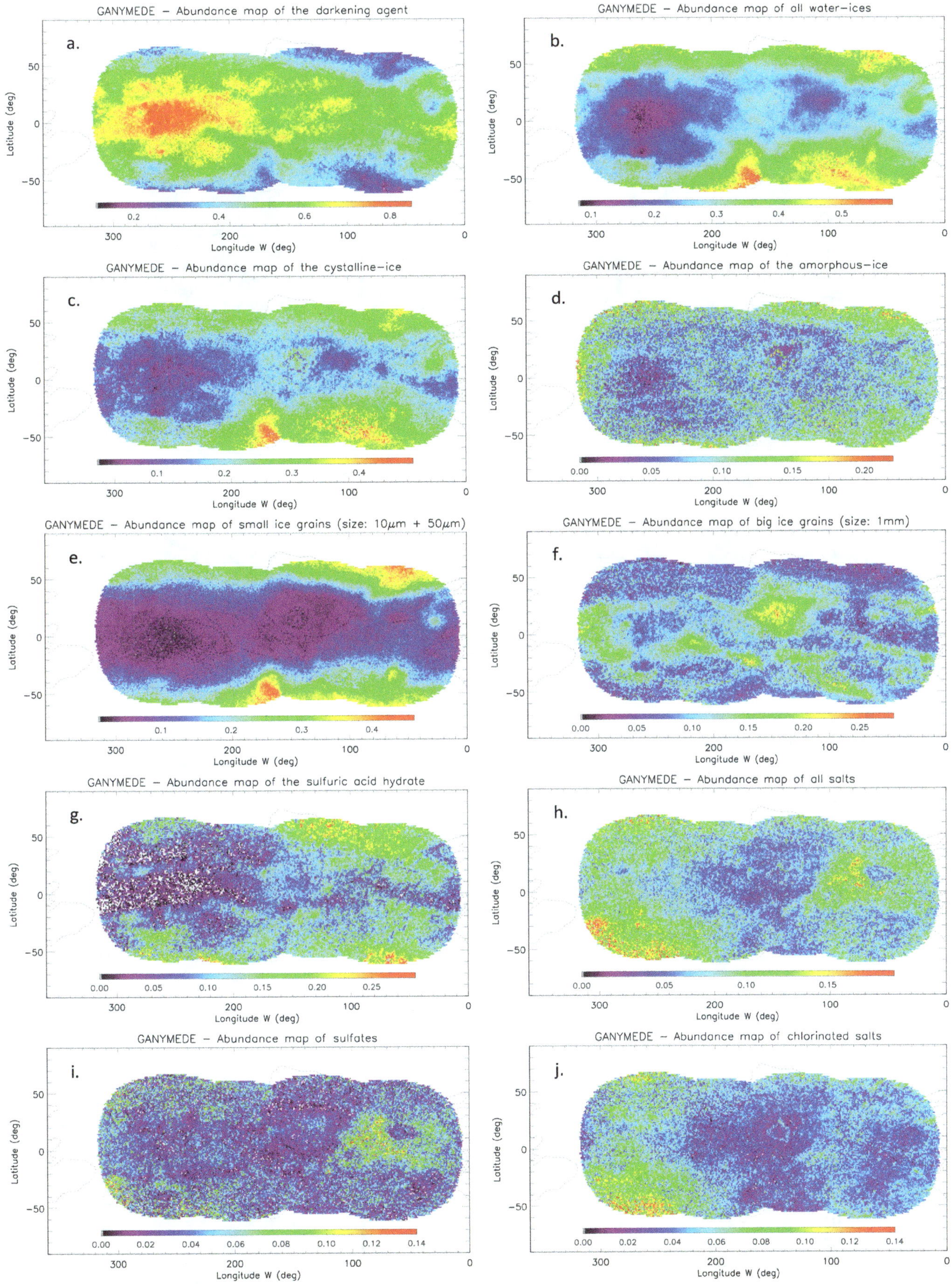

Figure 10.2 Abundance maps of the endmembers used in the modeling of SINFONI data: (a) the darkening agent, (b) H_2O ice, with H_2O ice separately shown for (c) crystalline, (d) amorphous form, (e) small ($\leq$50 μm), and (f) large grains, and the minor compounds (g) sulfuric acid hydrate and (h) salts in general, and separately shown for (i) sulfates and (j) chlorinate salts. Reprinted from *Icarus*, 333. N. Ligier, C. Parancias, J. Carter, F. Poulet, W. M. Calvin, T. A. Nordheim, C. Snodgrass, L. Ferellec. Surface composition and properties of Ganymede: Updates from ground-based observations with the near-infrared imaging spectrometer, © 2019, with permission from Elsevier.

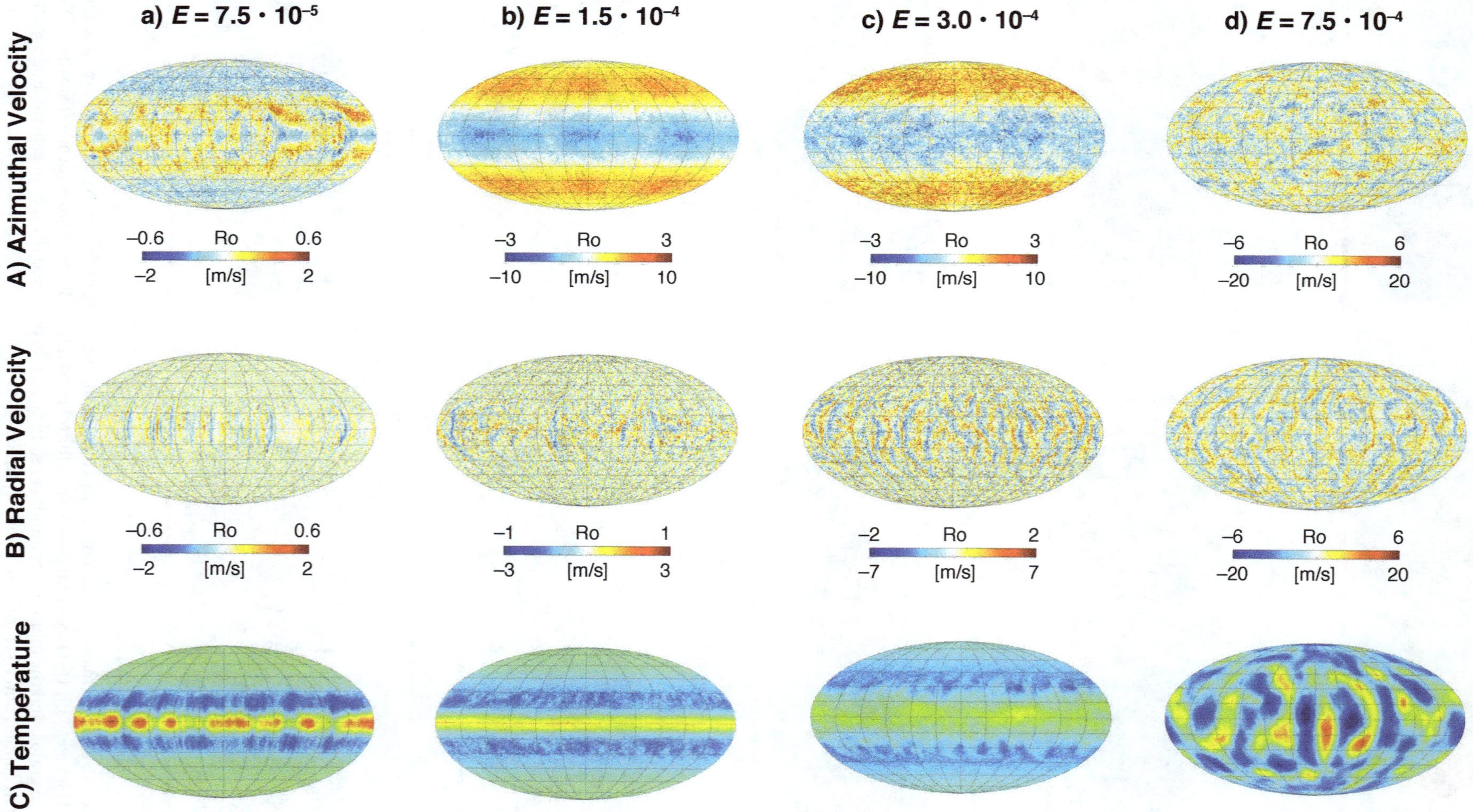

Figure 12.5 A) Azimuthal velocity, B) radial velocity, and C) temperature fields at mid-ocean depths for the four possible ocean convection regimes identified by Soderlund (2019), where the influence of rotation decreases from a) to d). The velocity fields are shown at an instantaneous snapshot in time, while the temperature field has been time-averaged. Black (white) indicates westward (eastward) zonal winds in A, black (white) indicates downwelling (upwelling) radial flows in B, and black (white) indicates cool (warm) fluid in C. Speeds are given in both dimensionless Rossby number units $Ro = U/\Omega D$ and dimensionalized units assuming rotation rate $\Omega = 1.0 \times 10^{-5}\ \mathrm{s}^{-1}$ and an ocean thickness of $D = 340$ km.

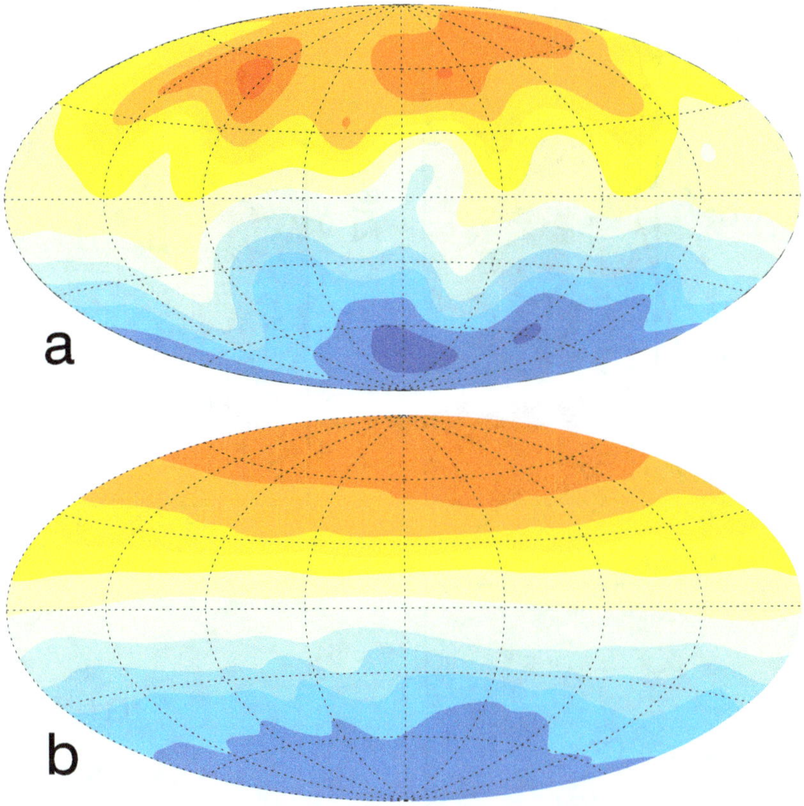

Figure 13.7 Snapshots of the radial magnetic field component for two dynamo models. (a) Case 14.0 without a stable top layer from Christensen (2015a), (b) case 14.4 with a stable layer (corrected as described in Christensen, 2015b). Red for positive B_r, blue for negative B_r. Gray-scale indicates value of $|B|$. See the text for more explanations.

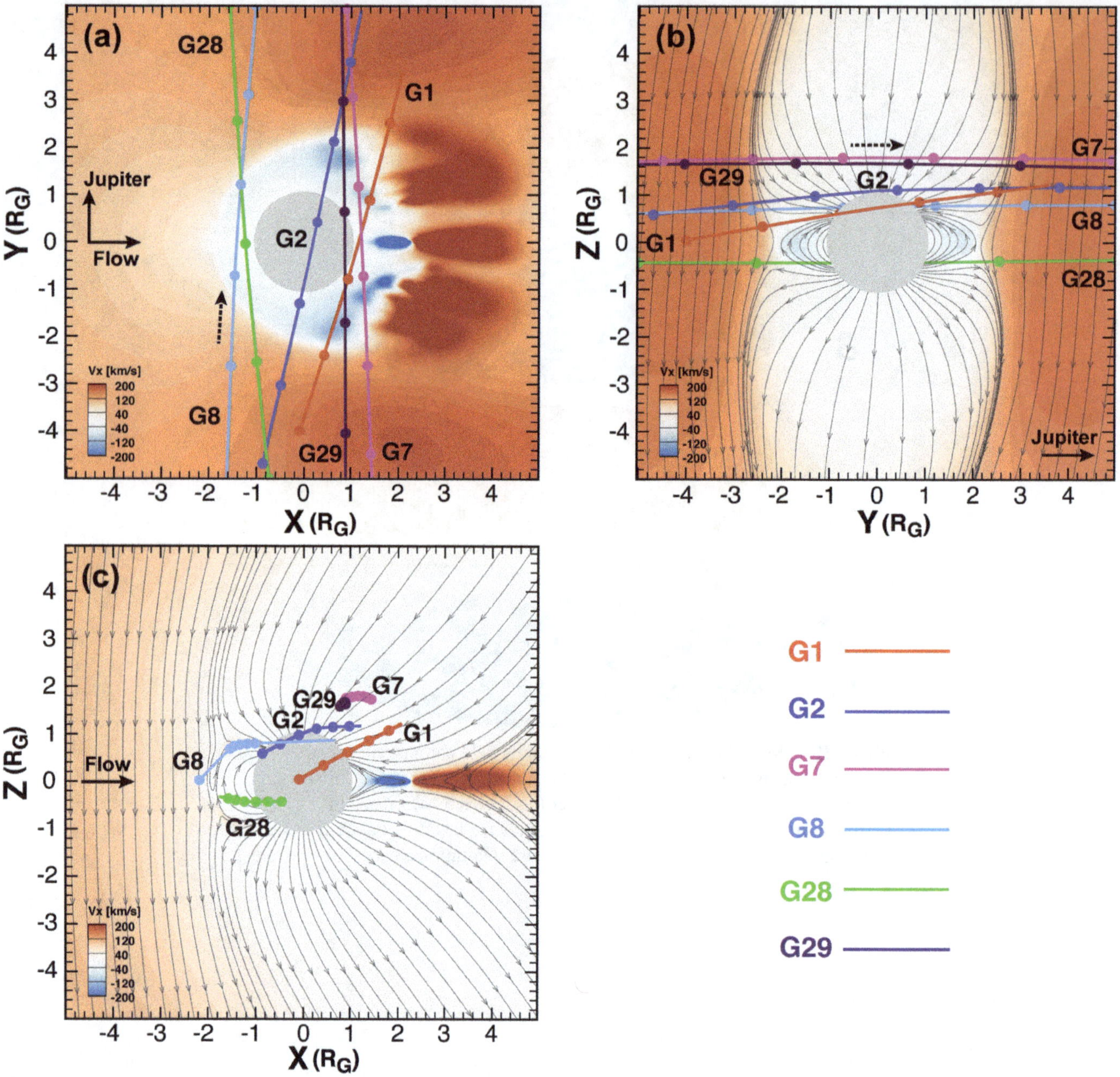

Figure 14.1 Trajectories of Galileo's six passes through Ganymede's magnetosphere plotted for the ~1 hour during which the highest-resolution data were acquired. Dots are separated by 10 minutes. The passes are projected into the (a) XY, (b) YZ, and (c) XZ planes of the GphiO coordinate system (X along the direction of corotating plasma flow at closest approach, Z parallel to Jupiter's spin axis, and Y pointing inward toward Jupiter. Panel (b) looks into the incident flow. The background is from the MHD simulation of Jia and colleagues (2009) for the Galileo G8 flyby with color representing the X-component of the flow velocity. Gray lines show the magnetic field lines, significantly distorted by the presence of the moon. The dashed arrows in panels (a) and (b) indicate the direction of the spacecraft motion, which is from -Y to +Y for all close encounters.

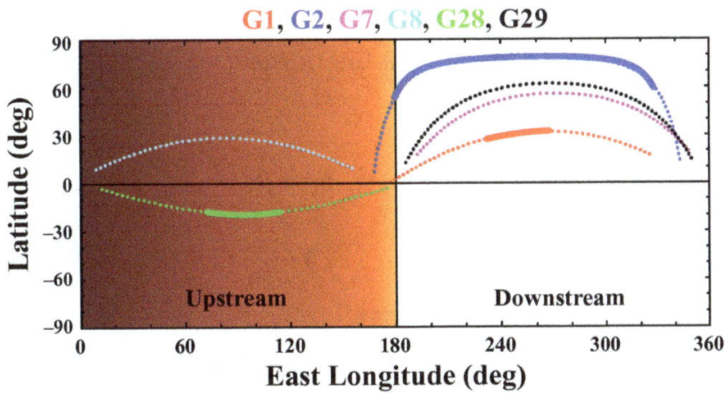

Figure 14.3 The latitude/longitude coverage of Ganymede by Galileo. The solid parts of the segments show the intervals when Galileo was within 1.4 RG from the center of Ganymede. Figure adapted by permission from Springer Nature: *Space Science Reviews*. Xianzhe Jia et al., Magnetic Fields of the Satellites of Jupiter and Saturn. © 2009.

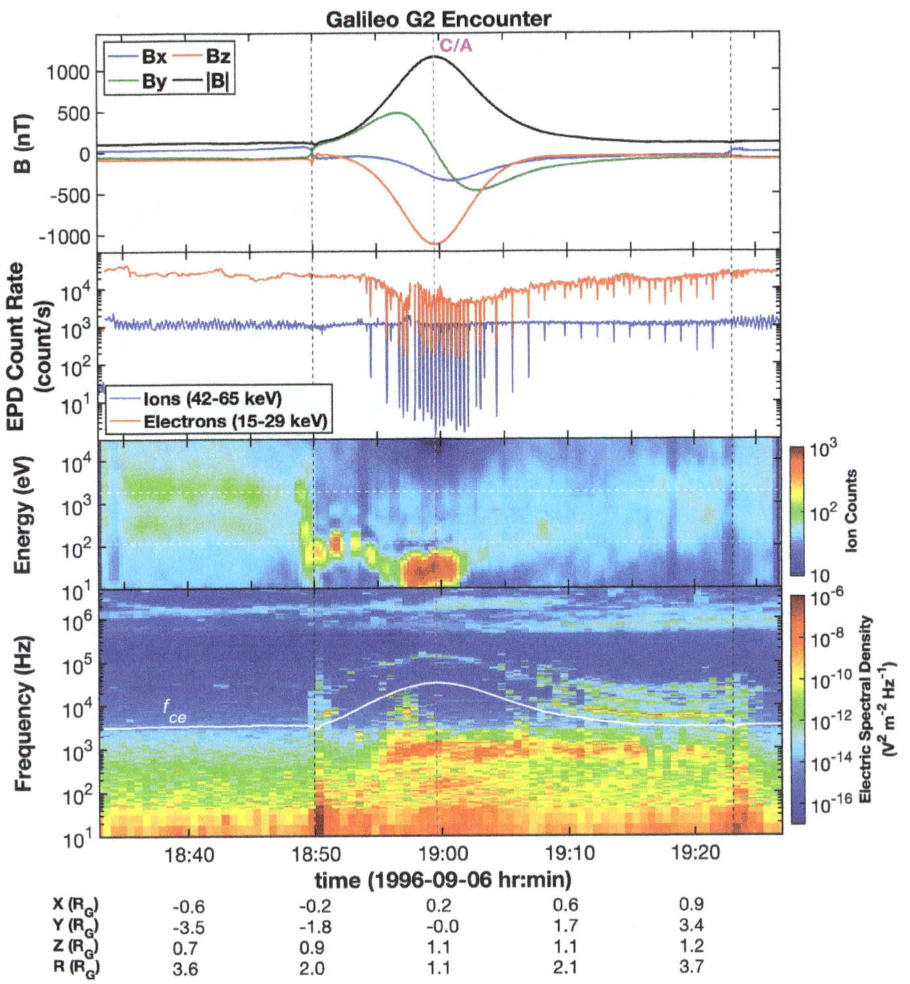

Figure 14.4 Combination of the MAG, EPD, PLS, and PWS data for the G2 encounter. The top panel shows the magnetic vector components and field strength in GphiO coordinates. The second panel shows the EPD counting rates for the ions (blue trace) from energy channel A1 (42–65 keV) and the electrons (red trace) from the energy channel E0 (15–29 keV), respectively. The third panel shows the ion energy spectrogram measured by the PLS. The color-coded ion count rate represents the maximum response from all available sensors at a given E/Q during one spacecraft spin period. For guidance, two white horizontal dashed lines at energies corresponding to the flow energy of ions with a convective flow speed of 150 km/s are plotted for M/Q = 16 and M/Q = 1, respectively. The bottom panel contains the PWS electric field spectrum with the electron gyrofrequency superimposed as white trace. The magenta vertical line marks closest approach (C/A), and the two gray vertical lines indicate the inbound and outbound magnetopause crossing identified based on the magnetic field data.

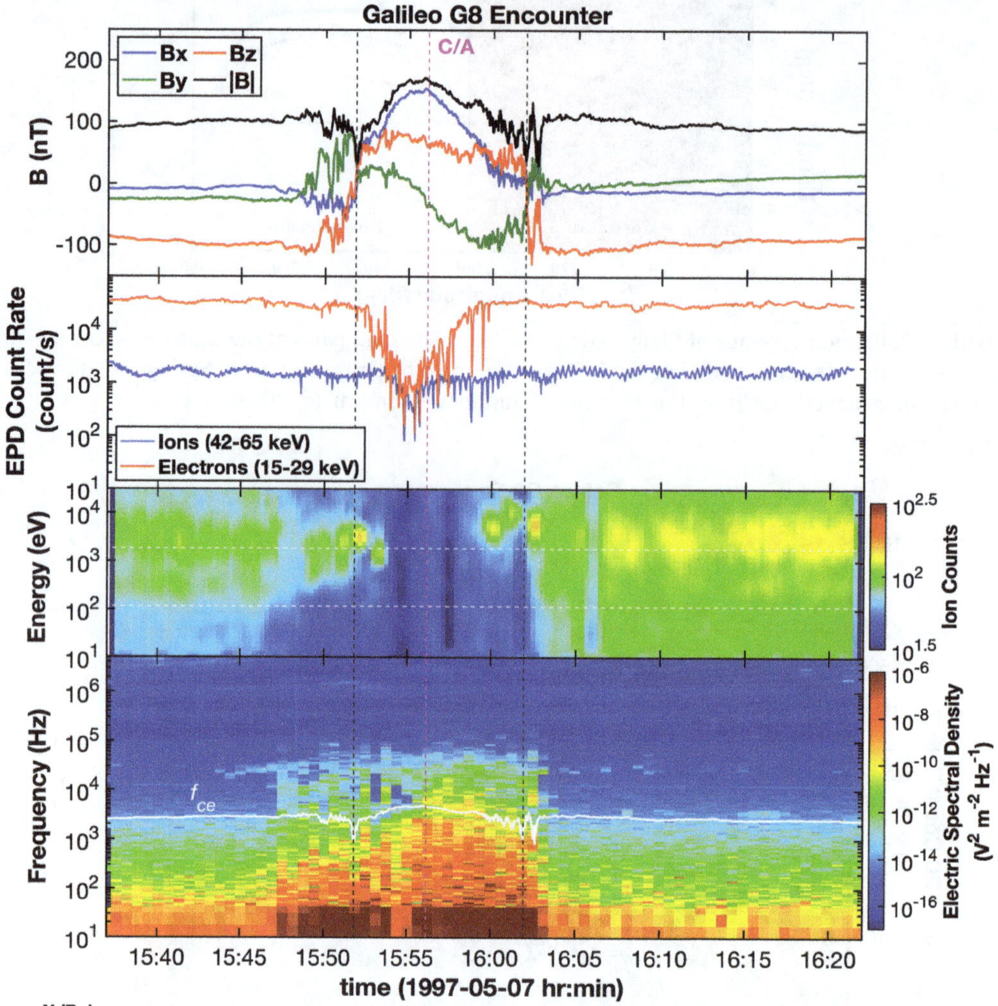

X (R_G)	-1.6	-1.5	-1.5	-1.4	-1.4	-1.3	-1.2	-1.1	-1.0
Y (R_G)	-3.1	-2.1	-1.1	-0.2	0.8	1.8	2.8	3.7	4.7
Z (R_G)	0.7	0.7	0.7	0.8	0.8	0.8	0.8	0.8	0.8
R (R_G)	3.5	2.7	2.0	1.6	1.8	2.3	3.1	4.0	4.9

Figure 14.5 Same as Figure 14.4 but for the G8 encounter.

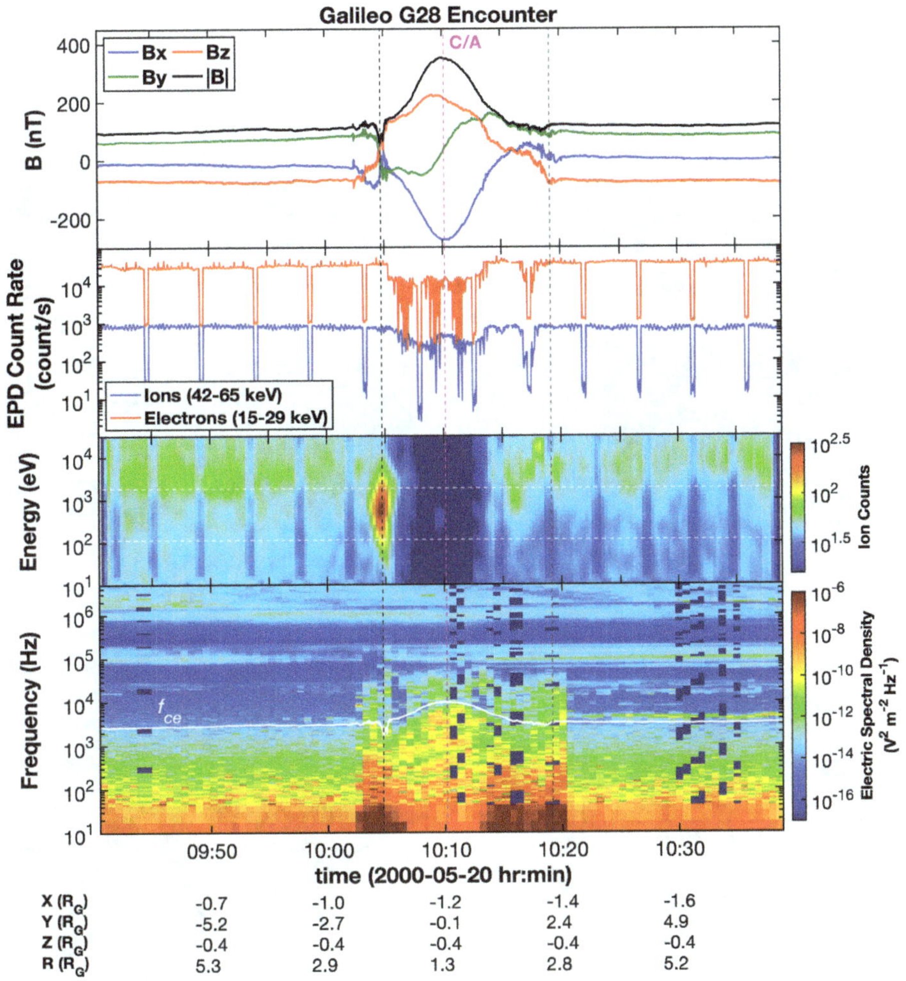

Figure 14.6 Same as Figure 14.4 but for the G28 encounter.

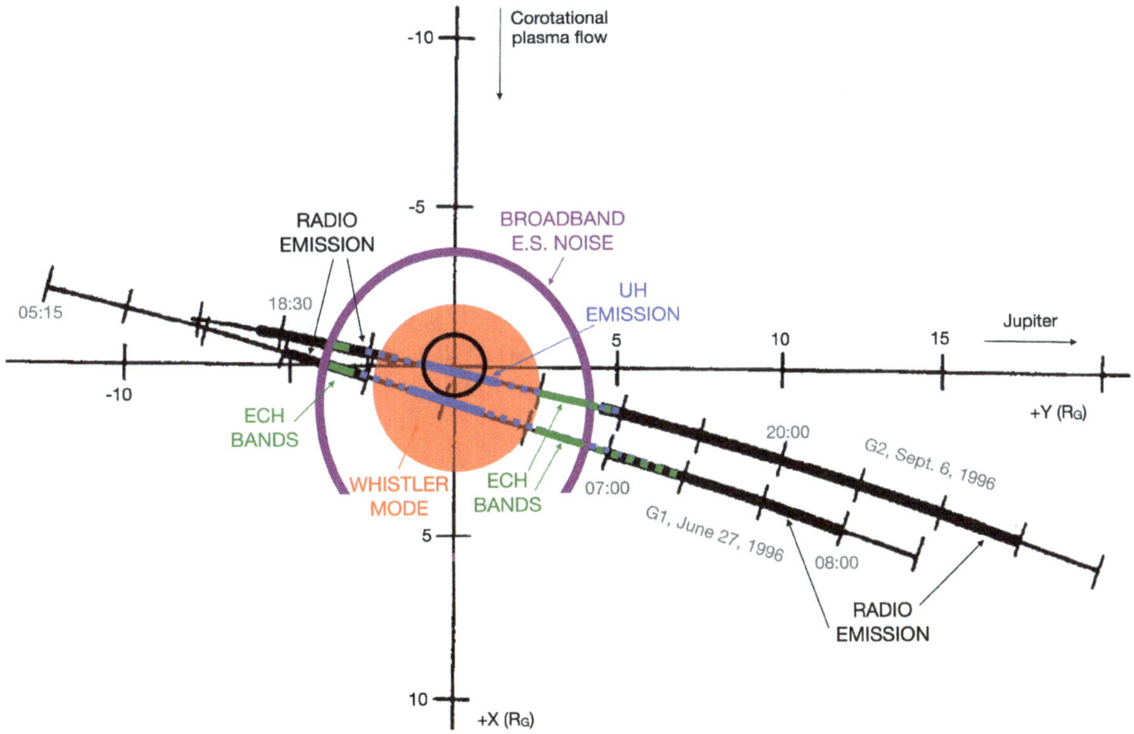

Figure 14.7 Schematic of the distribution of different types of plasma waves and radio emission in the region around Ganymede, deduced from observations during the first two Ganymede flybys. The black circle at the center represents the moon, Ganymede.

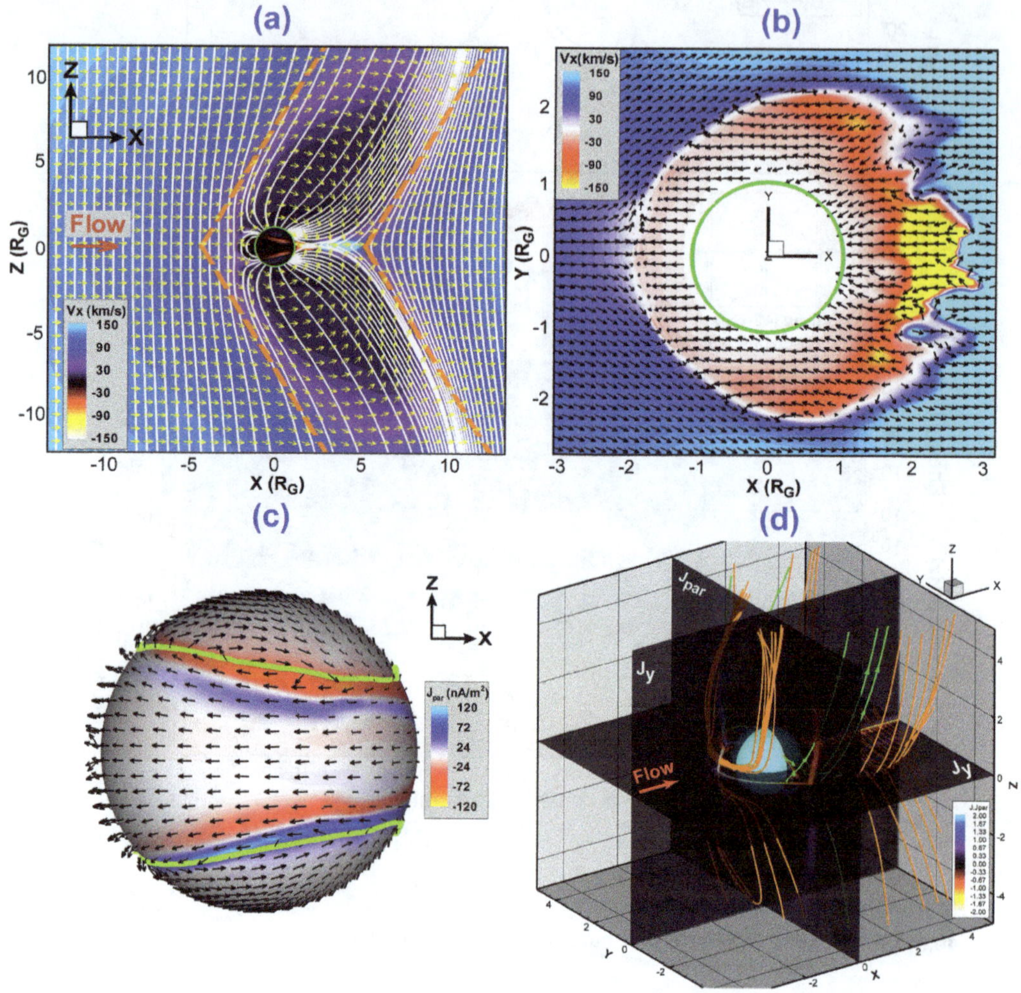

Figure 14.8 Global magnetospheric configuration and current systems extracted from MHD simulations. (a) Global plasma convection (colors: Vx contours; arrows: unit flow vectors) and magnetic field topology (white lines) in the XZ plane in GphiO coordinates. Orange dashed lines show the Alfvén characteristics for reference. (b) Convection pattern in the XY plane. Colors show the Vx contours whereas arrows show unit flow vectors. (c) Field-aligned current density along with unit flow vectors shown on a sphere of radius r= 1.08 R_G. The green curves show the flow separatrix between flows moving tailward and those moving toward the upstream. (d) 3D view of the current flowlines from the upstream flank side. Panels (a–c) reprinted by permission from John Wiley and Sons: *Journal of Geophysical Research: Space Physics*. Xianzhe Jia, et al., Properties of Ganymede's magnetosphere inferred from improved three-dimensional MHD simulations. © 2009. Panel (d) reprinted by permission from John Wiley and Sons: *Journal of Geophysical Research: Space Physics*. Xianzhe Jia, et al., Three-dimensional MHD simulations of Ganymede's magnetosphere. © 2008.

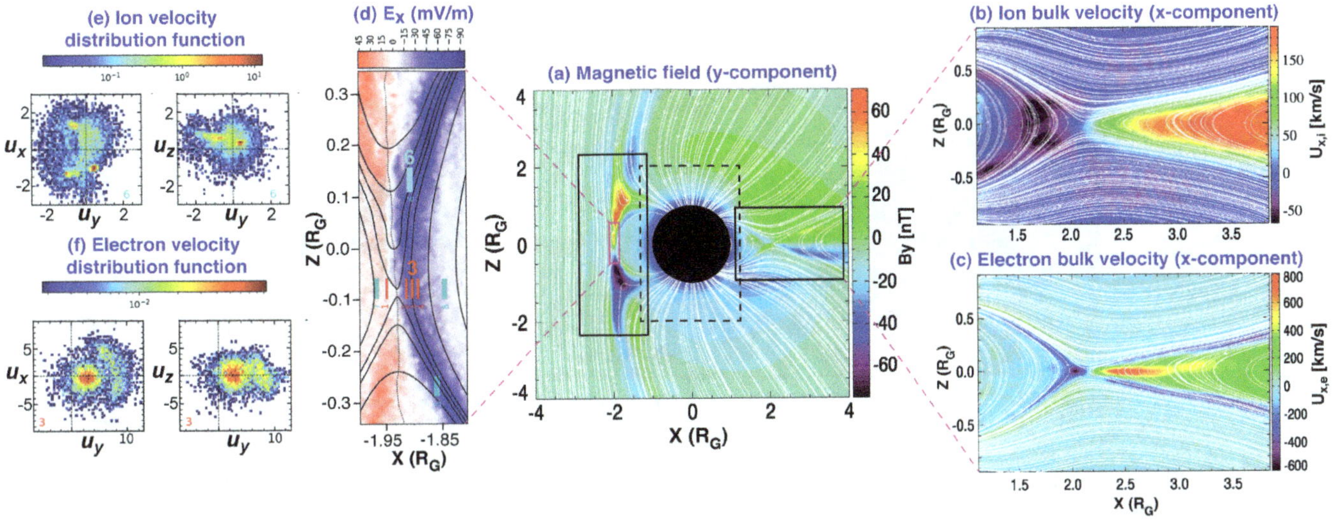

Figure 14.9 Reconnection-driven dynamics in Ganymede's magnetosphere extracted from MHD-EPIC simulations. (a) Contours of By overlaid with magnetic field lines (white lines) in the XZ plane in GphiO coordinates. The black solid and dashed boxes indicate the PIC regions embedded within the global MHD model. (b) and (c) Close-up views of the ion (b) and electron (c) flows (x-component of the bulk flow velocity) associated with reconnection in Ganymede's magnetotail as simulated by MHD-EPIC. (d) Close-up view of the upstream magnetopause reconnection, corresponding to the magenta box in (a). Colors indicate the x-component of the electric field, and black lines show magnetic field lines. (e) and (f) Normalized phase-space distribution functions in two cuts for ions and electrons, respectively. The velocity values shown along the axes are normalized to the ambient Alfvén speed. The ion distribution function shown here was extracted from location "6" labeled in (d), whereas the electron result shown was extracted from location "3" in (d). Panels (a–c) reprinted by permission from John Wiley and Sons: *Journal of Geophysical Research: Space Physics*. Gábor Tóth, et al., Extended magnetohydrodynamics with embedded particle-in-cell simulation of Ganymede's magnetosphere. © 2016. Panels (d-f) reprinted by permission from John Wiley and Sons: *Journal of Geophysical Research: Space Physics*. Hongyang Zhou, et al., Reconnection-Driven Dynamics at Ganymede's Upstream Magnetosphere: 3-D Global Hall MHD and MHD-EPIC Simulations. © 2020.

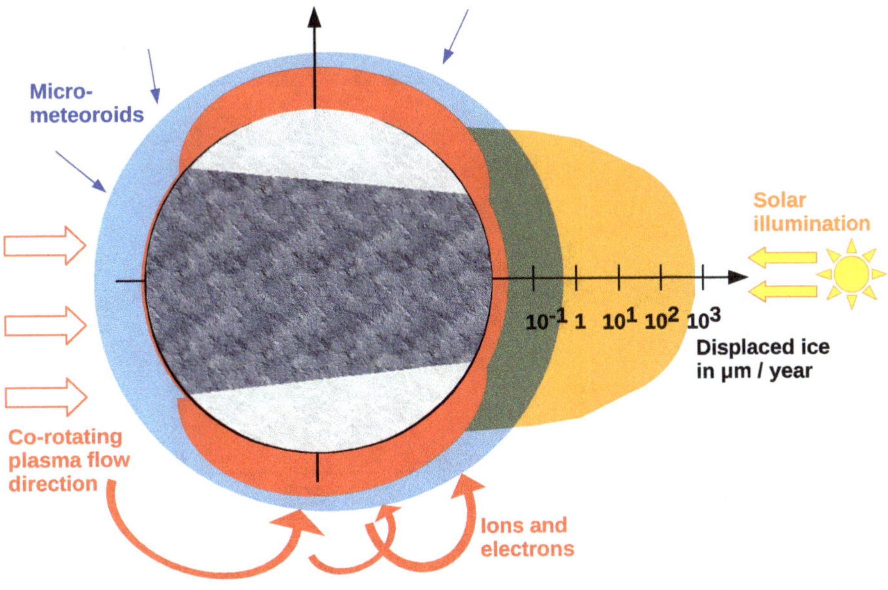

Figure 15.1 Sketch of water ice removal or turnover rates on Ganymede's surface in units of $\mu m\ yr^{-1}$. Orange shape: effect of thermal sublimation, blue: micrometeoroid impacts, red: irradiation-related rates (dominated by ion sputtering). Note the logarithmic scale of the rates (going from 0.1 to 1000 $\mu m\ yr^{-1}$). The sketch is centred on the equatorial plane of the anti-Jovian side. Ganymede's leading hemisphere with respect to Jupiter and the Jovian plasma is on the dayside to the right. The grey band and the white pole caps symbolize the surface areas of low and high albedo.

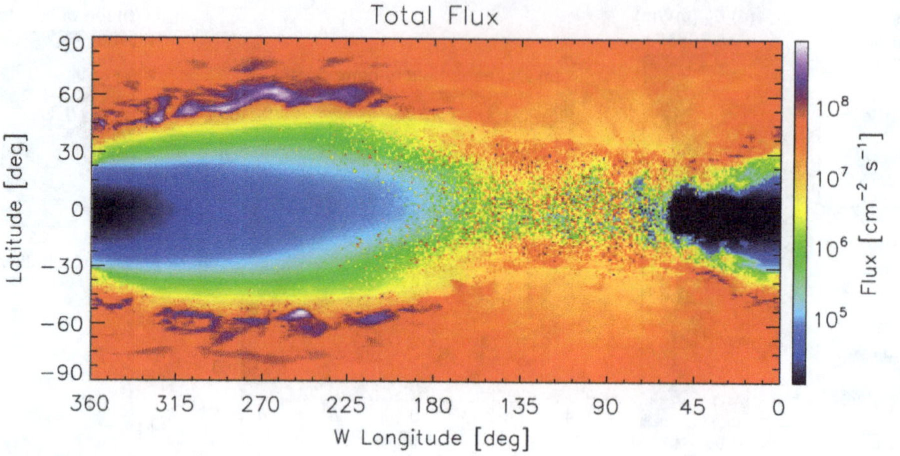

Figure 15.4 Total ion precipitation flux onto Ganymede's surface, summed over all ion species and energies, representative of G8 configurations when Ganymede was inside the Jovian plasma sheet. Leading hemisphere is centred at 90° W. Figure from Poppe et al. (2018).

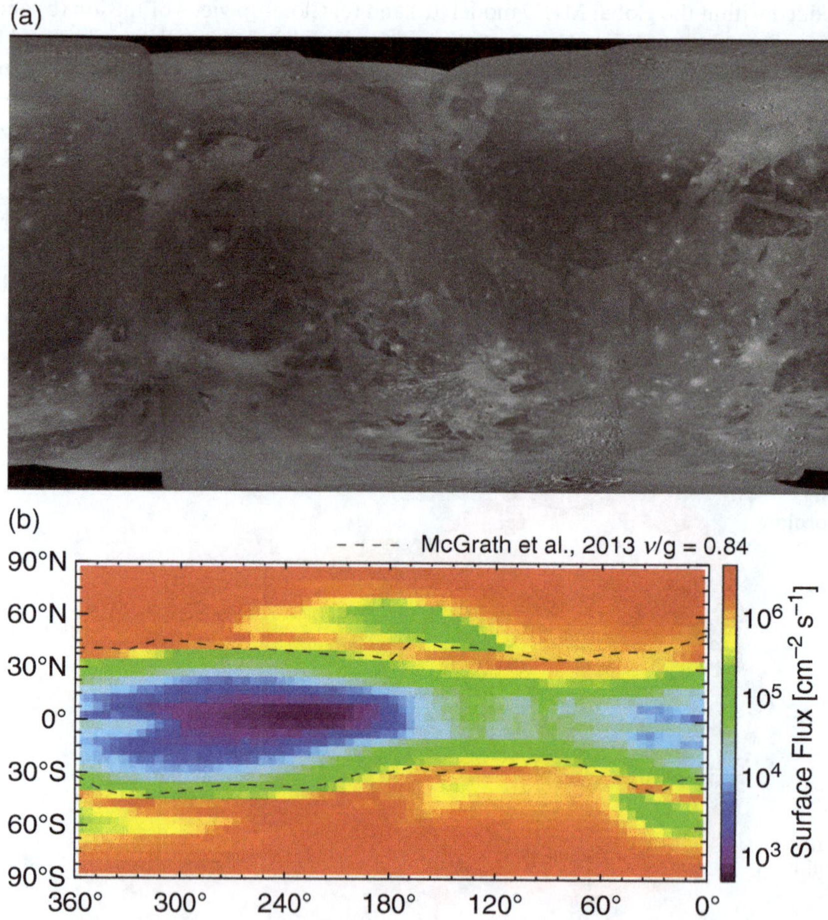

Figure 15.9 (a) Global mosaic of Ganymede using the best image quality and moderate resolution coverage supplied by Galileo Solid-State Imaging (SSI) and Voyager 1 and 2. The left hemisphere from 360° W to 180° W is the trailing hemisphere, and the right half is the leading hemisphere. Image credits: USGS Astrogeology Science Center, Batson (1984). (b) Total flux of precipitating energetic ions to the surface of Ganymede from the model by Fatemi, S., A. R. Poppe, K. K. Khurana, M. Holmström, and G. T. Delory (2016), On the formation of Ganymede's surface brightness asymmetries: Kinetic simulations of Ganymede's magnetosphere, *Geophys. Res. Lett.*, 43, 4745–4754. John Wiley and Sons. The dashed lines are the violet-to-green ratio of 0.84 from the composite image of Ganymede's surface by Khurana et al. (2007) showing the approximate boundaries of Ganymede's bright polar caps and dark low latitudes.

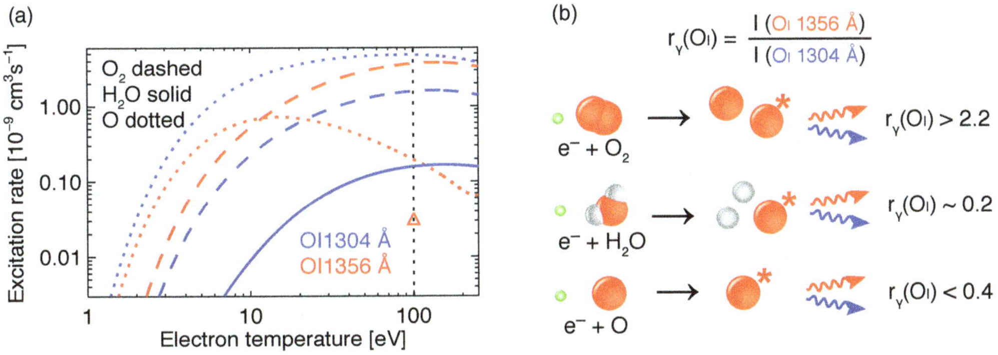

Figure 16.2 (a) Excitation rates for electron impact on O_2 (dashed), O (dotted) and H_2O (solid) for OI 1 356 Å (red) and OI 1 304 Å (blue) as a function of electron temperature. The rates are calculated from the cross sections (Kanik et al., 2001, 2003; Makarov et al., 2004), assuming Maxwellian temperature distribution for the electron population. For OI 1 356 Å from H_2O only a single cross-section measurement at 100 eV is available, and the red triangle (for the OI 1 356 Å rate) shows the 1 304 Å value scaled according to the relative cross section. (b) Schematic showing the approximate oxygen emission ratios for the three species (after Roth et al., 2021).

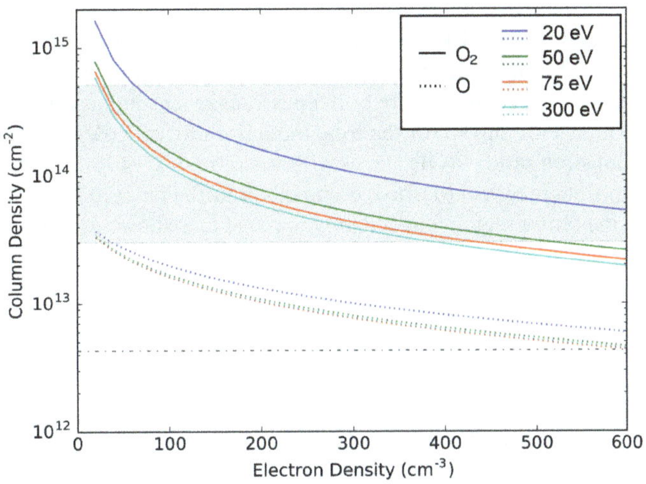

Figure 16.3 Global column densities of O_2 and O at Ganymede based on the oxygen emission intensities observed on the trailing hemisphere for a range of electron densities and temperatures. The O_2 column density is primarily constrained by a disk-average 1 356 Å intensity of 39 R, derived in Molyneux et al. (2018). The shaded region shows the range of O_2 column densities inferred from previous observations of Ganymede (Feldman et al., 2000). The dashed horizontal line indicates the maximum column density for an optically thin O atmosphere at a temperature of 150 K.

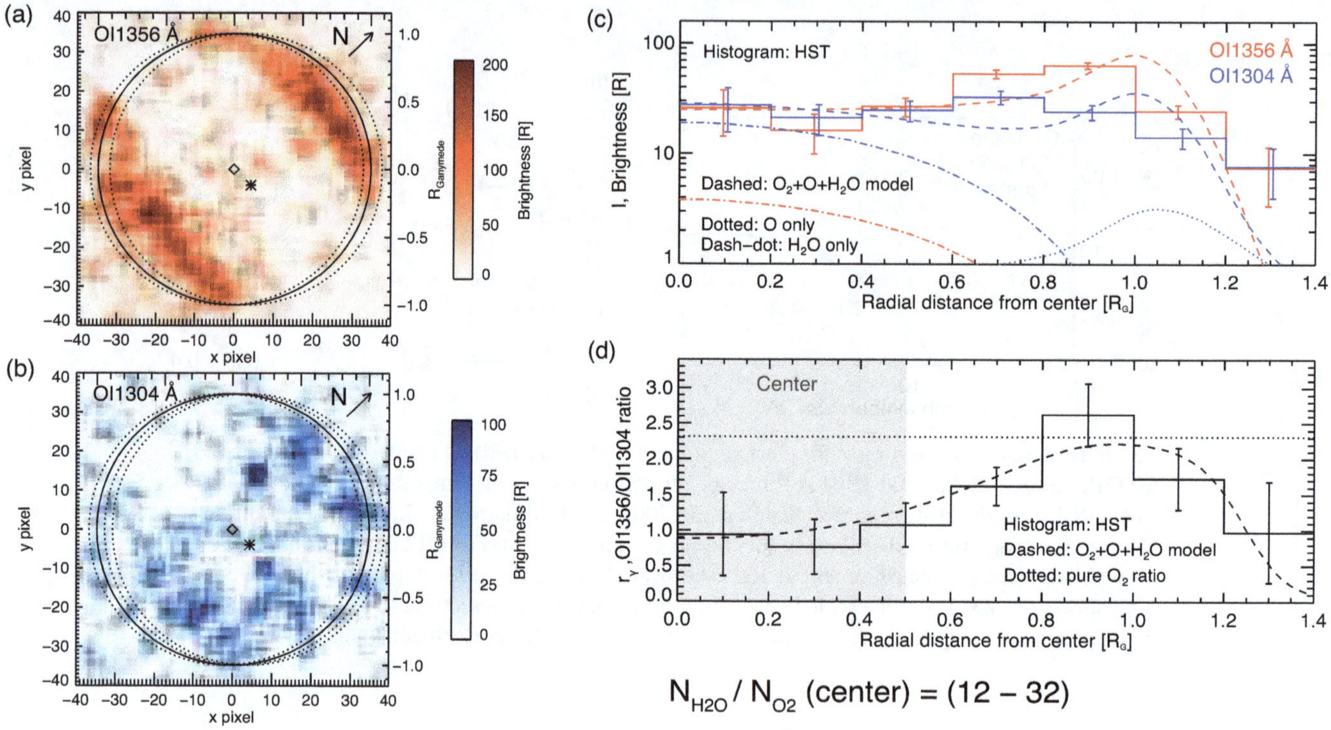

Figure 16.4 (a) and (b) HST/STIS images of the OI 1 356 Å and OI 1 304 Å emissions observing Ganymede's trailing hemisphere from Roth et al. (2021) (see their figure 3 for more details). (c) The radial brightness profiles derived from these OI 1 356 Å (red) and OI 1 304 Å (blue) images show that near the disk centre both emissions reach similar intensities. Simulated radial profiles for an assumed H_2O-O_2-O atmosphere model are shown for the total modelled intensity (dashed) and for individual contributions. (e) The profile of the observed oxygen emission ratio $r_\gamma(\mathrm{OI})$ agrees with the ratio of O_2 (dotted) only near the limb but is in good agreement with the O_2+O+H_2O atmosphere model (dashed) assumed by Roth et al. (2021) at all radial distances. The derived column density ratio N_{H_2O}/N_{O_2} in the centre region (shaded grey in panel e) is shown at the bottom.

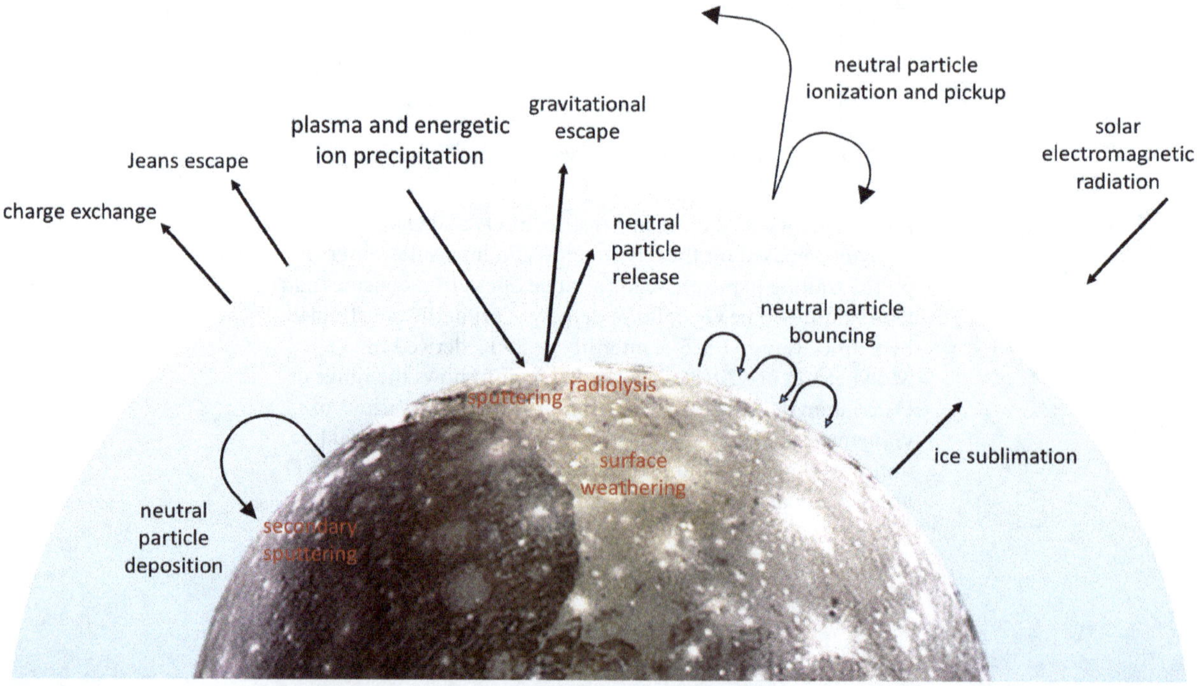

Figure 16.6 Main source and sink mechanisms for Ganymede's exosphere (from Plainaki et al. (2020a)). The moon's environment is strongly coupled with the Jovian magnetosphere through processes that determine the energy exchange between the two systems. Ganymede image credit: NASA/JPL/DLR.

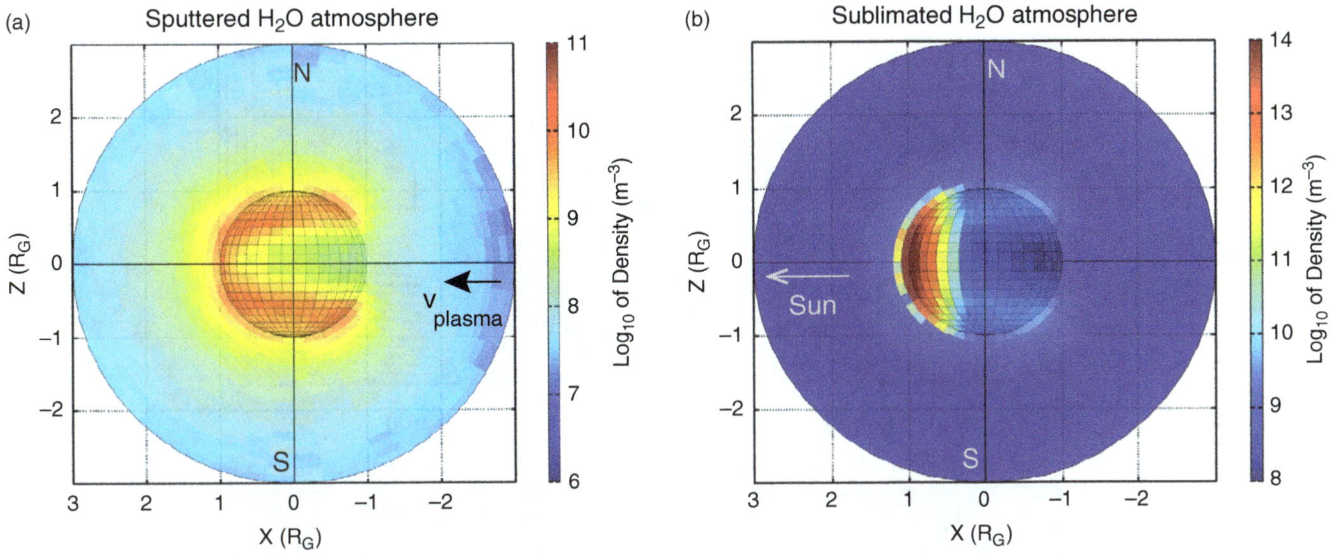

Figure 16.8 Projected atmospheric H_2O density distributions over the surface of Ganymede and in a plane through Ganymede's centre, where X is along the ambient plasma flow direction (shown by the black arrow) and Z is along Jupiter's spin axis (in units of Ganymede's radii, North 'N' is up), adapted from Plainaki et al. (2015). The Sun is to left, shown by the grey arrow. Panel a shows the H_2O that is produced from surface sputtering only, where H_2O is mostly confined to the regions near the open field lines. Panel b shows H_2O from sublimation only, with H_2O confined to the dayside. The peak sublimated density near the sub-solar points exceeds the maximum sputtered density by three orders of magnitude (note the different ranges of the color bars). Reprinted from *Icarus*, 245, Christina Plainaki, Anna Milillo, Stefano Massetti, Alessandro Mura, Xianzhe Jia, Stefano Orsini, Valeria Mangano, Elisabetta De Angelis, Rosanna Rispoli, The H2O and O2 exospheres of Ganymede: The result of a complex interaction between the jovian magnetospheric ions and the icy moon, 306–319, © 2015, with permission from Elsevier.

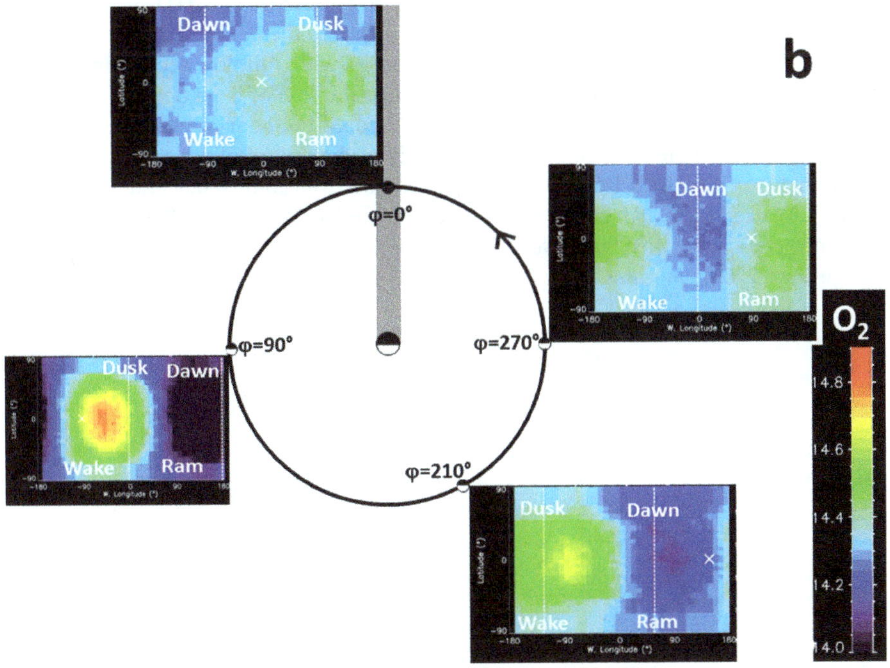

Figure 16.9 Simulated atmospheric evolution maps (latitude vs. W longitude) of zenith O_2 column densities at four Ganymede phase angles, from Leblanc et al. (2017). To follow diurnal behavior, the sub-solar point is indicated by an 'X' and dusk/dawn terminators are represented by the vertical dashed white lines. To follow plasma behavior, ram and wake are labeled. Ganymede's angular rotation vector Ω is counter-clockwise in the plane of Jovian co-rotation. Reprinted from *Icarus*, 293, F. Leblanc, A. V. Oza, L. Leclercq, C. Schmidt, T. Cassidy, R. Modolo, J. Y. Chaufray, R. E. Johnson, On the orbital variability of Ganymede's atmosphere, 185–198, © 2017, with permission from Elsevier.

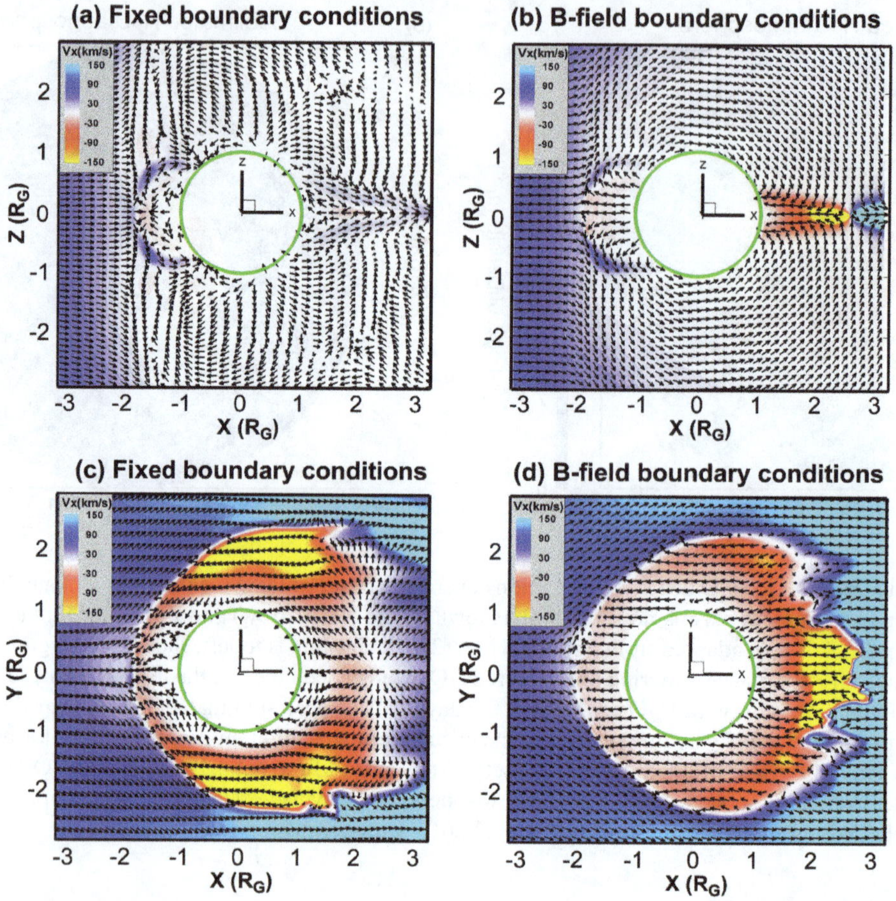

Figure 17.3 Comparisons of the plasma convection pattern from MHD simulations using different inner boundary conditions and applied to the G8 flyby. Contours of the X-component of the flow velocity (V_x) superimposed with unit flow vectors in the XZ plane (top row) and the XY plane (bottom row) for the case of fixed boundary conditions (panels (a) and (c)) and magnetic-field based boundary conditions (panels (b) and (d)). The circle in each panel shows the inner boundary of the MHD simulation ($r = 1.05\ R_G$). Reprinted with permission from John Wiley and Sons. Jia, X., R. J. Walker, M. G. Kivelson, K. K. Khurana, and J. A. Linker (2009), Properties of Ganymede's magnetosphere inferred from improved three-dimensional MHD simulations, *J. Geophys. Res.*, 114.

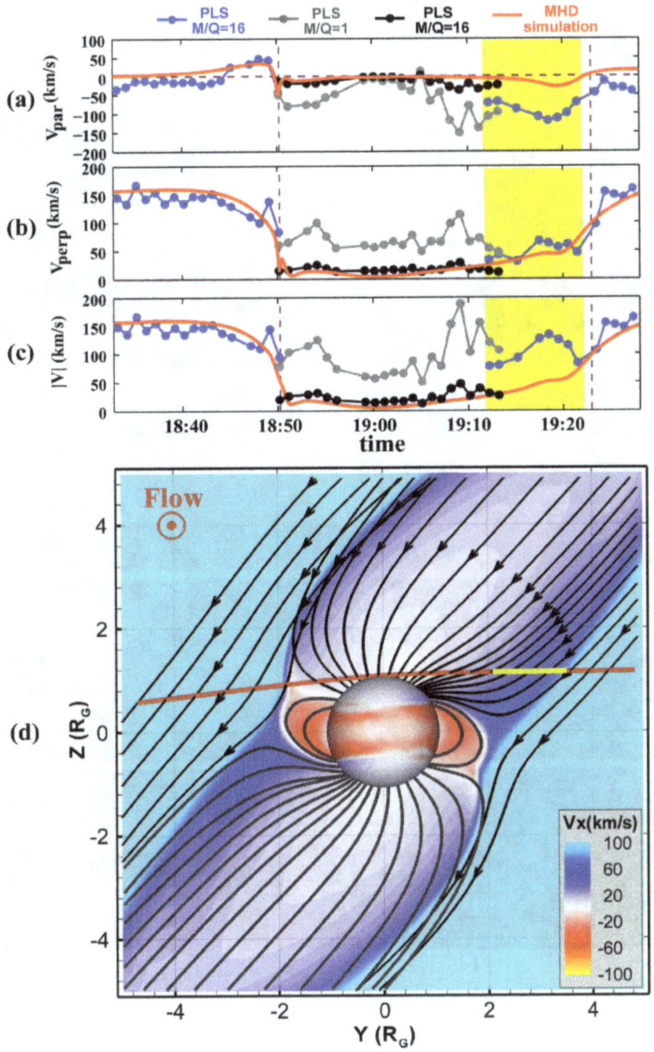

Figure 17.6 (a–c) Comparisons of the plasma bulk flow velocities from the MHD simulation of Jia et al. (2009) with the Galileo PLS observations for the G2 flyby. Panels (a) to (c) show the parallel, perpendicular velocity components with respect to the magnetic field and the total flow speed, respectively. In each panel, the blue dots represent bulk flows derived from the PLS data when the ion composition can be unambiguously determined as heavy ions with mass per charge (M/Q) = 16. The gray dots are the calculated moments assuming M/Q = 1, and the black dots represent bulk flows if M/Q = 16. The PLS moments are from Frank et al. (1997). Red traces are the bulk flow velocity from the MHD simulation. The two vertical dashed lines mark the inbound and outbound magnetopause crossings. (d) Colour contours of the modelled V_X in the YZ plane (perpendicular to the incident flow direction) and on a sphere of radius $r = 1.08\ R_G$, superposed with projection of simulated field lines. The black lines with arrows are field lines traced along the Galileo G2 trajectory (red trace), and the dark gray traces are arbitrarily chosen field lines within Ganymede's magnetosphere plotted for context. The locations where the PLS observed large field-aligned plasma flows are marked with yellow colour in all panels. Reprinted with permission from John Wiley and Sons. Jia, X., R. J. Walker, M. G. Kivelson, K. K. Khurana, and J. A. Linker (2009), Properties of Ganymede's magnetosphere inferred from improved three-dimensional MHD simulations, *J. Geophys. Res.*, 114.

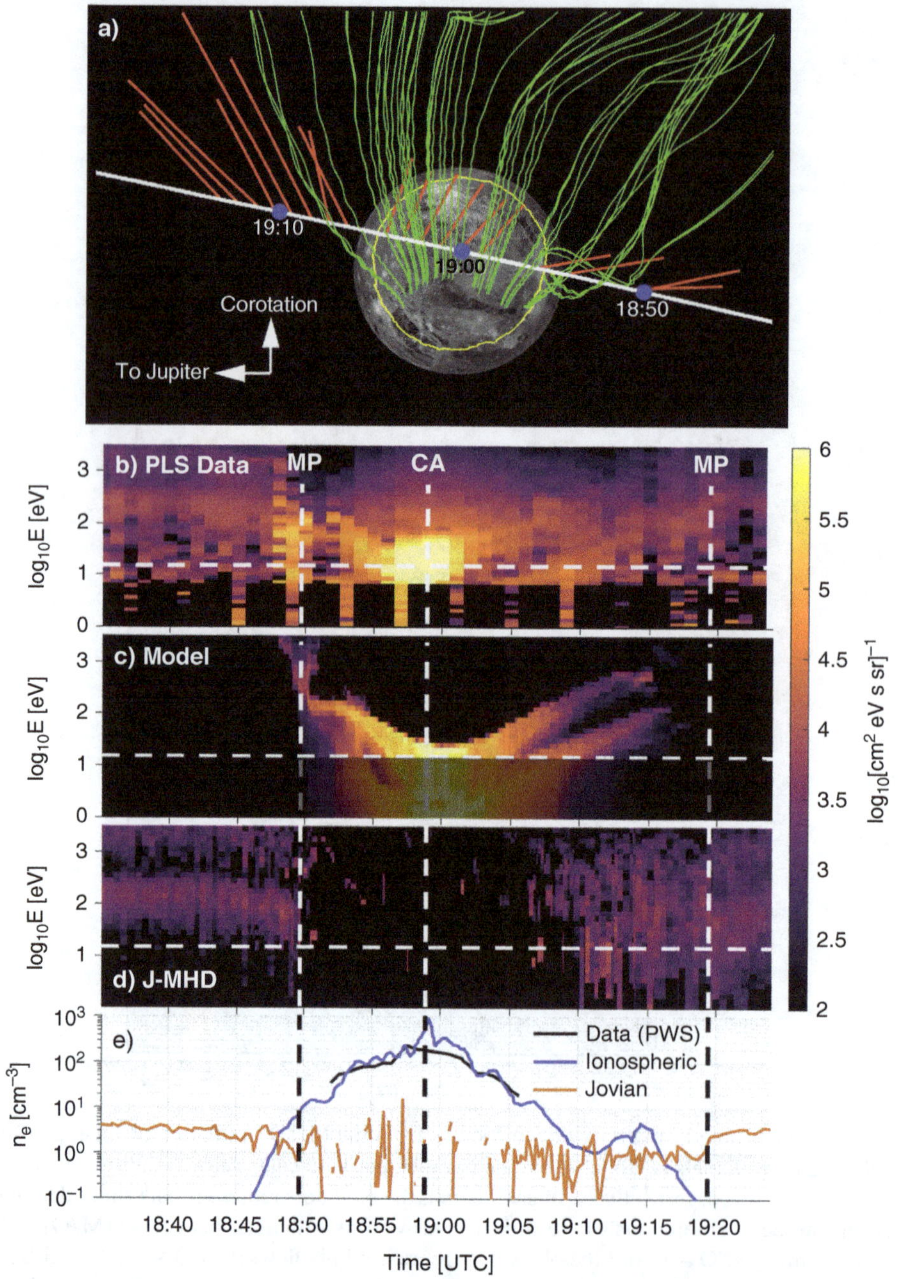

Figure 17.7 Comparison between Galileo observations and simulation output from the 3D test particle ion model by Carnielli et al. (2020a) for the G2 flyby. (a) White, bold line: Trajectory of the Galileo spacecraft. Red lines: Bulk velocity vectors as observed by the PLS instrument (Frank et al., 1997). Green curves: Selection of O_2^+ test particle trajectories which intersect the path of the spacecraft. Yellow circular line: Northern boundary between open and closed magnetic field lines. (b–d) Energy spectrograms along the Galileo trajectory, (b) as recorded by the PLS instrument and (c–d) generated by the test particle ion model applied to (c) the ionospheric population and (d) the Jovian magnetospheric population. The dashed vertical lines (white in panels (b–d) and black in panel (e)) indicate the times of magnetopause crossing (MP) and of closest approach (CA). The dashed, white horizontal lines in panels (b–d) show the upper energy limit below which the PLS dataset processing technique is considered as unreliable (see Carnielli et al. (2019) for more details). The simulation is driven by a number density for $O_2 \times 10$, compared to the 'dry' case from Leblanc et al. (2017), and an asymmetric electron-impact ionisation frequency (boosted case, Carnielli et al. (2020a)). The magnetic and electric fields are from the MHD model of Jia et al. (2009). (e) Electron number density along the Galileo trajectory, as a function of time, as recorded by the PWS instrument (black) (Eviatar et al., 2001) and as computed by the test particle ion model (blue). The simulated number density of the background Jovian ion density is shown in light brown. Panels (a–b, d) adapted from Carnielli et al. (2019). Reprinted from *Icarus*, 330, G. Carnielli, M. Galand, F. Leblanc, L. Leclercq, R. Modolo, A. Beth, H. L. F. Huybrighs, X. Jia, First 3D test particle model of Ganymede's ionosphere, 42–59, © 2019 with permission from Elsevier. Panels c, e adapted from Carnielli et al. (2020a). Reprinted from *Icarus*, 343, G. Carnielli, M. Galand, F. Leblanc, R. Modolo, A. Beth, X. Jia, Constraining Ganymede's neutral and plasma environments through simulations of its ionosphere and Galileo observations © 2020 with permission from Elsevier.

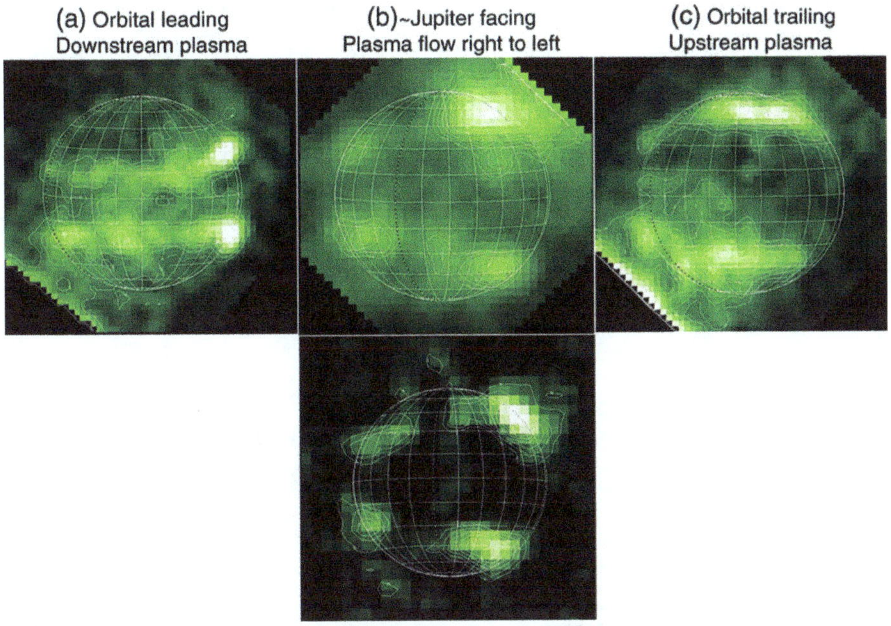

Figure 18.1 Ganymede's aurora seen from different directions. Both the upper and lower images in the central column (b) are Jupiter-facing but recorded during different observational dates. Figure from McGrath et al. (2013). Reprinted with permission from John Wiley and Sons. M. A. McGrath, X. Jia, K. Retherford, P. D. Feldman, D. F. Strobel, and J. Saur. Aurora on Ganymede. *J. Geophys. Res. (Space Physics)*, 118:2043–2054, © 2013.

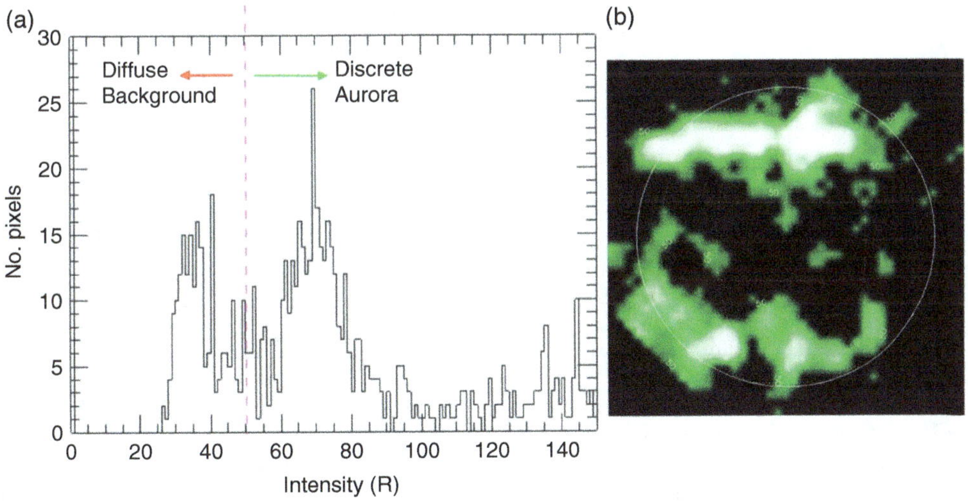

Figure 18.4 Bright versus faint or diffuse versus discrete aurora (Molyneux et al., 2018). The purple vertical dashed line in the histogram in panel (a) demarcates the threshold over which the emissions were considered bright as opposed to the diffuse background. The image in panel (b) only includes bright pixels. Pictures from Molyneux et al. (2018). Reprinted with permission from John Wiley and Sons. P. M. Molyneux, J. D. Nichols, N. P. N. P. Bannister, E. J. Bunce, J. T. Clarke, S. W. H. Cowley, J.-C. Gérard, D. Grodent, S. E. Milan, and C. Paty. Hubble space telescope observations of variations in Ganymede's oxygen atmosphere and aurora. *J. Geophys. Res. (Space Physics)*, 123(5):3777–3793 © 2018.

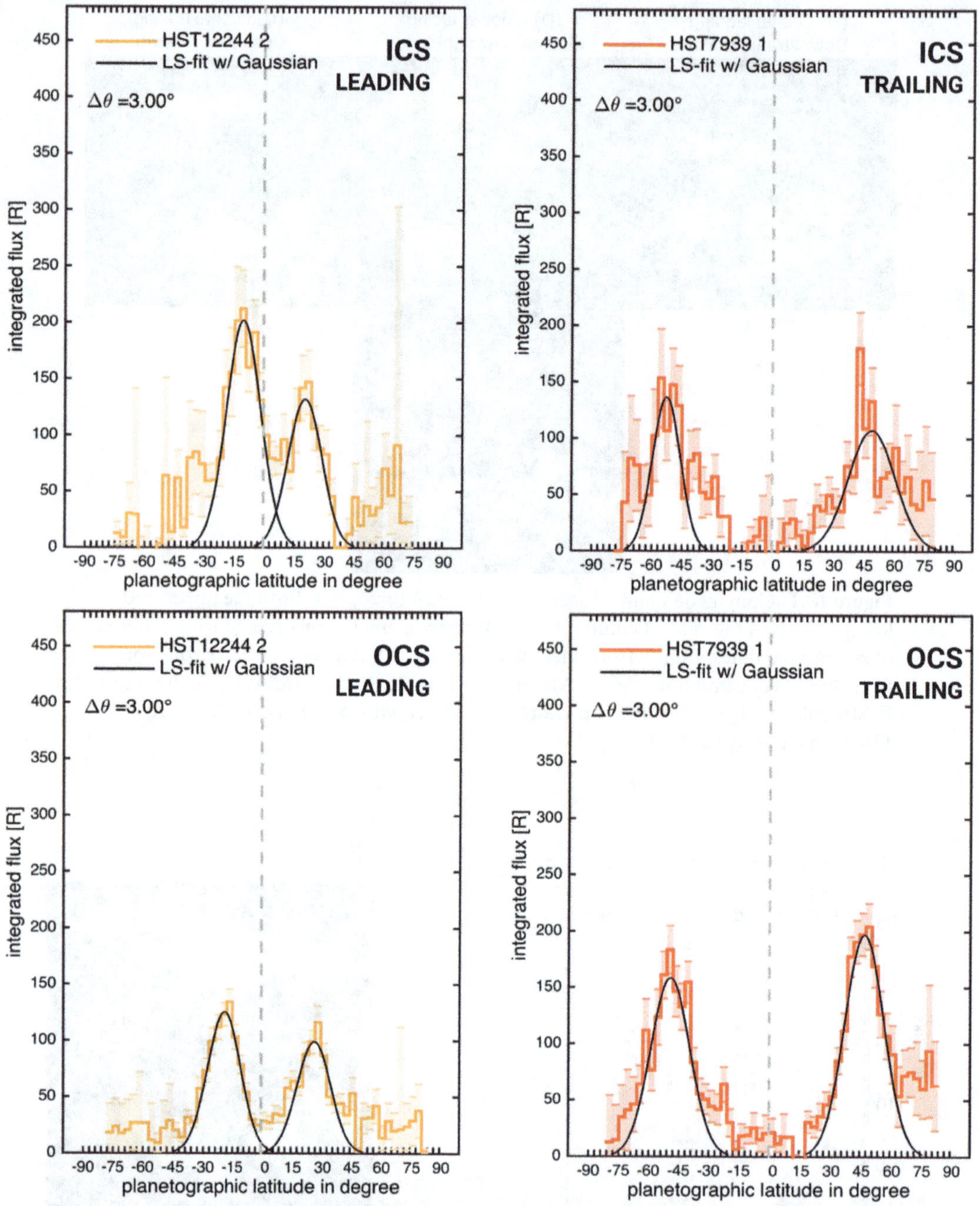

Figure 18.3 Integrated brightness as a function of latitude on both hemispheres when Ganymede is inside the dense plasma or current sheet (ICS) and above or below it (OCS) (Musacchio et al., 2017). Reprinted with permission from John Wiley and Sons. Fabrizio Musacchio, Joachim Saur, Lorenz Roth, Kurt D. Retherford, Melissa A. McGrath, Paul D. Feldman, and Darrel F. Strobel. Morphology of Ganymede's FUV auroral ovals. *J. Geophys. Res. (Space Physics)*, 122(3):2855–2876, © 2017.

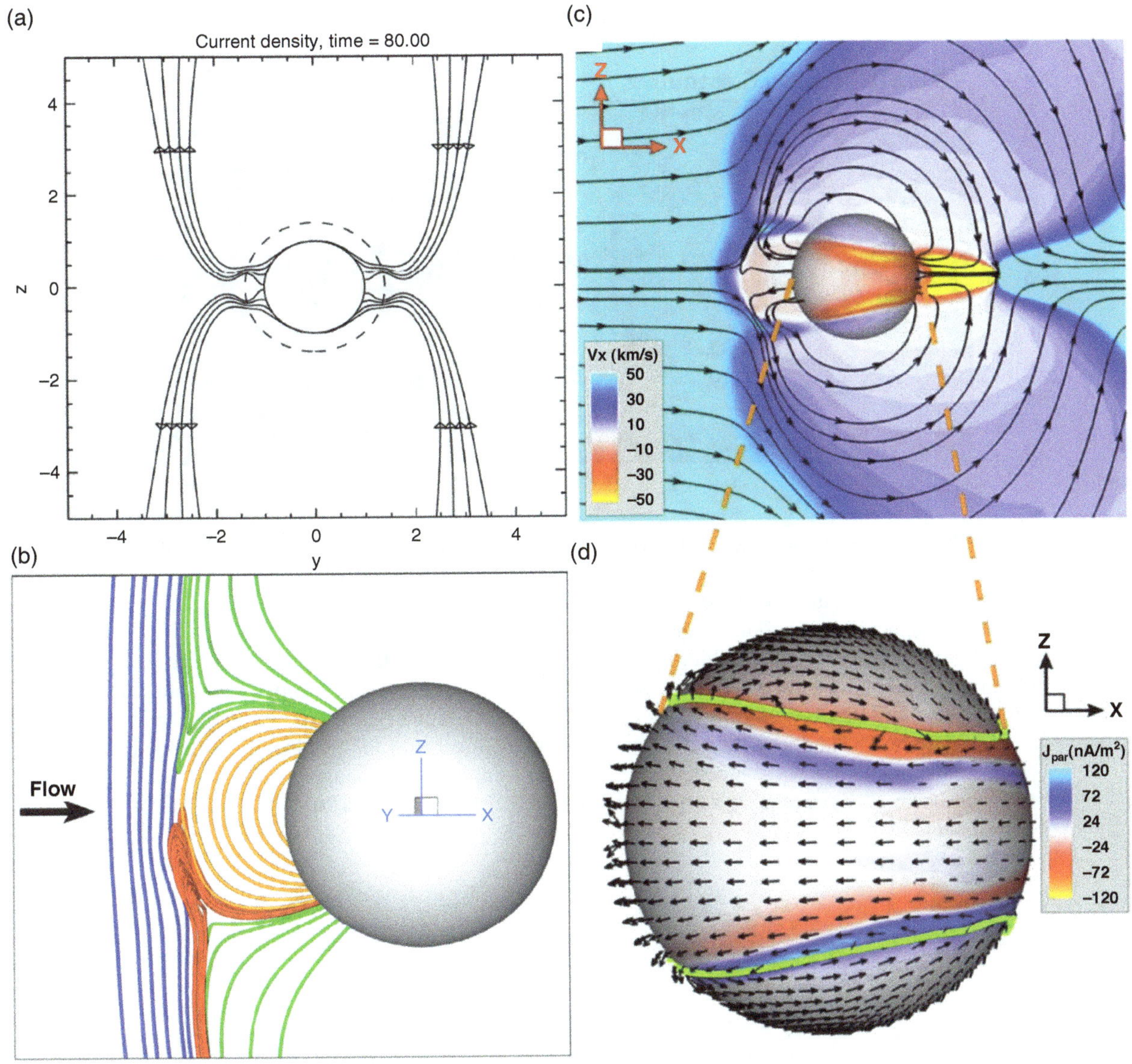

Figure 18.5 (a) field lines connected to the regions with the strongest ionospheric field-aligned currents in an MHD simulation (Ip and Kopp, 2002). The plane shown is perpendicular to the direction of the Jovian magnetospheric plasma flow around Ganymede. Reprinted with permission from John Wiley and Sons. Wing-Huen Ip and Andreas Kopp. Resistive MHD simulations of Ganymede's magnetosphere 2. Birkeland currents and particle energetics. *J. Geophys. Res. (Space Physics)*, 107(A12):SMP 42–1–SMP 42–7, © 2002. (b) Field line connectivity in an MHD simulation by Jia et al. (2010). Reprinted with permission from John Wiley and Sons. Xianzhe Jia, Raymond J. Walker, Margaret G. Kivelson, Krishan K. Khurana, and Jon A. Linker. Dynamics of Ganymede's magnetopause: Intermittent reconnection under steady external conditions. *J. Geophys. Res. (Space Physics)*, 115, © 2010. (c) Flow pattern in an MHD simulation (Jia et al., 2009). (d) Field-aligned current density at the ionosphere (Jia et al., 2009). Panels c, d reprinted with permission from John Wiley and Sons. Xianzhe Jia, Raymond J. Walker, Margaret G. Kivelson, Krishan K. Khurana, and Jon A. Linker. Properties of Ganymede's magnetosphere inferred from improved three-dimensional MHD simulations. *J. Geophys. Res.*, 114(A9):A09209, © 2009. The side of Ganymede shown in panels c and d is facing away from Jupiter.

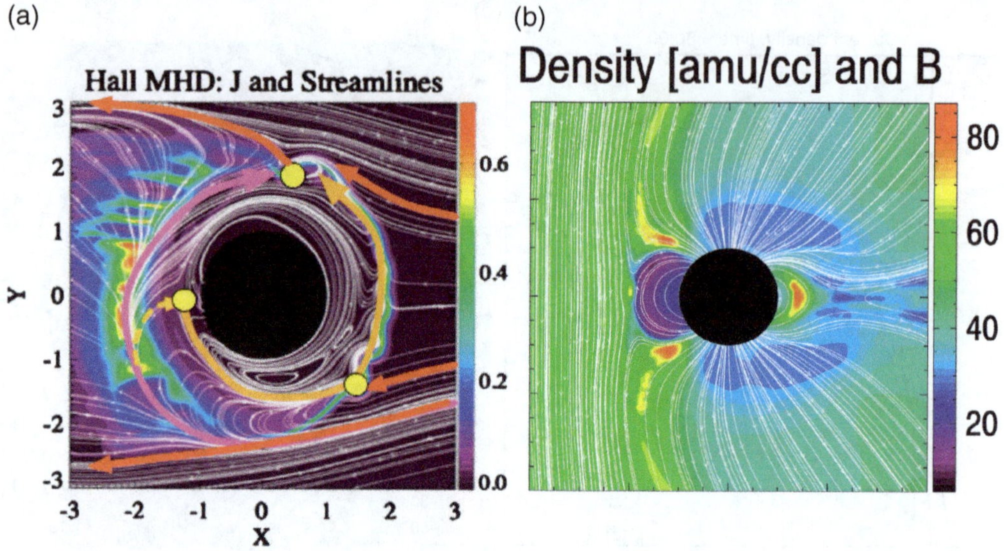

Figure 18.6 (a) Current densities and flow patterns in a Hall MHD simulation (Dorelli et al., 2015). Reprinted with permission from John Wiley and Sons. J. C. Dorelli, A. Glocer, G. Collinson., and G. Tóth. The role of the Hall effect in the global structure and dynamics of planetary magnetospheres: Ganymede as a case study. *J. Geophys. Res. (Space Physics)*, 120:5377–5392 © 2015. (b) Densities in a Hall MHD simulation with embedded iPIC patches (Tóth et al., 2016). Reprinted with permission from John Wiley and Sons. Gábor Tóth, Xianzhe Jia, Stefano Markidis, Ivy Bo Peng, Yuxi Chen, Lars K S Daldorff, Valeriy M. Tenishev, Dmitry Borovikov, John D. Haiducek, Tamas I. Gombosi, Alex Glocer, and John C. Dorelli. Extended magnetohydrodynamics with embedded particle-in-cell simulation of Ganymede's magnetosphere. *J. of Geophys. Res.: Space Physics*, 121(2):1273–1293, © 2016.

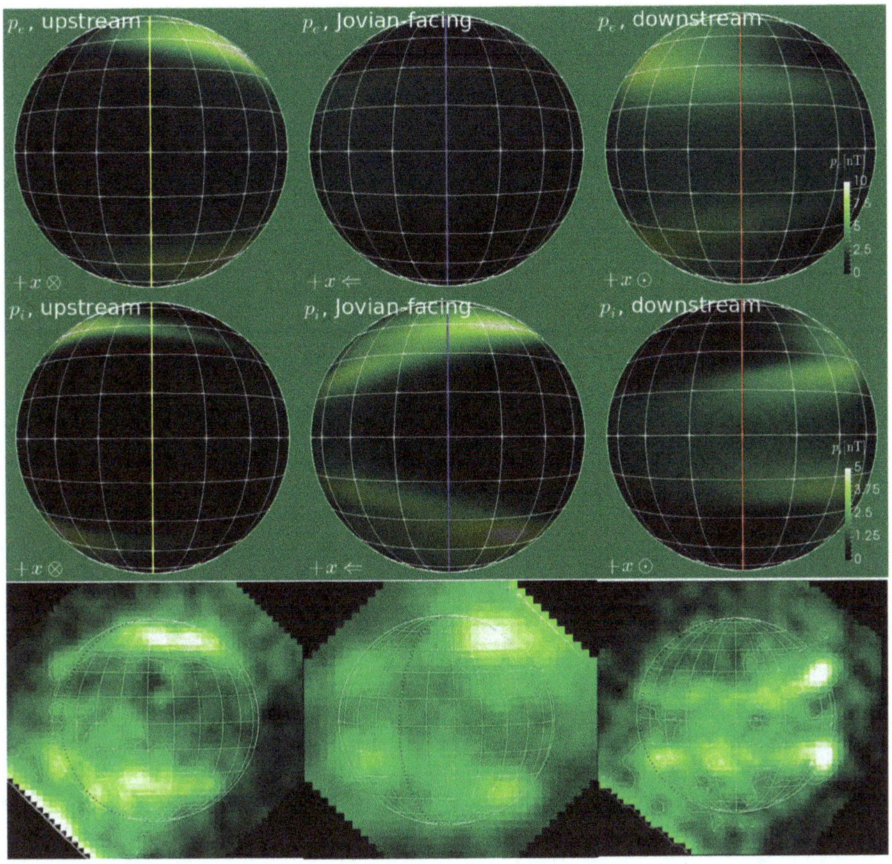

Figure 18.7 Scalar electron pressure (upper row) and scalar ion pressure (middle row) at 263 km altitude. The left column shows the upstream side; the middle column, the Jupiter-facing side; and the right column, the downstream side. The yellow, blue, and red curves mark longitudes 270° W, 0° W, 90° W, respectively. The coordinate system was defined by McGrath et al. (2013). The bottom row shows the light emissions observed by the Hubble Space Telescope (McGrath et al., 2013). The first two rows and the compilation of the compound figure were made by Wang et al. (2018). Reprinted with permission from John Wiley and Sons. Liang Wang, Kai Germaschewski, Ammar Hakim, Chuanfei Dong, Joachim Raeder, and Amitava Bhattacharjee. Electron physics in 3-D two-fluid 10-moment modeling of Ganymede's magnetosphere. *J. Geophys. Res. (Space Physics)*, 123(4):2815–2830 © 2018.

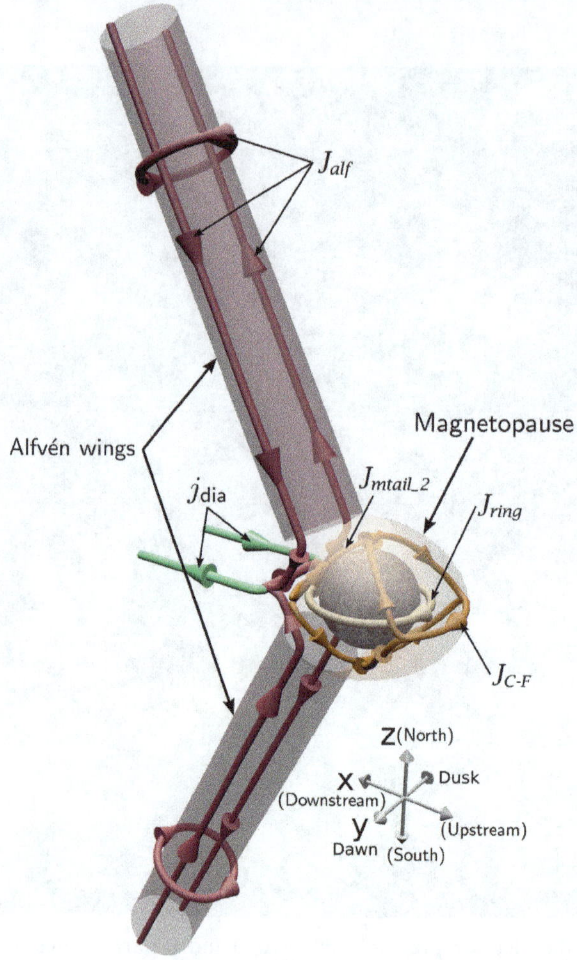

Figure 18.8 Three-dimensional diagram of the current systems for a sub-Alfvénic plasma interaction with a magnetised body (Ganymede-like), taken from (Vernisse et al., 2017). The magnetopause (Chapman–Ferraro) currents are shown in orange, Alvén wing currents are shown in purple, ring currents are in yellow, and the diamagnetic currents are in green.

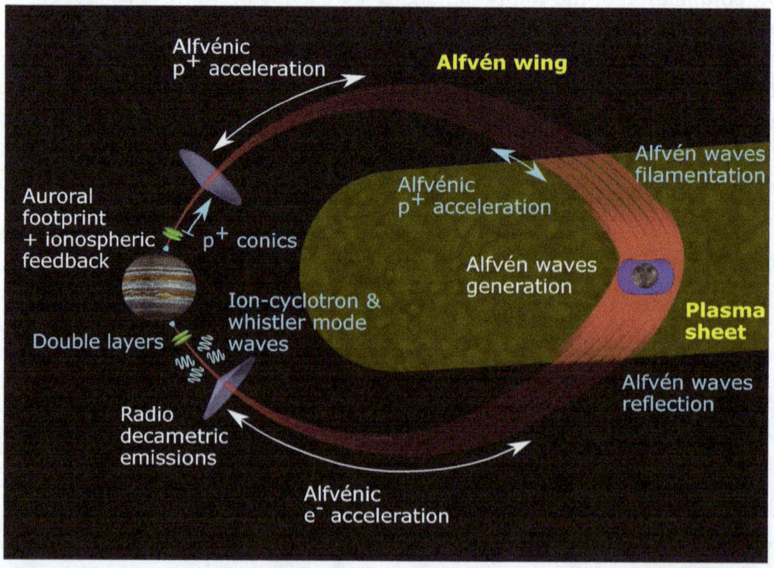

Figure 19.1 Schematic of the chain of processes taking place along the Alfvén wings (in red) generated by the interaction between the magnetosphere of Ganymede (in blue, not to scale) and that of Jupiter. The processes noted in white have already been observed at Ganymede and are the subject of the present chapter, while those noted in cyan have not been discovered yet, but are expected to exist because they have already been detected at Io.

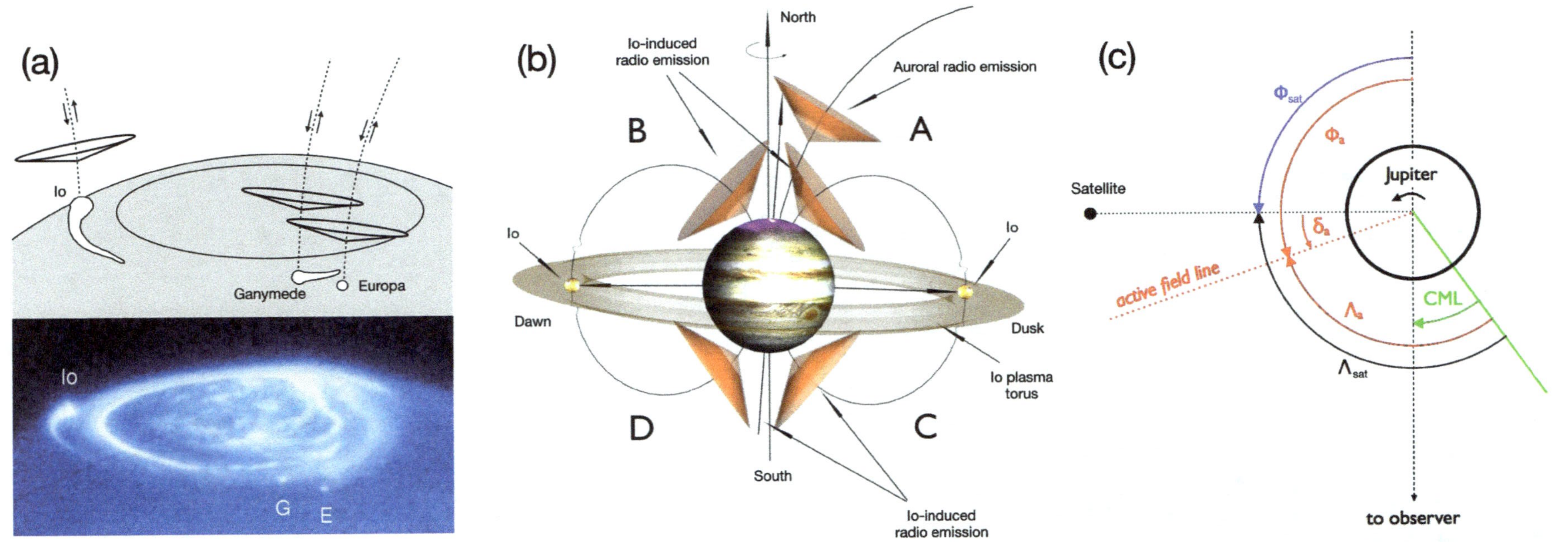

Figure 19.2 (a) Hubble Space Telescope UV image of northern Jovian auroral regions, showing the bright main auroral oval and the footprints of Io (and its tail-like, wake-induced emission), Ganymede, and Europa flux tubes. In the top part of the sketch, hollow conical beams of radio emission are displayed. They are produced above the UV hot spots by the energetic electrons precipitated along the satellite flux tubes or reflected upwards by magnetic mirroring. Similar radio emissions originate from sources distributed above the main oval. (b) Geometry and nomenclature of auroral and satellite-induced (here Io-induced) radio emissions. Radio emissions are produced along the displayed conical shells and thus are observable only when the source is near a limb of Jupiter. The magnetic field line connected to Io is sketched as the active, radio-emitting field line, but actually the active field line leads the instantaneous Io field line by several degrees (angle δ_a of panel (c)). (c) Definition of the different angles used to characterize the observing geometry. In green is the central meridian longitude (CML), which is the System III longitude of the observer. The satellite phase Φ_{sat} and its System III longitude Λ_{sat} are drawn in blue and black, respectively. Alfvén lead angle δ_a, phase, and phase Φ_a and longitude Λ_a of the so-called 'active' field line (in UV and radio) are shown in red. (a) From Clarke et al. (2002); Zarka (2007). Reprinted from *Planetary and Space Science*, 55(5). Philippe Zarka, Plasma interactions of exoplanets with their parent star and associated radio emissions,598-617 © 2007, with permission from Elsevier. (b & c) From M. S. Marques, P. Zarka, E. Echer, V. B. Ryabov, M. V. Alves, L. Denis and A. Coffre, *A&A*, 604, A17, 2017, reproduced with permission © ESO.

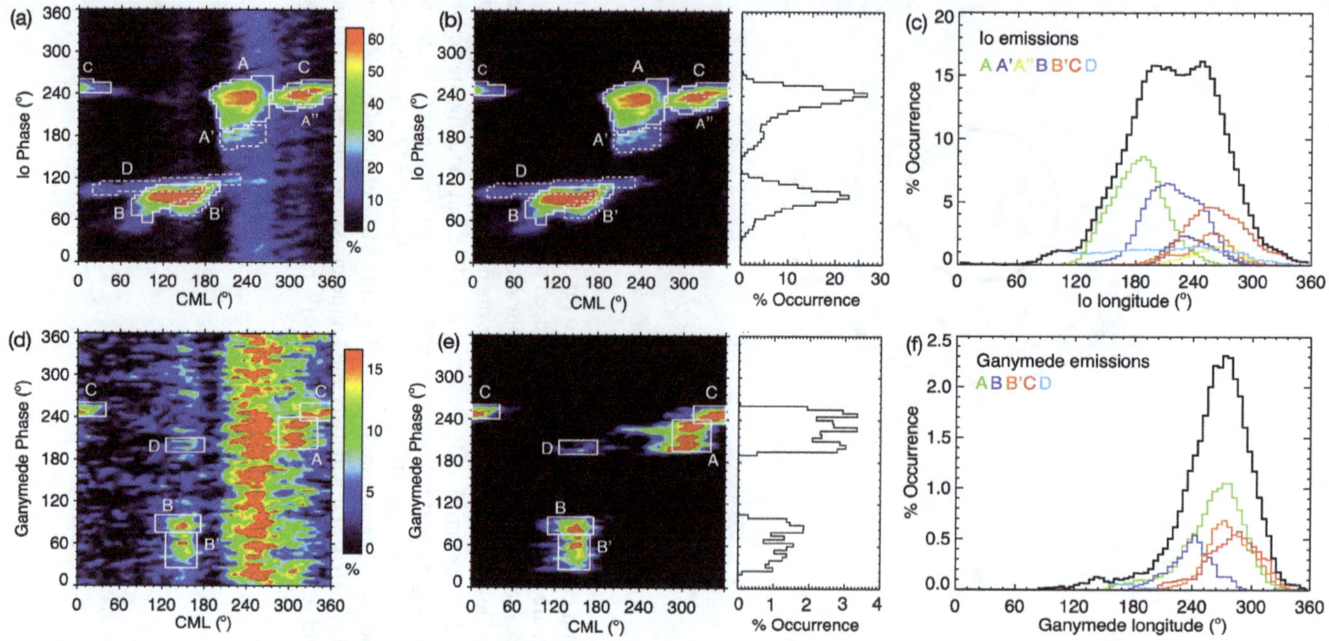

Figure 19.3 (a) Occurrence probability of Jovian radio emissions detected over 26 years (1990–2015) with the Nançay Decameter Array, displayed as 2D histograms as a function of planetary rotation (CML) and Io's orbital phase (Φ_{Io}) in 5° × 5° bins (smoothed via 1° interpolation). Regions of high occurrence labelled in white correspond to Io–Jupiter emissions (named Io-A, Io-B...). Vertical bands of emission covering restricted CML ranges at all Φ_{Io} correspond to non-Io emissions (auroral or induced by other satellites). (b) Occurrence probability of Io–Jupiter emissions only versus CML and Φ_{Io}. The profile integrated over all CML is displayed on the right-side panel. (c) Integrated occurrence probability of Io–Jupiter emissions versus Io's Jovicentric longitude ($\Lambda_{Io} = CML + 180° - \Phi_{Io}$). Components from panel (b) are identified by colours, and their sum is the black line. (d) Occurrence probability of non-Io emissions versus CML and Ganymede's orbital phase $\Phi_{Ganymede}$. Ganymede–Jupiter emissions show up within new regions of enhanced occurrence (white boxes), labelled A–D in reference to the non-Io components in which they have been identified. (e) Occurrence probability of Ganymede–Jupiter emissions only versus CML and $\Phi_{Ganymede}$. The profile integrated over all CML is displayed on the right-side panel. (f) Integrated occurrence probability of Ganymede–Jupiter emissions versus Ganymede's jovicentric longitude ($\Lambda_{Ganymede} = CML + 180° - \Phi_{Ganymede}$). Components from panel (e) are identified by colours, and their sum is the black line. From Zarka, P., Marques, M. S., Louis, C., Ryabov, V. B., Lamy, L., Echer, E. and Cecconi, B., *A&A*, 618, A84, 2018, reproduced with permission © ESO.

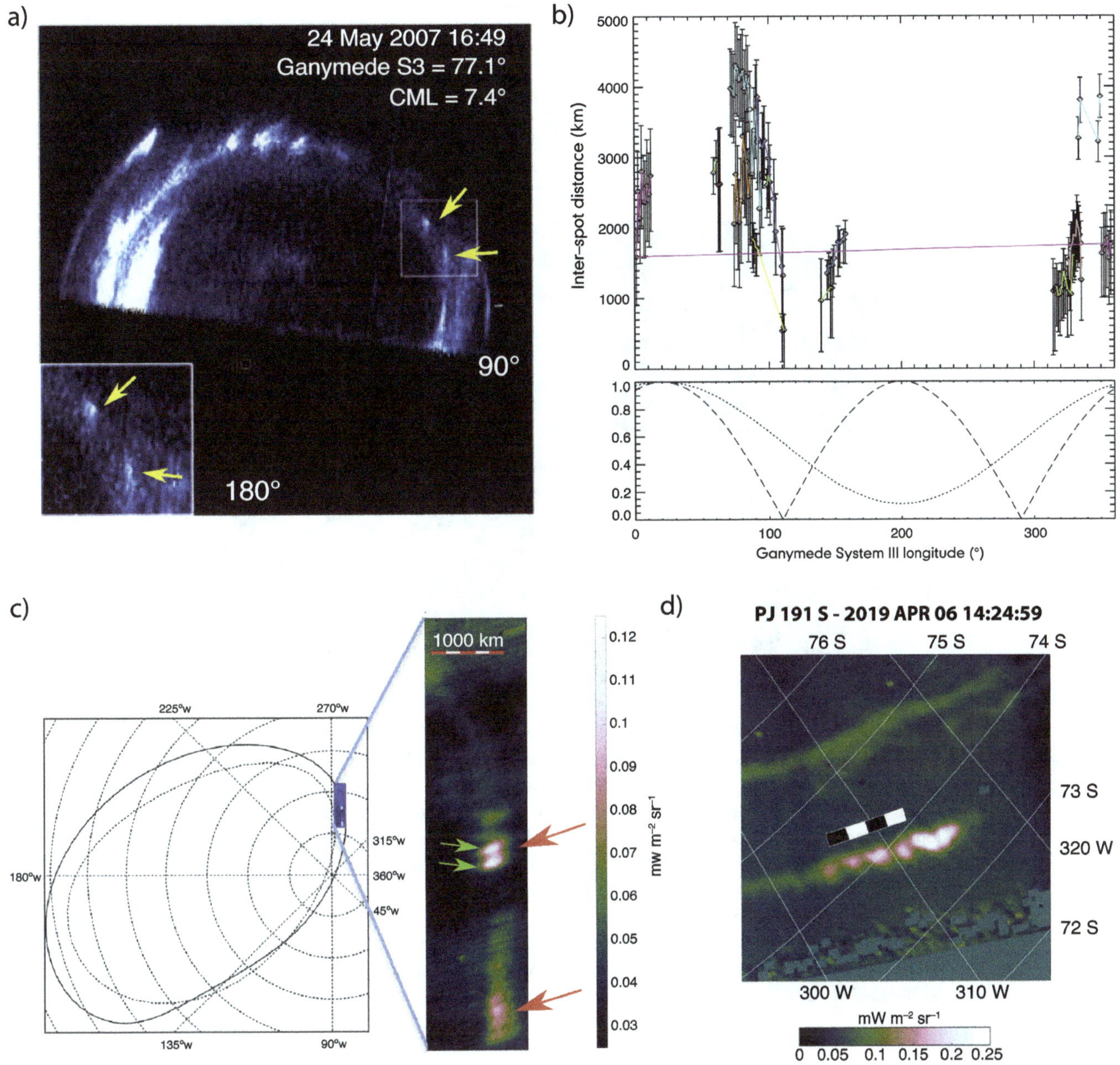

Figure 19.4 (a) Polar projections of an HST UV image acquired on 24 May 2007 at 16:40 UT. Two Ganymede footprint spots can be identified, as highlighted by the arrows. (From Bonfond et al. (2013)). (b) Inter-spot distance between the two spots of the Ganymede footprint. (top) The coloured lines connect points from the same HST orbit. The error bars assume a selection uncertainty of 1 pixel for the first spot and 2 pixels for the second one. (bottom) The long-dashed lines show the expected dependence of the distance for a trans-hemispheric electron beam spot (arbitrary units). In this case, the two spots merged as Ganymede crossed the centrifugal equator. The short-dashed line shows the expected behavior of the distance for a reflected Alfvén wing (RAW) spot. In this case, the minimum distance is expected when Ganymede is at its northernmost centrifugal latitude (∼200° System III longitude). (From Bonfond et al. (2013)). Panels a, b reprinted with permission from John Wiley and Sons. Bonfond, B., Hess, S., Bagenal, F., Gérard, J.-C., Grodent, D., Radioti, A., Gustin, J., and Clarke, J. T. The multiple spots of the Ganymede auroral footprint, *Geophys. Res. Lett.*, 40, 4977–4981 © 2013. (c) Infrared image of the Ganymede footprint spots from the JIRAM instrument on board Juno. We can see that each spot (red arrows) is actually formed from at least two sub-structures (green arrows). (From Mura et al. (2018)). AAPG © 2018. Mura, A., Adriani, A., Connerney, et al. Juno observations of spot structures and a split tail in Io-induced aurorae on Jupiter, *Science* 361(6404), 774 777. Reprinted by permission of the AAPG whose permission is required for further use. (d) Infrared images of the Ganymede footprint tail. The tail is also made of a string of sub-spots. (From Moirano et al. (2021)). Reprinted with permission from John Wiley and Sons. Moirano, A., Mura, A., Adriani, A., Dols, V., Bonfond, B., Waite, J. H., et al. (2021). Morphology of the auroral tail of Io, Europa, and Ganymede from JIRAM L-band imager. *Journal of Geophysical Research: Space Physics*, 126 © 2021.

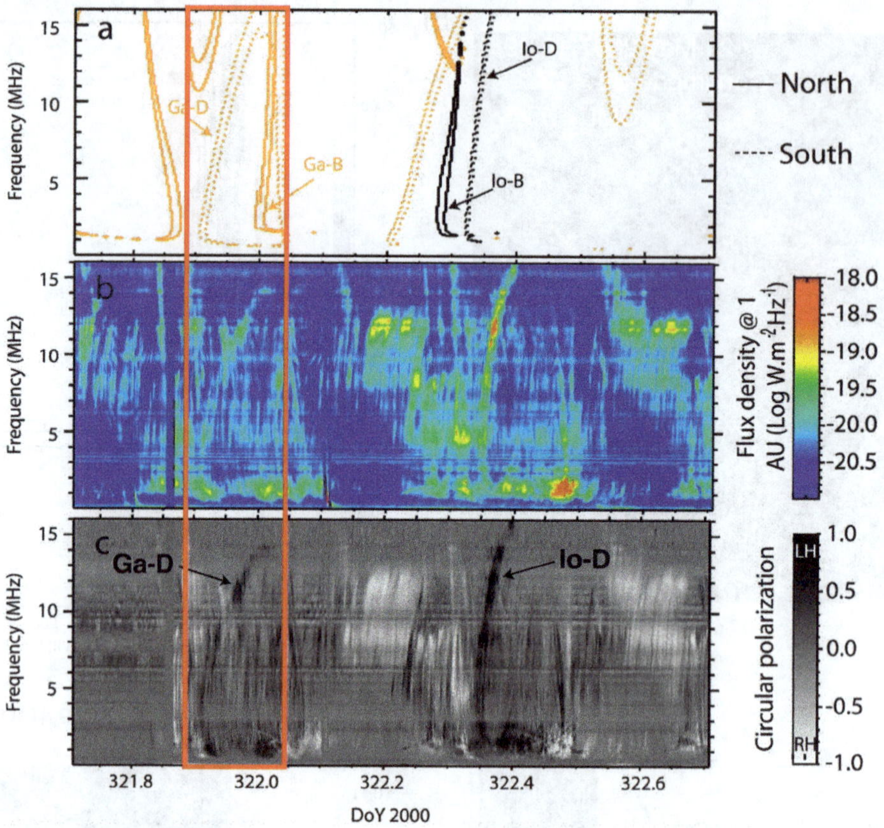

Figure 19.5 (a) ExPRES simulations of Io-induced (in black) and Ganymede-induced (in orange) radio arcs (solid line: northern emissions; dotted line, southern emissions) that should be detected by Cassini during its distant Jupiter flyby of 2000. (b) Dynamic spectrum of flux densities and (c) circular polarization measured by Cassini (LH = left-handed emission from the southern hemisphere; RH = right-handed emission from the northern one). A southern Ganymede D arc is clearly recognized. The background of nested smaller arcs is of auroral origin. From Louis et al. (2017b). Reprinted with permission from John Wiley and Sons. Louis, C. K., Lamy, L., Zarka, P., Cecconi, B., and Hess, S. L., Detection of Jupiter decametric emissions controlled by Europa and Ganymede with Voyager/PRA and Cassini/RPWS, *J. Geophys. Res. Space Physics*, 122, 9228–9247 © 2017.

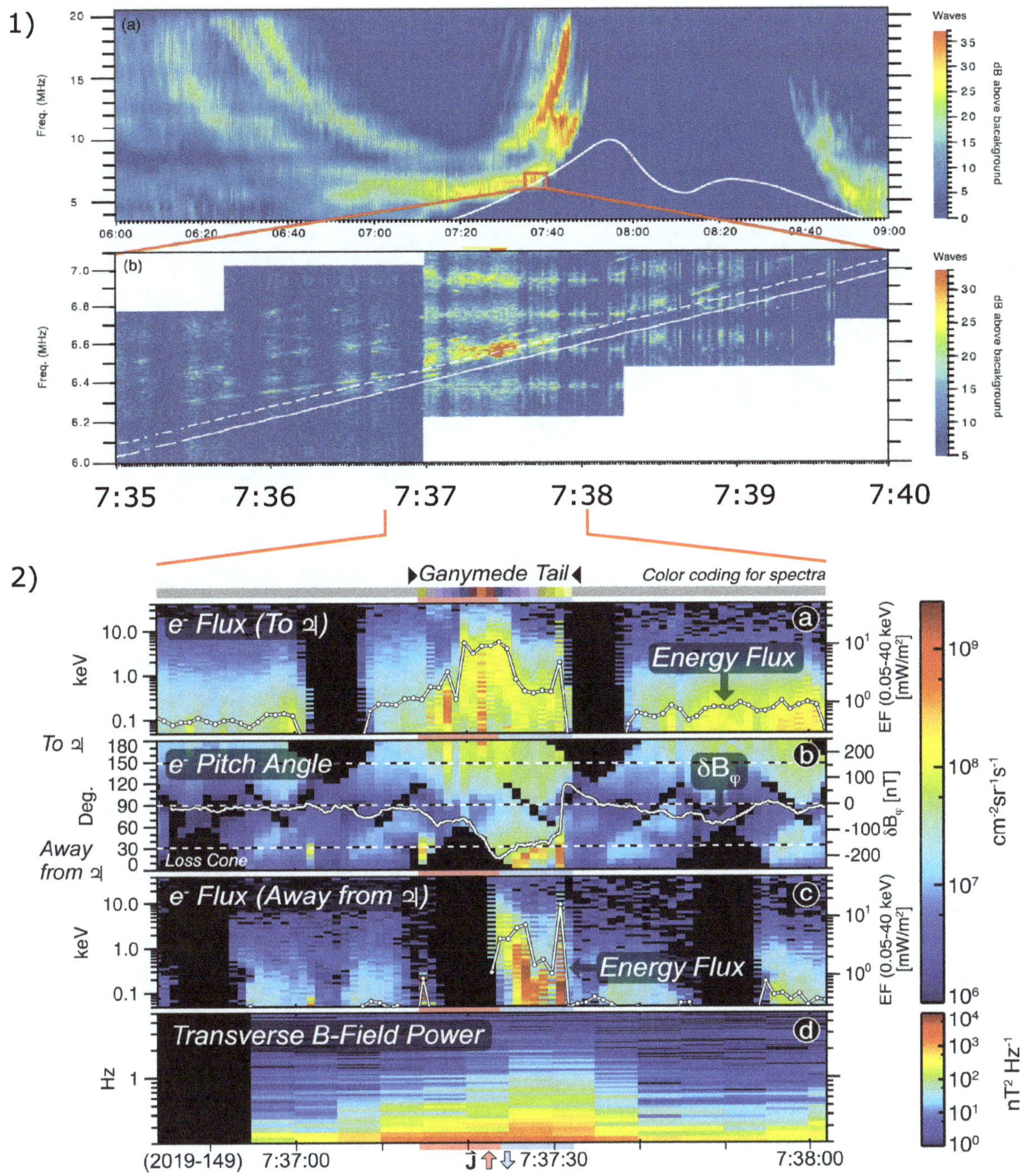

Figure 19.6 JADE, MAG and WAVES data during the Ganymede footprint tail flux tube encounter of 2019. (1a) Low-resolution Juno/Waves data and (1b) zoom at high resolution. The solid white line is the electron cyclotron frequency f_{ce} measured by Juno's magnetometer, and the dashed white line is $1.01 \times f_{ce}$ (adapted from Louis et al. (2020)). Reprinted with permission from John Wiley and Sons. Louis, C. K., Louarn, P., Allegrini, F., Kurth, W. S., & Szalay, J. R. Ganymede-induced decametric radio emission: In situ observations and measurements by Juno. *Geophysical Research Letters*, 47, © 2020. Panels (2a) and (2c) show the downward and upward electron differential energy flux (DEF) within the loss cone, respectively. Precipitating energy flux is overlaid on these panels with its separate axes on the right of the spectrograms. Panel (2b) shows electron pitch angles with δB_ϕ overlaid. Panel (2d) shows the transverse B field power spectral densities. Red/blue bars indicate approximate upward/downward current regions inferred from MAG data. (From Szalay et al. (2020a)). Reprinted with permission from John Wiley and Sons. Szalay, J. R., Allegrini, F., Bagenal, F., Bolton, S. J., Bonfond, B., Clark, G., et al. (2020). Alfvénic acceleration sustains Ganymede's footprint tail aurora. *Geophysical Research Letters*, 47, © 2020.

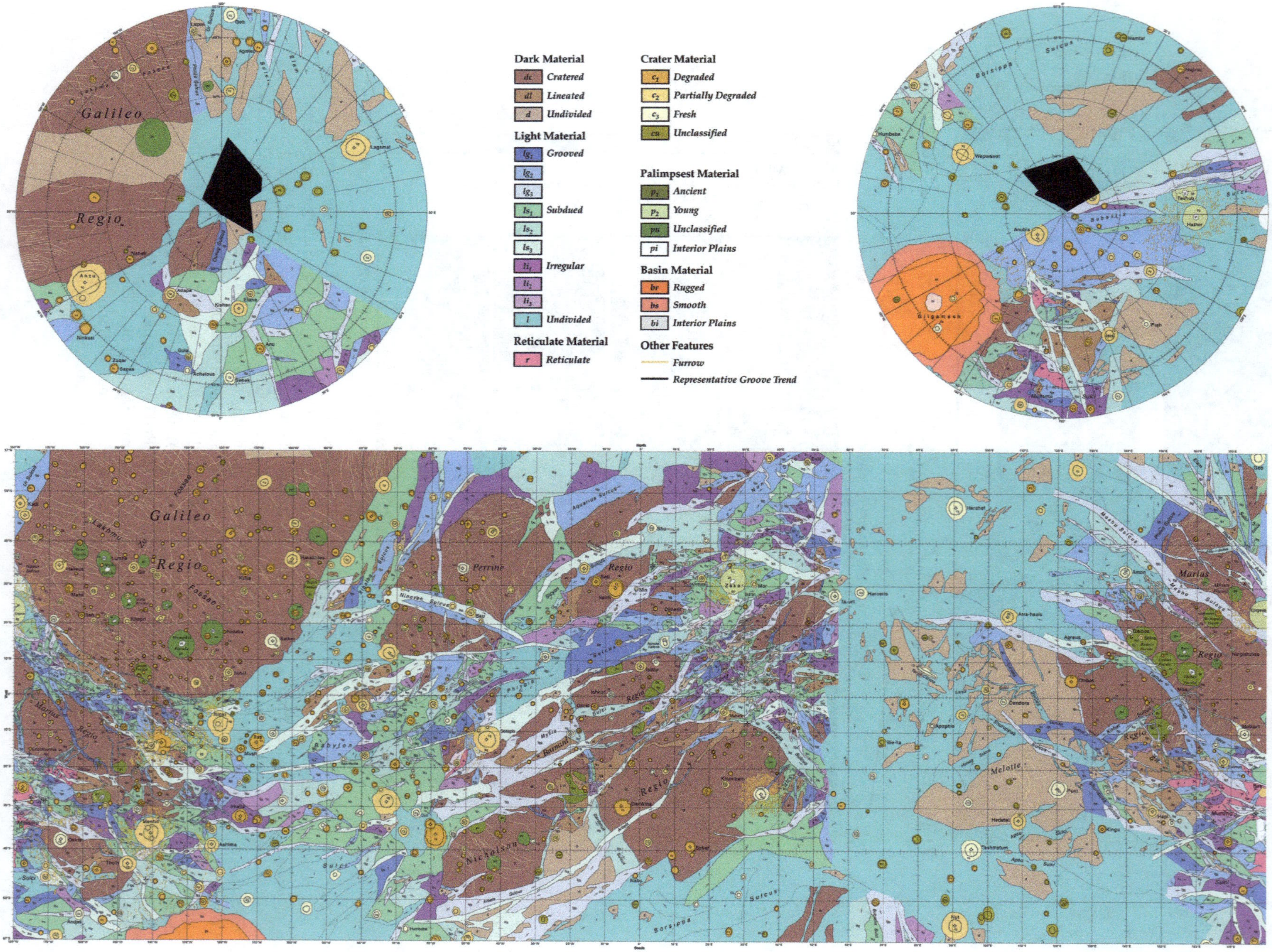

Figure A.3 Geological map of Ganymede.

15

Interactions between the Space Environment and Ganymede's Surface

ANDRÉ GALLI, AUDREY VORBURGER, PETER WURZ, MARINA GALAND, APURVA OZA, SHAHAB FATEMI, CHRISTINA PLAINAKI, AND ALESSANDRO MURA

15.1 INTRODUCTION

Because Ganymede lacks a substantial atmosphere, its surface is exposed to the harsh space environment in the Jovian magnetosphere. Generally speaking, the space environment interacts with and alters the surface of any airless icy moon via four processes:

1. Sputtering and radiolysis by particle irradiation
2. Thermal sublimation
3. Micrometeoroid impacts
4. Photo-stimulated desorption

These interaction processes are important both for the surface itself and as sources for the neutral atmosphere and the ionosphere of Ganymede. The latter two will be covered in detail in chapters 16 (atmosphere) and 17 (ionosphere). The focus of this chapter lies on the surface interaction processes themselves.

Most studies on Ganymede's surface alteration and atmosphere release processes concentrate on the first two processes. Photo-stimulated desorption and micrometeoroids are usually neglected as a source for neutral atmosphere models (Marconi, 2007; Khurana et al., 2007; Turc et al., 2014; Plainaki et al., 2015; Fatemi et al., 2016; Leblanc et al., 2017; Shematovich, 2016; Plainaki et al., 2020a,b). This approach was motivated by Marconi (2007) and Shematovich (2016), for example, who considered the widespread coverage of water-ice on the surface an indication that sublimation of water-ice and sputtering of water-ice by magnetospheric ions are the main sources for Ganymede's atmosphere. Shematovich (2016) also predicted sputtering would dominate in the polar regions, whereas sublimation should dominate in the equatorial regions near the subsolar point. We will come back to this question about relative importance of processes in the next section.

Section 15.2 of this chapter defines and describes the four interaction processes in more detail. Section 15.3 is dedicated to the most complex of the four processes, irradiation caused by charged particles. Section 15.4 discusses the role of the different interactions for some Ganymede surface features. Section 15.5 gives a compact summary of the chapter.

15.2 OVERVIEW OF SURFACE INTERACTION PROCESSES ON GANYMEDE

Before explaining the details of the four surface interactions, we attempt to compare the relative importance of irradiation (i.e., sputtering and radiolysis), sublimation, micrometeoroid impacts, and desorption on Ganymede's surface. This can be done in many different ways, depending on the specific science question at hand. Given the lack of accurate numbers and the sheer complexity of the interactions, we restrict ourselves to the case most investigated in recent surface and atmosphere: the erosion or turnover rate of water-ice on the surface of Ganymede, expressed as removed ice in μm yr^{-1} (in analogy to Cooper et al., 2001 and Johnson et al., 2004).

The results are summarized in Fig. 15.1 and in Table 15.1. Photo-stimulated desorption is negligible for water-ice removal compared to the three other processes everywhere on Ganymede's surface, whereas irradiation, sublimation, and micrometeoroid impacts may all turn out to be relevant at different specific surface areas. This can be recognized in Fig. 15.1 as follows: where T always remains below 120 K, micrometeoroid impacts and irradiation dominate. For 120 to 130 K (typically around 45° latitude during illumination) sublimation amounts to a comparable order of magnitude. Wherever T exceeds 130 K sometimes during the orbit, sublimation removes water-ice much faster than any other process. At the warmest region (150 K) about 600 μm yr^{-1} of water-ice would be removed compared to the 0.1 to 1 μm yr^{-1} removed by micrometeoroids and irradiation combined!

The total release rates related to irradiation are dominated by ion sputtering (see Section 15.3.2.1) and amount to 10^{26} to 10^{27} s^{-1} according to recent models by Carnielli et al. (2020), Poppe et al. (2018), and Plainaki et al. (2020a), with the exception of the ionospheric boost case by Carnielli et al. (2020); see Section 15.3.1). This order of magnitude translates into 0.01 to 0.1 μm yr^{-1} under the assumption that the ions precipitate onto the 50 per cent of Ganymede's surface at high latitudes, outside the closed magnetic field lines. The transition from closed to open magnetic field lines occurs at roughly 30° latitude on the leading, and roughly 45° on the trailing hemisphere as depicted in Fig. 15.1.

Figure 15.1 depicts the case that the illuminated hemisphere of Ganymede co-aligns with the leading hemisphere on the orbit around Jupiter. This is relevant because the precipitation of ions and electrons discriminates between upstream/trailing and downstream/leading hemisphere. When the trailing hemisphere is illuminated instead, the orange shape of erosion rates would flip around the vertical plot axis.

Figure 15.1 and Table 15.1 include only surface release or turnover rates. For the chemical evolution of Ganymede's surface, radiolysis and sputtering play a major role, too. Also keep in mind that the values in Table 15.1 apply to pure water-ice

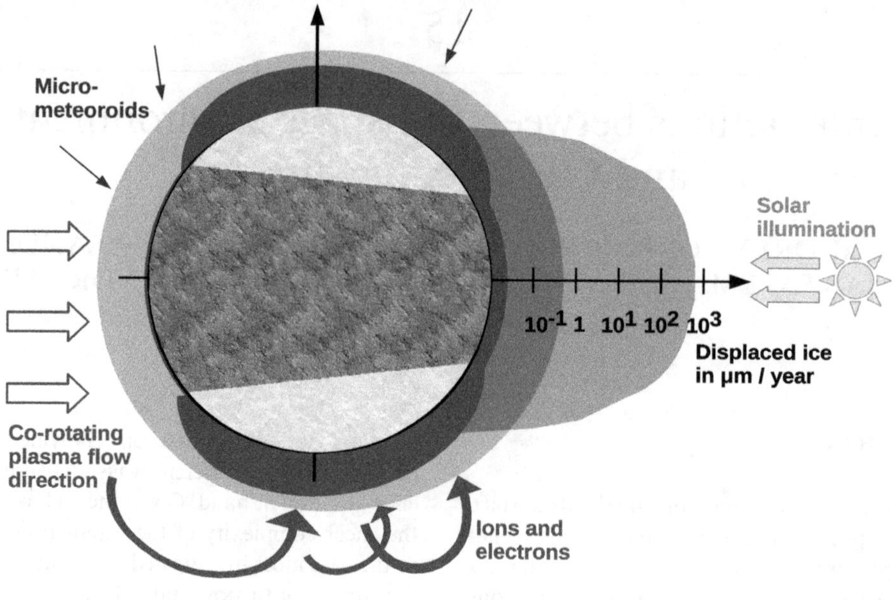

Figure 15.1 Sketch of water ice removal or turnover rates on Ganymede's surface in units of μm yr^{-1}. Orange shape: effect of thermal sublimation, blue: micrometeoroid impacts, red: irradiation-related rates (dominated by ion sputtering). Note the logarithmic scale of the rates (going from 0.1 to 1000 μm yr^{-1}). The sketch is centred on the equatorial plane of the anti-Jovian side. Ganymede's leading hemisphere with respect to Jupiter and the Jovian plasma is on the dayside to the right. The grey band and the white pole caps symbolize the surface areas of low and high albedo. (A black and white version of this figure will appear in some formats. For the colour version, please refer to the plate section.)

Table 15.1 *Water ice removal or turnover rates on Ganymede's surface in equatorial (Equ.) and polar regions in units of μm yr^{-1} (last three columns). References for ion irradiation: Carnielli et al. (2020); Poppe et al. (2018); Plainaki et al. (2020a), rates from electron irradiation are not quantifiable yet, but expected to be smaller. References for micrometeoroid impacts: Cooper et al. (2001); Johnson et al. (2004); Krüger et al. (2000). References for sublimation: Bohren and Albrecht (1998); Andreas (2007).*

Region	Temperature	Irradiation (ions)	Thermal sublimation	Meteoroid impacts
Equ., leading, dayside	125–150 K	0.025	0.1–600	0.6
Polar (lat. > 45°)	100–125 K	0.1	< 0.001–0.1	0.3
Equ., trailing, nightside	110 K	0.01	< 0.001	0.15

on Ganymede. For the other, more refractory materials present on the surface, we can assume that the erosion and turnover rate are dominated by micrometeoroids everywhere once the hydrated forms of salts and acids stable under irradiation have formed (see Section 15.3.2.2).

15.2.1 Sputtering and Radiolysis by Irradiation

Irradiation processes induced by charged particles are important for both the surface and the generation of the tenuous atmosphere of Ganymede (Marconi, 2007; Shematovich, 2016). For our purposes, we group irradiation processes into two broad categories: sputtering and radiolysis. Sputtering describes the ejection of surface species by an impacting energetic particle. The term radiolysis, on the other hand, designates the chemical alteration of surface species induced by the deposited energy of the particles. The sputtered ejecta may be the original molecules of the surface (e.g., H_2O), molecule or atomic fragments (e.g., OH, O), or radiolysis products (Teolis et al., 2017; Galli et al., 2018b). The detection of O_2 (Spencer et al., 1995) and O_3 (Noll et al., 1996) in the surface, for example, indicates radiolysis of water ice. Irradiation processes are the main topic of this chapter and are characterized in more detail in Section 15.3.

15.2.2 Thermal Sublimation

This is the surface alteration process that is easiest to quantify and understand in terms of theory and laboratory experiments. On Ganymede, thermal sublimation is most relevant at the equatorial regions, because they encounter more solar illumination and their surface is darker compared to polar regions (see Chapters 5 (Jaumann et al.) and 10 (Stephan et al.). The surface temperatures on Ganymede range between 100 K (polarregions) and 150 K (subsolar point) (Orton et al., 1996; Ligier et al.,2019) (see Chapter 7 for more details).

For a given surface temperature, we can predict the sublimation rate of water-ice, relying on work done by Bohren and Albrecht (1998) and Andreas (2007). The sublimation mass flux $S(T)$ in kg m^{-2} s^{-1} for pure water-ice of temperature T calculates to

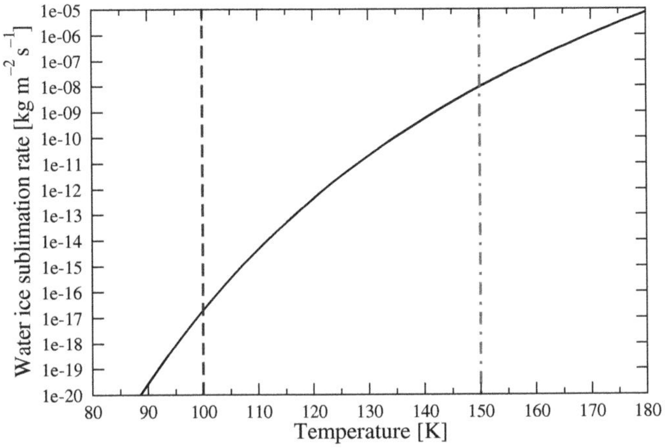

Figure 15.2 Water-ice sublimation rate (black) as a function of temperature, from the semi-empirical formula by Murphy and Koop (2005) and Andreas (2007). Note the logarithmic scale of the sublimation rate! Typical temperatures on Ganymede's surface are depicted by the blue, dashed line (night-side and polar regions, 100 K) and the red, dashed-dotted line (subsolar region, 150 K).

$$S(T) = p_{sat}(T) \left(\frac{m_W}{2\pi RT} \right)^{1/2}, \quad (15.1)$$

where R is the universal gas constant and m_w is the molecular mass of water. The saturation vapour pressure $p_{sat}(T)$ (in units of Pa) depends very strongly on temperature (in units of K) (Murphy and Koop, 2005):

$$p_{sat} = \exp\left(9.550426 - \frac{5723.265}{T} + 3.53068 \ln T - 0.00728332T\right). \quad (15.2)$$

Equation 15.2 accurately fits laboratory data over the temperature range of 110 to 273.15 K (Andreas, 2007). The corresponding sublimation rate $S(T)$ for water-ice is plotted in Fig. 15.2.

As a consequence of Equations 15.1 and 15.2, 1 cm of porous water ice with a density of 500 kg m^{-3} is sublimated within 16 billion years if said ice layer is kept at $T = 100$ K at Ganymede's polar regions (blue dashed line in Fig. 15.2). The same ice layer is removed within just 16 years if $T = 150$ K (red dashed-dotted line in Fig. 15.2)! Water-ice on Ganymede is thus very unstable at the dark, warm regions near the equator and very stable at the colder, brighter regions. This is illustrated by the pronounced erosion peak around the equator in Fig. 15.1 for water-ice. Non-volatile surface constituents (silicates, hydrated salts, etc.) are not much affected by thermal effects. Temperature-dependent sublimation at sunlit regions and subsequent cold-trapping of the released water in polar regions thus may cause the differentiation into bright and dark albedo regions on the surface (see Section 15.4.1). Because of the negative feedback between water-ice content and equilibrium surface temperature, this differentiation can also operate on smaller scales, leading to localized patches of pure water-ice persisting in equatorial regions where the surface temperatures on top of silicate-ice mixtures do not allow for permanent water-ice deposits. Indeed, surface temperature differences between two adjacent geological units can reach up to 25 K (Pappalardo et al., 2004; Stephan et al., 2020). This estimate from a geological perspective agrees with the sublimation curve in Fig. 15.2: Regions covered by pure water-ice cannot exceed surface temperatures of 128 K, otherwise these regions would lose more than 1 m of ice within less than 1 million years. Sublimation also leads to the formation of a transient H_2O atmosphere below altitudes of a few hundred kilometres at the equatorial regions around the subsolar point; see, for example, Marconi (2007), Turc et al. (2014), and Plainaki et al. (2015) and recent observations by Roth et al. (2021).

Other ices of interest for Ganymede, such as CO_2 (Mura et al., 2020) and O_2, are less stable than water-ice if they were to exist in pure form on the surface (Fray and Schmitt, 2009). On the other hand, volatile oxygen-bearing species produced by radiolysis may be trapped between water-ice molecules. In this form, CO_2, O_2, H_2O_2, HO_2, SO_2 and others may remain stable against sublimation until $T \approx 130 - 160$ K when the surrounding water-ice matrix sublimates or the water-ice transitions from amorphous ice to the cubic or hexagonal structure (Bar-Nun et al., 1985; Johnson et al., 2004; Bahr et al., 2001; Teolis et al., 2009; Laufer et al., 2017). O_2 molecules trapped at dangling H bonds inside the water-ice may outgas to the surface via thermal desorption at somewhat lower temperatures than required for full sublimation (Johnson et al., 2019). Hydrogen and noble gases cannot be permanently trapped in water-ice at any Ganymede surface temperatures (Bar-Nun et al., 1985).

15.2.3 Micrometeoroid Impacts

Micrometeoroid and meteoroid impact studies dedicated specifically to Ganymede are rare (e.g., Krüger et al., 2000; Miljkovic et al., 2012). Micrometeoroids are expected to release all surface species indiscriminately into the atmosphere by impact vaporization and ejection of dust grains. The ejection yield per impactor is very high (typically 10^4; see Krüger et al., 2000), but only a tiny fraction of the ejected grains leaves Ganymede entirely. Krüger et al. (2000) estimated, based on dust detector measurements onboard the Galileo spacecraft, that the majority of micrometeoroids hitting Ganymede's surface are interplanetary dust grains (predominantly from Jupiter-family comets according to Poppe (2016)) with a total mass flux of 30 g s^{-1} at the present. With a yield of 10^4, the authors estimated that 10^2 to 10^3 kg s^{-1} of Ganymede's surface material is released from the surface. Since most material returns to the surface eventually, the main effect of micrometeoroids on the surface is that they erode and turn over the surface layers, the so-called impact gardening. Cooper et al. (2001) and Johnson et al. (2004) estimated re-surfacing rates on Ganymede of 0.3–1.2 μm yr^{-1} ($\approx$ 0.3 μm yr^{-1} according to Krüger et al. (2000)) due to impact-gardening compared to 0.01–0.1 for ion sputtering, and 0.1 μm yr^{-1} for sublimation at $T = 125$ K (see Fig. 15.2). This implies that impact gardening is relevant compared with the previous processes (sputtering and sublimation) for specific areas on Ganymede's surface (see the summary in Section 15.5).

The average impact rate is about four times higher on the leading than on the trailing hemisphere, judging from the observed crater density (Dones et al., 2009) and the dynamical evolution model by Bottke et al. (2013). Micrometeoroid impacts therefore may have contributed to the uneven distribution of water-ice and the corresponding albedo differences of Ganymede's surface (see Section 15.4.1). One caveat is that the work by Bottke et al. (2013) addresses precipitation of dust fragments from disintegrated satellites early in the Jupiter system history rather than present-day micrometeoroid fluxes, which are dominated by interplanetary dust (Krüger et al., 2000; Poppe, 2016). The trend of micrometeoroids preferentially impacting the leading rather than the trailing hemisphere should hold for both cases, but the precipitation from disintegrated satellites may have resulted in an early mass influx rate to Ganymede of 10^5 to 10^6 g s^{-1} during the first 40 million years after the capture of irregular satellites (Bottke et al., 2013) compared with today's 30 g s^{-1}.

15.2.4 Photo-Stimulated Desorption

Neglecting photo-stimulated desorption or 'photosputtering' for Ganymede surface and atmosphere models is often motivated by references to Westley et al. (1995) and Johnson (1990). However, a more detailed examination by Johnson et al. (2004) for individual icy satellites indicated that the energy deposited by UV photons with sufficient energy to dissociate water-ice (8×10^8 keV cm^{-2} s^{-1} with $E > 6$ eV) exceeds the average energy flux deposited by charged particles (2×10^8 keV cm^{-2} s^{-1} ions and electrons with $E > 10$ keV (Cooper et al., 2001)) for Ganymede's equatorial regions. For the polar regions outside the closed magnetic field lines, the UV energy deposition is an order of magnitude lower than the energy flux deposited by charged particles. Neglecting photo-stimulated desorption for Ganymede surface alteration and release processes can be motivated nevertheless, but this has to include consideration of the yields for the various competing processes. Considering Lyman-alpha photons, the main constituent of the solar UV energy flux, the yield of released water molecules per photon is 0.005 to 0.01 for relevant water-ice temperatures of 80 and 120 K, respectively (Westley et al., 1995). This yield is several orders of magnitude lower than the sputter yield for ions from the Jovian magnetosphere and Ganymede's ionosphere (see Section 15.3.2.1).

15.3 IRRADIATION PROCESSES

The space environment of Ganymede is filled with three populations of charged particles: (1) The thermal plasma corotating with the magnetic field lines of Jupiter (Kivelson et al., 2004; Poppe et al., 2018), (2) energetic particles of the Jovian magnetosphere with tens of keV to MeV energy consisting of H^+, O^{2+}, and S^{3+} ions and electrons (Mauk et al., 2004; Allioux et al., 2013; Poppe et al., 2018; Plainaki et al., 2020a; Carnielli et al., 2020; Liuzzo et al., 2020 and chapter 14), and (3) the ions and electrons of Ganymede's own ionosphere (mostly O_2^+ for the ions; see Carnielli et al. (2020) and Chapter 17). The impacting particles penetrate Ganymede's surface at depths from hundreds of nanometres (slow ions) to metres (MeV electrons including their bremsstrahlung photon products) (Paranicas et al., 2002). In this process, particles can directly eject surface molecules (sputtering) and also trigger chemical reactions in the ice (radiolysis). In this section, we first characterize the particle precipitation for these surface processes (Section 15.3.1) and then proceed to the description of the irradiation processes taking place on the icy and non-icy surfaces of Ganymede (Sections 15.3.2.1 and 15.3.2.2).

15.3.1 Particle Precipitation on the Surface

The interactions between the Jovian plasma, Ganymede's intrinsic magnetic field, and the ionosphere make it very challenging to quantify the different particle populations and species bombarding Ganymede's surface. To compare the relative importance of particle populations, we must discern between the particle fluxes (where thermal plasma may dominate because of the higher number and density) and the deposited energy or sputtering rate or reaction rate per area, for which the energy of the particles and further energy-dependent parameters like the sputtering yield (see Section 15.3.2) must be taken into account. Consideration of particle precipitation at Ganymede must differentiate between the polar regions coinciding with open magnetic field lines versus the equatorial latitudes inside closed magnetic field lines and also between the leading versus the trailing hemisphere relative to the corotating Jovian plasma (see chapter 14).

The in situ measurements of the Galileo spacecraft in the 1990s allowed researchers to quantify the Jovian plasma environment for the first time. Based on these measurements, Cooper et al. (2001) and Mauk et al. (2004) compiled energy spectra of the main ions (H^+, O^{n+}, and S^{n+}) and electrons and estimated energy fluxes on the surfaces of Europa, Ganymede, and Callisto (see Fig. 15.3 for Ganymede). Cooper et al. (2001) predicted for Ganymede that electrons were the dominant contributor in terms of deposited energy in the polar regions, whereas energetic S^{3+} and O^{2+} dominated the energy flux in equatorial regions.

Based on recent modelling efforts (Poppe et al., 2018; Plainaki et al., 2020a; Carnielli et al., 2020), the ion precipitation on Ganymede's surface can be summarized as follows Energetic S^{3+} and O^{2+} from the Jovian magnetosphere are the

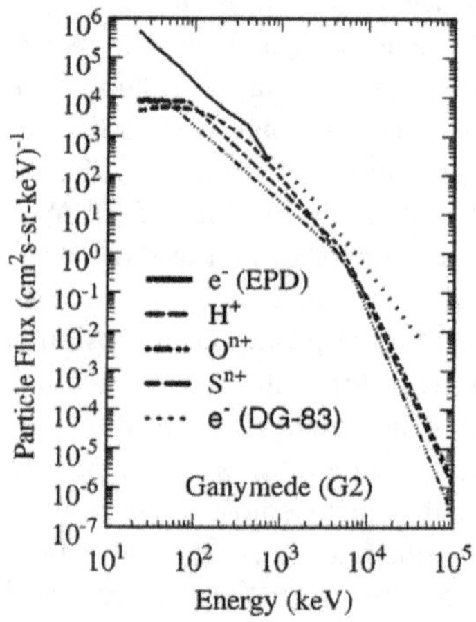

Figure 15.3 Energy spectra of the major energetic particle species in Jupiter's magnetosphere upstream of Ganymede, derived from Galileo measurements during the G2 flyby. Figure from Cooper et al. (2001). Reprinted from *Icarus*, 149.John F. Cooper, Robert E. Johnson, Barry H. Mauk, Henry B. Garrett, Neil Gehrels, Energetic Ion and Electron Irradiation of the Icy Galilean Satellites, 133–159, ©2001, with permission from Elsevier.

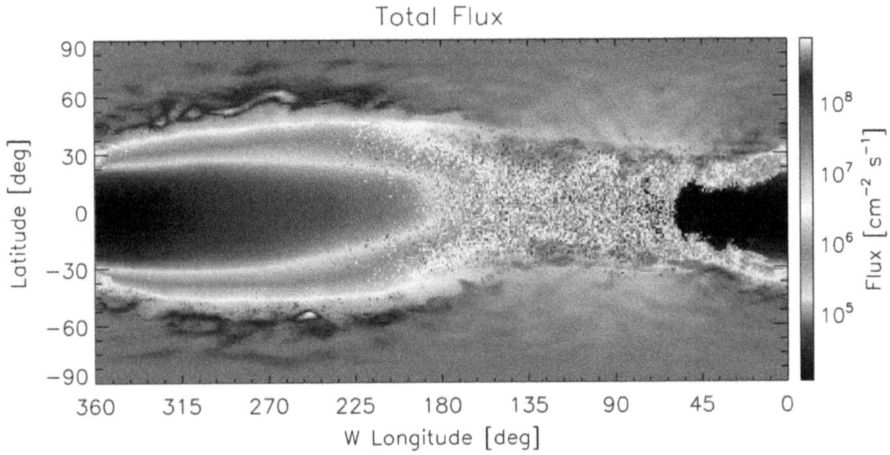

Figure 15.4 Total ion precipitation flux onto Ganymede's surface, summed over all ion species and energies, representative of G8 configurations when Ganymede was inside the Jovian plasma sheet. Leading hemisphere is centred at 90° W. Figure from Poppe et al. (2018). (A black and white version of this figure will appear in some formats. For the colour version, please refer to the plate section.)

dominant contributors, among all ion populations, in terms of deposited energy and triggered irradiation processes for most surface regions. The precipitating fluxes and sputtering rates of all energetic particle species decrease from polar regions to equatorial leading hemisphere to equatorial trailing hemisphere. This is illustrated by the map of combined ion precipitation flux in Fig. 15.4. The main reason for the polar equatorial dichotomy is Ganymede's dipole magnetic field, separating the surface into regions linked to open (polar regions) and regions linked to closed magnetic field lines (equatorial regions). The relative enhancement of precipitation on the leading versus the trailing equatorial regions is caused by Jovian plasma that overtakes Ganymede and then is dragged into and heated up in the magnetic field reconnection region downstream from the satellite, namely above Ganymede's leading side (Poppe et al., 2018).

The extent of these spatial differences in precipitation fluxes and sputtering rates seems to be model-dependent to some extent and may also depend on the importance of the ionosphere (Poppe et al., 2018; Plainaki et al., 2020a; Carnielli et al., 2020). Carnielli et al. (2020), for example, predict (see Fig. 15.5) S^{3+} to dominate among Jovian particle precipitation and the water-ice sputtering rate to be rather uniformly distributed across Ganymede's surface (for G2 flyby conditions). By comparison, Poppe et al. (2018) predict that neutral sputtering is reduced by factors of 2.5 and 10 in the leading and trailing hemispheres, respectively, with respect to the polar regions. The dynamics of the entry and circulation of the Jovian energetic ions inside Ganymede's magnetosphere, the formation and extent of the ionosphere, and the morphology of the ion precipitation onto the surface determine the variability of the water release. Plainaki et al. (2020a) have studied this variability for three specific Galileo flybys.

Energetic ions can be trapped inside Ganymede's magnetic field lines (see book Chapter 14). Contrary to the Jovian energetic particles, the energy-weighted precipitation pattern of these ionospheric ions peaks in the equatorial regions of the leading hemisphere. The relative importance of ionospheric ion precipitation relative to Jovian ions is hard to quantify because of the ill-constrained atmospheric densities. The contribution of ionospheric ions ranges between 10 and 70 per cent of the total water-ice release rate by all ion species (Carnielli et al., 2020).

Jupiter's electron environment was recently modelled by de Soria-Santacruz et al. (2016) and Jun et al. (2019) and the resulting precipitation of energetic Jovian electrons onto Ganymede's surface has been studied by Liuzzo et al. (2020). The precipitation pattern of energetic electrons (see Fig. 15.6) follows basically the one seen for energetic ions (see Fig. 15.4), with one qualitative difference: The minimum of energetic electron precipitation in equatorial regions (within closed field line regions) occurs near the anti-Jovian point (180° W longitude), in contrast to the sub-Jovian point at 0° for the ion precipitation minimum. This difference between positively and negatively charged particles is due to the particles' opposite drift direction within Ganymede's internal field (Liuzzo et al., 2020).

Open-closed field lines boundaries shield electrons below 40 MeV from the equatorial regions; above 100 MeV, electrons can hit Ganymede's surface anywhere. These equatorial electron fluxes are enhanced in the leading, sub-Jovian hemisphere (Liuzzo et al., 2020). Discussing relative contributions of particle precipitation in the polar regions, Liuzzo et al. (2020) find that the contribution to the number flux (and to the energy flux) of charged particles from energetic electrons exceeds the ion contribution by an order of magnitude and that the high-latitude regions are strongly irradiated to a depth of roughly 10 cm.

Low-energy electrons with $E \leq$ few keV from the thermal plasma have penetration depths of less than a micrometre (Johnson, 1990; Hand and Carlson, 2011), and their relevance to Ganymede surface processes has so far hardly been studied. Frank et al. (1997) derived, based on Galileo flyby measurements, that electron energy fluxes between 70 eV and 4.5 keV into the polar regions amounted to roughly 10^9 keV cm^{-2} s^{-1} at the time of flyby. This is five times less than the total energy deposited by energetic ions and electrons (Cooper et al., 2001). Neglecting low-energy (below roughly 10 keV) electron precipitation thus seems to be acceptable to first-order approximation, but future plasma models should cover this aspect, too, and also assess the importance of ionospheric electrons.

Now that we have set the scene with the plasma and energetic particle precipitation maps, we will discuss the reactions triggered by this precipitation.

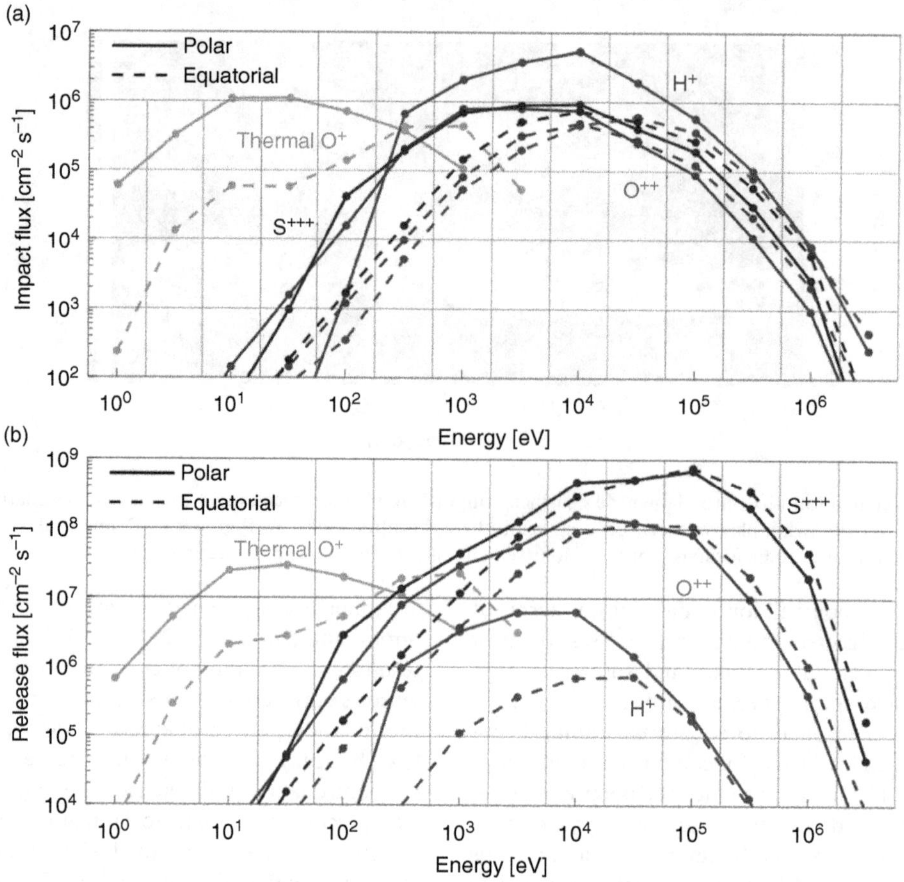

Figure 15.5 Impact (a) and release (b) fluxes as a function of impact energy for different Jovian ion species: O^+ from the thermal population (gold), energetic H^+ (red), energetic O^{2+} (green) and energetic S^{3+} (blue). The solid curves show values in the region of open magnetic field lines (north and south polar regions combined), while the dashed lines show values in the region of closed magnetic field lines. Figure adapted from Carnielli et al. (2020). Reprinted from *Icarus*, 351. G. Carnielli, M. Galand, F. Leblanc, R. Modolo, A. Beth, X. Jia, Simulations of ion sputtering at Ganymede ©2020, with permission from Elsevier.

Table 15.2 *Relative chemical abundances presumed for polar and equatorial surface regions, based on Ligier et al. (2019).*

	Water ice	Dark material	Sulphuric acid hydrate	Sulphates	Chlorinated salts
Equatorial	0.15–0.24	0.55–0.70	0.04–0.10	0.03–0.08	0.03–0.08
Polar	0.40–0.48	0.30	0.16	0.03	0.03–0.11

15.3.2 Description of Irradiation Processes

Beside regions rich in water ice, Ganymede's surface also shows areas covered with hydrated sulphuric acid ($H_2SO_4 * nH_2O$), salts, and an unidentified darkening agent hypothesized to contain hydrated silicates (Ligier et al., 2019). A simplified table of chemical abundances based on the near-infrared imaging spectrometer observations by Ligier et al. (2019) is shown in Table 15.2 (also refer to Chapters 10 and 11. We therefore will describe the irradiation processes known from theory and laboratory experiments, distinguishing between irradiation of water-ice (or other condensed gases) and non-icy target material of interest for Ganymede (hydrated sulphuric acid, salts, and potentially hydrated silicates).

15.3.2.1 Irradiation of Water-Ice

Irradiation of water-ice by charged particles is an important source process for the chemical and physical alteration of the surface and for the generation of Ganymede's atmosphere (see details in Chapter 16). Bearing in mind the wide variety of particles precipitating onto Ganymede, we attempt a general distinction before discussing the details for the different types of irradiation processes: energetic electrons dominate radiolysis rates of water molecules below the uppermost ice layers (depths of micrometres to centimetres), whereas energetic ions dominate the total mass loss rate from the icy surface. Finally, thermal plasma may be relevant because the implanted ions provide additional building blocks for chemical reactions in the ice.

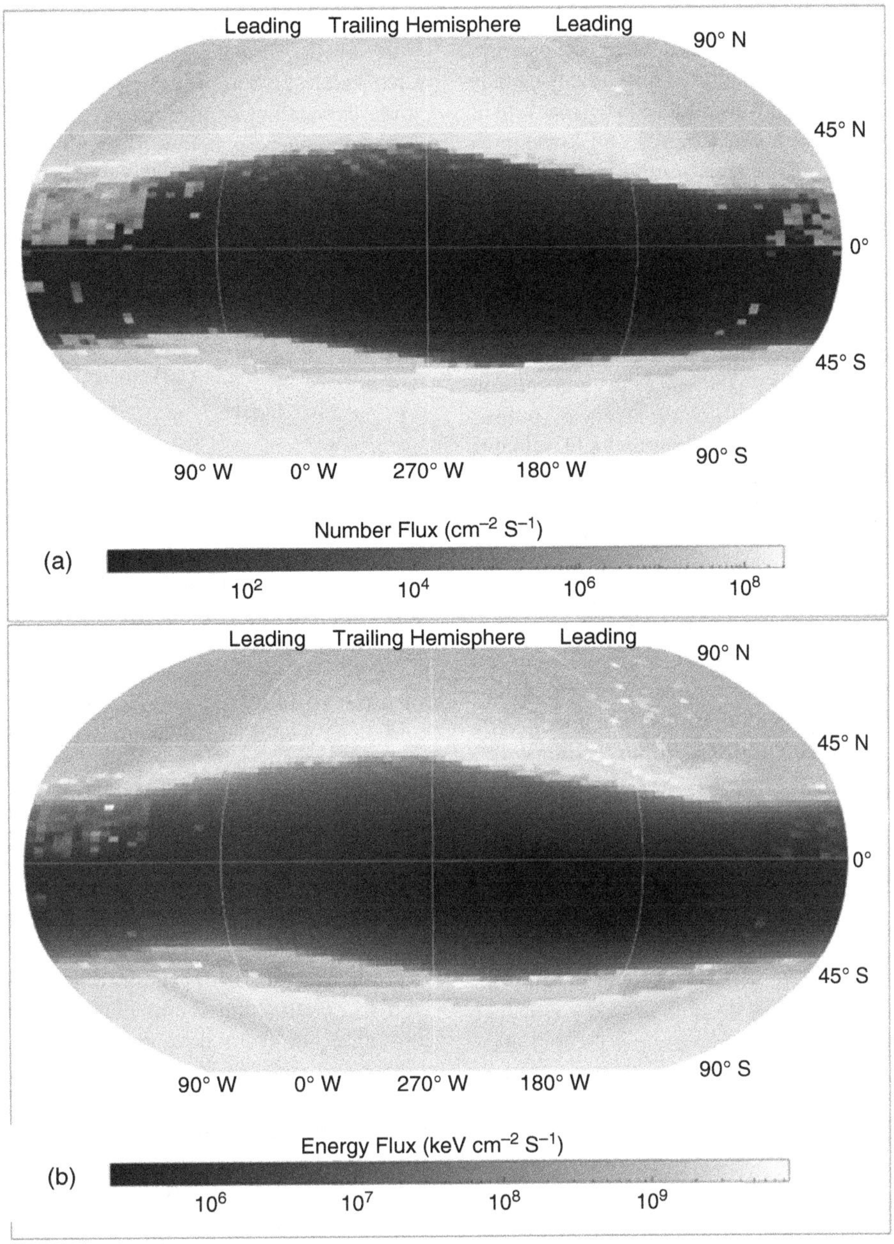

Figure 15.6 Energetic electron (a) number and (b) energy flux deposited onto Ganymede's surface averaged over a full Jovian synodic period. Figure taken from Liuzzo, L., Poppe, A. R., Paranicas, C., Nénon, Q., Fatemi, S., & Simon, S. (2020). Variability in the energetic electron bombardment of Ganymede. *Journal of Geophysical Research: Space Physics*, 125, John Wiley and Sons.

Electron Irradiation:

At ice temperatures relevant to Ganymede, most water molecules that are radiolyzed leave the ice via the reaction $H_2O \rightarrow H_2 + \frac{1}{2} O_2$ once irradiation has reached saturation levels (Petrik et al., 2006; Abdulgalil et al., 2017; Galli et al., 2018b). At much colder temperatures ($\simeq$10 K instead of $\simeq$100 K), radiolysis would produce and release more H_2O_2 than O_2 (Zheng et al., 2006). Little H_2O, compared with H_2 and O_2, is released by electron irradiation at temperatures relevant for Ganymede according to Abdulgalil et al. (2017), Teolis et al. (2017), and Galli et al. (2018b), whereas the ratio of released H_2O/O_2 reaches unity at such temperatures according to Davis et al. (2021). Some O_2 on the order of 1 per cent with respect to H_2O (Grieves and Orlando, 2005) and H_2O_2 (0.04 and 0.006 per cent by number relative to water at 80 K and 120 K, respectively (Hand and Carlson, 2011)) can remain trapped in irradiated water-ice, provided the temperature remains below 150 K (Zheng et al., 2006). Hydrogen is too volatile to be retained in the ice at Ganymede surface temperatures (Bar-Nun et al., 1985). Once O_2 builds up in the ice, electron irradiation will also form O_3 from O_2 (Sivaraman et al., 2007) but so far, to our knowledge, no O_3 was directly produced from irradiating water-ice with electrons in laboratory.

Electron irradiation can also amorphize crystalline water-ice (Baragiola et al., 2003; Loeffler et al., 2020), reduce microporosity (Behr et al., 2020), and lead to sintering (Howett et al., 2011, 2012; Ferrari and Lucas, 2016; Nordheim et al., 2017; Schaible et al., 2017). The first effect has been conjectured to be responsible for the observed amorphous water ice in the polar regions (Hansen and McCord, 2004; Liuzzo et al., 2020).

The total mass loss rate due to radiolysis and subsequent release of H_2, O_2 and so on. is sometimes described as sputter yield $Y(E)$ due to electron irradiation. This value is usually determined by measuring the electron-induced mass loss rate of ice deposited on a microbalance. For thermal and low-energy electrons, theory and experiments agree (Orlando and Sieger, 2003; Teolis et al., 2017; Galli et al., 2018b; Meier and Loeffler, 2020) that this yield increases with energy to reach a maximum or plateau around 1 keV. For higher energies, two conflicting interpretations of experiments exist: whereas consideration of electron stopping power and microbalance measurements with compact ice films at temperatures of 60 K indicate that $Y(E)$ decreases with energy above 1 keV (the solid curve in the top panel of Fig. 15.8), experiments with thick layers of porous water ice at 100 K indicated that the sputter yield remains roughly constant from 1 to 10 keV electron energy (Galli et al., 2018b). Moreover, the absolute sputter yields predicted or measured by Teolis et al. (2017), Galli et al. (2018b), Meier and Loeffler (2020) only agree within one order of magnitude. These open questions about the yield are maybe not crucial for our understanding of Ganymede's surface: The different studies do not disagree regarding the radiolysis rates triggered inside the irradiated ice, and all atmospheric source rates (with possible exceptions for H_2 and O_2) are dominated by energetic ions with their sputter yields $Y \approx 100$ to 1000 (Vorburger and Wurz, 2018).

The penetration depth of electrons in water-ice increases with energy in a way that can be approximated by a power law (Johnson, 1990; Hand and Carlson, 2011)

$$d \approx R_0 E^{\alpha}, \tag{15.3}$$

with E the electron energy in units of keV, $\alpha = 1.76$, and the depth $R_0 = 46$ nm for targets with density $\rho = 1$ g cm^{-3} at 1 keV. This implies a penetration depth of 1 cm for a 1 MeV electron, but such energetic electrons can cause radiolysis down to depths of $\sim$ 1 m because of the bremsstrahlung photons produced inside the ice (Paranicas et al., 2002).

Ion Irradiation:

Impacting ions directly eject H_2O molecules because of the larger energy deposition per unit path length of ions compared to electrons. But ions can initiate radiolysis, too, leading to chemical alteration of the ice and ejection of the newly formed species. Irradiation of water ice with H, O, and S ions can therefore result in the formation and/or ejection of H_2O, H_2, O_2, H_2O_2, and O_3 but the exact ratios in the irradiated ice and the ejecta are difficult to quantify and depend on many different parameters including at least ion energy, ion species, and ice temperatures (Haring et al., 1984; Cooper et al., 2005; Boduch et al., 2016).

Most results of ice sputtering experiments before 2010 were collected by Johnson and Liu (2010) in an online database. The impacting species include H, noble gases up to Xe, C, N, O, and F, with energies ranging from roughly 1 keV to 25 MeV. More recently, Muntean et al. (2016) and Galli et al. (2017) showed in laboratory experiments that singly and doubly charged ions produced the same sputtering yield on water-ice. The so-called potential sputtering by multiply charged ions (Aumayr and Winter, 2004) thus seems to be irrelevant for Ganymede's surface; in addition, the charge state of ions usually does not exceed 3 (see the previous section). On the other hand, Galli et al. (2017) found that O_2^+ molecular ions show a factor-of-2 higher sputter yield than expected from O^+ ions of the same total kinetic energy. Finally, Galli et al. (2018a) measured sulphur sputtering yields on water-ice for the first time under laboratory conditions.

For ion energies below 1 keV, the sputtering yield of ions in water ice can be described by a cascade of elastic collisions, whereas at higher energies, the so-called electronic sputtering dominates. Famá et al. (2008) derived a semi-empirical formula for the sputtering yield for the sum of both contributions, based on laboratory experiments with thin water-ice films:

$$\begin{aligned} Y(E, m_1, Z_1, \theta, T) = & \frac{1}{U_0}\left(\frac{3}{4\pi^2 C_0}\alpha S_n + \eta S_e^2\right) \\ & \times \left(1 + \frac{Y_1}{Y_0}\exp\left(-\frac{E_a}{kT}\right)\right)\cos^{-f}(\theta). \end{aligned} \tag{15.4}$$

Equation (15.4) quantifies the sputtering yield as a sum of elastic and electronic sputtering, described by the nuclear stopping power S_n and the electronic stopping power S_e. The sputtering yield depends on energy E, mass of impactor m_1, atomic number of impactor Z_1, the incidence angle θ from the surface normal, and temperature T. For U_0, the sublimation energy of water (0.45 eV) is assumed, $C_0 = 1.3$ Å^2, $E_a = 0.06$ eV, and $Y_1/Y_0 = 220$ are constants. The temperature-independent fraction in Equation (15.4) is due to the direct ejection of H_2O molecules. The temperature-dependent term with the activation energy E_a becomes dominant above $T = 120$ K and is due to the release of H_2 and O_2 (Johnson et al., 2004; Famá et al., 2008; Teolis et al., 2009). Water radicals inside the irradiated ice react to H_2 and O_2 (Cassidy et al., 2010; Galli et al., 2018b) and are then released by ion sputtering (O_2) or diffusion (H_2).

Cassidy et al. (2013) examined the data compiled by Johnson and Liu (2010) and found that the semi-empirical sputtering Equation (15.4) fits data well for energies below 100 keV. At higher energies, the formula by Johnson et al. (2009) for electronic sputtering is more accurate. This is illustrated in Fig. 15.7 for the three most common ion species H^+, O^+, and S^+ including some of the data from laboratory experiments. The yield curves peak in the energy range between 10 keV and 10 MeV. This fact, together with the yields (of the order of 10–1000 for 1 keV – 1 MeV O and S ions that form the bulk of energetic ions precipitating onto Ganymede, Fig. 15.5), explains why sputtering by energetic O and S dominates the surface erosion rate among all irradiation processes.

Experiments under laboratory conditions also showed that most of the removed water-ice is ejected as water molecules, but a fraction of the water-ice is radiolyzed by the ions into O_2, H_2, H_2O_2, and O_3 (Bahr et al., 2001; Teolis et al., 2006, 2009). The O_2/H_2O ratio of the released particles is not a fixed fraction of the total sputtering yield; it depends on both the energy of the impactor and the temperature of the ice. Below 100 K, the O_2 production upon Ar irradiation was found to be negligible, whereas some H_2O_2 was produced. At ice temperatures above 100 K, H_2O_2 production was suppressed whereas the O_2/H_2O ratio increased with temperature (Teolis et al., 2009). The relative contribution of O_2 to the total mass released from the ice was found to be highest (roughly 20 per cent) for ion energies around 10 keV (Teolis et al., 2017). The yields of H_2O and O_2 from water ice regolith, in the absence of thermal effect, are shown in Fig. 15.8 for all relevant impactors.

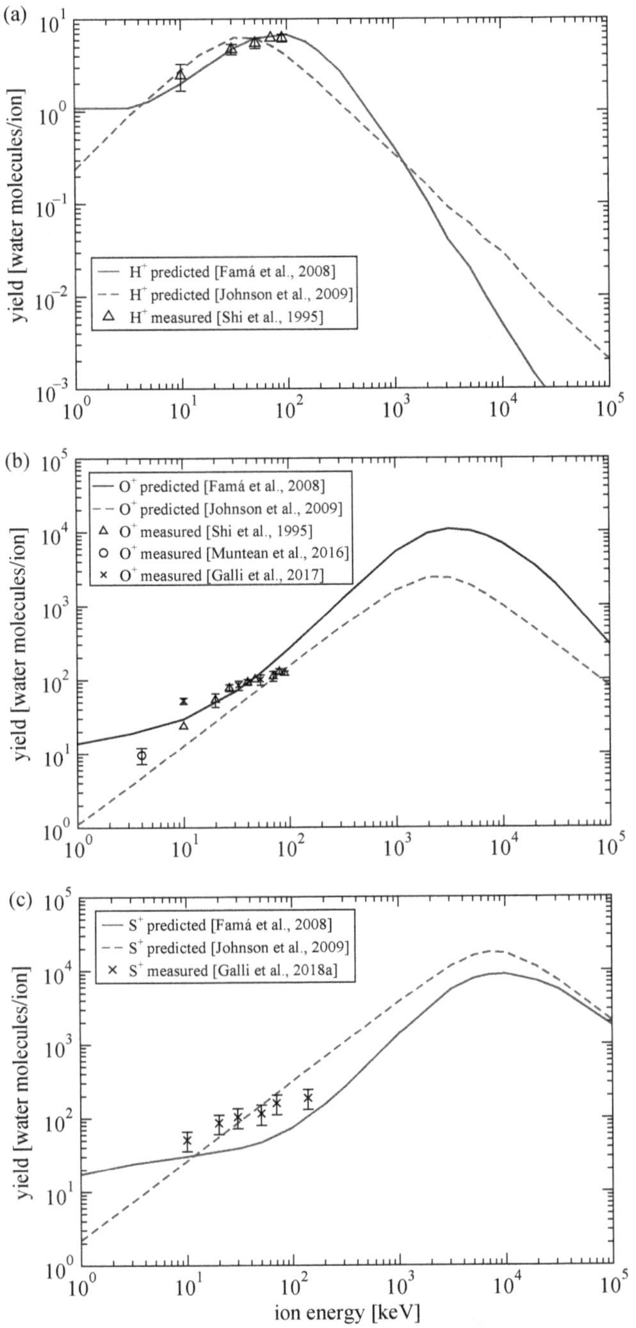

Figure 15.7 Ion sputtering yields for an ice temperature of 100 K and an incidence angle of 45° including semi-empirical formulae by Famá et al. (2008) (blue solid curves) and by Johnson et al. (2009) (red dashed curves) and some experiment results (Shi et al. (1995); Muntean et al. (2016); Galli et al. (2017, 2018b), black symbols) for H^+ (a), O^+ (b), and S^+ (c).

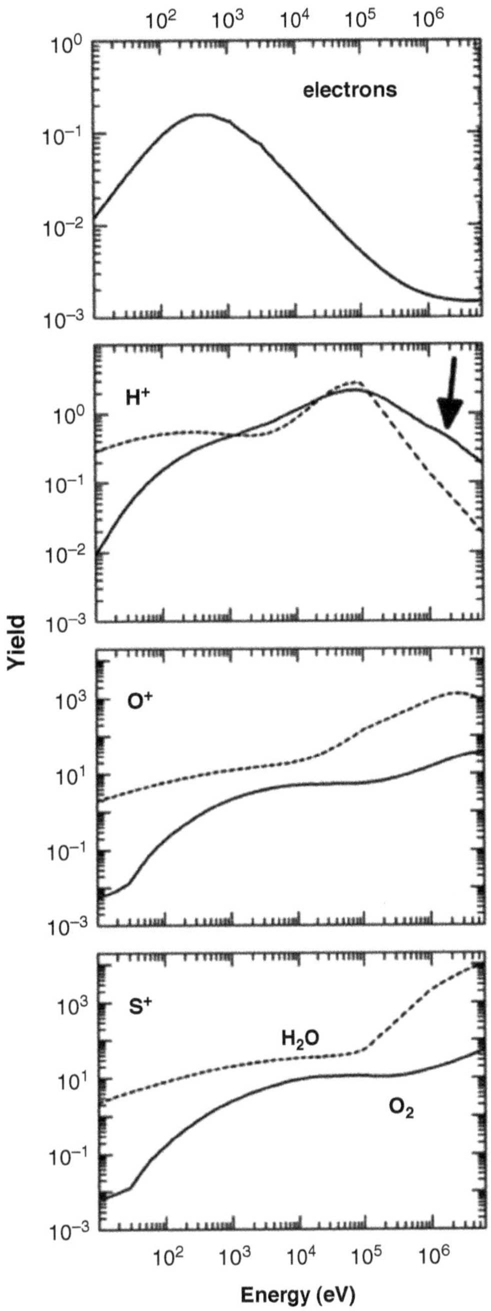

Figure 15.8 Predicted O_2 radiolysis yield (solid lines) and H_2O sputtering yield (dashed lines) in molecules per projectile, from a water-ice regolith versus energy E in the low-temperature limit ($\leq$ 80 K and $\leq$ 130 K for O_2 and H_2O, respectively). Arrow: model may be overestimating O_2 yields for high-energy H^+ ions. Reprint from Teolis, B. D., Plainaki, C., Cassidy, T. A., & Raut, U. (2017). Water ice radiolytic O_2, H_2, and H_2O_2 yields for any projectile species, energy, or temperature: A model for icy astrophysical bodies. *Journal of Geophysical Research: Planets*, 122, 1996–2012. John Wiley and Sons.

The penetration depths of ions increase with energy and decrease with ion mass. Sulphur and oxygen ions with energies between 10 and 100 keV reach typical ranges of tens to hundreds of nanometres in water-ice; 70 keV S^+ ions, for example, reach about 150 nm depth on average according to SRIM simulations (Galli et al., 2018a). Electrons of the same energy penetrate 500 times deeper in ice before being stopped (see Eq. (15.3)).

Finally, we also must consider that the impacting ions remain implanted in the water-ice. Energetic sulphur ions implanted in 80 K water-ice form hydrated sulphuric acid (H_2SO_4) at a high yield of 0.65 ± 0.1 molecules/ion under laboratory conditions (Strazzulla et al., 2007). The total flux of 10^6 cm^{-2} s^{-1} sulphur ions (see Fig. 15.5) thus results in the production rate of roughly 10^6 H_2SO_4 molecules cm^{-2} s^{-1} in the polar regions on Ganymede. As a result, the distribution of sulphur-bearing species on Ganymede's surface reflects the availability of Jovian sulphur ions and water-ice (see Section 15.4.2).

15.3.2.2 Irradiation of Non-watery Species

Both the chemical composition of the non-icy compounds (hydrated sulphuric acids, salts, silicates, etc.) and the interaction of these chemical compounds with energetic electrons and ions is far less well known than for water-ice targets.

The chemical reaction pathways and subsequent release processes of irradiated sulphuric acid or other sulphur-bearing compounds in water-ice have been investigated by Carlson et al. (1999, 2002), Moore et al. (2007), Loeffler et al. (2011), and Loeffler and Hudson (2012). In principle, ions and UV photons can decompose H_2SO_4 in the uppermost layer, but the high radiation stability of hydrated H_2SO_4 at temperatures $\simeq$100 K or higher means H_2SO_4 hydrates are more abundant than SO_2 and polymerized sulphur (Carlson et al., 1999; Loeffler and Hudson, 2012). Loeffler et al. (2011) observed for the 120–130 K temperature range that the ices become more stable with increasing hydration state of the sulphuric acid, which suggests that hydrates such as hemihexahydrate (H_2SO_4*6.5H_2O) and octahydrate (H_2SO_4*8H_2O) are stable on geological timescales. For the irradiation of heavily hydrated sulphuric acid on Ganymede's surface at temperatures of 100–150 K (see Section 15.2.2), we thus can assume that any potential new sulphur-bearing radiolysis product recombines to H_2SO_4*8H_2O or a similar hydrate and remains stable on geological timescales. The chemical composition of ejecta from irradiated hydrated sulphuric acid is unknown: based on ion sputtering experiments of H_2O ice, most ejecta should be entire H_2SO_4 molecules. However, ion irradiation of H_2SO_4 ices by MeV ions is able to produce also the radiolysis products SO_2, H_2O, S_2O_3, and H_3O, judging from reflectance spectra of the irradiated ice (Loeffler et al., 2011). If we assume that hydrated H_2SO_4 qualitatively behaves like H_2O ice (both are isolating condensed gases) upon irradiation, the ratio of ejected radiolyzed products to parent molecule abundances should follow Fig. 15.8 for water-ice: a large fraction ($\leq$ 90 per cent) of sputtered H_2SO_4 molecules would be ejected as H_2SO_4, and only the remaining 10 per cent would leave the ice as a combination of, for example, $SO_2 + H_2O + \frac{1}{2}\,O_2$, but the relative abundance among alternative pathways remains unknown. Not even the sputtering yield of H_2SO_4, unhydrated or hydrated, has ever been determined in laboratory experiments to our knowledge.

Regarding the irradiation of hydrated salts, our knowledge is even more limited. Only a few studies have been published, and the majority of them lack experimental data and are usually directed at the specific case of Europa's surface: Zolotov and Shock (2001); Brown and Hand (2013); McCord et al. (2001). Even the total sputtering yield of hydrated salts is unknown. Existing simulations may be of limited use to close this knowledge gap: the commonly used sputtering simulation SRIM (Ziegler et al., 2008) and its more comprehensive version SDTrimSP (Mutzke et al., 2009), for example, cannot simulate electronic sputtering nor the hydrate crystal structure. For anhydrous Na_2SO_4, sputtering experiments by Wiens et al. (1997) showed that the molecule usually fragments (the dominant Na-bearing species simply being the Na-atom) and the sputtering yield was akin to a metal or a silicate, that is, $Y \approx 0.1$ instead of the $Y \approx 100$ to 1000 for energetic ions irradiating H_2O or SO_2 ices (Johnson et al., 1984). McCord et al. (2001) reported that electron irradiation of anhydrous or hydrated epsomite released SO_2 as the only detectable species. In particular, the authors did not see O_2 or H_2. Hydrated salts thus should not be treated as salts plus independent water for radiolysis. Ion irradiation experiments with hydrated salts have not been performed, to the best of our knowledge. For future models and interpretation of Ganymede surface and atmosphere observations, more theoretical studies and laboratory experiments are required to study the sputtering yields and space weathering effects of hydrated sulphuric acid, hydrated salts, and possibly hydrated silicates (if they indeed constitute the darkening agent) at conditions relevant for Ganymede's surface.

15.4 SURFACE FEATURES CREATED BY THE SPACE ENVIRONMENT

To understand some characteristics of Ganymede's surface, we may have to refer to all of the space environment interaction processes listed at the beginning of this chapter. Here, we discuss three surface characteristics: the dark-bright albedo differences, the distribution of sulphur-bearing species, and the presence of oxygen and ozone. The reader is also referred to Chapters 5 (Jaumann et al.), 7 (Pappalardo et al.), and 10 (Stephan et al.).

15.4.1 The Dark-Bright Albedo Differences

One striking characteristic of Ganymede's surface is its differentiation into bright icy and dark, ice-poor regions (see upper panel in Fig. 15.9). This dark-bright dichotomy is most pronounced between the equatorial regions depleted of water ice and the polar ice-rich regions (Ligier et al., 2019; Mura et al., 2020). In addition, there is a less conspicuous longitudinal pattern in the surface albedo: the trailing hemisphere (180° to 360° W longitude, left part of the map) is darker on average than the leading hemisphere (Khurana et al., 2007). The nature and origin of the darkening agent concentrated on the trailing hemisphere is not known exactly; Bottke et al. (2013) and Ligier et al. (2019), for example, have proposed carbonaceous chondrite-like material or hydrated silicates. The dark material may represent the dusty remains of disintegrated smaller satellites from the early ages of the Jupiter system (Bottke et al., 2013). The authors argued that Ganymede's surface can be characterized as relatively old regions of dark terrains and younger cross-cutting lanes of bright, typically grooved terrain (Bottke et al., 2013; Pappalardo et al., 2004).

Researchers have been investigating mainly two processes to explain the dark-bright dichotomy: thermal sublimation of water ice and Ganymede's magnetic field concentrating Jovian plasma precipitation to the polar regions (see e.g. Spencer (1987) and Khurana et al. (2007)). Generally, the water-ice abundance and surface albedo correlate with the precipitation rates of energetic ions and electrons: from equatorial trailing to equatorial leading to polar regions in ascending order (see Section 15.3.1). Khurana et al. (2007) argued that reconnection of field lines allows for the redistribution of sputter-induced water ice at low latitudes on the leading hemisphere, and Fatemi et al. (2016) found a good correlation between the global surface albedo and the precipitation of Jovian energetic ions,

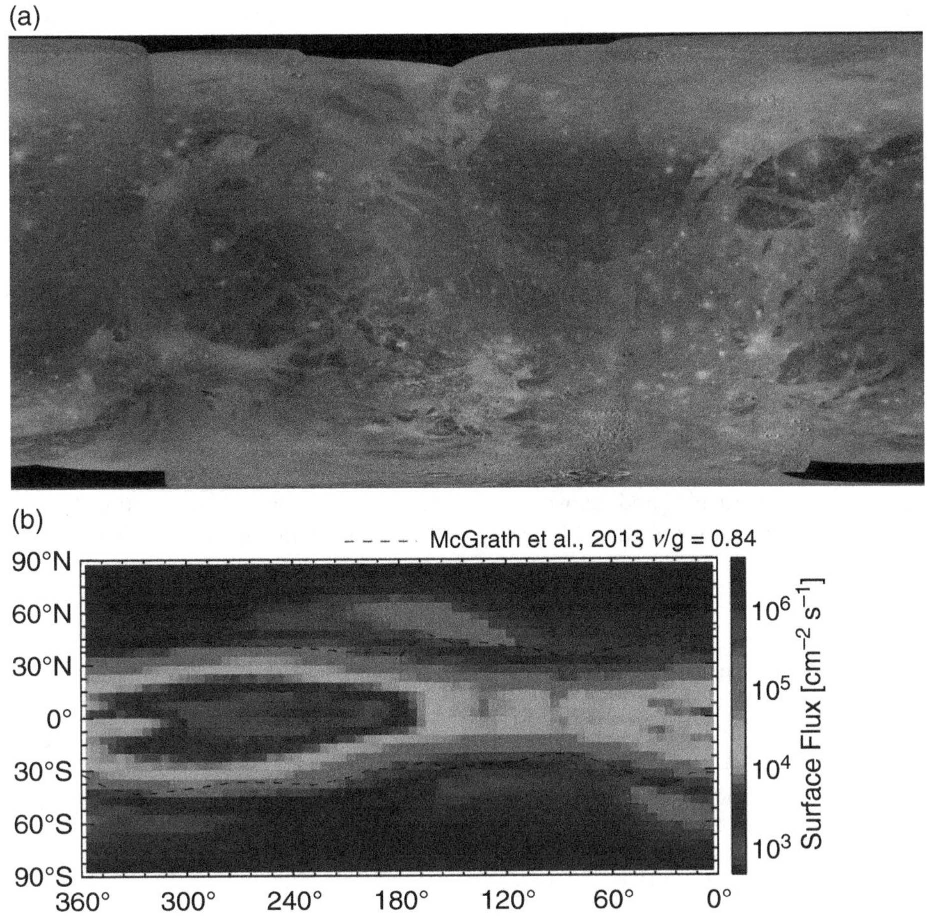

Figure 15.9 (a) Global mosaic of Ganymede using the best image quality and moderate resolution coverage supplied by Galileo Solid-State Imaging (SSI) and Voyager 1 and 2. The left hemisphere from 360° W to 180° W is the trailing hemisphere, and the right half is the leading hemisphere. Image credits: USGS Astrogeology Science Center, Batson (1984). (b) Total flux of precipitating energetic ions to the surface of Ganymede from the model by Fatemi, S., A. R. Poppe, K. K. Khurana, M. Holmström, and G. T. Delory (2016), On the formation of Ganymede's surface brightness asymmetries: Kinetic simulations of Ganymede's magnetosphere, *Geophys. Res. Lett.*, 43, 4745–4754. John Wiley and Sons. The dashed lines are the violet-to-green ratio of 0.84 from the composite image of Ganymede's surface by Khurana et al. (2007) showing the approximate boundaries of Ganymede's bright polar caps and dark low latitudes. (A black and white version of this figure will appear in some formats. For the colour version, please refer to the plate section.)

which is organized by Ganymede's magnetic field lines. This is illustrated in the lower panel of Fig. 15.9.

While irradiation processes are important for Ganymede's surface chemistry and erosion rates, they are likely insufficient to create the albedo dichotomy on their own. First, from the comparison of different precipitation models of Jovian and ionospheric ions (see Section 15.3.1) it is not clear if ion precipitation results in energy flux depositions or ice sputtering rates orders of magnitude stronger in polar regions compared to equatorial regions (see Fig. 15.5). Considerations may also have to include electron precipitation to obtain precipitation patterns strongly correlating with surface albedo (Liuzzo et al., 2020 Fig. 15.6). Second, the actual surface process of how impacting ions would cause a brightening is unclear: chemical reactions in water-ice triggered by ions (sulphur ions in particular) rather deplete the water-ice abundance to form other molecules. On Ganymede's neighbour Europa, the water ice-poor equatorial region (darker and yellowish, dominated by hydrated sulphuric acid and other hydrates) correlates with the maximum of ion precipitation, which occurs on the trailing hemisphere in the case of Europa (McEwen, 1986; Carlson et al., 1999). However, on Ganymede, the equatorial region most depleted of water-ice and highest in abundance of the darkening agent coincides with the minimum of ion precipitation on the trailing hemisphere (Ligier et al., 2019). Ion-induced sputtering also does not lead to a local increase of icy material per se because the relevant sputter yields for water-ice are orders of magnitude higher than those for non-ice species (see Section 15.3.2.1). Penetrating radiation (in particular due to energetic electrons), on the other hand, could brighten the surface by producing light-scattering defects in water ice to a lasting effect in polar regions, whereas in warmer regions above 100 K such radiation-induced defects are annealed rapidly (Johnson, 1997).

Besides these possible explanations related to irradiation, one should keep in mind (Section 15.2) that there are two additional interaction processes that regionally may rival or exceed irradiation in terms of erosion rates: sublimation and impact gardening. The spatial distribution of these processes must be compared to the particle precipitation patterns. Sublimation via its temperature dependence produces a strong latitudinal gradient, with the erosion rate of water-ice highest in the ice-poor equatorial regions. Since the water-ice sublimation rates there exceed the water-ice loss rates by any other process at any region by orders of magnitude, the thermal gradient must be the main reason for the latitudinal gradient of water-ice abundances. As a further argument for thermal effects, Ganymede and Callisto show similar latitudinal gradients of ice grain sizes and a dichotomy between bright polar and dark equatorial latitudes despite Callisto not possessing an intrinsic magnetic field (Stephan et al., 2020). Plainaki et al. (2020a) argued that a sputtering-assisted sublimation mechanism can explain why the trailing hemisphere low-latitude regions are darker and more depleted in H_2O ice than the leading equatorial region (Ligier et al., 2019). A clear attribution of space environment processes to surface properties becomes even more difficult when we also take into account micrometeoroid bombardment: it is expected to concentrate on the leading hemisphere (Bottke et al., 2013), too, thus coinciding with the equatorial gradient of particle precipitation. The gradient of albedo and ice content between the leading and trailing hemispheres may be caused by a combination of meteoroid impact gardening and ionospheric precipitation (Ligier et al., 2019), in analogy to what McEwen (1986) proposed for Europa's surface.

15.4.2 Sulphur-Bearing Species

The distribution of sulphur-bearing species looks like a well-understood phenomenon compared to the dark-bright differences. It is known (Fatemi et al., 2016; Poppe et al., 2018; Ligier et al., 2019; Plainaki et al., 2020a; Carnielli et al., 2020) that hardly any sulphur-bearing material is visible in equatorial trailing regions where sulphur ion precipitation is lowest, sulphates exist in equatorial leading hemisphere where precipitation is substantial and water-ice is depleted because of high temperatures, and sulphuric acid hydrates (but not sulphates) are abundant in polar regions where precipitation fluxes are high and water-ice is abundant because of the low temperatures.

While the occurrence of sulphates correlates with geomorphology (sulci) (Ligier et al., 2019) and thus might be related to an endogenic origin from the ocean (see Chapter 10), the hydrated sulphuric acid is most likely exogenic. The average abundance of hydrated sulphuric acid increases from equatorial trailing to equatorial leading to polar regions from 4 to 10 to 16 per cent, respectively (Ligier et al., 2019) and Table 15.2. This correlation with precipitation fluxes of energetic sulphur ions indicates that the sulphur-bearing species are the products of impacting S ions with the water-ice, in a similar way as postulated for the trailing hemisphere of Europa (McEwen, 1986; Carlson et al., 1999, 2002; Loeffler et al., 2011). Recent simulations by Plainaki et al. (2020a) of O^{2+} and S^{3+} ion precipitation fluxes agree with Ligier et al. (2019) who inferred the likely presence of sulphuric acid hydrate by matching the obtained spectra with a limited number of reference spectra.

The inferred abundances of sulphuric acid hydrates are consistent with ion fluxes and production rates known from experiments (see Section 15.3.2.1): given a production rate of 10^6 cm^{-2} s^{-1} H_2SO_4, 10 per cent of H_2O molecules in an initially pure water-ice surface are replaced by H_2SO_4 within 100 million years, which is compatible with typical surface ages of Ganymede (see Chapter 8).

15.4.3 Oxygen and Ozone at the Surface

The observed absorption lines of solid O_2 and O_3 in Ganymede's surface spectrum are usually taken as direct evidence of irradiation processes on the icy surface (Spencer et al., 1995; Calvin et al., 1996; Orlando and Sieger, 2003). The original detection of solid O_2 was reported based on absorption lines in the visible range reflectance spectrum at 577 and 627 nm (Spencer et al., 1995). This discovery came as a surprise because O_2 ice would sublimate immediately at Ganymede surface temperatures (Calvin et al., 1996). Subsequent laboratory experiments demonstrated that O_2 created by radiolysis in amorphous or crystalline water ice can be retained at temperatures close to Ganymede's dayside temperatures (Grieves and Orlando, 2005; Loeffler et al., 2006; Zheng et al., 2006). The O_2 abundance was inferred to be 0.1 to 1 per cent by volume relative to H_2O, and the O_2 densities appear similar to the liquid or solid γ-phase (Calvin et al., 1996). The O_2 band depths were found to be strongest at low latitudes on Ganymede's trailing side (Calvin et al., 1997). Absorption bands of solid O_2 were observed also on Europa's and Callisto's surfaces, albeit with a shallower absorption depth implying a O_2/H_2O ratio of the order of 0.1 per cent (Spencer et al., 2002).

On the trailing hemisphere of Ganymede, O_3 also was identified via the detection of the absorption band at 260 nm with the Hubble Space Telescope (Noll et al., 1996). The authors estimated a number density ratio of $[O_3]/[O_2] \approx 10^{-4}$ to 10^{-3} in Ganymede's surface. ozone can indeed be formed efficiently under laboratory conditions when water-ice is irradiated with heavy ions at ice temperatures exceeding 120 K, provided water-ice keeps recondensating onto the surface during the experiment (Teolis et al., 2006). Ozone can also be formed by irradiation of an H_2O-O_2 ice mixture with electrons (Jones et al., 2014).

In contrast to Europa (Carlson et al., 1999), H_2O_2 absorption lines so far were not discovered in Ganymede's surface reflectance spectra. This can be attributed to Ganymede's surface temperatures because the H_2O_2 radiolysis production in water-ice warmer than 100 K is known to become inefficient for electron and ion irradiation (Teolis et al., 2009; Hand and Carlson, 2011).

The O_2 and O_3 on Ganymede's surface cannot be a leftover from the creation time of Ganymede, given today's high temperatures and the temperatures immediately after formation (see Chapter 2). The explanation of a secondary creation via irradiation of water-ice is also plausible because the creation of O_2 and O_3 by radiolysis has been demonstrated under laboratory conditions. The spatial distribution of the O_2 and O_3, however, poses a puzzle: Spencer et al. (1995) and Calvin et al. (1996) argued that the concentration of the oxygen on Ganymede's trailing side suggests that it is generated by magnetospheric bombardment. There is a serious flaw in this argument, as we know now. The equatorial trailing side is the region with the lowest particle precipitation fluxes (see Section 15.3.1). This was not known or fully understood in the years 1995–7, as the discovery of Ganymede's magnetic field was reported only in 1996 (see Chapter 14). With the current knowledge of Jovian magnetosphere and particle precipitation we must reconcile the fact that the O_2 absorption bands seen on Ganymede's surface anti-correlate with irradiation levels (Bahr et al., 2001).

Moreover, the exact enriching and storage process of dense-phase O_2 in water-ice after radiolysis was unclear at the time of discovery (Calvin et al., 1997) and remains so to this day. (Calvin et al., 1996) favoured defect trapping or adsorption as the likely means to obtain the O_2 absorption lines, whereas Johnson and Jesser (1997) favoured gas-filled O_2 bubbles in the ice. Vidal et al. (1997), in turn, suggested that the O_2 molecules dissolve in the ice rather than aggregate in clusters or bubbles at Ganymede's surface temperatures. This theory was contradicted by subsequent observations finding the dense-phase or dimer oxygen form predominantly in equatorial and mid-latitude regions rather than in polar regions (Calvin et al., 1997). The abundance of oxygen appeared more dependent on latitude and longitude constraints than surface albedo. Calvin et al. (1997) thus argued that defect trapping is a more plausible process than physical adsorption to explain the solid O_2 absorption lines. In the experiment by Bahr et al. (2001), with energetic proton irradiation there was no indication of radiolytic O_2 bubbles as the source of the absorption bands of condensed O_2.

Finally, we must keep in mind that the O_2 absorption signal anti-correlates not only with particle precipitation fluxes, but also with water-ice abundance: at latitudes below 45°, the water-ice abundance (mostly in crystalline form) makes up only about 20 per cent (Ligier et al., 2019). The temperature dependence of the O_2 production rates in water-ice (Teolis et al., 2009) might be more important than the adverse trends of water-ice availability and irradiation doses, but the real explanation for the unexpected spatial distribution is currently unknown.

15.5 SUMMARY

This chapter has introduced all surface interaction processes at Ganymede. We have seen that irradiation, thermal sublimation, and micrometeoroid impacts can be relevant for different specific surface areas and surface alteration processes. In terms of water erosion or turnover rates, sublimation dominates in the warm region, and micrometeoroid impacts probably are generally more important than irradiation processes. The ions and electrons from the Jovian environment and the ionosphere, on the other hand, initiate chemical alterations in the topmost centimetres of ice that, over geological timescales, may also affect deeper layers depending on the turnover processes at work. All surface interaction processes combine to produce the intriguing surface features of Ganymede, such as the separate dark and bright regions, the distribution of sulphur-bearing species, and the surface reservoir of molecular oxygen and ozone. The surface alteration processes discussed here are also the sources for Ganymede's atmosphere, the topic of the next chapter.

ACKNOWLEDGEMENTS

A. Galli, A. Vorburger, and P. Wurz wish to thank the Swiss National Science Foundation for their financial support.

REFERENCES

Abdulgalil, A. G. M., Rosu-Finsen, A., Marchione, D. et al. 2017. Electron-promoted desorption from water ice surfaces: neutral gas-phase products. *ACS Earth and Space Chemistry* 1, 209–15.

Allioux, R., Louarn, P., André, N., 2013. Model of energetic populations at Ganymede, implications for an orbiter. *Advances in Space Research* 51, 1204–12.

Andreas, E. L., 2007. New estimates for the sublimation rate for ice on the Moon. *Icarus* 186, 24–30.

Aumayr, F., Winter, H., 2004. Potential sputtering. *Philosophical Transactions of the Royal Society of London* A 362, 77–102.

Bahr, D. A., Famfi, M., Vidal, R. A., Baragiola, R. A., 2001. Radiolysis of water ice in the outer solar system: sputtering and trapping of radiation products. *Journal of Geophysical Research* 106, 33,285.

Baragiola, R. A., 2003. Water ice on outer solar system surfaces: basic properties and radiation effects. *Planetary and Space Science* 51, 953–61. DOI: https://doi.org/10.1016/j.pss.2003.05.007.

Bar-Nun, A., Herman, G., Laufer, D., Rappaport, M. L., 1985. Trapping and release of gases by water ice and implications for icy bodies. *Icarus* 63, 317–32. DOI: https://doi.org/10.1016/0019-1035(85)90048-X.

Batson, R. M., 1984. *Voyager 1 and 2 Atlas of Six Saturnian Satellites (NASA-SP-474)*. Washington, DC: National Aeronautics and Space Administration (NASA).

Behr, P. R., Tribbett, P. D., Robinson, T. D., Loeffler, M. J., 2020. Compaction of porous H_2O ice via energetic electrons. *The Astrophysical Journal* 900, 147.

Boduch, P., Brunetto, R., Ding, J. J. et al. 2016. Ion processing of ices and the origin of SO_2 and O_3 on the icy surfaces of the icy jovian satellites. *Icarus* 277, 424–32.

Bohren, C. F., Albrecht, B. A., 1998. *Atmospheric Thermodynamics*. Oxford University Press, New York.

Bottke, W. F., Vokrouhlický, D., Nesvorný, D., Moore, J. M., 2013. Black rain: the burial of the Galilean satellites in irregular satellite debris. *Icarus* 223, 775–95.

Brown, M. E., Hand, K. P., 2013. Salts and radiation products on the surface of Europa. *The Astronomical Journal* 145, 110.

Calvin, W. M., Johnson, R. E., and Spencer, J. R., 1996. O_2 on Ganymede: spectral characteristics and plasma formation mechanisms. *Geophysical Research Letters* 23, 673–6.

Carlson, R. W., Anderson, M. S., Johnson, R. E., Schulman, M. B., Yavrouian, A. H., 2002. Sulfuric acid production on Europa: The radiolysis of sulfur in water ice. *Icarus* 157, 456–63.

Carlson, R. W., Johnson, R. E., Anderson, M. S., 1999. Sulfuric acid on Europa and the radiolytic sulfur cycle. *Science* 286, 97–9.

Carnielli, G., Galand, M., Leblanc, F. et al. 2020. Simulations of ion sputtering at Ganymede. *Icarus* 351, 113918.

Cassidy, T., Coll, P., Raulin, F. et al. 2010. Radiolysis and photolysis of icy satellite surfaces: Experiments and theory. *Space Science Reviews* 153, 299–315. DOI: https://doi:10.1007/s11214-009-9625-3.

Cassidy, T. A., Paranicas, C. P., Shirley, J. H. et al. Magnetospheric ion sputtering and water ice grain size at Europa. *Planetary and Space Science* 77, 64–73.

Cooper, J. F., Johnson, R. E., Mauk, B. H., Garrett, H. B., Gehrels, N., 2001. Energetic ion and electron irradiation of the icy Galilean satellites. *Icarus* 149, 133–59.

Cooper, P. D., Moore, M. H., Hudson, R. L., 2005. Ozone production in irradiated laboratory ices relevant to Europa and Ganymede. American Astronomical Society, DPS meeting #37, id.58.06, *Bulletin of the American Astronomical Society*, Vol. 37, p. 752.

Davis, M. R., Meier, R. M., Cooper, J. F., Loeffler, M. J., 2021. The contribution of electrons to the sputter-produced O_2 exosphere on Europa. *The Astrophysical Journal Letters* 908, L53.

de Soria-Santacruz, M., Garrett, H. B., Evans, R. W. et al. 2016. An empirical model of the high-energy electron environment at Jupiter. *Journal of Geophysical Research: Space Physics* 121, 9732–43. DOI: https://doi.org/10.1002/2016JA023059.

Dones, L., Chapman, C. R., McKinnon, W. B. et al. 2009. Icy satellites of Saturn: impact cratering and age determination. In Dougherty, M. K., Esposito, L. W., Krimigis, S. M. (eds.), *Saturn from Cassini-Huygens*. Springer, Dordrecht, pp. 613–35.

Famá, M., Shi, J., Baragiola, R. A., 2008. Sputtering of ice by low-energy ions. *Surface Science* 602, 156–61.

Fatemi, S., Poppe, A. R., Khurana, K. K., Holmström, M., Delory, G. T., 2016. On the formation of Ganymede's surface brightness asymmetries: kinetic simulations of Ganymede's magnetosphere. *Geophysical Research Letters* 43, 4745–54. DOI: https://doi:10.1002/2016GL068363.

Ferrari, C., Lucas, A., 2016. Low thermal inertias of icy planetary surfaces. *Astronomy & Astrophysics* 588, A133.

Frank, L. A., Paterson, W. R., Ackerson, K. L., Bolton, S. J., 1997. Low-energy electron measurements at Ganymede with the Galileo spacecraft: probes of the magnetic topology. *Geophysical Research Letters* 24, 2159. DOI: https://doi.org/10.1029/97GL01632.

Fray, N., Schmitt, B., 2009. Sublimation of ices of astrophysical interest: a bibliographic review. *Planetary and Space Science* 57, 2053–80.

Galli, A., Vorburger, A., Wurz, P., Cerubini, R., Tulej, M., 2018a. First experimental data of sulphur ions sputtering water ice. *Icarus* 312, 1. DOI: https://doi.org/10.1016/j.icarus.2018.04.029.

Galli, A., Vorburger, A., Wurz, P., Tulej, M., 2017. Sputtering of water ice films: a re-assessment with singly and doubly charged oxygen and argon ions, molecular oxygen, and electrons. *Icarus* 291, 36–45. DOI: https://doi.org/10.1016/j.icarus.2017.03.018.

Galli, A., Vorburger, A., Wurz, P. et al. 2018b. 0.2 to 10 keV electrons interacting with water ice: radiolysis, sputtering, and sublimation. *Planetary and Space Science* 155, 91. DOI: https://doi.org/10.1016/j.pss.2017.11.016.

Grieves, G. A., Orlando, T. M., 2005. The importance of pores in the electron stimulated production of D_2 and O_2 in low temperature ice. *Surface Science* 593, 180–6.

Hand, K. P., Carlson, R. W., 2011. H_2O_2 production by high-energy electrons on icy satellites as a function of surface temperature and electron flux. *Icarus* 215, 226–33. DOI: https://doi.org/10.1016/j.icarus.2011.06.031.

Hansen, G. B., McCord, T. B., 2004. Amorphous and crystalline ice on the Galilean satellites: a balance between thermal and radiolytic processes. *Journal of Geophysical Research* 109, E01012. DOI: https://doi.org/10.1029/2003JE002149.

Haring, R. A., Pedrys, R., Oostra, D. J., Haring, A., De Vries, A. E., 1984. Reactive sputtering of simple condensed gases by keV ions III: kinetic energy distributions. *Nuclear Instruments and Methods in Physics Research* B5, 483–8.

Howett, C. J. A., Spencer, J. R., Hurford, T. A., Verbiscer, A., Segura, M., 2012. PacMan returns: an electron-generated thermal anomaly on Tethys. *Icarus* 221, 1084–8.

Howett, C. J. A., Spencer, J. R., Schenk, P. et al. 2011. A high-amplitude thermal inertia anomaly of probable magnetospheric origin on Saturn's moon Mimas. *Icarus* 216, 221–6.

Johnson, R. E., 1990. Energetic Particle Interaction with Atmospheres and Surfaces. Springer, Berlin.

Johnson, R. E., 1997. Polar 'caps' on Ganymede and Io revisited. *Icarus* 128, 469–71.

Johnson, R. E., Burger, M. H., Cassidy, T. A. et al. 2009. Composition and detection of Europa's sputter-induced atmosphere. In Pappalardo, R. T., McKinnon, W. B., Khurana, K. K. (eds.), *Europa*. University of Arizona Press, Tucson, pp. 507–27.

Johnson, R. E., Carlson, R. W., Cooper, J. F. et al. 2004. Radiation effects on the surfaces of the Galilean satellites. In Bagenal, F. (ed.), *Jupiter: Atmosphere, Satellites and Magnetosphere*. University of Arizona Press, Tucson, pp. 485–512.

Johnson, R. E., Garrett, J. W., Boring, J. W., Barton, L. A., 1984. Erosion and Modification of SO_2 Ice by ion bombardment of the surface of Io. *Journal of Geophysical Research* 89, Supplement, B711.

Johnson, R. E., Jesser, W. A., 1997. O2/O3 Microatmospheres in the surface of Ganymede. *The Astrophysical Journal* 480, L79–L89.

Johnson, R. E. Liu, M., 2010. Sputtering of surfaces, sputtering data for H_2O ice. http://people.virginia.edu/~rej/h2o.html.

Johnson, R. E., Oza, A. V., Leblanc, F. et al. 2019. The origin and fate of O2 in Europa's ice: an atmospheric perspective. *Space Science Reviews* 215, 1–23.

Jones, B. M., Kaiser, R. I., Strazzulla, G., 2014. UV-VIS, infrared, and mass spectroscopy of electron irradiated frozen oxygen and carbon dioxide mixtures with water. *The Astrophysical Journal* 781, 85.

Jun, I., Garrett, H. B., Cassidy, T. A., Kim, W., Dougherty, L., 2019. Updating the Jovian electron plasma environment. *IEEE Transactions on Plasma Science* 47, 3915–22. DOI: https://doi.org/10.1109/TPS.2019.2901681.

Khurana, K. K., Pappalardo, R. T., Murphy, N., Denk, T., 2007. The origin of Ganymede's polar caps. *Icarus* 191, 193.

Kivelson, M. G., Bagenal, F., Kurth, W. S. et al. 2004. Magnetospheric interactions with satellites. In Bagenal, F., Dowling, T., McKinnon, W. K. (eds.), *Jupiter: The Planet, Satellites and Magnetosphere*. Cambridge University Press, Cambridge, Chapter 21.

Krüger, H., Krivov, A. V., Grün, E., 2000. A dust cloud of Ganymede maintained by hypervelocity impacts of interplanetary micrometeoroids. *Planetary and Space Science* 48, 1457–71.

Laufer, D., Bar-Nun, A., Greenberg, A. N., 2017. Trapping mechanism of O_2 in water ice as first measured by Rosetta spacecraft. *Monthly Notices of the Royal Astronomical Society* 469, S818. DOI: https://doi.org/10.1093/mnras/stx3359.

Leblanc, F. Oza, A. V., Leclercq, L. et al. 2017. On the orbital variability of Ganymede's atmosphere. *Icarus* 293, 185–98.

Ligier, N., Paranicas C., Carter, J. et al. 2019. Surface composition and properties of Ganymede: updates from ground-based observations with the near-infrared imaging spectrometer SINFONI/VLT/ESO. *Icarus* 333, 496.

Liuzzo, L., Poppe, A. R., Paranicas, C., Fatemi, S., Simon, S., 2020. Variability in the energetic electron bombardment of Ganymede. *Journal of Geophysical Research: Space Physics* 125, e2020JA028347.

Loeffler, M. J., Hudson, R. L., 2012. Thermal regeneration of sulfuric acid hydrates after irradiation. *Icarus* 219, 561–6.

Loeffler, M. J., Hudson, R. L., Moore, M. H., Carlson, R. W., 2011. Radiolysis of sulfuric acid, sulfuric acid monohydrate, and sulfuric acid tetrahydrate and its relevance to Europa. *Icarus* 215, 370–80.

Loeffler, M. J., Teolis, B. D., Baragiola, R.A., 2006. A model study of the thermal evolution of astrophysical ices. *The Astrophysical Journal* 639, L103.

Loeffler, M. J., Tribbett, P. D., Cooper, J. F., Sturner, S. J. 2020. A possible explanation for the presence of crystalline H_2O-ice on Kuiper Belt objects. *Icarus* 351, 113943. DOI: https://doi.org/10.1016/j.icarus.2020.113943.

Marconi, M. L., 2007. A kinetic model of Ganymede's atmosphere. *Icarus* 190, 155–74.

Mauk, B. H., Mitchell, D. G., McEntire, R. W. et al. 2004. Energetic ion characteristics and neutral gas interactions in Jupiter's magnetosphere. *Journal of Geophysical Research* 109, A09S12. DOI: https://doi.org/10.1029/2003JA010270.

McCord, T. B., Orlando, T. M., Teeter, G. et al. 2001. Thermal and radiation stability of the hydrated salt minerals epsomite, mirabilite, and natron under Europa environmental conditions. *Journal of Geophysical Research* 106, 3311–19.

McEwen, A. S., 1986. Exogenic and endogenic albedo and color patterns on Europa. *Journal of Geophysical Research*, 91, 8077–97.

Meier, R. M., and Loeffler, M. J., 2020. Sputtering of water ice by keV electrons at 60 K. *Surface Science* 691, 121509. DOI: https://doi.org/10.1016/j.susc.2019.121509.

Miljković, K., Hillier, J. K., Mason, N. J., Zarnecki, J. C., 2012. Models of dust around Europa and Ganymede. *Planetary and Space Science* 70, 20–7.

Moore, M. H., Hudson, R. L., Carlson, R. W., 2007. The radiolysis of SO_2 and H_2S in water ice: implications for the icy Jovian satellites. Icarus 189, 409–23.

Muntean, E. A., Lacerda, P., Field, T. A. et al. 2016. A laboratory study of water ice erosion by low-energy ions. *Monthly Notices of the Royal Astronomical Society* 462, 3361–7.

Mura, A., Adriani, A., Sordini, R. et al. 2020. Infrared observations of Ganymede from Juno/Jovian infrared auroral mapper. *Journal of Geophysical Research: Planets* 125, e2020JE006508.

Murphy, D. M., Koop, T., 2005. Review of the vapour pressures of ice and supercooled water for atmospheric applications. *Quarterly Journal of the Royal Meteorological Society* 131, 1539.

Mutzke A., Schneider R., Bizyukov, I. 2009. SDTrimSP-2D studies of the influence of mutual flux arrangement on erosion and deposition. *Journal of Nuclear Materials* 390, 115–18.

Noll, K. S., Johnson, R. E., Lane, A. L., Domingue, D. L., Weaver, H. A., 1996. Detection of ozone on Ganymede. *Science* 273, 341–3.

Nordheim, T. A., Hand, K. P., Paranicas, C. et al. 2017. The near-surface electron radiation environment of Saturn's moon Mimas. *Icarus* 286, 56–8.

Orlando, T. M., Sieger, M. T., 2003. The role of electron-stimulated production of O_2 from water ice in the radiation processing of outer solar system surfaces. *Surface Sciences* 528, 1–7.

Orton, G. S., Spencer, J. R., Travis, L. D., Martin, T. Z., Tamppari, L. K., 1996. Galileo photopolarimeter-radiometer observations of Jupiter and the Galilean satellites. *Science* 274, 389–91.

Pappalardo, R. T., Collins, G. C., Head, J. W., III, et al. 2004. Geology of Ganymede. In Bagenal, F., Dowling, T., McKinnon, W. K. (Eds.), *Jupiter: The Planet, Satellites and Magnetosphere*. Cambridge University Press, Cambridge, pp. 363–96.

Paranicas, C., Mauk, B. H., Ratliff, J. M., Cohen, C., Johnson, R. E., 2002. The ion environment near Europa and its role in surface energetics. *Geophysical Research Letters* 29, 5. DOI: https://doi.org/10.1029/2001GL014127.

Petrik, N. G., Kavetsky, A. G., Kimmel, G. A., 2006. Electron-stimulated production of molecular oxygen in amorphous solid water. *Journal of Physical Chemistry* B 110, 2723–31.

Plainaki, C., Massetti, S., Jia, X. et al. 2020a. Kinetic simulations of the jovian energetic ion circulation around Ganymede. *The Astrophysical Journal* 900, 74.

Plainaki, C., Milillo, A., Massetti, S. et al. 2015. The H_2O and O_2 exospheres of Ganymede: the result of a complex interaction between the jovian magnetospheric ions and the icy moon. *Icarus* 245, 306–19.

Plainaki, C., Sindoni, G., Grassi, D. et al. 2020b. Preliminary estimation of the detection possibilities of Ganymede's watervapor environment with MAJIS. *Planetary and Space Science* 191, 105004.

Poppe, A. R., 2016. An improved model for interplanetary dust fluxes in the outer Solar System. *Icarus* 264, 369–86.

Poppe, A. R., Fatemi, S., Khurana, K. K., 2018. Thermal and energetic ion dynamics in Ganymede's magnetosphere. *Journal of Geophysical Research: Space Physics* 123, 4614–37.

Roth, L., Ivchenko, N., Gladstone, G. R. et al. 2021. A sublimated water atmosphere on Ganymede detected from Hubble Space Telescope observations. *Nature Astronomy*, 5, 1043–51. DOI: https://doi.org/10.1038/s41550-021-01426-9.

Schaible, M. J., Johnson, R. E., Zhigilei, L. V., Piqueux, S., 2017. High energy electron sintering of icy regoliths: formation of the PacMan thermal anomalies on the icy Saturnian moons. *Icarus* 285, 211–23. DOI: https://org/10.1016/j.icarus.2016.08.033.

Shematovich, V. I., 2016. Neutral Atmosphere near the icy surface of Jupiter's moon Ganymede. Solar System Research 50, 262–80.

Shi, M., Baragiola, R. A., Grosjean, D. E. et al. 1995. Sputtering of water ice surfaces and the production of extended neutral atmospheres. *Journal of Geophysical Research* 100, 26387.

Sivaraman, B., Jamieson, C. S., Mason, N. J., Kaiser, R. I., 2007. Temperature-dependent formation of ozone in solid oxygen by 5 keV electron irradiation and implications for solar system ices. *The Astrophysical Journal* 669, 1414–21.

Spencer, J. R., 1987. Thermal segregation of water ice on the Galilean satellites. *Icarus* 69, 297–313.

Spencer, J. R., Calvin, W. M., Person, M. J., 1995. Charge-coupled device spectra of the Galilean satellites: molecular oxygen on Ganymede. *Journal of Geophysical Research* 100, 19049.

Spencer, J. R., Calvin, W. M., Person, M. J., 2002. Condensed O_2 on Europa and Callisto. *The Astronomical Journal* 124, 3400–3. DOI: https://doi.org/10.1086/344307.

Stephan, K., Hibbitts, C. A., Jaumann, R., 2020. H_2O-ice particle size variations across Ganymede's and Callisto's surface. *Icarus* 337, 113440.

Strazzulla, G., Baratta, G. A., Leto, G., Gomis, O., 2007. Hydrate sulfuric acid after sulfur implantation in water ice. *Icarus* 192, 623–8.

Teolis, B. D., Loeffler, M. J., Raut, U., Famá, M., Baragiola, R. A., 2006. Ozone synthesis on the icy satellites. *The Astrophysical Journal* 644, L141.

Teolis, B. D., Plainaki, C., Cassidy, T. A., Raut, U., 2017. Water ice radiolytic O_2, H_2, and H_2O_2 yields for any projectile species, energy, or temperature: a model for icy astrophysical bodies. *Journal of Geophysical Research: Planets* 122, 2017JE005285.

Teolis, B. D., Shi, J., Baragiola, R. A., 2009. Formation, trapping, and ejection of radiolytic O_2 from ion-irradiated water ice studied by sputter depth profiling. *The Journal of Chemical Physics* 130, 134704.

Turc, L., Leclercq, L., Leblanc, F., Modolo, R., Chaufray, J.-Y., 2014. Modelling Ganymede's neutral environment: a 3D test-particle simulation, *Icarus* 229, 157–69.

Vidal, R. A., Bahr, D., Baragiola, R. A., Peters, M., 1997. Oxygen on Ganymede: laboratory studies. *Science* 276, 1839–42.

Vorburger, A., Wurz, P., 2018. Europa's ice-related atmosphere: the sputter contribution. *Icarus* 311, 135–45.

Westley, M. S., Baragiola, R. A., Johnson, R. E., Baratta, G. A., 1995. Photodesorption from low-temperature water ice in interstellar and circumsolar grains. *Nature* 373, 405–7.

Wiens, R. C., Burnett, D. S., Calaway, W. F. et al. 1997. Sputtering products of sodium sulfate: implications for Io's surface and for sodium-bearing molecules in the Io torus. *Icarus* 128, 386–97.

Zheng, W.-J., Jewitt, D., Kaiser, R. I., 2006. Formation of hydrogen, oxygen, and hydrogen peroxide in electron-irradiated crystalline water ice. *The Astrophysical Journal* 639, 534–48.

Ziegler, J. F., Biersack, J. P., Ziegler, M. D., 2008. *SRIM – The Stopping and Range of Ions in Matter*. Vol. 5. SRIM Co., Chester, MD.

Zolotov, M. Y., Shock, E. L., 2001. Composition and stability of salts on the surface of Europa and their oceanic origin. *Journal of Geophysical Research* 106, 32815–27.

16

Ganymede's Tenuous Atmosphere

LORENZ ROTH, CHRISTINA PLAINAKI, APURVA V. OZA, AUDREY VORBURGER, ROBERT E. JOHNSON, PHILIPPA M. MOLYNEUX, KURT D. RETHERFORD, FRANÇOIS LEBLANC, STEFANO MASSETTI, ALESSANDRO MURA, AND PETER WURZ

16.1 INTRODUCTION

Years before a definitive detection of neutral gas abundance around Ganymede, studies suggested that an atmosphere consisting of water group species could be produced from sublimation. Based on an early but eventually incorrect detection of a significant atmosphere (Carlson et al., 1973), Yung and McElroy (1977) suggested that a significant atmosphere could be present, produced by photochemistry in a sublimated water vapor. Such an atmosphere would be dominated by O_2, which is gravitationally bound but does not recondense on the surface. During the Voyager 1 flyby, it was found that the plasma trapped in the Jovian magnetosphere was more intense and had a different composition than predicted.[1] This led a group at Bell Labs to carry out a series of experiments which showed that the icy satellite surfaces would be significantly eroded by the incident plasma ions (Lanzerotti et al., 1978). They called this process 'electronic sputtering' as, although it appeared to be related to a standard laboratory process for directly knocking atoms from a solid into the gas phase (called 'nuclear sputtering'), it was much more efficient and was a result of the electronic excitations produced in ice by the incident ions (Brown et al., 1978). The Voyager plasma ion flux combined with the laboratory measurements were used to show that, depending on the local surface temperature, the sputter production of gas phase species could dominate sublimation (Johnson et al., 1981). It was subsequently shown that the species ejected primarily consisted of H_2O, O_2, and H_2 with only small fractions of dissociation products (Brown et al., 1982). Because H_2 most easily escapes from the moons and H_2O readily condenses, O_2 is indeed likely to be the dominant atmospheric component over most of the icy satellite surfaces due to its gravitational confinement and the fact that it is directly produced from the surface (Johnson et al., 1981) rather than by photochemistry in a tenuous atmosphere (Yung and McElroy, 1977).

In this chapter we review the observational and theoretical studies related to Ganymede's atmosphere. An earlier review can be found in McGrath et al. (2004) but there have been crucial new insights and results since then, like the recent detection of H_2O. The chapter is structured as follows: Section 16.2 summarizes the observational studies. In Section 16.3, we elaborate on the role of the sources and sinks for Ganymede's atmosphere. Section 16.4 describes the various theoretical and modelling studies and their contributions to today's understanding. Finally, in Section 16.5, we summarize the current knowledge and open questions on Ganymede's atmosphere and describe the prospects and plans of the upcoming Jupiter ICy moons Explorer (JUICE) mission.

We also want to point to the most relevant related chapters in this book. Surface erosion processes and their role in constantly sustaining the atmosphere are discussed in Chapter 15. The ionosphere and how it is formed from the atmosphere through different ionization processes are described in Chapter 17. Both atmosphere and ionosphere are also crucial for the interaction of Ganymede and its magnetic field with the surrounding Jovian magnetosphere, presented in Chapter 14. Finally, Ganymede's aurora (Chapter 18) is closely related to the atmosphere, not only through the physical processes but also because observations of the electron-excited aurora are the primary evidence for the moon's bound atmosphere.

16.2 OBSERVATIONS

In this section we discuss the observational constraints on Ganymede's atmosphere. While there are observational means to indirectly constrain the neutral gas abundance in Ganymede's environment, for example, through plasma observations and simulations of the atmosphere–plasma interaction, the focus here will be on the direct detection of neutral species. Table 16.1 summarizes all observations of Ganymede's neutral atmosphere since the first detections and provides details on the instruments and set-ups.

16.2.1 Early Constraints

In their search for an atmosphere, Carlson et al. (1973) observed an occultation of a star (SAO 186800) by Ganymede from three locations in India, Java and Australia. The authors found a gradual increase above Ganymede's disk in the transit light curve and concluded that an atmosphere with surface pressure around 1 μbar was present on Ganymede (Carlson et al., 1973; Taylor, 1971). No atmospheric signature was found on Io in a similar observation, and the authors concluded that Ganymede's atmosphere should be denser than Io's (Taylor, 1971).

Almost a decade later, an occultation measurement by the Ultraviolet Spectrometer (UVS) on board the Voyager 1 spacecraft of the brighter star κ Centauri by Ganymede provided higher sensitivity, but no signatures of atmospheric absorption

[1] By 'plasma' we refer to charged particles of all energies throughout this chapter (including, e.g., keV and MeV ions).

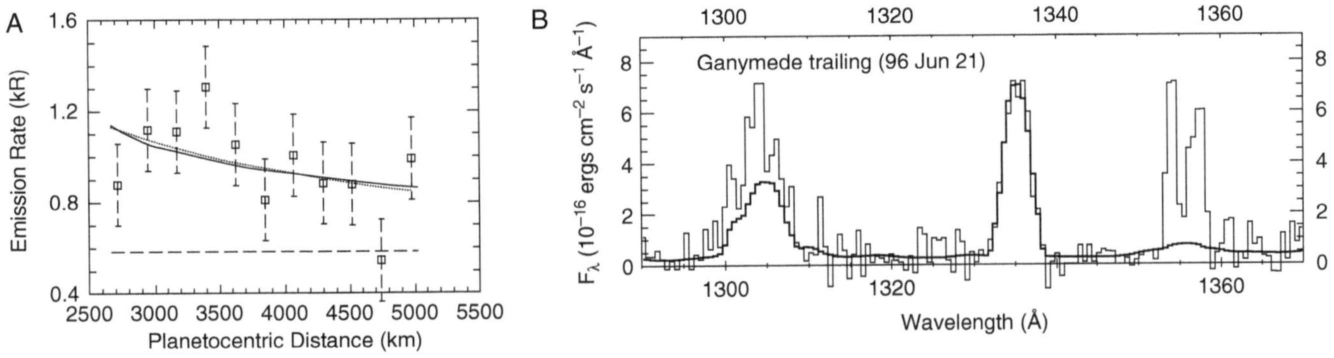

Figure 16.1 (a) First detection of Ganymede's H corona in observations of HI Lyman-α emissions above Ganymede's limb taken by Galileo UVS adapted from *Geophysical Research Letters*, 24(17), C. A. Barth, C. W. Hord, A. I. F. Stewart, W. R. Pryor, K. E. Simmons, W. E. McClintock, J. M. Ajello, K. L. Naviaux, J. J. Aiello. Galileo ultraviolet spectrometer observations of atomic hydrogen in the atmosphere of Ganymede, 185–198., ©1997, with permission from John Wiley and Sons. (b) First detection of Ganymede's bulk molecular atmosphere through emissions at the oxygen multiplets at 1304 Å and 1356 Å in a HST/GHRS spectrum (adapted from Hall et al., 1998, figure 1). The CII 1335 Å feature is used to adjust the surface-reflected solar spectrum.

were found (Broadfoot et al., 1981). A density upper limit of 1.5×10^9 cm^{-3} at the surface was inferred for a O_2 or H_2O atmosphere, which would convert to a column density of about 10^{16} cm^{-2}. This corresponds to a surface pressure of 2×10^{-5} μbar, almost five orders of magnitude lower than, and thus incompatible with, the value from Carlson et al. (1973). Later observations confirmed the tenuous nature of Ganymede's atmosphere, and the Voyager UVS upper limit still holds today.

16.2.2 First Detection of H Corona and Later Confirmations

The first positive detection of a neutral species around Ganymede was achieved by the ultraviolet spectrometer (UVS) on board the Galileo spacecraft in 1996. The spectrometer detected slowly decreasing H Lyman-α emissions in a scan from Ganymede's surface to 5 000 km altitude (Barth et al., 1997) (Figure 16.1a). The gradual emission profile and overall intensity were shown to be consistent with an extended corona of atomic hydrogen with maximum density at the surface of 1.5×10^4 cm^{-3}. The H abundance was suggested to originate from water vapor (H_2O) in the atmosphere (Barth et al., 1997).

Confirmation of extended H Lyman-α emissions around Ganymede and thus the presence of the H corona was provided by observations by the Space Telescope Imaging Spectrograph of the Hubble Space Telescope (HST/STIS) taken in 1998 (Feldman et al., 2000). The STIS was used in imaging spectroscopy mode, providing spatially resolved images at various wavelengths including hydrogen Lyman-α line and two far-UV oxygen multiplets. This STIS observation mode provided a series of important insights on Ganymede (cf. Sections 16.2.3 and 16.2.4) and was employed in several later observing campaigns (Table 16.1).

Alday et al. (2017) analyzed the extended H Lyman-α emissions in all HST/STIS observations taken between 1998 and 2014 (Table 16.1). Taking into account various contributions to the signal such as surface reflections scattered off the limb, they found a relatively stable H corona with surface densities in the range of $(5–8)\times10^3$ cm^{-3}. Systematic differences, for example between leading or trailing hemisphere or temporal changes, could not be identified. Alday et al. (2017) suggest that the H is produced in the atmosphere through dissociation of both H_2O (as first suggested by Barth et al. (1997)) and H_2, referring to the simulations from Marconi (2007).

16.2.3 Hubble Space Telescope Observations I: Atmospheric Oxygen Emissions

Far-UV atomic oxygen emissions near 1 304 Å and 1 356 Å from Ganymede's atmosphere were observed for the first time in a spectrum taken by the Goddard High Resolution Spectrograph of the Hubble Space Telescope (HST/GHRS) observing the moon's trailing hemisphere (Hall et al., 1998); see Figure 16.1b.

The relative brightness of the emissions from the spin-forbidden OI(^{5}S–^{3}P) 1 356 Å doublet and the optically allowed OI(^{3}S–^{3}P) 1 304 Å triplet was used as diagnostic of the excitation process. The derived OI 1356 Å/OI 1 304 Å oxygen ratio ($r_{\gamma O}$) of 1.3±0.3 was interpreted to relate to dissociative excitation of O_2 by electrons. The reasoning is that electron-impact excitation processes that involve other possible species in the atmosphere such as O or H_2O produce significantly brighter 1 304 Å emissions ($r_{\gamma O} < 1$). In addition, resonant scattering of solar emissions by atomic oxygen O contributes to the OI 1 304 Å brightness, but is absent at 1 356 Å (Hall et al., 1995). Electron-impact excitation of O_2 results in a larger OI 1 356 Å/OI 1 304 Å ratio of $r_{\gamma O} > 2.2$.

Figure 16.2 shows calculated excitation rates for a range of possible electron temperatures based on the laboratory cross sections and a schematic illustrating approximate oxygen emission ratios related to different constituents. The use of the oxygen ratio as diagnostic is further discussed in Section 16.2.4.

Before the detection at Ganymede, HST/GHRS observations had revealed oxygen FUV emissions at Europa, also with higher intensity at 1 356 Å than at 1 304 Å. The same diagnostic was used to derive O_2 in the atmosphere of Europa (Hall et al., 1995). Later on, the Cosmic Origins Spectrograph (COS), installed on HST in 2009, detected faint oxygen emissions also from Callisto, thanks to its superior sensitivity. The higher OI 1 356 Å intensity was again interpreted as originating from an O_2-dominated atmosphere (Cunningham et al., 2015).

The double-peak structure of the oxygen emissions in the GHRS spectrum on Ganymede (Figure 16.1, particularly at 1 356 Å) pointed to a non-uniform spatial distribution and was

Table 16.1 *Observations of Ganymede's neutral atmosphere and corona*

Facility/ program ID	Instrument	Year	Geometry/ hemisphere	Type	Comments	Ref.
Galileo	UVS	1996	above limb	FUV spectral scan	first detection of H through scattered solar Lyα	*a*
HST 6758	GHRS	1996	trailing	FUV spectrum	first detection of O_2	*b*
HST 7939	STIS	1998	trailing	spectral images	detection of auroral bands & of H_2O	*c,d,f,g,j*
HST 8224	STIS	2000	leading	spectral images	first images of leading hemisphere	*d,f,g*
HST 9296	STIS	2003	sub-Jovian	spectral images	first images of sub-Jovian hemisphere	*d*
New Horizons	Alice	2007	sub- & anti-Jovian	FUV spectra	observation of nightside atmosphere	Section 16.2.5
HST 10871	ACS	2007	sub-Jovian	FUV filter image	observation in eclipse	*d*
HST 12244	STIS	2010/11	leading	spectral images	temporal variability, H_2O	*e,f,g,j*
Herschel	HIFI	2011–13	leading & trailing	sub-mm spectrum	possible detection of H_2O absorption	*i*
HST 13328	STIS & COS	2014	leading & trailing	spectra & spectral images	first campaign to observe both hemispheres	*f,h*
HST 14634	STIS	2017	anti-Jovian	spectral images	first images of anti-Jovian hemisphere	
	COS	2018	sub-Jovian	high-res spectra	eclipse/egress observation/upper limits on O	*j*

[a]Barth et al. (1997), [b]Hall et al. (1998),[c]Feldman et al. (2000), [d]McGrath et al. (2013), [e]Saur et al. (2015), [f]Alday et al. (2017), [g]Musacchio et al. (2017), [h]Molyneux et al. (2018), [i]Hartogh et al. (2013), [j]Roth et al. (2021).

shown to be consistent with the emissions being confined to the polar regions (Hall et al., 1998). The first observations with HST/STIS in imaging spectroscopy mode from 1998 then provided confirmation that the oxygen OI 1 356 Å emissions on Ganymede's trailing hemisphere are indeed brightest near the north and south poles. Further STIS images of the orbital-leading and sub-Jovian hemispheres showed that the regions of brightest OI 1 356 Å emissions are roughly colocated with the open-closed-field-line-boundary (OCFB) of Ganymede's mini-magnetosphere (McGrath et al., 2013). The emission morphology is thus determined by the magnetic field topology and by the precipitation patterns of the electrons that excite the emissions in the atmosphere (see Chapter 17 for more details). Furthermore, Eviatar et al. (2001) showed that electrons must be accelerated to produce the observed intensities, but the acceleration mechanism and exact energies are unknown.

The auroral nature and particular morphology of the oxygen emissions enabled studies of various magnetospheric and plasma processes (Chapter 18; Saur et al., 2015; Musacchio et al., 2017). However, the derivation of neutral gas abundances from the electron-excited oxygen emissions is difficult given the uncertainty of the electron properties at the OCFB or inside and around Ganymede's magnetosphere. The first approximations assumed a homogeneous plasma with constant electron temperature and density to derive a line-of-sight O_2 column density averaged over the moon's disk around 10^{14} cm^{-2} to 10^{15} cm^{-2} (Hall et al., 1998; Feldman et al., 2000). Using the ionosphere model of Eviatar, Vasyliūnas and Gurnett (2001), Eviatar et al. (2001) found that electrons need to be accelerated locally such that some sub-population attains energies (temperatures) in the range of 75–300 eV to effectively produce the observed intensities with an atmosphere of $\sim 3\times10^{14}$ O_2 cm^{-2}.

These global column densities are derived from average intensities over the disk and neglect the spatial variations and peak intensities. In the case of Europa, the oxygen emissions are less confined to specific regions and are possibly excited by the thermal electron population in the plasma sheet (Hall et al., 1995; Saur et al., 1998; Roth et al., 2016). For this thermal population, estimates of density and temperature were made based on spacecraft measurements near Europa's orbit (Bagenal et al., 2015; Bagenal and Dols, 2020). There, the simplifying approach of relating a globally averaged intensity to a disk-average column density is probably more justified than in the

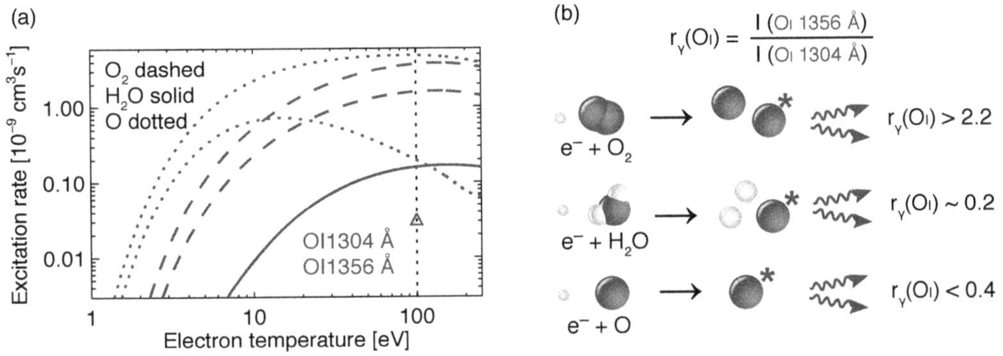

Figure 16.2 (a) Excitation rates for electron impact on O_2 (dashed), O (dotted) and H_2O (solid) for OI 1 356 Å (red) and OI 1 304 Å (blue) as a function of electron temperature. The rates are calculated from the cross sections (Kanik et al., 2001, 2003; Makarov et al., 2004), assuming Maxwellian temperature distribution for the electron population. For OI 1 356 Å from H_2O only a single cross-section measurement at 100 eV is available, and the red triangle (for the OI 1 356 Å rate) shows the 1 304 Å value scaled according to the relative cross section. (b) Schematic showing the approximate oxygen emission ratios for the three species (after Roth et al., 2021). (A black and white version of this figure will appear in some formats. For the color version, please refer to the plate section.)

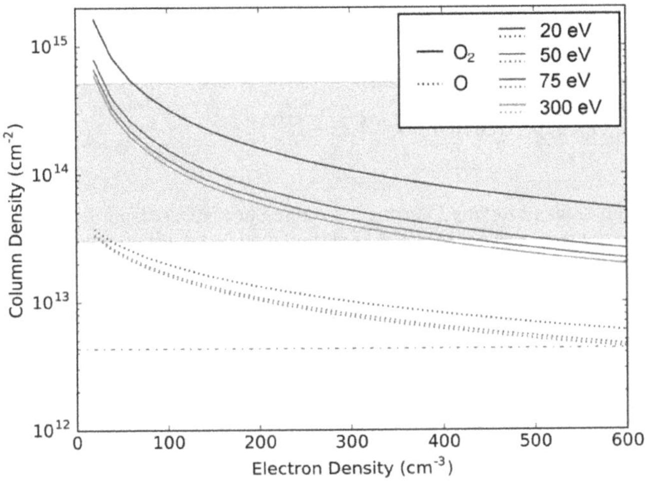

Figure 16.3 Global column densities of O_2 and O at Ganymede based on the oxygen emission intensities observed on the trailing hemisphere for a range of electron densities and temperatures. The O_2 column density is primarily constrained by a disk-average 1 356 Å intensity of 39 R, derived in Molyneux et al. (2018). The shaded region shows the range of O_2 column densities inferred from previous observations of Ganymede (Feldman et al., 2000). The dashed horizontal line indicates the maximum column density for an optically thin O atmosphere at a temperature of 150 K. (A black and white version of this figure will appear in some formats. For the color version, please refer to the plate section.)

case of Ganymede, where the system is likely highly dynamic and electron density and temperature can assume a wide range. Figure 16.3 shows the required column densities to explain the intensities observed by Molyneux et al. (2018) as a function of electron density for different electron temperatures, again assuming a homogeneous plasma environment.

If the atmospheric temperatures T are in the range of the surface temperatures of 80 K to 150 K (Orton et al., 1996), the nominal atmospheric scale height for O_2 is about 20 km. Assuming the actual effective atmospheric scale height of O_2 is between 10 km and 100 km, the estimated average O_2 column densities (10^{14} cm^{-2} to 10^{15} cm^{-2}) correspond to surface densities of 10^7 to 10^9 cm^{-3}. This is below, and thus consistent with, the upper limit on the molecular atmosphere from the Voyager UVS (Broadfoot et al., 1981).

16.2.4 Hubble Space Telescope Observations II: Oxygen Emission Ratio Studies

Hubble Space Telescope observations of Ganymede generally have a lower signal-to-noise ratio at 1 304 Å than at 1 356 Å because Ganymede's atmospheric OI 1 304 Å emissions have a lower intensity, while the background signal is higher due to both higher solar flux (leading to increased background from surface reflection) and an additional contribution from the Earth's geocorona at 1 304 Å. Therefore, most studies of the oxygen emissions and their morphology used primarily the 1 356 Å aurora. However, the relative intensity of the OI 1 304 Å to the OI 1 356 Å emissions (or the OI 1 356 Å/OI 1 304 Å ratio, $r_{\gamma O}$) can provide insights into both atmospheric and plasma processes. The ratio is widely different for different species such as O_2, H_2O or O, but also varies with the temperature of the exciting electrons as shown in Figure 16.2. Excitation of atomic oxygen is the most effective process at low temperatures ($T_e <$ 5 eV) and produces significantly more OI 1 304 Å emission. For electron temperatures near and above the O_2 dissociation threshold (roughly $T_e >$ 10 eV), excitation of O_2 becomes similarly efficient but produces more 1 356 Å emission. Dissociative excitation of H_2O has overall lower rates for producing oxygen emission, thus requiring larger abundances to reach similar intensities. A single measurement of the OI 1356 Å yield from impact of 100-eV electrons suggests a OI 1 356 Å/OI 1 304 Å ratio of $r_{\gamma O} \sim 0.2$ from H_2O.

The OI 1 356 Å/OI 1 304 Å ratio derived from earlier GHRS and STIS data revealed ratios between 1.2 and 3.2 (with large uncertainties), most consistent with an O_2 atmosphere (Hall et al., 1998; Feldman et al., 2000).

The first observations of Ganymede using HST's high-sensitivity Cosmic Origins Spectrograph (COS) and of both the leading and trailing hemispheres within a single HST campaign then revealed a systematic difference in the OI 1 356 Å/OI 1 304 Å ratio between the two hemispheres (Molyneux et al., 2018). On the sunlit leading side, the ratio of the globally average intensities was found to be $r_{\gamma O} = 2.7 \pm 0.5$, thus even larger than, but still consistent with, pure O_2 within uncertainty. The average ratio found for the sunlit trailing side of $r_{\gamma O} = 1.4 \pm 0.2$ was significantly lower, requiring a constituent other than O_2 to reduce $r_{\gamma O}$ to values below 2 (Figure 16.2b). Assuming this

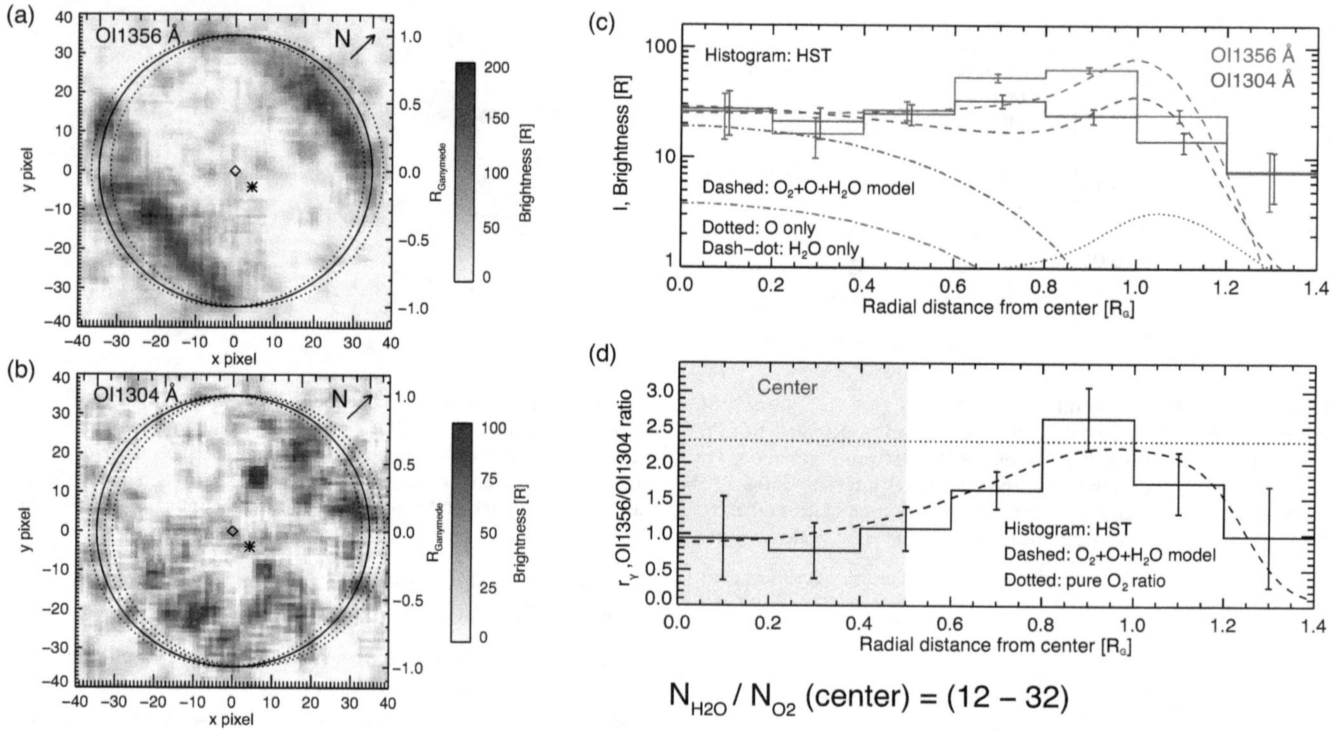

Figure 16.4 (a) and (b) HST/STIS images of the OI 1 356 Å and OI 1 304 Å emissions observing Ganymede's trailing hemisphere from Roth et al. (2021) (see their figure 3 for more details). (c) The radial brightness profiles derived from these OI 1 356 Å (red) and OI 1 304 Å (blue) images show that near the disk centre both emissions reach similar intensities. Simulated radial profiles for an assumed H_2O-O_2-O atmosphere model are shown for the total modelled intensity (dashed) and for individual contributions. (e) The profile of the observed oxygen emission ratio r_γ(OI) agrees with the ratio of O_2 (dotted) only near the limb but is in good agreement with the O_2+O+H_2O atmosphere model (dashed) assumed by Roth et al. (2021) at all radial distances. The derived column density ratio N_{H_2O}/N_{O_2} in the centre region (shaded grey in panel e) is shown at the bottom. (A black and white version of this figure will appear in some formats. For the color version, please refer to the plate section.)

constituent is atomic oxygen, O, Molyneux et al. (2018) showed that the required O column densities would put Ganymede's atmosphere in the optically thick range for OI 1 304 Å emission. This would imply that resonant scattering is an important contribution to the OI 1 304 Å intensity. Molyneux et al. (2018) also briefly discuss the morphology in the OI 1 304 Å images from individual exposures, mentioning large regions on the disk where the oxygen ratio is particularly low with $r_{\gamma O} < 1$.

Roth et al. (2021) tested the optical thickness of atomic oxygen atmosphere through a sequence of two exposures with COS before and after Ganymede enters into eclipse, but found no indication that resonant scattering contributes to the OI 1 304 Å emission. The upper limit on the O abundance imposed by this result implies the need for another species causing the measured $r_{\gamma O}$ ratios, which were found to be particularly low on the central trailing hemisphere; see Figure 16.4.

Roth et al. (2021) then showed that the observed oxygen emission ratios are consistent with an H_2O abundance around the central sunlit disks with H_2O/O_2 abundance ratios of 12–32 on the trailing side and 2–5 on the leading side. Around the sub-solar point, dissociative excitation of H_2O contributes significantly to the OI 1 304 Å emission, effectively reducing the oxygen ratio. We note that in all scenarios, the OI 1 356 Å emission almost exclusively originates from electron impact on O_2.

The higher H_2O/O_2 ratio above the warmer trailing hemisphere compared with the colder leading hemisphere and the spatial concentration at the sub-solar region make sublimation the most plausible source for this H_2O atmosphere. In addition, when assuming a column density for O_2 of $\sim 3\times10^{14}$ cm^{-2}, the resulting H_2O column densities are $\sim 10^{15}$ cm^{-2}, also consistent with expected abundances from ice sublimation at Ganymede's maximum surface temperature.

16.2.5 New Horizons P-Alice Observations

The New Horizons (NH) spacecraft observed Ganymede with the Pluto-Alice (P-Alice) instrument during its Jupiter encounter in 2007. The P-Alice UV imaging spectrograph spectrally images extreme- and far-ultraviolet light from 520 Å to 1 870 Å (Stern et al., 2008). P-Alice monitored the OI 1 356 Å and OI 1 304 Å emissions during two eclipse passages of Ganymede (denoted GEclipse01 and GEclipse02) from before ingress through after egress, viewing the sunlit sub-Jupiter (sub-observer W longitude ~9°, on February 27) and night-side anti-Jupiter (sub-observer W longitude ~155°, March 4), respectively.

Figure 16.5 shows that the disk-averaged oxygen OI 1 356 Å intensities measured in each event remain relatively stable over the ~200-minute time scale, with possible small increases of some 10 to 20 per cent from pre-ingress through eclipse to past egress. The stability of the OI 1 356 Å emissions, which almost exclusively originates from electron impact on O_2, suggests a stable O_2 atmosphere through the eclipse passage, assuming changes in electron excitation do not counterbalance changes in the atmosphere.

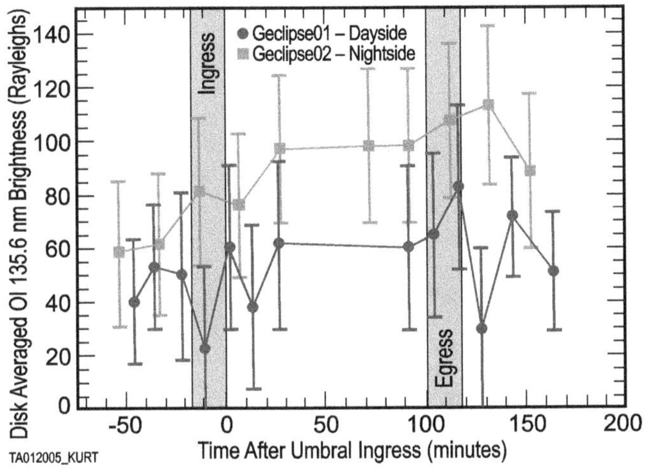

Figure 16.5 New Horizons Pluto-Alice observations of Ganymede's oxygen emissions obtained in 2007 provide the first views of nightside auroral emissions and generally demonstrate a stability of signals during the passage into and out of eclipse for both dayside (GEclipse01) and nightside (GEclipse02) hemispheres.

Moreover, the intensity in the second P-Alice eclipse event (GEclipse02), viewing primarily the nightside hemisphere (one week after the first event – GEclipse01 – that viewed the dayside), suggests that O_2 is similarly abundant on the nightside. The 50 per cent higher intensity of GEclipse02 compared to GEclipse01 might be attributed to the different Jovian magnetospheric plasma conditions at Ganymede. As HST observations are limited to the dayside, the P-Alice GEclipse02 data provide the first evidence for Ganymede's nightside atmosphere.

The analysis of the weaker oxygen OI 1 356 Å is hampered by the interfering surface reflections and possible extended cloud emissions. The average intensity for each event suggests OI 1 356 Å/OI 1 304 Å ratios between 1 and 3 on both days, consistent with previous results. The low signal-to-noise ratio during the second eclipse (GEclipse02) prevents further interpretation of the oxygen ratio with respect to the atmospheric mixing ratio on the nightside.

16.2.6 Other Atmosphere Observations

Already in 1995, Brown (1997) carried out long-slit spectroscopic observations of Ganymede in the search for emissions from sodium (Na) D1 and D2 lines at 5 896 Å and 5 890 Å, respectively. However, no emissions were detected, and an upper limit for the Na column density of 10^8 cm^{-2} was derived, more than an order of magnitude lower than the extended sodium corona discovered earlier at Europa (Brown and Hill, 1996).

Other noteworthy, yet unpublished observations were carried out by the Herschel Space Observatory (Hartogh et al., 2013). Herschel's Heterodyne Instrument for the Far Infrared (HIFI) measured absorption of the background continuum from Ganymede's surface at the 557 GHz water line. Interestingly, HIFI detected this absorption on the leading hemisphere but not on the trailing side, contrary to the results from HST (Roth et al., 2021). In addition, the derived column density for the leading hemisphere from Hartogh et al. (2013) (which we assume to be averaged over the disk of Ganymede) is about an order of magnitude lower than the value estimated based on the HST results for the same hemisphere.

16.3 SOURCES AND SINKS IN GANYMEDE'S ATMOSPHERE

Since the tenuous atmospheres of the icy moons are constantly replenished on timescales near or below their orbital periods, understanding the sources and sinks is key for understanding the abundance and distribution of the neutrals. In the absence of a direct detection of outgassing plumes, Ganymede's atmosphere is assumed to be produced from its icy surface through (i) solar irradiation (i.e., sublimation and photon-stimulated desorption), (ii) Jovian plasma irradiation (i.e., sputtering; this will be discussed later), and (iii) possibly micro-meteorite and dust grain impacts. Whereas this processing resembles that occurring at Europa, the presence of an intrinsic magnetic field on the plasma flow results in interesting spatial differences in the source rates. Therefore, comparisons between the two moons can help separate the relative importance of the sputtering sources. The most important source and sink mechanisms that act on Ganymede – sketched in Figure 16.6 – are described briefly in the following two sections. For a more detailed description of these, as well as the minor source processes, the reader is referred to Chapter 15 of this book.

16.3.1 Sources

As with other objects with thin atmospheres, the solar flux drives H_2O sublimation and thermal desorption of volatiles, namely release of non-condensable molecules trapped in Ganymede's regolith. Sublimation of H_2O is strongly temperature dependent. Ganymede's maximum surface temperature was measured by the Galileo photopolarimeter to reach ~150 K at the sub-solar point (Orton et al., 1996), though the maximum temperature of the water-ice may be significantly lower. At 150 K, the pure H_2O sublimation flux is equal to $\sim 1\times 10^{25}$ $m^{-2}s^{-1}$, whereas at 120 K, the sublimation flux reaches $\sim 4\times 10^{14}$ $m^{-2}s^{-1}$ (cf. Figure 15.2 in Chapter 15). The global sputter flux, on the other hand, was computed by the models that will be described shortly to be of the order of 10^{12}–10^{13} $m^{-2}s^{-1}$. Sublimation thus clearly dominates as a source process at Ganymede's sub-solar point. Away from the sub-solar point sublimation quickly drops off, and at the nightside temperatures of ~80 K the sublimation flux is negligible. Because the sublimation process is relatively well defined and modelling of it comparably simple, models usually do not focus on the implementation of this release process, but rather on the implementation of the more complex sputter release process. It is noteworthy, though, that even though most of the models that will be presented shortly put little emphasis on sublimation modelling, all of them agree on sublimated H_2O being the main exospheric constituent close to the sub-solar point.

The now extensive laboratory and plasma data sets confirm that the Jovian plasma bombardment is an important source of water products. Ganymede is embedded in the Jovian plasma, which consists of electrons, H^+, O^{n+}, and S^{n+}, as

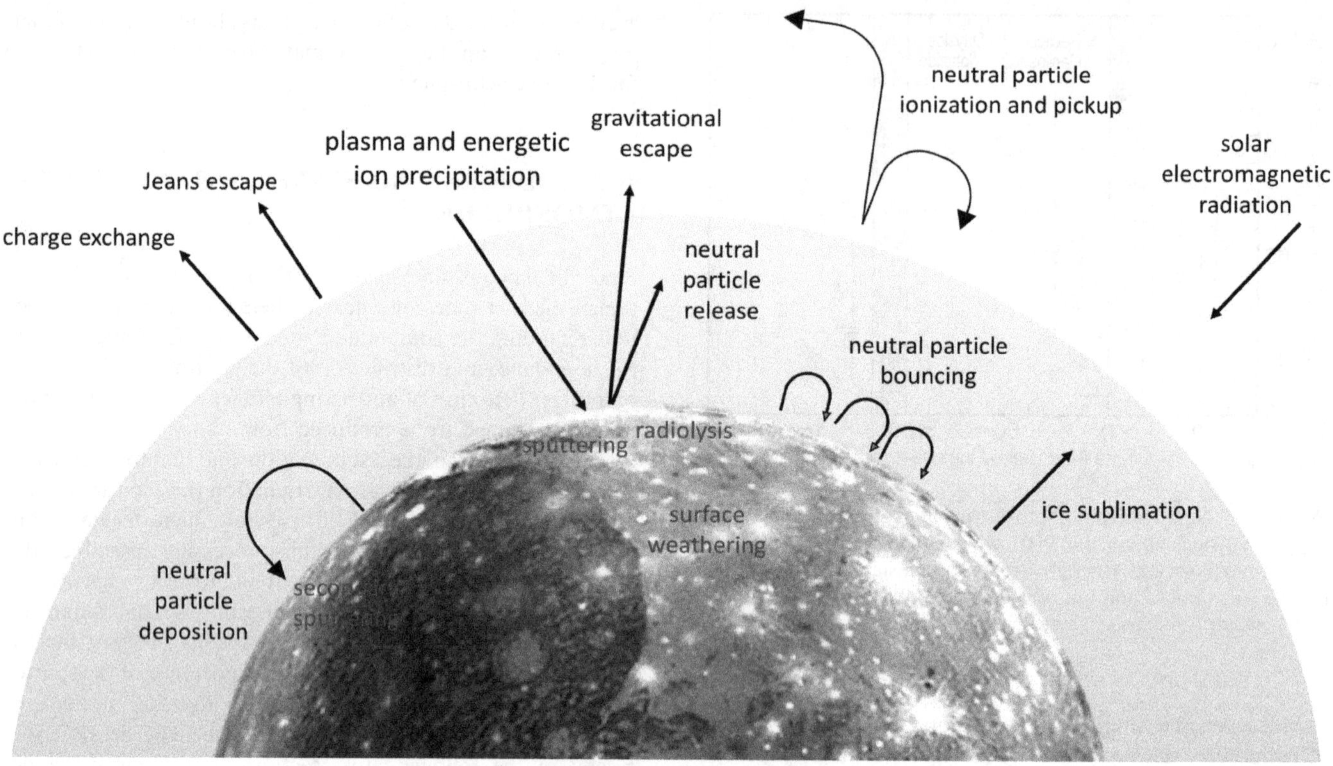

Figure 16.6 Main source and sink mechanisms for Ganymede's exosphere (from Plainaki et al. (2020a)). The moon's environment is strongly coupled with the Jovian magnetosphere through processes that determine the energy exchange between the two systems. Ganymede image credit: NASA/JPL/DLR. (A black and white version of this figure will appear in some formats. For the color version, please refer to the plate section.)

well as trace species with energies of a few eV to MeVs, and moves relative to Ganymede with bulk speeds equivalent to ~160eV/amu (Chapter 3.10). The source of sulphur and oxygen is primarily Iogenic with smaller sources of water products due to atmospheric escape from Europa and Ganymede fueling hydrogen-oxygen neutral tori (Smyth and Marconi, 2006; Marconi, 2007; Smith et al., 2019). Hence, plasma particles reaching Ganymede's surface can produce O_2, O_3, and H_2O_2 in the surface ice grains through a process called radiolysis [Spencer et al., 1995; Noll et al., 1996; Hendrix et al., 1999] and can brighten the polar regions (Johnson, 1997). Plasma bombardment leads to the ejection of H_2O, O_2, H_2, minor water dissociation products, and trace contaminants into the gas phase, a process often referred to as sputtering. Sputtering typically refers to the ejection produced by moment transfer collisions of incident ions with atoms and molecules in the surface of a solid (often referred to as knock-on collisions). Unlike metals [Wood, 1957], which are conductors, ices are weakly bound insulators and therefore can also undergo electronic sputtering, which has been shown to depend on the density of electronic excitations produced in the near-surface region (see figures 2.4 and 3.22b in Johnson (1990)) by impinging ions or electrons (Johnson et al., 2013).

The amount of each species ejected is typically given as a yield, Y, the number of ejected particles per incident particle. These yields depend on the incident plasma particle's energy, impact angle, mass, and charge, and on the temperature of the ice grains. Since the plasma varies temporally and Ganymede's field affects the spatial and energy distributions of particles reaching a surface that is porous and has contaminants, and since the incident plasma implants radicals and causes sintering of the grains, the parameter space of this interaction process is large. Therefore, observational and laboratory data are used synergetically in the modelling efforts.

16.3.2 Sinks

The atmospheric loss rates, as expected, are governed by interaction with solar photons and plasma particles, as well as by escape to space and loss to the regolith. Because of Ganymede's significant gravity, thermal escape is negligible except for the lightest neutrals (H, H_2; Marconi, 2007) or species energized in the gas phase by photon- and plasma-induced dissociation, as well as by charge exchange and momentum transfer collisions with the incident ions, which are also affected by Ganymede's magnetic field (e.g., Leblanc et al. (2017); Carnielli et al. (2020*a*)). Although these sinks exhibit considerable uncertainties, the physical processes are understood. However, the fallback onto the radiation-altered and very porous regolith (85 per cent at the surface: e.g., de Kleer et al. (2021)) and the further fate of the neutrals are not well understood.

Most of the gas phase ejecta return to the surface, where they can condense, react with the grains or permeate the porous regolith as molecules adsorbing and desorbing on the grains (e.g., Oza et al., 2019). The likely important role of the regolith as a potential sink is suggested, for example, by efforts to use the plasma observations to calculate the atmospheric density of O_2 at the icy moons Rhea [Teolis and Waite, 2016] and Ganymede [Leblanc et al., 2017]. Either extremely different models for the radiolytic production of O_2 are required or the fate of the returning O_2 in their regoliths is very different. Whereas there is considerable laboratory data on the production, much better models for the fate of gas phase species in the icy regoliths are required.

16.4 MODELLING

Much of our understanding of Ganymede's atmosphere is based on modelling efforts, which, very often, incorporate the outputs of extensive laboratory work (e.g., Johnson (1990); Johnson et al. (2004) and references therein). Although the modelling efforts presented in this section are clearly distinct from each other (applying different techniques, focusing on different plasma populations and neutral species, including different interaction processes, analyzing different atmospheric aspects, etc.), they all agree on the general structure of Ganymede's atmosphere: close to the sub-solar point, the atmosphere is dominated by sublimated H_2O molecules, whereas near the terminator and on the nightside the atmosphere is dominated by O_2 close to the surface and by H_2 at higher altitudes.

According to a model implementation by Shematovich (2016), Ganymede's atmospheric environment is characterized by a Knudsen number less than 0.1 up to altitudes of about 50 km, and by a Knudsen number in the range of 0.1–1 at altitudes between ~50 km and a few hundred kilometres. The Knudsen number compares the mean free path of the neutrals between collisions to the local scale of density variations. A Knudsen number (well) below unity indicates a collisional regime. Other modelling efforts (Marconi, 2007; Leblanc et al., 2017) have resulted in similar conclusions, with the atmosphere being collisional or quasi-collisional in a near-surface region near the sub-solar point, and collisionless elsewhere. As a result, some models treat the atmosphere as completely collisionless, while other studies account for neutral–neutral collisions.

Since Ganymede's atmosphere and the Jovian plasma environment are strongly intertwined, it is difficult to investigate one without the other. Atmospheric modelling efforts thus often focus on the Jovian plasma environment and its precipitation onto Ganymede's atmosphere and surface. Yet we remind the reader that sublimation is likely an equally important source but receives less attention solely due to its comparably simple nature.

In the discussion that follows, we first present modelling efforts of the effects of Ganymede's interaction with the magnetosphere on the surface sputtering source as well as the thermal atmosphere source, namely sublimation. Thereafter, we discuss the modelling studies of Ganymede's neutral atmosphere, highlighting new aspects that models included at the time of their introduction.

16.4.1 Modelling of the Sputtering Source

In order to understand where and how surface sputtering releases particles, one first has to understand where and how electrons and ions precipitate onto the moon. This is often accomplished through modelling. Space-based measurements of the environment around Ganymede (e.g., Cooper et al., 2001) have often served as an input for such numerical ion circulation models, which in addition make use of electric and magnetic fields from MHD simulations of Ganymede's environment.

Single-particle Monte Carlo simulations of the circulation and precipitation of energetic ions onto the moon can then indicate the surface regions where intense sputtering and radiolysis are expected (Carnielli et al., 2020*a*; Plainaki et al., 2015; Plainaki et al., 2020a; Plainaki, et al., 2020b). Such ion and electron trajectory computations have been performed for particles in a wide energy range (keV – MeV) for different configurations between Ganymede's magnetic field and Jupiter's plasma sheet: namely the moon above, inside and below the plasma sheet (e.g., Poppe et al. (2018); Plainaki et al. (2020a); Liuzzo et al. (2020); Carnielli et al. (2020*a*)). Such simulations have confirmed the existence of a shielded region close to the trailing hemisphere surface (due to the presence of an internal magnetic field) and enhancements in the ion flux at near-surface altitudes above the low-latitude leading hemisphere. In general, current modelling efforts agree in that ions precipitate predominantly at the moon's polar cap regions along open magnetic field lines, consistent with Ganymede's surface brightness map (Khurana et al., 2007).

As the Jovian plasma consists of electrons and different populations of ions, studies have been undertaken to determine which particles are most relevant for sputtering. According to these modelling efforts, even though the thermal ion precipitation flux of the corotating plasma is similar to the energetic ion precipitation flux (Carnielli et al., 2020*b*), it is the latter that primarily determines the atmosphere sources [Cooper et al., 2001; Gomis et al., 2004; Teolis et al., 2017; Plainaki et al., 2020a] due to the increase in sputter yield with energy in the relevant energy region, and it is also the latter that controls surface weathering (e.g., Johnson (1990); Shi et al. (1995); Khurana et al. (2007)).

Recently, Carnielli et al. (2020a) modelled precipitation of ionospheric ions (in addition to the Jovian plasma ions) onto Ganymede's surface to determine the ionosphere-induced sputter flux and its relative importance. The authors calculated that at least 10 per cent of the overall sputter flux could be due to ionospheric ions, and argued that this contribution could even become the dominant sputtering source under certain assumptions for not-well-constrained parameters such as the O_2 density and ionization frequency. The authors also found that an O_2 column density of $>10^{15}$ cm^{-2} (about 10 times higher than in other studies) is required to match the electron density profiles derived from Galileo measurements.

16.4.2 Modelling of Thermal Sources

Following the identification of energetic ions at Jupiter by Voyager 1 (Krimigis et al., 1977), a *sublimated* water-based atmosphere was suggested to transiently overwhelm the directly sputtered atmosphere predicted at Ganymede by Johnson et al. (1981) based on the equilibrium vapor pressure of H_2O ice. That is, while Europa experiences temperatures approaching ~135 K, equatorial temperatures on Ganymede approach ~150 K near noon, local time. For a collisionless atmosphere (Section 16.4.4) the canonical water column density is a product of the ballistic hop time and a sublimation flux Φ_{sub}, which can be expressed as $\Phi_{sub} = aT^{1/2}exp(-b/T)$, where a and b are empirical parameters constrained by vapor pressure experiments (Table 16.3).

Table 16.2 *Models of Ganymede's neutral atmosphere: *Collisionless and † Collisional. ‡ Sublimation refers to H_2O sublimation in this table. Rates can be found in Feistel and Wagner (2007) or Fray and Schmitt (2009). Ch-ex refers to charge exchange. * See also Plainaki et al. (2020a) ** Carnielli et al. (2019, 2020a,).*

Model	Species	Source	Loss	Novelty	Implications
Johnson et al. (1981)*	O_2, H_2O, H_2, H, OH	Sputtering, Sublimation‡	sticking, e-/photo-dissociation, escape	1st analytic Model	Prediction of O_2 and H_2O atmospheres on icy moons
Marconi (2007)†	O_2, H_2O, H_2, O, H, OH	Sputtering, Sublimation	sticking, e-/photo-dissociation, escape	2-D axisymmetric	Atmospheric profile; Latitudinal distributions: H_2O subsolar; O_2 elsewhere
Turc et al. (2014)*	O_2, H_2O, H_2, O, H, OH	Sputtering, Sublimation	e-/photo-dissociation, escape	3-D Model, Atm. collapse in eclipse	Collisional atm. at noon; H density underestimated
Plainaki et al. (2015)*	O_2, H_2O	Sputtering, Sublimation, Radiolysis	sticking, e- impact dissociation, ch-ex, escape	Plasma precipitation patterns*	Sputtered H_2O variability: leading/trailing; poles/equator; 2-component O_2 atm.
Shematovich (2016) †	O_2, H_2O, H_2, O, H, OH	Sputtering, Sublimation	e-/photo-dissociation, ch-ex	Collisions important	Suprathermal O
Leblanc et al. (2017) †	O_2, H_2O, H_2, O, H, OH	Thermal Radiolysis, Sputtering, Sublimation	sticking, e-/photo-dissociation, escape	3-D Rotational**	H_2O variability: Poles/Noon & leading/trailing; Dusk-over-Dawn O_2 Bulge
Oza et al. (2018)*	O_2	Outgassing, Thermal Radiolysis, Sputtering	e- impact dissociation	1-D Semi-Analytic Atm. Rotation Model	Dusk-over-Dawn O_2 Bulge; presence of near-infinite, outgassing O_2 reservoir
Vorburger et al. (2022)*	H_2O	Sputtering, Sublimation	sticking, e-/photo-dissociation, e-/photo-ionization, escape	Ion- and electron-sputtering, plasma precipitation patterns	Sublimation-dominated regions constrained, importance of $O^{+,++}$ and S^{+++} sputtering, electron-sputtering negligible

Along with the direct phase transition into sublimated water vapor, the desorption of the radiolytic O_2 that permeates the regolith (Oza et al., 2018; Johnson et al., 2019) and potential H_2O outgassing (Roth et al., 2014) as at Europa might also occur at Ganymede. Nevertheless, at present, observational evidence of a H_2O column at Ganymede remains to be confirmed against these predictions, presenting an avenue for characterization for future observers. In the following section we discuss simulations in an effort to characterize the preceding sources and sinks based on various assumptions, for example, collisional (Section 16.4.3), collisionless (Section 16.4.4), and rotation (Section 16.4.5).

16.4.3 Collisional and Quasi-Collisional Atmosphere Models

The 2-D axisymmetric kinetic model by Marconi (2007) used a multispecies approach to derive the atmospheric H_2O, O_2, H_2, O, H, and OH spatial structures. The model included implementation of a non-uniform surface temperature distribution, energetic plasma energy spectra as measured by the Galileo energetic particle detector when Galileo was near Ganymede, and energy-dependent sputter yields, which are based on laboratory studies. The authors found that near the sub-solar point the dominant component is sublimated H_2O vapor, whereas near the polar regions the main component is molecular oxygen

Table 16.3 *Pure H_2O ice grain sublimation flux parameters for a simplistic sublimation model.* Adapted from Oza (2017) Figure 3.5, *p. 93. We note that all experimental measurements to our knowledge of H_2O, summarized in Fray and Schmitt (2009), range from 130 to 273.15 K encouraging the need for more experiments at low temperatures.*

Model	a	b
	[$cm^{-2}\ s^{-1}$]	[K]
Johnson et al. (1981)	1.9×10^{32}	6 146
Marconi (2007)	1.1×10^{30}	5 737
Fray and Schmitt (2009)†	2.2×10^{32}	5 950
Leblanc et al. (2017)	1.9×10^{32}	8 500

† Fray and Schmitt (2009) use a more complicated formula for the sublimation flux, which we have interpolated to the simpler form for this comparison.

at low altitudes and molecular hydrogen at high altitudes, two species that do not condense easily and thus accumulate in the atmosphere (see Figure 16.7). A similar approach was also followed by Shematovich (2016) who used a kinetic model to investigate the near-surface atmosphere, taking collisions into account, and the extended atmosphere by including atmospheric sputtering and thermalization of supra-thermal atoms. The authors found that collisions caused the 10–100 km altitude to be populated by both O_2 and H_2O, and that inclusion of atmospheric sputtering and thermalization of supra-thermal atoms resulted in a substantial increase in the scale height of these species. These results were later confirmed by Leblanc et al. (2017) who used a 3D multi-species collisional approach (see the following discussion).

16.4.4 Collisionless Exosphere-Type Models

Several models have used a Monte-Carlo model approach to treat Ganymede's neutral gas environment as an exosphere, that is, neglecting collisions between atmospheric particles. Turc et al. (2014) applied a collisionless 3D model approach for simulating Ganymede's exosphere above the sunlit trailing hemisphere. The obtained exosphere confirmed the results that were obtained by the kinetic model of Marconi (2007), namely that sputtering and sublimation result in a strong dichotomy between the sub-solar region, where H_2O sublimation dominates, and the rest of the surface, where O_2 and H_2 dominate. In addition, the authors studied Ganymede's atmospheric response to time-varying parameters, namely Ganymede passing into the shadow of Jupiter and intrinsic variations in the Jovian plasma (for the results of these analyses see Section 16.4.5).

The exospheric model by Plainaki et al. (2015) was the first to account for the intensity and spatial distributions of the precipitating Jovian ions onto Ganymede's surface and to derive the induced surface-sputtered flux. They showed that sputtering alone would produce an H_2O distribution confined to mostly the polar regions and the low-latitude wake hemisphere, whereas sublimation releases H_2O in the sub-solar region; see Figure 16.8. The modelled O_2 exosphere comprises a nearly globally uniform and relatively dense part, one that is close to the surface and consists mainly of thermal molecules (molecules thermalized to surface temperature), and a less homogeneous part consisting of more energetic O_2 molecules resulting from direct surface sputtering, with a spatial distribution that depends on both the plasma surface impact and surface temperature. In addition, Plainaki et al. (2020a) investigated ion precipitation onto Ganymede's surface for different relative configurations between the moon's magnetic field and Jupiter's plasma sheet, that is, when Ganymede is above, close to the centre of, and below the Jovian plasma sheet, and found that the sputtered H_2O flux can easily change by an order of magnitude.

Finally, like Plainaki et al. (2015), Vorburger et al. (2022) accounted for the intensity and spatial distributions of the precipitating magnetospheric H^+, O^+, O^{++}, and S^{+++} ions onto Ganymede's surface, but also included electron precipitation fluxes in their analysis. The authors confirmed the H_2O distribution found earlier by Marconi (2007), Turc et al. (2014), and Plainaki et al. (2015), and showed that sputtering of the water atmosphere is mainly induced by the heavy ions, while protons and electrons play only a very minor role. In addition, their modelled H_2O atmosphere is in good agreement with the observations of a sublimated water component (Roth et al., 2021).

16.4.5 Atmosphere Models Including Orbital Variability

Leblanc et al. (2017) advanced the 3D model of Turc et al. (2014) in order to describe the orbital evolution of the atmosphere, by taking into account Jupiter's gravitational influence, the plasma upstream (trailing)/plasma wake (leading) orbital asymmetry, and Ganymede's rotation. They find that Ganymede's sublimated and sputtered atmosphere has a characteristic timescale (τ_i), determined primarily by electron-impact dissociation and ballistic flight times, which is similar to Ganymede's rotational period (τ_{orb}). This implies that Ganymede's atmosphere at a given orbital position is strongly dependent on the atmospheric source and sink processes occurring earlier in the orbit. In other words, in the absence of sources, the atmosphere is destroyed in less than one Ganymede day and is shaped by what happened earlier in the day. Figure 16.9 illustrates the consequences of the orbital evolution, most pronounced in O_2.

Such rotation-driven asymmetries are most apparent in the abundance of comparably long-lived volatiles like O_2. On Earth's Moon, non-condensable volatiles such as Ar exhibit a similar dusk-over-dawn asymmetry due to the balance of sources and sinks over an orbit. Hodges and Johnson [1968] tracked the latitudinal evolution of volatiles via density gradients. Recently Oza et al. (2018) used a simple analytic model to track the longitudinal evolution of the column density of a volatile. In this way, they examined the effect of atmospheric source properties on the longitudinal distribution of O_2

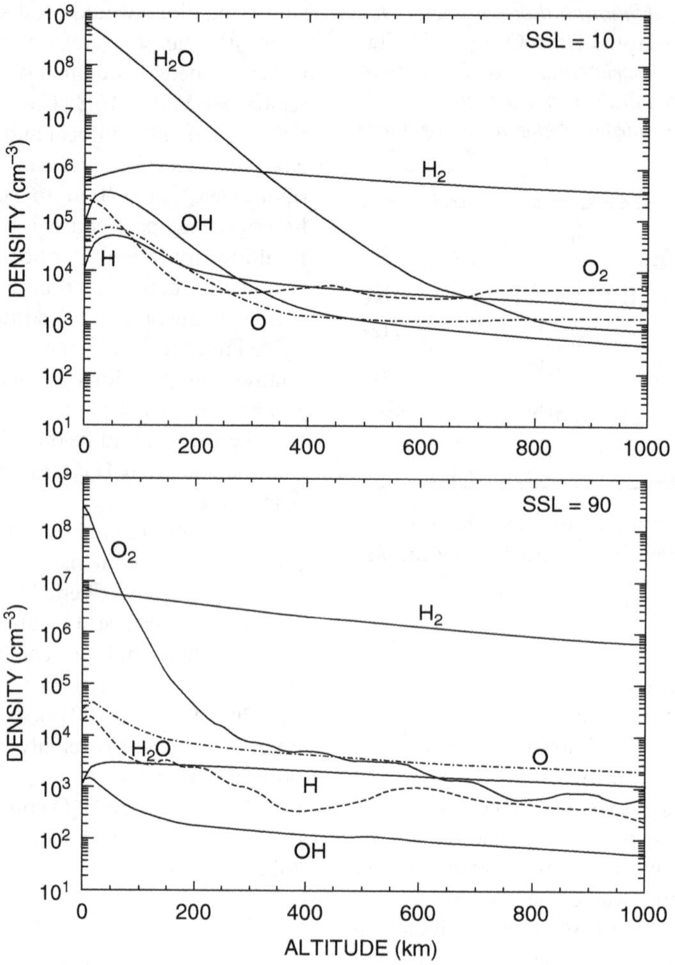

Figure 16.7 Atmospheric density profile at a sub-solar latitude (SSL) of 10° (a) and 90° (b), from Marconi (2007), suggesting a two-part atmosphere with a dominant sublimated H_2O atmosphere near the sub-solar point (upper panel) and a primary sputtering O_2 atmosphere elsewhere. This was later confirmed and supported by various models and observations. Reprinted from *Icarus*, 190(1), M.L. Marconi, A kinetic model of Ganymede's atmosphere, 155–174., ©2007, with permission from Elsevier.

in a surface-bounded atmosphere on tidally locked satellites, including Ganymede.

By modelling the atmospheric source and sink (Section 16.3) processes to assess the longitudinal and, therefore, time *evolution* of the atmospheric column density N in the form

$$\left(\frac{dN}{dt}\right) = Source - (Sink) \times N, \tag{16.1}$$

the atmospheric evolution parameter $\beta = 2\pi \frac{\tau_i}{\tau_{orb}}$, the ratio of the atmospheric lifetime to the diurnal timescale of the satellite, was shown to be critical. When β is very large, there is little longitudinal dependence, and when very small, the column is dominated by production. As β approaches unity, a dusk-over-dawn asymmetry can appear, similar to a thermal tide. This has been shown to account for the observed dusk/dawn enhancement in O_2 at Europa (Roth et al., 2016). That is, the radiolytically produced but returning O_2 permeates its porous regolith, providing a *thermal* source of O_2 that is much less sensitive to the local radiolytic production rate from the surface ice. A peak abundance near dusk local time is similarly expected for Ganymede (Leblanc et al., 2017; Oza et al., 2018) based on the estimated atmospheric residence time of O_2, which is similar to Ganymede's orbital period. In this picture, radiolysis-induced desorption of O_2 from its icy regolith is an important source at Ganymede as well as Europa (Johnson et al., 2019). Leblanc et al. (2017) also suggested that, even if the atmospheric O_2 molecules are preferentially ejected from the polar regions by sputtering, they will tend to accumulate in the regolith in the equatorial region as their many ballistic hops are affected by the centrifugal acceleration.

Several models (Plainaki et al., 2015; Leblanc et al., 2017) have suggested trailing-over-leading asymmetries of H_2O. Since H_2O freezes upon surface contact, it has a much shorter lifetime compared to O_2 and the distribution of H_2O is closely related to its source regions. Since Ganymede's trailing hemisphere has a lower albedo than the leading hemisphere (Spencer et al., 1989), the trailing hemisphere surface temperature is expected to be higher (Leblanc et al., 2017). Assuming the same level of contamination of the ice everywhere, highest sublimation flux is expected near the sub-solar point when the trailing hemisphere is illuminated, while the sublimation H_2O atmosphere is less dense on the sunlit leading hemisphere. The maximum sputtering flux, in contrast, is expected near the poles (Plainaki et al.,

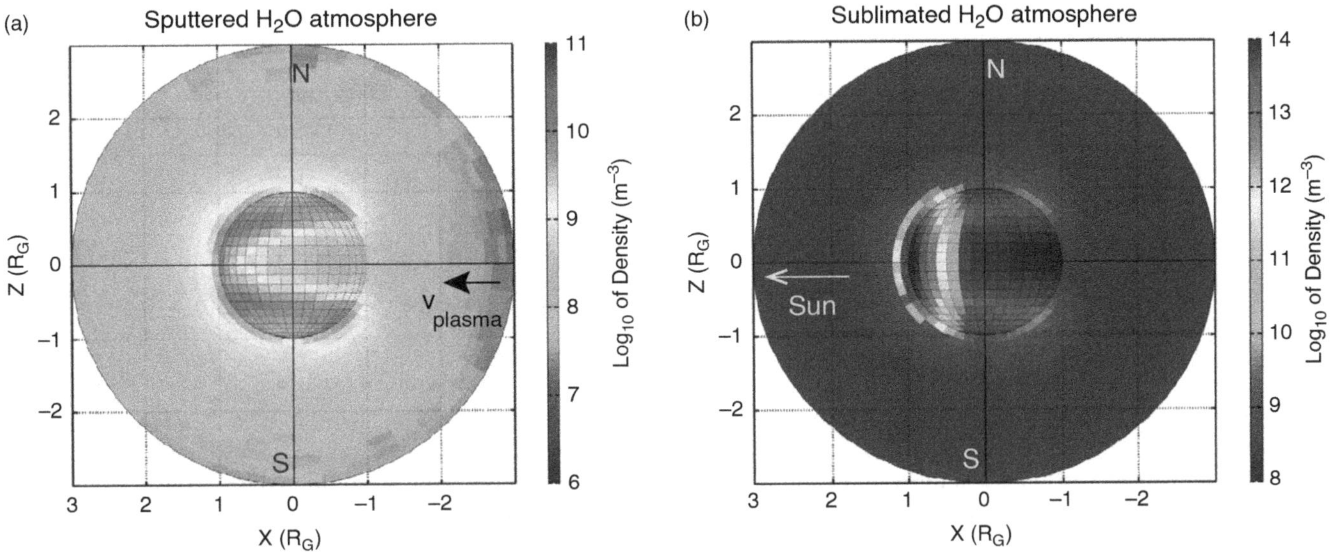

Figure 16.8 Projected atmospheric H_2O density distributions over the surface of Ganymede and in a plane through Ganymede's centre, where X is along the ambient plasma flow direction (shown by the black arrow) and Z is along Jupiter's spin axis (in units of Ganymede's radii, North 'N' is up), adapted from Plainaki et al. (2015). The Sun is to left, shown by the grey arrow. Panel a shows the H_2O that is produced from surface sputtering only, where H_2O is mostly confined to the regions near the open field lines. Panel b shows H_2O from sublimation only, with H_2O confined to the dayside. The peak sublimated density near the sub-solar points exceeds the maximum sputtered density by three orders of magnitude (note the different ranges of the color bars). Reprinted from *Icarus*, 245, Christina Plainaki, Anna Milillo, Stefano Massetti, Alessandro Mura, Xianzhe Jia, Stefano Orsini, Valeria Mangano, Elisabetta De Angelis, Rosanna Rispoli, The H_2O and O_2 exospheres of Ganymede: The result of a complex interaction between the jovian magnetospheric ions and the icy moon, 306–319, ©2015, with permission from Elsevier. (A black and white version of this figure will appear in some formats. For the color version, please refer to the plate section.)

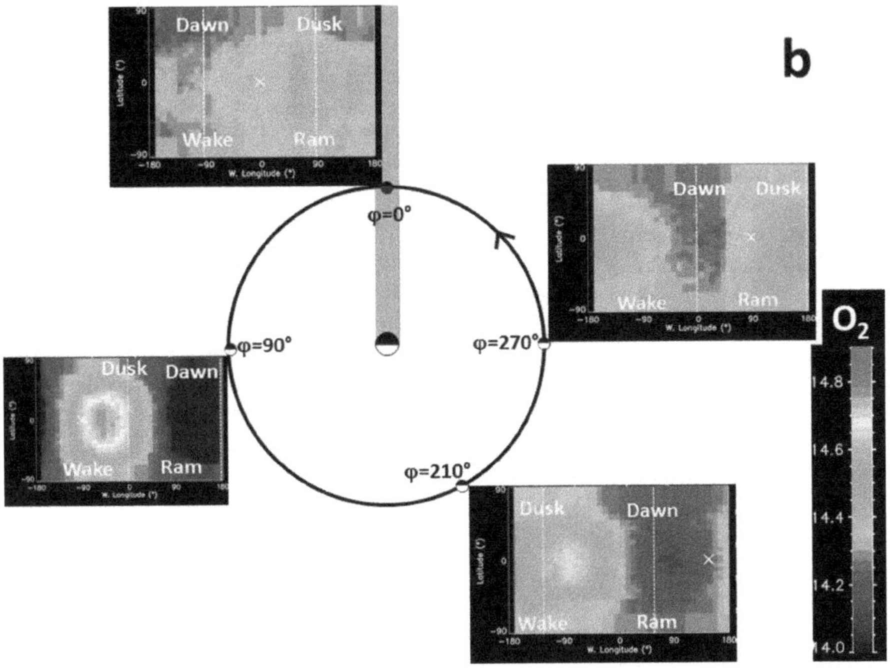

Figure 16.9 Simulated atmospheric evolution maps (latitude vs. W longitude) of zenith O_2 column densities at four Ganymede phase angles, from Leblanc et al. (2017). To follow diurnal behavior, the sub-solar point is indicated by an 'X' and dusk/dawn terminators are represented by the vertical dashed white lines. To follow plasma behavior, ram and wake are labeled. Ganymede's angular rotation vector Ω is counter-clockwise in the plane of Jovian co-rotation. Reprinted from *Icarus*, 293, F. Leblanc, A. V. Oza, L. Leclercq, C. Schmidt, T. Cassidy, R. Modolo, J. Y. Chaufray, R. E. Johnson, On the orbital variability of Ganymede's atmosphere, 185–198, ©2017, with permission from Elsevier. (A black and white version of this figure will appear in some formats. For the color version, please refer to the plate section.)

Table 16.4 *Surface density and column density ranges derived from atmospheric models. Values in parentheses are derived from observations, stated here for comparison.*

Species	Density [cm^{-3}]	Column Density [cm^{-2}]	Resident time [s] – limiting loss process†	References
H_2O (sub-solar)	2e8–4e9 (2e8–1e9)	6e15–2e16	3e2–1e3 – return to surface	a,b,c,d,e,f
H_2O (polar)	4e3–3e4	2e11–3e11		a,b,c,d,f
O_2	1e7–9e7 (1e8)	4e13–2e15 (3e13-1e15)	1e3–7e5 – dissociation/ionization	b,c,d,e
H_2	1e5–8e6	1e14–1e15	4e4–1e5 – escape	a,b,d,e
OH	~(1e2–1e4)	~(1e10–1e12)	~(1e3–1e5) – escape/return to surface	a,b,d
O	~1e4	~1e12	~1e4 – escape/return to surface	a,b,d
H	~5e3	~1e12	~1e4 – escape	a,b,d,e

†Resident time in the atmosphere and the relevant limiting process that determines the time scale, [a]Marconi (2007), [b]Turc et al. (2014),[c]Plainaki et al. (2015), [d]Shematovich (2016), [e]Leblanc et al. (2017), [f]Vorburger et al. (2022).

2015; Leblanc et al., 2017; Vorburger et al., 2022) (Figure 16.8). Depending mostly on the H_2O lifetime and surface thermal inertia, the dayside sublimation H_2O atmosphere should collapse or fade during an eclipse passage of Ganymede (Leblanc et al., 2017).

16.5 SUMMARY AND PERSPECTIVES

16.5.1 Summary of Observational and Modelling Work

Since the first detections in the mid-1990s and with the following increased model effort, substantial progress has been made in characterizing and understanding Ganymede's tenuous atmosphere.

On the observational side, the disk-integrated intensities of the OI 1 304 Å and OI 1 356 Å oxygen emissions from the atmosphere suggest that the majority of the total signal originates from dissociative excitation of molecular oxygen, O_2. At higher altitudes, namely in the extended exosphere, the observations are consistent with an increased mixing ratio of atomic oxygen to molecular oxygen.

However, interpreting the observations of the electron-excited oxygen emissions is generally limited by the fact that the signal is affected by the properties of both the neutral gas and the electrons as exciting species. Given the lack of knowledge about the distribution and temperature of the electrons (i.e., their phase space density), it is impossible to extract reliable constraints on absolute abundances or spatial distributions of individual molecules or atoms. Hence, the available data does not allow one to test hypotheses about asymmetries, inhomogeneities, or temporal variability in the atmospheric abundances.

The relative intensity of the OI 1 304 Å and OI 1 356 Å emissions is a diagnostic for mixing ratio. As such, it is less sensitive to the electron properties compared to the absolute intensities as a diagnostic for absolute abundance.

With a global OI 1 356 Å/OI 1 304 Å ratio that has been found to be around 2 (roughly between 1.3 and 2.7), electron excitation of O_2 is the overall primary emission source. The nightside observations by New Horizons as well as observations when Ganymede is eclipsed by Jupiter confirmed the presence of stronger OI 1 356 Å than OI 1 304 Å even in absence of solar illumination. However, it was found that the oxygen ratio systematically decreases from the limb region towards the central disk region, which coincides with the sub-solar region in HST images. The systematically changing ratio is consistent with an atmosphere that transitions from being O_2 dominated near the terminator (and on the nightside) to an H_2O-dominated atmosphere on the central dayside.

Simulation efforts generally produce results that are well in agreement with these observational constraints, providing additional context to improve the understanding. Table 16.4 gives rough ranges of the surface densities, column densities, as well as atmospheric resident times for all modelled water group species, derived and estimated from various models. We note that this is a simplification and some results might not be considered if they differ significantly from the usual ranges. The given numbers are mostly in agreement with the few observational constraints at hand. The simulations predict a number of characteristics, which remain to be confirmed by future missions and observational campaigns.

Given the various sources (sublimation, ion and electron sputtering, desorption) and their specific characteristics, all models predict an inhomogeneous and temporally variable (mostly on the timescale of Ganymede's orbit) atmosphere. Differences in the abundance and distribution of O_2 over an orbit are expected due to the changing geometry and in particular the change in the plasma stream direction and thus the related topology of Ganymede's mini-magnetosphere relative to the solar illumination. The presence and even dominance at higher altitudes of molecular hydrogen, H_2, is also consistently predicted. For the nightside, the simulations suggest an O_2-dominated atmosphere, since the sputtering and radiolysis source for O_2 has a comparably small dependence on surface temperature (and thus time of day) and the long lifetime of the O_2 molecules is similar to the orbital period. Molecular hydrogen is similarly expected to be globally abundant.

Various properties remain overall unclear. One example is the uncertainty of the surface ice properties (temperature, contamination level, porosity, reactions in the regolith) and of the resulting sublimation yield. Furthermore, until advanced spacecraft measurements rule this out, it remains a possibility that localized sources such as outgassing plumes like those on Enceladus (Hansen et al., 2020) and possibly Europa (Roth et al., 2014) are present on Ganymede.

16.5.2 Brief Comparison to Europa and Callisto

Ganymede's atmosphere can be expected to have many similarities to the atmospheres of Europa and Callisto. The main sources are likely charged particle sputtering and sublimation, as all three are embedded in Jupiter's (nearly) co-rotating magnetospheric plasma and their outermost layers consist predominantly of water-ice.

However, there are also differences in the properties of both the moons themselves and their environments. The magnetospheric plasma density and strength of Jupiter's magnetic field decreases from Europa out to Callisto by roughly two orders of magnitude, leading to differing plasma interactions (Kivelson et al., 2004) and particle sputtering efficiency and distribution. In addition, Ganymede has the internal magnetic field that affects the regions where charged particles can impinge onto the surface as discussed in detail previously. The difference in plasma environment also affects the atmospheric losses related to plasma collisions.

On the other hand, the characteristics of the icy surfaces also differ (Greeley et al., 2004; Pappalardo et al., 2004; Moore et al., 2004). Europa's surface is youngest with the highest ice content, and the moon's albedo is highest and surface temperatures thus the lowest. Callisto's surface is oldest and most contaminated, making it overall darker and likely warmer. Ganymede is again in between in terms of the surface properties like contamination, albedo, and temperature. These differences can lead to significantly different sublimation yields and also affect the sputtering yields of the water group species (Spencer, 1987).

The atmospheres of all three moons have been observed through the FUV oxygen emissions, confirming some of the expected similarities and differences. In all cases, the strongest contribution to the oxygen emissions is interpreted to originate from electron-impact dissociated excitation of O_2 (Hall et al., 1998; Cunningham et al., 2015). The properties of the exciting electrons are yet different: at Europa, the thermal plasma electrons are thought to excite the emissions (Roth et al., 2016). At Ganymede, the particular band shape of the emissions and the strong local intensities hint at electrons that are locally accelerated. Callisto's emissions are most consistent with airglow, meaning that low-energy photo-electrons excite the neutrals, while the magnetospheric electrons are effectively diverted around the moon (Strobel et al., 2002). The FUV signal near the disk centre that was related to a sublimated H_2O atmosphere at Ganymede (Roth et al., 2021) was similarly found at Europa but only on the darker trailing hemisphere (Roth, 2021). This is consistent with the expectation that sublimation is overall less effective at Europa than at Ganymede due to the lower surface temperature. The faint airglow signals from Callisto's atmosphere have not (yet) provided means to properly constrain the abundance of species other than O_2. The infrared detection of CO_2 in Callisto's atmosphere (Carlson, 1999) was possible due to a fortunate observing geometry during a flyby of the Galileo spacecraft. Carbon dioxide might be present in the atmospheres of Europa and Ganymede as well.

16.5.3 Future Observations

After more than 30 years in space, the Hubble Space Telescope is still operating and there are two new programs for FUV observations of Ganymede being carried out in 2021. In addition, the NASA Juno spacecraft performs two flybys at Ganymede the same year, with the possibility to provide new insights on Ganymede's atmosphere from its in situ and remote-sensing instruments. The long-awaited James Webb Space Telescope (JWST) was launched on 25 December 2021. The JWST has the potential to map different species in the atmosphere such as H_2O and CO_2 at infrared wavelengths. Ganymede is one of several targets in a granted JWST Early Release Science program, which aims at learning and testing JWST's capabilities to probe the Jupiter system.

Hence, hopes are high that we get more observational constraints on Ganymede's atmosphere in coming years. A large step forward is expected through the JUICE mission scheduled to arrive at Jupiter around 2030.

16.5.4 JUICE Investigation Plans

A major science objective of the JUICE mission is the characterization of Ganymede as a planetary object and possible habitat (Grasset et al., 2013), including a detailed investigation of Ganymede's tenuous atmosphere and its interaction with the Jovian magnetosphere. Both in situ and remote-sensing instruments will contribute measurements to constrain atmospheric composition, structure, temperature, and volatile content, and repeated observations over multiple flybys and throughout the Ganymede orbit mission phase will facilitate studies of atmospheric variability. In situ measurements by the Particle Environment Package (PEP) instrument suite will be used to determine the neutral composition at low altitudes, while remote-sensing instruments will map emission and absorption features of key atmospheric species in the ultraviolet, visible-NIR, and sub-mm spectral regions, respectively. Spectral images of reflected sunlight obtained will also allow surface composition and variability to be investigated, providing valuable information about atmospheric sources and sinks.

The JUICE UltraViolet imaging Spectrograph (UVS) will map neutral emissions of hydrogen (1 216 Å) and oxygen (1 304 Å and 1 356 Å) at spatial resolutions on the order of a few kilometers, allowing the distribution and column density of O, O_2, and H_2O in Ganymede's atmosphere to be derived. Limb stares will be used to investigate the vertical structure of the atmosphere and will also provide opportunities to search for fainter signatures of potential minor atmospheric species such as S, C, and Cl, which have prominent emissions in the 1 000–2 000 Å wavelength region. UVS will also measure O_2 and H_2O in absorption by observing the attenuation of stellar (or solar) light as the star (Sun) is occulted by Ganymede's atmosphere. Occultation spectra will also be assessed for the presence of additional minor species with distinctive absorption features in the 510–2 040 Å UVS bandpass, including CO_2, SO_2, and O_3. Complementary observations of the UV surface reflectance will help to constrain the sources and sinks of the atmosphere, since key volatile species including H_2O, CO_2, and SO_2 have characteristic reflectance features in the far ultraviolet (e.g. Hapke et al., 1981).

The Moons And Jupiter Imaging Spectrometer (MAJIS) is a compact visible and near-infrared imaging spectrometer covering the spectral range from 0.5 to 5.54 μm (Guerri et al., 2018; Langevin et al., 2018; Piccioni et al., 2019). The MAJIS system has the potential to observe the non-LTE photon emission from Ganymede's water vapor exosphere in the 2.4–3 μm range through limb measurements. A first preliminary evaluation of such a potential was performed by Plainaki, et al. (2020b), considering two representative observation cases with the instrument orbiting at an altitude of 5 000 km above the moon's surface: one at high vertical resolution (projected instantaneous field of view of ~1.1 km) and one at low vertical resolution (projected instantaneous field of view of ~10 km). The estimations showed that the instrument's sensitivity to the moon's water vapor environment reaches a SNR of 1 during limb observations with tangent at an altitude of 100 km at low latitudes (with higher SNR for lower altitudes). Although ion sputtering contributes in the release of water molecules in Ganymede's environment, the dominant exosphere generation mechanism is sublimation (Marconi, 2007; Plainaki et al., 2015; Shematovich, 2016). As a result, for the aforementioned geometry, the expected MAJIS sensitivity will allow one to measure Ganymede's sublimated-water exosphere.

The Submillimetre Wave Instrument (SWI) is a heterodyne spectrometer, consisting of two channels that measure spectra in wavelength ranges around 520 μm (530 GHz–625 GHz) and 250 μm (1080–1275 GHz). The SWI spectrometer will map the emission and absorption of the water isotopes $H_2^{16}O$, $H_2^{17}O$, $H_2^{18}O$, HDO, and the ortho-to-para ratio of gaseous water (a proxy for the conditions under which the water formed) as well as the CO molecule. Additionally, SWI can be tuned to 80 other molecules that partly may be present in plumes, characterizing evolutionary processes of the interior of the Galilean moons. Due to the high spectral resolution of SWI (>107), the exact position and shape of the molecular lines can be determined. This allows one to not only derive number densities of the molecules but other physical parameters like the transition region between collision- and non–collision-dominated regions in the moons' atmospheres as well as wind and temperature profiles. The capabilities of heterodyne spectrometers like SWI have been demonstrated for LTE and non-LTE conditions in one to three dimensions on comets (Hartogh et al., 2010, 2011b; Rezac et al., 2019), and the Enceladus torus (Hartogh et al., 2011a) and simulations for Ganymede have been recently published (Yamada et al., 2018; Wirström et al., 2020).

The Neutral and Ion Mass (NIM) spectrometer of the PEP instrument suite is a compact, high-resolution, time-of-flight mass spectrometer designed to measure thermal (<5 eV) atmospheric neutrals and ions in the mass range 1–1 000 u/q, with a mass resolution (M/ΔM) of 1 100 and a detection threshold of 1 cm^{-3} (~ 10^{-16} mbar) [Föhn et al. 2021]. The NIM spectrometer's design results in a very high dynamic range, covering more than six decades in a single setup and about 12 decades using different gains and integration times. In addition, NIM can take measurements at a high cadence, with one complete mass spectrum becoming available every 5 s at closest approach. The NIM spectrometer will perform the first direct sampling (in situ measurements) of Ganymede's neutral gas environment, determining atmospheric properties with a high degree of confidence, because direct sampling (in contrast to remote sensing) is not affected by the physical properties of the material to be analyzed. It is foreseen that NIM may measure Ganymede's sputtered and sublimated bound atmosphere and possible extended neutral clouds whenever JUICE is closer to Ganymede's surface than 100 000 km, and continuously during the Ganymede orbits. At near-surface altitudes, NIM may be able to characterize atmospheric O_2 bulges, which might occur at certain local times or longitudes (Leblanc et al., 2017; Oza et al., 2018). In addition, NIM's atmospheric measurements will allow determination of the chemical and isotopic composition, density, temperature, and scale height of the neutral atmosphere. The NIM spectrometer will be able to resolve important possible trace species such as Mg, Al, Si, Ca, and Na, and allow identification of isotopes for various elements, depending on their abundance.

Set into the larger context, the JUICE measurements will improve the understanding of the atmosphere of the largest planetary and only known 'magnetic' moon. It will thereby allow more general and systematic comparisons to atmospheres of other bodies in the solar system and beyond.

REFERENCES

Alday, J., Roth, L., Ivchenko, N. et al. (2017), 'New constraints on Ganymede's hydrogen corona: analysis of Lyman-α emissions observed by HST/STIS between 1998 and 2014', *Planet. Space Sci.* **148**, 35–44.

Bagenal, F. and Dols, V. (2020), 'The space environment of Io and Europa', *J. Geophys. Res.* **125**(5), e27485.

Bagenal, F., Sidrow, E., Wilson, R. J. et al. (2015), 'Plasma conditions at Europa's orbit', *Icarus* **261**, 1–13.

Barth, C. A., Hord, C. W., Stewart, A. I. F. et al. (1997), 'Galileo ultraviolet spectrometer observations of atomic hydrogen in the atmosphere at Ganymede', *Geophys. Res. Lett.* **24**(17), 2147–50.

Broadfoot, A. L., Sandel, B. R., Shemansky, D. E. et al. (1981), 'Overview of the Voyager ultraviolet spectrometry results through Jupiter encounter', *J. Geophys. Res.* **86**(A10), 8259–84.

Brown, M. E. (1997), 'NOTE: a search for a sodium atmosphere around Ganymede', *Icarus* **126**(1), 236–8.

Brown, M. E. and Hill, R. E. (1996), 'Discovery of an extended sodium atmosphere around Europa', *Nature* **380**(6571), 229–31.

Brown, W. L., Lanzerotti, L. J. and Johnson, R. E. (1982), 'Fast ion bombardment of ices and its astrophysical implications', *Science* **218**(4572), 525–31.

Brown, W. L., Lanzerotti, L. J., Poate, J. M. and Augustyniak, W. M. (1978), 'Sputtering of ice by MeV ions', *Phys. Rev. Lett.* **49**(15), 1027–30.

Carlson, R. W. (1999), 'A tenuous carbon dioxide atmosphere on Jupiter's moon Callisto', *Science* **283**(5400), 820–1.

Carlson, R. W., Bhattacharyya, J. C., Smith, B. A. et al. (1973), 'An atmosphere on Ganymede from its occultation of SAO 186800 on 7 June 1972', *Science* **182**(4107), 53–5.

Carnielli, G., Galand, M., Leblanc, F. et al. and Jia, X. (2019), 'First 3D test particle model of Ganymede's ionosphere', *Icarus* **330**, 42–59.

Carnielli, G., Galand, M., Leblanc, F. et al. (2020*a*), 'Constraining Ganymede's neutral and plasma environments through simulations of its ionosphere and Galileo observations', *Icarus* **343**, 113691.

Carnielli, G., Galand, M., Leblanc, F. et al. (2020*b*), 'Simulations of ion sputtering at Ganymede', *Icarus* **351**, 113918.

Cooper, J. F., Johnson, R. E., Mauk, B. H., Garrett, H. B. and Gehrels, N. (2001), 'Energetic ion and electron irradiation of the icy Galilean satellites', *Icarus* **149**(1), 133–59.

Cunningham, N. J., Spencer, J. R., Feldman, P. D. et al. (2015), 'Detection of Callisto's oxygen atmosphere with the Hubble Space Telescope', *Icarus* **254**, 178–89.

de Kleer, K., Butler, B., de Pater, I. et al. (2021), 'Ganymede's surface properties from millimeter and infrared thermal emission', *Planet. Sci.* **2**(1), 5–24.

Eviatar, A., Strobel, D. F., Wolven, B. C. et al. (2001), 'Excitation of the Ganymede ultraviolet aurora', *Astrophys. J.* **555**, 1013–19.

Eviatar, A., Vasyliūnas, V. M. and Gurnett, D. A. (2001), 'The ionosphere of Ganymede', *Planet. Space Sci.* **49**(3–4), 327–36.

Feistel, R. and Wagner, W. (2007), 'Sublimation pressure and sublimation enthalpy of H_2O ice Ih between 0 and 273.16 K', *Geochim. Cosmochim. Acta* **71**(1), 36–45.

Feldman, P. D., McGrath, M. A., Strobel, D. F. et al. (2000), 'HST/STIS ultraviolet imaging of polar aurora on Ganymede', *Astrophys. J.* **535**, 1085–90.

Fray, N. and Schmitt, B. (2009), 'Sublimation of ices of astrophysical interest: a bibliographic review', *Planet. Space Sci.* **57**(14–15), 2053–80.

Gomis, O., Satorre, M. A., Strazzulla, G. and Leto, G. (2004), 'Hydrogen peroxide formation by ion implantation in water ice and its relevance to the Galilean satellites', *Planet. Space Sci.* **52**(5–6), 371–8.

Grasset, O., Dougherty, M. K., Coustenis, A. et al. (2013), 'JUpiter ICy moons Explorer (JUICE): an ESA mission to orbit Ganymede and to characterise the Jupiter system', *Planet. Space Sci.* **78**, 1–21.

Greeley, R., Chyba, C. F., Head, I. J. W. et al. (2004), 'Geology of Europa', in Bagenal, F., Dowling, T. E., and McKinnon, W. B., eds., *Jupiter: The Planet, Satellites and Magnetosphere*, Cambridge University Press, pp. 329–62.

Guerri, I., Fabbri, A., Tommasi, L. et al. (2018), The optical design of the MAJIS instrument on board of the JUICE mission, in Mazuray, L., Wartmann, R. and Wood, A. P., eds., *Optical Design and Engineering VII*, Vol. 10690 of Society of Photo-Optical Instrumentation Engineers (SPIE) Conference Series. https://ui.adsabs.harvard.edu/abs/2018SPIE10690E..1LG.

Hall, D. T., Feldman, P. D., McGrath, M. A. and Strobel, D. F. (1998), 'The far-ultraviolet oxygen airglow of Europa and Ganymede', *Astrophys. J.* **499**(5), 475–81.

Hall, D. T., Strobel, D. F., Feldman, P. D., McGrath, M. A. and Weaver, H. A. (1995), 'Detection of an oxygen atmosphere on Jupiter's moon Europa', *Nature* **373**(6516), 677–9.

Hansen, C. J., Esposito, L. W., Colwell, J. E. et al. (2020), 'The composition and structure of Enceladus' plume from the complete set of Cassini UVIS occultation observations', *Icarus* **344**, 113461.

Hapke, B., Wells, E., Wagner, J. and Partlow, W. (1981), 'Far-UV, visible, and near-IR reflectance spectra of frosts of H_2O, CO_2, NH_3 and SO_2', *Icarus* **47**(3), 361–7.

Hartogh, P., Bockelée-Morvan, D., Rezac, L. et al. (2013), Detection and characterization of Ganymede's and Callisto's water atmospheres, in 'Presented at the International Symposium "The Universe Explored by Herschel" 15–18 October 2013, ESTEC, Netherlands'.

Hartogh, P., Crovisier, J., de Val-Borro, M. et al. (2010), 'HIFI observations of water in the atmosphere of comet C/2008 Q3 (Garradd)', *A&A* **518**, L150.

Hartogh, P., Lellouch, E., Moreno, R. et al. (2011a), 'Direct detection of the Enceladus water torus with Herschel', *A&A* **532**, L2.

Hartogh, P., Lis, D. C., Bockelée-Morvan, D. et al. (2011b), 'Ocean-like water in the Jupiter-family comet 103P/Hartley 2', *Nature* **478**(7368), 218–20.

Hendrix, A. R., Barth, C. A. and Hord, C. W. (1999), 'Ganymede's ozone-like absorber: observations by the Galileo ultraviolet spectrometer', *J. Geophys. Res.* **104**(E6), 14169–78.

Hodges Jr., R. R. and Johnson, F. S. [1968]. 'Lateral transport in planetary exospheres', *J. Geophys. Res. Space Phys.* 73(23), 7307–7317. DOI: https://doi.org/10.1029/JA073i023p07307.

Johnson, R. E. (1990), *Energetic Charged-Particle Interactions with Atmospheres and Surfaces*. Springer. DOI: https://doi.org/10.1007/978-3-642-48375-2.

Johnson, R. E. (1997), 'Polar "caps" on Ganymede and Io revisited', *Icarus* **128**(2), 469–71.

Johnson, R. E., Carlson, R. W., Cassidy, T. A. and Fama, M. (2013), 'Sputtering of Ices', in Gudipati, M., Castillo-Rogez, J. (eds.), *The Science of Solar System Ices*. Astrophysics and Space Science Library, vol. 356. DOI: https://doi.org/10.1007/978-1-4614-3076-6_17.

Johnson, R. E., Carlson, R. W., Cooper, J. F. et al. (2004), Radiation effects on the surfaces of the Galilean satellites, in Bagenal, F., Dowling, T. E. and McKinnon, W. B., eds., 'Jupiter: The Planet, Satellites and Magnetosphere', Cambridge University Press, pp. 485–512.

Johnson, R. E., Lanzerotti, L. J., Brown, W. L. and Armstrong, T. P. (1981), 'Erosion of Galilean satellite surfaces by Jovian magnetosphere particles', *Science* **212**(4498), 1027–30.

Johnson, R. E., Oza, A. V., Leblanc, F. et al. (2019), 'The origin and fate of O_2 in Europa's ice: an atmospheric perspective', *Space Sci. Rev.* **215**(1), 20.

Kanik, I., Johnson, P. V., Das, M. B., Khakoo, M. A. and Tayal, S. S. (2001), 'Electron-impact studies of atomic oxygen: I. Differential and integral cross sections; experiment and theory', *J. Phys. B: At. Mol. Opt. Phys.* **34**, 2647–65.

Kanik, I., Noren, C., Makarov, O. P. et al. (2003), 'Electron impact dissociative excitation of O_2: 2. Absolute emission cross sections of the OI(130.4 nm) and OI(135.6 nm) lines', *J. Geophys. Res.* **108**(E11), 5126.

Khurana, K. K., Pappalardo, R. T., Murphy, N. and Denk, T. (2007), 'The origin of Ganymede's polar caps', *Icarus* **191**(1), 193–202.

Kivelson, M. G., Bagenal, F., Kurth, W. S. et al. (2004), 'Magnetospheric interactions with satellites', in Bagenal, F., Dowling, T. E., and McKinnon, W. B., eds., *Jupiter: The Planet, Satellites and Magnetosphere*, Cambridge University Press, pp. 513–36.

Krimigis, S. M., Armstrong, T. P., Axford, W. I. et al. (1977), 'The Low Energy Charged Particle (LECP) experiment on the Voyager Spacecraft', *Space Sci. Rev.* **21**(3), 329–54.

Langevin, Y., Piccioni, G., Filacchione, G., Poulet, F. and Dumesnil, C. (2018), 'MAJIS, the VIS-IR imaging spectrometer of JUICE', in *IPM 2018, International Workshop on Instrumentation for Planetary Missions, Technische Universität Berlin*, p. E1. https://ui.adsabs.harvard.edu/abs/2018ipm..confE...1L

Lanzerotti, L. J., Brown, W. L., Poate, J. M. and Augustyniak, W. M. (1978), 'On the contribution of water products from Galilean satellites to the Jovian magnetosphere', *Geophys. Res. Lett.* **5**(2), 155–8.

Leblanc, F., Oza, A. V., Leclercq, L. et al. (2017), 'On the orbital variability of Ganymede's atmosphere', *Icarus* **293**, 185–98.

Liuzzo, L., Poppe, A. R., Paranicas, C. et al. (2020), 'Variability in the energetic electron bombardment of Ganymede', *J. Geophys. Res.* **125**(9), e28347.

Makarov, O. P., Ajello, J. M., Vatti Palle, P. et al. (2004), 'Kinetic energy distributions and line profile measurements of dissociation products of water upon electron impact', *J. Geophys. Res.* **109**(A18), 9303.

Marconi, M. L. (2007), 'A kinetic model of Ganymede's atmosphere', *Icarus* **190**, 155–74.

McGrath, M. A., Jia, X., Retherford, K. et al. (2013), 'Aurora on Ganymede', *J. Geophys. Res.* **118**, 2043–54.

McGrath, M. A., Lellouch, E., Strobel, D. F., Feldman, P. D. and Johnson, R. E. (2004), Satellite atmospheres, *in* Bagenal, F., Dowling, T. E., and McKinnon, W. B., eds., *Jupiter: The Planet, Satellites and Magnetosphere*, Cambridge University Press, pp. 457–83.

Molyneux, P. M., Nichols, J. D., Bannister, N. P. et al. (2018), 'Hubble Space Telescope observations of variations in Ganymede's oxygen atmosphere and aurora', *J. Geophys. Res.* **123**(5), 3777–93.

Moore, J. M., Chapman, C. R., Bierhaus, E. B. et al. (2004), Callisto, in F. Bagenal, T. E. Dowling and W. B. McKinnon, eds., *Jupiter: The Planet, Satellites and Magnetosphere*, Cambridge University Press, pp. 397–426.

Musacchio, F., Saur, J., Roth, L. et al. (2017), 'Morphology of Ganymede's FUV auroral ovals', *J. Geophys. Res.* **122**(3), 2855–76.

Noll, K. S., Johnson, R. E., Lane, A. L., Domingue, D. L. and Weaver, H. A. (1996), 'Detection of ozone on Ganymede', *Science* **273**(5273), 341–3.

Orton, G. S., Spencer, J. R., Travis, L. D., Martin, T. Z. and Tamppari, L. K. (1996), 'Galileo photopolarimeter-radiometer observations of Jupiter and the Galilean satellites', *Science* **274**(5286), 389–91.

Oza, A. (2017), Detection and dynamics of satellite exospheres, Thesis, Université Pierre et Marie Curie – Paris VI. https://tel.archives-ouvertes.fr/tel-01687043.

Oza, A. V., Johnson, R. E. and Leblanc, F. (2018), 'Dusk/dawn atmospheric asymmetries on tidally-locked satellites: O_2 at Europa', *Icarus* **305**, 50–5.

Oza, A. V., Leblanc, F., Johnson, R. E. et al. (2019), 'Dusk over dawn O_2 asymmetry in Europa's near-surface atmosphere', *Planet. Space Sci.* **167**, 23–32.

Pappalardo, R. T., Collins, G. C., Head, J. W. et al. (2004), Geology of Ganymede, in F. Bagenal, T. E. Dowling and W. B. McKinnon, eds., *Jupiter: The Planet, Satellites and Magnetosphere*, Vol. 1, pp. 363–96.

Piccioni, G., Tommasi, L., Langevin, Y. et al. (2019), Scientific goals and technical challenges of the MAJIS imaging spectrometer for the JUICE mission, in '2019 IEEE 5th International Workshop on Metrology for AeroSpace (MetroAeroSpace)', IEEE, pp. 318–23.

Plainaki, C., Massetti, S., Jia, X. et al. (2020a), 'Kinetic simulations of the Jovian energetic ion circulation around Ganymede', *ApJ* **900**(1), 74.

Plainaki, C., Milillo, A., Massetti, S. et al. and Rispoli, R. (2015), 'The H_2O and O_2 exospheres of Ganymede: the result of a complex interaction between the Jovian magnetospheric ions and the icy moon', *Icarus* **245**, 306–19.

Plainaki, C., Sindoni, G., Grassi, D. et al. (2020b), 'Preliminary estimation of the detection possibilities of Ganymede's water vapor environment with MAJIS', *Planet. Space Sci.* **191**, 105004.

Poppe, A. R., Fatemi, S. and Khurana, K. K. (2018), 'Thermal and energetic ion dynamics in Ganymede's magnetosphere', *J. Geophys. Res.* **123**(6), 4614–637.

Rezac, L., Zhao, Y., Hartogh, P. et al. (2019), 'Three-dimensional analysis of spatial resolution of MIRO/Rosetta measurements at 67P/Churyumov-Gersimenko', *A&A* **630**, A34.

Roth, L. (2021), 'A stable H_2O atmosphere on Europa's trailing hemisphere from HST images', *Geophys. Res. Lett.* 48(20), e2021GL094289. DOI: https://doi.org/10.1029/2021GL094289.

Roth, L., Ivchenko, N., Gladstone, G. R. et al. (2021), 'A sublimated water atmosphere on Ganymede detected from Hubble Space Telescope observations', *Nat. Astron.* **5**, 1043–51. DOI: https://doi.org/10.1038/s41550-021-01426-9.

Roth, L., Saur, J., Retherford, K. D. et al. (2014), 'Transient water vapor at Europa's south pole', *Science* **343**(6167), 171–4.

Roth, L., Saur, J., Retherford, K. D. et al. (2016), 'Europa's far ultraviolet oxygen aurora from a comprehensive set of HST observations', *J. Geophys. Res.* **121**, 2143–70.

Saur, J., Duling, S., Roth, L. et al. (2015), 'The search for a subsurface ocean in Ganymede with Hubble Space Telescope observations of its auroral ovals', *J. Geophys. Res.* **120**, 1715–37.

Saur, J., Strobel, D. F. and Neubauer, F. M. (1998), 'Interaction of the Jovian magnetosphere with Europa: constraints on the neutral atmosphere', *J. Geophys. Res.* **103**(E9), 19947–62.

Shematovich, V. I. (2016), 'Neutral atmosphere near the icy surface of Jupiter's moon Ganymede', *Sol. Syst. Res.* **50**(4), 262–80.

Shi, M., Baragiola, R. A., Grosjean, D. E. et al. (1995), 'Sputtering of water ice surfaces and the production of extended neutral atmospheres', *J. Geophys. Res.* **100**(E12), 26387–96.

Smith, H. T., Mitchell, D. G., Johnson, R. E., Mauk, B. H. and Smith, J. E. (2019), 'Europa neutral torus confirmation and characterization based on observations and modeling', *ApJ* **871**(1), 69.

Smyth, W. H. and Marconi, M. L. (2006), 'Europa's atmosphere, gas tori, and magnetospheric implications', *Icarus* **181**, 510–26.

Spencer, J. R. (1987), 'Icy Galilean satellite reflectance spectra: less ice on Ganymede and Callisto?', *Icarus* **70**(1), 99–110.

Spencer, J. R., Calvin, W. M. and Person, M. J. (1995), 'CCD spectra of the Galilean satellites: molecular oxygen on Ganymede', *J. Geophys. Res.* **100**(E9), 19049–56.

Spencer, J. R., Lebofsky, L. A. and Sykes, M. V. (1989), 'Systematic biases in radiometric diameter determinations', *Icarus* **78**(2), 337–54.

Stern, S. A., Slater, D. C., Scherrer, J. et al. (2008), 'ALICE: the ultraviolet imaging spectrograph aboard the New Horizons Pluto-Kuiper Belt Mission', *Space Sci. Rev.* **140**(1–4), 155–87.

Strobel, D. F., Saur, J., Feldman, P. D. and McGrath, M. A. (2002), 'Hubble Space Telescope space telescope imaging spectrograph search for an atmosphere on Callisto: a Jovian unipolar inductor', *Astrophys. J.* **581**, L51–L54.

Taylor, G. E. (1971), 'Occultation of Beta Scorpii C by Io on May 14, 1971', *Nature* **234**(5329), 405–6.

Teolis, B. D., Plainaki, C., Cassidy, T. A. and Raut, U. (2017), 'Water ice radiolytic O_2, H_2, and H_2O_2 yields for any projectile species, energy, or temperature: a model for icy astrophysical bodies', *J. Geophys. Res.* **122**(10), 1996–2012.

Teolis, B. D., and Waite, J. H. [2016], 'Dione and Rhea seasonal exospheres revealed by Cassini CAPS and INMS', *Icarus* **272**, 277–89. DOI: https://doi.org/10.1016/j.icarus.2016.02.031.

Turc, L., Leclercq, L., Leblanc, F., Modolo, R. and Chaufray, J. Y. (2014), 'Modelling Ganymede's neutral environment: a 3D test-particle simulation', *Icarus* **229**, 157–169.

Vorburger, A., Fatemi, S., Galli, A. et al. (2022), '3D Monte-Carlo simulation of Ganymede's water exosphere', *Icarus* **375**, 114810. DOI: https://doi.org/10.1016/j.icarus.2021.114810.

Wirström, E. S., Bjerkeli, P., Rezac, L., Brinch, C. and Hartogh, P. (2020), 'Effect of the 3D distribution on water observations made with the SWI. I. Ganymede', *A&A* **637**, A90.

Wood, J. A. (1957), 'Sputtering and the drag coefficient in the regime of free molecular flow', *AJ* **62**, 40.

Yamada, T., Rezac, L., Larsson, R. et al. (2018), 'Solving non-LTE problems in rotational transitions using the Gauss–Seidel method and its implementation in the Atmospheric Radiative Transfer Simulator', *A&A* **619**, A181.

Yung, Y. L. and McElroy, M. B. (1977), 'Stability of an oxygen atmosphere on Ganymede', *Icarus* **30**(1), 97–103.

17

The Ionosphere of Ganymede

MARINA GALAND, GIANLUCA CARNIELLI, AND XIANZHE JIA

17.1 INTRODUCTION

Besides sublimation of its icy surface and bombardment of micro-meteorites, Ganymede is undergoing irradiation by energetic, charged particles (see Chapter 15). All of these processes lead to the formation of a tenuous atmosphere, (see Chapter 16), which is primarily composed of water, released through sublimation and sputtering of its surface, and of O_2 and H_2, released after forming through radiolysis following particle irradiation of its surface (see Chapter 15). Ionisation of these neutral species and of their dissociated products through

- absorption of the solar Extreme UltraViolet (EUV) radiation in the sunlit tenuous atmosphere, and
- impact from energetic (a few 10s to a few keV) electrons, such as electrons from Jupiter's magnetosphere which access Ganymede's magnetosphere primarily through the polar regions and electrons energised within Ganymede's magnetosphere,

produces electrons and ions all the way to the moon's surface and forms an **ionosphere**, as illustrated in Figure 17.1. The major ion species O_2^+ is affected by charge exchange over the first 100 km, while transport affects ionospheric densities over all altitudes (Carnielli, 2019).

This ionospheric plasma is immersed and evolves in a unique electromagnetic environment. Ganymede is indeed the only known moon in the solar system to generate its own magnetic field (see Chapter 13). The field environment around Ganymede results from the interaction of this internal magnetic field with the sub-Alfvénic, near-corotating plasma in Jupiter's magnetosphere (see Chapter 14). While Jovian charged particles have access to Ganymede's magnetosphere and some reach its surface (see Chapters 14 and 15), the plasma within 1 R_G (equatorial region) to 2 R_G (polar regions) is primarily of ionospheric origin (see Section 17.4.1).

The ionosphere plays a critical role in the coupling between Ganymede's surface, tenuous atmosphere, and space environment. It is, therefore, essential to assess Ganymede's ionosphere (formation, 3D distribution, variability) in order to understand Ganymede's neutral, plasma, and field environments and its interaction with the Jovian magnetosphere within which it is embedded. In particular, the quantitative assessment of Ganymede's ionosphere is critical for the following:

- for evaluating electric conductivity and currents and for modelling the electric and magnetic field environment around Ganymede with implications for the interaction of the moon with the Jovian magnetosphere (see Section 17.3.2 and Chapter 14)
- for estimating sources of the tenuous atmosphere, as energised ionospheric plasma irradiates Ganymede's surface (see Section 17.4.2 and Chapter 15)
- for constraining the atmospheric, neutral densities which are poorly known, while their knowledge is critical for planning the JUpiter ICy moons Explorer (JUICE) spacecraft operation during its low-altitude orbits around Ganymede (see Sections 17.4.1 and 17.5.2 and Chapter 16)
- in preparation for the JUICE mission for example, assessing the operability of plasma and particle instruments (see Section 17.5.2)
- once the JUICE magnetic field and plasma dataset are acquired, for interpreting the observations and detecting a subsurface salty ocean (see Section 17.5.2 and Chapter 14).

Despite the critical role the ionosphere plays, there is only a very limited number of observations of the ionospheric plasma. These observations were primarily taken for a similar Jupiter's solar local time of Ganymede and when the moon was located above the Jovian plasma sheet (see Section 17.2). Hence, currently we have to rely heavily on models to describe the plasma environment. In magnetospheric models which self-consistently treat the plasma and fields, the ionosphere is often very simplified, treated partly or fully as a fluid, and assumed to be spherically symmetric. It is only recently that the first 3D magnetospheric and ionospheric models, treating kinetically, as required, the plasma in Ganymede's environment, have been developed (see Sections 17.3.2 and 17.3.3). To enhance the science return, electron, ion, and field measurements from the Galileo spacecraft have been compared to simulation outputs (see Section 17.4.1).

The radius of Ganymede is taken to be 1 R_G = 2,634 km. The Cartesian coordinate system used for discussions in this chapter is called GPhiO, which is centred on Ganymede and fixed in Ganymede's frame. The X axis points in the corotation direction of the Jovian plasma and of Ganymede's motion, Y points towards Jupiter and Z along the moon's spin axis. Though the Jovian plasma is slightly sub-corotating at Ganymede's orbit, as it is orbiting Jupiter at about 10 to 20 per cent below rigid corotation, it is still significantly faster than the moon. As a result:

- a **wake** is formed on the leading side of Ganymede, and
- the **ram** is located on the trailing side of Ganymede.

Finally, we refer to

- the regions where the magnetic field lines of Ganymede's magnetosphere connect to the Jovian field lines, as **polar** (or **polar cap**). So, with respect to Ganymede, the magnetic field lines are said to be **open**;
- lower latitudes corresponding to the region where Ganymede's magnetic field lines are closed, as **equatorial**;
- the open/closed magnetic field line boundary as **separatrix**.

We first review electron/wave and ion measurements obtained in Ganymede's magnetosphere in Sections 17.2.1 and 17.2.2,

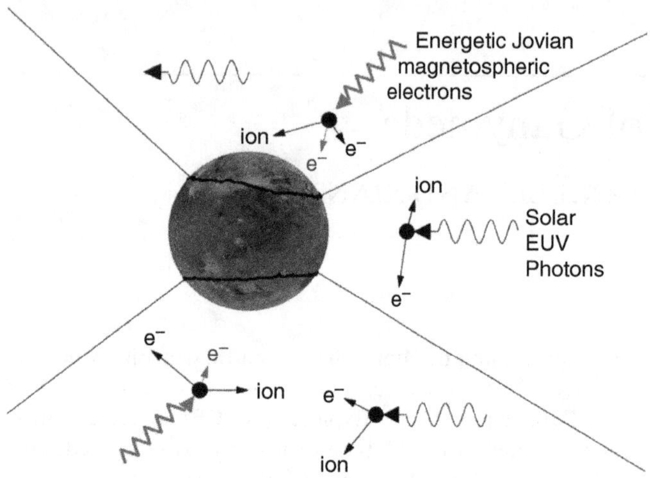

Figure 17.1 Creation of an ionosphere around Ganymede through the ionisation of atmospheric neutral species (black dots). Two ionisation processes dominate: photo-ionisation by EUV solar radiation (sun on the right of the figure) and impact by electrons from the Jovian magnetosphere. The latter primarily occurs in the polar regions, poleward of the separatrices (open/closed magnetic field line boundaries), shown as black, thick lines on Ganymede's surface. Contribution from electrons produced within Ganymede's near environment has not yet been assessed.

respectively. Next, in Section 17.3 we present different types of models: (1) early ionospheric models (see Section 17.3.1), (2) the ionospheric component of 3D magnetospheric models (see Section 17.3.2), and (3) a 3D multi-species, kinetic ionospheric model (see Section 17.3.3). In Section 17.4 we discuss what we have learned so far through a comprehensive comparison of simulation output with a Galileo plasma wave, particle, and field dataset (see Section 17.4.1) and we review the first estimation of the contribution of ionospheric ion bombardment of the moon's surface as a source for the tenuous atmosphere (see Section 17.4.2). Finally, in Section 17.5 we highlight key findings from ionospheric studies and provide a list of outstanding questions (see Section 17.5.1); we also propose how ionospheric models could be used in preparing for the JUICE mission and interpreting anticipated JUICE observations (see Section 17.5.2).

17.2 OBSERVATIONS OF GANYMEDE'S IONOSPHERE

The amount of measurements of Ganymede's ionosphere is very limited to date. It primarily relies on the low-altitude Galileo flybys over Ganymede, G1 and G2, presented in Table 17.1, during which a clear signature of ionospheric plasma was detected.

17.2.1 Galileo Observations of the Electron Population

Ganymede's ionosphere was first detected by upper hybrid resonance frequency measurements from the Galileo/Plasma Wave Spectrometer (PWS) with an accuracy of 6 per cent in electron number density along the G1 and G2 trajectories (Gurnett et al., 1996; Eviatar et al., 2001). The method to derive the electron density is discussed in Gurnett et al. (1996). The electron density profiles as a function of radial distance from the centre of the moon are shown for the inbound (solid) and outbound (dashed) parts of the trajectory in Figure 17.2. The closest approach of Galileo to the moon was reached during G2 at an altitude of 261 km (0.1 R_G), where a value of about 250 cm^{-3} for the electron density was derived from PWS measurements. This is well below the upper limit of 4 000 cm^{-3} which was inferred from remote-sensing observations from Galileo radio occultations (Kliore, 1998).

Though there is some variability between the electron density profiles in Fig. 17.2, they exhibit a relatively similar slope around 600 km. The departure from the common slope during the G2 inbound, with a steeper profile, is related to too limited a dataset to be considered as real (Eviatar et al., 2001). Using the value of 600 km for extrapolating towards lower altitudes provides an electron density of 400 cm^{-3} at the surface. The scale height below the CA altitude may, however, be smaller, as seen in Fig. 17.4 (a). The electron density measured by PWS around CA is primarily of ionospheric origin (see Section 17.4.1). There is currently no information from observations of the electron temperature in Ganymede's ionosphere. Close to the surface where the neutral densities are the highest, part of the electron population may be thermalised. However, at higher altitudes, the atmosphere is not dense enough to thermalise the electrons; intense electric and magnetic fields combined with wave interaction may accelerate and heat the electrons.

Galileo/Energetic Particle Detector (EPD) measured the electron energy spectra above 15 keV during the G2 flyby (Williams et al., 1998). These observations attest to the presence of energetic electrons within Ganymede's magnetosphere. They are consistent with Jovian electrons on magnetic field lines connected to Ganymede in the polar region; some of the electrons become trapped between the moon and the near-Jupiter mirror point as they convect across Ganymede's magnetosphere. At such large energies, the population is indeed anticipated to be mostly from Jupiter's magnetosphere and not to interact with Ganymede's thin atmosphere, though some energetic electrons may reach the surface and contribute to sputtering (Liuzzo et al., 2020).

17.2.2 Galileo Observations of the Ion Population

During the G2 encounter, the Galileo Plasma Analyzers (PLS) detected cold (a few eV) and dense plasma outflowing from Ganymede's ionosphere across the northern polar cap (Frank et al., 1997). The observed densities of the cold plasma peak near CA and fall off with distance from the surface, indicating an ionospheric origin. Due to the lack of direct measurement of the ion mass per charge as well as various challenges associated with measuring very cold plasma population, there is ambiguity in deriving the bulk plasma properties based on the raw PLS measurements. The outflowing ionospheric ions were first identified by Frank et al. (1997) as protons, which implies a maximum plasma density of $\sim$75 cm^{-3} near the CA altitude of 261 km and an average plasma convection speed of $\sim$70 km s^{-1}.

In addition to thermal plasma observations from the PLS, the Galileo EPD measured the fluxes of energetic particles at Ganymede, and anisotropies in the measured energetic ion

Table 17.1 *Characteristics of the G1 and G2 flybys by the Galileo spacecraft: date, local time (LT) of Ganymede with respect to Jupiter, the region crossed relative to Ganymede, and coordinated Universal Time (UTC) and altitude at closest approach (CA). During these flybys, Galileo flew towards Jupiter and Ganymede was located above the centre of the plasma sheet.*

Flyby	Date	LT (hh:mm)	Region crossed relative to Ganymede	CA UTC (hh:mm:ss)	CA altitude (km)
G1	27 June 1996	11:15	Central wake	06:29:07	831 km
G2	06 Sept 1996	10:46	Near north pole	18:59:34	261 km

Note: A LT of 00:00 with respect to Jupiter corresponds to Ganymede being at the centre of Jupiter's shadow.

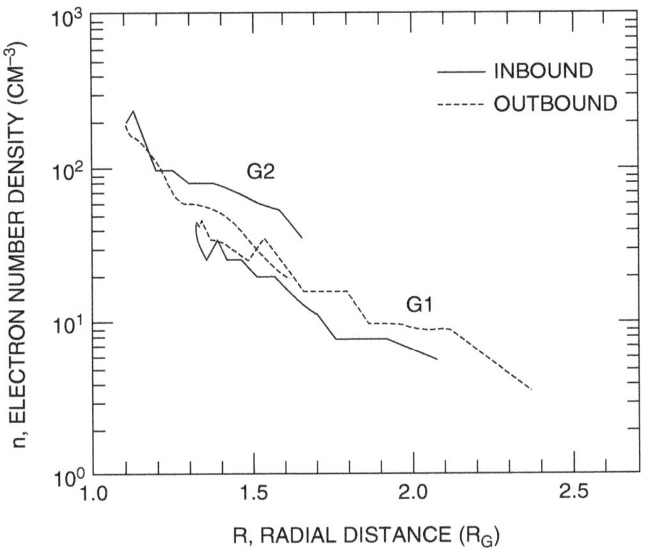

Figure 17.2 Electron number density measured by the PWS instrument on board Galileo, as a function of the radial distance from Ganymede's centre (expressed in Ganymede's radius, R_G, 1 R_G = 2 634 km), along the Galileo trajectory, for two Galileo flybys of the moon, G1 and G2. Inbound and outbound profiles are shown with solid and dashed lines, respectively. Reprinted from *Planetary and Space Science*, 49, Aharon Eviatar, Vytenis M. Vasyliunas, Donald A. Gurnett, The ionosphere of Ganymede, 327–336, ©2001 with permission from Elsevier.

count rates due to spacecraft spin modulation have been used to derive plasma flow velocity. During the G2 encounter, the Galileo EPD shows no detectable flow anisotropies inside Ganymede's magnetosphere, resulting in an upper limit of ~25 to 45 km s^{-1} for the convection speed (Williams et al., 1997), which is significantly lower than the plasma convection speed derived from the PLS data assuming the outflowing ionospheric plasma to be dominated by protons. The discrepancies between the ionospheric plasma density and velocity originally derived from the PLS data and the total electron density from PWS (Section 17.2.1), and plasma flow speed from EPD led to a reinterpretation of the G2 flyby data by Vasyliūnas and Eviatar (2000), who pointed out that the observed ionospheric ions are not protons, but instead oxygen ions (O^+). The resulting plasma density and convective flow speed should differ by factors of 4 and 1/4 from those calculated by Frank et al. (1997), respectively, which gives a maximum plasma density of ~300 cm^{-3} and a bulk flow speed of ~18 km s^{-1}. Consequently, the derived PLS moments assuming oxygen ions are more consistent with the electron density measurements from PWS (Gurnett et al., 1996; Eviatar et al., 2001) and the flow measurements inferred from the EPD (Williams et al., 1997) during this pass. The conclusion that the ionospheric outflow is composed mainly of heavy ions is further supported by global magnetosphere and ionosphere simulations (Jia et al., 2009; Zhou et al., 2019; Carnielli et al., 2020a), combined with the recent re-analysis of the PLS data for G2 (Collinson et al., 2018; Carnielli et al., 2020a) (see Section 17.4.1).

Recently, Collinson et al. (2018) analysed the PLS data from the G1 flyby that passed through the mid-latitude region of the magnetosphere on the downstream side. They showed that near the closest approach on this pass, there was also cold (~10 eV) and dense plasma outflowing from Ganymede's ionosphere. Assuming the measured ions are oxygen ions, the maximum plasma density observed is ~240 cm^{-3} and the average plasma flow speed is ~17 km s^{-1}, both of which are comparable to those measured on the G2 flyby, though the CA distance on G1 (831 km) is larger than on G2 (261 km).

17.3 MODELS OF GANYMEDE'S IONOSPHERE

17.3.1 Early Ionospheric Models

Eviatar et al. (2001) were the first to propose an ionospheric model for Ganymede. They solved the rate equations at a fixed location for both ion and neutral species under steady-state conditions. In the polar regions, O_2 was the only species released from the moon's surface, while O was produced in the tenuous atmosphere through dissociation of O_2. Assuming photochemical equilibrium, O_2^+ density was derived from the balance between its production by electron-impact ionisation and its loss through electron–ion dissociative recombination, and was found to be the major ion species. Over the G2 flyby to which they compared their results, transport drives the ion densities. O^+ density was derived from the balance between its electron-impact rate from ionisation of O and its escape rate. O^+ is, however, predominantly produced from dissociative ionisation of O_2 (Carnielli, 2019). Over the equatorial region, they considered ionisation of H_2O, followed by the reaction of H_2O^+ with H_2O to produce H_3O^+, which then recombines with an electron. As a result, they suggested that H_3O^+ was minor and the major ion was O^+ produced from ionisation of O. O^+ may, however, be primarily produced through dissociative ionisation of water. In addition, transport dominates over chemical loss for molecular ions over the altitude range probed. Finally, Eviatar et al. (2001) disregarded the presence of protons due to the escape of H from Ganymede's environment. Nevertheless, H^+ is produced through dissociative ionisation of H_2O and is then expected to be present in Ganymede's ionosphere.

Cessateur et al. (2012) calculated the ion production rates resulting from ionisation by solar EUV radiation. They took into account photo-absorption of the solar radiation by the background tenuous atmosphere, which was based on the atmospheric model from Marconi (2007): H_2O dominates in the first 200 km above the surface in the equatorial region, while O_2 dominates in the first few tens of kilometres in the polar regions. Molecular hydrogen is the major species at higher altitudes. They neglected electron-impact ionisation. Assuming photochemical equilibrium and applying the simulation to the Galileo G2 polar flyby near CA (see Table 17.1), they found that H_2^+ is dominating. However, transport plays a key role at that altitude (~260 km), so photochemical equilibrium does not hold. In addition, H_2^+ is very reactive and is most likely to be chemically lost through ion–neutral reactions rather than ion–electron dissociative recombination (as they assumed). Finally, H_2^+ is too light to be the major ion species during the G2 flyby, as attested by the analysis of the Galileo/PLS ion observations (see Section 17.4.1).

These early ionospheric models are not suitable for describing Ganymede's environment due to the combination of a tenuous atmosphere with intense electric and magnetic fields (see Chapter 14). A 3D treatment of the plasma environment where the ions can be transported and accelerated is required, as presented in Sections 17.3.2 and 17.3.3.

17.3.2 Ionosphere in 3D Magnetospheric Models

A number of global simulations have been developed for Ganymede to characterise the large-scale structure and dynamics of the moon's magnetosphere (see Chapter 14). The existing magnetosphere simulations not only adopted different approaches in treating the plasma interaction with electromagnetic fields based on magnetohydrodynamics (MHD), hybrid or coupled fluid-kinetic methods, but also differed significantly in how they treat Ganymede's ionosphere. For example, the hybrid simulation by Fatemi et al. (2016), where ions were modelled as kinetic particles and electrons were modelled as a massless fluid, did not include an ionosphere in their model, but instead treated Ganymede as a plasma absorber. In this case, there is no source of ionospheric plasma allowed in the simulation.

Most of the other magnetosphere simulations attempted to include the effects of Ganymede's ionosphere in a relatively simplified manner, mainly through imposing boundary conditions at the simulation inner boundary near the moon's surface. For instance, the first global MHD simulation developed for Ganymede by Kopp and Ip (2002) was adapted from their model for Titan, in which a mass-loading term was added to the MHD continuity equation to represent a plasma source provided by Ganymede's ionosphere. Although the mass-loading rate used in their Ganymede simulation is much smaller than that used for Titan, consistent with Ganymede having a much weaker atmosphere and ionosphere, the plasma flow velocity was set to be zero at the simulation inner boundary. One consequence of this flow boundary condition is that the global convection inside Ganymede's magnetosphere is nearly stagnant, and such a flow pattern is inconsistent with the flow measurements from Galileo (see Section 17.2.2).

Paty and Winglee (2004) adapted a multi-fluid MHD model to Ganymede in which the plasmas originating from Jupiter's magnetosphere and Ganymede's ionosphere were treated as separate fluids. Ganymede's ionosphere in the multi-fluid MHD model was introduced through prescribing a density profile near the simulation inner boundary with a surface density of 5 200 ions cm^{-3} assuming a 4:1 ratio of O^+ to H^+ ions and a scale height of 125 km for both ion species (Paty et al., 2008). Both the prescribed surface density and scale height are larger than those inferred from the Galileo PWS data (see Section 17.2.1). The temperatures of the ionospheric ions at the simulation inner boundary were fixed in time and were chosen to vary as a function of latitude, changing smoothly from the pole (0.1 eV) to the equator (6.0 eV). The prescribed plasma density profile was then held fixed throughout the simulation so that the ionosphere behaves as a constant source of plasma for the magnetosphere. Furthermore, to account for collisional effects in the ionosphere, the resistivity in the model was set to 3 800 ohm-meters right at the simulation inner boundary corresponding to the base of the ionosphere, and zero everywhere else in the simulation domain. Another model for Ganymede's magnetosphere, the hybrid model developed by Leclercq et al. (2016), adopted an approach similar to that used in Paty et al. (2008) in treating the ionosphere, namely prescribing a density profile for the ionosphere near the moon's surface and then maintaining the profile by continuously injecting new ions into the domain during the simulation.

Jia et al. (2008) applied a resistive MHD model to Ganymede that takes into account the electrical conductivity of various regions relevant to the magnetospheric interaction. These regions include (1) the moon's interior composed of a conducting core and a resistive mantle, (2) the ionosphere with finite conductance, and (3) the magnetosphere that contains highly conducting plasma, except in regions such as current sheets where plasma resistivity may become significant enough to enable magnetic reconnection to occur. Similar to the models discussed previously, the ionosphere in the Jia et al. MHD model was represented as a finite-conductivity, thin layer populated with cold and dense plasma. The simulation inner boundary was placed at $r = 1.05\ R_G$, under the assumption that the ionospheric conductance peaks at the altitude of $0.05\ R_G$, or ~130 km. At this boundary, the plasma density and temperature are uniformly specified with fixed values (550 amu cm^{-3} for mass density and 20 eV for temperature) such that the ionosphere behaves as a plasma reservoir. The resistivity in the ionosphere layer increases from the background value used far from Ganymede to the value assumed at the moon's surface in order to mimic the finite conductance of the ionosphere.

In their initial simulations, Jia et al. (2008) set the height-integrated Pedersen conductance in the ionospheric layer to be 2 S, corresponding to a weakly conducting ionosphere, but assumed a high conductance for the moon's interior, resulting in a high total conductance when summed over the moon's interior and ionosphere. Correspondingly, the plasma flow velocity was set to be zero at the simulation inner boundary (referred to as the "fixed" boundary condition), similar to the condition used in the Kopp and Ip (2002) simulation, which is a reasonably

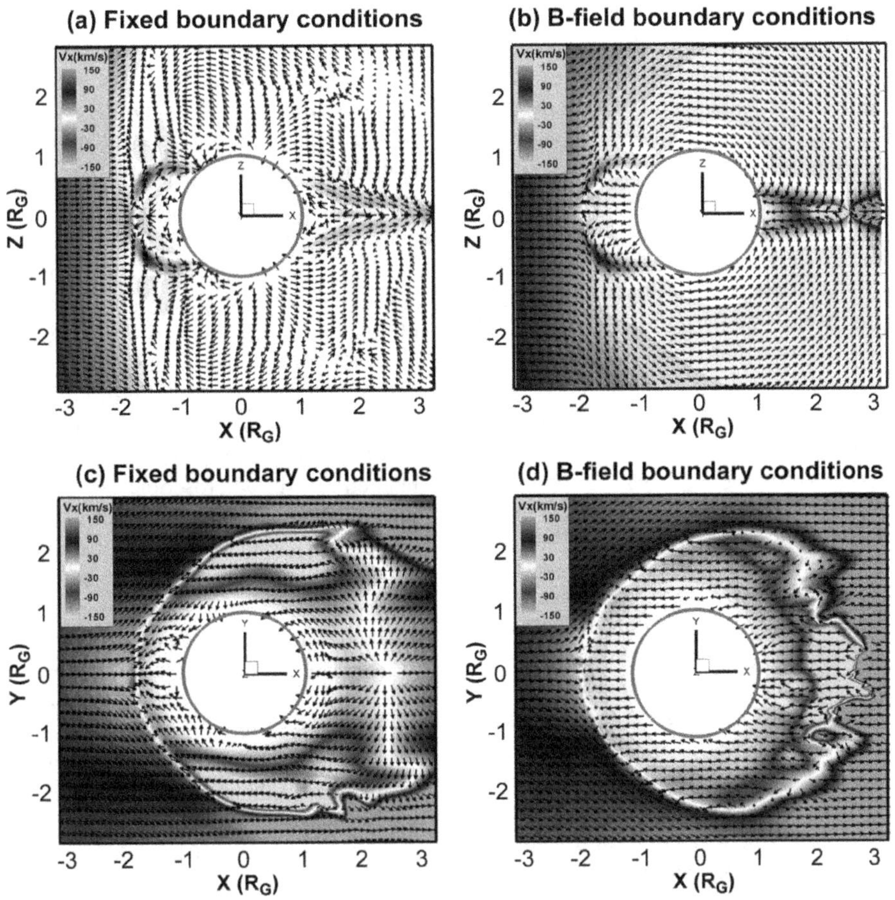

Figure 17.3 Comparisons of the plasma convection pattern from MHD simulations using different inner boundary conditions and applied to the G8 flyby. Contours of the X-component of the flow velocity (V_X) superimposed with unit flow vectors in the XZ plane (top row) and the XY plane (bottom row) for the case of fixed boundary conditions (panels (a) and (c)) and magnetic-field based boundary conditions (panels (b) and (d)). The circle in each panel shows the inner boundary of the MHD simulation ($r = 1.05\ R_G$). Reprinted with permission from John Wiley and Sons. Jia, X., R. J. Walker, M. G. Kivelson, K. K. Khurana, and J. A. Linker (2009), Properties of Ganymede's magnetosphere inferred from improved three-dimensional MHD simulations, *J. Geophys. Res.*, 114. (A black and white version of this figure will appear in some formats. For the colour version, please refer to the plate section.)

good approximation for a highly conducting obstacle. As an example, Figures 17.3 (a) and 17.3 (c) show the plasma convection pattern extracted from a simulation for the Galileo G8 flyby using the fixed boundary condition. Both figures indicate that the flow speed in general is very small close to the inner boundary and throughout the entire polar cap. The latter contains open magnetic field lines connected to Jupiter's magnetospheric field and thus is expected to have considerable plasma flows convecting in the direction from upstream to downstream. While there are some flows moving downstream inside the polar cap, there is no coherent large-scale convection pattern consistent with expectations. Jia et al. (2008) further showed that the size of the magnetosphere produced in this model is smaller than that inferred from the Galileo observations. It was suggested that these discrepancies can be attributed to the fixed flow boundary condition and to the high conductance assumed for the moon's interior, which is applicable to an obstacle with high conductance but does not provide the freedom of studying the magnetosphere under different ionospheric conductances.

Later, Jia et al. (2009) improved their MHD model by implementing new inner boundary conditions that self-consistently model the coupling between the magnetosphere and the ionosphere. In addition to adjusting the resistivity profile for the moon's interior such that the moon's mantle is not a significant contributor to the total conductance, the main modification to the model was to construct a flow boundary condition at the inner boundary, taking into account the magnetic field (referred to as the 'B-field' boundary condition). In particular, the 'B-field' boundary condition sets the component of the plasma flow velocity perpendicular to the local magnetic field flows at the boundary). Figures 17.3 (b) and 17.3 (d) present the convection pattern extracted from the same G8 flyby simulation but using the 'B-field' boundary condition. In contrast to the results shown in Figures 17.3 (a) and (c), the plasma convection in Figures 17.3 (b) and 17.3 (d) exhibits a well-organised coherent pattern consistent with that expected for a Dungey-cycle dominated magnetosphere. Overall, the 'B-field' boundary condition provides a more convincing physical picture of magnetospheric convection than the fixed boundary condition. It was also found that the MHD model with the new boundary condition produces magnetic configurations and plasma conditions very consistent with the Galileo observations for multiple passes (Jia et al., 2009). As will be discussed in Section 17.3.3, the electric and magnetic fields output from global magnetosphere models have been used to trace charged particles around Ganymede to understand the formation and behavior of Ganymede's ionosphere. As such, having a realistic description of the magnetospheric fields, such as that provided by the Jia et al.

(2009) MHD model with appropriate boundary conditions, is crucial for obtaining a high-fidelity model for the ionosphere.

The importance of proper treatment of the coupling of Ganymede's magnetosphere with its ionosphere and surface has also been emphasised in the MHD modeling study by Duling et al. (2014). In contrast to the numerical approach adopted by Jia et al. (2008, 2009), Duling et al. (2014) developed an analytical method for setting the boundary conditions at the surface based on decomposition of the magnetic field into poloidal and toroidal components. By implementing the boundary conditions into a single-fluid MHD model and testing them for Ganymede, Duling et al. demonstrated that their approach is suitable for coupling the magnetosphere with a non-conducting body, which is a reasonably good approximation for treating the icy surface of Ganymede. Furthermore, Ganymede's ionosphere was produced and maintained in the MHD model through various mass-loading terms added to the MHD equations, an approach that is in some respects similar to that used by Kopp and Ip (2002) but includes more source and loss terms to account for collisional effects important for the ionosphere. The ionospheric plasma was generated through ionisation of a prescribed neutral atmosphere that is spherically symmetric and composed of molecular oxygen (O_2). The height-integrated Pedersen conductance of the ionosphere used in their model is 12.5 S, higher than the 2 S used in the Jia et al. (2009) model but still close to the lower end of available estimates for Ganymede's ionospheric conductance (Kivelson et al., 2004; Eviatar et al., 2001). The MHD model was performed for different Galileo flybys of Ganymede, and the model comparisons with the Galileo magnetic field data also show a satisfactory agreement for multiple flybys.

Recent modeling efforts on Ganymede's magnetosphere have been largely focused on including non-ideal MHD effects in global simulations, such as the Hall MHD model by Dorelli et al. (2015), the ten-moment MHD model by Wang et al. (2018), and the MHD with Embedded Particle-in-Cell (MHD-EPIC) model by Tóth et al. (2016). While all of these models included additional physics beyond ideal MHD that is important for understanding kinetic effects present in Ganymede's magnetosphere, they adopted a relatively simple approach in setting the inner boundary conditions near the moon's surface, in which case the moon essentially was treated as a plasma absorber and there was no source of plasma provided by the ionosphere. Zhou et al. (2019, 2020) improved the MHD-EPIC model by implementing the same method and boundary conditions as proposed by Jia et al. (2009) for coupling the magnetosphere with the moon's interior and ionosphere. It was demonstrated that, with appropriately selected boundary conditions, the improved MHD-EPIC model not only is able to capture the kinetic physics crucial for understanding reconnection-driven dynamics at Ganymede but also provides a realistic global representation of Ganymede's magnetosphere that is consistent with the observations (Zhou et al., 2019, 2020).

17.3.3 3D Multi-Species Ion Test-Particle Model

Magnetospheric hybrid (Leclercq et al., 2016) and Particle-in-Cell (PiC) (Zhou et al., 2019) models, which self-consistently compute particles and fields while treating (at least) ions kinetically, have been applied to Ganymede (see Section 17.3.2). However, their treatment of the ionosphere is limited (e.g., ion density profile imposed (hybrid), no mass loading (PiC)). A 3D kinetic, ionospheric model is a required, complementary tool to assess in detail the ionospheric population.

17.3.3.1 Approach

Previously published models of Ganymede's ionosphere had very limiting assumptions (see Section 17.3.1). In order to address the need for a more realistic ionosphere, Carnielli et al. (2019) developed the first 3D kinetic model of Ganymede's ionosphere. It is a 3D multi-species, test-particle code (Carnielli, 2019) which computes the trajectories of newly born ions through a 3D environment with prescribed electric and magnetic fields. Microscopic quantities, such as energy distribution, and macroscopic quantities, such as number density, bulk velocity, and average kinetic energy, can then be derived for each ion species. The ion test-particle code has also been applied to the Jovian ions, present in Jupiter's magnetosphere and penetrating Ganymede's magnetosphere (Carnielli et al., 2020b). The Jovian population is, however, not discussed in the current chapter, which focuses on the ionospheric population. For further information on the Jovian ions which are irradiating Ganymede's surface, the reader should refer to Chapter 15.

To describe the ionospheric population around Ganymede, the first step is to create the ions. Two processes ionising the atmospheric neutrals (dominated by O_2, H_2O, and H_2) have been considered: photo-ionisation by solar EUV radiation over the sunlit tenuous atmosphere and electron-impact ionisation (currently reduced to Jovian energetic electrons over the polar, open magnetic field line regions due to the lack of information on the electron population within Ganymede's magnetosphere). The tenuous atmosphere of Ganymede is of low enough density (e.g., 2.4×10^{14} O_2/cm^2 and 1.4×10^{13} H_2O/cm^2 average radial column densities from Leblanc et al. (2017)) so that it is optically thin to solar radiation in the EUV wavelength (Carnielli, 2019). In this case, the mean energy of the produced electrons is around 15 eV for O_2 and H_2O (and even lower for H_2), which is close to the ionisation threshold. Hence the contribution of photoelectron-impact ionisation is negligible compared to photo-ionisation. Values for photo-ionisation frequencies, $\nu^{\gamma}_{n,i}$, of the main neutral species n, yielding the production of an ion species i, at a heliocentric distance of 5.46 AU are given in Table 17.2.

The second source of ionisation is the energetic electrons from the Jovian magnetosphere. The electron-impact ionisation frequency, $\nu^{e}_{n,i}$, is given by

$$\nu^{e}_{n,i} = \int_{E^{th}_{n,i}}^{E_{max}} I^{e}(E)\,\sigma^{e}_{n,i}(E)\,dE, \tag{17.1}$$

where $E^{th}_{n,i}$ is the threshold energy for the ionisation ($n \rightarrow i$; see Table 17.2), E_{max}, the highest energy considered (here 10^6 eV), I^e, the electron differential flux, and $\sigma^{e}_{n,i}$, the electron-impact cross section for the neutral n and produced ion species i. Due to the lack of data within Ganymede's magnetosphere, the electron differential flux was derived from Voyager and Galileo electron measurements taken in Jupiter's magnetosphere near

Table 17.2 *Characteristics of ionisation processes for the main neutral species at Ganymede. The threshold wavelength (column 3) and energy (column 4) are given for the single, non-dissociative ionisation of a neutral species (column 1) and produced ion species (column 2). The associated photo-ionisation frequency by solar EUV radiation and electron-impact frequency by the Jovian electrons are given in columns 5 and 6, respectively.*

Neutral species n	Ion species i	Threshold wavelength (nm)	Threshold energy (eV)	Photo-ionisation frequency $\nu_{n,i}^{\gamma}$ (s^{-1})	Electron-impact ionisation frequency $\nu_{n,i}^{e}$ (s^{-1})
H_2	H_2^+	80.4	15.43	0.23×10^{-8}	3.02×10^{-8}
H_2O	H_2O^+	98	12.6	1.13×10^{-8}	4.25×10^{-8}
O_2	O_2^+	102.7	12.07	1.75×10^{-8}	9.05×10^{-8}

Note: The photo-ionisation frequency has been calculated at a heliocentric distance of 5.46 AU for low solar activity (Carnielli et al., 2019). The electron flux used to describe the energetic electrons near Ganymede is based on Voyager/PLS, Galileo/EPD, and Galileo/PLS particle measurements (see Scudder et al. (1981); Paranicas et al. (1999); Cooper et al. (2001)) taken in the Jovian magnetosphere near or extrapolated to the orbit of Ganymede outside of the Jovian central plasma sheet (Carnielli et al., 2019).

(or extrapolated to) the orbital distance of Ganymede, away from Jupiter's central plasma sheet (Carnielli et al., 2019). In the region of closed magnetic field lines, the energetic electron population is expected to be dominated by photoelectrons, and electron-impact ionisation based on the Jovian magnetospheric dataset is only included over the polar regions where Ganymede's magnetic field lines connect to the Jovian field lines. Values for electron-impact ionisation for the main neutral species are given in Table 17.2. At Ganymede, energetic electrons are more efficient at ionising the main atmospheric neutrals than solar EUV radiation.

The total production rate, $S_{i,\mathrm{cell_{exo}}}$, of the ion species i (O_2^+, O^+, H_2^+, H^+, H_2O^+, OH^+) in a cell of the atmospheric grid in Ganymede's ionosphere is given by

$$S_{i,\mathrm{cell_{exo}}} = \sum_n (\nu_{n,i}^{\gamma} + \nu_{n,i}^{e})\, n_{n,\mathrm{cell_{exo}}}, \tag{17.2}$$

where $n_{n,\mathrm{cell_{exo}}}$ is the number density of the neutral species n in the cell of the atmospheric grid. The sum is over all neutral species which can produce the ion species i through ionisation.

The 3D maps of the atmospheric neutral densities used as input of the 3D ionospheric model are output from a simulation of the atmospheric model from Leblanc et al. (2017) in the low sublimation rate scenario consistent with Herschel observation (Hartogh et al., 2013) (referred hereafter as the 'dry' case). This simulation matches the ratio of the H_2O column densities observed on the leading and trailing hemispheres (Hartogh et al., 2013). The O_2 column density values are in agreement with that estimated by Hall et al. (1998). The major neutral species are O_2, H_2, and H_2O. Combining the low sublimation rate with the fact that H_2O is sticky, and hence absorbed by the surface, means that the O_2 and H_2 which do not easily stick to the surface, accumulate in the tenuous atmosphere and dominate over H_2O. Their distributions over latitude and longitude are smoother than the distribution of H_2O. Molecular oxygen is confined close to the surface, compared with H_2 and H_2O, as it is heavier and hence ejected with a lower speed from the surface. O_2 to H_2O number density ratio is higher in the equatorial region than in the polar regions. This characteristic is peculiar to this atmospheric setting. Other scenarios may be occurring, such as having H_2O dominating in the subsolar region as observed recently (Roth et al., 2021) ('wet' case). The reader should refer to Chapters 15 and 16 for further insight on Ganymede's tenuous atmosphere and its source.

In the test-particle model, the ion species are represented by macroparticles, which are created at a random position in every cell of the grid used to represent the tenuous atmosphere. At creation, they are assigned a weight (which is equal to the number of particles represented by the macroparticle) and a velocity. The initial weight of the macroparticle of species i produced in a given cell of the atmospheric grid is defined as

$$W_0 = S_{i,\mathrm{cell_{exo}}}\, V_{\mathrm{cell_{exo}}}\, dt, \tag{17.3}$$

where $V_{\mathrm{cell_{exo}}}$ is the volume of the atmospheric cell and dt, the timestep. To assume an incident velocity of zero for the macroparticle or a more realistic value based on the bulk velocity of the parent neutral n (with a spreading due to the thermal velocity of n and weighted by $S_{i,cell_{exo}}$) does not change the results due to the presence of an intense electromagnetic field (Carnielli et al., 2019).

Once the ions are created, they are transported through the electromagnetic environment via the Lorenz force, until they leave the simulation domain through an outer boundary or by impacting the moon's surface (Carnielli et al., 2019). The 3D electric field and magnetic field maps are output from a global MHD (Jia et al., 2009) or hybrid (Leclercq et al., 2016) model. They are kept fixed during the ionospheric simulations. While in the hybrid model ionospheric and Jovian ions are separated, it is not the case in the MHD model. As a result, using the fields from the MHD model means that the dynamics in Ganymede's ionosphere reflects primarily those arising from the coupling between the magnetosphere and ionosphere.

There is no feedback of the kinetically treated ions on the electric and magnetic fields in the approach presented here. Nevertheless, it was checked that the tangential components of the ion bulk velocity from the test-particle code and from the magnetospheric code are consistent with each other. The Debye length is between 1 m and 20 m, assuming an electron temperature of 5 eV (Gilet, 2019). Large-scale fields, obtained from the MHD model, dominate over short-lived microscopic fields. Finally, localised electric fields may exist due to local and transient charge separation, which would require future modelling work.

O_2^+ and O^+ can undergo charge-exchange with slow O_2 along their trajectories, which decreases the initial weight of the macroparticle (Carnielli et al., 2020a). Including this additional process allows one to assess the production of energetic O_2; however, it does not affect significantly the ion population, as the loss through charge exchange remains modest. Three-dimensional density maps of energetic O_2 have been obtained by applying the test particle model to the hot O_2 produced through charge exchange, turning off the Lorentz force (Carnielli et al., 2020a).

The timestep is adapted along the trajectory, but checks were made that no significant difference with a fixed timestep was found in the final results. The timestep is chosen such that it fulfills the following criteria (Carnielli, 2019): (1) the distance travelled by a macroparticle between two consecutive steps is less than the ionospheric grid cell to ensure that it does not skip a cell; (2) the travelled distance within one timestep is not larger than the spatial resolution of the grid on which the electromagnetic field has been computed; (3) the distance is less than 1/20 of the particle's instantaneous gyroradius to describe the ion trajectory realistically.

Each simulation corresponds to one given ion species. The simulation domain over which the ion trajectories are computed is centred around Ganymede and extends spherically up to 6.7 R_G. Macroscopic quantities, moments of the ion distribution, are calculated on a spherical grid defined by $100 \times 90 \times 180$ cells in the radial, polar, and azimuthal directions, respectively. The cell width in the radial direction increases with altitude. The polar and azimuthal cells have an angular width of 2°. In order to have statistically robust simulations, it was found that a number, $N_{\text{stat,cell}_{\text{exo}}}$, of 25 test particles injected initially per atmospheric cell was sufficient when comparing moments (Carnielli, 2019). The first two moments, number density and bulk velocity in a cell of the ionospheric grid, are computed as follows:

$$n_i = \frac{\sum_j W_j}{V_{\text{cell}_{\text{iono}}} N_{\text{stat,cell}_{\text{exo}}}}, \tag{17.4}$$

$$\vec{u}_i = \frac{\sum_j W_j \vec{v}_j}{\sum_j W_j}, \tag{17.5}$$

where $V_{\text{cell}_{\text{iono}}}$ is the volume of the ionospheric grid cell and $\vec{v}_j$ is the velocity of a test particle inside the ionospheric grid cell. The index j loops over all test particles which were present in the ionospheric grid cell over at least one timestep during their simulated lifetime and, for such a test particle, it loops over all timesteps during which the test particle was inside the cell of the ionospheric grid.

The thermal energy, $E_{th,i}$, and the average kinetic energy, $E_{K,i}$, associated with moments of order 2, for ion species i in a cell of the ionospheric grid, are defined as:

$$E_{th,i} = \frac{3}{2} k_B T_i = \frac{m_i}{2} \left(\frac{\sum_j W_j |\vec{v}_j|^2}{\sum_j W_j} - |\vec{u}_i|^2 \right), \tag{17.6}$$

$$E_{K,i} = \frac{m_i}{2} \frac{\sum_j W_j |\vec{v}_j|^2}{\sum_j W_j} = \frac{1}{2} m_i |\vec{u}_i|^2 + E_{th,i}, \tag{17.7}$$

where T_i is the temperature of the ion species i. $E_{th,i}$ corresponds to the velocity dispersion around the mean, $\vec{u}_i$, while $E_{K,i}$ corresponds to the arithmetic average of the kinetic energy within the ionospheric grid cell, hence the sum of the kinetic energy associated with the bulk plasma and its thermal energy. While for H^+ the total energy is primarily thermal, for O_2^+, which has higher inertia, convection (energy associated with the bulk motion) dominates.

17.3.3.2 Main Modelling Outcome

We present several results from simulations of the 3D multi-species test-particle code, introduced in Section 17.3.3.1 and applied to the ionospheric population:

- The averaged number density in three typical regions around Ganymede, for each ion species considered;
- 2D maps of the first three moments of the major ion, O_2^+.

The simulations have been applied to the Galileo G2 flyby conditions (see Table 17.1). Ganymede was at 10:00 LT with respect to Jupiter (solar phase angle of 330°), which is an approximation for G2 (see Table 17.1), but photo-ionisation represents a minor ion source (see Table 17.2). The 3D atmospheric density maps for O_2, H_2, H_2O, O, H, and OH are taken from Leblanc et al. (2017) ('dry' case; see Section 17.3.3.1). The 3D electric and magnetic field configuration is from an MHD simulation applied to G2 (see Jia et al. (2009) and Fig. 17.5).

Ion composition. Fig. 17.4 presents the number density profiles in altitude for the ion species considered. The profiles are each the average of the ion number density over a typical region (Carnielli, 2019):

- Region I (panel a) in the polar northern and southern Alfvén wings, where Ganymede's magnetic field lines connect with those from Jupiter (see Chapter 14);
- Region II (panel b) in the sunlit equatorial region in the corotation direction of the Jovian magnetospheric plasma;
- Region III (panel c) in the equatorial region in the trailing hemisphere in part under darkness.

In all three regions, O_2^+ is the dominant ion species ('dry' atmospheric case). Molecular oxygen is the dominant neutral at low altitudes where most of the ionisation occurs and has a high ionisation frequency. Molecular hydrogen is the second most abundant neutral species. Hence the most produced ion species, over all altitudes, latitudes, and longitudes, are O_2^+ (67 per cent of total ion production globally), followed by H_2^+, O^+ (produced primarily through dissociative ionisation of O_2), and H_2O^+ (all three around 10 per cent), and finally H^+ and OH^+ (around 2 per cent). As O_2 is mainly confined close to the surface: (1) the O_2^+ and O^+ densities decrease more sharply with increasing altitude than the densities of the other ion species, whose parent neutrals have more spread density profiles in altitude; (2) O_2^+ and O^+ primarily impact the moon's surface (see Section 17.4.2): only 22 to 24 per cent escape the near-Ganymede environment (not shown). For the other ion species, as their parent neutral density profiles are more spread in altitude, they are more likely to escape compared with the oxygen ions: 50 per cent for H_2^+ and 74 per cent for H_2O^+ escape instead of impacting the moon's surface.

The highest ionospheric densities are found in the Alfvén wings (Region I) where the production, driven primarily by electron impact in the polar regions, is peaking and where

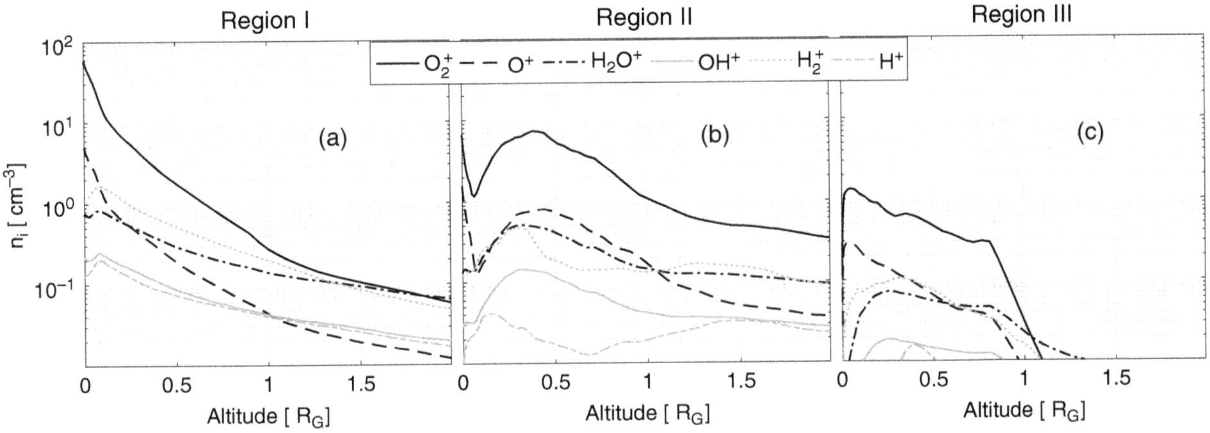

Figure 17.4 Ion number densities in Ganymede's ionosphere as a function of altitude (a) in the polar Alfvén wings, (b) in the sunlit equatorial region in the Jovian plasma corotating direction, and (c) in the equatorial region in the trailing hemisphere in large part under darkness. The profiles were generated from the test particle ion model by Carnielli et al. (2019). The neutral atmospheric model from Leblanc et al. (2017) ('dry' case) and the electric and magnetic fields from Jia et al. (2009) were used for these simulations. Ganymede was located at 10 a.m. in Jupiter's solar local time. A comparison between the modelled and Galileo/PWS ionospheric densities is presented in Fig. 17.7 (see Section 17.4.1). (Credit: Adapted from Carnielli (2019))

the local electric field is weaker, such that the plasma undergoes slow acceleration. Regions II and III, where ions are here locally produced through photo-ionisation, are also associated with high densities close to the surface, at least for the oxygen ions. This results from ions trapped in the closed magnetic field line region ($<1\ R_G$) and undergoing an $\vec{E} \times \vec{B}$ drift motion around the moon with typical energies of a few keV. This contributes to the formation of partial radiation belts. Ions coming from the polar region also contribute to the peak seen just below 0.5 R_G in Region II. H_2O^+ becomes dominant in Region I only beyond 2 R_G from the surface. As Region II experiences a higher O_2/H_2O density ratio than in the polar region, O_2^+ remains dominant even beyond 2 R_G. In Region III beyond 1 R_G altitude, the ion densities decrease sharply. This region located outside of Ganymede's magnetosphere, whose upstream magnetopause standoff distance is typically $\sim 1\ R_G$ above the surface (Jia et al., 2009), is not accessible to ionospheric ions; only ions produced locally are present. This is in contrast to Region II where the density profiles in altitude have a large scale height at high altitudes: a significant fraction of ionospheric ions escape. In Region I, the scale height of the ionospheric profiles is lower than in Region II, as ions, when escaping through the Alfvén wings, undergo steady acceleration starting from very low flow speed values close to the surface. Outside of Ganymede's magnetosphere, the escaping ion population reaches bulk speed similar to the speed of the Jovian plasma with which they merge, that is, around 140 km s^{-1}.

First three moments of O_2^+. Having established that O_2^+ is the major ion species (in the 'dry' atmospheric case), let us examine the spatial distribution of its various velocity moments. Figure 17.5 (a) shows the trajectory of an O_2^+ ion (grey, thick line) through the magnetic field environment (black). The trajectory was generated by the test-particle code. In this particular example, the O_2^+ ion, produced close to the moon (at radial distance of 1.005 R_G), is escaping the simulation volume through the northern Alfvén wing. Accumulating such individual trajectories for 5×10^7 macroparticles, it is possible to calculate the ion moments. The 2D map of the number density (panel b), bulk velocity (panel c), and average kinetic energy (panel d), derived from Eqs. (17.4), (17.5), and (17.7), are shown in the YZ plane in GPhiO coordinate system.

Firstly, it is apparent that the number density (see Fig. 17.5 (b)) of the ionospheric plasma is not spherically symmetric around Ganymede, an assumption usually made in magnetospheric models. While the ions are mainly produced close to the moon where O_2 is confined, the ion population extends along the open magnetic field lines in the Alfvén wings, well visible over the polar regions. It also extends in the direction of corotation of the Jovian plasma where it merges with the Jovian ions (Carnielli et al., 2019). Three regions can be identified, as seen in Fig. 17.5 (b):

- Region 1 (dark grey/black) of low density mainly present outside the Alfvén wings and away from the moon ($n_{O_2^+} < 10^{-2}$ cm^{-3});
- Region 2 (lighter grey) with higher densities, located in the polar regions inside the Alfvén wings away from the moon, except for a small (dark) region over the southern polar region which corresponds to Region 1; it is also present in the low-latitude, closed field line region;
- Region 3 (light grey) with most intense densities, in the polar regions inside the Alfvén wings close to the moon ($n_{O_2^+} > 1$ cm^{-3}); it results from high ion production combined with weak electric field.

The boundary between Region 1 and Region 2 corresponds to the outer boundary of the Alfvén wings intersecting the YZ plane in polar regions and to the magnetopause in the equatorial region. At low latitudes, the ions trapped on the closed magnetic field lines and mirroring between the two hemispheres appear clearly through these croissant-type light-grey shapes (Region 2). The asymmetry in the O_2^+ density between (−Y, anti-Jovian direction) and (+Y, Jovian-facing direction) results from the solar illumination: the (+Y) equatorial region is in darkness where no ions are produced (Region 1 very close to the moon, Region 2 otherwise), while the (−Y) equatorial region is sunlit (Region 2). Region 3 is present at low latitudes in the leading hemisphere (not shown).

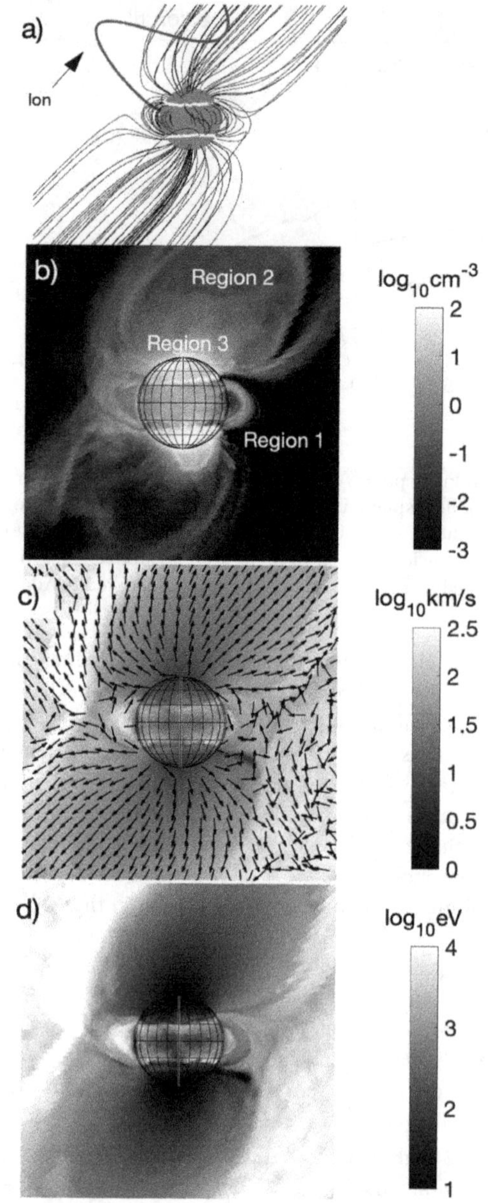

Figure 17.5 Examples of outputs generated by the 3D test particle ion model by Carnielli et al. (2019): (a) trajectory (grey, thick line) of an O_2^+ ion produced close to the moon's surface. The ion is travelling through the magnetic field lines (black) produced by the MHD model from Jia et al. (2009). The separatrix is shown in white. (b–d) 2D maps of the first three moments (0–2) for O_2^+: (b) number density, (c) bulk speed (with arrows showing the direction of the ion flow), and (d) average kinetic energy. All maps are shown in the YZ plane: (+Y) points to the right towards Jupiter and (+Z) points up along the moon's rotation axis. Ganymede is located at 10 a.m. in Jupiter's solar local time. The neutral atmospheric model from Leblanc et al. (2017) was used for these simulations (non-boosted case). See text for information on Regions 1, 2, and 3. Panels (b–d) adapted from Carnielli et al. (2019). Reprinted from *Icarus*, 330, G. Carnielli, M. Galand, F. Leblanc, L. Leclercq, R. Modolo, A. Beth, H. L. F. Huybrighs, X. Jia, First 3D test particle model of Ganymede's ionosphere, 42–59, ©2019 with permission from Elsevier.

Figure 17.5 (c) presents a 2D map of the bulk flow speed (greyscale) and flow direction (arrows) of the O_2^+ ions. The ions are escaping Ganymede's magnetosphere through the Alfvén wings. They are also escaping in the leading hemisphere (not shown) in the direction of corotation of the Jovian plasma. On average, within Ganymede's magnetosphere, the ion bulk speed in the Alfvén wings (1–10 km s^{-1}) is lower than the bulk speed of the Jovian magnetospheric plasma (140 km s^{-1}), while the ion bulk speed is faster in the Alfvén wing boundaries (>1 000 km s^{-1}) and in the closed magnetic field line regions corresponding to the return flux tubes from tail reconnection on the downstream side. The electric field is significantly weaker inside the Alfvén wings compared to their boundaries where currents accelerate ions. These currents flow along the Jovian magnetic field lines all the way to Jupiter's ionosphere (see Chapter 18). The asymmetry seen in the bulk speed where lower values are found in the Jupiter-facing hemisphere (+Y) results from the drift motion of the ions around the moon, which is primarily in the clockwise direction when looking over the northern pole. The drift motion from the (+Y) to the (−Y) implies larger acceleration for the ions reaching the anti-Jupiter hemisphere (as attested by checking individual ion trajectories, not shown here) (Carnielli, 2019).

Figure 17.5 (d) shows a 2D map of the average kinetic energy of the O_2^+ ions. The lower energies are found in the Alfvén wings (1–100 eV), while the highest are observed at the Alfvén wing boundaries (>10^4 eV) and inside the closed magnetic field line region (keV). The boundaries between these regions are less well defined for O_2^+ than for lighter ions, due to its higher mass. Energisation occurs close to the magnetopause boundary (near 1 R_G) where strong currents are present as a part of the Alfvén wing current system formed due to Ganymede's magnetospheric interaction (e.g., Jia et al. (2009) and Chapter 14). O_2^+ has its maps of bulk velocity (Fig.17.5 (c)) and average kinetic energy (Fig.17.5 (d)) that exhibit very similar features. The average kinetic energy for this heavy ion is dominated by dynamics of the flow. The thermal energy (defined as Eq.(17.6)), which traces the dispersion in the velocity around the average, is significantly lower than the kinetic energy. For lower-mass ions, the thermal energy becomes increasingly important compared to the convective energy (Carnielli, 2019).

17.4 WHAT HAVE WE LEARNED SO FAR?

Beyond the direct analysis of the ionospheric quantities from simulations (see Sections 17.3.2 and 17.3.3.2), we can gain essential insights on the field, plasma, and neutral environment of Ganymede from multi-instrument analysis of the moon's ionosphere observed by the Galileo spacecraft (see Section 17.4.1). Furthermore, it is now possible to assess the contribution of the ionospheric ion impact on the moon's surface as a source for the moon's tenuous atmosphere (see Section 17.4.2).

17.4.1 Comparison between Galileo Observations and 3D Simulations during the G2 Flyby

We present comparisons between observations from the Galileo spacecraft (see Section 17.2) and output from magnetospheric (see Section 17.3.2) and ionospheric (see Section 17.3.3) models, in terms of ion bulk velocity, ion energy spectrograms, and electron density. This comparison is applied to the G2 flyby of Ganymede (see Table 17.1). The goal is to validate the model description of the electromagnetic environment, to identify regions where Jovian versus ionospheric ions dominate, to get insight on ionospheric composition, and to constrain the poorly known neutral, atmospheric densities.

Electromagnetic Environment and Ion Bulk Speed and Composition During the G2 encounter, the Galileo spacecraft came closest to Ganymede's surface at a CA distance of ~261 km. As such, the Galileo magnetic field and plasma data acquired during this flyby provide critical datasets for validating global magnetosphere models (see Section 17.3.2). Some modeling studies focused on just magnetic field comparison (e.g., Fatemi et al., 2016; Duling et al., 2014), while a few other studies also compared the plasma data (e.g., Paty et al., 2008; Jia et al., 2009, 2010; Leclercq, 2015; Zhou et al., 2019). As discussed in Section 17.3.2, many of the published magnetosphere models either included no ionosphere or adopted a rather simplified treatment of the ionosphere. Therefore, it is important to understand how well a global model represents the reality through comparisons with available observations. Here, we use the results from the MHD model of Jia et al. (2009) to illustrate how such data-model comparisons can help validate global magnetosphere models and provide context valuable for interpreting observations.

As detailed in Jia et al. (2009), the MHD simulation is able to reproduce the observed magnetic field during multiple Galileo flybys with high fidelity, including both the vector components and the field strength as well as the locations of the magnetopause boundary. Such successful comparisons validate the simulated magnetic field along the Galileo trajectory. In addition to the magnetic field, Jia et al. (2009) also compared the simulated plasma properties with the published Galileo PLS data from the G2 flyby to gain further insight into the plasma environment around Ganymede. An example of the plasma data comparison is provided in Figure 17.6, which shows the plasma bulk flow velocity (panels a–c) as well as the simulated magnetic field topology (panel d) that provides a global context for this flyby. As shown in panel (d), during this flyby the Galileo spacecraft traversed Ganymede's northern polar cap that contains open magnetic field lines connecting Ganymede's ionosphere with Jupiter's through which the Jovian plasma can gain access to Ganymede's surface, consistent with Galileo observations (see Chapter 14). The magnetopause boundary can be readily identified based on the magnetic field topology shown in the figure, that is, the separatrix between external field lines and open or closed field lines. Similar large-scale structure of the magnetosphere was also obtained in other global simulations (Fatemi et al., 2016; Duling et al., 2014; Leclercq et al., 2016; Zhou et al., 2019). Due to lack of direct ion composition measurements, two sets of derived plasma velocity data based on different assumptions of the ion composition are included in Figure 17.6 for comparison. Considering that an MHD model mainly describes the convective flow velocity transverse to the magnetic field, the flow velocity shown in this figure has been decomposed into components perpendicular and parallel to the magnetic field to identify potential data-model discrepancies.

In panels (a–c), the two vertical lines mark the inbound and outbound magnetopause crossings as identified from the magnetic field data. Outside of Ganymede's magnetosphere, the observed ion spectrogram from PLS (see Figure 17.7 discussed hereafter in this section) contains two bands of enhanced

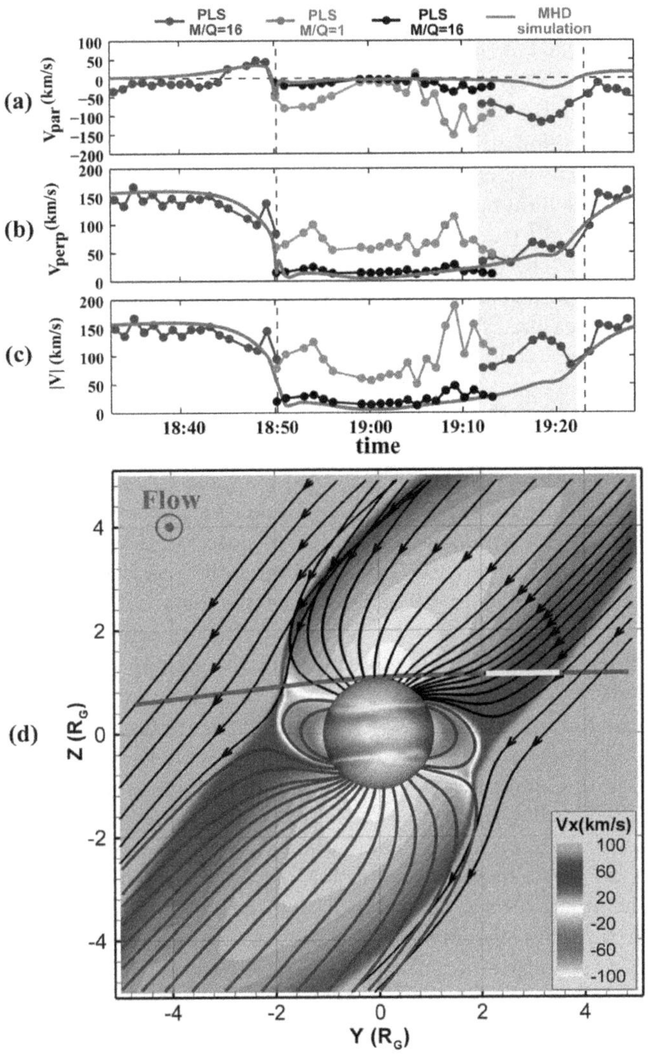

Figure 17.6 (a–c) Comparisons of the plasma bulk flow velocities from the MHD simulation of Jia et al. (2009) with the Galileo PLS observations for the G2 flyby. Panels (a) to (c) show the parallel, perpendicular velocity components with respect to the magnetic field and the total flow speed, respectively. In each panel, the blue dots represent bulk flows derived from the PLS data when the ion composition can be unambiguously determined as heavy ions with mass per charge (M/Q) = 16. The gray dots are the calculated moments assuming M/Q = 1, and the black dots represent bulk flows if M/Q = 16. The PLS moments are from Frank et al. (1997). Red traces are the bulk flow velocity from the MHD simulation. The two vertical dashed lines mark the inbound and outbound magnetopause crossings. (d) Colour contours of the modelled V_x in the YZ plane (perpendicular to the incident flow direction) and on a sphere of radius $r = 1.08\ R_G$, superposed with projection of simulated field lines. The black lines with arrows are field lines traced along the Galileo G2 trajectory (red trace), and the dark gray traces are arbitrarily chosen field lines within Ganymede's magnetosphere plotted for context. The locations where the PLS observed large field-aligned plasma flows are marked with yellow colour in all panels. Reprinted with permission from John Wiley and Sons. Jia, X., R. J. Walker, M. G. Kivelson, K. K. Khurana, and J. A. Linker (2009), Properties of Ganymede's magnetosphere inferred from improved three-dimensional MHD simulations, *J. Geophys. Res.*, 114. (A black and white version of this figure will appear in some formats. For the colour version, please refer to the plate section.)

ion counts, which were suggested to correspond to different ion species present in the sub-corotating ambient plasma of Jupiter's magnetosphere (Frank et al., 1997). The two distinct peaks in the observed spectrogram allows the ambient plasma composition to be unambiguously determined as heavy ions with mass per charge (M/Q) = 16 and light ions with M/Q = 1. The resultant bulk flow speed of the ambient plasma is around 150 km s^{-1} (see Fig. 17.6 (c)), which is in close agreement with the flow speed determined by analysis of the anisotropies of energetic ion data from the Galileo EPD (Williams et al., 1997). The measured flow speed was used to prescribe the upstream boundary conditions for the MHD simulation, and as shown by Fig. 17.6 (c), the modeled flow velocity in general agrees very well with the PLS observations outside of the magnetosphere.

Inside of the magnetosphere, the PLS data shows that the ambient Jovian ions are largely absent around CA, and instead, cold ions outflowing from Ganymede's ionosphere dominate the particle population in the region close to the moon's surface (Frank et al., 1997). There is only one noticeable energy peak in the observed PLS spectrogram (see Figure 17.7). Therefore, assumptions must be made about the ion composition in order to derive plasma moments, such as plasma density and bulk velocity. It was originally assumed by Frank et al. (1997) that the ions observed inside of the magnetosphere were protons, and the corresponding bulk velocities are shown as gray dots in Figures 17.6 (a–c). In Figure 17.6 (c), the averaged flow speed inside the magnetosphere (between 18:51 and 19:13 UT) assuming protons is around 70 km s^{-1}, which is significantly higher than the upper limit of 25–45 km s^{-1} for the convective flow speed as determined from the EPD data (Williams et al., 1997). Vasyliūnas and Eviatar (2000) suggested that the observed cold ions around CA should instead be heavy ions, most likely oxygen ions with M/Q = 16. The resulting bulk flow velocities assuming oxygen ions are indicated by the black dots in Figures 17.6 (a–c). In the scenario of heavy ions, the averaged bulk flow speed inside the magnetosphere is reduced by a factor of 4 to around 20 km s^{-1}, which sits comfortably below the upper limit determined from the EPD data.

After the inbound magnetopause crossing at ~18:50 UT, the MHD simulation predicts a significant slowing down of the plasma flow. Inside the magnetosphere, the plasma flow speed is, in general, quite small, which is on the order of about 15 km s^{-1} and reaches the minimum value of ~6 km s^{-1} at CA. In this region, the simulated flow velocity agrees much better with the calculated flow velocity for oxygen ions than with that calculated for protons. During the outbound magnetopause crossing at ~9:23 UT, the model predicts the flow speed gradually increases to the ambient flow speed, a trend that is consistent with the PLS observations. Overall, the MHD model results are more consistent with the scenario that the cold ionospheric ions observed during the G2 flyby near CA were oxygen ions, thereby confirming the conclusion of Vasyliūnas and Eviatar (2000). The same conclusion was reached in another modeling study of the G2 flyby by Zhou et al. (2019) using a coupled fluid-kinetic simulation model (MHD-EPIC), which predicts that the convection speed in Ganymede's polar cap is significantly smaller than that derived from the PLS data assuming protons. The dominance of oxygen ions, in particular O_2^+ along the G2 flyby around CA, and the agreement between Galileo/PLS and calculated oxygen-ion bulk velocities were derived using the multi-species test-particle model (Carnielli et al., 2019, 2020a). This further supports the MHD and MHD-EPIC findings.

The comparison presented in Figure 17.6 shows large deviations between the model and the PLS data in some regions. Specifically, between 19:12 and 19:21 UT, the MHD simulation underpredicts the flow speed compared to the PLS data. The velocity components shown in Figures 17.6 (a) and 17.6 (b) reveal that the discrepancy largely lies in the parallel component of the flow velocity. The magnetic field topology in the MHD simulation (Figure 17.6 (d)) indicates that during this interval the spacecraft were on open magnetic field lines (see Fig. 17.7(a)) near the magnetopause boundary where field-aligned currents are present (Jia et al., 2009). The observed strong parallel flows (of the order of 100 km s^{-1}) could have arisen from intense field-aligned currents and the associated parallel electric fields, which are not fully captured by an MHD model.

In contrast, Figure 17.6 (b) shows that the perpendicular flow velocity from the MHD simulation is quite consistent with that calculated from the PLS data during the entire close flyby when the observed ionospheric ions are assumed to be oxygen ions. Since the convective electric field depends on the perpendicular component of the plasma bulk velocity, the satisfactory data-model agreement for both the perpendicular flow velocity and the vector magnetic field (see Jia et al. (2009) for details) makes it appropriate to apply the electromagnetic field's output from the MHD simulation to drive test particle simulations, which are discussed in detail below.

Ion Energy Distribution and Ionospheric Composition Further validation of the electromagnetic environment in which the ions evolve is possible through comparison of the shape of the ion spectrograms between measurements and simulations. From the Galileo/PLS dataset (see Section 17.2.2), we have access not only to macroscopic quantities, such as ion bulk velocity, as discussed earlier in this section, but also to microscopic quantities. The ion spectrogram acquired by Galileo/PLS during the G2 flyby is shown in Fig. 17.7 (b) (usable only above 15 eV, identified with a horizontal, dashed white line). Its shape is sensitive to the electromagnetic environment and to ion composition. Paty et al. (2008) undertook the first comparison between the Galileo/PLS spectrograms and simulation results using a multi-fluid MHD model. As ions were treated in a fluid way, a Maxwellian distribution was assumed for each ion species in order to produce ion spectrograms. The ion species considered in the model were Jovian magnetospheric O^+, ionospheric O^+, and ionospheric H^+ (including a few per cent of Jovian magnetospheric H^+). Paty et al. (2008) found that around CA of the G2 flyby the dominant ion species above 10 eV was ionospheric H^+, while ionospheric O^+ was confined to lower energies. Multi-species kinetic ion models, such as hybrid (Leclercq et al., 2016) and test-particle (Carnielli et al., 2019) models, are inherently more suitable to assess microscopic ion quantities. Leclercq (2015) simulated Jovian O^+ and H^+, and ionospheric O^+, using a hybrid model. When Galileo was within Ganymede's ionosphere, they found that ionospheric O^+ reached energies above 15 eV and was detectable by Galileo/PLS, in contrast to the earlier findings from Paty et al. (2008).

Figure 17.7 (c) and 17.7 (d) show the simulated ion spectrograms applied to ionospheric and Jovian ions, respectively, obtained from the 3D multi-species test particle model (see Section 17.3.3.1). The ionospheric spectrogram (see Fig. 17.7 (c)) is the sum of the contributions from O_2^+, O^+, H_2O^+, H_2^+, H^+,

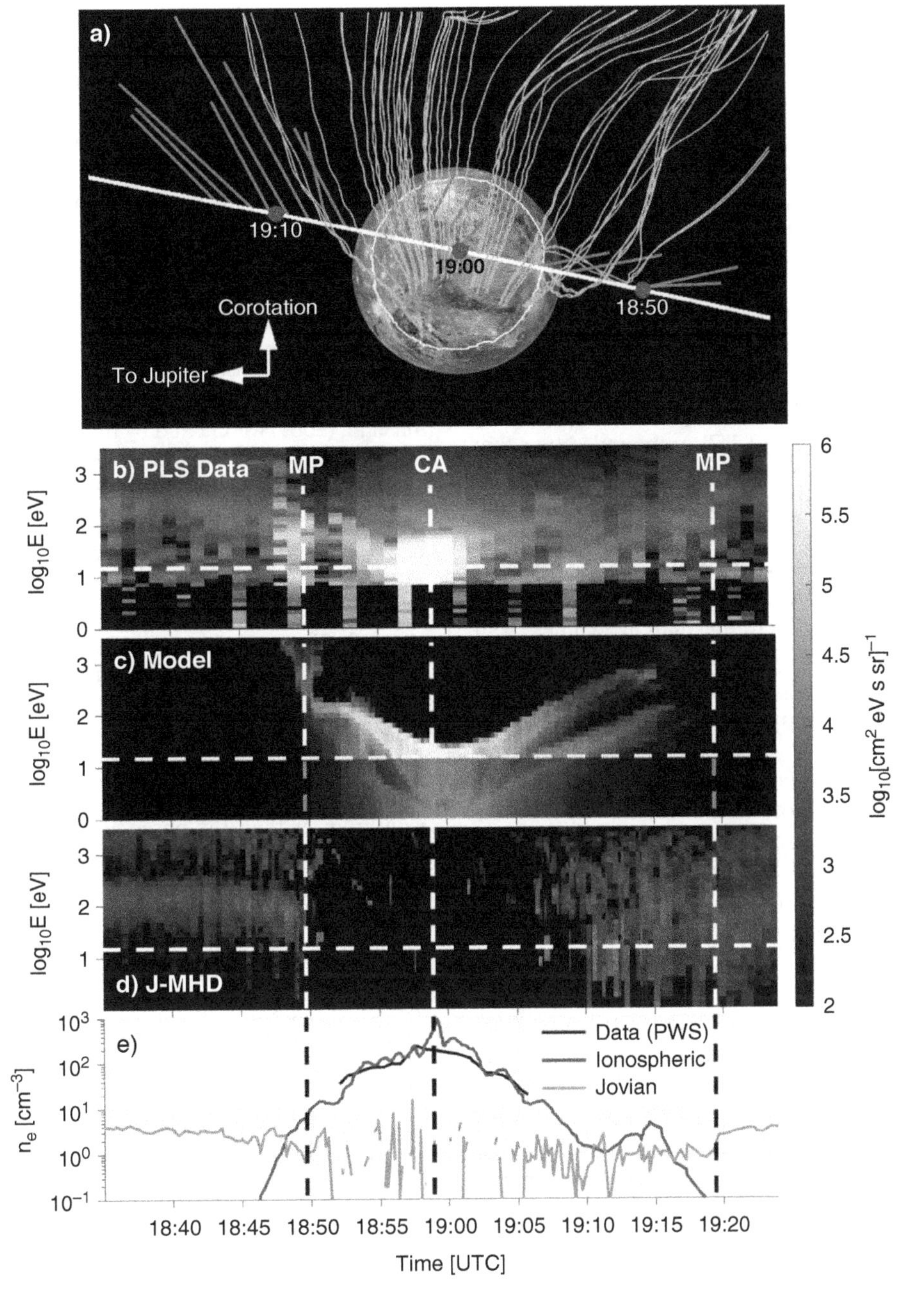

Figure 17.7 Comparison between Galileo observations and simulation output from the 3D test particle ion model by Carnielli et al. (2020a) for the G2 flyby. (a) White, bold line: Trajectory of the Galileo spacecraft. Red lines: Bulk velocity vectors as observed by the PLS instrument (Frank et al., 1997). Green curves: Selection of O_2^+ test particle trajectories which intersect the path of the spacecraft. Yellow circular line: Northern boundary between open and closed magnetic field lines. (b–d) Energy spectrograms along the Galileo trajectory, (b) as recorded by the PLS instrument and (c–d) generated by the test particle ion model applied to (c) the ionospheric population and (d) the Jovian magnetospheric population. The dashed vertical lines (white in panels (b–d) and black in panel (e)) indicate the times of magnetopause crossing (MP) and of closest approach (CA). The dashed, white horizontal lines in panels (b–d) show the upper energy limit below which the PLS dataset processing technique is considered as unreliable (see Carnielli et al. (2019) for more details). The simulation is driven by a number density for $O_2 \times 10$, compared to the 'dry' case from Leblanc et al. (2017), and an asymmetric electron-impact ionisation frequency (boosted case, Carnielli et al. (2020a)). The magnetic and electric fields are from the MHD model of Jia et al. (2009). (e) Electron number density along the Galileo trajectory, as a function of time, as recorded by the PWS instrument (black) (Eviatar et al., 2001) and as computed by the test particle ion model (blue). The simulated number density of the background Jovian ion density is shown in light brown. Panels (a–b, d) adapted from Carnielli et al. (2019). Reprinted from *Icarus*, 330, G. Carnielli, M. Galand, F. Leblanc, L. Leclercq, R. Modolo, A. Beth, H. L. F. Huybrighs, X. Jia, First 3D test particle model of Ganymede's ionosphere, 42–59, ©2019 with permission from Elsevier. Panels c, e adapted from Carnielli et al. (2020a). Reprinted from *Icarus*, 343, G. Carnielli, M. Galand, F. Leblanc, R. Modolo, A. Beth, X. Jia, Constraining Ganymede's neutral and plasma environments through simulations of its ionosphere and Galileo observations ©2020 with permission from Elsevier. (A black and white version of this figure will appear in some formats. For the colour version, please refer to the plate section.)

and OH^+ created in Ganymede's ionosphere. Along the trajectory of G2 over the northern, polar regions, the main ion species was found to be O_2^+ which represents 60 to 80 per cent of the ion species. All the other ion species represent less than 20 per cent. The only exception is outbound from 19:05 UT during which the model shows that O_2^+ represents only 40 to 60 per cent of the total ionospheric species, while O^+ peaks up to 35 per cent (Carnielli et al., 2019). The ionospheric spectrogram (see Fig. 17.7 (d)) is the sum of the contributions from thermal O^+, energetic H^+, O^{2+}, and S^{3+} coming from Jupiter's magnetosphere.

Comparing Fig. 17.7 (c–d) with Fig. 17.7 (b) confirms the earlier findings that the ionospheric ions dominate over the Jovian ions from the inbound magnetopause (MP) crossing (around 18:50 UT) to about 19:12 UT outbound. Note that only the ionospheric ions which have undergone acceleration are observed, as below 15 eV the PLS dataset is no longer reliable. Jovian ions dominate in Jupiter's magnetosphere, before the MP crossing inbound and after the MP crossing outbound (19:20 UT). Both ion populations are present in Ganymede's polar magnetospheric region where magnetic field lines from Ganymede connect with those from Jupiter. The ionospheric and Jovian ion densities (see Fig. 17.7.e) are of similar magnitude between 19:10 UT and the outbound MP crossing. During this period, the region crossed by Galileo is further away from the moon, well out in the Alfvén wings. Ionospheric ions are escaping to the Jovian magnetosphere, while Jovian ions are accessing Ganymede's near environment.

Focusing on the 18:50–19:15 UT period during which the ionospheric population is significant, various structures are present above the usable PLS range of 15 eV in both the Galileo/PLS spectrogram (Fig. 17.7 (b)) and the simulated ionospheric spectrogram (Fig. 17.7 (c)). They include a wing decreasing in energy from 100 eV to 10 eV towards CA where a maximum is found when Galileo was closest to the moon, followed outbound by two wings increasing in energy. Assessing the contribution of different ion species to the ionospheric spectrogram shows that the intense wing decreasing from 100 eV to 10 eV inbound is composed primarily of O_2^+ ions; the contribution at lower energy, shortly after the MP crossing, is due to lower-mass ions. The intense peak near CA is also driven by O_2^+. In the simulation, it extends from a few eV to 20 eV. The peak present around 1 eV in the ionospheric simulations near CA is due to H_2^+ (Carnielli, 2019). Outbound, the higher-energy wing is generated by heavy, O_2^+ ions, while the lower-energy wing is due to lighter, H_2^+ ions. Both locally produced ions and ions accelerated by the electric field from their source region contribute to the wings. This result highlights the necessity to apply a kinetic treatment to analyse the PLS spectrograms. While there is general agreement in the structure of the spectrograms seen in Fig. 17.7 (b–c), there are some discrepancies: the peak near CA is extending to energies higher in the PLS dataset than in the simulations. This may be due to the spacecraft potential which has not been taken into account in the simulations, as it is unknown, though to which low-energy particles are most sensitive.

The very good agreement between simulations and Galileo/PLS ion observations, applied to the G2 flyby in terms of bulk velocity and ion distribution (shape in the ion spectrogram), demonstrates that

- the electromagnetic field simulated by fluid and kinetic models provides a reasonable description of the environment encountered by the ionospheric ions from birth to the region crossed by Galileo;
- close to Ganymede the dominant ion population is of ionospheric origin;
- Ganymede's ionosphere is dominantly made of heavy ions consistent with O_2^+, with contribution of lighter ions, such as H_2^+. This concludes a long-standing debate on the ionospheric composition in Ganymede's ionosphere (Carnielli et al., 2019).

Ionospheric Density. Figures 17.7 (a) and 17.7 (c–e) were generated from simulations driven by the electromagnetic field from Jia et al. (2009) for G2 and by the neutral, atmospheric densities from Leblanc et al. (2017) ('dry' case), except for a boosted O_2 tenuous atmosphere (density increased by a factor of 10 from the baseline) and the boosted electron-ionisation frequencies (partially enhanced from the baseline shown in Table 17.2). Using the baseline inputs does not affect the ion bulk velocity (see Fig. 17.7 (a)) and ion distribution (spectrogram shape) shown in Figure 17.7 (c). This, however, highlights that the baseline tenuous atmosphere seems to be not dense enough. This is confirmed by the comparison between the Galileo/PWS electron density measured in Ganymede's ionosphere around CA (see Fig. 17.7 (e), Section 17.2.1) and the simulated ionospheric densities from the baseline inputs (not shown), which are significantly lower than the observations (see Carnielli et al. (2019) for further details).

How to reconcile Galileo/G2 observations and simulations in terms of magnitude of the ion distribution and ionospheric densities? Carnielli et al. (2020a) explored several directions. First, they included the effect of collisions. The most likely collisional process for the major ionospheric ion, O_2^+, which has been accelerated by intense fields and is hence fast ('*f*'), is charge exchange with slow ('*s*') atmospheric O_2:

$$O_{2,f}^+ + O_{2,s} \rightarrow O_{2,s}^+ + O_{2,f}.$$

This produces a slow O_2^+ and hence could increase the ionospheric densities. However, the test-particle simulations show that the ion is quickly accelerated again and the ionospheric densities are not significantly affected by the inclusion of charge-exchange processes.

Next, Carnielli et al. (2020a) assessed the required enhancement in neutral, atmospheric densities to reach an agreement between observed and simulated ionospheric densities. As the major ion species is O_2^+, only O_2 density was increased. A factor-of-10 increase applied to O_2 density everywhere in the simulation domain is required to reach an agreement along the outbound part of the Galileo trajectory (see Fig. 17.7 (e) after CA). Molecular oxygen density is poorly known. There is only one column density measurement of O_2 available (Hall et al., 1998). As for neutral, atmospheric models, they are limited by the poor knowledge of the ejection yields through sputtering or radiolysis (see Chapter 15). An uncertainty of a factor of 10 in the yields, necessary to boost similarly the O_2 densities, is not unrealistic. This would bring the mean O_2 column density to a factor of only 2 higher than the upper limit derived from Hall et al. (1998).

Even with a boosted O_2 tenuous atmosphere, the simulated ionospheric densities still remain too low by a factor of about 4 compared to the observed values, along the inbound part of the trajectory (see Carnielli et al. (2020a) for further details). It is unlikely that the O_2 density would exhibit such a strong asymmetry between inbound and outbound which would imply a further increase in the anti-Jovian hemisphere at high latitudes. Longitudinal asymmetry features were found in neutral, atmospheric simulations, but they are located in the equatorial regions and do not exceed a factor of 2 (Leblanc et al., 2017).

Another input parameter, which is not currently well constrained, is the electron-impact ionisation frequency. In the baseline simulations of the test-particle model, the values are reduced to a constant value, based on Voyager and Galileo particle measurements at (or extrapolated to) Ganymede's orbit (see Table 17.2). Such an asymmetry in the electron-impact ionisation frequency is motivated by the increase seen in the Galileo/EPD electron fluxes in the 15–29 keV range (Williams et al., 1998) before CA (Carnielli et al., 2020a). Even though the electrons responsible for ionising the thin, tenuous atmosphere are significantly less energetic, an increase in the high-energy electron population may attest to an increase over a larger energy range. This is supported by the presence of intense currents near the separatrix and the proximity of the magnetopause to the moon during inbound. By boosting the O_2 density by a factor of 10 and the electron-impact frequency by a factor of 4 in the open magnetic field line region around the sub-solar longitude, the region crossed by Galileo inbound, the magnitudes in the ionospheric density (see Fig. 17.7 (e)) and in ion spectrograms (see Fig. 17.7 (b–c)) from Galileo observations, and test-particle simulations reach a close agreement. The multi-instrument analysis between Galileo plasma observations and an ionospheric, kinetic model highlights that the neutral, atmospheric densities, in particular for O_2 which is the major contributor to the ionosphere (for the 'dry' atmospheric scenario), are greater than anticipated, by a factor as large as 10 during the G2 flyby.

17.4.2 Surface Irradiation by Ionospheric Ions

Besides ice sublimation of the sunlit surface and impact of micro-meteoroids, a key source of Ganymede's thin atmosphere is the irradiation of its surface by energetic charged particles (see Chapter 3.2). Irradiation leads to (1) sputtering of Ganymede's surface, that is, direct ejection of atoms or molecules primordial to Ganymede's surface (e.g., H_2O), and (2) radiolysis from water ice, that is, production of new species, such as O_2, O_3, and H_2 through chemical reactions between water radicals in the ice (Teolis et al., 2017), followed by their release from the surface through sputtering (e.g., O_2) or diffusion (e.g., H_2). Hereafter the term 'release' is used to refer to atoms and molecules released through either sputtering or radiolysis.

So far, only sublimation of water ice and irradiation of Ganymede's surface by Jovian ions have been considered as a primary source of neutrals in atmospheric models of Ganymede (Marconi (2007); Turc et al. (2014); Leblanc et al. (2017); Plainaki et al. (2015), and Chapter 3.3). While the former (along with thermal degassing also considered by Shematovich (2016)) generates an H_2O-dominated tenuous atmosphere (Marconi, 2007), the latter also leads to radiolysis and induces an O_2-dominated tenuous atmosphere (Leblanc et al., 2017). The irradiating Jovian ion population that has been considered to assess release rates includes Jovian thermal, corotating ions, such as H^+ and O^+, and Jovian, energetic ions, such as H^+, O^{2+}, and S^{3+} (Ip et al., 1997; Cooper et al., 2001; Poppe et al., 2018; Plainaki et al., 2020; Carnielli et al., 2020b).

The first quantitative estimation of impact rates, related to ionospheric ions impacting the moon's surface, and of the release rates, induced by these impacts and leading to the release of neutral species, was recently undertaken by Carnielli et al. (2020b) using the 3D multi-species test-particle model presented in Section 17.3.3.1. The impact and release fluxes at Ganymede's surface are calculated on a spherical grid of radius 1 R_G with an angular separation of 2° between the cells: 90 cells along the polar angle θ and 180 cells along the azimuthal angle ϕ. The release flux, $I_{\text{rel},i}$ induced by irradiation by ion species i, is derived from the impact flux, $I_{\text{imp},i}$, of ion species i as follows:

$$I_{i,n}^{\text{rel}}(E,\theta,\phi) = Y_{i,n}(E) \times I_i^{\text{imp}}(E,\theta,\phi), \tag{17.8}$$

where E is the energy of the impacting ion i and $Y_{i,n}(E)$ is the energy-dependent release yield of species n from impact by ion i on water-ice, expressed in molecules per ion. The yield $Y_{i,n}$ increases with increasing energy E_{imp} of the impacting ions: as expected, the impact by an ion of higher energy causes a larger number of particles to be released from the surface. It is taken from Famá et al. (2008) for $E_{\text{imp}} < 100$ keV and from Johnson et al. (2009) for impact energies above. It depends on the angle of incidence of the impacting ion assumed to be 45° (where 0° corresponds to normal to the surface) and on the surface temperature map taken from Leblanc et al. (2017) spreading from 80 K to 150 K. The total impact and release rates are calculated by integrating over the full energy range of the impacting energies and over the polar and azimuthal angles to cover the whole surface of Ganymede, as well as summing over all impacting ion species (O_2^+, O^+, H_2O^+, H_2^+, H^+, and OH^+) and released species (Carnielli et al., 2020b).

Figure 17.8 shows the total ionospheric impact flux (a), the average impact energy (b), and the total induced release flux (c) on Ganymede's surface. These fluxes include the contribution from all ionospheric species considered, though O_2^+ dominates in all regions: it represents between 67 and 78 per cent of the total ionospheric impact flux depending on the regions and between 86 and 89 per cent of the total release rate induced by ionospheric ions (Carnielli et al., 2020b). This is not surprising because O_2^+ is the most abundant ion species produced close to the surface where its mother molecule is confined and it is the most abundant species present in the ionosphere (see Section 17.3.3.2).

The impact flux map, shown in Fig. 17.8 (a), is strongly linked to the ion dynamics (Carnielli et al., 2020b). Most of the ionospheric impact occurs in the equatorial region in the leading hemisphere, centred in 0° longitude.

- In the equatorial region, the flow close to the surface is predominantly radial over the leading and trailing hemispheres. Over the former, the flow is inward, especially close to the separatrix, and leads to impact on the surface. Over the latter, the flow is outward and ions escape ultimately through

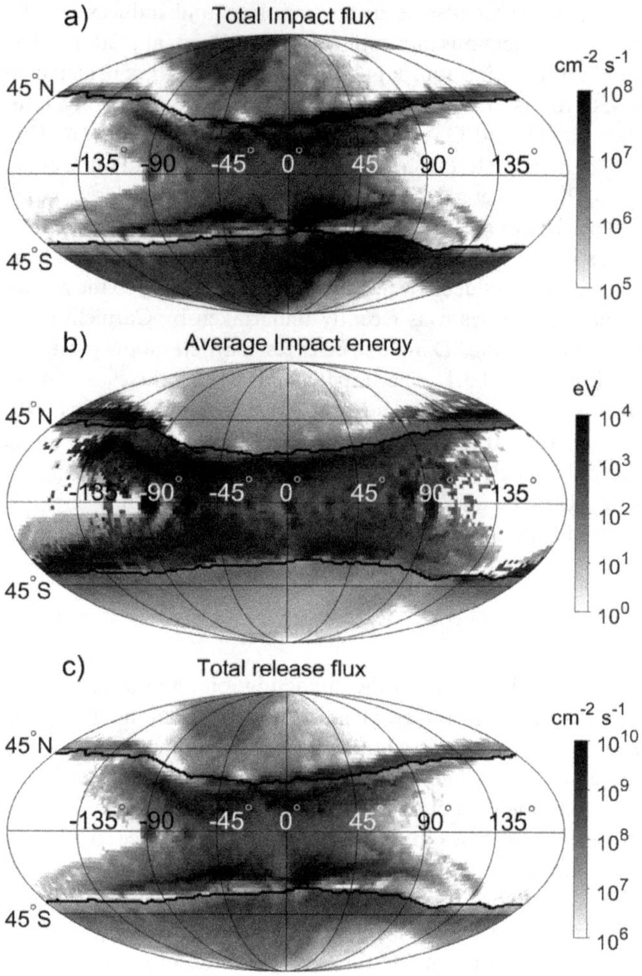

Figure 17.8 Maps induced by the irradiation of ionospheric ions on the surface of Ganymede: (a) total ionospheric impact flux, (b) average impact energy of ionospheric ions, and (c) total release flux of neutral species. All surface maps which include the contribution from all ionospheric species have been produced by the 3D test particle ion model from Carnielli et al. (2020b) for G2 conditions. The boundaries between open and closed magnetic field lines (separatrices) are plotted in black. The 0° longitude corresponds to the bisector of the leading hemisphere and 90° is towards Jupiter. The atmospheric, neutral model from Leblanc et al. (2017) ('dry' case) and the electric and magnetic fields from Jia et al. (2009) were used for these simulations (non-boosted case). Adapted from Carnielli et al. (2020b). Reprinted from *Icarus*, 351, G. Carnielli, M. Galand, F. Leblanc, R. Modolo, A. Beth, X. Jia, Simulations of ion sputtering at Ganymede ©2020 with permission from Elsevier.

the magnetopause boundary. In the Jovian-facing and anti-Jovian hemispheres the ion flow is more tangential and ions convect towards the trailing hemisphere.

- Over the northern polar region, the impact flux is low, as most ions close to the surface convect downstream to low latitudes (tangential flow towards the leading hemisphere) or escape through the Alfvén wing in the anti-Jovian hemisphere close to the separatrix where the flow is outward. Some of the ions which drift downstream reach the reconnection region and are accelerated back to the moon. They either impact the surface, contributing further to the impact flux at low latitudes, or drift around the moon and escape through the Alfvén wings (Carnielli et al., 2019).
- Over the southern polar region, the impact rate is greater than over the northern one and is more evenly spread over latitudes and longitudes. This is explained by the radial flow more inward over the southern than northern polar regions. Hence, while ions still convect downstream, those close enough to the surface impact it. Another reason for the hemispheric asymmetry over the polar regions is the ion flow close to the separatrix in the leading hemisphere: ions flow in the anti-Jovian direction over the northern polar region, while ions flow in the Jovian direction over the southern polar region.

The asymmetry between the polar regions is driven by that in the electromagnetic environment induced by the 45° tilt in the Jovian magnetic field during G2 (see Fig. 17.6 (d)), as apparent by the asymmetry in the separatrices (Carnielli et al., 2020b).

The average impact energy map (see Fig. 17.8 (b)) is primarily representative of the main ion species, O_2^+, whose thermal energy is negligible compared with convective energy (see Section 17.3.3.1). Ions move relatively slowly close to the surface over the polar regions, which explains the low impact energy found there. Over the equatorial region, the plasma includes both local, low-energy ions and higher-energy ions coming originally from the polar regions and convecting towards the downstream reconnection region before being accelerated back towards the moon. The highest speeds are reached near the separatrices. The strong flow shear occurring between open and closed magnetic field lines induces strong field-aligned currents that accelerate ionospheric ions towards the surface (Carnielli et al., 2020b).

The release flux map (see Fig. 17.8 (b)) is sensitive to both the impact flux and impact energy (see Eq. (17.8)). The most intense release fluxes are found in the equatorial region, especially close to the separatrices where impact flux and impact energy are very large. Larger-impact fluxes are observed in the southern polar region near the separatrix, but are associated with low-impact energy. Carnielli et al. (2020b) found that unlike the polar regions that are dominantly impacted by thermal, Jovian ions, the equatorial region is impacted by both Jovian, energetic ions and ionospheric ions. The latter impact more dominantly the leading side of the moon as a result of the global convection pattern. Such an asymmetry is consistent with Herschel observations from which a higher amount of water, by an order of magnitude, was derived for the leading hemisphere compared to the trailing hemisphere (Hartogh et al., 2013).

To summarise these findings, Figure 17.9 presents the total impact flux (grey) and total release flux (black) summed over all ionospheric ions, as a function of the impact energy, for four different regions on Ganymede's surface under the Galileo G2 flyby conditions. It clearly illustrates that more energetic impacting ions are more efficient at releasing molecules from the surface: the release to impact flux ratio increases with increasing impact energy. Figure 17.9 (a) highlights the asymmetry between the polar (solid) and equatorial (dashed) regions. Ions with low energy ($<$100 eV) primarily impact the polar, open-magnetic field line regions, where the electric field is weak. Ions with higher energies dominantly impact the equatorial region. While this region is undergoing impact by local, low-energy ions, it is also affected by ions accelerated efficiently by intense currents at the magnetopause before returning towards the moon and impacting its surface. When summing over the

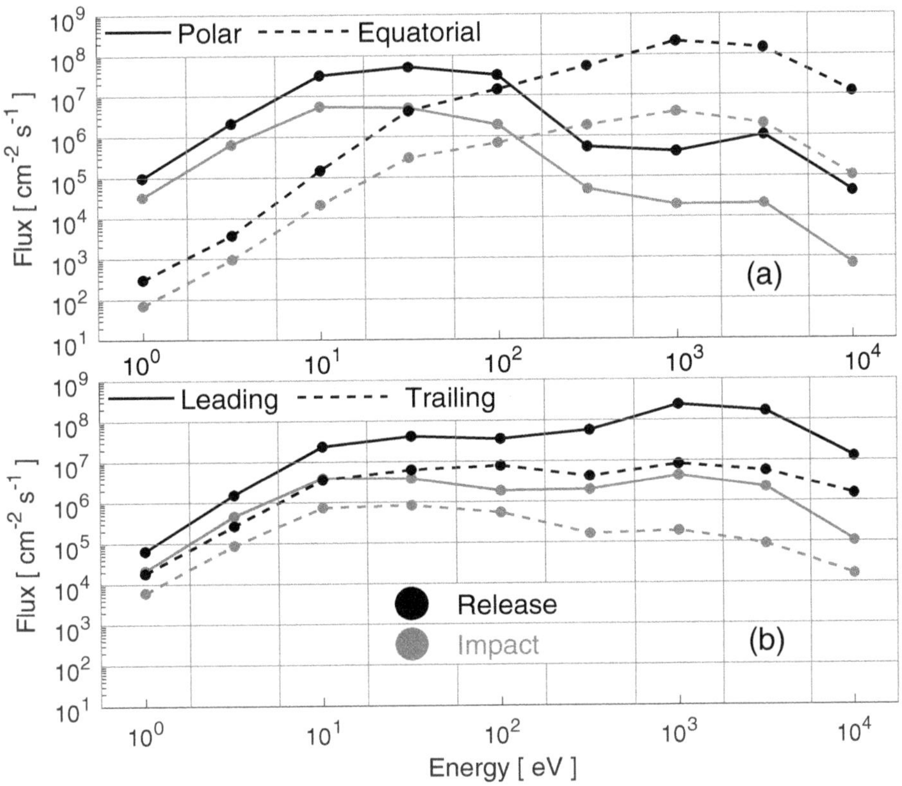

Figure 17.9 Energy distribution of the ionospheric ions impacting on Ganymede's surface, expressed in terms of impact flux (grey) and release flux of neutral species (black) (a) over the polar regions (solid lines), where magnetic field lines are open, and the equatorial region (dashed lines), where magnetic field lines are closed, (b) over the leading (solid lines) and trailing (dashed lines) hemispheres. The profiles, which include the contribution from all ionospheric species, have been produced by the 3D test particle ion model from Carnielli et al. (2020b) for G2 conditions. The neutral, atmospheric model from Leblanc et al. (2017) ('dry' case) and the electric and magnetic fields from Jia et al. (2009) were used for these simulations (non-boosted case). Adapted from Carnielli et al. (2020b). Reprinted from *Icarus*, 351, G. Carnielli, M. Galand, F. Leblanc, R. Modolo, A. Beth, X. Jia, Simulations of ion sputtering at Ganymede ©2020 with permission from Elsevier.

whole energy range, the polar impact flux dominates over the equatorial impact flux by a factor of 1.6; however, the release flux is almost four times higher in the equatorial region, where impact energies are higher, than in the polar regions (Carnielli et al., 2020b).

Figure 17.9 (b) highlights the asymmetry between leading (solid) and trailing (dashed) hemispheres. While a wide range of impact energies contribute to both, the asymmetry is larger, more than one order of magnitude from a few hundreds of eV, of both the impact and release fluxes over the leading hemisphere compared to the trailing hemisphere. This is explained by the fact that here over the leading hemisphere, the ion flow close to the surface is predominantly radial, oriented towards the surface, while over the trailing hemisphere, it is mainly radially outwards. When summing over all energies, the release flux is more than 15 times higher over the leading than the trailing hemispheres as a result of a higher impact flux (factor 6.3) and higher impact energies over the leading hemisphere (see Figure 17.9 (b)). Finally, it should be noted that, while for the G2 configuration the leading hemisphere is partly sunlit, the finding that the dominant impact and release rates are occurring over the leading hemisphere is not affected by the local time of Ganymede with respect to Jupiter and the related change in the atmospheric, neutral densities (Carnielli (2019), 2019).

The total ionospheric ion impact and release rates are compared with those from the Jovian, thermal O^+, and energetic ions (H^+, O^{2+}, S^{3+}) (see Table 17.3). All rates have been derived from the same 3D multi-species test particle model under the same G2 flyby conditions (Carnielli et al., 2020b). In the baseline case, while the ionospheric contribution to impact rate is similar to the Jovian, energetic contribution at a level of $\sim$ 45 per cent, it decreases down to 11 per cent in terms of release rate: the ionospheric ions are less energetic than the bulk of the Jovian, suprathermal ions, whose impact fluxes peak around 10^4 eV. In the boosted case which uses inputs consistent with the Galileo observations during G2 (see Section 17.4.1), the ionospheric ions are the main source of the atmosphere with a release rate representing 70 per cent of the total rate through ion irradiation. The reader should refer to Chapter 3.2 for further discussion on impact and release rates of Ganymede's surface as a source for its tenuous atmosphere.

17.5 PERSPECTIVES

17.5.1 Brief Highlight and Outstanding Questions

Our current understanding of Ganymede's ionosphere, which has largely benefited from the outcome of Galileo multi-instrument analysis of the ionospheric plasma, is summarised

Table 17.3 *Contribution of different ion populations from the Jovian magnetosphere and Ganymede's ionosphere to the impact and release rates.*

	Baseline		Boosted case	
Species	Total impact rate (s^{1})	Total release rate (s^{1})	Total impact rate (s^{1})	Total release rate (s^{1})
All (Jovian + Ionosph.)	21.9×10^{24}	25.1×10^{26}	196.3×10^{24}	74.4×10^{26}
Species	Contribution to impact rate (%)	Contribution to release rate (%)	Contribution to impact rate (%)	Contribution to release rate (%)
Jovian, thermal ions	9.2	2.4	1.0	0.8
Jovian, energetic ions	47.2	86.8	5.3	29.3
Ionospheric	43.6	10.8	93.7	69.9

Note: All values are derived from simulations applied for G2 conditions (Carnielli et al., 2020b). For the boosted case, the O_2 density has been increased everywhere by a factor of 10 ('dry' atmospheric scenario), and the electron-impact ionisation frequency by a factor of 4, in the polar regions in the anti-Jovian hemisphere, compared with the baseline case, in order to explain the Galileo/PWS ionospheric density and Galileo/PWS ion energy spectra observed along the G2 trajectory (see Section 17.4.1 and Carnielli et al. (2020a)).

Table 17.4 *Summary of the highlights of Ganymede's ionosphere in terms of our current understanding and implication of its analysis through Galileo multi-instrument observations and the use of kinetic modelling.*

Topic	Highlight	Relevant table, figure, or section
Formation of an ionosphere	• Ionosphere primarily produced from impact of atmospheric neutrals by Jovian electrons (having assumed a uniform frequency ν^e)	Table 17.2 in Section 17.3.3.1
Ionospheric composition	• O_2^+: major species, followed by H_2^+, O^+, and H_2O^+ (simulations for 'dry' atmosphere)	Fig. 17.4 in Section 17.3.3.2
	• O_2^+: consistent with comparison with Galileo/PLS in terms of ion energy spectrograms and ion bulk velocity during G2	Fig 17.6 and Fig. 17.7 in Section 17.4.1
Dynamics and spatial distribution of ionospheric plasma	• Non-uniform, structured ionosphere (e.g., Alfvén wing, equatorial trailing, equatorial leading)	Fig. 17.5 in Section 17.3.3.2
	• Significant influence from transport and acceleration processes driven by magnetospheric interaction	Section 17.4.1
Plasma present in Ganymede's magnetosphere	• Ionospheric plasma dominates close to the moon, even if Jovian plasma is present (G2)	Fig. 17.7 in Section 17.4.1
Electromagnetic field environment	• Validation during G2 flyby	Fig. 17.6 in Section 17.4.1
Atmospheric densities	• Enhancement by a factor of 10, at least for O_2 ('dry' case), during G2 flyby (multi-instrument analysis)	Section 17.4.1
Irradiation of the moon's surface by ionospheric ions	• First estimate of impact/release rates by ionospheric ions (mainly O_2^+)	Sections 17.3.3.2 and 17.4.2, Table 17.3
	• Ionospheric contribution to total release rate up to 70 per cent (G2 case)	Section 17.4.2
	• Largest impact flux in the equatorial, leading hemisphere and southern polar region; highest release rates in the equatorial region, near separatrices over the leading hemisphere (G2 case)	Fig. 17.8 and Fig. 17.9 in Section 17.4.2

Table 17.5 *Outstanding questions of relevance to Ganymede's ionosphere, its modelling, and its impact on its environment (not exhaustive).*

Outstanding questions
Energetic electrons (10eV–10 keV): • What are the spatial and energy distributions of electrons inside Ganymede's magnetosphere? • What and how significant are the processes accelerating, heating, and cooling them? • How variable are these distributions with solar and magnetospheric conditions? • How is the ionisation frequency map affected by a more realistic estimate of the energetic electrons? • What are the consequences in terms of ionospheric densities and impact rates? • Are electrons, produced within Ganymede's ionosphere or of Jovian origin, energetic enough to contribute significantly to the surface weathering? This question extends over a larger energy range, up to hundreds of MeV.
Dust: • Is dust present in the ionosphere, such as water-based dust as observed at Enceladus? • If so, does it have any impact on the ionosphere?
Magnetosphere–ionosphere coupling: • What is the effect of a more realistic ionosphere (e.g., asymmetric ionosphere, multi-species) on the simulated magnetospheric environment? • How different are the 3D maps of ionospheric moments with solar, magnetospheric, and orbital conditions?
Atmospheric densities: • Is the increase in O_2 densities by a factor of 10 ('dry' case), derived from G2 analysis, also found during other Galileo flybys? How different is the outcome of the analysis for a 'wet' atmosphere? • What is the source of inconsistency between the total release rate assumed in atmospheric models and that estimated from kinetic, ion simulations of ionospheric and Jovian ions (Carnielli et al., 2020b)? • What is the implication for atmospheric densities of the analysis of the auroral emissions, when energetic electron spectral maps in Ganymede's ionosphere become available? • What is the effect, on the 3D atmospheric, neutral distribution, of ionospheric ions which impact Ganymede's surface, especially in the equatorial, leading region?

in Table 17.4. Sections and figures, where key findings are discussed and illustrated, are given.

Keeping in mind what we have learned so far, some questions still remain outstanding, while new ones have been raised. A non-exhaustive list of key outstanding questions is provided in Table 17.5.

17.5.2 Plans in Support of the JUICE Mission

In preparation for the JUICE mission, 3D modelling of Ganymede's ionosphere is critical:

- for assessing the operability of plasma instruments, such as Radio and Plasma Wave Investigation (RPWI) and configuring most suitable operation modes (Rauch et al., 2017; Gilet, 2019; Carnielli, 2019)
- for estimating the spacecraft potential which affects the lower limit of the energy range over which electrons and ions will be able to be measured by the Particle Environment Package (PEP)
- for constraining the tenuous atmosphere, which may be denser by a factor as high as 10 than what atmospheric models predicted (see Section 17.4.1, 'dry' case); this has crucial implications for planning JUICE operation during its nine-month phase in orbit around Ganymede; circular orbits are planned all the way down to 200 km.

Ionospheric models will play an essential part in interpreting the anticipated JUICE observations of Ganymede's space environment, and in improving our understanding of Ganymede's ionosphere in terms of source, distribution, variability, and contribution to the atmospheric generation, to the field environment, and, more generally, to the coupling between Ganymede's and Jupiter's magnetospheres. More specifically, ionospheric models will be critical:

- for interpreting electron measurements in Ganymede's ionosphere, such as density and temperature from RPWI and density from the Gravity and Geophysics of Jupiter and the Galilean Moons (3GM) radio science instrument (see Section 17.4.1),

- for interpreting measured energy and phase distributions of energetic electrons, ions, and hot oxygen from PEP; hot neutrals are produced from charge exchange of energetic ions with the cold tenuous atmosphere (see Section 17.3.3.1 and Carnielli et al. (2020b)),
- for interpreting the ion composition observed from the Neutral Ion Mass (NIM) spectrometer (see Section 17.3.3.2), though only a small subset of the ionospheric population would be detectable to NIM which is capable of measuring ions only below 5 eV,
- for assessing the contribution of the ionospheric population as a source of the tenuous atmosphere, for input to atmospheric, neutral models in light of NIM neutral gas observations,
- for assessing electric conductivity and currents and for removing the ionospheric contribution from the Jupiter magnetometer (J-MAG) dataset in order to infer constraints on the internal magnetic field and to detect and characterise putative subsurface salty ocean (see Chapter 14),
- for interpreting the auroral emissions observed by the Ultra-Violet imaging Spectrograph (UVS) induced by energetic electrons whose 3D distribution will be critical to estimate (see Chapter 18).

At the same time, NIM observations of the neutral composition and density and J-MAG observations of the magnetic field, combined with atmospheric and magnetospheric models, respectively, will contribute to building a 3D picture of the atmospheric, neutral density distribution and of the electromagnetic field environment to be used as input to ionospheric models. A very fruitful approach will be multi-instrument analysis of JUICE observations, using physics-based ionospheric, atmospheric, and magnetospheric models, to link the gas, plasma, particle, field, and ultraviolet measurements together, enhance the science return from the JUICE mission, and reveal the complex and closely coupled environment around this intrinsically magnetised, plasma-diverse, and ionospheric-irradiated moon, which is Ganymede.

ACKNOWLEDGEMENTS

Work at Imperial College London was supported by STFC of UK under grant ST/W001071/1. X. Jia acknowledges support from NASA through contract #NNN13AA07T through the Applied Physics Laboratory at Johns Hopkins University and Early Career Fellow Startup Grant 80NSSC20K1286.

REFERENCES

Carnielli, G. 2019. Kinetic modelling of the plasma environment around Ganymede. PhD thesis Imperial College London, DOI: https://doi.org/10.25560/78695.

Carnielli, G., Galand, M., Leblanc, F. et al. 2019. First 3D test particle model of Ganymede's ionosphere. *Icarus*, **330**(Sept.), 42–59.

Carnielli, G., Galand, M., Leblanc, F. et al. 2020a. Constraining Ganymede's neutral and plasma environments through simulations of its ionosphere and Galileo observations. *Icarus*, **343**(June), 113691.

Carnielli, G., Galand, M., Leblanc, F. et al. 2020b. Simulations of ion sputtering at Ganymede. *Icarus*, **351**(Nov.), 113918.

Cessateur, G., Lilensten, J., Barthélémy, et al. 2012. Photoabsorption in Ganymede's atmosphere. *Icarus*, **218**(1), 308–19.

Collinson, G., Paterson, W. R., Bard, C. et al. 2018. New results from Galileo's first flyby of Ganymede: reconnection-driven flows at the low-latitude magnetopause boundary, crossing the cusp, and icy ionospheric escape.. *Geophysical Research Letters*, **45**(8), 3382–92.

Cooper, J. F., Johnson, R. E., Mauk, B. H., Garrett, H. B., and Gehrels, N. 2001. Energetic ion and electron irradiation of the icy Galilean satellites. *Icarus*, **149**(1), 133–59.

Dorelli, J. C., Glocer, A., Collinson, G., and Tóth, G. 2015. The role of the Hall effect in the global structure and dynamics of planetary magnetospheres: Ganymede as a case study. *Journal of Geophysical Research (Space Physics)*, **120**(7), 5377–92.

Duling, S., Saur, J., and Wicht, J. 2014. Consistent boundary conditions at nonconducting surfaces of planetary bodies: applications in a new Ganymede MHD model. *Journal of Geophysical Research (Space Physics)*, **119**(6), 4412–40.

Eviatar, A., Vasyliūnas, V. M., and Gurnett, D. A. 2001. The ionosphere of Ganymede. *Planetary and Space Science*, **49**(3–4), 327–36.

Famá, M., Shi, J., and Baragiola, R. A. 2008. Sputtering of ice by low-energy ions. *Surface Science*, **602**(1), 156–61.

Fatemi, S., Poppe, A. R., Khurana, K. K., Holmström, M., and Delory, G. T. 2016. On the formation of Ganymede's surface brightness asymmetries: kinetic simulations of Ganymede's magnetosphere. *Geophysical Research Letters*, **43**(10), 4745–54.

Frank, L. A., Paterson, W. R., Ackerson, K. L., and Bolton, S. J. 1997. Outflow of hydrogen ions from Ganymede. *Geophysical Research Letters*, **24**(17), 2151–4.

Gilet, N. 2019. Modélisation et calibration des sondes à impédance mutuelle: application à la sonde MIP à bord de Rosetta et préparation de BepiColombo et JUICE. *PhD thesis Université Orléans*, Dec.

Gurnett, D. A., Kurth, W. S., Roux, A., Bolton, S. J., and Kennel, C. F. 1996. Evidence for a magnetosphere at Ganymede from plasma-wave observations by the Galileo spacecraft. *Nature*, **384**(6609), 535–7.

Hall, D. T., Feldman, P. D., McGrath, M. A., and Strobel, D. F. 1998. The far-ultraviolet oxygen airglow of Europa and Ganymede. *Astrophysical Journal*, **499**(1), 475–81.

Hartogh, P., Bockelée-Morvan, D., Rezac, L. et al. 2013. Detection and Characterisation of Ganymede's and Callisto's Water Atmospheres. *International Symposium 'The universe explored by Herschel' 15–18 October 2013, ESTEC, Netherlands*, http://herschel.esac.esa.int/TheUniverseExploredByHerschel/presentations/13a--1720_HartoghP.pdf.

Ip, W. H., Williams, D. J., McEntire, R. W., and Mauk, B. 1997. Energetic ion sputtering effects at Ganymede. *Geophysical Research Letters*, **24**(21), 2631–4.

Jia, X., Walker, R. J., Kivelson, M. G., Khurana, K. K., and Linker, J. A. 2010. Dynamics of Ganymede's magnetopause: intermittent reconnection under steady external conditions. *Journal of Geophysical Research (Space Physics)*, **115**(Dec.), 12202.

Jia, X., Walker, R. J., Kivelson, M. G., Khurana, K. K., and Linker, J. A. 2008. Three-dimensional MHD simulations of Ganymede's magnetosphere. *Geophysical Research Letters*, **113**(A6), A06212.

Jia, X., Walker, R. J., Kivelson, M. G., Khurana, K. K., and Linker, J. A. 2009. Properties of Ganymede's magnetosphere inferred from improved three-dimensional MHD simulations. *Geophysical Research Letters*, **114**(A9), A09209.

Johnson, R. E., Burger, M. H., Cassidy, T. A. et al. 2009. *Composition and Detection of Europa's Sputter-Induced Atmosphere*. Pages 527–48 in Pappalardo, R. T., McKinnon, W. B., and Khurana,

K. (eds.), *Europa*. University of Arizona Press, Project MUSE muse.jhu.edu/book/57223.

Kivelson, M. G., Bagenal, F., Kurth, W. S. et al. 2004. *Magnetospheric Interactions with Satellites*. Cambridge, UK: Cambridge University Press. Pages 513–36.

Kliore, A. J. 1998. *Satellite Atmospheres and Magnetospheres*. Dordrecht: Springer Netherlands. Pages 1065–9.

Kopp, A., and Ip, W.-H. 2002. Resistive MHD simulations of Ganymede's magnetosphere 1. Time variabilities of the magnetic field topology. *Journal of Geophysical Research (Space Physics)*, **107**(A12), 1490.

Leblanc, F., Oza, A. V., Leclercq, L. et al. 2017. On the orbital variability of Ganymede's atmosphere. *Icarus*, **293**(Sept.), 185–98.

Leclercq, L. 2015. De l'exosphère á la magnétosphère des objets planétaires faiblement magnétisés: optimisation de modélisations parallélisées pour une application à Ganymède. UVSQ/Université Paris-Saclay, https://hal–insu.archives–ouvertes.fr/tel–0125619 1v2.

Leclercq, L., Modolo, R., Leblanc, F., Hess, S., and Mancini, M. 2016. 3D magnetospheric parallel hybrid multi-grid method applied to planet-plasma interactions. *Journal of Computational Physics*, **309**(Mar.), 295–313.

Liuzzo, L., Poppe, A. R., Paranicas, C. et al. 2020. Variability in the energetic electron bombardment of Ganymede. *Journal of Geophysical Research (Space Physics)*, **125**(9), e28347.

Marconi, M. L. 2007. A kinetic model of Ganymede's atmosphere. *Icarus*, **190**(1), 155–74.

Paranicas, C., Paterson, W. R., Cheng, A. F. et al. 1999. Energetic particle observations near Ganymede. *Geophysical Research Letters*, **104**(A8), 17459–70.

Paty, C., Paterson, W., and Winglee, R. 2008. Ion energization in Ganymede's magnetosphere: using multifluid simulations to interpret ion energy spectrograms. *Journal of Geophysical Research (Space Physics)*, **113**(A6), A06211.

Paty, C., and Winglee, R. 2004. Multi-fluid simulations of Ganymede's magnetosphere. *Geophysical Research Letters*, **31**(24), L24806.

Plainaki, C., Milillo, A., Massetti, S. et al. 2015. The H_2O and O_2 exospheres of Ganymede: the result of a complex interaction between the jovian magnetospheric ions and the icy moon. *Icarus*, **245**(Jan.), 306–19.

Plainaki, C., Massetti, S., Jia, X. et al. 2020. Kinetic simulations of the Jovian energetic ion circulation around Ganymede. *Astrophysical Journal*, **900**(1), 74.

Poppe, A. R., Fatemi, S., and Khurana, K. K. 2018. Thermal and energetic ion dynamics in Ganymede's magnetosphere. *Journal of Geophysical Research (Space Physics)*, **123**(6), 4614–37.

Rauch, J. L., Henri, P., Wahlund, J. E. et al. 2017 (Sept.). Electron Density Measurement on JUICE Mission by Mutual Impedance Technique: MIME Instrument as a Part of RPWI Consortium. *Pages EPSC2017–518 of: European Planetary Science Congress*.

Roth, L., Ivchenko, N., Gladstone, G. R. et al. 2021. A sublimated water atmosphere on Ganymede detected from Hubble Space Telescope observations. *Nature Astronomy*, **5**(July), 1043–51.

Scudder, J. D., Sittler, E. C., and Bridge, H. S. 1981. A survey of the plasma electron environment of Jupiter: a view from voyager. *Geophysical Research Letters*, **86**(A10), 8157–79.

Shematovich, V. I. 2016. Neutral atmosphere near the icy surface of Jupiter's moon Ganymede. *Solar System Research*, **50**(4), 262–80.

Teolis, B. D., Plainaki, C., Cassidy, T. A., and Raut, U. 2017. Water ice radiolytic O_2, H_2, and H_2O_2 yields for any projectile species, energy, or temperature: a model for icy astrophysical bodies. *Journal of Geophysical Research (Planets)*, **122**(10), 1996–2012.

Tóth, G., Jia, X., Markidis, S. et al. 2016. Extended magnetohydrodynamics with embedded particle-in-cell simulation of Ganymede's magnetosphere. *Journal of Geophysical Research (Space Physics)*, **121**(2), 1273–93.

Turc, L., Leclercq, L., Leblanc, F., Modolo, R., and Chaufray, J. Y. 2014. Modelling Ganymede's neutral environment: a 3D test-particle simulation. *Icarus*, **229**(Feb.), 157–69.

Vasyliūnas, V. M., and Eviatar, A. 2000. Outflow of ions from Ganymede: a reinterpretation. *Geophysical Research Letters*, **27**(9), 1347–9.

Wang, L., Germaschewski, K., Hakim, A. et al. 2018. Electron physics in 3-D two-fluid 10-moment modeling of Ganymede's magnetosphere. *Journal of Geophysical Research (Space Physics)*, **123**(4), 2815–30.

Williams, D. J., Mauk, B. H., McEntire, R. W. et al. 1997. Energetic particle signatures at Ganymede: implications for Ganymede's magnetic field. **24**(17), 2163–6.

Williams, D. J., Mauk, B., and McEntire, R. W. 1998. Properties of Ganymede's magnetosphere as revealed by energetic particle observations. *Geophysical Research Letters*, **103**(A8), 17523–34.

Zhou, H., Tóth, G., Jia, X., Chen, Y., and Markidis, S. 2019. Embedded kinetic simulation of Ganymede's magnetosphere: improvements and inferences. *Geophysical Research Letters*, **124**(7), 5441–60.

Zhou, H., Tóth, G., Jia, X., and Chen, Y. 2020. Reconnection-driven dynamics at Ganymede's upstream magnetosphere: 3-D global Hall MHD and MHD-EPIC simulations. *Journal of Geophysical Research (Space Physics)*, **125**(8), e28162.

18

Ganymede's Aurora

HERBERT GUNELL, LORENZ ROTH, SHAHAB FATEMI, CHUANFEI DONG, AND LIANG WANG

18.1 AURORAS

We begin this chapter with a discussion of the nature of auroras and how the auroras at Ganymede can be compared to those on Earth, where the phenomenon has been studied the most. On both Ganymede and Earth the aurora is externally driven. On Earth the energy ultimately comes from the relative motion between the planet and the solar wind, and on Ganymede the energy source is the relative motion between the moon and the plasma in the Jovian magnetosphere. This is another reason for comparing Ganymede to Earth rather than to the nearby Jupiter, whose aurora is driven by processes within Jupiter's own magnetosphere.

What is an aurora? The phenomenon was first observed on Earth, where anyone would recognise it, particularly the bright discrete auroral arcs. When light emissions are seen at other bodies, a definition is needed to determine whether or not we may classify them as aurora. A definition is not a law of nature, and reasonable minds may differ on the question of classification. We hold as suitable the proposition that extraterrestrial auroras should look like auroras on Earth and be caused by the same kind of process. The definition we shall adopt here is that an aurora is the light emissions caused by the impact of charged particles from the magnetosphere that precipitate on an atmosphere along the lines of a planetary magnetic field (Knudsen et al., 2021). This distinguishes aurora from airglow, which is caused by way of local chemistry, and the requirement of a magnetic field sets the aurora apart from other unstructured emissions. It is the magnetic field that leads to the formation of arcs and the large-scale organisation into auroral ovals. For comparison we may test this definition by applying it to Mars. Light emissions from the cusps of the Martian crustal magnetic field (Bertaux et al., 2005) are auroras according to our definition, whereas emissions caused by neutral particles (Deighan et al., 2018) or by solar energetic particles (Schneider et al., 2015) are not. One example of an emission on Earth that is not an aurora according to our definition is STEVE (Strong Thermal Emission Velocity Enhancement), as that is caused by protons moving horizontally, that is to say, not by field-aligned precipitating particles (Gallardo-Lacourt et al., 2021). Turning now to our object of interest, Ganymede is itself magnetised (Chapter 13) and is thereby surrounded by a small magnetosphere within Jupiter's magnetosphere (Chapter 14). As we shall see shortly, the emissions come from Ganymede's atmosphere (Chapter 16), where they form circumpolar bands in regions of vertical magnetic field. They peak in regions with a vertical magnetic field, where models predict the strongest field-aligned current, indicating that the emissions are in agreement with our definition of aurora.

On Earth the aurora is found in ovals around the magnetic poles. The poleward edge of this oval reaches the polar cap boundary, that is to say, the boundary between closed and open field lines. Discrete auroral arcs can at times reach locations near this boundary, and there is an Alfvén Poynting flux going down to the ionosphere and powering this aurora (Keiling et al., 2002); see also the review by Karlsson et al. (2020). Following the magnetic field lines on the nightside of the planet, this boundary traces the plasma sheet boundary layer all the way to the region of the tail where reconnection takes place (e.g. Akasofu et al., 2010). However, the vast majority of the aurora is emitted from the part of the ionosphere that is on closed field lines (e.g. Karlsson et al., 2020). Even during substorms, when bright emissions appear poleward of where the strongest emissions occur, these are equatorward of the open–closed field line boundary (Mauk and Bagenal, 2012).

On Ganymede there is also an open–closed field line boundary, but the magnetospheric configuration differs significantly from that of Earth (Kivelson et al., 1996). The relative motion of Ganymede and the magnetospheric plasma is slower than the Alfvén speed, and therefore there is no bow shock at Ganymede. On the other hand, Alfvén wings emanate from its polar region. These Alfvén wings connect the moon to Jupiter (see also Chapter 14 or Kivelson et al., 2002; Neubauer, 1998). This sub-Alfvénic interaction of Jupiter's magnetospheric plasma with Ganymede's magnetosphere influences the shape and structure of Ganymede's aurora. It leads to an asymmetry, with emissions at higher latitudes on the upstream and lower latitudes on the downstream side.[1] As the moon is magnetically connected to Jupiter, the field-aligned potential drops associated with the Alfvén wings accelerate electrons that cause auroras on that planet (e.g. Hess and Delamere, 2012). When the Juno spacecraft passed above the tail of Ganymede's footprint on Jupiter's ionosphere, it was found that Alfvénic acceleration was the dominating electron energisation process (Szalay et al., 2020). Interaction between Ganymede and Jupiter is treated in Chapter 19.

While there are many sub-categories of aurora at Earth, the two main kinds are the discrete arcs and the diffuse aurora.

[1] The speed of Ganymede in its orbit around Jupiter is $\sim 11\,\mathrm{km\,s^{-1}}$, and the co-rotating plasma of Jupiter's magnetosphere moves at $\sim 150\,\mathrm{km\,s^{-1}}$. Thus, Ganymede is overtaken by the surrounding plasma, and the upstream direction is opposite that of Ganymede's orbital motion.

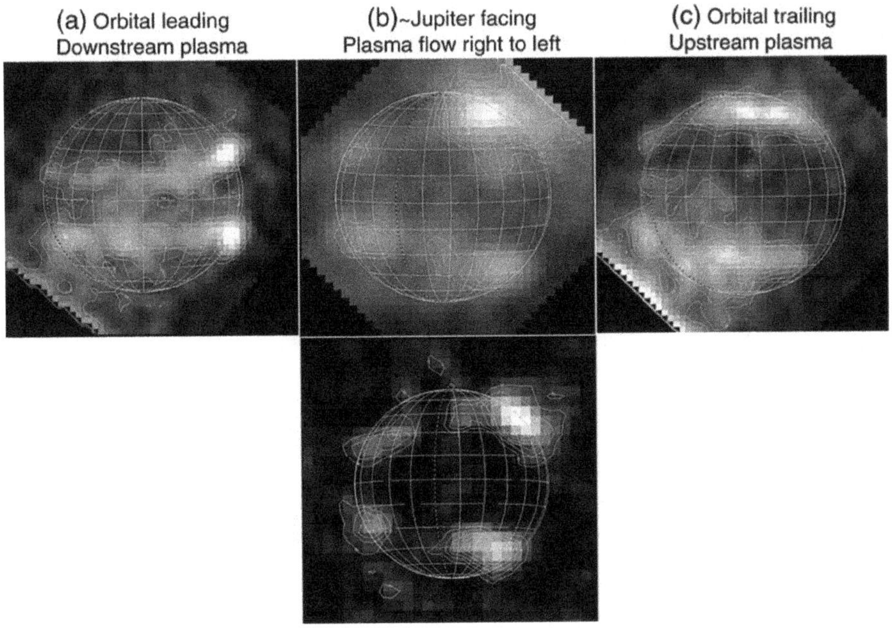

Figure 18.1 Ganymede's aurora seen from different directions. Both the upper and lower images in the central column (b) are Jupiter-facing but recorded during different observational dates. Figure from McGrath et al. (2013). Reprinted with permission from John Wiley and Sons. M. A. McGrath, X. Jia, K. Retherford, P. D. Feldman, D. F. Strobel, and J. Saur. Aurora on Ganymede. *J. Geophys. Res. (Space Physics)*, 118:2043–2054, ©2013. (A black and white version of this figure will appear in some formats. For the color version, please refer to the plate section.)

The discrete arcs are brighter and more localised, but the total precipitated energy of the diffuse aurora, integrated over the auroral oval, is higher than that of the discrete arcs (Newell et al., 2009). In discrete arcs, electrons are accelerated downward by electric fields that are parallel to the magnetic field (e.g. Lysak et al., 2020). In the diffuse aurora, however, there is no indication of any such acceleration; instead, electrons precipitate after waves have scattered them into the loss cone (e.g. Nishimura et al., 2020). What drives an auroral arc is not yet completely understood. However, an electric generator mechanism is required to provide the potential drop that accelerates the auroral electrons. A number of generator processes were reviewed recently by Borovsky et al. (2020). The Juno spacecraft, observing Jupiter's aurora, found electron and ion distributions indicative of discrete acceleration by both electrostatic potential drops and broadband or stochastic processes. The energy flux was larger for the broadband than for the discrete acceleration processes (Mauk et al., 2017). Juno also observed charged particle distributions indicative of both broadband and discrete acceleration on the same field line in ways that are not known from Earth (Mauk et al., 2018). For Ganymede even less is known today due to the scarcity of in situ measurements. In this chapter, we review the observations that have been made of the aurora at Ganymede, numerical modelling relevant to the aurora, and finally we discuss the open questions and how these can be addressed in coming years.

18.2 OBSERVATIONS

Observations of the Ganymede's auroral emissions serve also as the primary observational diagnostic tool for the moon's atmosphere, although an atmosphere has also been detected via starlight occultation measurements (Carlson et al., 1973), and an ionosphere was detected through radio occultations by the Galileo spacecraft (Kliore, 1998). For an overview on all aurora observations, we refer to reader to Table 1 of Chapter 16 in this book. All listed observations except for the Galileo UVS data include the far-ultraviolet oxygen aurora, discussed here.

First Detection

The first detection of emissions from Ganymede's oxygen atmosphere, which now generally are called 'Ganymede's aurora,' were achieved through observations by the Goddard High Resolution Spectrograph (GHRS) of the Hubble Space Telescope (HST) (Hall et al., 1998). The obtained spectrum (see Figure 1 in Chapter 16) showed atmospheric emissions at the atomic oxygen multiplets near 1 304 Å and 1 356 Å on top of the surface reflectance signal, similar to the emissions detected on Europa (Hall et al., 1995). The GHRS data did not provide spatial resolution to map the emissions, and Hall et al. (1998) used the term 'airglow' in their study of the observations of the two moons (Europa and Ganymede). 'Airglow' is often used for emissions that originate from energising processes within an atmosphere (e.g., recombination). However, the clear double-peak structure of the OI1 356 Å emissions in the Ganymede spectrum allowed Hall et al. (1998) already to speculate about the auroral nature of the Ganymede emissions in the sense of being governed by electron precipitation along magnetic field lines. Indeed, the auroral nature of Ganymede's emissions was confirmed later on, which is in contrast to the neighbouring Callisto, where the faint FUV emissions are thought to originate from photo-electrons ('airglow') (Cunningham et al., 2015).

Static Aurora Morphology

The first images of the oxygen emissions were taken when HST's Space Telescope Imaging Spectrograph (STIS) observed the plasma upstream (orbital trailing) hemisphere of Ganymede in 1998 (Feldman et al., 2000). Indeed, they disclosed an auroral

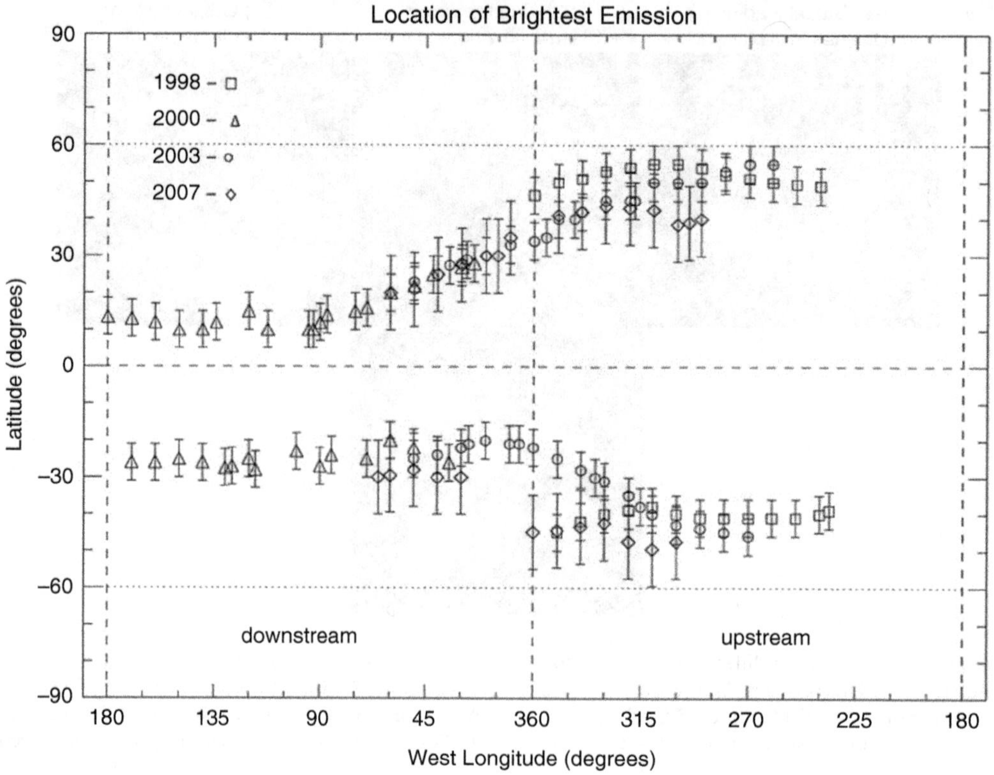

Figure 18.2 Positions in latitude and longitude of the peak of the emission bands. The different symbols represent the four different observation campaigns. Longitude 0°, or equivalently 360°, is at the centre of the Jupiter-facing hemisphere; 90° is facing downstream and 270° upstream. Picture from McGrath et al. (2013). Reprinted with permission from John Wiley and Sons. M. A. McGrath, X. Jia, K. Retherford, P. D. Feldman, D. F. Strobel, and J. Saur. Aurora on Ganymede. *J. Geophys. Res. (Space Physics)*, 118:2043–2054, ©2013.

type of topology with the brightest emissions clustered near the high latitudes; see Figure 18.1(c).

Further STIS imaging of the orbital leading hemisphere revealed different oxygen emission morphology with longer bands much closer to the equator (Figure 18.1a). The images of Ganymede's sub-Jovian hemisphere provide an intermediate vantage point, viewing both the high-latitude upstream bands and low-latitude downstream bands, with a possible gap in between (Figure 18.1b).

A comparison to magnetohydrodynamic (MHD) simulations showed that the regions of the brightest OI 1 356 Å emissions are roughly collocated with the Open–Closed Field Line Boundary (OCFB) of Ganymede's mini-magnetosphere (McGrath et al., 2013). On the orbital leading side, which is also the plasma downstream or wake side, the band-like emissions are close to the equator, as the magnetosphere is stretched. The high-latitude emissions observed on the trailing side, which is also the plasma upstream side, are consistent with a compressed magnetosphere pushing the OCFB further to the poles.

Projections on a latitude-longitude map (Figure 18.2) suggest that the bright aurora bands can get as close as 10° to the equator on the downstream side and are centred at latitudes 50° N/S on the upstream side. Musacchio et al. (2017) used another approach to quantify the latitude as well as the brightness of the aurora bands by integrating along the horizontal axes in the images; see Figure 18.3. The peaks of resulting brightness profiles lie between 9° and 28° latitude on the downstream side and upstream close to 50° N/S. Moreover, the study by Musacchio et al. (2017) quantifies an asymmetry in the latitude of the northern and southern bands. Relating this asymmetry in the band latitudes directly to the orientation of Ganymede's dipole magnetic moment, they derive a more westward orientation than the estimations based on in situ measurement of the magnetic field (Kivelson et al., 2002).

The integrated brightness of the 1 356 Å emissions in the auroral bands is roughly between 100 R and 200 R (Figure 18.3) and is generally similar in the more polar upstream (trailing) side and equatorial downstream (leading) side auroral bands. Local peak intensities inside the auroral bands reach ~300 R (Feldman et al., 2000; McGrath et al., 2013). The changes in location and brightness (between upper and lower panels in Figure 18.3) are due to changes in the plasma environment and are discussed below.

Away from the bright aurora bands, there is a background intensity in the oxygen emissions across the entire disk of about 20–40 R (Feldman et al., 2000; McGrath et al., 2013; Musacchio et al., 2017; Molyneux et al., 2018) (Figure 18.3, away from peaks). In order to distinguish the emissions from the brighter auroral bands, they are sometimes called 'diffuse' emissions (Figure 18.4). There is, however, no observational constraint on the generating processes and, for example, on the involved energies of the exciting electron, which would allow a clear separation between different types of auroral emissions across Ganymede. The images of the fainter 1 304 Å emissions revealed a morphology generally similar to the 1 356 Å emission images (Feldman et al., 2000; Molyneux et al., 2018) with intensities of roughly half the intensity of the emissions discussed in the previous paragraph. An analysis of the relative intensities of the two oxygen emissions across the moon disk in images showed

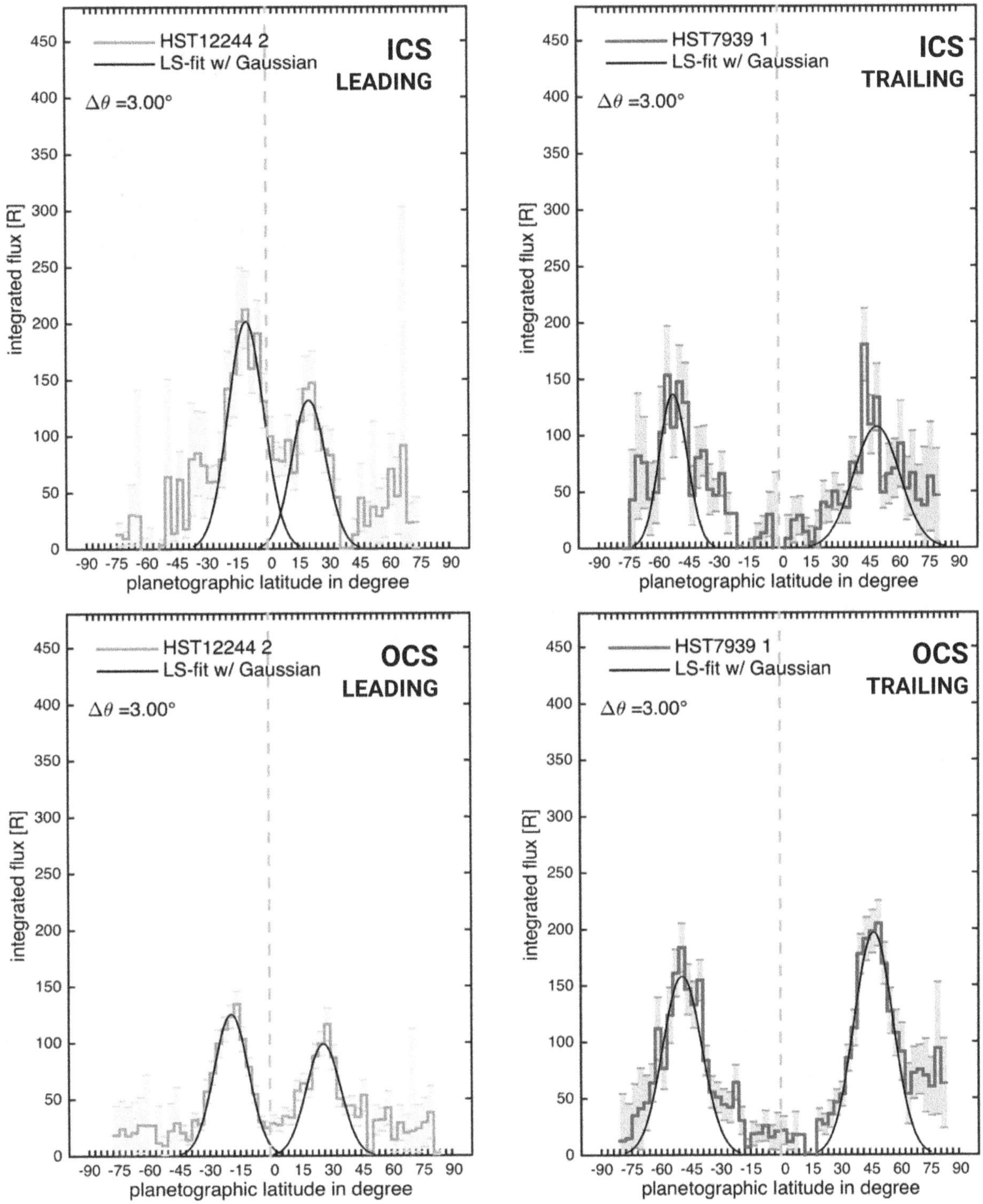

Figure 18.3 Integrated brightness as a function of latitude on both hemispheres when Ganymede is inside the dense plasma or current sheet (ICS) and above or below it (OCS) (Musacchio et al., 2017). Reprinted with permission from John Wiley and Sons. Fabrizio Musacchio, Joachim Saur, Lorenz Roth, Kurt D. Retherford, Melissa A. McGrath, Paul D. Feldman, and Darrel F. Strobel. Morphology of Ganymede's FUV auroral ovals. *J. Geophys. Res. (Space Physics)*, 122(3):2855–2876, ©2017. (A black and white version of this figure will appear in some formats. For the color version, please refer to the plate section.)

that the 1 304 Å emissions become relatively stronger towards the sub-solar point (Roth et al., 2021). This change in the relative emissions originates from contributions of water vapor aurora (dissociative excitation of H_2O) in this region due to sublimation of the ice surface.

Besides the HST observations, Ganymede's aurora was observed by the Pluto-Alice spectrograph on board the New Horizons spacecraft in 2007. The OI1 356 Å intensity measured during two passages of Ganymede through eclipse of Jupiter is presented in Chapter 16. The stability through eclipse as well

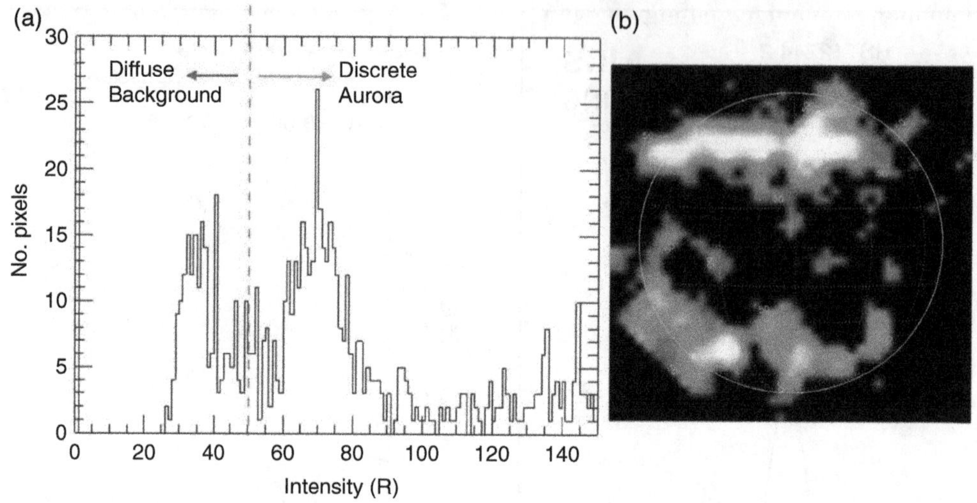

Figure 18.4 Bright versus faint or diffuse versus discrete aurora (Molyneux et al., 2018). The purple vertical dashed line in the histogram in panel (a) demarcates the threshold over which the emissions were considered bright as opposed to the diffuse background. The image in panel (b) only includes bright pixels. Pictures from Molyneux et al. (2018). Reprinted with permission from John Wiley and Sons. P. M. Molyneux, J. D. Nichols, N. P. N. P. Bannister, E. J. Bunce, J. T. Clarke, S. W. H. Cowley, J.-C. Gérard, D. Grodent, S. E. Milan, and C. Paty. Hubble space telescope observations of variations in Ganymede's oxygen atmosphere and aurora. *J. Geophys. Res. (Space Physics)*, 123(5):3777–3793 ©2018. (A black and white version of this figure will appear in some formats. For the color version, please refer to the plate section.)

as the intensity of the nightside during the second observation confirm that the emissions are independent of solar illumination and thus their auroral nature.

Collinson et al. (2018) analysed ion velocity distributions recorded by the Galileo Plasma Subsystem (PLS) during the first flyby of Ganymede by the Galileo spacecraft. The authors found downgoing field-aligned ions at the time the spacecraft crossed the OCFB. No electron observations were available, as the electron sensors had been damaged by the harsh environment.

Time–Variability

Although the global appearance of the auroral bands is considered to be relatively stable (McGrath et al., 2013), there are variations of the location of the bands on the timescale of Jupiter's rotation and possibly also temporal variability in the local intensities.

In a dedicated observation campaign, Saur et al. (2015) monitored the auroral bands on the downstream side over half a period of Jupiter's rotation during two visits in 2010 and 2011. They show that the locations of the auroral bands oscillate due to the periodic oscillation of Jupiter's time-varying magnetic field relative to Ganymede but find that the angle defining the amplitude of this oscillation is only $2.0\pm1.3°$. Using resistive MHD simulations, they estimated the expected oscillation of the OCFB to be at least 4.5°. Saur et al. (2015) concluded that a process to reduce the aurora oscillation is required, such as magnetic induction in a sub-surface ocean. The authors assumed that the observed optical emissions coincided with the lines where the OCFB intersects Ganymede's surface, and as discussed in Section 18.4, this assumption is not yet observationally confirmed. However, for the conclusion of Saur et al. (2015) to hold, collocation of the OCFB and the emissions is not necessarily required. What is necessary is that the two have the same angular oscillation amplitude. To completely confirm this, simultaneous global optical observations and in situ plasma measurements are needed.

Further analysing the variability over the Jupiter rotation, Musacchio et al. (2017) showed that when Ganymede moves from the denser current sheet ("inside current sheet" or ICS) to more dilute plasma regions above and below ("outside current sheet" or OCS), the emission bands move to lower latitudes on the upstream side and to higher latitudes on the downstream side, as evidenced by comparing the top and bottom rows of Figure 18.3. These shifts are in accordance with the excepted change of the OCFB in response to plasma density (and thus plasma pressure) changes.

Musacchio et al. (2017) further analyse the evolution of the intensity on average on the disk as well as in the bands over a rotation of Jupiter. On the leading side, both average and band brightness are higher by 30 to 40 per cent when Ganymede is in the plasma sheet centre compared to far outside the plasma sheet (compare the top and bottom panels in Figure 18.3 left). The authors explain this increase by the increased upstream ram pressure, which possibly generates larger magnetic stresses, resulting in larger electric currents near the open–closed field line boundary. On the trailing side (top and bottom panels in Figure 18.3 right), the opposite trend is observed with a decrease by ~20 per cent from the outside-current-sheet (OCS) images to the inside-current-sheet (ICS) images. This is suggested as being caused by a shift of the auroral bands towards the downstream hemisphere (not captured in the image) when the upstream ram pressure is higher (inside the current sheet).

Using later STIS and COS observations from 2014, Molyneux et al. (2018) confirmed these differing brightness changes on the leading and trailing hemispheres. Molyneux et al. (2018) note furthermore that the measured standard deviations of the brightness between individual exposures indicate that there is more temporal variation in the emissions on the leading hemisphere than on the trailing hemisphere. Systematic changes in the background emissions away from the oval were not found (Musacchio et al., 2017).

18.3 MODELLING

Modelling of Ganymede's magnetosphere has been performed in the past, using analytical, fluid, and hybrid models. Most of this modelling has not focused on the aurora but rather on the larger-scale physics of Ganymede's magnetosphere. Photochemistry has been used to model dayglow emissions from Ganymede (Cessateur et al., 2016), but that phenomenon is different from the aurora. In this chapter, the analytical models in the next paragraph were used to estimate the auroral emissions themselves, whereas the vast majority of the fluid and hybrid models described later in the chapter do not include photochemistry. The modelled magnetic field has been compared to magnetic field data obtained during the Galileo flybys, and improvements in modelling have through the years led to a better ability to reproduce the observed fields. The magnetic field configuration and the physics of the magnetosphere are closely linked to the processes causing the auroral emissions. In this section, we review some of the Ganymede magnetospheric modelling that has been done, while concentrating on the auroral aspects of it.

Analytical Models

Eviatar et al. (2000) compared electron distributions observed by the Galileo spacecraft to analytical models, and it was found that, due to an energy-dependent injection mechanism, mostly high-energy electrons are trapped on the closed field lines in Ganymede's magnetosphere. Electron spectrometer energy channels up to 884 keV were considered. Further analytical modelling of electrons, ions, and neutrals led to the conclusion that there is enough power in the relative motion of Ganymede through Jupiter's magnetosphere to drive the aurora. However, due to the low density of the plasma surrounding Ganymede, an acceleration mechanism in near-Ganymede space is required to achieve the observed auroral intensities (Eviatar et al., 2001b,a). It was concluded that, although both are possible, acceleration by a magnetic field–aligned electric field is more likely than stochastic acceleration by waves (Eviatar et al., 2001a). Much later, observations by the Juno spacecraft revealed that stochastic acceleration was more important than static electric fields in causing the aurora on Jupiter (Mauk et al., 2017, 2018). Whether similar discoveries will take place at Ganymede remains to be seen.

Fluid Models

In global MHD modelling of Ganymede interacting with the plasma of the Jovian magnetosphere, a magnetopause forms upstream of the moon. As Jupiter's and Ganymede's magnetic fields are oppositely oriented, reconnection takes place in that region, and that is seen in resistive MHD models. These models also reproduce Alfvén wings over Ganymede's polar regions (Ip and Kopp, 2002; Kopp and Ip, 2002). Furthermore, resistive MHD models predict that field-aligned currents run between the region around the OCFB and the Jovian magnetosphere (Ip and Kopp, 2002; Kopp and Ip, 2002), as illustrated in panel (a) of Figure 18.5.

Jia et al. (2008) used resistive MHD to model Ganymede's interaction with the ambient Jovian magnetosphere and compared the results to the Galileo magnetic field measurements. This was improved in a later paper by including models of anomalous resistivity in the reconnection region and an improved ionospheric boundary condition (Jia et al., 2009). The enhanced resistivity layer takes on the role of small-scale physics that is not explicitly included in the MHD model, and in this way better agreement with observations is reached on a global scale. We see in panel (b) of Figure 18.5 that field-aligned currents coming from the Alfvén wings reach Ganymede's atmosphere also in this model (Jia et al., 2010), although along a more complicated path than in the earlier simulations by Ip and Kopp (2002) (those shown in panel (a) of Figure 18.5). The flow pattern is shown on the right-hand side of the figure, and in the lower right panel bands of field-aligned currents are seen to run across the globe at latitudes approximately corresponding to the location of the auroral ovals. Both upward and downward current regions are seen next to each other, as would be expected in an auroral current circuit. The maxima of the field-aligned current density in these resistive MHD models are located on the sides of Ganymede facing Jupiter and away from Jupiter. This should be compared to the configuration of the auroral emissions shown in Figure 18.1. There, the emission maxima can be found on the sides leading and trailing the moon in the co-rotating Jovian magnetospheric flow.

The use of Hall MHD to model Ganymede's interaction leads to a magnetospheric configuration that differs from that of resistive MHD. Panel (a) of Figure 18.6 shows the presence of significant plasma flows inside Ganymede's magnetosphere (Dorelli et al., 2015). On Earth magnetospheric plasma flows are one of the processes that generate field-aligned potential drops that in turn accelerate the auroral electrons. As pointed out by Dorelli et al. (2015), to adequately represent field-aligned acceleration and electron precipitation into the ionosphere would require a kinetic plasma model. However, the presence of the magnetospheric plasma flows suggests the possibility of an auroral current circuit similar to that of discrete auroral arcs on Earth. Furthermore, the field-aligned currents in the Hall MHD simulations connect to Ganymede's ionosphere at the sides leading and trailing in the co-rotating flow, which corresponds well to where the emissions were observed.

Tóth et al. (2016) used a combined approach: a global Hall MHD simulation where the upstream and tail reconnection regions were modelled, using an embedded implicit particle in cell code. This could be said to be a hybrid model in that it combines two different approaches to plasma modelling, but it is different from the hybrid models described shortly. The implicit particle in the cell model treats both electrons and ions as particles, but it does not resolve the Debye length and the plasma period. A density map produced by this simulation is shown in panel (b) of Figure 18.6, where the density maximum inside the magnetosphere on the tail, right-hand side, is a feature that is not seen in models based on Hall MHD alone. The plasma density in this region of Ganymede's magnetosphere could influence the auroral current circuits. Zhou et al. (2019) developed this technique further, including an electrical model of Ganymede's interior. From the auroral perspective, the

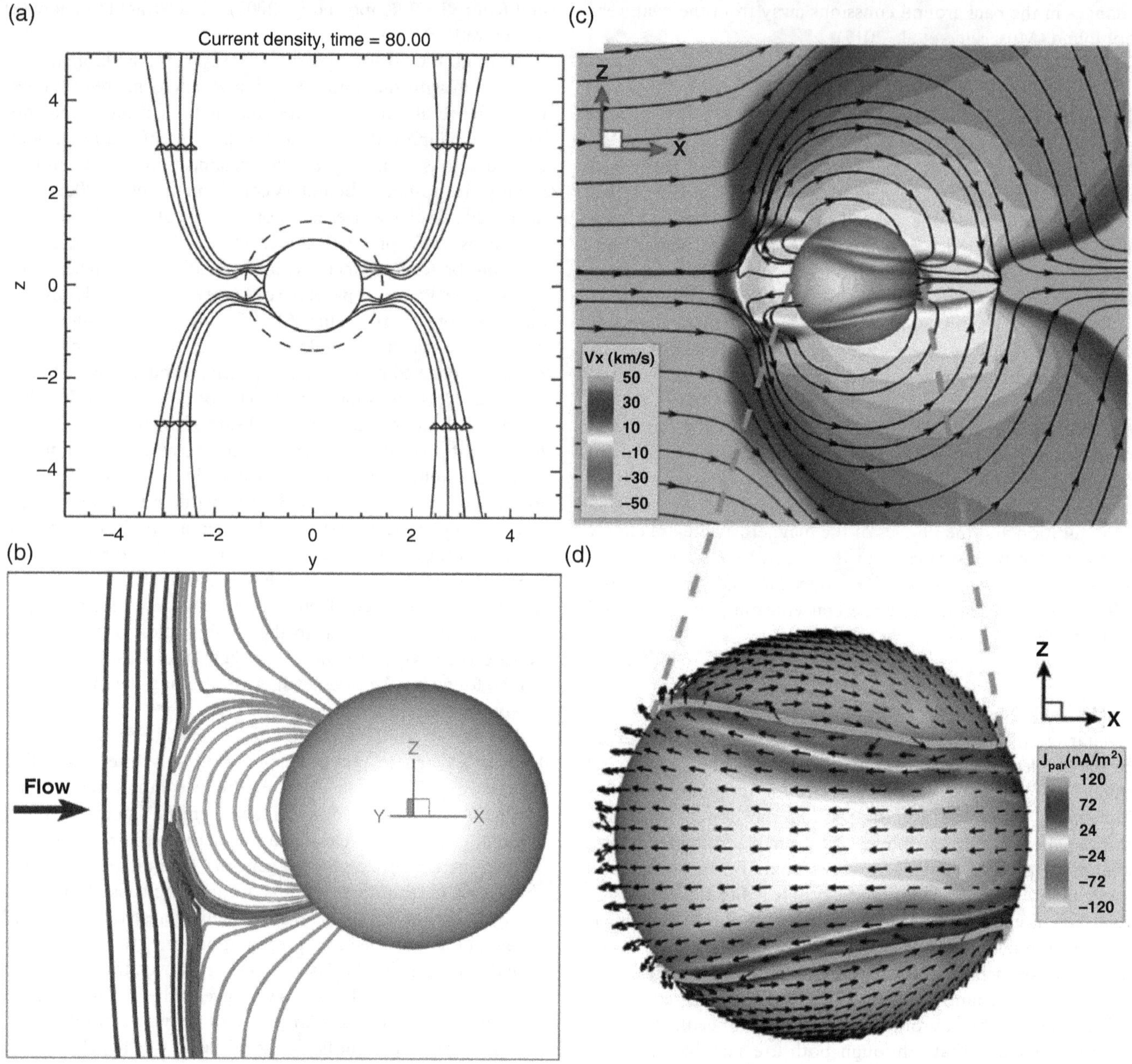

Figure 18.5 (a) field lines connected to the regions with the strongest ionospheric field-aligned currents in an MHD simulation (Ip and Kopp, 2002). The plane shown is perpendicular to the direction of the Jovian magnetospheric plasma flow around Ganymede. Reprinted with permission from John Wiley and Sons. Wing-Huen Ip and Andreas Kopp. Resistive MHD simulations of Ganymede's magnetosphere 2. Birkeland currents and particle energetics. *J. Geophys. Res. (Space Physics)*, 107(A12):SMP 42–1–SMP 42–7, ©2002. (b) Field line connectivity in an MHD simulation by Jia et al. (2010). Reprinted with permission from John Wiley and Sons. Xianzhe Jia, Raymond J. Walker, Margaret G. Kivelson, Krishan K. Khurana, and Jon A. Linker. Dynamics of Ganymede's magnetopause: Intermittent reconnection under steady external conditions. *J. Geophys. Res. (Space Physics)*, 115, ©2010. (c) Flow pattern in an MHD simulation (Jia et al., 2009). (d) Field-aligned current density at the ionosphere (Jia et al., 2009). Panels c, d reprinted with permission from John Wiley and Sons. Xianzhe Jia, Raymond J. Walker, Margaret G. Kivelson, Krishan K. Khurana, and Jon A. Linker. Properties of Ganymede's magnetosphere inferred from improved three-dimensional MHD simulations. *J. Geophys. Res.*, 114(A9):A09209, ©2009. The side of Ganymede shown in panels c and d is facing away from Jupiter. (A black and white version of this figure will appear in some formats. For the color version, please refer to the plate section.)

relevant result of the study was the estimate that the energy flux carried by the particles from the reconnection regions down to the atmosphere could account for 40 per cent of the auroral emissions. Since the regions ofspace that were modelled using the particle in cell code did not reach down to auroral altitudes, this was based on an assumption of constant flux and a calculation of the loss cone. Thus, the possible influence of an acceleration region further below was not taken into account.

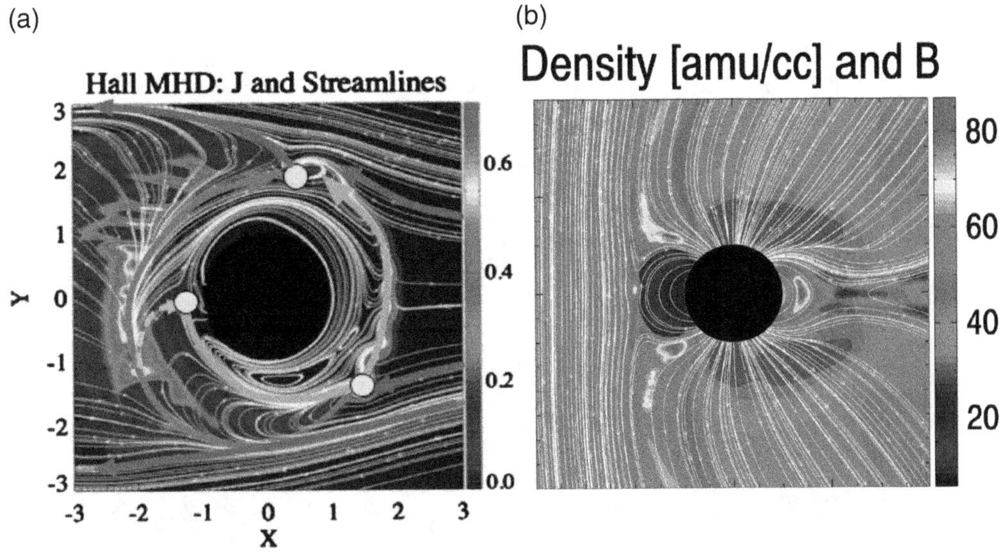

Figure 18.6 (a) Current densities and flow patterns in a Hall MHD simulation (Dorelli et al., 2015). Reprinted with permission from John Wiley and Sons. J. C. Dorelli, A. Glocer, G. Collinson., and G. Tóth. The role of the Hall effect in the global structure and dynamics of planetary magnetospheres: Ganymede as a case study. *J. Geophys. Res. (Space Physics)*, 120:5377–5392 ©2015. (b) Densities in a Hall MHD simulation with embedded iPIC patches (Tóth et al., 2016). Reprinted with permission from John Wiley and Sons. Gábor Tóth, Xianzhe Jia, Stefano Markidis, Ivy Bo Peng, Yuxi Chen, Lars K S Daldorff, Valeriy M. Tenishev, Dmitry Borovikov, John D. Haiducek, Tamas I. Gombosi, Alex Glocer, and John C. Dorelli. Extended magnetohydrodynamics with embedded particle-in-cell simulation of Ganymede's magnetosphere. *J. of Geophys. Res.: Space Physics*, 121(2):1273–1293, ©2016. (A black and white version of this figure will appear in some formats. For the color version, please refer to the plate section.)

Effects similar to those of Hall MHD were also seen in multi-fluid simulations that model more than one ion species (Paty and Winglee, 2004, 2006). The multi-fluid approach has also been combined with an auroral brightness model (Payan et al., 2015). Wang et al. (2018) employed a ten-moment multi-fluid model to study Ganymede–Jovian magnetosphere interaction. A ten-moment model means that it keeps track of ten variables related to the moments of the distribution function for each species: the density, three components of the momentum, and six elements of the pressure tensor. The two fluids modelled represent protons and electrons with a mass ratio of $m_p/m_e = 25$. The results were that high-speed jets flowed from the reconnection regions towards the surface of Ganymede. Figure 18.7 shows electron (top row) and ion (middle row) scalar pressures mapped onto a sphere 263 km above the surface of the moon. The bottom row shows the emissions observed by the HST (McGrath et al., 2013) for comparison. There is a partial overlap between the electron pressure peak in the northern hemisphere (upper left panel) on the side facing diagonally upstream and away from Jupiter and the light emission shown in the lower left panel. However, apart from that partial overlap, regions with observed light emissions and enhanced simulated electron pressure are disjoint. For the ions, the emission peaks overlap partially with the simulated pressure peaks on the Jupiter-facing and downstream sides, but there are also regions where the emissions and ion pressures do not coincide. This emphasises the need for kinetic modelling of the electrons down to ionospheric altitudes.

Hybrid Models

At the time of this writing, very few hybrid-kinetic models (particle ions and fluid electrons) have been applied to study the Jovian plasma interaction with Ganymede (Fatemi et al., 2016; Leclercq et al., 2016) or Ganymede-like bodies (Vernisse et al., 2017).

Fatemi et al. (2016) showed the global structure of Ganymede's magnetosphere when Ganymede is inside and outside the Jovian plasma sheet. They found a good agreement between their simulations and Galileo magnetic field observations during the six close encounters of Ganymede. They also calculated the flux of the Jovian ions precipitating to the surface of Ganymede and found an excellent correlation between the ion precipitation flux and Ganymede's surface brightness patterns. They showed that Ganymede's closed magnetic field lines at low latitudes considerably limit the access of the Jovian ions to the surface, while the open field lines over the polar caps facilitate the access of plasma to high latitudes. Leclercq et al. (2016) showed the global structure of Ganymede's magnetosphere as a planetary application of a newly developed method in their hybrid model. The global structure of Ganymede's magnetosphere obtained from their model agrees with previous MHD simulations (e.g. Jia et al., 2008, 2009; Dorelli et al., 2015) and the hybrid simulation by Fatemi et al. (2016), but they did not compare their results with any previous observations.

Precipitating charged particles carry currents, and therefore the current system is relevant to auroral physics (see Baumjohann et al., 2010, for a review of currents in magnetospheres). Vernisse et al. (2017) applied a hybrid model of plasma to study plasma interaction with magnetised bodies in super-Alfvénic and sub-Alfvénic plasma flow. Their general intention was to investigate the morphology of the current systems and their closure around magnetised bodies. An example of the current systems around a magnetised body (Ganymede-like) that interacts with a sub-Alfvénic plasma flow is shown in Figure 18.8. The major difference between the modelled body and

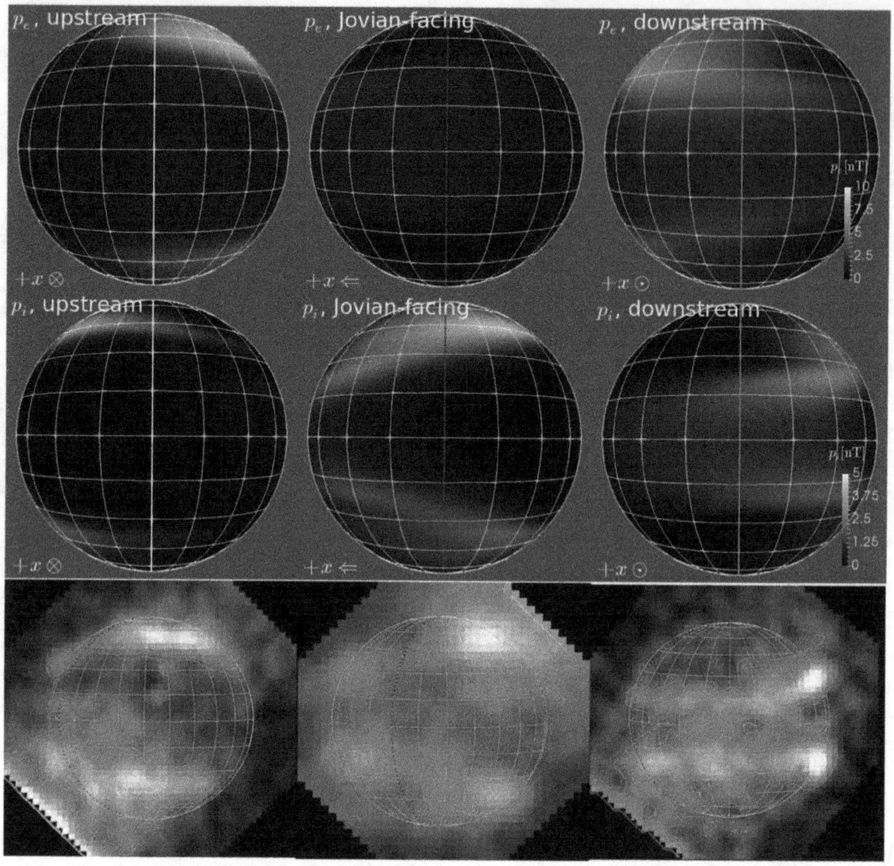

Figure 18.7 Scalar electron pressure (upper row) and scalar ion pressure (middle row) at 263 km altitude. The left column shows the upstream side; the middle column, the Jupiter-facing side; and the right column, the downstream side. The yellow, blue, and red curves mark longitudes 270°W, 0°W, 90°W, respectively. The coordinate system was defined by McGrath et al. (2013). The bottom row shows the light emissions observed by the Hubble Space Telescope (McGrath et al., 2013). The first two rows and the compilation of the compound figure were made by Wang et al. (2018). Reprinted with permission from John Wiley and Sons. Liang Wang, Kai Germaschewski, Ammar Hakim, Chuanfei Dong, Joachim Raeder, and Amitava Bhattacharjee. Electron physics in 3-D two-fluid 10-moment modeling of Ganymede's magnetosphere. *J. Geophys. Res. (Space Physics)*, 123(4):2815–2830 ©2018. (A black and white version of this figure will appear in some formats. For the color version, please refer to the plate section.)

Ganymede is the orientation of the magnetic dipole moment. Vernisse et al. (2017) have used a southward oriented dipole moment interacting with a southward background magnetic field, while the Ganymede dipole orientation is primarily northward. However, there are similarities between the generated currents in the magnetosphere of the modelled body and those in the magnetosphere of Ganymede. For example, the magnetopause current forms at very close distances to the body and mainly closes to the Alfvén wing currents. However, the ring current, shown in Figure 18.8, has not been observed in Ganymede's magnetosphere.

18.4 SUMMARY AND OUTLOOK

What do we know? What do we not know? And how can we find out? Observations by the Hubble Space Telescope have shown emissions of ultraviolet light in bands around Ganymede similar to the auroral ovals on Earth. Observations by the Galileo spacecraft made during flybys of Ganymede showed that this moon has an intrinsic magnetic field and a small magnetosphere of its own inside Jupiter's magnetosphere. Furthermore, it is from the auroral observations that we know that Ganymede has an oxygen atmosphere.

Numerical modelling has helped paint a picture of the physical processes behind the aurora at Ganymede. As the plasma flow around the moon is subsonic, no bow shock is formed; instead, there are Alfvén wings above the polar regions, and the Jovian magnetosphere plasma interacts directly with the magnetopause, where reconnection occurs. The relative motion of Ganymede and the surrounding plasma provides the energy that drives the aurora. However, to cause the atmosphere to emit light there must be currents that reach down in the atmosphere and processes that accelerate the electrons downward. In early MHD models, field-aligned currents from the Alfvén wings connected directly to the ionosphere at the open–closed field line boundary. Later, models including more of the physics showed more complicated current paths, including plasma flows inside Ganymede's magnetosphere. Such flows may generate field-aligned potential drops that can cause electron acceleration on closed field lines, and in turn auroral emissions equatorward of the OCFB. On Earth,this is where the majority

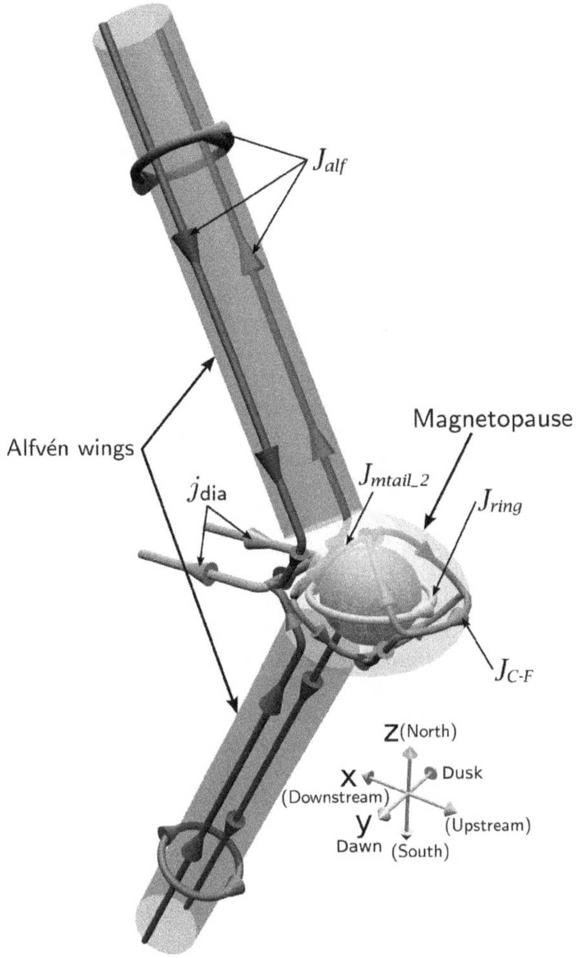

Figure 18.8 Three-dimensional diagram of the current systems for a sub-Alfvénic plasma interaction with a magnetised body (Ganymede-like), taken from (Vernisse et al., 2017). The magnetopause (Chapman–Ferraro) currents are shown in orange, Alvén wing currents are shown in purple, ring currents are in yellow, and the diamagnetic currents are in green. (A black and white version of this figure will appear in some formats. For the color version, please refer to the plate section.)

of the auroral emissions occur. To simulate these acceleration processes would require kinetic models that can resolve electron scales and that reach all the way down to ionospheric altitudes.

This picture of the physics of Ganymede's aurora is consistent with the observations that have been made, but we cannot say that the observations confirm the picture. We do not have simultaneous observations of the auroral emissions and the magnetic field configuration yet. We do not know from observations what proportion of the aurora is located at the OCFB and how much is generated away from it. We have models of where the field-aligned currents are, but we have no direct observations of them. We do not know how the auroral electrons are accelerated or even whether the aurora is dominated by processes similar to discrete arcs or diffuse aurora, both of which are common at Earth. At this time, the physical processes behind Ganymede's aurora are not well constrained by observations.

The upcoming JUpiter ICy moons Explorer (JUICE) mission from the European Space Agency aims at putting a spacecraft in orbit around Ganymede (Grasset et al., 2013). This will shed some light upon the outstanding issues. Simultaneously observing emissions, magnetic fields, and charged particles should be sufficient to answer the question of where the emissions appear in relation to the different parts of the magnetosphere. Distribution functions of electrons on the field lines that connect to the aurora will provide information about acceleration processes and pitch angle scattering. Observations of cross-field plasma flows in the equatorial regions of Ganymede's magnetosphere could be used together with auroral particle observations in the polar regions to put constraints on the parameter space spanned by auroral generators and the auroral current circuit as a whole. To completely understand both the current circuit and the emissions, one also needs to know the properties of the atmosphere and ionosphere as discussed in Chapters 16 and 17. All these observations can subsequently be used to put constraints on simulations, both by providing realistic initial and boundary conditions and by telling us which physical processes need to be included to obtain results that are consistent with observations. In this way, Ganymede auroral modelling will advance to the next level and illuminate the auroral processes of a magnetosphere inside a magnetosphere. This, in turn, will push the boundaries of our understanding of magnetospheres in general.

ACKNOWLEDGEMENTS

H. G. was supported by the Swedish National Space Agency (SNSA), grant 108/18. L. R. appreciates the support from the Swedish National Space Agency (SNSA) through grant 154/17 and the Swedish Research Council (VR) through grant 2017-04897. S. F. acknowledges support from the Swedish National Space Agency, grants 179/18 and 115/18, and the Swedish Research Council, grant 2018-03454.

REFERENCES

Akasofu, S.-I., Lui, A. T. Y., and Meng, C.-I., Importance of auroral features in the search for substorm onset processes. *Journal of Geophysical Research: Space Physics*, 115(A8), 2010. DOI: https://doi.org/10.1029/2009JA014960.

Baumjohann, W., Blanc, M., Fedorov, A. and Glassmeier, K.-H. Current systems in planetary magnetospheres and ionospheres. *Space Science Reviews*, 152(1–4):99–134, May 2010. DOI: https://doi.org/10.1007/s11214-010-9629-z.

Bertaux, J.-L., Leblanc, F., Witasse, O. et al. Discovery of an aurora on Mars. *Nature*, 435(7403):790–4, June 2005. DOI: https://doi.org/10.1038/nature03603.

Borovsky, J. E., Birn, J., Echim, M. M. et al. Quiescent discrete auroral arcs: a review of magnetospheric generator mechanisms. *Space Science Reviews*, 216(1):1, 2020. DOI: https://doi.org/10.1007/s11214-019-0619-5.

Carlson, R. W., Bhattacharyya, J. C., Smith, B. A. et al. An atmosphere on Ganymede from its occultation of SAO 186800 on 7 June 1972. *Science*, 182(4107):53–5, 1973. ISSN 0036-8075. DOI: https://doi.org/10.1126/science.182.4107.53.

Cessateur, G., Barthelemy, M., and Peinke, I. Photochemistry-emission coupled model for Europa and Ganymede. *Journal of Space Weather and Space Climate*, 6(27):A17, March 2016. DOI: https://doi.org/10.1051/swsc/2016009.

Collinson, G., Paterson, W. R., Bard, C. et al. New results from Galileo's first flyby of Ganymede: reconnection-driven flows at the low-latitude magnetopause boundary, crossing the cusp, and icy ionospheric escape. *Geophysical Research Letters*, 45(8):3382–92, 2018. DOI: https://doi.org/10.1002/2017GL075487.

Cunningham, N. J. Spencer, J. R. Feldman, P. D. et al. Detection of Callisto's oxygen atmosphere with the Hubble Space Telescope. *Icarus*, 254:178–89, 2015.

Deighan, J., Jain, S. K., Chaffin, M. S. et al. Discovery of a proton aurora at Mars. *Nature Astronomy*, 2:802–7, July 2018. DOI: https://doi.org/10.1038/s41550-018-0538-5.

Dorelli, J. C., Glocer, A., Collinson, G., and Tóth, G. The role of the Hall effect in the global structure and dynamics of planetary magnetospheres: Ganymede as a case study. *Journal of Geophysical Research: Space Physics*, 120(7):5377–92, July 2015. DOI: https://doi.org/10.1002/2014JA020951.

Eviatar, A., Strobel, D. F., Wolven, B. C. et al. Excitation of the Ganymede ultraviolet aurora. *The Astrophysical Journal*, 555(2): 1013–19, July 2001a. DOI: https://doi.org/10.1086/321510.

Eviatar, A., Vasyliūnas, V. M., and Gurnett, D. A. The ionosphere of Ganymede. *Planetary and Space Science*, 49(3–4):327–36, March 2001b. DOI: https://doi.org/10.1016/S0032-0633(00)00154-9.

Eviatar, A., Williams, D. J., Paranicas, C. et al. Trapped energetic electrons in the magnetosphere of Ganymede. *Journal of Geophysical Research*, 105(A3):5547–54, March 2000. DOI: https://doi.org/10.1029/1999JA900450.

Fatemi, S., Poppe, A. R., Khurana, K. K., Holmström, M., and Delory, G. T. On the formation of Ganymede's surface brightness asymmetries: kinetic simulations of Ganymede's magnetosphere. *Geophysical Research Letters*, 43(10):4745–54, 2016. DOI: https://doi.org/10.1002/2016GL068363.

Feldman, P. D., McGrath, M. A., Strobel, D. F. et al. HST/STIS ultraviolet imaging of polar aurora on Ganymede. *The Astrophysical Journal*, 535(2):1085–90, June 2000. DOI: https://doi.org/10.1086/308889.

Gallardo-Lacourt, B., Frey, H. U., and Martinis, C. Proton aurora and optical emissions in the subauroral region. *Space Science Reviews*, 217(1):10, February 2021. DOI: https://doi.org/10.1007/s11214-020-00776-6.

Grasset, O., Dougherty, M. K., Coustenis, A. et al. JUpiter ICy moons Explorer (JUICE): an ESA mission to orbit Ganymede and to characterise the Jupiter system. *Planetary and Space Science*, 78: 1–21, April 2013. DOI: https://doi.org/10.1016/j.pss.2012.12.002.

Hall, D. T., Feldman, P. D., McGrath, M. A., and Strobel, D. F. The far-ultraviolet oxygen airglow of Europa and Ganymede. *The Astrophysical Journal*, 499(1):475–81, May 1998. DOI: https://doi.org/10.1086/305604.

Hall, D. T., Strobel, D. F., Feldman, P. D., McGrath, M. A., and Weaver, H. A. Detection of an oxygen atmosphere on Jupiter's moon Europa. *Nature*, 373(6516):677–9, February 1995.

Hess, S. L. G. and Delamere, P. A. *Satellite-Induced Electron Acceleration and Related Auroras*. Washington DC American Geophysical Union Geophysical Monograph Series, 197:295–304, January 2012. DOI: https://doi.org/10.1029/2011GM001175.

Ip, W.-H. and Kopp, A. Resistive MHD simulations of Ganymede's magnetosphere 2. Birkeland currents and particle energetics. *Journal of Geophysical Research: Space Physics*, 107(A12):SMP 42–1–SMP 42–7, 2002. DOI: https://doi.org/10.1029/2001JA005072.

Jia, X., Walker, R. J., Kivelson, M. G., Khurana, K. K., and Linker, J. A. Three-dimensional MHD simulations of Ganymede's magnetosphere. *Journal of Geophysical Research: Space Physics*, 113: A06212, June 2008. DOI: https://doi.org/10.1029/2007JA012748.

Jia, X., Walker, R. J., Kivelson, M. G., Khurana, K. K., and Linker, J. A. Properties of Ganymede's magnetosphere inferred from improved three-dimensional MHD simulations. *Journal of Geophysical Research*, 114(A9):A09209, September 2009. DOI: https://doi.org/10.1029/2009JA014375.

Jia, X., Walker, R. J., Kivelson, M. G., Khurana, K. K., and Linker, J. A. Dynamics of Ganymede's magnetopause: intermittent reconnection under steady external conditions. *Journal of Geophysical Research: Space Physics*, 115(A12):A12202, December 2010. ISSN 0148-0227. DOI: https://doi.org/10.1029/2010JA015771.

Karlsson, T., Andersson, L., Gillies, D. M. et al. Quiet, discrete auroral arcs—observations. *Space Science Reviews*, 216(1):16, February 2020. DOI: https://doi.org/10.1007/s11214-020-0641-7.

Keiling, A., Wygant, J. R., Cattell, C. et al. Correlation of Alfvén wave poynting flux in the plasma sheet at 4–7 RE with ionospheric electron energy flux. *Journal of Geophysical Research: Space Physics*, 107(A7):SMP 24–1–SMP 24–13, 2002. DOI: https://doi.org/10.1029/2001JA900140.

Kivelson, M. G., Khurana, K. K., Russell, C. T. et al. Discovery of Ganymede's magnetic field by the Galileo spacecraft. *Nature*, 384(6609):537–41, December 1996. DOI: https://doi.org/10.1038/384537a0.

Kivelson, M. G., Khurana, K. K., and Volwerk, M. The permanent and inductive magnetic moments of Ganymede. *Icarus*, 157:507–22, June 2002. DOI: https://doi.org/10.1006/icar.2002.6834.

Kliore, A. J. Satellite atmospheres and magnetospheres. In J. Andersen, editor, *Highlights of Astronomy: As Presented at the XXIIIrd General Assembly of the IAU, 1997*, pages 1065–9. Springer Netherlands, Dordrecht, 1998. ISBN 978-94-011-4778-1. DOI: https://doi.org/10.1007/978-94-011-4778-1_138.

Knudsen, D. J., Borovsky, J. E., Karlsson, T., Kataoka, R., and Partamies, N. Editorial: topical collection on auroral physics. *Space Science Reviews*, 217(1):19, February 2021. DOI: https://doi.org/10.1007/s11214-021-00798-8.

Kopp, A. and Ip, W.-H. Resistive MHD simulations of Ganymede's magnetosphere 1. time variabilities of the magnetic field topology. *Journal of Geophysical Research: Space Physics*, 107(A12):SMP 41–1–SMP 41–5, 2002. DOI: https://doi.org/10.1029/2001JA005071.

Leclercq, L., Modolo, R., Leblanc, F., Hess, S., and Mancini, M. 3d magnetospheric parallel hybrid multi-grid method applied to planet–plasma interactions. *Journal of Computational Physics*, 309(6609):295–313, 2016. DOI: https://doi.org/10.1016/j.jcp.2016.01.005.

Lysak, R., Echim, M., Karlsson, T. et al. Quiet, discrete auroral arcs: acceleration mechanisms. *Space Science Reviews*, 216(5):92, July 2020. DOI: https://doi.org/10.1007/s11214-020-00715-5.

Mauk, B. and Bagenal, F. Comparative auroral physics: Earth and other planets. *Washington DC American Geophysical Union Geophysical Monograph Series*, 197:3–26, January 2012. DOI: https://doi.org/10.1029/2011GM001192.

Mauk, B. H., Haggerty, D. K., Paranicas, C. et al. Discrete and broadband electron acceleration in Jupiter's powerful aurora. *Nature*, 549(7670):66–9, September 2017. DOI: https://doi.org/10.1038/nature23648.

Mauk, B. H., Haggerty, D. K., Paranicas, C. et al. Diverse electron and ion acceleration characteristics observed over Jupiter's main aurora. *Geophysical Research Letters*, 45(3):1277–85, 2018. DOI: https://doi.org/10.1002/2017GL076901.

McGrath, M. A., Jia, X., Retherford, K. et al. Aurora on Ganymede. *Journal of Geophysical Research: Space Physics*, 118(5):2043–54, May 2013. DOI: https://doi.org/10.1002/jgra.50122.

Molyneux, P. M., Nichols, J. D., Bannister, N. P. et al. Hubble space telescope observations of variations in Ganymede's oxygen atmosphere and aurora. *Journal of Geophysical Research: Space Physics*, 123(5):3777–93, 2018. DOI: https://doi.org/10.1029/2018JA025243.

Musacchio, F., Saur, J., Roth, L. et al. Morphology of Ganymede's FUV auroral ovals. *Journal of Geophysical Research: Space Physics*, 122 (3):2855–76, 2017. DOI: https://doi.org/10.1002/2016JA023220.

Neubauer, F. M. The sub-Alfvénic interaction of the Galilean satellites with the Jovian magnetosphere. *Journal of Geophysical Research: Planets*, 103(E9):19843–66, 1998.

Newell, P. T., Sotirelis, T., and Wing, S. Diffuse, monoenergetic, and broadband aurora: the global precipitation budget. *Journal of Geophysical Research: Space Physics*, 114:A09207, September 2009. DOI: https://doi.org/10.1029/2009JA014326.

Nishimura, Y., Lessard, M. R., Katoh, Y. et al. Diffuse and pulsating aurora. *Space Science Reviews*, 216(1):4, January 2020. DOI: https://doi.org/10.1007/s11214-019-0629-3.

Paty, C. and Winglee, R. Multi-fluid simulations of Ganymede's magnetosphere. *Geophysical Research Letters*, 31(24):L24806, 2004. ISSN 0094-8276. DOI: https://doi.org/10.1029/2004GL021220.

Paty, C. and Winglee, R. The role of ion cyclotron motion at Ganymede: magnetic field morphology and magnetospheric dynamics. *Geophysical Research Letters*, 33(10), May 2006. ISSN 00948276. DOI: https://doi.org/10.1029/2005GL025273.

Payan, A. P., Paty, C. S., and Retherford, K. D. Uncovering local magnetospheric processes governing the morphology and variability of Ganymede's aurora using three-dimensional multifluid simulations of Ganymede's magnetosphere. *Journal of Geophysical Research: Space Physics*, 120(1):401–413, January 2015. ISSN 21699380. DOI: https://doi.org/10.1002/2014JA020301.

Roth, L., Ivchenko, N., Gladstone, G. R. et al. A sublimated water atmosphere on Ganymede detected from Hubble Space Telescope observations. *Nature Astronomy*, July 2021. DOI: https://doi.org/10.1038/s41550-021-01426-9.

Saur, J., Duling, S., Roth, L. et al. The search for a subsurface ocean in Ganymede with Hubble Space Telescope observations of its auroral ovals. *Journal of Geophysical Research: Space Physics*, 120(3):1715–37, March 2015. DOI: https://doi.org/10.1002/2014JA020778.

Schneider, N. M., Deighan, J. I., Jain, S. K. et al. Discovery of diffuse aurora on Mars. *Science*, 350(6261):0313, November 2015. DOI: https://doi.org/10.1126/science.aad0313.

Szalay, J. R., Allegrini, F., Bagenal, F. et al. Alfvénic acceleration sustains Ganymede's footprint tail aurora. *Geophysical Research Letters*, 47(3):e2019GL086527, 2020. DOI: https://doi.org/10.1029/2019GL086527.

Tóth, G., Jia, X., Markidis, S. et al. Extended magnetohydrodynamics with embedded particle-in-cell simulation of Ganymede's magnetosphere. *Journal of Geophysical Research: Space Physics*, 121(2):1273–93, February 2016. ISSN 21699402. DOI: https://doi.org/10.1002/2015JA021997.

Vernisse, Y., Riousset, J. A., Motschmann, U., and Glassmeier, K.-H. Stellar winds and planetary bodies simulations: magnetized obstacles in super-Alfvénic and sub-Alfvénic flows. *Planetary and Space Science*, 137:40–51, 2017. DOI: https://doi.org/10.1016/j.pss.2016.08.012.

Wang, L., Germaschewski, K., Hakim, A. et al. Electron physics in 3-D two-fluid 10-moment modeling of Ganymede's magnetosphere. *Journal of Geophysical Research: Space Physics*, 123(4):2815–30, 2018. DOI: https://doi.org/10.1002/2017JA024761.

Zhou, H., Tóth, G., Jia, X., Chen, Y., and Markidis, S. Embedded kinetic simulation of ganymede's magnetosphere: improvements and inferences. *Journal of Geophysical Research: Space Physics*, 124(7):5441–60, 2019. DOI: https://doi.org/10.1029/2019JA026643.

19

Electrodynamic Coupling between Ganymede and the Jovian Ionosphere

BERTRAND BONFOND AND PHILIPPE ZARKA

19.1 INTRODUCTION

Having conductive interiors and/or ionospheres, the Galilean satellites are in electrodynamic interaction with the fast-rotating Jovian magnetic field and magnetosphere in which they are embedded. In the inner Jovian magnetosphere, within the orbit of Ganymede, the magnetic pressure of the rotating flow is larger than its ram pressure (except in the near-equatorial current sheet at Ganymede's orbit) and the flow is sub-Alfvénic (Kivelson and Bagenal, 2014). The flow–obstacle interactions generate both local and distant effects. In this chapter, we do not focus on the local interaction, which is covered in Chapter 3.1 (Kivelson); rather we will zoom out and explore the outcome of this interaction further down the disturbed magnetic field lines. There, the electrodynamic satellite–Jupiter interaction takes the form of transverse perturbations of the Jovian magnetic field propagating away from the satellite along Jovian magnetic field lines at the Alfvén velocity, in both hemispheres, down to the Jovian ionosphere. These perturbations are called Alfvén waves and their envelopes Alfvén wings, primarily theorized by Neubauer (1980, 1998).

As these waves propagate along the magnetic field lines, they initiate a long chain of processes and generate a variety of signatures associated with the long-range interaction, as summarized in Figure 19.1. Some of them, noted in white, have already been observed at Ganymede, while others are expected to take place because they have been identified at Io (in cyan). Ganymede's magnetosphere forms a large obstacle for Jupiter's magnetospheric plasma and magnetic field rotating with the planet. Its motion relative to the plasma launches large-scale Alfvén waves, which then undergo some turbulent filamentation, and partial reflections at the Alfvén velocity gradients (either at the plasma sheet boundary or at the Jovian ionosphere). These waves ultimately enter a regime (mostly inertial Alfvén waves) in which they can accelerate charged particles (Section 19.4). Part of them hit the atmosphere, exciting and ionizing neutral atoms and molecules, which generates specific auroral emissions, called the Ganymede footprint (Section 19.2), as they de-excite. Another part of the accelerated electrons bounces due to the magnetic mirror effect. The resulting upgoing electron distribution is far from Maxwellian, as it lacks the electrons with small pitch angle which have precipitated into the atmosphere, causing the feature called a loss-cone in the velocity space $(v_{\parallel}, v_{\perp})$. This feature notably displays a positive gradient of the electron distribution function (or phase space density $f(v_{\parallel}, v_{\perp})$) towards increasing perpendicular velocities $v_{\perp}$. This positive gradient is the free energy source of the Cyclotron-Maser Instability (CMI) mechanism (Zarka, 1998; Hess, Mottez and Zarka, 2007; Hess et al., 2008), through which the perpendicular energy of electrons in cyclotron motion around the magnetic field lines is directly and collectively transferred to electromagnetic radio waves having this same cyclotron frequency ($f_{ce} = eB/2\pi m_e$). The corresponding electrons are said to be *resonant* and the wave is hugely amplified, either causing the electron distribution to relax towards a Maxwellian one, or trapping resonant electrons in the wave electric field, and in both cases quenching the instability (Le Quéau, 1988). Radio emissions are further detailed in Section 19.3.

Because Io is closer to Jupiter, where the magnetic field magnitude is larger than at Ganymede (~2 000 nT at Io compared to ~100 nT at Ganymede), and embedded into the dense plasma torus stemming from the moon's outstanding volcanism, all the outcomes of the moon–magnetosphere interaction are magnified there. Hence many of the phenomena related to the moon–magnetosphere interactions were first unveiled at Io before they were found to be also applicable at Ganymede. These distant consequences are key to studying these interactions because they often are the only way to get information in absence of a spacecraft directly measuring the local environment.

For example, it was the intense radio emissions related to Io that first revealed that this moon was playing such an important role in the Jovian magnetosphere. Over half a century ago, Bigg (1964) noticed that the recently discovered and variable Jovian decametre (DAM) radio emission showed maximum occurrence at specific orbital phases of Io (Φ_{Io}) around Jupiter as seen from a terrestrial observer (~90° and ~240° from the anti-observer's position – Figures 19.3a and 19.2b,c).

The picture grew in complexity as spacecraft brought new pieces of information on the Jovian system, starting from the simpler unipolar inductor model (Goldreich and Lynden-Bell, 1969), to the Alfvén wing theory (Neubauer, 1980) and then all its subsequent developments, involving the propagation and reflections of the Alfvén waves (Gurnett and Goertz, 1981; Jacobsen et al., 2007, 2010; Hinton et al., 2019), their turbulent filamentation (Chust et al., 2005), or their capability to accelerate particles (Jones and Su, 2008; Hess et al., 2010, 2013; Damiano et al., 2019).

The arrival of Juno around Jupiter and the traversal of the field lines connected to the Io footprint at high latitude offered a brand new perspective on the richness of the wave–particle interactions resulting from the moon–magnetosphere

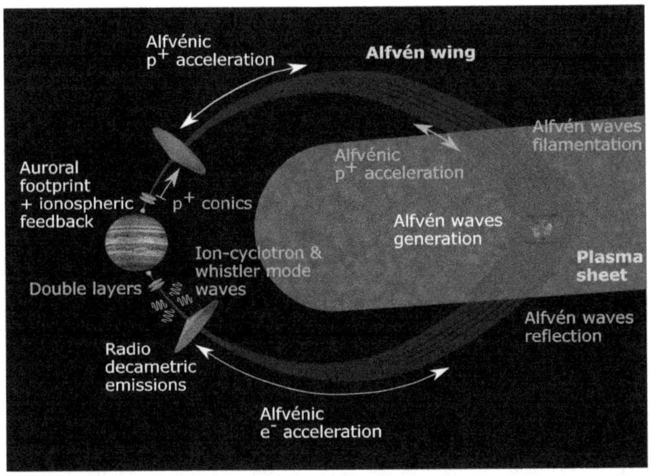

Figure 19.1 Schematic of the chain of processes taking place along the Alfvén wings (in red) generated by the interaction between the magnetosphere of Ganymede (in blue, not to scale) and that of Jupiter. The processes noted in white have already been observed at Ganymede and are the subject of the present chapter, while those noted in cyan have not been discovered yet, but are expected to exist because they have already been detected at Io. (A black and white version of this figure will appear in some formats. For the colour version, please refer to the plate section.)

couplings. The importance of Alfvén waves undergoing a turbulent cascade and accelerating electrons along the magnetic field lines through a broad range of energies was confirmed by in situ particle and electromagnetic wave observations (Szalay et al., 2018; Sulaiman et al., 2020). However, a greater surprise came from the finding of proton beams and conics, originating from three different regions along the Alfvén wing (Szalay et al., 2020a; Clark et al., 2020). Finally, ion-cyclotron waves and whistler-mode waves excited by the field-aligned electrons are also observed (Sulaiman et al., 2020).

An important question arose, however. Were these phenomena specific to Io's case, since this moon is the most volcanically active body of the solar system and the main source of material in the magnetosphere, or could they be extended to other moons, such as Europa and its induced magnetosphere related to its sub-surface ocean or – and this is the focus of our present interest – to Ganymede and its permanent intrinsic magnetic field.

Io lacks a permanent magnetic dipole. So, its non-local interaction with the Jovian magnetic field consists of (1) its steady Alfvén wings, which result from the stacking of Jupiter's magnetic field upstream of Io and its subsequent deflection around the satellite (Saur et al., 2004), and (2) a dense plasma wake that is rapidly re-accelerated downstream (Hinson et al., 1998).

The existence of an intrinsic magnetic field implies some modifications of this picture for Ganymede (Chapter 3.1). The obstacle in Jupiter's rotating magnetic field is not Ganymede's body or ionosphere but its magnetosphere, which is of 2–3 R_G radius (R_G stands for Ganymede's radius, ~2 634 km). The orientation of Ganymede's internal field, antiparallel to Jupiter's, leads to a favourable orientation of interacting magnetic fields resulting in reconnection at Ganymede's upstream and downstream magnetopauses and subsequent particle energization. A signature of this reconnection is the double loss-cone distributions (i.e. in both directions along the field line) observed by Galileo on magnetic field lines connected to both Ganymede and Jupiter (Williams et al., 1997). At high energies (keV–tens of keV), particles can bounce several times between mirror points close to Ganymede and close to Jupiter before the magnetic field line drifts across Ganymede's magnetosphere, and then be temporarily trapped and exhibit double loss-cone distributions, whereas at low energies, the larger pitch angle diffusion fills the loss-cones at some distance from the mirror points (Williams and Mauk, 1997).

The plasma energized by magnetic reconnection at the magnetopause also generates local electrostatic and electromagnetic emissions (Gurnett et al., 1996; Kurth et al., 2000). It is much more developed than at Io, as described in chapter 14.

Beyond a few R_G downstream of Ganymede, one finds Alfvén wings similar to Io's (Jia et al., 2008) as well as an extended plasma wake (Kivelson et al., 1998). The Alfvén wings carry a current of about 0.5 MA (Lavrukhin and Alexeev, 2015) driven by a transverse electric potential drop of about 80 kV across Ganymede's magnetosphere (Zhou et al., 2020) (versus a current ~1 MA driven by a 400 kV potential drop across Io's ionosphere). The Alfvén wings' field-aligned currents close in Ganymede's and Jupiter's ionospheres and in Ganymede's magnetotail or plasma wake (Jia et al., 2008).

For solar wind–planet interactions, the energy release is often sporadic (substorms) as a result of the variable character of the Solar wind and of magnetic flux storage in the magnetotail. But at Ganymede, immersed in Jupiter's magnetosphere, the upstream magnetic field conditions are essentially steady, with a slowly rocking orientation of the magnetic field external to Ganymede's magnetosphere. As a consequence, steady reconnection was anticipated at Ganymede's magnetopause, not a bursty, substorm-like one (Kivelson et al., 2004). These expectations were challenged by the reanalysis of Galileo plasma data recorded near Ganymede, showing plasma flows accelerated by time-variable magnetic reconnection consistent with Dungey-type substorms (Collinson et al., 2018).

Another source of variability of the interaction comes from the fact that Ganymede does not orbit permanently in a dense plasma environment like Io in its torus, but it periodically crosses the Jovian plasma sheet. Its plasma environment is thus very variable and induces a variable travel time of the Alfvénic perturbations between Ganymede and Jupiter. Bonfond et al. (2013) found a maximum longitudinal shift of 13° between the multiple spots of the Ganymede northern UV auroral footprint and Ganymede's instantaneous longitude, quite comparable to Io's lead angle up to ~15° in the north and ~8° in the south (Bonfond et al., 2017; Hinton et al., 2019).

19.2 AURORAL FOOTPRINTS

Discovery of the Ganymede Footprint

While the first detection of a satellite auroral footprint, Io's, took place in the infrared domain from a ground-based telescope (Connerney et al., 1993), all the subsequent first detections happened in the ultraviolet (UV) domain. The Europa and Ganymede footprints were first discovered with the Hubble Space Telescope (HST) (Clarke et al., 2002). Then came the discovery of the Enceladus footprint with the Ultraviolet Imaging Spectrograph (UVIS) on board Cassini (Pryor et al., 2011). Finally, a tentative identification of a Callisto footprint

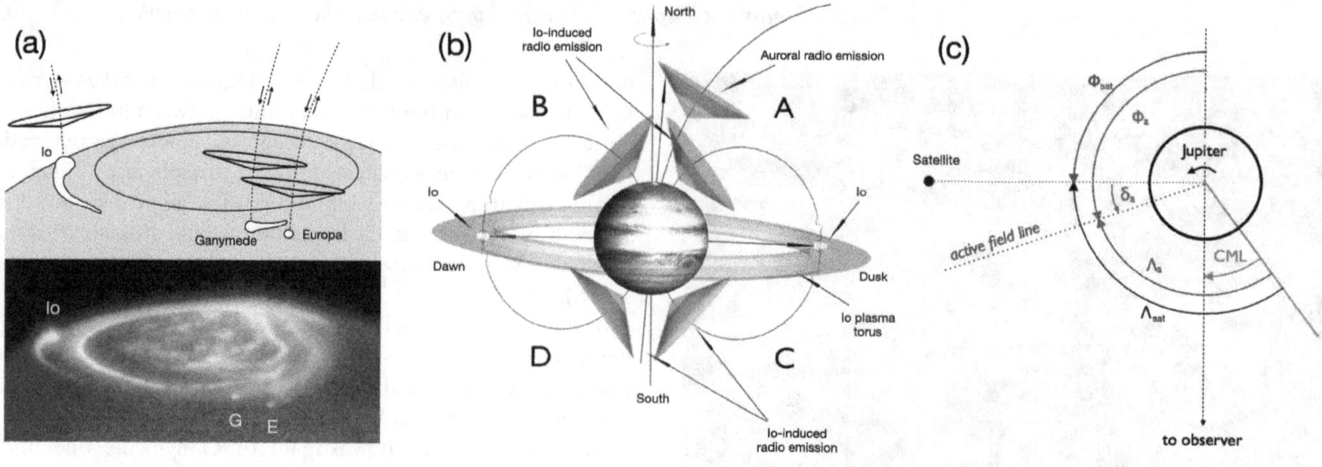

Figure 19.2 (a) Hubble Space Telescope UV image of northern Jovian auroral regions, showing the bright main auroral oval and the footprints of Io (and its tail-like, wake-induced emission), Ganymede, and Europa flux tubes. In the top part of the sketch, hollow conical beams of radio emission are displayed. They are produced above the UV hot spots by the energetic electrons precipitated along the satellite flux tubes or reflected upwards by magnetic mirroring. Similar radio emissions originate from sources distributed above the main oval. (b) Geometry and nomenclature of auroral and satellite-induced (here Io-induced) radio emissions. Radio emissions are produced along the displayed conical shells and thus are observable only when the source is near a limb of Jupiter. The magnetic field line connected to Io is sketched as the active, radio-emitting field line, but actually the active field line leads the instantaneous Io field line by several degrees (angle δ_a of panel (c)). (c) Definition of the different angles used to characterize the observing geometry. In green is the central meridian longitude (CML), which is the System III longitude of the observer. The satellite phase Φ_{sat} and its System III longitude Λ_{sat} are drawn in blue and black, respectively. Alfvén lead angle δ_a, phase, and phase Φ_a and longitude Λ_a of the so-called 'active' field line (in UV and radio) are shown in red. (a) From Clarke et al. (2002); Zarka (2007). Reprinted from *Planetary and Space Science*, 55(5). Philippe Zarka, Plasma interactions of exoplanets with their parent star and associated radio emissions, 598–617 ©2007, with permission from Elsevier. (b & c) From M. S. Marques, P. Zarka, E. Echer, V. B. Ryabov, M. V. Alves, L. Denis and A. Coffre, *A&A*, 604, A17, 2017, reproduced with permission © ESO. (A black and white version of this figure will appear in some formats. For the colour version, please refer to the plate section.)

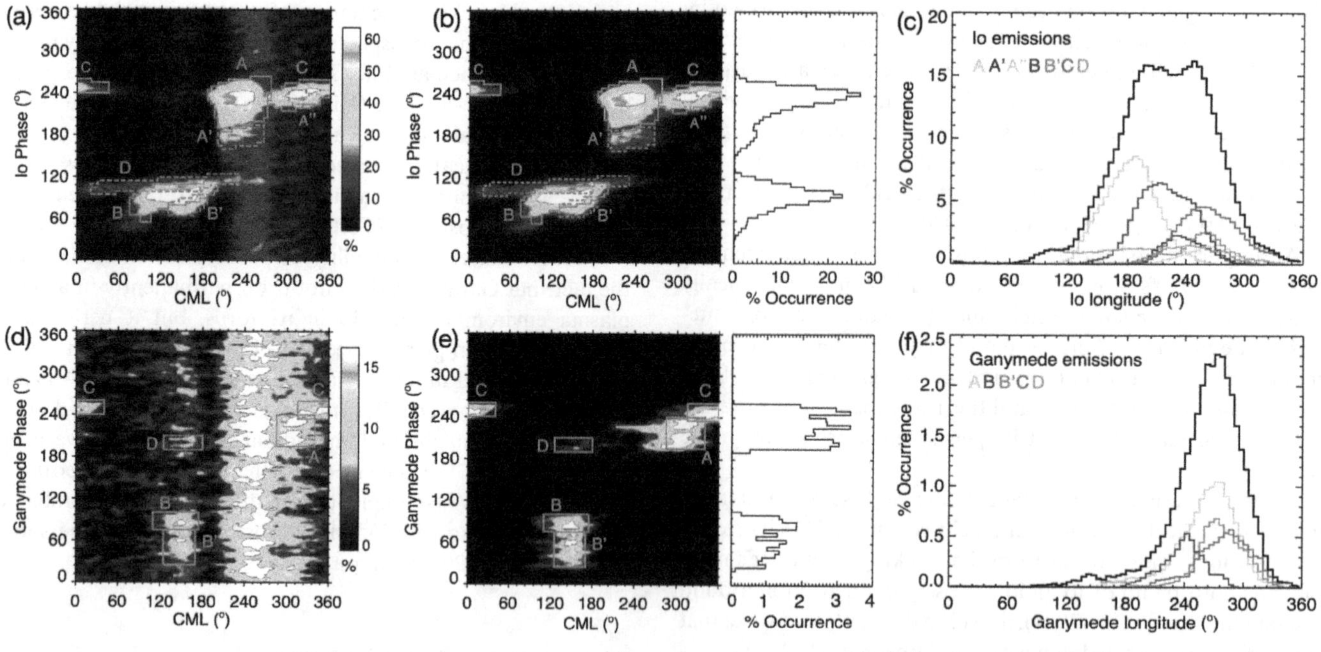

Figure 19.3 (a) Occurrence probability of Jovian radio emissions detected over 26 years (1990–2015) with the Nançay Decameter Array, displayed as 2D histograms as a function of planetary rotation (CML) and Io's orbital phase (Φ_{Io}) in 5° × 5° bins (smoothed via 1° interpolation). Regions of high occurrence labelled in white correspond to Io–Jupiter emissions (named Io-A, Io-B...). Vertical bands of emission covering restricted CML ranges at all Φ_{Io} correspond to non-Io emissions (auroral or induced by other satellites). (b) Occurrence probability of Io–Jupiter emissions only versus CML and Φ_{Io}. The profile integrated over all CML is displayed on the right-side panel. (c) Integrated occurrence probability of Io–Jupiter emissions versus Io's Jovicentric longitude ($\Lambda_{Io} = CML + 180° - \Phi_{Io}$). Components from panel (b) are identified by colours, and their sum is the black line. (d) Occurrence probability of non-Io emissions versus CML and Ganymede's orbital phase $\Phi_{Ganymede}$. Ganymede–Jupiter emissions show up within new regions of enhanced occurrence (white boxes), labelled A–D in reference to the non-Io components in which they have been identified. (e) Occurrence probability of Ganymede–Jupiter emissions only versus CML and $\Phi_{Ganymede}$. The profile integrated over all CML is displayed on the right-side panel. (f) Integrated occurrence probability of Ganymede–Jupiter emissions versus Ganymede's jovicentric longitude ($\Lambda_{Ganymede} = CML + 180° - \Phi_{Ganymede}$). Components from panel (e) are identified by colours, and their sum is the black line. From Zarka, P., Marques, M. S., Louis, C., Ryabov, V. B., Lamy, L., Echer, E. and Cecconi, B., *A&A*, 618, A84, 2018, reproduced with permission © ESO. (A black and white version of this figure will appear in some formats. For the colour version, please refer to the plate section.)

was reported, based on HST UV images as well (Bhattacharyya et al., 2018). The Ganymede and Europa footprints have since also been identified in the infrared domain around 3.4 μm with the Jupiter InfraRed Auroral Mapper (JIRAM) camera on board Juno (Mura et al., 2017).

Both H and H_2 UV auroral emissions on one hand and H_3^+ infrared emissions on the other hand arise from the precipitation of charged particles into Jupiter's H and H_2 atmosphere. The UV emissions result from the de-excitation of atomic or molecular hydrogen after impact by precipitating particles (electrons or ions) or secondary electrons. Conversely, auroral infrared emissions are thermal emissions of H_3^+ ions which are themselves an indirect product of the electron precipitation followed by charge transfer (see review in Badman et al., 2015):

$$e^- + H_2 \rightarrow H_2^+ + 2e^-,$$

$$H_2^+ + H_2 \rightarrow H_3^+ + H.$$

Usually, H_3^+ emissions are observed around 3.4 μm because the strong absorption from CH_4 molecules located at lower altitude than most of the aurora offer a high contrast between the auroral emissions and the planetary background at this wavelength.

Morphology of the Ganymede Footprint

The first detections associated only one auroral spot with the Ganymede footprint (Clarke et al., 2002). On HST images, this spot can be fitted with an ellipse of area $\sim 5 \times 10^5$ km^2. When mapped back into the equatorial plane along magnetic field lines, such a surface corresponds to a disk with 8–20 times Ganymede's radius (R_G = 2 634 km), thus corresponding to the size of Ganymede's magnetosphere rather than Ganymede itself (Grodent et al., 2009).

Using their characteristic motion on polar projections fixed in System III (the longitude system fixed with the magnetic field), it was found that, in some cases, at least two spots could be associated with the footprint of Ganymede (Bonfond et al., 2013) (Figure 19.4). The spacing between these two spots increases up to a maximum of about 4 000 km and then decreases systematically as a function of the position of Ganymede with respect to the plasma sheet, and they seem to merge when Ganymede is close to the central region of the sheet. This evolution of the inter-spot distances suggests that one spot corresponds to the main Alfvén wing (MAW) and the second one to the trans-hemispheric electron beam (TEB), that is, generated by electrons accelerated in the opposite hemisphere (see Figure 19.4 b).

The inter-spot distance also varies on timescales larger than Jupiter's rotation. Observations carried out in 2007 in similar Ganymede longitude ranges, but a few weeks apart, showed that this distance can vary by a factor of 2. Since the distance between spots is directly related to the Alfvén propagation time, such variations were attributed to an increase of plasma density in the Jovian magnetosphere that took place in the first month of 2007 (Bonfond et al., 2013). Indeed, several other observations, such as the brightening of Jupiter's sodium nebula (Yoneda et al., 2009), the decrease of the hectometric (HOM) radio emissions unrelated to the solar wind fluctuations (Yoneda et al., 2013), the increased occurrence rate of large plasma injection auroral signatures and the expansion of the main auroral oval over three months (Bonfond et al., 2012) all indirectly suggest that the plasma input from Io increased from February to June 2007.

In addition to the spots, an extended auroral tail (≥24°) can also sometimes be seen in the downstream direction along the Ganymede footpath (Bonfond et al., 2017). As with many features of the Ganymede footprint, this tail is harder to detect in UV images than Io's footprint tail. Moreover, even in similar geometrical configurations, the Ganymede footprint UV tail has only been identified in a handful of cases. These rare detections indicate that parameters other than just the position of Ganymede in the plasma sheet and the viewing geometry impact its apparent brightness (such as the plasma energy distribution, composition, density, the magnetic field strength, etc.). However, the existence of tails in both Ganymede's and Europa's footprints indicates that lessons learned from Io's footprint most likely also apply for the others and vice versa. Two main ideas have been proposed to explain the Io footprint tail. The first one involves a steady current loop related to the reacceleration of the stagnant plasma in Io's wake (Hill and Vasyliunas, 2002; Delamere et al., 2003; Su et al., 2003; Ergun et al., 2006; Matsuda et al., 2012), while the second involves the increasingly intricate reflection pattern of the Alfvén waves downstream of the satellite (Jacobsen et al., 2007; Bonfond et al., 2017). The broad energy distribution inferred from remote observations (Bonfond et al., 2009) and observed directly by Juno favours the second (Szalay et al., 2018). It is noteworthy that the existence of a footprint tail at Ganymede, where no significance mass loading takes place, indicates that this process is not a necessary ingredient to the formation of the tail.

Infrared observations from the JIRAM instrument on board Juno offer an unprecedented spatial resolution, down to 15 km/pixel (Mura et al., 2018; Moirano et al., 2021). They not only showed similar features in infrared H_3^+ emissions than in the UV, such as the pair of spots followed by the extended tail (Mura et al., 2017), but they also unveiled further details of the footprint morphology. In particular, and similarly to the Io footprint spots, each one of the two Ganymede footprint spots appears to be formed of a pair of smaller auroral dots separated by 170 km (Figure 19.4c) (Mura et al., 2018). One possible explanation is that these dots correspond to the front and tail of Ganymede's magnetosphere, where magnetic reconnection takes place. However, the presence of similar substructures at Io's footprint rather suggests another (and not mutually exclusive) origin, possibly related to Jupiter's ionosphere or to the electron acceleration process. This conclusion is further strengthened by the finding that the Ganymede footprint tail (as well as Io's and Europa's) was also made of sub-dots separated by ~270 km (Figure 19.4d). Contrary to the larger spots discussed earlier, these sub-dots appear fixed with the planet rather than following the motion of the satellite (Moirano et al., 2021). Because of this behaviour, as well as the size and spacing of the spots, Moirano et al. (2021) suggest that these features result from the ionosphere feedback instability, in which a localized increase of ionospheric conductivity in the presence of background ionospheric electric field enhances the ionospheric currents. In this scenario, the current enhancement closes through field-aligned currents at the conductivity gradient through secondary Alfvén waves, which further increases the precipitating electron flux and the conductivity, closing the feedback loop.

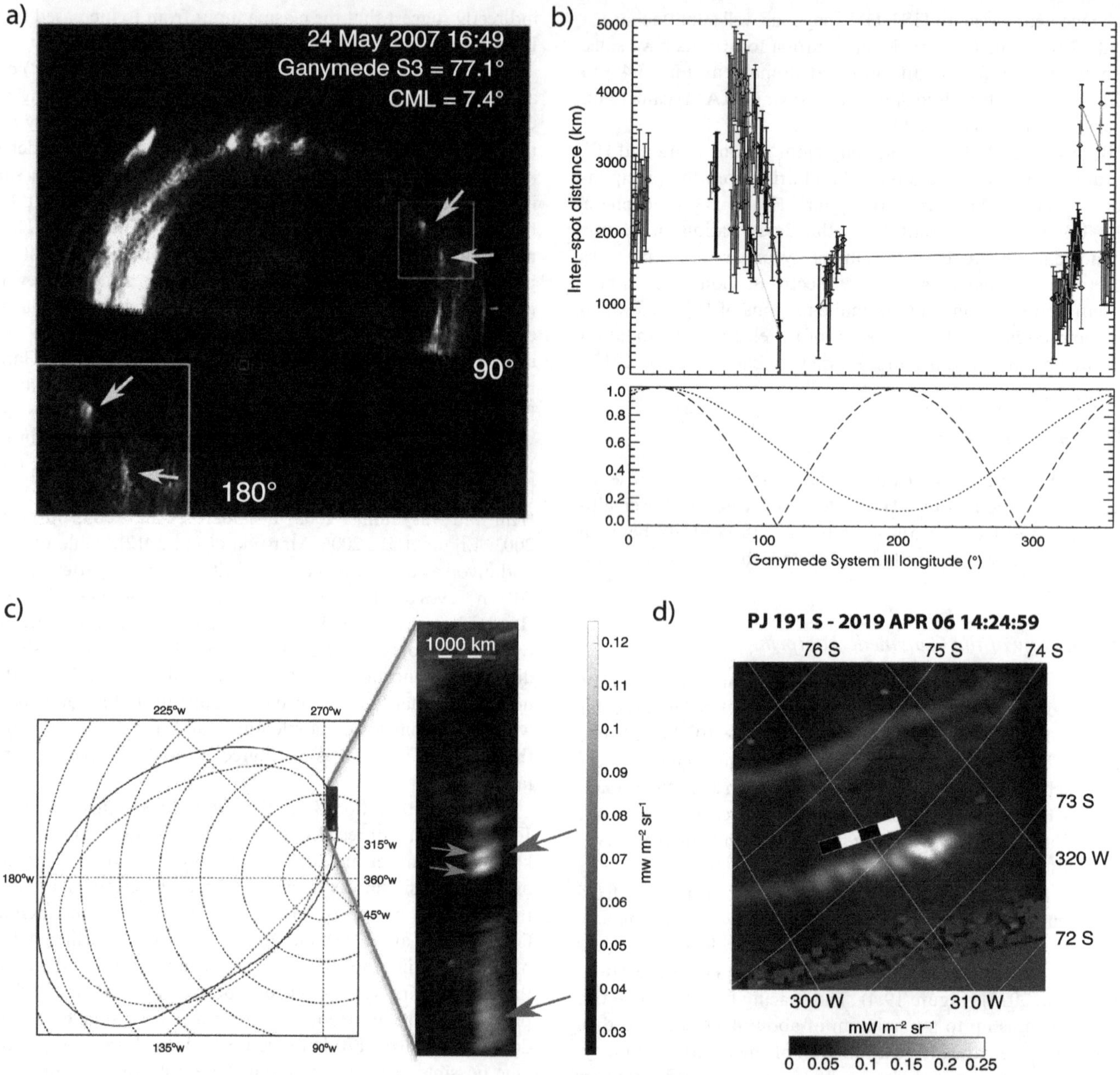

Figure 19.4 (a) Polar projections of an HST UV image acquired on 24 May 2007 at 16:40 UT. Two Ganymede footprint spots can be identified, as highlighted by the arrows. (From Bonfond et al. (2013)). (b) Inter-spot distance between the two spots of the Ganymede footprint. (top) The coloured lines connect points from the same HST orbit. The error bars assume a selection uncertainty of 1 pixel for the first spot and 2 pixels for the second one. (bottom) The long-dashed lines show the expected dependence of the distance for a trans-hemispheric electron beam spot (arbitrary units). In this case, the two spots merged as Ganymede crossed the centrifugal equator. The short-dashed line shows the expected behaviour of the distance for a reflected Alfvén wing (RAW) spot. In this case, the minimum distance is expected when Ganymede is at its northernmost centrifugal latitude (~200° System III longitude). (From Bonfond et al. (2013)). Panels a, b reprinted with permission from John Wiley and Sons. Bonfond, B., Hess, S., Bagenal, F., Gérard, J.-C., Grodent, D., Radioti, A., Gustin, J., and Clarke, J. T. The multiple spots of the Ganymede auroral footprint, *Geophys. Res. Lett.*, 40, 4977–4981 ©2013. (c) Infrared image of the Ganymede footprint spots from the JIRAM instrument on board Juno. We can see that each spot (red arrows) is actually formed from at least two sub-structures (green arrows). (From Mura et al. (2018)). AAPG ©2018. Mura, A., Adriani, A., Connerney, et al. Juno observations of spot structures and a split tail in Io-induced aurorae on Jupiter, *Science* 361(6404), 774–777. Reprinted by permission of the AAPG whose permission is required for further use. (d) Infrared images of the Ganymede footprint tail. The tail is also made of a string of sub-spots. (From Moirano et al. (2021)). Reprinted with permission from John Wiley and Sons. Moirano, A., Mura, A., Adriani, A., Dols, V., Bonfond, B., Waite, J. H., et al. (2021). Morphology of the auroral tail of Io, Europa, and Ganymede from JIRAM L-band imager. *Journal of Geophysical Research: Space Physics*, 126 ©2021. (A black and white version of this figure will appear in some formats. For the colour version, please refer to the plate section.)

The Brightness of the Ganymede Footprint

On HST UV images acquired at slant angle, the brightness of the Ganymede footprint can reach up to 180 kR (Wannawichian et al., 2010). The total power emitted by the Ganymede footprint's main spot typically ranges between 1 and 6 GW in the H_2 UV Lyman and Werner bands. However, depending on the auroral background emissions, it can also become so dim that it is no longer identifiable (Bonfond et al., 2017a). The precipitated energy is approximately 10 times higher than the energy of the resulting UV emissions (Gustin et al., 2012). Assuming that 10 per cent of the Poynting flux carried by the Alfvén waves is converted into electron acceleration, this emitted UV power is in the highest part or slightly higher than the range predicted by the models of satellite–magnetosphere interaction (Saur et al., 2013). Because the moon–magnetosphere interaction depends on the local plasma density, the emitted power is the highest when Ganymede is in the center of the plasma sheet and decreases outside of it (Grodent et al., 2009). However, it also varies at two shorter timescales: in the range of 10–40 minutes and in the range of two to three minutes. The first one probably corresponds to Ganymede's traversal time of plasma bubbles with properties differing from the surrounding environment, such as the plasma injections typically observed at such radial distances (Mauk et al., 1997). A detailed analysis of some HST sequences showed the Ganymede footprint disappearing as it traversed the auroral signature of such a plasma injection, further adding confidence into this explanation (Bonfond et al., 2017a). A similar behaviour had already been reported for the Io footprint (Bonfond et al., 2012), and Hess, Bonfond and Delamere (2013) argued that this disappearance was most probably due to the increased electron density at high latitude, which limited the efficiency of the parallel electron acceleration by inertial Alfvén waves. The two- to three-minute timescale could either be related to the recurrence time of bursty magnetic reconnection at the front of Ganymede's magnetosphere, in accordance with the MHD simulations (Jia et al., 2008; Zhou et al., 2019, 2020), or to the quasi-periodic formation, upward migration and disappearance of electric potential drops of a few hundred volts at altitudes around 0.2 R_J above Jupiter's surface. The latter scenario, which is not mutually exclusive with the first one, is proposed in analogy with similar structures found at the Io footprint from observations of fast-drifting radio S-bursts (Hess et al., 2009) and could explain the two- to three-minute fluctuations of the UV brightness of the Io, Europa and Ganymede footprints (Bonfond et al., 2007; Grodent et al., 2009; Bonfond et al., 2017a).

19.3 RADIO EMISSIONS

Induced Radio Emissions

The electrons accelerated in the Alfvén wings to energies of a few keV to tens of keV give rise, close to Jupiter, to intense radio emission via the Cyclotron-Maser Instability (CMI) mechanism (Zarka, 1998; Hess, Mottez and Zarka, 2007; Hess et al., 2008). The emission is produced close to the local electron cyclotron frequency, which can reach 40 MHz in Jupiter's ionosphere (Connerney et al., 2018), and it is beamed at a large angle from the magnetic field in the source (Figures 19.2a,b). The part above 10 MHz (corresponding to the decametre – DAM – wavelength range) can propagate through the Earth's ionosphere and be detected by ground-based antenna arrays, as done by Bigg (1964) and others (e.g. Lamy et al., 2017). Accumulated observations revealed for the Io–Jupiter interaction four islands of enhanced emission occurrence in the Φ_{Io}–CML (Central Meridian Longitude = observer's Jovian System III longitude) plane (Figure 19.3a,b). Those correspond to two physical sources, near both Io Flux Tube (IFT) northern and southern footprints, seen on the eastern and western limbs of the planet because CMI emission is beamed nearly perpendicular to the IFT field lines (Marques et al., 2017) (Figure 19.2b). The high-occurrence islands are not symmetrical around $\Phi_{Io} = 180°$ due to Alfvén wave propagation through Io's torus to Jupiter's ionosphere, which induces a time-variable angular shift (so-called lead angle) that depends on the position of Io in the plasma torus (Figures 19.2c).

Radio Searches

The result from Bigg (1964) demonstrated that existence of a satellite-induced radio emission could be statistically established by building emission occurrence versus the satellite phase Φ, or in the Φ–CML plane. Many observers subsequently searched for the statistical evidence of the interaction of Jupiter's magnetic field with the other Galilean moons plus Amalthea via ground-based DAM observations $\geq$10 MHz, at the Universities of Florida, Chile (narrow-band observations distributed between 15 and 28 MHz), and Colorado (swept-frequency spectrograph 7.6–41 MHz). Lebo et al. (1965) and Bigg (1966) mentioned marginal effects of Europa and Ganymede, but those were not confirmed by subsequent studies involving up to 18 years of accumulated observations (1957–75), that found no other effect than Io's (Dulk, 1967; Kaiser and Alexander, 1973; St. Cyr, 1985). The instruments used had low sensitivity (minimum detectable flux density $\sim 10^4$ Jy = 10^{-22} $\mathrm{Wm^{-2}Hz^{-1}}$).

The Galileo plasma wave instrument provided several years of continuous observations (1995–2002) in the hectometre range (2.1–5.6 MHz), that were statistically analyzed in the same way. Low-significance occurrence peaks were found in the $\Phi_{Ganymede}$–CML plane by Menietti et al. (1998a). Including 10 months of Cassini observations, Hospodarsky et al. (2001) found a similar result as well as a marginal dependence on $\Phi_{Callisto}$. Higgins (2007) obtained analog results from the analysis of Voyager 1 and 2 data in the range of 2.1–5.8 MHz. Besides these statistical studies, radio emission induced by Ganymede was tentatively identified through an occultation by Ganymede's body itself (Kurth et al., 1997a) and radio direction-finding by Galileo (Menietti et al., 1998b), and it was proposed as the possible source of rare circular patches observed in Ulysses radio dynamic spectra (Kaiser and MacDowall, 1998). But none of these interpretations was unique and hence convincing.

Radio emissions produced by Ganymede's magnetosphere itself were also discovered at low frequencies ($\leq$100 kHz) by Galileo (Gurnett et al., 1996; Kurth et al., 1997b) (see also Chapter 3.1). These local radio emissions are not generated by the CMI, due to a f_{pe}/f_{ce} ratio > 0.3 close to Ganymede, too high for the CMI to develop, but by conversion of electrostatic waves.

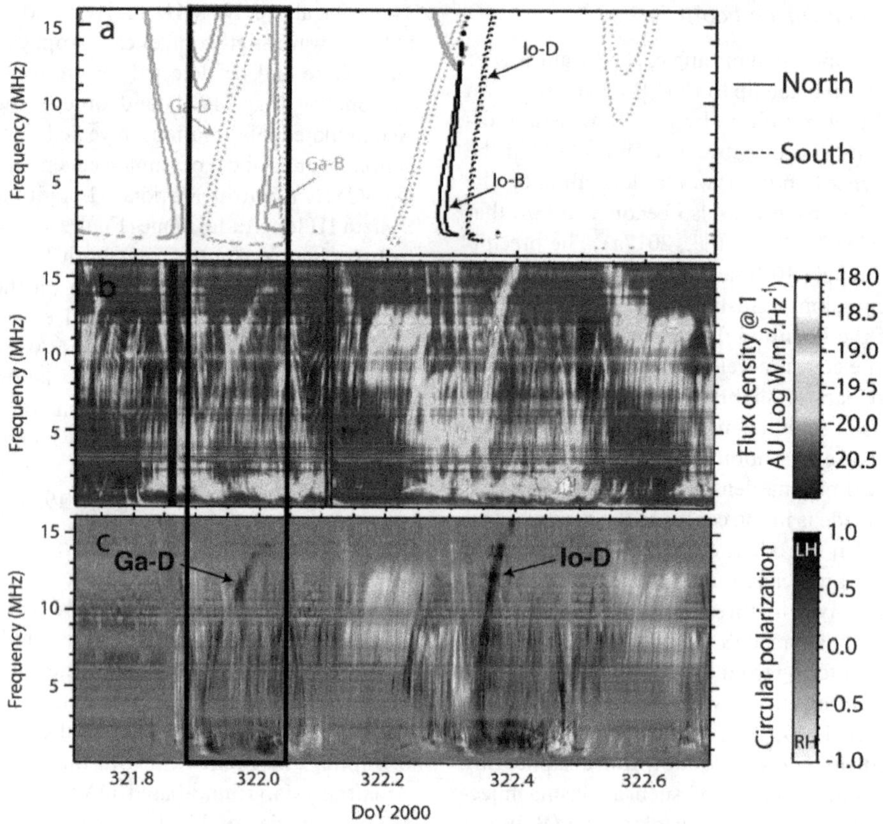

Figure 19.5 (a) ExPRES simulations of Io-induced (in black) and Ganymede-induced (in orange) radio arcs (solid line: northern emissions; dotted line, southern emissions) that should be detected by Cassini during its distant Jupiter flyby of 2000. (b) Dynamic spectrum of flux densities and (c) circular polarization measured by Cassini (LH = left-handed emission from the southern hemisphere; RH = right-handed emission from the northern one). A southern Ganymede D arc is clearly recognized. The background of nested smaller arcs is of auroral origin. From Louis et al. (2017b). Reprinted with permission from John Wiley and Sons. Louis, C. K., Lamy, L., Zarka, P., Cecconi, B., and Hess, S. L., Detection of Jupiter decametric emissions controlled by Europa and Ganymede with Voyager/PRA and Cassini/RPWS, *J. Geophys. Res. Space Physics*, 122, 9228–9247 ©2017. (A black and white version of this figure will appear in some formats. For the colour version, please refer to the plate section.)

Radio Detections

Three recent studies detected independently and unambiguously the DAM emissions induced by Ganymede in Jupiter's magnetic field.

Louis et al. (2017a) was based on simulations by the ExPRES code (published in (Louis et al., 2019)). This code, based on CMI physics, predicts the shape of radio arcs in the time-frequency plane for a selected observer's position or trajectory relative to Jupiter. For satellite–Jupiter interactions, that results in an 'active' magnetic flux tube (where electron acceleration takes place – Figure 19.2c) and thus a radio source of limited size; the predicted arcs are well defined and locally isolated in the time-frequency plane rather than being drowned in a series of nested arcs as is the case for auroral DAM (Figure 19.5). ExPRES simulations were compared to Voyager and Cassini radio observations in (Louis et al., 2017a,b), leading to the identification of ~100 Ganymede–Jupiter radio arcs around $\Phi_{Ganymede} \sim 100°$ and 260°, in broad CML ranges. These ranges of $\Phi_{Ganymede}$ and CML correspond to Ganymede's Jovian longitude of $\Lambda_{Ganymede} = 160°–300°$ in the northern hemisphere and $\Lambda_{Ganymede} = 0°–100°$ in the southern hemisphere ($\Lambda_{Ganymede} = CML + 180° - \Phi_{Ganymede}$). Radio arcs cover the spectral ranges ~1–30 MHz in the north and ~1–16 MHz in the south. Their duration is ~40–60 minutes. The typical energy of the electrons in ExPRES simulations matching the observations is a few keV.

Zarka et al. (2018) performed a statistical study similar to those of the earlier radio searches described previously, but this time applied to a much longer database, consisting of 26 years of daily observations of Jupiter with the Nançay Decameter Array in the range of 10–40 MHz, with a sensitivity ~1.5×10^3 Jy, that is, ~7 times better than earlier observations (Marques et al., 2017). With this database, islands of enhanced occurrence clearly showed up in the $\Phi_{Ganymede}$–CML plane (Figures 19.3d,e), with a signal-to-noise ratio of up to 13σ. These islands, strongly reminiscent of Io–Jupiter ones, cover the ranges of Ganymede phases $65° \pm 40°$ and $225° \pm 35°$ at CML around 155° and 310° respectively, and gather 360 Ganymede–Jupiter emissions. They correspond to a range of Ganymede's longitude $220° \pm 50°$ (Figure 19.3f), very similar to the range of Io's longitudes for which Io-DAM emission is detected (Figure 19.3c). Northern emissions reach 33 MHz and southern ones 27 MHz. The difference in the CML and $\Phi_{Ganymede}$ ranges found by Louis et al. (2017a) can be explained by the different spectral ranges covered by the observations and by the slightly different Jovian latitudes of the observers. The fact that Ganymede–Jupiter emissions do not reach frequencies as high as the Io–Jupiter ones (~40 MHz) is due to the fact that the northern footprint of the Ganymede flux tube (latitude ~75°) lies northward of the northern high-amplitude magnetic anomaly at the surface of Jupiter crossed by the northern Io flux tube footprint (latitude ~66°), implying that lower electron cyclotron frequencies are reached at Ganymede's flux tube footprint.

The third study is the in situ exploration by Juno of Ganymede's wake, described in the next section.

19.4 IN SITU OBSERVATIONS ASSOCIATED WITH THE GANYMEDE FOOTPRINT

Juno's orbit and instrument suite are particularly well suited to combine remote-sensing observations of the auroral emissions and in situ measurements of the particles and waves giving rise to these emissions. In particular, on 29 May 2019, during Juno's 20th perijove operations, between 07:37:14 and 07:37:32, the spacecraft crossed Ganymede's footpath (i.e. the mapping of Ganymede's orbit along the magnetic field lines) only 8° downstream of the footprint's main spot, as observed from the Juno UltraViolet Spectrograph (Juno-UVS). All in situ instruments showed clear signatures of the crossing (Szalay et al., 2020a). For example, the magnetometer data recorded evidence of both significant field-aligned currents (in the form of deviations of the azimuthal magnetic field component δB_ϕ) and strong Alfvénic activity with Poynting fluxes ~100 mW/m^2 (see Figure 19.6(2d)). On the other hand, the JADE (Jovian Auroral Distributions Experiment) particle instrument recorded field-aligned enhancement of the electron flux in both directions (see Figure 19.6(2a), (2b) and (2c)). The precipitating electrons' energy flux reached 11 mW/m^2, which is ~10 per cent of the Poynting energy flux. The electron energy distribution did not show a peaked feature, which would be expected from acceleration by a discrete quasi-static electric field, but a broadband enhancement in the range from 0.5 to 40 keV, compatible with an Alfvénic acceleration process.

From the crossing time, the inferred width of the whole footprint tail is 660 km; however, the current system is highly structured, and smaller-scale (~50 km) sub-structures also are apparent in the JADE data. Moreover, the pitch angle distribution shows that the electron acceleration is bi-directional since a significant flux of upwards-moving electrons was observed. Finally, JADE measurements also revealed upward electron conics in downward currents, which further bolsters the explanation involving particle acceleration by inertial Alfvén waves near the Jovian ionosphere. These findings, as well as the discovery of a similar accelerated broadband electron distribution in the Europa footprint tail (Allegrini et al., 2020), further strengthen the idea that similar processes are at play for all footprints.

During the same event, Juno also crossed the source region of the decametric radio emissions related to the Ganymede footprint tail (Louis et al., 2020, Figure 19.6 1). Measurements by Juno/Waves and JADE showed that these decametric emissions are produced slightly (0.5–2.1%) above the local f_{ce} and beamed at 76° to 83° from the magnetic field. These results are fully consistent with their generation by the loss-cone-driven CMI (as opposed to a shell-driven one that would generate perpendicular waves). The electrons triggering these emissions, likely accelerated by Alfvén waves, have an energy of 4–15 keV. The size of the radio source region was at least 250±50 km wide perpendicular to the magnetic field. It is spread along thousands of kilometres along the field lines, different frequencies being emitted at different altitudes, where $f \sim f_{ce}$; as a rule of thumb, 1 kHz bandwidth corresponds to ~1 km extent along Jovian high-latitude field lines.

19.5 WHAT ELECTROMAGNETIC EMISSIONS FROM SATELLITE FOOTPRINTS TEACH US ON MAGNETOSPHERIC PHYSICS

The Ganymede Footprint as a Landmark in the Jovian Magnetosphere

The satellite footprints are extremely valuable landmarks in the aurora, as they directly connect via magnetic field lines the associated moons at their orbital distance to their ionospheric conjugates. The location of the Io footprint has thus been used as a constraint to increase the accuracy of the internal magnetic field models. For example, the VIP4 (Voyager, Io, Pioneer, 4th order) magnetic field model uses infrared observations of the Io footpath location to complement the in situ magnetic measurements from the Pioneer and Voyager spacecraft and increase the model accuracy in the polar regions (Connerney et al., 1998). However, an analysis of the footpaths of Io, Europa and Ganymede based on HST UV observations in the northern hemisphere showed that the three contours diverge in the region centered around 100° System III longitude (Grodent et al., 2008). The peculiar shape of the different footpaths could be reproduced by adding a small localized dipole magnetic field to the global multipolar magnetic field model, interpreted as indicative of a localized magnetic anomaly. The VIPAL (Voyager, Io, Pioneer, Anomaly, Longitudes) model was based on a larger set of UV observations, using both the latitude and the longitude of the Io footprint's main spot (rather than just the footpath location) to improve the model's accuracy, especially in the magnetic anomaly region (Hess and Delamere, 2012). Finally, the ISaAC (In Situ and Auroral Constrains) model (Hess et al., 2017) used the same technique, but also accounting for the location of the Europa and Ganymede footprints to further refine the model. In the polar regions, the result was indeed remarkably close to the later results from the JRM09 model, which is derived solely from the highly accurate magnetic field measurements from Juno's first nine orbits (Connerney et al., 2018). Juno's measurements also confirmed the presence and location of the magnetic anomaly in the northern polar region. It should be noted, however, that only Juno's measurements could identify the larger magnetic anomaly often named the 'Big Blue Spot', which is located much closer to the equator (Moore et al., 2017; Connerney et al., 2018), where footprint locations provide no useful constraints.

The size of the contour of the main auroral emissions at Jupiter can change from one Jovian rotation to another, superimposed onto long-term trends over a few months (Bonfond et al., 2012). It is, however, challenging to infer whether these changes are related to variations of the radial distance from which these auroras originate, or to the variable stretching of

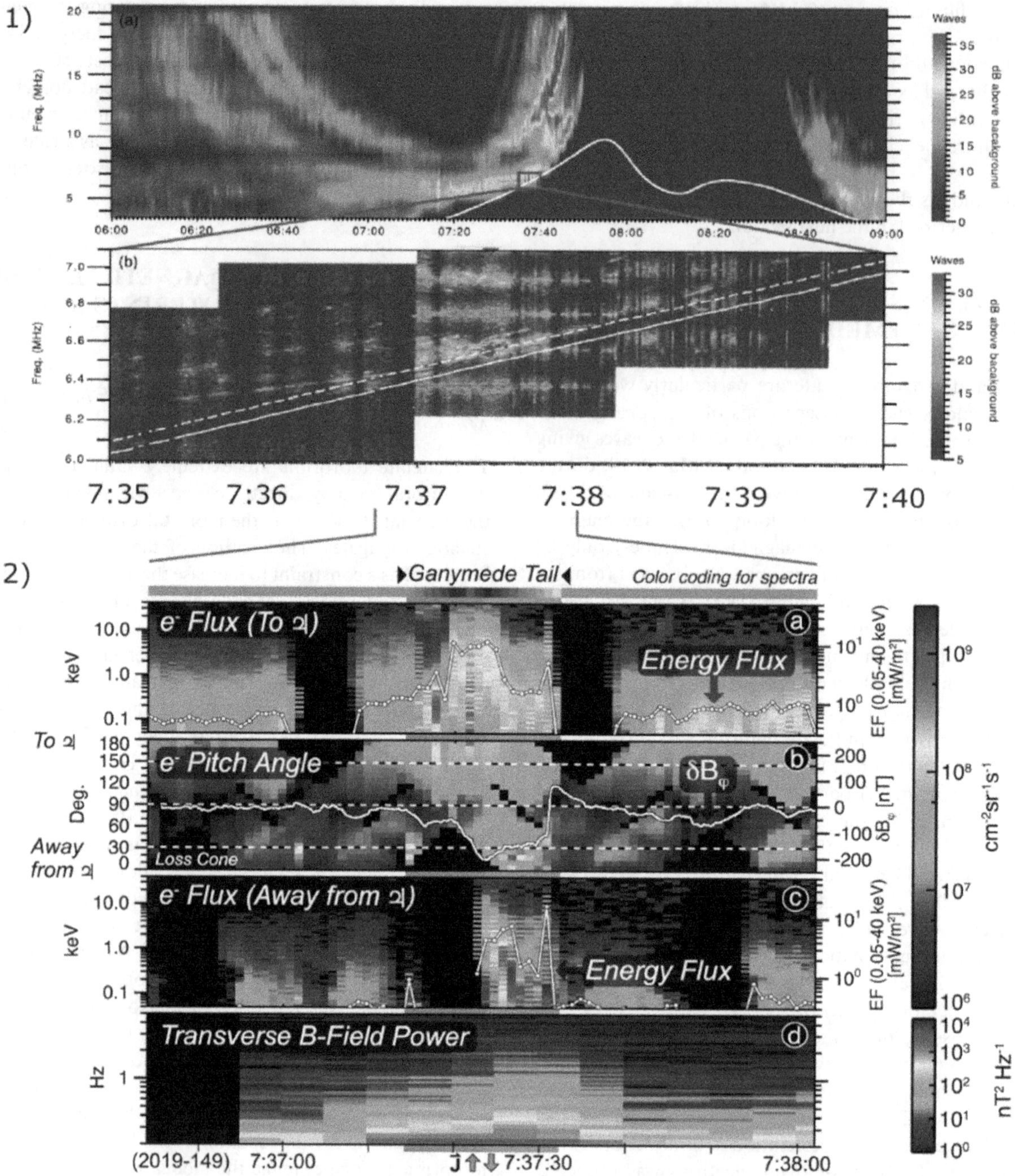

Figure 19.6 JADE, MAG and WAVES data during the Ganymede footprint tail flux tube encounter of 2019. (1a) Low-resolution Juno/Waves data and (1b) zoom at high resolution. The solid white line is the electron cyclotron frequency f_{ce} measured by Juno's magnetometer, and the dashed white line is $1.01 \times f_{ce}$ (adapted from Louis et al. (2020)). Reprinted with permission from John Wiley and Sons. Louis, C. K., Louarn, P., Allegrini, F., Kurth, W. S., & Szalay, J. R. Ganymede-induced decametric radio emission: In situ observations and measurements by Juno. *Geophysical Research Letters*, 47, ©2020. Panels (2a) and (2c) show the downward and upward electron differential energy flux (DEF) within the loss cone, respectively. Precipitating energy flux is overlaid on these panels with its separate axes on the right of the spectrograms. Panel (2b) shows electron pitch angles with δB_ϕ overlaid. Panel (2d) shows the transverse B field power spectral densities. Red/blue bars indicate approximate upward/downward current regions inferred from MAG data. (From Szalay et al. (2020a)). Reprinted with permission from John Wiley and Sons. Szalay, J. R., Allegrini, F., Bagenal, F., Bolton, S. J., Bonfond, B., Clark, G., et al. (2020). Alfvénic acceleration sustains Ganymede's footprint tail aurora. *Geophysical Research Letters*, 47, ©2020. (A black and white version of this figure will appear in some formats. For the colour version, please refer to the plate section.)

the magnetic field lines. At Ganymede's distance from Jupiter, the influence on the mapping of the current sheet's magnetic field, which distorts the magnetic field radially, is much larger than at Io's. Grodent et al. (2008b) indeed noted that, while the location of the Io footpath remained remarkably stable through time, the magnetic latitude of the Ganymede footpath could move by as much as 2.4° and the main emissions by ~3°. Such a shift could be reproduced by modifying the current sheet thickness from $5R_J$ to $2.5R_J$ in the VIP4 model (Connerney et al., 1998), which uses the current sheet model of Connerney (1981). Moreover, on at least one occasion, the Ganymede footprint was seen inside the main emissions instead of outside

(Bonfond et al., 2012). Not only did the main emissions expand equatorward down to Ganymede, but the Ganymede footprint itself had moved out by 0.5°. This unique observation indicated that both the stretching of the magnetic field lines and the radial distance of the region mapping to the main emissions can change through time.

Finally, magnetic field models based on multi-polar developments of the internal field and an axisymmetric representation of the current sheet, such as VIP4, VIPAL, ISaAC or JRM09, are increasingly inaccurate beyond 30 R_J. In particular, local time effects become increasingly important beyond this distance (Khurana, 1997). An alternative method to field tracing models is based on the flux equivalence principle (Vogt et al., 2011, 2015). Starting from a distance where the mapping is known, the iterative construction of this mapping model consists of finding the ionospheric counterpart of an elemental area in the equatorial magnetosphere by equating the magnetic flux in the two regions. Again, the footpath of Ganymede serves as a reliable reference point, from which contours at increasingly large distances in the magnetosphere are progressively mapped into the ionosphere.

Radio Emissions Durations

The duration of Io–Jupiter DAM emissions was found to be statistically twice that of auroral DAM emissions (Marques et al., 2017; Zarka et al., 2018). This can be understood because auroral DAM is controlled by Jupiter's rotation of period ~10 h whereas Io–Jupiter DAM is primarily controlled by Io's orbital motion of period ~42 h, combined with a narrow radio beaming and a steady Alfvén wings system fixed (at first order) relative to Io. Ganymede's orbital period being four times that of Io, one might have expected, in the case of steady magnetic reconnection and Alfvén wings attached to that moon, a duration of Ganymede–Jupiter DAM emissions statistically longer than for Io. Zarka et al. (2018) developed a method for comparing the broad distributions of durations of these radio emissions, and not only their moments. They showed that the duration of Ganymede–induced DAM radio bursts, between ~10 min. and ~3 h 30, was statistically 1.7 times shorter than that of Io–Jupiter ones, and only ~1.2 times longer than auroral DAM bursts. This implies that Ganymede–Jupiter interaction is dominated by Jupiter's rotation.

One possible explanation is that the efficiency of the reconnection between Jupiter and Ganymede magnetic fields varies with Jupiter's rotation. Analyzing an analytical criterion for reconnection onset, Kaweeyanun et al. (2020) found that reconnection may occur anywhere on Ganymede's magnetopause in an unpredictable, disordered way, the average reconnection rate being controlled by the ambient Jovian field orientation and hence driven by Jupiter's rotation. Magnetohydrodynamics simulations (Jia et al., 2009, 2010) and MHD-Hall simulations (Zhou et al., 2019, 2020) also showed that magnetic reconnection is intrinsically intermittent, involving flux ropes and flux transfer events at timescales down to 10–100 s near the upstream magnetopause, even for constant external conditions. But these timescales are much shorter than the duration of Ganymede–induced DAM radio bursts.

Another explanation is that the conditions permitting CMI emission at Ganymede's footprints (for example, the magnetic field topology favouring the existence of a loss cone) exist in a range of longitudes more restricted than those for Io. This is indeed what is suggested by Figures 19.3c and 19.3f. Combined with the synodic period of Jupiter relative to Ganymede being shorter than the synodic period of Jupiter relative to Io, this explains the statistically shorter durations of Ganymede-induced radio bursts.

This, and the persistence of the UV footprints of Ganymede, suggests that the remote electrodynamic interaction between Ganymede and the Jovian magnetic field is rather steady, and consequently that the electrons responsible for the electromagnetic footprint emissions are likely accelerated by Alfvén waves rather than by reconnection.

Radio Emissions Energetics

Kurth et al. (2000) qualitatively compared the strength of the local and distant radio emissions and plasma waves at the four Galilean moons, noting that it is strongest at Io and 'intermediate' at Europa and Ganymede. Zarka (2007) extended this comparison to the power emitted in UV footprints and induced CMI radio emissions, still poorly constrained at that time. Grodent et al. (2009) better quantified the power emitted in the UV by Ganymede's footprint (0.2–1.5 GW). Following their statistical detection of Ganymede-induced radio emissions, Zarka et al. (2018) measured their intensity (marginally lower than Io-induced ones) and their emitted power (~15× lower than for Io-induced radio emissions). This strengthened the radio-magnetic scaling law proposed in (Zarka et al., 2001; Zarka, 2007) and generalized in figure 7 of (Zarka et al., 2018) that relates the emitted radio power to the intercepted Poynting flux (or magnetic energy flux) in all interactions involving a magnetized plasma flow (sub- or super-Alfvénic) and an obstacle (magnetized or not). According to this scaling law, the dissipated electromagnetic power writes in all cases

$$P_{dissipated}(W) \simeq \epsilon(V_{flow}B^2_{\perp flow}/\mu_o)\pi R^2_{obstacle}, \tag{19.1}$$

with $B_{\perp flow}$ the flow's magnetic field component perpendicular to the flow direction in the obstacle's frame, and an efficiency $0 < \epsilon \leq 1$ ($\epsilon \simeq M_A$, the Alfvén Mach number, for a sub-Alfvénic flow). Following Equation (19.1), the dissipated power (from the intercepted Jovian magnetic field) in the Alfvén wings is similar for Europa and Ganymede, Europa being closer to Jupiter but Ganymede having a much larger cross section due to its magnetosphere, and about one order of magnitude smaller than in Io's Alfvén wings. Furthermore, it was found in (Zarka et al., 2001; Zarka, 2007) that the emitted radio power resulting from the flow–obstacle interaction follows the relation

$$P_{radio}(W) \simeq \beta(V_{flow}B^2_{\perp flow}/\mu_o)\pi R^2_{obstacle} = (\beta/\epsilon)P_{dissipated}, \tag{19.2}$$

with an efficiency factor $\beta = 2 - 10 \times 10^{-3}$. Subsequent works explored the theoretical foundations of this radio-magnetic scaling law and found that it only provides order-of-magnitude estimates (see (Zarka, 2020) and references therein), but as it seems to hold over >10 orders of magnitude, it remains adapted to predictions and analyses of populations. In particular, the Ganymede–Jupiter interaction provides a useful model for studying star–planet plasma interactions in which a magnetized hot Jupiter interacts with its magnetized parent star, and also possibly some pulsar–planet interactions.

19.6 SUMMARY AND PERSPECTIVES

Summary

Thanks to the data collected by Galileo during its six close flybys of Ganymede in 1995–2000, and by HST, Juno and ground-based radio telescopes, the electrodynamic coupling between Ganymede and the Jovian ionosphere thus seems to be fairly well described and understood at first order. Despite the very different nature of the local interaction close to the satellites, the processes taking place further away from them appear common to all moons, suggesting that there is a universal physics at play, applicable to other systems as well, even beyond our solar system. Among these processes are the prominent role of Alfvén waves, not only to carry the electric current but also to accelerate the electrons along the field lines in both directions. Another common characteristic is the presence of short length-scale features and short timescale variations of the footprints, even if the characteristics of these behaviours are not yet fully elucidated. Moreover, the characteristics of the radio decametric emissions at both Io and Ganymede demonstrate the importance of the cyclotron-maser instability, and the power of the decametric emissions at Ganymede helped validate the scaling law between the emitted radio power and the intercepted Poynting flux. Finally, the location of the Ganymede auroral footprint has also served as a useful landmark to both constrain the internal magnetic field models before the arrival of Juno and to map auroral features into the magnetosphere.

Open Questions and Perspectives

In addition to its use as a reference point in the magnetosphere, a careful analysis of the magnetic footprint of Ganymede could even help us to get one step further: documenting the changes in the Jovian magnetosphere. The footprint could be a tool to monitor the state of the magnetosphere, by studying both the latitudinal position of the spots and their spacing. The first one is related to the azimuthal currents in the plasma sheet, while the second varies directly with the Alfvén propagation time, which depends on the plasma density and the magnetic field strength. However, to achieve such an objective, a calibration of the relationship between auroral and magnetospheric parameters remains to be performed.

The coordinates of the high-occurrence islands in the $\Phi_{Ganymede}$–CML plane (Figures 19.3d,e), as well as their extension to nearly all CML and different Ganymede phases at low frequencies (Louis et al., 2017b), remain to be quantitatively explained. Another open question concerning radio emissions is the existence of Ganymede-induced S-bursts and the spatial structure of electric fields and electrons' acceleration along the Ganymede flux tube (e.g., (Hess, Zarka and Mottez, 2007; Hess et al., 2009) for Io).

It would be interesting to search in radio and UV data for direct signatures of the intermittent reconnection between Jupiter's and Ganymede's magnetic fields, and more broadly to correlate the occurrence of radio and UV emissions at all timescales. However, this quest will not be simple, because short-timescale (a few minutes) variations of the footprint brightness, which could result from those bursty reconnections, have also been identified at Io and Europa, where reconnection is not expected (Bonfond et al., 2017a).

These two processes are not mutually exclusive and will thus be difficult to disentangle. Furthermore, the sub-structure of the Ganymede footprint spots could also be interpreted as a signature of the reconnection sites at the front and back of Ganymede's magnetosphere (Mura et al., 2018). However, here again, a similar spatial pattern has been identified at Io's footprint, which also calls for a common explanation.

Many pieces of the scenario proposed to explain the footprint spots' multiplicity, such as the measurements of strong Alfvén waves and bi-directional electron beams with a broad energy distribution, have been confirmed by in situ measurements. However, a clear demonstration that the electrons accelerated away from Jupiter in one hemisphere can actually precipitate in the opposite one remains to be found.

Finally, one of the most unexpected findings of Juno regarding the satellite footprints was the discovery of proton beams during Io's Alfvén wings crossings, with three different acceleration regions identified, at altitudes between 0.9 and 2.5 R_J at the torus boundary and very close to Jupiter at altitudes around 0.16 R_J (Szalay et al., 2020b; Clark et al., 2020). The two former populations probably arise from Alfvénic acceleration, while the third possibly stems from interactions with ion-cyclotron waves (Sulaiman et al., 2020). The first indications that a similar process also takes place at Ganymede need to be confirmed (Szalay et al., 2020a). Similarly, energetic proton depletion found in Io's wake (Paranicas et al., 2019) probably has a counterpart at Ganymede, even if it hasn't been found yet.

While the Juno mission is essentially dedicated to Jupiter itself, the European JUpiter ICy moons Explorer (JUICE) mission, which was launched on 14 April 2023, will further investigate Ganymede and its surrounding space environment (Grasset et al., 2013). After a first phase of the mission orbiting around Jupiter and flying by Europa, Ganymede and Callisto, the spacecraft is planned to insert into Ganymede's orbit. This mission profile will allow it to connect directly for the first time in situ measurements of particles and fields near Ganymede (e.g. signatures of reconnection), or observations of the incoming Jovian magnetospheric plasma via energetic neutral atoms (ENAs) imaging, to simultaneous multi-wavelength remote-sensing observations of Ganymede's auroral footprint on Jupiter. Such observations will be instrumental to discriminate the many processes involved in the electrodynamic interaction and answer the questions listed earlier.

REFERENCES

Allegrini, F., Gladstone, G. R., Hue, V. et al. (2020), 'First report of electron measurements during a Europa footprint tail crossing by Juno', *Geophysical Research Letters* **47**(18), e2020GL089732. eprint: https://agupubs.onlinelibrary.wiley.com/doi/pdf/10.1029/2020GL089732. **URL:** https://agupubs.onlinelibrary.wiley.com/doi/abs/10.1029/2020GL089732.

Badman, S. V., Branduardi-Raymont, G., Galand, M. et al. (2015), 'Auroral processes at the giant planets: energy deposition, emission mechanisms, morphology and spectra', *Space Science Reviews* **187**(1), 99–179. **URL:** https://doi.org/10.1007/s11214-014-0042-x.

Bhattacharyya, D., Clarke, J. T., Montgomery, J. et al. (2018), 'Evidence for auroral emissions from Callisto's footprint in HST UV images', *Journal of Geophysical Research: Space Physics* **123**(1),

364–73. **URL:** https://agupubs.onlinelibrary.wiley.com/doi/abs/10.1002/2017JA024791.

Bigg, E. K. (1964), 'Influence of the Satellite Io on Jupiter's decametric emission', *Nature* **203**(4949), 1008–10. **URL:** www.nature.com/articles/2031008a0.

Bigg, E. K. (1966), 'Periodicities in Jupiter's decametric radiation', *Planetary and Space Science* **14**(8), 741–58.

Bonfond, B., Gérard, J.-C., Grodent, D. and Saur, J. (2007), 'Ultraviolet Io footprint short timescale dynamics', *Geophysical Research Letters* **34**, L06201, doi:https://doi.org/10.1029/2006GL028765. **URL:** https://agupubs.onlinelibrary.wiley.com/doi/abs/10.1029/2006GL028765.

Bonfond, B., Grodent, D., Badman, S. V. et al. (2017a), 'Similarity of the Jovian satellite footprints: spots multiplicity and dynamics', *Icarus* **292**, 208–17. **URL:** www.sciencedirect.com/science/article/pii/S0019103516304547.

Bonfond, B., Grodent, D., Gérard, J.-C. et al. (2009), 'The Io UV footprint: location, inter-spot distances and tail vertical extent', *Journal of Geophysical Research: Space Physics* **114**, A07224, doi:https://doi.org/10.1029/2009JA014312. **URL:** https://agupubs.onlinelibrary.wiley.com/doi/abs/10.1029/2009JA014312.

Bonfond, B., Grodent, D., Gérard, J.-C. et al. (2012), 'Auroral evidence of Io's control over the magnetosphere of Jupiter', *Geophysical Research Letters* **39**, L01105, doi:https://doi.org/10.1029/2011GL050253. **URL:** https://agupubs.onlinelibrary.wiley.com/doi/abs/10.1029/2011GL050253.

Bonfond, B., Hess, S., Bagenal, F. et al. (2013), 'The multiple spots of the Ganymede auroral footprint', *Geophysical Research Letters* **40**(19), 4977–81. **URL:** https://agupubs.onlinelibrary.wiley.com/doi/abs/10.1002/grl.50989.

Bonfond, B., Saur, J., Grodent, D. et al. (2017), 'The tails of the satellite auroral footprints at Jupiter', *Journal of Geophysical Research: Space Physics* **122**(8), 7985–96. **URL:** https://agupubs.onlinelibrary.wiley.com/doi/abs/10.1002/2017JA024370.

Chust, T., Roux, A., Kurth, W. S. et al. (2005), 'Are Io's Alfvén wings filamented? Galileo observations', *Planetary and Space Science* **53**(4), 395–412. **URL:** www.sciencedirect.com/science/article/pii/S003206330400159X.

Clark, G., Mauk, B. H., Kollmann, P. et al. (2020), 'Energetic proton acceleration associated with Io's footprint tail', *Geophysical Research Letters* **47**(24), e2020GL090839. eprint: https://agupubs.onlinelibrary.wiley.com/doi/pdf/10.1029/2020GL090839. **URL:** https://onlinelibrary.wiley.com/doi/abs/10.1029/2020GL090839.

Clarke, J. T., Ajello, J., Ballester, G. et al. (2002), 'Ultraviolet emissions from the magnetic footprints of Io, Ganymede and Europa on Jupiter', *Nature* **415**(6875), 997–1000. **URL:** www.nature.com/articles/415997a.

Collinson, G., Paterson, W. R., Bard, C. et al. (2018), 'New results from Galileo's first flyby of Ganymede: reconnection-driven flows at the low-latitude magnetopause boundary, crossing the cusp, and icy ionospheric escape', *Geophysical Research Letters* **45**(8), 3382–92. eprint: https://agupubs.onlinelibrary.wiley.com/doi/pdf/10.1002/2017GL075487. **URL:** https://agupubs.onlinelibrary.wiley.com/doi/abs/10.1002/2017GL075487.

Connerney, J. E. P. (1981), 'The magnetic field of Jupiter: a generalized inverse approach', *Journal of Geophysical Research: Space Physics* **86**(A9), 7679–93. **URL:** https://agupubs.onlinelibrary.wiley.com/doi/abs/10.1029/JA086iA09p07679.

Connerney, J. E. P., Acuña, M. H., Ness, N. F. and Satoh, T. (1998), 'New models of Jupiter's magnetic field constrained by the Io flux tube footprint', *Journal of Geophysical Research: Space Physics* **103**(A6), 11929–9. **URL:** https://agupubs.onlinelibrary.wiley.com/doi/abs/10.1029/97JA03726.

Connerney, J. E. P., Baron, R., Satoh, T. and Owen, T. (1993), 'Images of excited $H3^+$ at the foot of the lo flux tube in Jupiter's atmosphere', *Science* **262**(5136), 1035–38. **URL:** https://science.sciencemag.org/content/262/5136/1035.

Connerney, J. E. P., Kotsiaros, S., Oliversen, R. J., Espley, J. R. et al. (2018), 'A new model of Jupiter's magnetic field from Juno's first nine orbits', *Geophysical Research Letters* **45**(6), 2590–6. **URL:** https://agupubs.onlinelibrary.wiley.com/doi/abs/10.1002/2018GL077312.

Damiano, P. A., Delamere, P. A., Stauffer, B., Ng, C.-S. and Johnson, J. R. (2019), 'Kinetic simulations of electron acceleration by dispersive scale Alfvén waves in Jupiter's magnetosphere', *Geophysical Research Letters* **46**(6), 3043–51. eprint: https://agupubs.onlinelibrary.wiley.com/doi/pdf/10.1029/2018GL081219. **URL:** https://agupubs.onlinelibrary.wiley.com/doi/abs/10.1029/2018GL081219.

Delamere, P. A., Bagenal, F., Ergun, R. and Su, Y.-J. (2003), 'Momentum transfer between the Io plasma wake and Jupiter's ionosphere', *Journal of Geophysical Research: Space Physics* **108**(A6). **URL:** https://agupubs.onlinelibrary.wiley.com/doi/abs/10.1029/2002JA009530.

Dulk, G. A. (1967), 'Lack of effects of satellites Europa, Ganymede, Callisto, and Amalthea on the decametric radio emission of Jupiter', *The Astrophysical Journal* **148**, 239.

Ergun, R. E., Su, Y.-J., Andersson, L. et al. (2006), 'S bursts and the Jupiter ionospheric Alfvén resonator', *Journal of Geophysical Research: Space Physics* **111**, A06212, doi: https://doi.org/10.1029/2005JA011253. **URL:** https://agupubs.onlinelibrary.wiley.com/doi/abs/10.1029/2005JA011253.

Goldreich, P. and Lynden-Bell, D. (1969), 'Io, a Jovian unipolar inductor', *The Astrophysical Journal* **156**, 59–78.

Grasset, O., Dougherty, M. K., Coustenis, A. et al. (2013), 'JUpiter ICy moons Explorer (JUICE): an ESA mission to orbit Ganymede and to characterise the Jupiter system', *Planetary and Space Science* **78**, 1–21. **URL:** www.sciencedirect.com/science/article/pii/S0032063312003777.

Grodent, D., Bonfond, B., Gérard, J.-C. et al. (2008), 'Auroral evidence of a localized magnetic anomaly in Jupiter's northern hemisphere', *Journal of Geophysical Research: Space Physics* **113**, A09201, doi:https://doi.org/10.1029/2008JA013185. **URL:** https://agupubs.onlinelibrary.wiley.com/doi/abs/10.1029/2008JA013185.

Grodent, D., Bonfond, B., Radioti, A. et al. (2009), 'Auroral footprint of Ganymede', *Journal of Geophysical Research: Space Physics* **114**, A07212, doi:https://doi.org/10.1029/2009JA014289. https://agupubs.onlinelibrary.wiley.com/doi/abs/10.1029/2009JA014289.

Grodent, D., Gérard, J.-C., Radioti, A., Bonfond, B. and Saglam, A. (2008b), 'Jupiter's changing auroral location', *Journal of Geophysical Research: Space Physics* **113**, A01206, doi:https://doi.org/10.1029/2007JA012601. eprint: https://agupubs.onlinelibrary.wiley.com/doi/pdf/10.1029/2007JA012601. **URL:** https://agupubs.onlinelibrary.wiley.com/doi/abs/10.1029/2007JA012601.

Gurnett, D. A. and Goertz, C. K. (1981), 'Multiple Alfven wave reflections excited by Io: Origin of the Jovian decametric arcs', *Journal of Geophysical Research* **86**(A2), 717–22.

Gurnett, D. A., Kurth, W. S., Roux, A., Bolton, S. J. and Kennel, C. F. (1996), 'Evidence for a magnetosphere at Ganymede from plasma-wave observations by the Galileo spacecraft', *Nature* **384**(6609), 535–7. **URL:** www.nature.com/articles/384535a0.

Gustin, J., Bonfond, B., Grodent, D. and Gérard, J.-C. (2012), 'Conversion from HST ACS and STIS auroral counts into brightness, precipitated power, and radiated power for H2 giant planets', *Journal of Geophysical Research: Space Physics* **117**, A07316, doi:https://doi.org/10.1029/2012JA017607. **URL:** https://agupubs.onlinelibrary.wiley.com/doi/abs/10.1029/2012JA017607.

Hess, S., Bonfond, B., Bagenal, F. and Lamy, L. (2017), *A Model of the Jovian Internal Field Derived from In-Situ and Auroral Constraints*. Austrian Academy of Sciences Press. **URL:** https://orbi.uliege.be/handle/2268/221222.

Hess, S. and Delamere, P. (2012), Satellite-Induced Electron Acceleration and Related Auroras, in A. Keiling, E. Donovan, F. Bagenal and T. Karlsson, eds., *Auroral Phenomenology and Magnetospheric Processes: Earth and Other Planets*, American Geophysical Union. **URL:** http://onlinelibrary.wiley.com/doi/10.1029/2011GM001175/summary.

Hess, S. L. G., Bonfond, B., Chantry, V. et al. (2013), 'Evolution of the Io footprint brightness II: modeling', *Planetary and Space Science* **88**, 76–85. **URL:** www.sciencedirect.com/science/article/pii/S0032063313002109.

Hess, S. L. G., Bonfond, B. and Delamere, P. A. (2013), 'How could the Io footprint disappear?', *Planetary and Space Science* **89**, 102–10. **URL:** www.sciencedirect.com/science/article/pii/S0032063313002195.

Hess, S. L. G., Delamere, P., Dols, V., Bonfond, B. and Swift, D. (2010), 'Power transmission and particle acceleration along the Io flux tube', *Journal of Geophysical Research: Space Physics* **115**, A06205, doi:https://doi.org/10.1029/2009JA014928. https://agupubs.onlinelibrary.wiley.com/doi/abs/10.1029/2009JA014928.

Hess, S., Mottez, F. and Zarka, P. (2007), 'Jovian S burst generation by Alfvén waves', *Journal of Geophysical Research: Space Physics* **112**(A11), A11212.

Hess, S., Mottez, F., Zarka, P. and Chust, T. (2008), 'Generation of the Jovian radio decametric arcs from the Io flux tube', *Journal of Geophysical Research: Space Physics* **113**(A3), A03209.

Hess, S., Zarka, P. and Mottez, F. (2007), 'Io Jupiter interaction, millisecond bursts and field-aligned potentials', *Planetary and Space Science* **55**(1–2), 89–99.

Hess, S., Zarka, P., Mottez, F. and Ryabov, V. B. (2009), 'Electric potential jumps in the Io-Jupiter flux tube', *Planetary and Space Science* **57**(1), 23–33. **URL:** www.sciencedirect.com/science/article/pii/S0032063308003358.

Higgins, C. A. (2007), 'Satellite control of Jovian 2–6 MHz radio emission using Voyager data', *Journal of Geophysical Research: Space Physics* **112**, A05213, doi:https://doi.otg/10.1029/2006JA012100. eprint: https://agupubs.onlinelibrary.wiley.com/doi/pdf/10.1029/2006JA012100. **URL:** https://agupubs.onlinelibrary.wiley.com/doi/abs/10.1029/2006JA012100.

Hill, T. W. and Vasyliunas, V. M. (2002), 'Jovian auroral signature of Io's corotational wake', *Journal of Geophysical Research* **107**(A12), 1464.

Hinson, D. P., Kliore, A. J., Flasar, F. M. et al. (1998), 'Galileo radio occultation measurements of Io's ionosphere and plasma wake', *Journal of Geophysical Research* **103**(A12), 29343–58.

Hinton, P. C., Bagenal, F. and Bonfond, B. (2019), 'Alfvén wave propagation in the Io plasma torus', *Geophysical Research Letters* **46**(3), 1242–9. eprint: https://agupubs.onlinelibrary.wiley.com/doi/pdf/10.1029/2018GL081472. **URL:** https://agupubs.onlinelibrary.wiley.com/doi/abs/10.1029/2018GL081472.

Hospodarsky, G. B., Christopher, I. W., Menietti, J. D. et al. (2001), Control of Jovian radio emissions by the Galilean moons as observed by Cassini and Galileo, in H. O. Rucker, M. L. Kaiser and Y. Leblanc, eds., *Planetary Radio Emissions V*, pp. 155–64.

Jacobsen, S., Neubauer, F. M., Saur, J. and Schilling, N. (2007), 'Io's nonlinear MHD-wave field in the heterogeneous Jovian magnetosphere', *Geophysical Research Letters* **34**, L10202, doi:https://doi.org/10.1029/2006GL029187.

Jacobsen, S., Saur, J., Neubauer, F. M. et al. (2010), 'Location and spatial shape of electron beams in Io's wake', *Journal of Geophysical Research* **115**(A14), A04205.

Jia, X., Walker, R. J., Kivelson, M. G., Khurana, K. K. and Linker, J. A. (2009), 'Properties of Ganymede's magnetosphere inferred from improved three-dimensional MHD simulations', *Journal of Geophysical Research* **114**, A09209, doi:https://doi.org/10.1029/2009JA014375.

Jia, X., Walker, R. J., Kivelson, M. G., Khurana, K. K. and Linker, J. A. (2008), 'Three-dimensional MHD simulations of Ganymede's magnetosphere', *Journal of Geophysical Research* **113**(A12), 6212.

Jia, X., Walker, R. J., Kivelson, M. G., Khurana, K. K. and Linker, J. A. (2010), 'Dynamics of Ganymede's magnetopause: intermittent reconnection under steady external conditions', *Journal of Geophysical Research: Space Physics* **115**(A14), A12202.

Jones, S. T. and Su, Y.-J. (2008), 'Role of dispersive Alfvén waves in generating parallel electric fields along the Io-Jupiter fluxtube', *Journal of Geophysical Research* **113**(A12), 12205.

Kaiser, M. L. and Alexander, J. K. (1973), 'Periodicities in the Jovian decametric emission', *The Astrophysical Journal Letters* **14**, 55.

Kaiser, M. L. and MacDowall, R. J. (1998), 'Jovian radio "bullseyes" observed by Ulysses', *Geophysical Research Letters* **25**(16), 3113–16.

Kaweeyanun, N., Masters, A. and Jia, X. (2020), 'Favorable conditions for magnetic reconnection at Ganymede's upstream magnetopause', *Geophysical Research Letters* **47**(6), e2019GL086228. eprint: https://agupubs.onlinelibrary.wiley.com/doi/pdf/10.1029/2019GL086228. **URL:** https://agupubs.onlinelibrary.wiley.com/doi/abs/10.1029/2019GL086228.

Khurana, K. K. (1997), 'Euler potential models of Jupiter's magnetospheric field', *Journal of Geophysical Research: Space Physics* **102**(A6), 11295–306. **URL:** https://agupubs.onlinelibrary.wiley.com/doi/abs/10.1029/97JA00563.

Kivelson, M. G. and Bagenal, F. (2014), Chapter 7 – Planetary Magnetospheres, in T. Spohn, D. Breuer and T. V. Johnson, eds., *Encyclopedia of the Solar System* (Third Edition), Elsevier, Boston, pp. 137–57. **URL:** www.sciencedirect.com/science/article/pii/B9780124158450000074.

Kivelson, M. G., Bagenal, F., Kurth, W. S. et al. (2004), Magnetospheric interactions with satellites, in *Jupiter: The Planet, Satellites and Magnetosphere*, pp. 513–36.

Kivelson, M. G., Warnecke, J., Bennett, L. et al. (1998), 'Ganymede's magnetosphere: magnetometer overview', *Journal of Geophysical Research: Planets* **103**(E9), 19963–72. eprint: https://agupubs.onlinelibrary.wiley.com/doi/pdf/10.1029/98JE00227. **URL:** https://agupubs.onlinelibrary.wiley.com/doi/abs/10.1029/98JE00227.

Kurth, W. S., Bolton, S. J., Gurnett, D. A. and Levin, S. (1997a), 'A determination of the source of Jovian hectometric radiation via occultation by Ganymede', *Geophysical Research Letters* **24**(10), 1171–4.

Kurth, W. S., Gurnett, D. A. and Menietti, J. D. (2000), 'The influence of the Galilean satellites on radio emissions from the Jovian system', *Washington DC American Geophysical Union Geophysical Monograph Series* **119**, 213–25.

Kurth, W. S., Gurnett, D. A., Roux, A. and Bolton, S. J. (1997b), 'Ganymede: a new radio source', *Geophysical Research Letters* **24**(17), 2167–70. eprint: https://agupubs.onlinelibrary.wiley.com/doi/pdf/10.1029/97GL02249. **URL:** https://agupubs.onlinelibrary.wiley.com/doi/abs/10.1029/97GL02249.

Lamy, L., Zarka, P., Cecconi, B. et al. (2017), 1977–2017: 40 years of decametric observations of Jupiter and the Sun with the Nancay Decameter Array, in G. Fischer, G. Mann, M. Panchenko and P. Zarka, eds., *Planetary Radio Emissions VIII*, pp. 455–66.

Lavrukhin, A. S. and Alexeev, I. I. (2015), 'Aurora at high latitudes of Ganymede', *Astronomy Letters* **41**(11), 687–92. **URL:** https://doi.org/10.1134/S1063773715110043.

Le Quéau, D. (1988), Planetary radio emissions from high magnetic latitudes: the 'Cyclotron-Maser' theory, in H. O. Rucker, S. J. Bauer and B. M. Pedersen, eds., *Planetary Radio Emissions II*, Austrian Academy of Sciences Press, Graz, Austria, pp. 381–98. **URL:** www.austriaca.at:8080/1523-6inhalt?frames=yes.

Lebo, G. R., Smith, A. G. and Carr, T. D. (1965), 'Jupiter's decametric emission correlated with the longitudes of the first three Galilean satellites', *Science* **148**(3678), 1724–25. **URL:** https://science.sciencemag.org/content/148/3678/1724.

Louis, C. K., Hess, S. L. G., Cecconi, B. et al. (2019), 'ExPRES: an Exoplanetary and Planetary Radio Emissions Simulator', *Astronomy & Astrophysics* **627**, A30.

Louis, C. K., Lamy, L., Zarka, P., Cecconi, B. and Hess, S. L. G. (2017a), 'Detection of Jupiter decametric emissions controlled by Europa and Ganymede with Voyager/PRA and Cassini/RPWS', *Journal of Geophysical Research: Space Physics* **122**(9), 9228–47. eprint: https://agupubs.onlinelibrary.wiley.com/doi/pdf/10.1002/2016JA023779. **URL:** https://agupubs.onlinelibrary.wiley.com/doi/abs/10.1002/2016JA023779.

Louis, C. K., Louarn, P., Allegrini, F., Kurth, W. S. and Szalay, J. R. (2020), 'Ganymede-induced decametric radio emission: in situ observations and measurements by Juno', *Geophysical Research Letters* **47**(20), e2020GL090021. eprint: https://agupubs.onlinelibrary.wiley.com/doi/pdf/10.1029/2020GL090021. **URL:** https://agupubs.onlinelibrary.wiley.com/doi/abs/10.1029/2020GL090021.

Louis, C., Lamy, L., Zarka, P. et al. (2017b), Simulating Jupiter-satellite decametric emissions with ExPRES: a parametric study, in G. Fischer, G. Mann, M. Panchenko and P. Zarka, eds., *Planetary Radio Emissions VIII*, pp. 59–72.

Marques, M. S., Zarka, P., Echer, E. et al. (2017), 'Statistical analysis of 26 yr of observations of decametric radio emissions from Jupiter', *Astronomy & Astrophysics* **604**, A17.

Matsuda, K., Terada, N., Katoh, Y. and Misawa, H. (2012), 'A simulation study of the current-voltage relationship of the Io tail aurora', *Journal of Geophysical Research: Space Physics* **117**(A16), 10214.

Mauk, B. H., Williams, D. J. and McEntire, R. W. (1997), 'Energy-time dispersed charged particle signatures of dynamic injections in Jupiter's inner magnetosphere', *Geophysical Research Letters* **24**(23), 2949–2952. eprint: https://agupubs.onlinelibrary.wiley.com/doi/pdf/10.1029/97GL03026. **URL:** https://agupubs.onlinelibrary.wiley.com/doi/abs/10.1029/97GL03026.

Menietti, J. D., Gurnett, D. A., Kurth, W. S. and Groene, J. B. (1998a), 'Control of Jovian radio emission by Ganymede', *Geophysical Research Letters* **25**(23), 4281–4. eprint: https://agupubs.onlinelibrary.wiley.com/doi/pdf/10.1029/1998GL900112. **URL:** https://agupubs.onlinelibrary.wiley.com/doi/abs/10.1029/1998GL900112.

Menietti, J. D., Gurnett, D. A., Kurth, W. S., Groene, J. B. and Granroth, L. J. (1998b), 'Galileo direction finding of Jovian radio emissions', *Journal of Geophysical Research* **103**(E9), 20001–10.

Moirano, A., Mura, A., Adriani, A. et al. (2021), 'Morphology of the auroral tail of Io, Europa, and Ganymede from JIRAM L-Band Imager', *Journal of Geophysical Research: Space Physics* **126**(9), e2021JA029450. eprint: https://onlinelibrary.wiley.com/doi/pdf/10.1029/2021JA029450. **URL:** https://onlinelibrary.wiley.com/doi/abs/10.1029/2021JA029450.

Moore, K. M., Bloxham, J., Connerney, J. E. P., Jørgensen, J. L. and Merayo, J. M. G. (2017), 'The analysis of initial Juno magnetometer data using a sparse magnetic field representation', *Geophysical Research Letters* **44**, 4687–93.

Mura, A., Adriani, A., Altieri, F. et al. (2017), 'Infrared observations of Jovian aurora from Juno's first orbits: main oval and satellite footprints', *Geophysical Research Letters* **44**, 5308–16.

Mura, A., Adriani, A., Connerney, J. E. P. et al. (2018), 'Juno observations of spot structures and a split tail in Io-induced aurorae on Jupiter', *Science* **361**(6404), 774–7. **URL:** http://science.sciencemag.org/content/361/6404/774.

Neubauer, F. M. (1980), 'Nonlinear standing Alfven wave current system at Io - Theory', *Journal of Geophysical Research* **85**, 1171–8.

Neubauer, F. M. (1998), 'The sub-Alfvénic interaction of the Galilean satellites with the Jovian magnetosphere', *Journal of Geophysical Research: Planets* **103**(E9), 19843–66. eprint: https://agupubs.onlinelibrary.wiley.com/doi/pdf/10.1029/97JE03370. **URL:** https://agupubs.onlinelibrary.wiley.com/doi/abs/10.1029/97JE03370

Paranicas, C., Mauk, B. H., Haggerty, D. K. et al. (2019), 'Io's effect on energetic charged particles as seen in Juno data', *Geophysical Research Letters* **46**(23), 13615–20. eprint: https://agupubs.onlinelibrary.wiley.com/doi/pdf/10.1029/2019GL085393. **URL:** https://agupubs.onlinelibrary.wiley.com/doi/abs/10.1029/2019GL085393.

Pryor, W. R., Rymer, A. M., Mitchell, D. G. et al. (2011), 'The auroral footprint of Enceladus on Saturn', *Nature* **472**(7343), 331–3. **URL:** www.nature.com/articles/nature09928.

Saur, J., Grambusch, T., Duling, S., Neubauer, F. M. and Simon, S. (2013), 'Magnetic energy fluxes in sub-Alfvénic planet star and moon planet interactions', *Astronomy & Astrophysics* **552**, A119. **URL:** www.aanda.org/articles/aa/abs/2013/04/aa18179-11/aa18179-11.html.

Saur, J., Neubauer, F. M., Connerney, J. E. P., Zarka, P. and Kivelson, M. G. (2004), Plasma interaction of Io with its plasma torus, in *Jupiter. The Planet, Satellites and Magnetosphere*, pp. 537–60. Citation Key Alias: saurPlasmaInteractionsIo.

St. Cyr, O. C. (1985), Jupiter's decameter and kilometer emissions: satellite effects and long term periodicities, PhD thesis, University of Florida, Gainesville.

Su, Y.-J., Ergun, R. E., Bagenal, F. and Delamere, P. A. (2003), 'Io-related Jovian auroral arcs: Modeling parallel electric fields', *Journal of Geophysical Research: Space Physics* **108**(A2). **URL:** https://agupubs.onlinelibrary.wiley.com/doi/abs/10.1029/2002JA009247.

Sulaiman, A. H., Hospodarsky, G. B., Elliott, S. S. et al. (2020), 'Wave-particle interactions associated with Io's auroral footprint: evidence of Alfvén, ion cyclotron, and whistler modes', *Geophysical Research Letters* **47**(22), e2020GL088432. eprint: https://agupubs.onlinelibrary.wiley.com/doi/pdf/10.1029/2020GL088432. **URL:** https://agupubs.onlinelibrary.wiley.com/doi/abs/10.1029/2020GL088432.

Szalay, J. R., Allegrini, F., Bagenal, F. et al. (2020a), 'Alfvénic acceleration sustains Ganymede's footprint tail aurora', *Geophysical Research Letters* **47**(3), e2019GL086527. eprint: https://agupubs.onlinelibrary.wiley.com/doi/pdf/10.1029/2019GL086527. **URL:** https://agupubs.onlinelibrary.wiley.com/doi/abs/10.1029/2019GL086527.

Szalay, J. R., Bagenal, F., Allegrini, F. et al. (2020b), 'Proton acceleration by Io's Alfvénic interaction', *Journal of Geophysical Research: Space Physics* **125**(1), e2019JA027314. eprint: https://agupubs.onlinelibrary.wiley.com/doi/pdf/10.1029/2019JA027314. **URL:** https://agupubs.onlinelibrary.wiley.com/doi/abs/10.1029/2019JA027314.

Szalay, J. R., Bonfond, B., Allegrini, F. et al. (2018), 'In situ observations connected to the Io footprint tail aurora', *Journal of Geophysical Research: Planets* **123**(11), 3061–77. **URL:** https://agupubs.onlinelibrary.wiley.com/doi/abs/10.1029/2018JE005752.

Vogt, M. F., Bunce, E. J., Kivelson, M. G. et al. (2015), 'Magnetosphere-ionosphere mapping at Jupiter: quantifying the effects of using different internal field models', *Journal of Geophysical Research: Space Physics* **120**(4), 2584–99. **URL:** https://agupubs.onlinelibrary.wiley.com/doi/abs/10.1002/2014JA020729.

Vogt, M. F., Kivelson, M. G., Khurana, K. K. et al. (2011), 'Improved mapping of Jupiter's auroral features to magnetospheric sources', *Journal of Geophysical Research: Space Physics* **116**, A03220, doi:https://doi.org/10.1029/2010JA016148. **URL:** https://agupubs.onlinelibrary.wiley.com/doi/full/10.1029/2010JA016148.

Wannawichian, S., Clarke, J. T. and Nichols, J. D. (2010), 'Ten years of Hubble Space Telescope observations of the variation of the Jovian satellites' auroral footprint brightness', *Journal of Geophysical Research: Space Physics* **115**, A02206, doi:https//doi.org/10.1029/2009JA014456. **URL:** https://agupubs.onlinelibrary.wiley.com/doi/abs/10.1029/2009JA014456.

Williams, D. J. and Mauk, B. (1997), 'Pitch angle diffusion at Jupiter's moon Ganymede', *Journal of Geophysical Research* **102**(A11), 24283–302.

Williams, D. J., Mauk, B. H., McEntire, R. W. et al. (1997), 'Energetic particle signatures at Ganymede: implications for Ganymede's magnetic field', *Geophysical Research Letters* **24**(17), 2163–6.

Yoneda, M., Kagitani, M. and Okano, S. (2009), 'Short-term variability of Jupiter's extended sodium nebula', *Icarus* **204**(2), 589–96. **URL:** www.sciencedirect.com/science/article/pii/S0019103509003157.

Yoneda, M., Tsuchiya, F., Misawa, H. et al. (2013), 'Io's volcanism controls Jupiter's radio emissions', *Geophysical Research Letters* **40**(4), 671–5. **URL:** https://agupubs.onlinelibrary.wiley.com/doi/abs/10.1002/grl.50095.

Zarka, P. (1998), 'Auroral radio emissions at the outer planets: observations and theories', *Journal of Geophysical Research: Planets* **103**(E9), 20159–94. **URL:** https://agupubs.onlinelibrary.wiley.com/doi/abs/10.1029/98JE01323.

Zarka, P. (2007), 'Plasma interactions of exoplanets with their parent star and associated radio emissions', *Planetary and Space Science* **55**(5), 598–617.

Zarka, P. (2020), Star-planet interactions in the radio domain: prospect for their detection, in H. J. Deeg and J. A. Belmonte, eds., *Handbook of Exoplanets*, Springer International Publishing, Cham, pp. 1–16. **URL:** https://doi.org/10.1007/978-3-319-30648-3_22-2.

Zarka, P., Marques, M. S., Louis, C. et al. (2018), 'Jupiter radio emission induced by Ganymede and consequences for the radio detection of exoplanets', *Astronomy & Astrophysics* **618**, A84. **URL:** www.aanda.org/articles/aa/abs/2018/10/aa33586-18/aa33586-18.html.

Zarka, P., Treumann, R. A., Ryabov, B. P. and Ryabov, V. B. (2001), 'Magnetically-driven planetary radio emissions and application to extrasolar planets', *Astrophysics and Space Science* **277**, 293–300.

Zhou, H., Tóth, G., Jia, X. and Chen, Y. (2020), 'Reconnection-driven dynamics at Ganymede's upstream magnetosphere: 3-D Global Hall MHD and MHD-EPIC simulations', *Journal of Geophysical Research: Space Physics* **125**(8), e28162.

Zhou, H., Tóth, G., Jia, X., Chen, Y. and Markidis, S. (2019), 'Embedded kinetic simulation of Ganymede's magnetosphere: improvements and inferences', *Journal of Geophysical Research: Space Physics* **124**(7), 5441–60.

Part IV

ASTROBIOLOGY

20

Habitability of Ganymede: Conditions for Sustainability of Life and Geochemical Biosignatures

JULIAN CHELA-FLORES

20.1 INTRODUCTION

This chapter focuses on two questions that are related but different, yet overlap significantly. Firstly, as implied by the title, habitability of Ganymede is considered as a measure of its potential to develop and maintain environments that are hospitable to life as we know it. Secondly, as implied by the subtitle, we discuss biogenicity, which requires the identification of reliable biosignatures. As Ganymede is a confirmed ocean world (Hendrix et al., 2018), its habitability is inserted in the wider context of exploration of ocean worlds (Hand et al., 2020). Aspects of habitability and biogenesis should be highlighted from the start. We will return to these topics in this chapter.

Firstly, for Ganymede, careful analysis has demonstrated the existence of a frozen brine, possibly emanating from a subsurface ocean (McCord et al., 2001a, b; 2002). Secondly, there are many places on Europa's icy surface where there are claims of the existence of endogenous materials. For instance, in Castalia Macula (0°N, 225°W) there is evidence of chemical elements emanating from the subsurface ocean (Prockter and Schenk, 2005).

Exploring the tenuous atmosphere of the icy moons, which will be possible for Ganymede in the short term, can yield information on surface components (most surficial trace species will be present in the extended atmosphere. A sensitive ion and/or neutral mass spectrometer in orbit would be able to identify surface components (Johnson et al., 2009) and also obtain insights into the oceanic chemical elements.

Sulphur is highly fractionated by microbes. Consequently, biogenically processed sulphur can be easily distinguished from mantle sulphur (Ireland, 2013). Microorganisms that consume sulphate while producing sulphide as a waste product have been confirmed to be among the oldest terrestrial life forms (Shen et al., 2001; Shen and Buick, 2004). This is the reason why we choose to focus on stable isotope geochemistry of sulphur (Hoefs, 2009). In view of recent progress in instrumentation, including what has been selected in the payloads of forthcoming missions, biogeochemistry will provide a valuable additional biosignature (cf. Section 20.3).

The high-pressure ice layer is not a barrier to water–rock interactions, a topic discussed another chapter of this book (Kalousova et al., 2022). The phenomenon of oceanic water and rock interacting raises the possibility of habitability conditions existing on Ganymede, which on Earth are known to support life. The terrestrial analogy for the emergence of life is supported by the suggestion that hydrothermal vents are likely to be sites for the origin and evolution of life, providing the necessary pathways for the evolution of free-living organisms (Baross and Hoffman, 1985). For long periods of time during Ganymede's history, water that had been generated in contact with the rocky core has been transported to the overlying ocean (Choblet et al., 2017).

A robust type of geochemical biosignature is needed to distinguish molecule preservation. In view of the intense radiation at the surface of Ganymede, we would prefer to search for biosignatures of biogenic molecules with a higher level of immunity to destruction than confining ourselves simply to biosignatures of biomolecules (nucleic acids, amino acids, lipids and phospholipids). Inorganic S-bearing compounds that have been isotopically fractionated offer an alternative (cf. Section 20.3). This is especially important, given that sulphur compounds have been found on Ganymede's surface, for instance, $MgSO_4$ hydrates (McCord et al., 1997; 1998; 2001a, b; Ahrens, 2022). The forthcoming JUpiter ICy moons Explorer (JUICE) mission and the Europa Clipper will carry instruments that could potentially investigate S isotopes in sufficient detail.

Terrestrial biota is a preliminary point of reference for considering eventual biogenicity elsewhere in the solar system, including Ganymede. For staying alive on the Ganymede ocean, an energy source is needed, as well as mainly four chemical elements: carbon, hydrogen, oxygen and nitrogen (CHON), making up some 99.9 per cent of life as we know it (Schopf, 1999); but phosphorus and, significantly, sulphur also are biogenic elements.

Biogenesis on Earth is well documented for an early origin of life (Dodd et al., 2017). The primordial solar system may have had various simultaneous non-terrestrial episodes of life's emergence; the search for them is one of the main objectives of space exploration with robotic missions. This is a simpler task than learning how life first emerged. The focus of this chapter will be especially on habitability and the closely associated question of biogenesis on Ganymede (Chela-Flores, 2010).

On the Galilean moons, both habitability and biogenesis have been discussed frequently (Chela-Flores and Kumar, 2008; Phillips and Pappalardo, 2014; Chela-Flores et al., 2015: this last publication will serve as a basis for the present discussion). Due to the remarkable evolution of instrumentation since the Galileo mission, the origin of life will be tested again in forthcoming missions.

The radiation to which the surficial molecules of Europa and Ganymede are exposed is so strong (Kivelson et al., 2004; Kivelson, 2022) that regardless of whether they are biogenic in origin, any organics in existence would have very short lifetimes. Relevant biomolecules could be identified if they were protected below the surface. Both Europa and Ganymede have strong radiation environments at the surface, but the doses and the exposure time of target materials (age of terrains) are different at each moon (Johnson et al., 2004).

To discuss the specific case of Ganymede's habitability, we should recall that the radiation environment constrains the strategies of future missions. It is necessary to discuss the effect of Ganymede's magnetosphere on the molecules' preservation. It is evident that a number of surficial spectral characteristics are likely determined by radiolysis caused by the impacting energetic ions and electrons trapped in the Jovian magnetosphere. This will contribute to changes in the molecules on the surface. Other changes affecting these molecules are due to the addition of materials of astrobiological interest that have been delivered by cometary impacts (Pierazzo and Chyba, 2002) and subsequently modified by radiolysis and photolysis (Dalton et al., 2003). These environmental effects on Ganymede's surface are discussed in Section 20.3.

Less than a year after the arrival of the Galileo spacecraft in the Jovian System, NASA organized the Europa Ocean Conference at San Juan Capistrano, California (Nash, 1996). There were already preliminary discussions related to extra-terrestrial oceans (Pappalardo et al., 1996), submarine silicate volcanism (Shoemaker, 1996), biological activity (Bada, 1996) and habitability (Chela-Flores, 1996; Lorenz and Lunine, 1996). At an early stage of the mission there were great expectations for detecting biosignatures on Europa rather than on Ganymede. The reason was that at that time, before more data was retrieved from Ganymede, Europa was considered to be the most enigmatic of the Galilean satellites (Squyres, 1996). More importantly, in Capistrano for the first time there were suggestions for novel instrumentation to probe the unprecedented challenges presented by unexpected oceanographic questions.

Since that time, science instruments have gone through a gradual and continuous process of improvement. To start with, the possibility of exploring the habitability of a new ocean world was suggested with a submersible called a hydrobot, coupled to an ice-breaker, called a cryobot (Trowell et al., 1996; Horvath et al., 1997).

Direct ocean exploration had not been raised for Ganymede, firstly since the existence of such an ocean was not immediately evident. When this idea dawned upon the astrobiological community, both the thickness and the age of Ganymede's icy surface were factors that militated against such unrealistic submersible exploration. We are now aware that the icy surface could be up to 80 km thick (Hussmann et al., 2006).

Yet, in the presence of the first extra-terrestrial ocean world, we argued then as now that the cryobot-hydrobot proposal has the merit of highlighting the need for developing instrumentation capable of probing ocean worlds. This need remains even at the time of writing, since now there are almost 20 candidates for ocean worlds in the solar system, some of which are confirmed oceans (Hendrix et al., 2018). In the case of the smaller of the icy Galilean moons, emphasis today has shifted to the possibility of placing a lander at least on Europa (Hand et al., 2017), possibly close to the icy surface from which plumes emerge (Roth et al., 2014). Sadly, no plumes are known for Ganymede.

Another possibility for inferring the nature of subsurface oceans without a cryobot-hydrobot couple would be to search for molecules on the icy surfaces that may reveal their oceanic origin (cf. Section 20.1). This approach provides indirect and possibly more feasible ways to probe for the presence of life on ocean worlds.

A subsequent early step towards improved instrumentation for ocean worlds was later suggested: the penetrators (Gowen et al., 2011). Although these instruments have not yet found space in approved space missions, we argued then as now that they are an attractive low-cost alternative or complement to landers. They could add significantly to understanding habitability of the Galilean moons (Kömle et al., 2018; Romero-Wolf et al., 2015). Penetrators are a good example of the early stages in the evolution of instrumentation for the exploration of the solar system. The advantages of testing the habitability of Ganymede with new instrumentation will be discussed in Section 20.5.

20.2 EXPLORING HABITABILITY OF GANYMEDE AND OTHER OCEAN WORLDS

The Galileo spacecraft was launched on 18 October 1989. Six years later, on 7 December 1995, Galileo was subjected to a manoeuvre that inserted it in a trajectory to tour the Galilean moons. (The mission ended in 2003.) On 27 June 1996, there was a close encounter with Ganymede. Its data indicates that the largest moon in the solar system possesses important prerequisites to be considered a habitable world. Some of Galileo's major discoveries were as follows:

(i) Galileo detected induced magnetic fields (Khurana et al., 1998; Kivelson et al., 2002). Magnetic data provides sufficient evidence that the icy Galilean moons have salty hidden oceans underneath, as well as thin atmospheres above.
(ii) Galileo provided some images of surface characteristics. Especially remarkable were the sulphur patches, ridges and the chaotic terrain (Pappalardo et al., 2004).
(iii) With Galileo, uncertainty remained regarding the depth and composition of the Europan icy crust (Spohn and Schubert, 2003).

We should pay special attention to the Ganymede-related mission of the European Space Agency (ESA) – The JUpiter ICy moons Explorer (JUICE) will spend at least three years probing Jupiter and its ocean worlds, not only Ganymede, but also Europa and Callisto. JUICE's payload will carry out several scientific experiments. Its main focus will be on Ganymede. It will be the first time that a moon in the outer solar system would have a permanent orbiter (Grasset et al., 2013). The main JUICE science objectives for Ganymede deal with the chemistry essential to life, including the composition of the non–water-ice materials.

One of the components of the payload will be the Particle Environment Package (PEP), which will be relevant for the search of biosignatures. It has a direct flight heritage from Cassini, Mars Express, Venus Express, Rosetta, New Horizons and Chandrayaan (Barabash et al., 2013). PEP has several significant aspects:

(i) It includes a neutral and ion gas mass spectrometer (Wurz et al., 2018) and has the accuracy needed for testing biogenicity by means of S isotope measurements. PEP addresses some of the scientific objectives of the JUICE mission.

Table 20.1 *A selection of current instrumentation relevant for biogeochemical mass spectrometry measurements for the detection of biomarkers. The information concerns the exploration of the Jovian system with JUICE and Clipper.*

Miniaturized instruments (JUICE)	**Miniaturized instruments (Europa Clipper)**
Particle Environment Package (PEP)[1,2].	
	The MAss Spectrometer for Planetary EXploration/Europa (MASPEX)[3]
	SUrface Dust Analyzer (SUDA)[4].

[1] Barabash et al., 2013.
[2] Meyer et al., 2017
[3] Brockwell et al., 2016.
[4] Kempf et al., 2014.

(ii) Together with the Neutral Gas and Ion Mass Spectrometer (NGIMS), it will characterize the Ganymedean exosphere (McGrath et al., 2004; Roth et al., 2022).
(iii) It is in a privileged position to measure isotopic composition of major species. Among such candidates hopefully are hydrated magnesium sulphate that may have been sputtered to orbital altitudes. However, as yet it is not certain that ejected hydrated sulphates can reach the altitude to be detected by PEP at the height of the JUICE orbits.
(iv) Radiolytic products such as SO_2 can be detected by PEP. We recall that Europa's atmosphere is normally considered as an extension of its surface (Johnson et al., 2009). Besides, if chemical elements in the exosphere are of endogenic origin, as, for instance, sulphur compounds, their source must be regions having a young surface, where the upwelling of subsurface material may occur. This raises the possibility that the observed chemical elements may be from the ocean (Leblanc et al., 2002).

Europa Clipper is an upcoming NASA orbital mission designed to investigate exclusively the habitability of Europa, not of Ganymede. Instead, the larger moon's habitability will be a concern of the JUICE mission. Europa Clipper has the capability to measure biogenic stable S-isotope fractionation: the MAss Spectrometer for Planetary EXploration/Europa (MASPEX) will be included in the Europa Clipper scientific payload (Pappalardo et al., 2013). It was designed to determine the composition of the surface and subsurface ocean by measuring Europa's tenuous atmosphere and any surface material ejected into space (see Table 20.1).

As a part of the Europa Clipper mission, the SUrface Dust Analyzer (SUDA) will measure the composition of small solid particles ejected from the icy surface of Europa (Kempf et al., 2012). The SUDA instrument will not sample Europa's surface directly. Instead, it will sample only Europa's exosphere. The type of biosignature that would be identifiable, concerning the objective of characterizing habitability, could be biogeochemical biosignatures. Indeed, the abundance of the ejected surficial sulphur is sufficient for the accuracy available to SUDA and the detection of sulphur compounds is feasible. For example, a neutral mass spectrometer orbiting at a height of 100 km above the surface is capable of performing such measurements (Cassidy et al. 2009) as well as retrieving data on globally averaged densities 110–600 $cm^{-3}$of SO_2at the orbital height.

At the time of writing, a lander remains a mission concept, in spite of the fact that its relevance has long been advocated (Phillips and Chyba, 2001). In any case it would be valuable to include a mass spectrometer in its payload (Pitesky et al., 2020).

Laplace-Posadka (English: Laplace-lander) is a proposal in conceptual phase of the Russian Space Agency (Roscosmos). Originally it should have consisted of an orbiter and a lander, but at the time of writing the lander has not been approved. A spacecraft would orbit Ganymede (Martynov et al., 2017). From the point of view of Ganymede's habitability, the initial planning for Laplace-P would have been in an optimal position for evaluating the nature of the non–water-ice materials of dark terrains, in which sulphur compounds are present with their potential implications for determining biogeochemical biosignatures (Ligier et al., 2019).

The outermost ice-I shell is expected to be too thick for a future type of hydrobot to be inserted by means of a cryobot for directly probing the Ganymede ocean, which is about 80 km thick (Hussmann et al., 2006). One may argue that the thickness of the ice shell could prevent the exposure of biomolecules of extant life on the icy surface, but we have already pointed out the possibility that endogenous processes may induce subsurface ocean organics or key inorganic elements relevant for biology, such as sulphur, to reach the Ganymedean surface.

Hopefully an eventual success of the lander Laplace-P, or subsequent ones by other space agencies, would offer the opportunity of verifying the 'ground truth' of the ocean biomolecules that could have migrated to the dark terrains earlier in Ganymede's history. However, if the lander is not eventually approved by Roscosmos, these molecules may still be probed independently from orbit.

In biogeochemical studies mass spectrometers have unique capabilities that can be applied to study habitability and, perhaps more importantly, life detection on icy worlds. In this regard, two of the forthcoming missions to the Jovian system have been provided with relevant instrumentation (cf. Table 20.1). Possible candidates for biota on the icy worlds are extremophiles, which account for a substantial portion of life on Earth.

It is worth noting that remarkable variations of microbial sulphate reduction have been discovered in pure culture experiments (Sim et al., 2011). These experiments revealed a large S mass-dependent fractionation of $\delta^{34}S$ of up to −70‰. This biogenic phenomenon is exclusively observed in a handful of natural environments on Earth, which raises the question as to whether a Europan or a Ganymedean biota, would be composed of sulphate reducers and disproportionators. This would be significant, as they could give rise to measurable geochemical biomarkers. The mass spectrometer approved for both Europa Clipper and JUICE missions would be useful if we intend to identify this kind of biota on both Galilean moons, Ganymede and Europa.

A fundamental development in instrumentation that has been achieved is the ability to carry out remote analysis (Ireland, 2013). The source of biogeochemical biosignatures on the Galilean moons could be isotopic fractionation of biological processes. We should evaluate the instrumentation that can serve to estimate biogeochemical measurements, with regard to the relative advantages of remote space observations in contrast to in situ measurements. Fortunately, there is a long history of instrumentation in previous and current missions, and so highly miniaturized devices have been developed (Riedo et al., 2013a). With them, measurements of relevant chemical elements can be performed, in principle, with detection sensitivity better than parts per million. Consequently, such data for investigations in planetary science are feasible.

Mass spectrometers also have advantages, as they can be dedicated to remote space observation. One example is given by instrumentation for the investigation of isotopic fractionation caused by biological activity, at least at a semi-quantitative level (Tulej et al., 2011). In addition, for in situ measurements, mass spectrometric analyses of elemental and isotopic compositions can be performed using an fs-laser ablation ion source (Riedo et al., 2012, 2013b). The results of these analyses indicate that under certain conditions, the measurements of isotope abundances can be conducted with an accuracy at the per mill level for isotope concentrations higher than 100 ppm. In such cases, measurements of isotope abundances could be used in remote space observation search for biogeochemical biosignatures on Ganymede and Europa.

20.3 STABLE ISOTOPE GEOCHEMISTRY AS A GUIDE TO HABITABILITY AND BIOGENICITY

Instrumentation for the detection of life is not exclusively developed with the biomolecules of cell biology (amino acids, nucleic acids, lipids and polysaccharides) in mind. Payloads of past or current missions with astrobiologic objectives usually consider not just one, but a battery of possible signatures of life and analytical techniques, from isotopic signatures to chemical products of life metabolism. Out of the possible signatures of life, we choose to search for the effect of metabolic alteration of the environment, not as an alternative to carbon/liquid water-based life, but rather keeping in mind a less restrictive search for other life that may not be close to life as we know it. Our experience with reliable primordial microfossils that are known to be close to the root of the tree of life suggests that we should focus on biogeochemical alteration of the environment rather than search for specific molecules that we are familiar with on Earth. With all the advantages of the paleobiological record, the reliability of biosignatures is guaranteed by the well-established experimental techniques that are available, such as Raman imagery (Schopf and Kudryavtsev, 2009).

In any case, granted that the search for isotopically fractionated, S-bearing compounds may be included in the payloads of current and future missions for exploration of the solar system, even in this case additional biosignatures would still be needed to determine whether their source is due to a *Second Genesis* (McKay, 2001) – a term that has been extended beyond its original context of the emergence of life on Mars and Europa (Chela-Flores, 2009).

The remarkable advantage of searching for geochemical biosignatures is that they are more robust after death than the alternative option of searching for a statistically anomalous distribution of biotic organic molecules. To ascertain that biological organics are distinct from organic material of non-biological origin is a complex problem (McKay, 2008). While they are alive, organisms maintain their original distribution of organics that is different from distribution of organics from abiotic sources. For instance, a notable example of abiotic sources is the 70 amino acids present in the Murchison meteorite, well over the 20 amino acids found in biological proteins (Kvenvolden et al., 1970). In this sense, the biogeochemical perspective may be an attractive and more reliable complementary biosignature, or it may even be a valid alternative for the search of the emergence of life independent of searching for cellular biomolecules (Johnson et al., 2018).

Of all the biogenic elements, sulphur has the most relevant isotopic fractionation for the detection of traces of biogenic activity (Kaplan, 1975; Chela-Flores, 2006). Sulphur consists largely of two stable isotopes, ^{32}S and ^{34}S, which have a natural abundance of 95.02 per cent and 4.21 per cent, respectively, exceeding by far the abundance of other sulphur isotopes ^{33}S and S (Ryu et al., 2006). Bacterial sulphate reduction results in the depletion of ^{32}S in the remaining dissolved sulphate and the enrichment of ^{32}S in sulphide as a waste product. For this reason, it is reasonable to consider the redistribution of the primordial isotopic mixtures in terms of the following function:

$$\delta^{34}S = [(^{34}S/^{32}S)_{sa}/(^{34}S/^{32}S)_{st} - 1] \times 10^3 [‰, \mathrm{CDM}]. \tag{20.1}$$

For simplicity, the function in Eq. (20.1) will be referred to as the delta-34 parameter, or simply the delta parameter. Its value is close to zero when the sample coincides with the corresponding value of the Canyon Diablo meteorite (CDM), a troilite (FeS) that was found in the Barringer crater, north of Phoenix, Arizona. The delta parameter allows a comparison of a sample (sa) with the standard (st) CDM. The relevant terms are the dominant sulphur isotope (^{32}S) and the one next in abundance (^{34}S). In fact, $(^{34}S/^{32}S)_{st}$ coincides with the average terrestrial fraction of the two most abundant isotopes of sulphur. In a given sample we obtain a positive value of the delta-34 parameter when the quotient $(^{34}S/^{32}S)$sa exceeds the value of the standard Canyon Diablo meteorite.

There is a time-honoured and overwhelming amount of data supporting the view that metabolic pathways of sulphur bacteria have enzymes that preferentially select the isotope ^{32}S over ^{34}S (Harrison and Thode, 1957; Kaplan and Rittenberg, 1964). As pointed out previously, this will be reflected in the habitats that are depleted of ^{34}S. In other words, where sulphur microbes are present in aqueous environments, the value of the delta-^{34}S parameter will have characteristically large negative values. In addition, in the case of sulphur mass-dependent fractionations (S-MDF), advances in the understanding of the metabolism of sulphate-reducing bacteria lead to the conclusion that for the largest S-MDF, bacterial sulphate reduction can be of the order of –70‰ (Wortmann et al. 2001; Brunner et al. 2005). In the case of application biogeochemical biosignatures to Ganymede, it is worth recalling that JUICE will not measure the surface directly, but instead will measure the isotopic ratio of some elements in the exosphere (Roth et al., 2022).

It has been shown quantitatively that radiolysis and photolysis are, in fact, occurring on the surfaces of the Galilean satellites (Johnson et al., 2004). Environmental effects on Ganymede, such as radiolysis or photolysis, have to be kept in mind in our search for biosignatures. Atmospheric photochemical reactions may result in mass-independent fractionation (S-MIF) (Thamdrup, 2007). Abiotic fractionation pathways for sulphur, such as hydrolysis, may be ruled out, since only minor sulphur isotope fractionation accompanies hydrolysis (Smith, 2000).

We can also consider other external environmental effects on Ganymedean ice chemistry. They are more significant than just identifying biosignatures by producing changes in chemical elements. For instance, surficial sulphur, which is our main concern, has been considered in radiolysis and photolysis of icy satellite surfaces. Their effect can produce chemical reactions on sulphur compounds. This phenomenon, in turn, leads to other sulphur compounds that still remain on the surface (Cassidy et al., 2010). For those S compounds remaining on the icy surface, it would be possible to test for sulphur biogeochemical biosignatures.

Cycles of fractionation accompanying the bacterial disproportionation of S^0, followed by sulphide oxidation, can generate large depletions of many marine sulphides (Canfield and Thamdrup, 1994). Sulphur disproportionating bacteria (SDB) have been found to grow autotrophically, requiring not only S^0 but also basic inorganic nutrients and a sulphide scavenger (Thamdrup et al., 1993). Under the assumption that the Ganymede ocean is close to euxinicity due to a community of SDBs, biogeochemical biosignatures would then be measurable, if chemical elements were to emerge on the Ganymedean surface. But, in spite of Ganymede's thick and ancient surface, we have seen that such events cannot be ruled out (Section 20.1).

The effect of high pressure on possible biogenicity has been discussed: ocean worlds could host life under layers of high-pressure ice (Emspak, 2016), which can form in oceans at depths of around 170 kilometres, as in the case of Ganymede (Sauer et al., 2015). Given a source of heat from the planet's interior, it is not altogether impossible that under high pressures water can remain liquid under the ice long enough for life to take hold (Noack et al., 2016).

On the other hand, besides the possible effect of surficial sulphur distribution (for instance, the previously mentioned high pressure), a major factor that should be considered is external implantation of sulphur on the surfaces of the icy Galilean moons. They would be expected to produce a more uniform distribution, provided that the source was from the extraordinary volcanic activity of Io (Carlson et al., 1999, 2002). The cases of Europa and Ganymede are of mutual interest, but should be discussed separately:

(i) **Europa**. Higher spatial resolution in the ultraviolet suggests that the main role of sulphur from Io and the Jovian magnetosphere is as an agent in sputtering erosion (Fanale et al., 1999). The colouration is potentially due to radiation. Coloured materials can be endogenous or exogenous. As an endogenous input, colouration can be due to sulphur in the frequent cases of the Europa linea and mottled terrain. However, the origin of the colour of the non–water-ice materials is still under discussion (Hand and Carlson, 2015). The only remaining argument is the possible exogenous implantation in the dark trailing hemisphere.

(ii) **Ganymede**. There is Galileo evidence for a conducting water-rich layer situated deep underneath Ganymede's surface (Kivelson et al., 2002). This discovery suggests entertaining the possibility that some of Ganymede's surficial sulphur may be endogenous, as it happened on Europa. However, this challenging possibility that would help in the search for biogeochemical biosignatures requires further analysis, especially due to the differences between these two Galilean moons.

Once the materials in a primordial planetary mantle (on the Earth) or in a satellite's internal silicate nucleus (for example, on Ganymede) had entered their corresponding geochemical cycles, their initial isotope mixtures began to be redistributed. The Earth's upper mantle and crust are believed to reflect the isotopic distribution patterns of chondritic meteorites (Libby, 1971).

Models of Ganymede suggest that a type of chondrite can carry sufficient amounts of water, carbon compounds and sulphur to stand as a good model of the planetesimals, including LL chondrites that gave rise to the proto-Galilean satellites (Kuskov and Kronrod, 2001; Schubert et al., 2004; Hussmann et al., 2022). This suggests that collisions with the proto-satellite planetesimals of this composition would have carried with them sufficient amounts of water to account for an ocean. Other models of the origin of the Galilean satellites have been discussed independently (Kargel et al., 2000). Right from the very beginning of the Galileo Mission, the surfaces of Europa and other icy Galilean satellites were studied by spectroscopic means (Noll et al., 1995). Measurements have provided some evidence for various chemical elements on their surfaces including sulphur: firstly, on Europa (Table 20.2 and Carlson et al., 1999, 2002) and, secondly, on Ganymede (Ligier et al., 2019).

Sulphur fractionation is a possibility that could be tested in a significant case, as we would be in the presence of a

Table 20.2 *Possible interpretation of the non–water-ice constituents on the icy surface of Ganymede, according to the Galileo NIMS measurements.*

New absorption features (μm)	Possible non–water-ice constituents
3.50	H_2O_2
3.88	C_2H_5SH (mercaptan)
4.05	SO_2
4.25	CO_2
4.57	CN_2 (cyanogen)

From McCord et al., 1998.

large depletion of ^{34}S, if we were faced with a euxinic Ganymede ocean. This possibility has a well-understood terrestrial counterpart in the Black Sea, known to the early Romans as *Pontus Euxinus*. Its waters are consequently called 'euxinic'. Sulphur-reducing populations have been shown to be capable of turning seas such as the Black Sea into dark, sulphur-laden waters (Grice et al., 2005). Naturally occurring sulphides in sediments and in euxinic waters can be depleted in ^{34}S by as much as −70‰. Such large depletions are measurable with current miniaturized mass spectrometers, the inclusion of which is within the possibilities of payloads for future missions.

In a hypothetical Ganymedean euxinic ocean, there is a repeated cycle of sulphide oxidation to elemental sulphur, followed by a reaction in which a single compound is simultaneously oxidized and reduced, a phenomenon called disproportionation (Habicht and Canfield, 2001). These microbes can generate large fractionations that go well beyond the so-called Rees upper bound (Canfield and Thamdrup, 1994; Rees, 1973). If sulphur fractionations (by sulphate reducers and disproportionators) are present on the icy surfaces of the Galilean moons, then detecting large surficial fractionations would offer possible biosignatures, to be complemented with other data that may be available. Such biosignatures could even be traces of a putative Ganymedean euxinic ocean.

Probing the extreme hypothetical Ganymedean euxinic ocean may be within the capability of JUICE. Despite the magnetic shielding of Ganymede, its surface has spectral features resembling those on Callisto, and perhaps those on Europa (Johnson et al., 2004). Therefore, sulphur ion implantation might account for the hydrated material on the trailing side in the form of hydrated sulphuric acid. Measurements could prove, or rule out, the biogenicity implied by the euxinicity of a global Ganymede ocean, since a frozen brine has possibly emanated from a subsurface ocean (McCord and co-workers, 2001a, b, 2002). This information would add a further constraint to habitability conditions.

20.4 ARE HABITABILITY AND BIOGENICITY PROPERTIES OF GANYMEDE?

From the Voyager images, the most evident features of Ganymede were an icy surface with some dark and bright terrains. Unfortunately, the highest resolution of the Voyager images was only about 1 km per pixel for Voyager 1 and up to 500 m/pixel for Voyager 2 (Pappalardo et al., 2004). Yet, these images were only an initial step in deciphering the dark and bright terrains of the Ganymedean surface.

Even though Ganymede is a confirmed ocean world, thus having an important ingredient of habitability, much remains to be known, as it is not enough to establish this Jovian moon as a *biogenic* ocean world. In earlier times both Ganymede's surface and its interior may have been moulded by a resonance with other Galilean moons. The Voyagers attempted to give us information on this phenomenon, but sadly the data for Ganymede was lost.

To assess the likelihood of the ocean–surface interactions on Ganymede, consider that its surface appears to be ancient, since it is dominated by impact cratering and the slow decay of ice sublimating into the surrounding vacuum, leaving behind dark dust. Yet, two-thirds of Ganymede's surface was rejuvenated during a dramatic event at some point in the middle of the solar system's history by intense tectonism and possibly cryovolcanism (Collins and Johnson, 2014). Hence, in spite of its present appearance, past ocean–surface interactions are not to be excluded (cf. Section 20.1).

To understand what motivated the evolution of post-Galileo instrumentation, the presence of liquid oceans underneath the surfaces of the three Galilean icy moons is of prime importance. The orbital resonance of the inner three Galilean satellites was first pointed out in 1771 by French scientist Pierre-Simon Laplace. But it was only much later that Peale, Cassen and Reynolds realized that the Laplace resonance provided profound insight into the thermal state of Io (Peale et al., 1979). In addition, possibly more significant from the point of view of habitability, this resonance, largely responsible for internal heating, has provided an understanding of why since the early evolution of the solar system, oceans may have originated and persevered on the icy moons of the Jovian system.

These considerations lead us to the general question of habitability. Ganymede, Callisto and Europa may have oceans underneath their icy crust. In the case of the largest Galilean satellite, a subsurface ocean would be located between layers of near-surface and high-pressure ices: the fluid region could be between two ice sheets (cf. Kaoulosova et al., 2022). It was suggested that this liquid is a brine rich in $MgSO_4$ (Vance et al., 2014) and that some of the brine material reached the surface of Ganymede over the history of its evolution (McCord et al., 2001a, b). This result follows from their analysis of water absorption bands in the Galileo Near Infrared Mapping Spectrometer (NIMS) spectra for the non-ice materials that have emerged from the ocean.

The relevance of surficial contamination by sulphur-bearing materials has been discussed in more detail for Europa than for Ganymede. The presence of sulphur on Europa is patchy. For instance, sulphur is a prominent chemical element in the water-ice surface in the area north of the equatorial region, between 0° and 30° N, and between the longitudes 240° and 270°. These patches were discovered by the Galileo mission (McCord et al., 1998, 1999).

It should be underlined that the surficial distribution of non–water-ice components on Ganymede is not well understood, mainly because the Galileo NIMS data was subject to a downlink restriction, providing data only in the range of 0.7 to 5.2 μm. Sulphur is present, providing an opportunity to question whether there are biogeochemical biosignatures. The spatial distribution of the darkening surficial chemical elements is not related to the Jovian magnetospheric bombardment (Ligier et al., 2019). Well-identified chemical elements are mostly detected on bright-grooved terrains. To ascertain their identity, a better fit to the measurements is provided by sulphuric acid hydrate and salts, most likely sulphates. Endogenous processes, such as freezing of upwelling fluids going through the ice shell, may explain the distribution of chemical elements on bright-grooved terrains.

Although there are significant differences between Ganymede and Europa, it is instructive to return to Europa, where there is valuable evidence of endogenous materials emerging from the ocean lying under the surface. This is necessary in order to explain the phenomenon observed in Castalia Macula (cf. Section 20.1).

Ganymede has three main layers: a sphere of metallic iron at the centre, the core, which generates a magnetic field (Andersen et al., 1996); a spherical shell of rock (mantle) surrounding the core; and a hydrosphere. This last layer is about 800 km thick, composed of a three-layered structure. From the surface downward through the ice mantle, there are concentric shells of ice I, briny liquid water, and high pressure (HP) polymorphs ice III, V, and VI – the phases present depend on the thermal gradient (Durham et al., 1997; Hussmann et al., 2022). The bulk of this ice mantle, unlike that on Ganymede's neighbouring moon Europa, decouples the ocean from the silicate interior. The main consequence of this would seem to imply that there is little or no water interaction with silicates. Yet, this is not the complete picture (Kalousova et al., 2022).

We consider, in turn, the contact we know of an icy moon's silicate floor with the ocean above. Firstly, for Europa, evidence for endogenous salts on its surface suggests that a water-silicate is formed on its ocean, which interacts with its silicate seafloor. For assessing habitability of Europa this contact is critical (Hand and Carlson, 2015). Such a direct contact with the silicate nucleus is fundamental in origin-of-life scenarios, where hydrothermal vents at the seafloor are sources of nourishment for early microbial communities (Baross and Hoffman, 1985; Reysenbach and Cady, 2001).

Secondly, a question arises that is relevant for both habitability and biogenicity: when we consider the interface of Ganymede's silicate seafloor and the Ganymede ocean, does its ice mantle decouple the ocean from the silicate interior? The Ganymede ocean contains multiple phases of high-pressure ice. The bulk of the HP ice mantle decouples the ocean from the silicate interior, indicating little or no water interaction with silicates, unlike the preceding case of Europa. A key question in the habitability of Ganymede – the decoupling of core and water ice above – has been reviewed in this book (Kalousova et al., 2023).

However, solid-state convection in the HP ice mantle, together with meltwater generation, migration and extraction into the ocean, may provide a suitable mechanism for the exchange through this seeming barrier. In light of other recent studies of fluids within and under the ices, these HP ice shells are not a barrier to water–rock interactions (Kalousova et al., 2018). Indeed, it had been speculated that during long periods of time in this moon's history, water generated in contact with the rocky core had been transported to the subsurface ocean above (Choblet et al., 2017).

Several questions in comparative planetology are relevant for understanding the habitability of Ganymede: are there analogies on Earth that may help us understand the transport of nutrients for ice-covered oceans? Which biosignatures could facilitate life detection in extreme environments? Which would be the most robust biosignatures in such conditions?

Antarctica's McMurdo Dry Valley Lakes (DVL) have long been of interest to astrobiology (Doran et al., 2010). Biotopes are known to thrive under difficult constraints in these lakes, including living in environments that are permanently covered by ice. There are some significant differences to keep in mind, however, such as the fact that the DVLs differ from the Ganymede ocean, since in the DVLs the lake floor is not in contact with HP water ice.

However, in spite of the evident differences, the DVLs could orient us on how to plan experiments on Ganymede and Europa. They are windows to glance at analogies with biological events for ancient life that thrived in watery bodies submerged under surficial ice. For example, Lake Untersee in a dry area in East Antarctica is of special interest, as it is ice covered year-round (Andersen et al., 2011). It supports cyanobacteria in extreme environmental conditions deep inside the lake. Many of the structures on the lake bottom look like fossils in the Archean rock record from about 3 thousand million years (Gyrs) before the present (BP), because its waters harbour microbialites formed by layers of cyanobacteria.

The McMurdo DVLs discussed in the previous Section 20.2 represent only one environment for life's emergence that may throw some light on not only the habitability of Ganymede but also the hypothetical biogenicity of its subsurface ocean. Other analogues are the following:

(i) In the Antarctic there is an additional significant analogue. One example is provided by a surficial feature – Blood Falls – suggesting the presence of microbial mats underneath the Taylor glacier. The name is due to the feature's resemblance to a blood-red waterfall at the glacier's extreme end. Isotopic measurements of sulphate, water, carbonate, ferrous iron and gene analyses imply that a microbial consortium facilitates a sulphur cycle that may have analogies elsewhere in the solar system (Mikucki et al., 2009). This is especially relevant to Ganymede, which displays a variety of chemical elements on its dark terrains, as well as on Europa, where there are the Galileo's sulphur patches that have already been discussed. These stains are suggestive of chemosynthetic products of metabolism (Chela-Flores, 2010). Some of the biomolecules of the internal ocean may betray the possible biogenic interior, a daring hypothesis that is still subject to refutation by future feasible tests. Contrary to the situation on Ganymede, in Antarctica the biosignatures are easy to detect.

(ii) In the Canadian Arctic at Ellesmere Island, where sulphur patches, like those on the surface of Europa, are accumulating on glacial ice lying over saline springs rich in sulphide and sulphate (Gleeson et al., 2012), there is another major Europan analogue. Here we are dealing with a valley that has been given the name of 'Borup Fiord Pass', where minerals accumulate on glacial ice. This valley lies over saline springs rich in sulphide and sulphate that make it all the way to the icy, patchy surface. This additional traffic of surficial sulphur provides a terrestrial laboratory to test the instrumentation that may be used on a future mission to Europa or Ganymede. What has been followed up in Ellesmere Island is the process of biomineralization (Gleeson et al., 2010). This interesting biogenic process takes place when biominerals are generated as a result of interactions of microbial life and its environment. In chemosynthesis this is a well-understood process (Konhauser, 2007).

20.5 HABITABILITY AND BIOGENICITY OF GANYMEDE AND OTHER OCEAN WORLDS

In conjecturing the presence of a Ganymede oceanic population of biological sulphate reducers and disproportionators, we confront a hypothesis that is subject to experimental refutation, due to advances in miniaturizing mass spectrometers (Tulej et al., 2011). More precisely, a biotope is an area of uniform environmental conditions with a population of organisms forming a biological community. Speculations on the existence of biotopes in Ganymede or Europa are not idle exercises.

In addition, on Europa the presence of oxygen has been explained by photochemistry (Hall et al., 1995). The case of oxygen on Ganymede is also well documented. Optical reflectance measurements of Ganymede have demonstrated the presence of oxygen on its exosphere (Vidal et al., 1997) following the discovery of two weak absorption bands in the visible (at 5 773 and 6 275 Å) spectrum, but not on that of other icy satellites (Spencer et al., 1995). These are signatures of interacting pairs of O_2 molecules (double transitions in adjacent molecules). In addition, a strong absorption band was found in the near ultraviolet (UV), which suggested the presence of condensed ozone (Noll et al., 1996).

A hypothetical Ganymedean biota would have the effect of fractionating the sulphate that reaches the high-albedo regions ('bright terrains'), whose spectra were observed by the IR-Galileo spectrometer. Just as the geological record of stable sulphur isotopes is an archive of insights into Earth's history (Detmers et al., 2001), we can expect the same with the coming era of exploration of the Jupiter system.

We have raised the possibility of testing for the habitability of the icy Galilean moons with the Europa Clipper and JUICE. As explained previously, for this purpose we may take advantage of the S isotope analysis in conjunction with other approaches for planetary exploration. By these means we extend these ideas to question whether biogenesis may have occurred on Ganymede.

Rocks of Archean age (older than 2.5 Gyrs BP) provide the best evidence of early terrestrial metabolic processes. Their study allows reconstruction of the biogeochemical cycle for sulphur since the origin of life on Earth. The imprint of the possibly biogenic signature of the surficial sulphur would be retained in the dust cloud that surrounds both Europa and Ganymede (Kruger et al., 2003; Taylor et al., 2007).

In recent years we have witnessed significant progress in the miniaturization of mass spectrometers for various applications (Tulej et al., 2011; Wiesendanger et al., 2017; 2018a, b). Mass spectrometer instruments were developed specifically for space applications. Remote-sensing spectroscopic methods can be used to obtain further information in addition to what is known at present about atmospheric composition (Roth et al., 2022). On the other hand, in situ mass spectrometry provides the ability to identify a variety of atmospheric constituents across a wide range of concentrations. It is fortunate that mass spectrometry is included in mass payloads of Europa Clipper (Waite et al., 2015) and JUICE (Meyer et al., 2017). In the case of the ESA mission, implications for further insights into Ganymede will be possible.

The question of oxygen on Europa and Ganymede – one of the possible requirements for their habitability – has been discussed in the past. For this reason, much research has been devoted to exoatmospheric studies. The Hubble Space Telescope found evidence for a tenuous oxygen exoatmosphere on Ganymede (Hall et al., 1998).

The tenuous atmospheres on both Ganymede and Europa have a similar origin (Roth et al., 2022). Surficial water ice is split into hydrogen and oxygen by radiation, with the hydrogen then being lost quicker, due to its lower atomic mass.

Care has been taken to notice that there are very significant limitations in appealing to comparative planetology as a guide for the Ganymede ocean. The main reasons are due to the particular geophysical characteristics and apparent geological inactivity of Ganymede's crust at present (Pappalardo et al., 2022). Analogies are useful for aqueous geochemical approaches, but implied by the extreme pressure on Ganymede's ocean is a major difference with life that is presently thriving under the Dry Valley lakes (cf. Section 20.2).

There remain common challenges for determining the habitability of Titan, as there are for Ganymede. Titan has some similarities with the slightly larger Ganymede. For example, Titan is 5 150 km in diameter; it is composed of about 45 per cent water and 55 per cent rocky materials. In view of these parameters, Titan is in the same size class with Ganymede's 5 268 km and has an analogous composition of 60 per cent water and 40 per cent rocky materials (Greely et al. 2013; Showman and Malhotra, 1999). They are both candidate ocean worlds (Hendrix et al., 2018).

Titan may have liquid water from an impact, which could be preserved under a frozen isolation layer (Artemivia and Lunine, 2003). The case for a Ganymede ocean has been discussed extensively in this chapter. However, there are many differences between the two moons, including their atmospheres, geophysical structure and surfaces.

Biogenicity, which has been discussed for Ganymede, but for different reasons, is not excluded on Titan. As organisms on Earth combine molecular oxygen with organic compounds to get energy, organisms on Titan might induce the reaction of molecular hydrogen with organic materials, such as acetylene. Hence, on Titan methanogens could consume hydrogen, acetylene and ethane (McKay and Smith, 2005).

A second analogy is that sulphur may be present on Ganymede (cf. Section 20.3) but also on Titan. This follows from the low temperature at Titan's tropopause. In this case the maximum hydrogen sulphide (H_2S) mole fraction is expected to be present in the stratosphere of Titan from internal sources, for instance from cryovolcanism (Hickson et al., 2014). The hydrogen sulphide abundance is extremely low (about 10^{-15}), so most sulphur would be from external sources. However, on Ganymede sulphur could be endogenous (cf. Section 20.1). For Titan an analogous discussion for endogenicity of sulphur has also been raised (Dubois et al., 2020). Then, for both giant moons isotopic fractionation could eventually provide a possible test for biosignatures. At the present time, specific measurements for Titan are not feasible. Fortunately, there has been renewed interest, with the proposal of various mission concepts and related instrumentation.

Table 20.3 *Proposals for the exploration of the outer solar system, including Titan.*

Solar system bodies	Proposed missions	Proposer/Approved
Enceladus and Titan	Explorer of Enceladus and Titan	Proposed by ESA and NASA[1]
Titan	Dragonfly to Titan	Approved by NASA. Postponed until 2027 [2,3]

[1] Mitri et al., 2018.
[2] Barnes et al., 2017.
[3] Turtle et al., 2017.

All of these new efforts would add to our understanding of habitability of Titan. Consequently, clarifying comments on the present status of some of the mission proposals are timely at this stage (Table 20.3).

In addition, it should be noticed that since ELSAH was presented at the New Frontiers 4 and the New Frontiers 5 programme, this proposed mission will be able to undertake a new selection of targets. On the other hand, Explorer of Enceladus and Titan was a mission that had been proposed to the ESA M5 Cosmic Vision programme, but sadly it was not preselected.

20.6 SUMMARY AND CONCLUSIONS

Following the end of the Galileo mission, habitability was extensively reviewed (Bagenal et al., 2004). Since then there have been some changes in our understanding of habitability of Ganymede, such as new observations (2017–19) on Ganymede's atmosphere in eclipse and in transit of Jupiter from recent Hubble programs (Roth et al., 2022).

The main focus in this chapter has been on two related but different overlapping questions: habitability and biogenicity. For this purpose, biogeochemical biosignatures had already been proposed for other ocean worlds of the outer solar system beyond Ganymede (Chela-Flores, 2017).

Firstly, an ocean over a silicate nucleus (Greeley et al., 2004) makes the first ocean world – Europa – a candidate for testing biogenicity (Pappalardo et al., 2013). We have endeavoured to show that tests are possible for Ganymede's habitability, but even more, tests are possible for Ganymede's inhabitability as well. In other words, the Ganymede ocean is a second possibility for testing not only for habitability but also for the possible presence of another evolutionary pathway of autochthonous life with the approved payload of the JUICE mission. We should underline that habitability of a confirmed global ocean is not enough to establish the biogenicity candidature of Ganymede.

We have singled out a hypothetical biotope in the Ganymedean ocean – its euxinicity – that could be tested with the PEP instrument in the payload of the JUICE mission, since a frozen brine possibly has emanated from a subsurface ocean. This information would constrain the habitability conditions.

At present, however, some challenges remain for developing instrumentation that can overcome the effects of the strong radiation of the Jovian system. This is especially significant concerning the detection efficiency of the instruments to be used in the payload of the missions that are to focus on Ganymede (Tulej et al., 2016). As the exoatmosphere can be considered to be an extension of the surface of these icy worlds (Coustenis et al., 2010), some orbital probes may contribute to the search for geochemical biosignatures that could even have the subsurface ocean as their source.

Finally, tests of biogenicity on Ganymede are feasible on particles in the exoatmosphere that may arise from confirmed surficial non-water elements, which in turn could be indirect tests of the ocean chemical elements. If sulphur compounds have been biogenically processed, this phenomenon would add one possible biogeochemical biomarker that would need to be complemented with others.

ACKNOWLEDGEMENTS

The author would like to thank Dr. Christopher P. McKay for his generous relevant comments, and corrections for an earlier stage of the manuscript. His reading has greatly benefited the present chapter. We are also equally grateful for the careful, relevant and extensive reading of a second reviewer that was wisely selected by the editors of this book.

REFERENCES

Ahrens, C., Solomonidou, A., Ligier, N. et al. (2022). This book, Chapter 2.6, Physical chemistry and thermal evolution of ices at Ganymede.

Andersen, D. T., Sumner, D. Y., Hawes, I. et al. (2011). Discovery of large conical stromatolites in Lake Untersee, Antarctica. *Geobiology*, 9 (3), 280–93.

Anderson, J. D., Lau, E. L., Sjogren, W. L. et al. (1996). Gravitational constraints on the internal structure of Ganymede. *Nature*, 384(6609), 541–3.

Artemivia, N. and Lunine, J. (2003). Cratering on Titan: impact melt, ejecta, and the fate of surface organics. *Icarus,* 164(2), 471–80.

Bada, J. (1996). Europa as an analogue of an ice-covered primitive Earth. In E. Nash, ed., *The Europa Ocean Conference at San Juan Capistrano Research Institute*. San Juan Capistrano, CA, pp. 11–12.

Bagenal, F., Dowling, T. E. and McKinnon, W. B. (2004). *Jupiter: The Planet, Satellites and Magnetosphere*. Cambridge Planetary Science Series. Cambridge: Cambridge University Press.

Barabash, S., Wurz, P., Brandt, P. et al. (2013). Particle Environment Package (PEP). *European Planetary Science Congress*, 8–13 September, London, Vol. 8.

Barnes, J. W., Turtle, E. P., Trainer, M. G. and Lorenz, R. (2017). Dragonfly: exploring Titan's surface with a New Frontiers relocatable lander. *American Astronomical Society, DPS meeting #49, id.219.02.*

Baross, J. A. and Hoffman, S. E. (1985). Submarine hydrothermal vents and associated gradient environments as sites for the origin and evolution of life. *Origins of Life and Evolution of the Biosphere*, 15, 327–45.

Brockwell, T. G., Meech, K., Pickens, K. et al. (2016). The mass spectrometer for planetary exploration (MASPEX). *IEEE Aerospace Conference*, pp. 1–17, DOI: https://doi.org/10.1109/AERO.2016.7500777.

Brunner, B., Bernasconi, S. M., Kleikemper, J. and Schroth, M. H. (2005). A model for oxygen and sulfur isotope fractionation in sulfate during bacterial sulfate reduction processes. *Geochimica et Cosmochimica Acta*, 69, 4773–85.

Canfield, D. and Thamdrup, B. (1994). The production of ^{34}S-depleted sulfide during bacterial disproprtionation of elemental sulfur. *Science*, 266(5193), 1973–5.

Carlson, R. W., Anderson, M. S., Johnson, R. E. et al. (2002). Sulfuric acid production on Europa: the radiolysis of sulfur in water ice. *Icarus*, 157, 456–63.

Carlson, R. W., Johnson, R. E. and Anderson, M. S. (1999). Sulfuric acid on Europa and the radiolytic sulfur cycle. *Science*, 286, 97–9.

Cassidy, T. A., Coll, P., Raulin, F. et al. (2010). Radiolysis and photolysis of icy satellite surfaces: experiments and theory. *Space Science Reviews*, 153(1–4), 299–315.

Cassidy, T. A., Johnson, R. E. and Tucker, O. J. (2009). Trace constituents of Europa's atmosphere. *Icarus*, 201, 182–90.

Chela-Flores, J. (1996). Habitability of Europa: possible degree of evolution of Europan biota. In *The Europa Ocean Conference.* San Juan Capistrano Research Institute: San Juan Capistrano, CA, pp. 21–21a. users.ictp.it/~chelaf/ss29.html.

Chela-Flores, J. (2006). The sulphur dilemma: are there biosignatures on Europa's icy and patchy surface? *International Journal of Astrobiology*, 5(1), 17–22. users.ictp.it/~chelaf/ss238.

Chela-Flores, J. (2009). *A Second Genesis: Stepping-stones towards the Intelligibility of Nature*. Singapore: World Scientific Publishers.

Chela-Flores, J. (2010). Instrumentation for the search of habitable ecosystems in the future exploration of Europa and Ganymede. *International Journal of Astrobiology*, 9(2), 101–8. users.ictp.it/~chelaf/jcf_IJA_2010.pdf.

Chela-Flores, J. (2017). Forum article: instrumentation for testing whether the icy moons of the gas and ice giants are inhabited. *Astrobiology*, 17(10), 958–61.

Chela-Flores, J., Cicuttin, A., Crespo, M. L. and Tuniz, C. (2015). Biogeochemical fingerprints of life: earlier analogies with polar ecosystems suggest feasible instrumentation for probing the galilean moons. *International Journal of Astrobiology*, 14(3), 427–34. users.ictp.it/~chelaf/IJA2015.pdf.

Chela-Flores, J. and Kumar, N. (2008). Returning to Europa: can traces of surficial life be detected? *International Journal of Astrobiology*, 7(3–4), 263–69. users.ictp.it/~chelaf/JCFKumar.pdf.

Choblet, G., Tobie, G., Sotin, C. et al. (2017). Heat transport in the high-pressure ice mantle of large icy moons. *Icarus*, 285, 252–62.

Collins, G. and Johnson, T. V. (2014). Ganymede and Callisto. In T. Spohn, D. Breuer, and T. V. Johnson, eds., *Encyclopedia of the Solar System* (3rd. ed.), Amsterdam, Elsevier, pp. 813–29.

Coustenis, A., Tokano, T., Burger, M. H. et al. (2010). Atmospheric/exospheric characteristics of icy satellites. *Space Science Reviews*, 153, 155–84.

Dalton, J. B., Mogul, R., Kagawa, H. K. et al. (2003). Near-infrared detection of potential evidence for microscopic organisms on Europa. *Astrobiology*, 3(3), 505–29.

Detmers, J., Bruanchert, V., Habich, K. S. and Kuever, J. (2001). Diversity of sulfur isotope fractionations by sulfate-reducing prokaryotes. *Applied and Environmental Microbiology*, 67(2), 888–94.

Dodd, M. S., Papineau, D., Grenne, T. et al. (2017). Evidence for early life in Earth's oldest hydrothermal vent precipitates. *Nature*, 543(7643), 60–4.

Doran, P. T., Berry Lyons, W. and McKnight, D. M. (2010). *Life in Antarctic Deserts and Other Cold Dry Environments.* Cambridge: Cambridge University Press.

Dubois, D., Oza, A. and Radebaugh, J. (2020). The sulfur conundrum on Titan: cryovolcanism, outgassing, and chemical impact on the atmosphere. AAS Division of Planetary Science meeting #52, id. 218.04. *Bulletin of the American Astronomical Society*, Vol. 52, No. 6 e-id 2020n6i218p04.

Durham, W. B., Kirby, S. H. and Stern, L. A. (1997). Creep of water ices at planetary conditions: a compilation. *Journal of Geophysical Research*, 102, 16293–302.

Emspak, J. (2016). Ocean worlds could host life under layers of high-pressure ice, *New Scientist Magazine*, issue 3078, published 14 June 2016. www.newscientist.com/article/2093631-ocean-worlds-could-host-life-under-layers-of-high-pressure-ice/.

Fanale, F. P., Granahan, J. C., McCord, T. B. et al. (1999). *Galileo's* multiinstrument spectral view of Europa's surface composition. *Icarus*, Issue 139(E7), 179–88.

Gleeson, D., Pappalardo, R. T., Grasby, S. E. et al. (2010). Characterization of a sulfur-rich, Arctic spring site and field analog of Europa using hyperspectral data. *Remote Sensing of Environment*, 114(6), 1297–311.

Gleeson, D. F., Pappalardo, R. T., Anderson, M. S. et al. (2012). Biosignature detection at an Arctic analog to Europa, *Astrobiology*, 12(E7), 1–16.

Gowen, R. A., Smith, A., Fortes, A. D. et al. (2011). Penetrators for in situ sub-surface investigations of Europa. *Advances in Space Research*, 48(4), 725–42. users.ictp.it/~chelaf/Adv_Space_Res_3.pdf.

Grasset, O., Dougherty M. K., Coustenis, A. et al. (2013). JUpiter ICy moons Explorer (JUICE): an ESA mission to orbit Ganymede and to characterise the Jupiter system. *Planetary and Space Science*, 78, 1–21.

Greeley, R. (2013). *Planetary Geomorphology*. New York: Cambridge University Press.

Greeley, R., Chyba, C. F., Head III, J. W. et al. (2004). Geology of Europa. In F. Bagenal, T. E. Dowling, W. B. McKinnon, eds., *Jupiter: The Planet, Satellites and Magnetosphere* (Cambridge Planetary Science, Vol. 1). Cambridge: Cambridge University Press, pp. 329–62.

Grice, K., Cao, C., Love, G. D. et al. (2005). Photic zone euxinia during the Permian-Triassic superanoxic event. *Science*, 307(5710), 706–9.

Habicht, K. S. and Canfield, D. E. (2001). Isotope fractionation by sulfate-reducing natural populations and the isotopic composition of sulfide in marine sediments. *Geology*, 29(6), 555–8.

Hall, D. T., Feldman, P. D., McGrath, M. A. and Strobel, D. F. (1998). The far-ultraviolet oxygen airglow of Europa and Ganymede. *The Astrophysical Journal*, 499(1), 475–81.

Hall, D. T., Strobel, D. F., Feldman, P. D. et al. (1995). Detection of an oxygen atmosphere on Jupiter's moon Europa. *Nature*, Issue No. 373(6516), 677–81.

Hand, K. P. and Carlson, R. W. (2015). Europa's surface color suggests an ocean rich with sodium chloride. *Geophysical Research Letters*, 42(9), 3174–8.

Hand, K. P., Murray, A. E. and Garvin, J. B. (2017). *Europa Lander Study 2016, Report*. https://europa.nasa.gov/resources/58/europa-lander-study-2016-report/.

Hand, K. P., Sotin, C., Hayes, A. et al. (2020). On the habitability and future exploration of ocean worlds. *Space Science Reviews*, Issue 216(5), 95.

Harrison, A. G. and Thode, H. G. (1957). The kinetic isotope effect in the chemical reduction of sulphate. *Transactions of the Faraday Society*, 53, 1648–51.

Hendrix, A. R., Hurford, T. A., Barge, L. M. et al. (2018). The NASA roadmap to ocean worlds. *Astrobiology*, 19(1), 1–27.

Hickson, K. M., Loison, J. C., Cavalié, T. et al. (2014). The evolution of infalling sulfur species in Titan's atmosphere. *Astronomy and Astrophysics*, 572, A58, DOI: https://doi.org/10.1051/0004-6361/201424703.

Hoefs, J. (2009). *Stable Isotope Geochemistry*. Berlin: Springer.

Horvath, J., Carsey, F., Cutts, J. et al. (1997). Searching for ice and ocean biogenic activity on Europa and Earth. In Hoover, R. B., ed., *Instruments, Methods and Missions for Investigation of Extraterrestrial Microorganisms: Proceedings of the Society of Photo-Optical Instrumentation Engineers*, 3111, pp. 490–500. users.ictp.it/~chelaf/searching_for_ice.html.

Hussmann, H., Schubert, G. and Frank Sohl, F. (2022). This book, Chapter 2.1, Internal structure of Ganymede.

Hussmann, H., Sohl, F. and Spohn, T. (2006). Subsurface oceans and deep interiors of medium-sized outer planet satellites and large trans-Neptunian objects. *Icarus*, 185(1), 258–73.

Ireland, T. R. (2013). Recent developments in isotope-ratio mass spectrometry for geochemistry and cosmochemistry. *Review of Scientific Instruments*, 84(1), 011101-21.

Johnson, R. E., Burger, M. H., Cassidy, T. A. et al. (2009). Composition and detection of Europa's sputter-induced atmosphere. In R. T. Pappalardo, W. B. McKinnon and K. K. Khurana, eds., *Europa*. Tucson, AZ: University of Arizona Press, pp. 507–27.

Johnson, R. E., Carlson, R. W., Cooper, J. F. et al. (2004). Radiolysis and photolysis on the Galilean satellites. In *Jupiter*, Cambridge: Cambridge University Press.

Johnson, S. S., Anslyn, E. V., Graham, H. V. et al. (2018). Fingerprinting non-terran biosignatures. *Astrobiology*, 18(7), 1–8.

Kalousova, K., Soderlund, K. and Solomonidou, A. (2022). This book, Chapter 2.7, Ganymede's hydrosphere (HP ice mantle, liquid ocean, outer ice I crust).

Kalousova, K., Sotin, C., Choblet, G. et al. (2018). Two-phase convection in Ganymede's high-pressure ice layer – Implications for its geological evolution. *Icarus*, 299, 133–47.

Kaplan, I. R. (1975). Stable isotopes as a guide to biogeochemical processes. *Proceedings of the Royal Society of London. Series B*, Issue 189(1095), 183–211.

Kaplan, I. R. and Rittenberg, S. C. (1964). Microbiological fractionation of sulphur isotopes. *Microbiology*, 34(2), 195–212.

Kargel, J. S., Kaye, J. Z., Head, J. W., III et al. (2000). Europa's crust and ocean: origin, composition, and the prospects for life. *Icarus*, 148(1), 226–65.

Kempf, S., Altobelli, N., Briois, C. et al. (2014). SUDA: a dust mass spectrometer for compositional surface mapping for a mission to Europa. *EPSC Abstracts* 9, EPSC2014-229, European Planetary Science Congress.

Kempf, S., Sram, R., Grun, E. et al. (2012). Linear high resolution dust mass spectrometer for a mission to the Galilean satellites. *Planetary and Space Science*, 65(1), 10–20.

Khurana, K. K., Kivelson, M. G., Stevenson, D. J. et al. (1998). Induced magnetic fields as evidence for subsurface oceans in Europa and Callisto, *Nature*, 395(6704), 777–80.

Kivelson, M. (2022). This book, Chapter 3.1, Ganymede's magnetic field and magnetosphere.

Kivelson, M. G., Bagenal, F. Kurth, W. S. et al (2004). Magnetospheric interactions with satellites. In F. Bagenal, T. E. Dowling, W. B. McKinnon, eds., *Jupiter: The Planet, Satellites and Magnetosphere* (Cambridge Planetary Science, Vol. 1). Cambridge: Cambridge University Press.

Kivelson, M. G., Khurana, K. K. and Volwerk, M. (2002). The permanent and inductive magnetic moments of Ganymede. *Icarus*, 157(2), 507.

Kömle, N. I., Tiefenbacher, P., Weiss, P. and Bendiukova, A. (2018). Melting probes revisited – Ice penetration experiments under Mars surface pressure conditions. *Icarus*, 308, 117–27.

Konhauser, K. (2007). *Introduction to Geomicrobiology*. Oxford: Blackwell Publishing.

Kruger, H., Krivov, A. V., Sremcevi, M. and Grün, E. (2003). Impact-generated dust clouds surrounding the Galilean moons. *Icarus*, 164, 170–87.

Kuskov, O. L. and Kronrod, V. A. (2001). Core sizes and internal structure of the Earth's and Jupiter's satellites. *Icarus*, 151(2), 204–27.

Kvenvolden, K. A., Lawless, J., Pering, K. et al. (1970). Evidence for extraterrestrial amino-acids and hydrocarbons in the Murchison meteorite. *Nature*, 228(5275), 923–6.

Leblanc, F., Johnson, R. E. and Brown, M. E. (2002). Europa's sodium atmosphere: an ocean source? *Icarus*, 159(1), 132–44.

Libby, W. F. (1971). Terrestrial and meteorite carbon appear to have the same isotopic composition. *Proceedings of the National Academy of Sciences of the United States of America*, 68(2), 377.

Ligier, N., Paranicas, C., Carter, J., et al. (2019). Surface composition and properties of Ganymede: updates from ground-based observations with the near-infrared imaging spectrometer SINFONI/VLT/ESO. *Icarus*, 333(2), 496–515.

Lorenz, R. D and Lunine, J. L. (1996). Light and heat under the snow: Europa habitability revisited. In *The Europa Ocean Conference*. San Juan Capistrano Research Institute: San Juan Capistrano, CA, pp. 48–9.

Martynov, M. B., Merkulov, P. V., Lomakin, I. V. et al. (2017). Advanced Russian mission Laplace-P to study the planetary system of Jupiter: scientific goals, objectives, special features and mission profile. *Solar System Research*, 51(7), 555–62.

McCord, T. B., Carlson, R. W., Smythe, W. D., et al. (1997). Organics and other molecules in the surfaces of Callisto and Ganymede. *Science*, 278(5336), 271–5.

McCord, T. B., Hansen, G. B., Clark, R. N. et al. (1998). Non-water-ice constituents in the surface material of the icy Galilean satellites from the Galileo near-infrared mapping spectrometer investigation. *Journal of Geophysical Research: Planets*, 103(E4), 8603–26.

McCord, T. B., Hansen, G. B., Matson, D. L. et al. (1999). Hydrated salt minerals on Europa's surface from the Galileo near-infrared mapping spectrometer (NIMS) investigation. *Journal of Geophysical Research: Planets*, 104(E5), 11827–51.

McCord, T. B., Hansen, G. B. and Hibbitts, C. A. (2001a). Hydrated salt minerals on Ganymede's surface: evidence of an ocean below. *Science*, 292(E5), 1523–25.

McCord, T. B., Orlando, T. M., Teeter, G. et al. (2001b). Thermal and radiation stability of the hydrated salt minerals epsomite, mirabilite, and natron under Europa environmental conditions. *Journal of Geophysical Research: Planets*, 106, 3311–20.

McCord, T. B., Teeter, G., Hansen, G. B. et al. (2002). Brines exposed to Europa surface conditions. *Journal of Geophysical Research*, 107(E1), 1029.

McGrath, M. A., Lellouch, E., Strobel, D. F. et al. (2004). Satellite atmospheres. In F. Bagenal, T. E. Dowling, W. B. McKinnon, eds., *Jupiter: The Planet, Satellites and Magnetosphere* (Cambridge Planetary Science, Vol. 1). Cambridge: Cambridge University Press, pp. 457–83.

McKay, C. P. (2001). The search for a second genesis of life in our solar system. In J. Chela-Flores, T. Owen, and F. Raulin, eds., *The First Steps of Life in the Universe*. Dordrecht: Kluwer Academic Publishers, pp. 269–77.

McKay, C. P. (2008). An approach to searching for life on Mars, Europa, and Enceladus. *Space Science Reviews*, 135, 49–54.

McKay, C. P. and Smith, H. D. (2005). Possibilities for methanogenic life in liquid methane on the surface of Titan. *Icarus*, 178(1), 274–6.

Meyer, S., Tulej, M. and Wurz, P. (2017). Mass spectrometry of planetary exospheres at high relative velocity: direct comparison of open- and closed-source measurements. *Geoscientific Instrumentation, Methods and Data Systems*, 6(1), 1–8.

Mikucki, J. A., Pearson, A., Johnston, D. T. et al. (2009). A contemporary microbially maintained subglacial ferrous 'ocean'. *Science*, 324(5925), 397–400.

Mitri, G., Postberg, F., Soderblom, J. M. et al. (2018). Explorer of Enceladus and Titan (E2 T): investigating ocean worlds' evolution and habitability in the solar system. *Planetary and Space Science*, 155, 73–90.

Nash, D., ed. (1996). *The Europa Ocean Conference*. San Juan Capistrano Research Institute: San Juan Capistrano, CA.

Noack, L., Höning, D., Rivoldini, A. et al. (2016). Water-rich planets: how habitable is a water layer deeper than on Earth? *Icarus*, 277, 215–36.

Noll, K. S., Johnson, R. E., Lane, A. L., Domingue, D. L. and Weaver, H. A. (1996). Detection of ozone on Ganymede. *Science*, 273(5273), 341–3.

Noll, K .S., Weaver, H. A. and Gonnella, A. M. (1995). The albedo spectrum of Europa from 2200 Angstrom to 3300 Angstrom. *Journal of Geophysical Research*, 100(E9), 19057–9.

Pappalardo, R. T., Collins, G. C., Head, J. W., III, et al. (2004). Geology of Ganymede. In F. Bagenal, T. E. Dowling, W. B. McKinnon, eds., *Jupiter: The Planet, Satellites and Magnetosphere* (Cambridge Planetary Science, Vol. 1). Cambridge: Cambridge University Press, 363–96.

Pappalardo, R., Head, J., Greely, R. and the Galileo Solid State Imaging Team. (1996). A Europa ocean? The (circumstantial) geological evidence. *The Europa Ocean Conference.* San Juan Capistrano Research Institute: San Juan Capistrano, CA, pp. 59–60.

Pappalardo, R., McCord, T. B. and Schenk, P. M. (2022). This book. Chapter 2.2, Ganymede's geology.

Pappalardo, R. T., Vance, S., Bagenal, F. et al. (2013). Science potential from a Europa lander. *Astrobiology*, 13(8), 740–73.

Peale, S. J., Cassen, P. and Reynolds, R. T. (1979). Melting of Io by tidal dissipation. *Science*, 203, 892–4.

Phillips, C. B. and Chyba, C. F. (2001). Europa: prospects for an ocean and life. In J. Chela-Flores, T. Owen and F. Raulin, eds., *The First Steps of Life in the Universe*. Dordrecht: Kluwer Academic Publishers, pp. 25–34.

Phillips, C. B. and Pappalardo, R. T. (2014). Europa Clipper mission concept. *Eos, Transactions, American Geophysical Union*, 95(20), 165–7.

Pierazzo, E. and Chyba, C. F. (2002). Cometary delivery of biogenic elements to Europa. *Icarus*, 157(1), 120–7.

Pitesky, J. K., Hand, P., Phillips, C. B. et al. (2020). The Europa Lander mission concept: *in situ* exploration of an ocean world. AGU Fall Meeting 2020, San Francisco, P051-01.

Prockter L. and Schenk P. M. (2005) Origin and evolution of Castalia Macula, an anomalous young depression on Europa. *Icarus*, 177(2), 305–26.

Rees, C. E. (1973). A steady-state model for sulphur isotope fractionation in bacterial reduction processes. *Geochimica et Cosmochimica Acta*, 37(5), 1141–62.

Reysenbach, A. L. and Cady, S. (2001). Microbiology of ancient and modern hydrothermal systems. *Trends in Microbiology*, 9(2), 79–86.

Riedo, A., Bieler, A., Neuland, M., Tulej, M. and Wurz, P. (2013a). Performance evaluation of a miniature laser ablation time-of-flight mass spectrometer designed for in situ investigations in planetary space research. *Journal of Mass Spectrometry*, 48(1), 1–15.

Riedo, A., Meyer, S., Heredia, B. et al. (2012). Highly accurate isotope composition measurements by a miniature laser ablation mass spectrometer designed for in situ investigations on planetary surfaces. *Planetary and Space Science*, 87, 1–13.

Riedo, A., Neuland, M., Meyer, S., Tulej, M. and Wurz, P. (2013b). Coupling of LMS with fs-laser ablation ion source: elemental and isotope composition measurements. *Journal of Analytical Atomic Spectrometry*, 28(8), 1256–69.

Romero-Wolf, A., Vance, S., Maiwald, F. et al. (2015). A passive probe for subsurface oceans and liquid water in Jupiter's icy moons. *Icarus*, 248, 463–77.

Roth, L., Plainaki, C., Wurz, P. et al. (2022). This book, Chapter 3.3, Ganymede's tenuous atmosphere.

Roth, L., Saur, J., Retherford, K. D. et al. (2014). Transient water vapor at Europa's south pole. *Science*, 343(6167), 171–4.

Ryu, J., Zierenbergb, R. A. and Dahlgrena, R. A. (2006). Sulfur biogeochemistry and isotopic fractionation in shallow groundwater and sediments of Owens Dry Lake, California. *Chemical Geology*, 229(4), 257–2.

Sauer, J., Duling, S., Roth, L. et al. (2015). The search for a subsurface ocean in Ganymede with Hubble Space Telescope observations of its auroral ovals. *Journal of Geophysical Research: Space Physics*, DOI: https://doi.org/10.1002/2014JA020778.

Schopf, J. W. (1999). *Cradle of Life: The Discovery of Earth's Earliest Fossils*. Princeton: Princeton University Press.

Schopf, J. W. and Kudryavtsev, A. B. (2009). Confocal laser scanning microscopy and Raman imagery of ancient microscopic fossils. *Precambrian Research*, 173(1–4), 39–49.

Schubert, G., Anderson, J. D., Spohn, T. and McKinnon, W. B. (2004). Interior composition, structure and dynamics of the Galilean satellites. In F. Bagenal, T. E. Dowling and W. B. McKinnon, eds., *Jupiter: The Planet, Satellites and Magnetosphere* (Cambridge Planetary Science, Vol. 1). Cambridge: Cambridge University Press, 281–306.

Shen, Y. and Buick, R. (2004). The antiquity of microbial sulfate reduction. *Earth-Science Reviews*, 64(3–4), 243–72.

Shen, Y., Buick, R. and Canfield, D. E. (2001). Isotopic evidence for microbial sulphate reduction in the early Archaean era. *Nature*, 410(6824), 77–81.

Shoemaker, E. M. (1996). The age of Europa's surface. In *The Europa Ocean Conference.* San Juan Capistrano Research Institute: San Juan Capistrano, CA, pp. 65–6. users.ictp.it/~chelaf/Shoemaker.pdf.

Showman, A. P. and Malhotra, R. (1999). The Galilean satellites. *Science*, 286(5437), 77–84.

Sim, M. S., Bosak, T. and Ono, S. (2011). Large sulfur isotope fractionation does not require disproportionation. *Science*, 333(6038), 74–7.

Smith, J. W. (2000). Isotopic fractionations accompanying sulfur hydrolysis, *Geochemical Journal*, 34, 95–9.

Spencer, J. R., Calvin, W. M. and Person, M. J. (1995). CCD spectra of the Galilean satellites: molecular oxygen on Ganymede. *Journal of Geophysical Research*, 100, 19049–56.

Spohn, T. and Schubert, G. (2003). Oceans in the icy Galilean satellites of Jupiter? *Icarus*, 161(2), 456–67.

Squyres, S. W. (1996). What we thought we 'knew' about Europa. In *The Europa Ocean Conference.* San Juan Capistrano Research Institute: San Juan Capistrano, CA, pp. 67–8.

Taylor, E. A., Ball, A. J., Barber, S. J. et al. (2007). A combined dust impact detector and ion trap mass spectrometer for a Europa orbiter. *Geophysical Research Abstracts*, 9, 10928, SRef-ID: 1607-7962/gra/EGU2007-A-10928, European Geosciences Union.

Thamdrup, B. (2007). New players in an ancient cycle. *Science*, 317(1), 1508–9.

Thamdrup, B., Finster, K., Hansen, J. W. and Bak, F. (1993). *Applied Environmental Microbiology*, 59(5844), 101.

Trowell, S., Wild, J., Horvath, J. et al. (1996). Through the Europan ice: advanced lander mission options In *The Europa Ocean Conference*. San Juan Capistrano Research Institute: San Juan Capistrano, CA, p. 76. users.ictp.it/~chelaf/Trowell1996.pdf.

Tulej, M., Iakovleva, M., Leya, I. and Wurz, P. (2011). A miniature mass analyser for in-situ elemental analysis of planetary material–performance studies. *Analytical and Bioanalytical Chemistry*, 399(6), 2185–200.

Tulej, M., Wurz, P., Meyer, S., et al. (2016). Experimental investigation of the radiation shielding efficiency of a MCP detector in the radiation environment near Jupiter's Moon Europa: nuclear instruments and methods. *Physics Research Section B*, 383, 21–37.

Turtle, E. P., Barnes, J. W., Trainer, M. G. et al. (2017). Dragonfly: exploring Titan's prebiotic organic chemistry and habitability. *Lunar and Planetary Science Conference.*

Vance, S., Bouffard, M., Choukroun, M. and Sotin, C. (2014). Ganymede's internal structure including thermodynamics of magnesium sulfate oceans in contact with ice. *Planetary and Space Science*, 96, 62–70.

Vidal, R. A., Bahr, D., Baragiola, R. A. and Peters, M. (1997). Oxygen on Ganymede: laboratory studies. *Science*, 276(5320), 1839–42.

Waite, J. H., Jr., Brockwell, T., McGrath, M. A. et al. (2015). MASPEX Europa. *American Geophysical Union Fall Meeting*, Abstract P13E-06.

Wiesendanger, R., Tulej, M., Riedo, A. et al. (2017). Improved detection sensitivity for heavy trace elements using a miniature laser ablation ionisation mass spectrometer. *Journal of Analytical Atomic Spectrometry*, 32(11), 2182–8.

Wiesendanger, R., Tulej, M., Grimaudo, V. et al. (2018a). A method for improvement of mass resolution and isotope accuracy for laser ablation time-of-flight mass spectrometers. *Journal of Chemometrics*, DOI: https://doi.org/10.1002/cem.3081.

Wiesendanger, R., Wacey, D., Tulej, M. et al. (2018b). Chemical and optical identification of micrometer-sized 1.9 billion-year-old fossils by combining a miniature laser ablation ionization mass spectrometry system with an optical microscope. *Astrobiology*, 18(8), 1071–80.

Wortmann, U. G., Bernasconi, S. M. and Bottcher, M. E. (2001). Hypersulfidic deep biosphere indicates extreme sulfur isotope fractionation during single-step microbial sulfate reduction. *Geology*, 29(7), 647–50.

Wurz, P., Meyer, S., Galli, A. et al. (2018). The Neutral Gas and Ion Mass spectrometer of the PEP experiment on the JUICE mission. *EGU General Assembly Conference*, 20, Abstract 10091.

Appendix A

Maps of Ganymede

THOMAS ROATSCH, ELKE KERSTEN, GEOFFREY COLLINS, AND WES PATTERSON

A.1 INTRODUCTION

In 1979 Voyager 1 and 2 arrived at Jupiter to acquire 490 Narrow Angle Camera (NAC) and Wide Angle Camera (WAC) images of Ganymede's surface with resolutions from 470 m/pxl to 20 km/pxl. The Galileo spacecraft, with its Solid State Imaging (SSI) camera onboard, entered orbit around Jupiter in 1995 and took 149 images (<20 km/pxl) of Ganymede during 15 flybys. Taken together these images cover almost the entire surface of the Jovian moon. Today these images are the foundation of our understanding of the geological processes that formed Ganymede's surface, albedo changes across the surface, crater distribution, and much more. Later observations of Ganymede during the Cassini flyby in 2000 (Porco et al., 2003) and during the Juno orbits around Jupiter (Mura et al., 2013) did not reach high enough resolution to improve the global imaging coverage.

A.2 GLOBAL MOSAICS

The first regional mosaics of Ganymede with a resolution of about 1 km/pxl were assembled soon after the Voyager flybys (Smith et al., 1979). The first controlled global mosaic with combined imaging from Voyager and Galileo was published by Becker et al. (2001). This mosaic has a resolution of 1 km/pxl and was the standard mosaic for many years. Schenk (2010) also created a global mosaic of Ganymede with a similar resolution and published it in the Atlas of the Galilean Satellites where he also included global color mosaics for the first time.

Kersten et al. (2021) recently calculated a new controlled global mosaic of Ganymede with a resolution of 359 m/pxl using the best available Voyager and Galileo images. During the stepwise mosaicking process images with resolutions better than 100 m/pxl to worse than 10 km/pxl have been integrated. The major difference between the mosaics with 1 km/pxl spatial resolution and this one is the integration of very high-resolution Galileo images at some areas. The final mosaic is shown in different projections in Figures A.1 and A.2. The full resolution mosaic is available at the DLR JANUS web page in GeoTIFF format: https://janus.dlr.de, and at the ESA Guest Storage Facility: https://www.cosmos.esa.int/web/psa/dlr_ganymede_voyager-ga-lileo-_v1.0.

All named mosaics used the rotational parameters as defined by International Astronomical Union (IAU) in Archinal et al. (2011). Crater Anat defining the longitude system was always rectified to 232° East (https://planetarynames.wr.usgs.gov/Feature/251).

A.3 GEOLOGICAL MAPPING

Imaging data from the Voyager flybys covered two separated regions on opposite sides of Ganymede, showing similar types of terrain features on the subjovian and antijovian hemispheres. The regionally separated nature of the Voyager data led to an effort to produce several regional geological maps at the 1:5M scale (Guest et al., 1988; Murchie and Head, 1989; Croft et al., 1990; Lucchitta et al., 1992; De Hon et al., 1994; Wagner et al., 1995; Wilhelms, 1990; Underwood et al., 1997). All of the regional maps took the approach of separating Ganymede into two main categories of geologic units, light material and dark material, but the details of the subcategories within light and dark material varied by region.

Galileo imaging data filled in most of the gaps in the Voyager data set, resulting in 94% coverage at better than 3.5 km/pxl. This global data set allowed the construction of the first global map of light and dark material units (Collins et al., 2013), as well as a global survey of large craters, though the image resolution over most of the trailing hemisphere and parts of the leading hemisphere are still too poor to distinguish most tectonic features and material subunits in these areas.

The approach to mapping the global geology of Ganymede (Patterson et al., 2010) was similar to that used for other planetary global maps (e.g., Tanaka et al., 1994; Greeley et al., 2000), where map units are defined as those that can be distinguished and delimited on the basis of observable physical features (Wilhelms, 1990). This enables the maximum amount of geological information to be clearly conveyed by the map, with a minimum of confusion and extraneous symbols. The global map of Ganymede also adopted the convention started by Shoemaker et al. (1982) of mapping light material unit boundaries that enclose unified patterns of grooves that are distinct from neighboring light materials. High-resolution Galileo images have revealed that the boundaries between adjacent polygons of light material generally occur at the edge of a group of crosscutting faults, and do not represent the transition to a compositionally distinct material.

For the global geologic map of Ganymede (Fig. A.3), dark material has been subdivided into three units cratered, lineated, and undivided while light material has been subdivided into four units – grooved, subdued, irregular, and undivided. Two other basic geologic units are also recognized reticulate material and impact material. Impact material encompasses crater, palimpsest, and basin materials. Palimpsest material is further subdivided into four units: three of these are distinguished by their stratigraphic relationship with light material units and the fourth is an interior plains unit. Five units comprise the crater

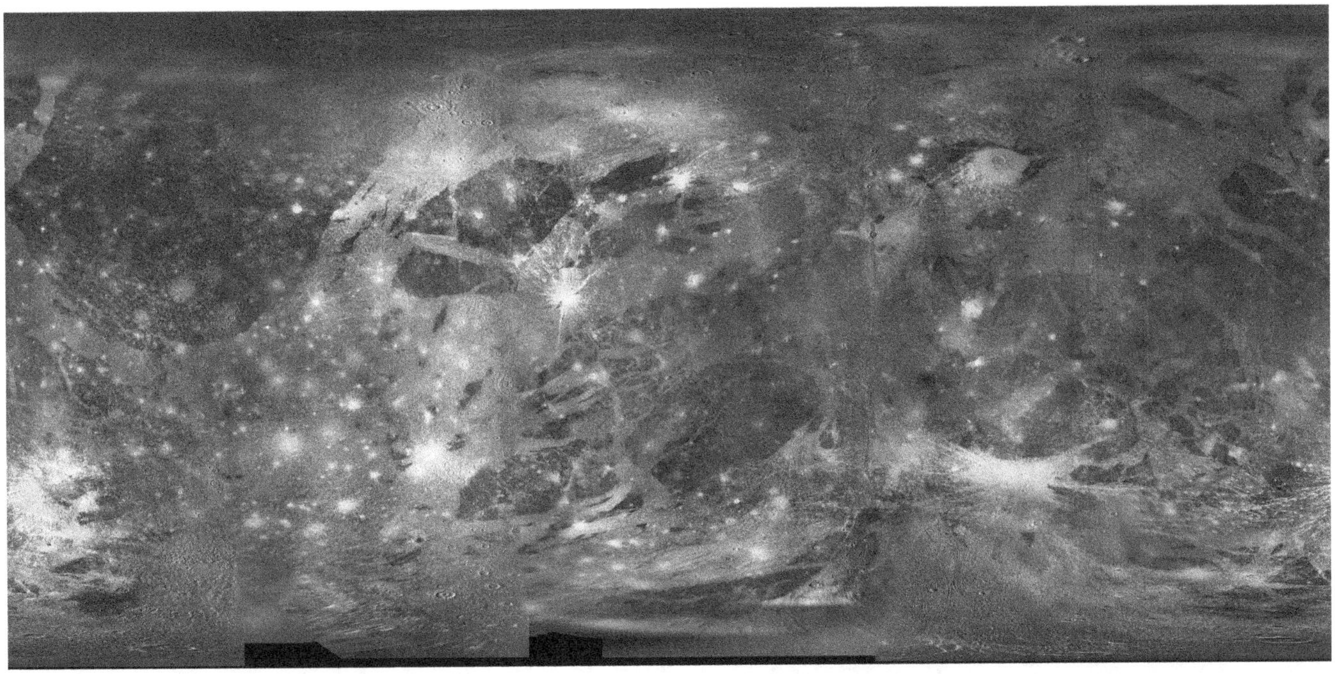

Figure A.1 Global Ganymede mosaic from Kersten et al. (2021). The mosaic is in equidistant projection centered at 0° longitude.

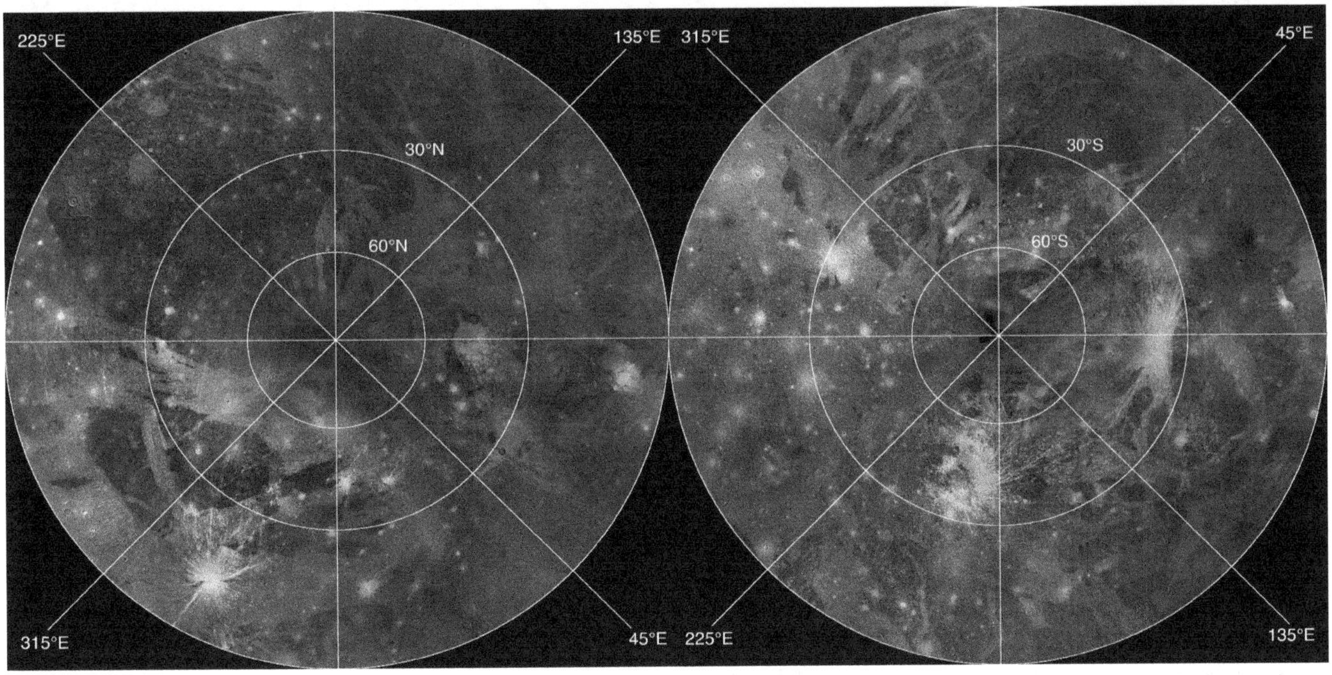

Figure A.2 Global Ganymede mosaic from Kersten et al. (2021). The mosaic is in stereographic projection centered at 90°/−90°.

materials – fresh, partially degraded, degraded, unclassified, and ejecta – and two units comprise the basin material – rugged and smooth. A key to these unit types is provided in Figure A.3 see Collins et al. (2013) for full unit descriptions.

The global correlation of map units in the Collins et al. (2013) geological map was systematized into three geologic ages for Ganymede. The earliest age, the Nicholsonian, is characterized by the formation of dark terrain, furrow systems, and most palimpsests. The middle age, the Harpagian, is defined by the formation of light material covering most of the surface by tectonically breaking up and/or cryovolcanically flooding the dark material. The most recent age, the Gilgameshan, comes after the formation of light material and is characterized by the formation of impact craters and the Gilgamesh basin.

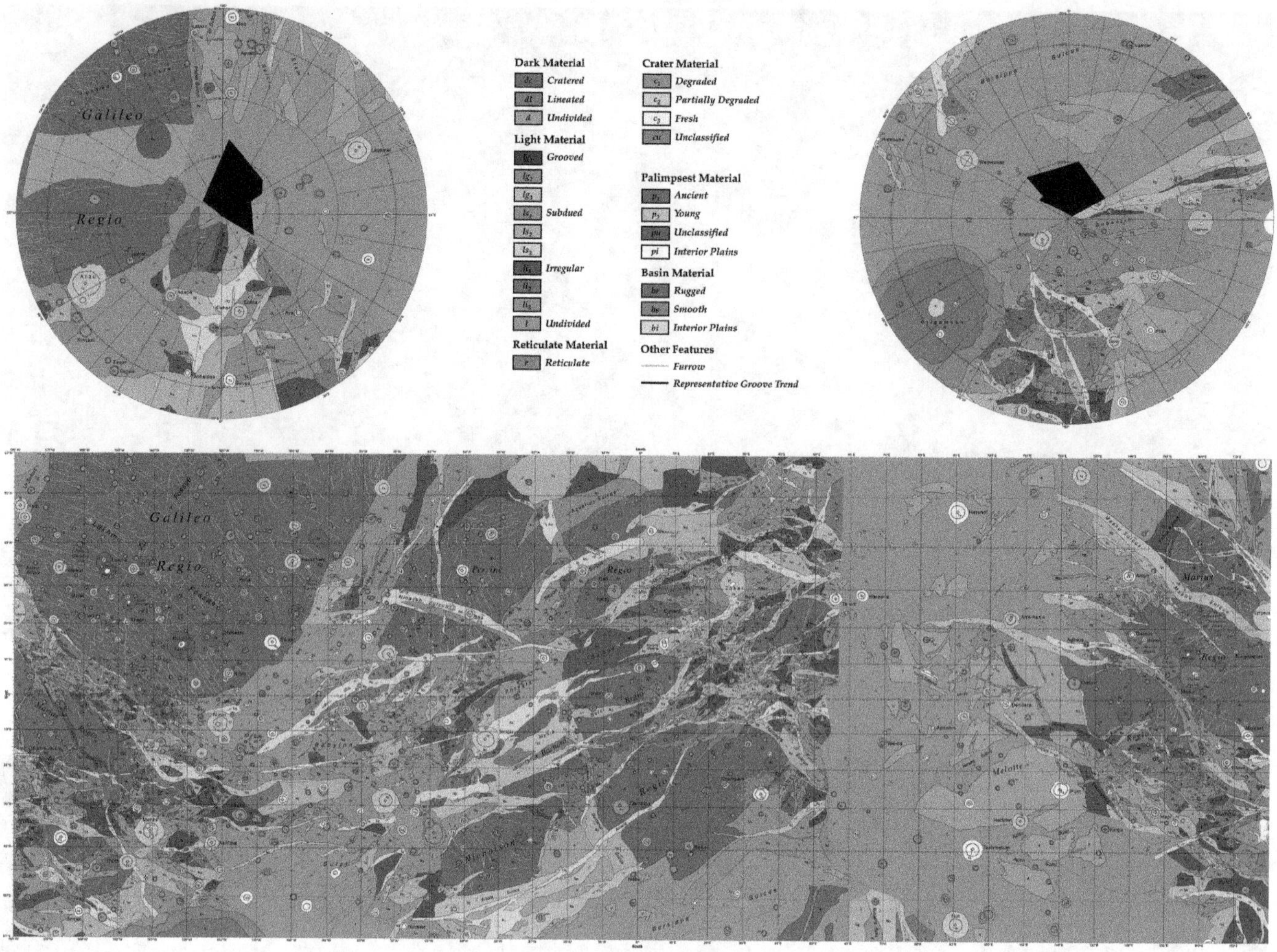

Figure A.3 Geological map of Ganymede. (A black and white version of this figure will appear in some formats. For the color version, please refer to the plate section.)

A.4 OUTLOOK

In the early 2030s the JUpiter ICy moons Explorer (JUICE) will enter the Jovian system to investigate Jupiter and three of its large moons, Ganymede, Europa, and Callisto. The mission will culminate in an eight-month orbital tour around Ganymede (Grasset et al., 2013). In this phase the onboard JANUS camera will acquire high-resolution global coverage of Ganymede's surface with spatial resolutions of 400 to 70 m during the GCO5000 (Ganymede Circular Orbit with an altitude of about 5000 km) in four different filters. This will allow the calculation of new global mosaics in these four filters with a resolution of at least 400 m/pxl.

REFERENCES

Archinal, B. A., A'Hearn, M. F., Bowell, E., Conrad, A., Consolmagno, G. J., Courtin, R., Fukushima, T., Hestroffer, D., Hilton, J. L., Krasinsky, G. A., Neumann, G., Oberst, J., Seidelmann, P. K., Stooke, P., Tholen, D. J., Thomas, P. C., and Williams, I. P. [2011], 'Report of the IAU Working Group on Cartographic Coordinates and Rotational Elements: 2009', *Celest. Mech. Dyn. Astr.* Vol. 109, 101–135, doi:10.1007/s10569-010-9320-4.

Becker, T. L., Archinal, B. A., Colvin, T., Davies, M., Gitlin, A., Kirk, R. L., and Weller, L. [2001], 'Final Digital Global Maps of Ganymede, Europa, and Callisto', 32nd Lunar and Planetary Science Conference, Lunar and Planetary Institute, Houston, TX.

Collins, G. C., Patterson, G. W., Head, J. W., Pappalardo, R. T., Prockter, L. M., Lucchitta, B. K., and Kay, J. P. [2013], 'Global Geological Map of Ganymede', *US Geol. Surv. Map* 3237.

Croft, S., Cassachia, R., and Strom, R. [1990], 'Geologic Map of the Tiamat Sulcus (Jg-9) Quadrangle of Ganymede', *US Geol. Surv. Map* I-1548.

De Hon, R. A., Leith, A. C., and McKinnon, W. B. [1994], 'Geologic Map of the Hathor Region (Jg-15) of Ganymede', *US Geol. Surv. Map* I-2388.

Grasset, O., Dougherty, M. K., Coustenis, A., Bunce, E. J., Erd, C., Titov, D., Blanc, M., Coates, A., Drossart, P., Fletcher, L. N., Hussmann, H., Jaumann, R., Krupp, N., Lebreton, J.-P., Prieto-Ballesteros, O., Tortora, P., Tosi, F., and Van Hoolst, T. [2013]. 'JUpiter ICy Moons Explorer (JUICE): An ESA Mission to Orbit Ganymede and to Characterise the Jupiter System', *Planetary and Space Science*, Vol. 78, 1–21, doi:10.1016/S0032063312003777.

Greeley, R., Figueredo, P. H., Williams, D. A., Chuang, F. C., Klemaszewski, J. E., Kadel, S. D., Prockter, L. M., Pappalardo, R. T., Head, J. W., Collins, G. C., Spaun, N. A., Sullivan, R. J., Moore, J. M., Senske, D. A., Tufts, B. R., Johnson, T. V., Belton, M. J. S., and Tanaka, K. L. [2000], 'Geologic Mapping of Europa', *J. Geophys. Res.*, Vol. 105, 22559–22578, doi:10.1029/1999JE001173.

Guest, J. E., Bianchi, R., and Greeley, R. [1988], 'Geologic Map of the Uruk Sulcus (Jg-8) Quadrangle of Ganymede', *US Geol. Surv. Map* I-1934.

Kersten, E., Zubarev, A. E., Roatsch, Th., and Matz, K.-D. [2021], 'Controlled Global Ganymede Mosaic from Voyager and Galileo Images', *Planetary and Space Science*, Vol. 206, 105310, doi:10.1016/j.pss.2021.105310.

Lucchitta, B. K., Barnes, C. W., Glotfelty, M. F. [1992], 'Geologic Map of the Memphis Facula (Jg-7) Quadrangle of Ganymede', *US Geol. Surv. Map* I-2289.

Mura, A., Adriani, A., Sordini, R., Sindoni, G., Plainaki, C., Tosi, F., Filacchione, G., Bolton, S., Zambon, F., Hansen, C. J., Ciarniello, M., Brooks, S., Piccioni, G., Grassi, D., Altieri, F., Migliorini, A., Moriconi, M. L., Noschese, R., and Cicchetti, A. [2020], 'Infrared Observations of Ganymede from Juno/JIRAM', *JGR Planets*, Vol. 125, Issue 12, doi:10.1029/2020JE006508.

Murchie, S. L., and Head, J. W. [1989], 'Geologic Map of the Philus Sulcus (Jg-4) Quadrangle of Ganymede', *US Geol. Surv. Map* I-1966.

Patterson, G. W., Collins G. C., Head, J. W., Pappalardo, R. T., Prockter, L. M., Lucchitta, B. K., and Kay, J. P. [2010], 'Global Geological Mapping of Ganymede', *Icarus*, Vol. 207, 845–867.

Porco, C. C., West, R. A, McEwen, A., Del Genio, A. D., Ingersoll, A. P., Thomas, P., Squyres, S., Dones, L., Murray, C. D., Johnson, T. V., Burns, J. A., Brahic, A., Neukum, G., Veverka, J., Barbara, J. M., Denk, T., Evans, M., Ferrier, J. J., Geissler, P., Helfenstein, P., Roatsch, Th., Throop, H., Tiscareno, M., and Vasavada, A. R. [2003], 'Cassini Imaging of Jupiter's Atmosphere, Satellites, and Rings', *Science* Vol. 299, Issue 5612, 1541–1547, doi:10.1126/science.1079462.

Schenk, P. [2010], *Atlas of the Galilean Satellites*, Cambridge University Press, Cambridge.

Shoemaker, E. M. and Wolfe, R. F. [1982], 'Cratering Time Scales for the Galilean Satellites', in D. Morrison (ed.), *Satellites of Jupiter*, University of Arizona Press, Tucson, pp. 277–339.

Smith, B.A., Soderblom, L. A., Johnson, T. V., Ingersoll, A. P., Collins, S. A., Shoemaker, E. M., Hunt, G. E., Masursky, H., Carr, M. H., Davies, M. E., Cook 2nd, A. F., Boyce, J., Danielson, G. E., Owen, T., Sagan, C., Beebe, R. F., Veverka, J., Strom, R. G., McCauley, J. F., Morrison, D., Briggs, G. A., and Suomi, V. E. [1979], 'The Jupiter System through the Eyes of Voyager 1', *Science*, Vol. 204, No. 4396, 951–972. doi:10.1126/science.204.4396.951.

Tanaka, K. L., Moore, H. J., Schaber, G. G., Chapman, M. G., Stofan, E. R., Campbell, D. B., Davis, P. A., Guest, J. E., McGill, G. E., Rogers, P. G., Saunders, R. S., and Zimbelman, J. R. [1994], *The Venus Geologic Mappers' Handbook*, USGS Open File Report 94-438.

Underwood, J. R., Casacchia, R., Woronow, A., and Teeling, M. J. [1997], 'Geologic Map of the Galileo Regio (Jg-3) Quadrangle of Ganymede', *US Geol. Surv. Map* I-2534.

Wagner, R. J., Jaumann, R., and Neukum, G. P. [1995], 'Geologic Map of the Etana Region (Jg-1) of Ganymede', *US Geol. Surv. Map* I-2497.

Wilhelms, D. E. [1990], 'Geologic Mapping', in R. Greeley and R. M. Batson (eds.), *Planetary Mapping*, Cambridge University Press, New York, pp. 208–260.

Wilhelms, D. E. [1997], 'Geologic Map of the Osiris (Jg-12) and Apsu (Jg-13) Quadrangles of Ganymede', *US Geol. Surv. Map* I-2422.

Appendix B

Ganymede Parameters

In this appendix we have gathered together (and slightly adapted) most tables from the chapters in this book, to give quick access to information about Ganymede. The order of the tables does not follow the order of the chapters in the book, but are bundled in certain categories, where a table from Chapter 11 "Physical Chemistry and Thermal Properties of Ices at Ganymede" by Ahrens et al. can be followed by on from Chapter 15 "Interactions between the Space Environment and Ganymede's Surface" by Galli et al., because they both give information about the icy surface.

Most of the references in the captions have been taken out to save some space, and the original number of the table has been put into the caption in order for the reader to know where to find the details of said table.

B.1 GANYMEDE'S PHYSICAL PARAMETERS AND GALILEO'S FLYBYS

Table B.1 *Ganymede's physical parameters*

Name	Ganymede (Jupiter III)
Discovery	7 January 1610 by Galileo Galilei (but see Chapter 1)
Mass of Ganymede M_G	1.4818×10^{23} kg
Mean radius R_G	2631.2 ± 1.7 km
Mean density ρ_G	1942.0 ± 4.8 kg/m^3
Surface Gravity	1.428 m/s^2 (0.146 g)
Escape Velocity	2.741 km/s
Moment of inertia	0.3115 ± 0.0028
J_2	$(127.53 \pm 2.9) \times 10^{-6}$
C_{22}	$(38.26 \pm 0.87) \times 10^{-6}$
Fluid Love-number k_f	0.804 ± 0.018
Semi-major axis	1,070,400 km / 14.97 R_J
Excentricity	0.0013
Orbital Period	7.155 days
Inclination	0.20° to Jupiters equator 2.214° to ecliptic
Geometric Albedo	0.44
Mass of Jupiter M_J	1.8986×10^{27} kg
grav. constant G	6.67259×10^{-11} m^{3}kg^{-1}s^{-2}
GM_G	9887.83 ± 0.03 km^{3}s^{-2}

Table B.2 *Galileo's encounters with Ganymede (Chapter 14, Kivelson et al.)*

Pass	Date UT	Dist. Jupiter (RJ)	LT (hours)	Jupiter Latitude	SIII east long	Alt. at C/A	Gn Lat.	GN Lon.	Region
G1	06/27/96	14.9	11.2	0.13	185.1	838.0 km	30.4	246.7	Central
	06:29:07	15.2	11/3						Wake
G2	09/06/96	14.8	10.7	0.19	202.0	264.4 km	79.3	236.4	Polar
	06:29:07	15.2	10.9						
G7	04/05/97	14.8	19.7	0.23	340.2	3104.9 km	55.8	270.4	Central
	07:09:58	15.2	19.8						Wake
G8	05/07/97	14.8	8.0	0.07	73.3	1606.2 km	28.3	84.8	Upstream
	15:56:10	15.1	8.1						
G28	05/20/00	14.7	0.7	–0.10	8.22	808.7 km	–19.0	92.4	Upstream
	10:10:10	15.3	0.9						
G29	12/28/00	14.7	23.9	0.23	139.5	2337.5 km	62.2	269.0	Polar
	08:25:27	15.3	30.0						
C9G	06/26/97	15.16	8.04	–0.16	300.1	30.3 RG	–0.0	261.2	GN Dist.
	16:00 – 19:00					79,750 km			Wake
E12G	12/15/97	15.01	6.75	–0.12	349.8	5.47 RG	–5.8	266.1	GN Dist.
	08:02 – 12:00					14,300 km			Wake

B.2 GANYMEDE'S ROTATION

Table B.3 *Libration forcing and response amplitudes for an interior model consisting of three layers of homogeneous density. Periods are given in years (a) or days (d). (Chapter 3, Stark et al.)*

Period	Phase	Forcing amplitude	Response amplitude	Response amplitude at equator	Difference forcing and response
575.34359 a	159.808°	0.00309°	0.00309°	141.9 m	0.00000°
137.71359 a	127.748°	0.00280°	0.00280°	128.9 m	0.00000°
30.90433 a	340.031°	0.00078°	0.00078°	35.7 m	0.00000°
26.39823 a	326.086°	0.00089°	0.00089°	40.8 m	0.00000°
11.86294 a	122.630°	0.01324°	0.01325°	609.3 m	0.00001°
5.63872 a	289.313°	0.00097°	0.00098°	44.9 m	0.00000°
1.32951 a	265.581°	0.00547°	0.00594°	272.9 m	0.00046°
1.31980 a	67.453°	0.01048°	0.01138°	523.3 m	0.00090°
1.26630 a	282.887°	0.00867°	0.00952°	437.8 m	0.00085°
50.14182 d	190.809°	0.00083°	0.00489°	225.0 m	0.00406°
12.52325 d	217.225°	0.00540°	−0.00037°	17.2 m	−0.00503°
7.16008 d	36.203°	0.00082°	< 1″	<12.8 m	-
7.15559 d	254.678°	0.16379°	−0.00352°	162.0 m	−0.16027°
7.15481 d	56.549°	0.08838°	−0.00190°	87.4 m	−0.08648°
7.05093 d	142.132°	0.06893°	−0.00144°	66.1 m	−0.06749°
6.94928 d	209.587°	0.00081°	< 1″	<12.8 m	-
6.26162 d	74.436°	0.01845°	−0.00030°	13.9 m	−0.01815°
4.17442 d	111.597°	0.00130°	< 1″	<12.8 m	-
3.52546 d	104.268°	0.00108°	< 1″	<12.8 m	-
2.35031 d	66.341°	0.00125°	< 1″	<12.8 m	-

Table B.4 *Amplitudes, frequencies, periods, and phases (phases at the J1950.0 epoch) of the orbit precession of Ganymede with respect to its Laplace plane. The Laplace plane has a node of* 138.27696° *and an inclination of* 0.0977821° *with respect to the equatorial plane of Jupiter (Chapter 3, Stark et al.)*

j	i_j (deg)	$\dot{\Omega}_j$ (rad/year)	Period (years)	γ_j (deg)	ε_j (deg)	ε_j^{el} for $k_2 = 0.067$ (deg)	ε_j^{el} for $k_2 = 0.5$ (deg)
1	0.18258	−0.0456245	−137.715	52.9125	0.03096	0.03340	0.06812
2	0.04026	−0.0111624	−562.888	203.898	0.00148	0.00158	0.00287
3	−0.01652	−0.207903	−30.2218	−78.2478	−0.03217	−0.03944	0.08596
4	0.00180	1.05936	5.93112	180.311	−0.00139	−0.00141	−0.00156
5	0.00029	−0.529633	−11.8633	−121.96	−0.00071	−0.00065	−0.00042
6	0.00023	0.529663	11.8626	−37.1349	−0.00015	−0.00015	−0.00018
7	0.00021	1.58901	3.95416	122.814	−0.00017	−0.00017	−0.00019
8	0.00016	−9.22056	−0.681432	−89.6335	−0.00016	−0.00016	−0.00016

B.3 CRATERING ON GANYMEDE

Table B.5 *Summary of Ganymede crater dimension scaling trends (Chapter 8, Kirchoff et al.)*

Dimension	Equation/ value	Diameter Range
Depth d	$d = 0.2D$	< 1.9(0.5) km
	$d = 0.227D^{0.446}$	5–50 km
Rim height H_r	$H_r = 0.021D^{1.22}$	< 9 km
	$H_r = 0.032D^{0.72}$	9–51 km
Wall Slope S	$S = 173.41(d/D)^{0.84}$	4–50 km
Continuous ejecta radius e_R	$E_r = 2.74D^{0.793}$	
Largest secondary crater diameter (D_s) about a primary (D)	$D_s = 0.12D^{0.805}$	
Central peak height H_{cp}	$H_{cp} = 0.0004D^{2.557}$	< 15 km
	$H_{cp} = 0.0018D^{1.54}$	12–50 km
Central pit Diameter D_p	$D_p = 0.005D^{0.89}$	4–150 km
Central pit depth d_p	$d_p = 0.0016D^{1.42}$	< 100 km
Dome width D_d	$D_d = 0.1D$	50–70 km
	$D_d = 0.25D$	> 70 km
Dome Height H_d	$H_d = 0.025(Dd)^{1.2}$	Dome widths of 5–30 km

Table B.6 *Measured cumulative crater density on Ganymede (Chapter 8, Kirchoff et al.)*

Terrain	$N(D > 8\text{km})$ per 10^6 km^2	$N(D > 30\text{km})$ per 10^6 km^2
Bright Terrain	62 ± 57	8 ± 7
Galileo Regio	246 ± 7	25 ± 2
Perrine Regio	764 ± 38	37 ± 7
Marius Regio	964 ± 84	28 ± 13
Nicholson Regio	512 ± 114	26 ± 26
Nicholson Regio at	25000 ± 800	
high resolution $N(D > 1\text{km})$	10000 ± 830	

Table B.7 *Absolute Model Ages (AMAs) computed by the given references for various terrains on Ganymede. Outer solar system models are identified by S&W82, Z98, and Z03; lunar extrapolation model is identified by N97 (Chapter 8, Kirchoff et al.)*

Terrain	AMA (Ga)	Model	Citation
Galileo Regio	3.8	S&W82	Passey and Shoemaker (1982)
	3.8–4.2	N97	Prockter et al. (1998)
	4.2	N97	Neukum (1997)
Nichelson Regio	3.9	S&W82	Passey and Shoemaker (1982)
	4.1	N97	Wagner et al. (2018)
	2.87–4.56	Z03	Wagner et al. (2018)
Gilgamesh	3.5	S&W82	Passey and Shoemaker (1982)
	0.7–2.0	Z98	Zahnle, Dones and Levison (1998)
	0.8	Z03	Zahnle, Dones and Levison (2003)
	0.11–2.13	Z98	Wagner et al. (1999)
	0.55–3.14	Z03	Wagner et al. (2018)
Grooved Terrain	3.4	S&W82	Shoemaker et al. (1982)
Leading Hemisphere	0.5–2	Z98	Zahnle et al. (1998)
Grooved Terrain	3.6	S&W82	Shoemaker et al. (1982)
Trailing Hemisphere/	0.5–3	Z98	Zahnle et al. (1998)
Harpagian Sulcus	3.85	N97	Wagner et al. (2018)
	0.93–3.97	Z03	Wagner et al. (2018)
Grooved Terrain	3.8	S&W82	Shoemaker et al. (1982)
South Pole			
Western Equatorial Basin	1–2	Z98	Zahnle, Dones and Levison (1998)
Uruk Sulcus	3.7	N97	Neukum (1997)
Osiris	0.26	N97	Wagner et al. (1999)
	0.003–0.01	Z98	Wagner et al. (1999)
	0.3–1.4	N97	Wagner et al. (2018)
	0.03–0.24	Z03	Wagner et al. (2018)

B.4 GANYMEDE'S ICE

Table B.8 *Temperature and pressure ranges for non-H2O volatile species observed on Ganymede (Chapter 11, Ahrens et al.)*

Species	Solid Phase	Temperature Range (K)	Pressure (bar)
O_2	α	20–23.78	$(1.49 \pm 0.03) \times 10^{-3}$
	β	23.78–43.77	
	γ	43.77–54.33	
O_3		< 79.6	$(5.6 \pm 0.6) \times 10^{-6}$
CO_2		40–194.7	5.185 ± 0.005
		194.7–216.58	
H_2S	I	126.2–187.57	0.229 ± 0.008
	II	103.6–126.2	
	III	< 103.6	
SO_2		15–197.63	0.0167 ± 0.0001

Table B.9 *List of ice compounds and relevant Young's (E) or bulk moduli (B). (Chapter 11, Ahrens et al.)*

Ice Species	Modulus Type	Temperature (K)	Modulus (GPa)
Ice I_h	E	< 273	0.009–0.0112
	E	90	0.1–0.5
	E	100	0.2–0.9
CO_2	E	80	13.12
CH_4 clathrate SI	E	< 273	8.4
	B	< 273	8.76
CH_4 clathrate SII	E	< 273	8.2
	B	< 273	8.48
NH_3-H_2O	E	90	0.85
	E	100	0.1
Ice III	B	250	9.761
Ice VI	B	300	20.27

Table B.10 *Physical property comparison between water ice I_h phase and clathrate hydrates based on cage structures (CS)–I and II. (Chapter 11, Ahrens et al.)*

Physical Property	Ice I_h	Clathrate CS I	Clathrate CS II
Dielectric constant (at 273 K)	94	58	58
Young's modulus (at 268 K) [GPa]	8–9.5	8.5	8.5
Poisson's ratio	0.331	0.31403	0.31119
Thermal Expansion (at 200 K) [10^{-6}/K]	56	77	52
Thermal Conductivity (at 263 K) [W/m K]	2.18	0.51	0.50

Table B.11 *Water ice removal or turnover rates on Ganymede's surface in equatorial (Equ.) and polar regions (Chapter 11,)*

Region	Temperature	Irradiation (ions) $\mu m\ yr^{-1}$	Thermal sublimation $\mu m\ yr^{-1}$	Meteoroid impact $\mu m\ yr^{-1}$
Equ., leading, dayside	125 ... 150 K	0.025	0.1 ... 600	0.6
Polar (lat. > 45°)	100 ... 125 K	0.1	< 0.001 ... 0.1	0.3
Equ., trailing, nightside	110 K	0.01	< 0.001	0.15

Table B.12 *Relative chemical abundances presumed for polar and equatorial surface regions (Chapter 15, Galli et al.)*

	Water ice	Dark material	Sulphuric acid hydrate	Sulfates	Chlorinated salts
Equatorial	0.15 ... 0.24	0.55 ... 0.70	0.04 ... 0.10	0.03 ... 0.08	0.03 ... 0.08
Polar	0.40 ... 0.48	0.30	0.16	0.03	0.03 ... 0.11

Table B.13 *Possible interpretation of the non-water-ice constituents on the icy surface of Ganymede, according to the Galileo NIMS measurements. (Chapter 20, Chela-Flores)*

New absorption features (μm)	Possible non-water-ice constituents
3.50	H_2O_2
3.88	C_2H_5SH (mercaptan)
4.05	SO_2
4.25	CO_2
4.57	CN_2 (cyanogen)

B.5 GANYMEDE'S MAGNETIC FIELD AND PLASMA

Table B.14 *Plasma and field properties near Ganymede (adapted from Chapter 14, Kivelson et al.)*

Property Symbol (units)	Typical & Range of Values
Magnetic Field Magnitude B_0 (nT)	95 ± 25
Magnetic Pressure $B_0^2/2\mu_0$ (nPa)	4 ± 2
Vertical Field B_θ (nT)	80 ± 8
Radial Field B_r (nT)	0 ± 90
Azimuthal Field B_ϕ (nT)	0 ± 15
Corotation Speed V_{co} (km/s)	187
Azimuthal Flow Speed V_{az} (km/s)	150 ± 20
V_{az}/V_{co}	$80\% \pm 10\%$
Ganymede's Orbital Speed V_{orb} (km/s)	11
$V_{az} - V_{orb}$ (km/s)	140 ± 20

Table B.15 *Plasma and field properties near Ganymede continued (adapted from Chapter 14, Kivelson et al.)*

Property Symbol (units)	Highest magnetic latitude	Intermediate magnetic latitude	Near magnetic equator
Magnetic Field Magnitude B_0 (nT)	120	95	70
Total Electron Density N_e (cm-3)	3	6	10
Cold Electron Temperature T_e (eV)	60	40	30
Hot Electron Density Percentage N_{eh}	10%	10%	10%
Hot Electron Temperature T_{eh} (eV)	200	200	200
Proton Density N_p (cm^{-3})	0.5	0.8	1.2
Proton Temperature T_p (eV)	300	200	100
Heavy Ion Density N_{hea} (cm^{-3})	1.7	3.5	6
Heavy Ion Temperature T_{hea} (eV)	100	75	50
Hot Heavy Ion Density N_{heavyh}	20%	15%	10%
Hot Heavy Ion Temperature T_{heah} (eV)	3000	2000	1500
Heavy Ion Mass/Charge A/Z (amu/q)	19/1.2	24/1.7	25/1.6
Ion Mass Density (amu cm^{-3})	48	85	160
Total Ion Thermal Pressure $P_{ion} = P_p + P_{hea} + P_{heah} = \Sigma nkT$ (nPa)	0.25	0.25	0.20
Energetic Ion Pressure P_{ener} (nPa)	2.5	2.5	2.5
Electron Pressure P_e (nPa)	0.04	0.05	0.08
Total Plasma Pressure P_{tot} (nPa)	2.8	2.8	2.8
Plasma-$\beta = P_{tot}/(B^2/2\mu_0)$	0.5	0.7	1.4
Alfvén Speed V_A (km/s)	380	220	120
Sound Speed C_s (km/s)	110	95	85
Alfvén Mach Number M_A	0.4	0.6	1
Magnetosonic Mach Number M_{ms}	0.4	0.6	1
Ram Pressure P_{ram} (nPa)	0.6	2	4

Table B.16 *Fitted parameters for Ganymede's internal field (Chapter 14, Kivelson et al.)*

Fitted dipole plus quadrupole moments g_i^j and h_i^j (in nT).							
g_1^0	g_1^1	h_1^1	g_2^0	g_2^1	g_2^2	h_2^1	h_2^2
-711.0 ± 4.2	46.8 ± 1.0	22.3 ± 1.9	0.9 ± 2.6	27.0 ± 1.2	-0.4 ± 1.7	1.8 ± 2.2	-11.0 ± 0.8
Fitted dipole moment g_i^j and h_i^j (in nT) and induction efficiency α.							
g_1^0	g_1^1	h_1^1	α				
-716.8 ± 1.6	49.3 ± 0.7	22.2 ± 1.1	0.84 ± 0.018				

B.6 GANYMEDE'S ATMOSPHERE

Table B.17 *Observations of Ganymede's neutral atmosphere and corona (Chapter 16, Roth et al.)*

Facility program ID	Instrument	Year	Geometry hemisphere	Type	Comments
Galileo	UVS	1996	above limb	FUV spectral scan	1st detection of H through scattered solar Lyα
HST 6758	GHRS	1996	trailing	FUV spectrum	1st detection of O_2
HST 7939	STIS	1998	trailing	spectral images	detection of auroral bands & of H_2O
HST 8224	STIS	2000	leading	spectral images	1st images of leading hemisphere
HST 9296	STIS	2003	sub-Jovian	spectral images	1st images of sub-Jovian hemisphere
New Horizons	Alice	2007	sub- & anti-Jovian	FUV spectra	observation of nightside atmosphere
HST 10871	ACS	2007	sub-Jovian	FUV filter image	observation in eclipse
HST 12244	STIS	2010/11	leading	spectral images	temporal variability, H_2O
Herschel	HIFI	2011-2013	leading & trailing	sub-mm spectrum	possible detection of H_2O absorption
HST 13328	STIS & COS	2014	leading & trailing	spectra & spec. images	1st campaign to observe both hemispheres
HST 14634	STIS	2017	anti-Jovian	spectral images	1st images of anti-Jovian hemisphere
	COS	2018	sub-Jovian	high-res spectra	eclipse/egress observation upper limits on O

Table B.18 *Models of Ganymede's neutral atmosphere (collisionless) (Chapter 16, Roth et al.)*

Model	Species	Source	Loss	Novelty	Implications
Johnson et al. (1981)	O_2, H_2O, H_2, H, OH	Sputtering, Sublimation†	sticking, e-/photo-dissociation, escape	1st analytic Model	Prediction of O_2 and H_2O atmospheres on icy moons
Marconi (2007)‡	O_2, H_2O, H_2, O, H, OH	Sputtering, Sublimation	sticking, e-/photo-dissociation, escape	2-D axisymmetric	Atmospheric profile; Latitudinal distributions: H_2O subsolar; O_2 elsewhere
Turc et al. (2014)	O_2, H_2O, H_2, O, H, OH	Sputtering, Sublimation	e-/photo-dissociation, escape	3-D Model, Atm. collapse in eclipse	Collisional atm. at noon; H density underestimated
Plainaki et al. (2015)	O_2, H_2O	Sputtering, Sublimation, Radiolysis	sticking, e- impact dissociation, charge exchange, escape	Plasma precipitation patterns*	Sputtered H_2O variability: leading/trailing; poles/equator; 2-component O_2 atm.
Shematovich (2016)†	O_2, H_2O, H_2, O, H, OH	Sputtering, Sublimation	e-/photo- dissociation, charge exchange	Collisions important	Suprathermal O
Leblanc et al. (2017)†	O_2, H_2O, H_2, O, H, OH	Thermal Radiolysis, Sputtering, Sublimation	sticking, e-/photo dissociation, escape	3-D Rotational**	H_2O variability: Poles/Noon & leading/trailing; Dusk-over-Dawn O_2 Bulge
Oza et al. (2018)	O_2	Outgassing, Thermal Radiolysis, Sputtering	e- impact dissociation	1-D Semi-Analytic Atm. Rotation Model	Dusk-over-Dawn O_2 Bulge; presence of near-infinite, outgassing O_2 reservoir
Vorburger et al. (2021)	H_2O	Sputtering, Sublimation	sticking, e-/photo-dissociation, e-/photo-ionization, escape	Ion- and electron-sputtering, plasma precipitation patterns	Sublimation-dominated regions constrained, importance of $O^{+,++}$ and S^{+++} sputtering, electron-sputtering negligible

†Sublimation refers to H_2O sublimation in this table.
‡& Collisional

Table B.19 *Surface density and column density ranges derived from atmospheric models. Values in parentheses are derived from observations, stated here for comparison (Chapter 16, Roth et al.)*

Species	Density [cm^{-3}]	Column Density [cm^{-2}]	Resident time [s] *limiting loss process*†
H_2O (sub-solar)	$2 \times 10^8 - 4 \times 10^9$ ($2 \times 10^8 - 1 \times 10^9$)	$6 \times 10^{15} - 2 \times 10^{16}$	$3 \times 10^2 - 1 \times 10^3$ *return to surface*
H_2O (polar)	$4 \times 10^3 - 3 \times 10^4$	$2 \times 10^{11} - 3 \times 10^{11}$	$3 \times 10^2 - 1 \times 10^3$ *return to surface*
O_2	$1 \times 10^7 - 9 \times 10^7$ (1×10^8)	$4 \times 10^{13} - 2 \times 10^{15}$ ($3 \times 10^{13} - 1 \times 10^{15}$)	$1 \times 10^3 - 7 \times 10^5$ *dissociation/ionization*
H_2	$1 \times 10^5 - 8 \times 10^6$	$1 \times 10^{14} - 1 \times 10^{15}$	$4 \times 10^4 - 1 \times 10^5$ *escape*
OH	$\sim(1 \times 10^2 - 1 \times 10^4)$	$\sim(1 \times 10^{10} - 1 \times 10^{12})$	$\sim(1 \times 10^3 - 1 \times 10^5)$ *escape/return to surface*
O	$\sim 1 \times 10^4$	$\sim 1 \times 10^{12}$	$\sim 1 \times 10^4$ *escape/return to surface*
H	$\sim 5 \times 10^3$	$\sim 1 \times 10^{12}$	$\sim 1 \times 10^4$ *escape*

†resident time in the atmosphere and the relevant limiting process that determines the time scale.

B.6 GANYMEDE'S IONOSPHERE

Table B.20 *Characteristics of ionisation processes for the main neutral species at Ganymede (Chapter 17, Galand et al.)*

Neutral species	Ion species	Threshold wavelength (nm)	Threshold energy (eV)	Photo-ionisation frequency (s^{-1})	Electron-impact ionisation frequency (s^{-1})
H_2	H_2^+	80.4	15.43	0.23×10^{-8}	3.02×10^{-8}
H_2O	H_2O^+	98	12.6	1.13×10^{-8}	4.25×10^{-8}
O_2	O_2^+	102.7	12.07	1.75×10^{-8}	9.05×10^{-8}

Table B.21 *Contribution of different ion populations from the Jovian magnetopshere and Ganymede's ionosphere to the impact and release rates. (Chapter 17, Galand et al.)*

	Baseline	
Species	Total impact rate (s^{1})	Total release rate (s^{1})
All (Jovian + Ionosph.)	21.9×10^{24}	25.1×10^{26}
Species	Contribution to impact rate (%)	Contribution to release rate (%)
Jovian, thermal ions	9.2	2.4
Jovian, energetic ions	47.2	86.8
Ionospheric	43.6	10.8
	Boosted case	
Species	Total impact rate (s^{1})	Total release rate (s^{1})
All (Jovian + Ionosph.)	196.3×10^{24}	74.4×10^{26}
Species	Contribution to impact rate (%)	Contribution to release rate (%)
Jovian, thermal ions	1.0	0.8
Jovian, energetic ions	5.3	29.3
Ionospheric	93.7	69.9

INDEX

For EU product safety concerns, contact us at Calle de José Abascal, 56–1°,
28003 Madrid, Spain or eugpsr@cambridge.org.

www.ingramcontent.com/pod-product-compliance
Lightning Source LLC
LaVergne TN
LVHW082110200926
848484LV00008B/1938

* 9 7 8 1 1 0 8 8 3 2 9 5 3 *